Concrete Segmental Bridges

Concrete Segmental Bridges

Theory, Design, and Construction to AASHTO LRFD Specifications

Dongzhou Huang, Ph.D., P.E.
President, American Bridge Engineering Consultants
Chief Engineer, Atkins North America
Professor, Fuzhou University

Bo Hu, PhD., P.E., P. Eng.,
Associate Technical Director, COWI

CRC Press
Taylor & Francis Group
Boca Raton London New York

CRC Press is an imprint of the
Taylor & Francis Group, an **informa** business

CRC Press
Taylor & Francis Group
6000 Broken Sound Parkway NW, Suite 300
Boca Raton, FL 33487-2742

First issued in paperback 2021

ISBN-13: 978-1-4987-9900-3 (hbk)
ISBN-13: 978-1-03-217576-8 (pbk)
DOI: 10.1201/9780429485473

Library of Congress Cataloging-in-Publication Data

A catalog record for this book has been requested.

Visit the Taylor & Francis Web site at
http://www.taylorandfrancis.com

and the CRC Press Web site at
http://www.crcpress.com

Dedication

The first author is eternally grateful for his lovely wife, Yingying, without whose patience and support this book would not have been possible.

Contents

Preface

Segmental concrete bridges have become a preferred choice for major transportation projects throughout the world because they meet the prevailing need for expedited construction with minimal traffic disruption, lower life cycle costs, appealing aesthetics, and the adaptability to accommodate curved roadway alignments. Although there is an abundance of literature pertaining to the design and construction of concrete segmental bridges, most of it is focused on construction and is directed toward experienced concrete segmental bridge designers. However, it is important for bridge engineers and university seniors with little or no experience in designing complex segmental bridges to have a comprehensive, well-written book that can be used as a reference tool for both design and for self-study. This book is intended to assist all levels of practicing bridge engineers and university seniors in fully understanding concrete segmental bridge design techniques, analytical methods, design specifications, theory, construction methods, and common industry practices. It presents comprehensive design theories, practical analysis, and key construction methods used for segmental bridge design in a systematic, easy-to-follow manner while highlighting the importance of key design checks. The design principles presented are further reinforced with a design example based on the AASHTO LRFD design specifications for each of the main segmental bridge types. The examples encompass both hand solutions as well as computer modeling techniques that are currently used in practice.

Chapter 1 is intended to familiarize readers with the basic concepts needed to design concrete segmental bridges. First, the types of materials and their mechanical properties used in concrete segmental bridges are introduced. Then, the basic components, the types of concrete segmental bridges, and their typical construction methods are briefly discussed. At the end of this chapter, the selection of different types of segmental bridges, bridge general design procedures, and bridge aesthetics are discussed.

It is especially useful for bridge designers to clearly understand bridge design philosophy, types and magnitudes of bridge loadings, and the safety and reliability characteristics required during the service life of a bridge. In Chapter 2, the different types of highway bridge design loadings and their analytical methods are introduced. Then, the basic theoretical background of the AASHTO LRFD method is discussed. Finally, design limits, load combinations, load, and resistance factors are presented.

Though bridge designers are formally trained in the theories of structural design and analysis, it requires considerable knowledge and skill to recall, utilize, and implement these theories in bridge design. For this reason, Chapter 3 first presents the basic theory of prestressing concrete structures and the design theories of bending, shear, torsion, and compression. Then, the current AASHTO design equations are briefly presented and interpreted.

Concrete segmental bridges are generally statically indeterminate structures, and their analytical methods are less familiar to bridge engineers than those of simply supported bridges. To help readers fully understand the design theory and behavior of concrete segmental bridges, detailed bridge analytical theories are given in Chapter 4. First, the basic analytical theory of indeterminate structures is briefly reviewed. Then, the longitudinal analysis of segmental bridges is discussed, including secondary forces due to post-tensioning, temperature, support deformation, shrinkage, and creep. Finally, the transverse analysis of segmental box girder bridges is presented.

The design of span-by-span concrete segmental bridges is discussed in Chapter 5. The common details, elements, and analytical methods of concrete segmental bridges are presented, including the different types of segments, tendon layouts, diaphragms, as well as blister design and analysis. At the end of the chapter, a design example is given that is generated based on an existing segmental bridge. It is recommended that this chapter be understood prior to reading about the design of other types of segmental bridges.

In Chapter 6, the design of balanced cantilever segmental bridges is presented. The distinguishing design characteristics and details of this type of bridge are first presented. Then, the design theory and detailing of curved segmental bridges are discussed. At the end of the chapter, an example of a balanced cantilever segmental bridge design based on an existing bridge is provided.

Designs of incrementally launched segmental bridges, post-tensioned spliced girder bridges, arch bridges, cable-stayed bridges, and substructures are addressed in Chapters 7, 8, 9, 10, and 11, respectively. In each of these chapters, the distinguishing design theories, details, and behaviors of each individual bridge type are presented. For each type of bridge and substructure, a design example is generated based on an existing bridge.

In Chapter 12, some typical construction methods in concrete segmental bridge construction are presented. These include typical segment cast methods, geometrical control, construction tolerances, and post-tensioning tendon installation and grouting.

The *AASHTO LRFD Bridge Design Specifications*, 7th edition, 2014, is used throughout the text as the design code. The English system of measuring units used in the book is consistent with the code and the bridge building industry in the United States. For the convenience of those accustomed to the metric system, a table, "Unit Conversion Factors for English and Metric Units," is provided.

Because of the tremendous work effort involved in writing this book, I invited my colleague Dr. Bo Hu, Ph.D., P.E., P.Eng., to be a co-author to finish this book. I wrote Chapters 1 to 11 and part of Chapter 12, while Dr. Hu wrote part of Chapter 12, performed literature reviews, and drew most of the final figures contained in this book.

Though significant efforts have been taken to eliminate errors, there may be some minor oversights. Readers are encouraged to e-mail me any corrections and/or recommendations that may improve the next edition of this book at dhuang.abe@gmail.com.

Dongzhou Huang, Ph.D., P.E.
August 30, 2019
Tampa, Florida, United States

Acknowledgments

I would like to gratefully acknowledge the Florida Department of Transportation (FDOT), American Segmental Bridge Institute (ASBI), Journal of Structural Engineering International, International Association for Bridge and Structural Engineering (IABSE), American Association of State Highway and Transportation Officials (AASHTO), DYWIDAG, FREYSSINET, TENSA, and D.S. Brown for their permission to use many of the drawings contained in this book. I would also like to express my deepest appreciation to Mr. William R. Cox, Dr. Andy Richardson, Mr. Jose S. Rodriguez, Mr. Scott C. Arnold, Mr. Thomas A. Andres, Mr. Yicheng Huang, Mr. Teddy S. Theryo, and Mr. Uma Pallapolu for their valuable comments, insightful suggestions, and assistance.

I am also deeply indebted to my mentors Professor Guohao Li, Professor Haifan Xiang, and Professor Dong Shi for their gracious support and distinguished supervision during my pursuit of my master's and Ph.D. degrees at Tongji University.

SPECIAL ACKNOWLEDGMENTS

The design examples contained in Chapters 7, 9, and 10 have been revised by myself based on the following design examples and writers whose contributions are highly valued and sincerely appreciated:

- "Design Example of Incremental Launched Segmental Bridge" provided by Prof. Chunsheng Wang and Dr. Peijie Zhang, School of Highway, Chang'an University, China.
- "Design Example of Arch Bridge" provided by Prof. Baochun Chen, Dr. Xiaoye Luo, Fuzhou University; Mr. Zenghai Lin, Mr. Guisong Tu, Institute of Guangxi Communication Planning Surveying and Designing Co., Ltd; and Dr. Jianjun Wang, Guangxi Road and Bridge Engineering Group Co., Ltd, China.
- "Design Example of Cable-Stayed Segmental Bridge" provided by Mr. Xudong Shen, and Mr. Fangdong Chen, Planning and Design Institute of Zhejiang Province, Inc., China.

Dongzhou Huang
August 30, 2019

Authors

Dr. Dongzhou Huang, Ph.D., P.E., is president of American Bridge Engineering Consultants and chief engineer of Atkins North America. He has been a professor and visiting professor at Tongji University, Fuzhou University, and Florida International University for more than 20 years. He has been engaged in bridge engineering for over 40 years and has published more than 100 technical papers and books on bridge design, dynamic and stability analysis, practical analysis methods, capacity evaluations of different types of bridges, including box girder, curved girder, cable-stayed, truss, and arch bridges—over 60 of which have appeared in peer-reviewed journals. He has worked in two bridge design firms and been extensively involved in the design, analysis, and construction of different types of long-span and complex bridges, including span-by-span constructed segmental bridges, precast and cast-in-place cantilever segmental bridges, cable-stayed bridges, arch bridges, and spliced concrete I-girder bridges. He is an editor/board member of *Journal of Structural Engineering International*, IABSE, and former associate editor of *Journal of Bridge Engineering*, ASCE. He is a fellow of IABSE.

Dr. Bo Hu, Ph.D., P.E., P.Eng., is an associate technical director in the Edmonton office of COWI. After receiving his bachelor's and master's degrees in bridge engineering at Tongji University in 1995 and 1998, respectively, he obtained his Ph.D. in civil engineering from the University of Delaware in 2006. He has been engaged in bridge engineering for 20 years with research and design experiences in various bridge types, seismic behaviors of bridge structures, seismic design methodology for bridge structures, and development of innovative structures. He is a professional engineer in the United States and Canada and has been playing leading technical and project roles in a variety of large bridge projects in North America, including segmental concrete girder bridges, extradosed concrete segmental bridges, and cable-stayed concrete segmental bridges.

Unit Conversion Factors

SI to English

Length
1 mm = 0.0394 in.
1 cm = 0.394 in.
1 m = 3.281 ft
1 m = 39.37 in.
1 km = 0.622 mile

Area
$1 \text{ mm}^2 = 0.00155 \text{ in.}^2$
$1 \text{ m}^2 = 10.76 \text{ ft}^2$
$1 \text{ m}^2 = 1.196 \text{ yd}^2$

Volume
$1 \text{ cm}^3 = 0.0610 \text{ in.}^3$
$1 \text{ m}^3 = 35.31 \text{ ft}^3$
$1 \text{ m}^3 = 1.564 \text{ yd}^3$

Moment of Inertia
$1 \text{ mm}^4 = 2.40 \times 10^{-6} \text{ in.}^4$
$1 \text{ cm}^4 = 2.40 \times 10^{-2} \text{ in.}^4$
$1 \text{ m}^4 = 2.40 \times 10^{6} \text{ in.}^4$

Mass
1 kg = 2.205 lb
1 Mg = 1.102 ton (2000 lb)
1 tonne (metric) = 1.102 ton (2000 lb)

Force
1 N = 0.2248 lb
1 MN = 224.8 kips

Stress
1 MPa = 145 psi

$1 \text{ N/mm}^2 = 145 \text{ psi}$
$1 \text{ MN/m}^2 = 145 \text{ psi}$

English to SI

1 in. = 25.4 mm
1 in. = 2.54 cm
1 ft = 0.3048 m
1 in. = 0.0254 m
1 mile = 1.609 km

$1 \text{ in.}^2 = 645.2 \text{ mm}^2$
$1 \text{ ft}^2 = 0.0929 \text{ m}^2$
$1 \text{ yd}^2 = 0.835 \text{ m}^2$

$1 \text{ in.}^3 = 16.387 \text{ cm}^3$
$1 \text{ ft}^3 = 0.0283 \text{ m}^3$
$1 \text{ yd}^3 = 0.765 \text{ m}^3$

$1 \text{ in.}^4 = 416,200 \text{ mm}^4$
$1 \text{ in.}^4 = 41.62 \text{ cm}^4$
$1 \text{ in.}^4 = 0.4162 \times 10^{-6} \text{ m}^4$

1 lb = 0.454 kg
1 ton (2000 lb) = 907.2 kg

1 lb = 4.448 N
1 kip = 4.448 kN

$1 \text{ psi} = 6.896 \text{ kPa} \left(\frac{kN}{m^2}\right)$
$1 \text{ ksi} = 6.895 \text{ MN/m}^2$

Principal Notations

A	=	cross-sectional area
	=	area of object facing wind
	=	fatigue constant
A_b	=	effective net area of bearing plate
A_c	=	area of concrete section
A_{cp}	=	total area enclosed by outside perimeter of cross section
A_{cs}	=	effective cross-sectional area of strut
A_{cv}	=	area of concrete considered to be engaged in interface shear transfer
$ADDT$	=	average number of trucks per day in one direction over design life of bridge
A_g	=	gross area of section
AI	=	dynamic response due to accidental release or application of a precast segment load
A_l	=	required total longitudinal reinforcement
A_o	=	area enclosed by shear flow path
A_{ps}	=	area of prestressing steel
A_{psb}	=	area of bonded prestressing steel
A_{psu}	=	area of unbonded prestressing steel
A_s	=	area of non-prestressed tension reinforcement
A_s'	=	area of compression reinforcement
A_{st}	=	total area of longitudinal non-prestressed reinforcement
A_v	=	area of transverse reinforcement within distance s
A_{vf}	=	area of interface shear reinforcement crossing shear plane
a	=	depth of equivalent rectangular stress block
	=	vehicle acceleration
B	=	bridge width
b	=	beam width
	=	flange width on each side of web
	=	width of compression face of member
	=	half of bridge width
b_c	=	length of top cantilever flange of box girder
b_e	=	effective flange width
	=	lateral dimension of effective bearing area
b_{em}	=	effective flange width for interior portions of span
b_{es}	=	effective flange width at supports
BL	=	blast loading
BR	=	vehicular braking force
	=	base rate of vessel aberrancy
b_t	=	web spacing
c	=	damping ratio
	=	cohesion factor
	=	distance from extreme compression fiber to nuetral axis
$\bar{c}$	=	generalized damping ratio
C	=	total resulting compression force
C_A	=	modification factor for object area and direction of wind
C_D	=	drag coefficient
CE	=	vehicular centrifugal force
CT	=	vehicular collision force
CV	=	vessel collision force

CEQ	=	specialized construction equipment
CLE	=	longitudinal construction equipment load
CLL	=	distributed construction live load
CR	=	creep effects
	=	force effects due to creep
cgc	=	center of gravity of concrete section
cgs	=	center of gravity of prestressing steel
c_t	=	distance from cgc to top fiber
c_b	=	distance from cgc to bottom fiber
D	=	external diameter of circular member
D_r	=	internal diameter of pot
DC	=	weight of supported structure
	=	dead load of structural components and nonstructural attachments
DD	=	downdrag force
$DIFF$	=	differential load
DW	=	superimposed dead load
	=	dead load of wearing surfaces and utilities
DWT	=	dead weight tonnage of vessel
d	=	depth of pier
	=	depth of beam
d_{burst}	=	distance from anchorage device to centroid of bursting force
d_c	=	thickness of concrete cover measured from extreme tension fiber to center of flexural reinforcement
d_e	=	effective depth from extreme compression fiber to centroid of tensile force in tensile reinforcement
d_o	=	girder depth
	=	depth of super-structure
d_p	=	distance from extreme compression fiber to centroid of prestressing tendons
d_s	=	distance from extreme compression fiber to centroid of non-prestressed tensile reinforcement
d'_s	=	distance from extreme compression fiber to centroid of compression reinforcement
d_v	=	effective shear depth
E	=	modulus of elasticity of material
EH	=	horizontal earth pressure load
EL	=	miscellaneous locked-in force effects resulting from construction process, including jacking apart of cantilevers in segmental construction
ES	=	earth surcharge load
EV	=	vertical pressure from dead load of earth fill
EQ	=	earthquake load
FR	=	friction load
E_c	=	modulus of elasticity of concrete
E'_c	=	effective modulus of elasticity
E_{ci}	=	modulus of elasticity of concrete at transfer
E_{eq}	=	equivalent modulus of elasticity of cable
EI	=	flexural stiffness
$E_p = E_{ps}$	=	modulus of elasticity of prestressing steel
E_s	=	modulus of elasticity of steel reinforcement
E_{28}	=	modulus of elasticity at 28 days
$E_\phi(t_n, t_{n-1})$	=	effective modulus of elasticity

e	=	eccentricity
	=	base of natural logarithms
	=	correction factor for distribution
e_g	=	distance between centers of gravity of beam and deck
F	=	force effect
	=	braking force
F_a	=	site factor for short-period range of acceleration response spectrum
F_h	=	wind-induced horizontal force
F_{pga}	=	site factor at zero-period on acceleration response spectrum
F_v	=	site factors in long-period range
	=	wind-induced lift force
$F_{u\text{-}out}$	=	out-of-plane force per unit length of tendon due to horizontal curvature
$F_{v\text{-}out}$	=	out-of-plane force per unit length of tendon due to vertical curvature
F_w	=	total axial force acting on web due to post-tensioning force
f	=	angle of internal friction
	=	cycle per second
	=	normal stress
	=	design rise of arch
f_a	=	normal compression stresses
f^t	=	top fiber stress
f^b	=	bottom fiber stress
f_b	=	stress in anchor plate at section taken at edge of wedge hole or holes
	=	normal stress due to bending moment
f_{burst}	=	bursting stress
f_{bv}	=	shear stress due to bending
f'_c	=	compressive strength of concrete
f_{ca}	=	concrete allowable compression stress
f'_{ci}	=	concrete compressive strength at time of prestressing for pretensioned members and at time of initial loading for non-prestressed members
	=	design concrete strength at time of application of tendon force
f_{cpe}	=	compressive stress in concrete due to effective prestress forces only at extreme fiber of section where tensile stress is caused by externally applied loads
f_{cr}	=	flexural cracking stress of concrete beam
f_{ct}	=	concrete tensile stress
f'_{cu}	=	factored concrete compressive stress
$\{F_e\}$	=	nodal forces vector of finite element
f_{max}	=	maximum principal stress in the web, compression positive
f_{pc}	=	unfactored compressive stress in concrete after prestress losses have occurred
f_{pe}	=	effective stress in prestressing steel after losses
f_{pi}	=	prestressing steel stress immediately prior to transfer
	=	initial prestressing stress
f_{pR}	=	remaining prestressing in strands
f_{ps}	=	average stress in prestressing steel when the nominal resistance of member is required
f'_{ps}	=	compression stress in prestressing steel
f_{pu}	=	tensile strength of prestressing steel
f_{py}	=	yield strength of prestressing steel
f_r	=	modulus of rupture of concrete
f_s	=	stress in steel
	=	maximum average stress of bearing
f_{ss}	=	calculated tensile stress in mild steel reinforcement at the service limit state

f_s'	=	stress in the non-prestressed compression reinforcement at nominal flexural resistance
f_t	=	tensile strength of concrete
f_{ta}	=	concrete allowable tensile stress
f_{tD}	=	tensile stress caused by dead loads
f_{tL}	=	tensile stress caused by live loads
f_v	=	shear stress due to shear force
$f_{\omega v}$	=	shear stress due to warping torsion
f_ω	=	normal stress due to warping torsion
$f_x(x)$	=	nominal density function
f_y	=	minimum yield strength of reinforcement
f_y'	=	minimum yield strength of compression reinforcement
f_0	=	average compression stress in the section
	=	clear rise of arch
G	=	shear modulus of inertia
g	=	gravitational acceleration
	=	live load distribution factor representing the number of design lanes
$g_{Exterior}^M$	=	load distribution factor for moment in exterior beam
$g_{Interior}^M$	=	load distribution factor for moment in interior beam
$g_{Exterior}^S$	=	load distribution factor for shear in exterior beam
$g_{Interior}^S$	=	load distribution factor for shear in interior beam
H	=	horizontal component of cable force or arch rib
	=	height of wall
	=	average annual ambient relative humidity
h	=	overall thickness or depth of member
	=	notational size of member (mm)
	=	height of pylon
h_b	=	distance between centroidal axis and bottom fiber of section
h_f	=	compression flange depth
h_t	=	distance between centroidal axis and top fiber of section
	=	tower/pylon height
	=	total thickness of elastomeric bearing
h_{rt}	=	elastomeric thickness
h_0	=	girder depth
IC	=	ice load
IE	=	dynamic load from equipment
IM	=	dynamic load allowance
I	=	moment of inertia
I_g	=	moment of inertia of gross concrete section about centroidal axis
I_s	=	moment of inertia of longitudinal reinforcement about centroidal axis
I_ω	=	warping moment of inertia
$I_{\bar\omega}$	=	distortional warping constant
J	=	St. Venant torsional inertia
K_g	=	longitudinal stiffness parameter
K_1	=	fraction of concrete strength available to resist interface shear
	=	correction factor for source of aggregate
K_2	=	limiting interface shear resistance
K	=	wobble friction coefficient
K_D	=	stiffness of box section against distortion
K_u	=	average longitudinal stiffness
K_v	=	average lateral stiffness

$[K]$	=	global structural stiffness matrix
$[K_g]$	=	global geometric matrix
$[k]$	=	stiffness matrix of finite element
$[k_g]$	=	element geometrical matrix
$[k_\phi]$	=	creep stiffness matrix of finite element
k	=	lateral stiffness
	=	effective length factor
	=	factor used in calculation of distribution factor for multibeam bridges
$[\underline{k}]$	=	stiffness matrix of finite element
$\overline{k}$	=	generalized stiffness
k_b	=	lower kern point
k_c	=	factor for effect of volume-to-surface ratio for creep
	=	ratio of maximum concrete compressive stress to design compressive strength of concrete
k_f	=	factor for effect of concrete strength
k_{hc}	=	humidity factor for creep
k_{hs}	=	humidity factor for shrinkage
k_s	=	factor for effect of volume-to-surface ratio of component
k_{td}	=	time development factor
k_x	=	vertical bending curvature
k_y	=	lateral bending curvature about y-axis
k_t	=	up kern point
k_z	=	curvature of beam deflection
	=	torsion curvature about z-axis
L	=	length of bearing pad
	=	span length
	=	total bridge length
$[L]$	=	coordinate transformation matrix
LL	=	vehicular live load
$LLDF$	=	live load distribution factor
L_c	=	critical length of yield line failure pattern
L_{max}	=	maximum launching length
LS	=	live load surcharge
L_P	=	length of plastic hinge
l	=	unbraced length of horizontally curved girder
	=	span length
l_e	=	effective tendon length
l_{ps}	=	tendon length
l_u	=	unsupported length of compression member
	=	one-half of length of arch rib
l_0	=	distance between two spring lines
	=	length of launching nose
M	=	bending moment
M_w	=	warping torsion lateral moment
M_{1b}	=	smaller end moment on compression member due to factored gravity loads that result in no appreciable sidesway
M_{2b}	=	moment on compression member due to factored gravity loads that result in no appreciable sidesway calculated by conventional first-order elastic frame analysis
M_{2s}	=	moment on compression member due to factored lateral or gravity loads that result in sidesway

M_d	=	total unfactored dead load moment
M_{cr}	=	cracking moment
M_{dnc}	=	total unfactored dead load moment acting on monolithic or noncomposite section
$M_{Dt}(x)$	=	final dead load moment at time t in final structural system
$M_{D1}(x)$	=	initial moment due to dead load in first stage of structure
$M_{D2}(x)$	=	moment due to dead load in second stage of structure
M_{Df}	=	final moment in continuous girder
M_{Dca}	=	moment calculated based on cantilever girder system
M_{Dco}	=	moment calculated based on continuous girder system
M_i	=	bending moment resulting from force effect or superimposed deformation
M_{max}^r	=	maximum positive moment in rear zone
M_{min}^r	=	minimum negative moment in rear zone
M_n	=	nominal flexural resistance
M_r	=	factored flexural resistance
M_{rx}	=	uniaxial factored flexural resistance of section in direction of x-axis
M_{ry}	=	uniaxial factored flexural resistance of section in direction of y-axis
M_t	=	transverse moment
M_u	=	factored moment at section
M_{ux}	=	factored applied moment about x-axis
M_{uy}	=	factored applied moment about y-axis
M_ω	=	bi-moment
$M_{\tilde{\omega}}$	=	distortional bi-moment
M_1	=	smaller end moment at strength limit state due to factored loads acting on compression member
M_2	=	larger end moment at strength limit state due to factored loads acting on compression member
$m_i(x)$	=	moment due to unit force
m	=	lumped mass
$\bar{m}$	=	generalized mass
N	=	number of identical prestressing tendons
	=	axial force
N_{cr}	=	buckling axial force of arch
$[N]$	=	shape function of element
N_b	=	number of beams or girders
N_c	=	number of cells in concrete box girder
N_L	=	number of design lanes
N_u	=	factored axial force
n	=	modular ratio = E_s/E_c or E_p/E_c
P	=	axle load
$\{P\}$	=	global load matrix
P_B	=	base wind pressure
P_c	=	permanent net compressive force
p_c	=	length of outside perimeter of section
p_D	=	design wind pressure
P_e	=	Euler buckling load
P_i	=	wheel loads
p	=	coefficient of traffic in single lane
p_e	=	equivalent uniform static seismic loading per unit length of bridge that is applied to represent primary mode of vibration
PGA	=	peak seismic ground acceleration coefficient on rock

PL	=	pedestrian live load
P_n	=	nominal axial resistance
P_o	=	nominal axial resistance of section at 0.0 eccentricity
P_{ps}	=	prestressing force
P_r	=	factored axial resistance
	=	factored splitting resistance in pretensioned anchorage zones provided by reinforcement in end of pretensioned beams
P_{rx}	=	factored axial resistance determined on basis that only eccentricity e_y is present
P_{rxy}	=	factored axial resistance in biaxial flexure
P_{ry}	=	factored axial resistance determined on basis that only eccentricity e_x is present
PS	=	secondary forces from post-tensioning for strength limit states
	=	total prestress forces for service limit states
P_u	=	factored applied axial force
Q	=	total factored load
	=	moment of area
Q_i	=	force effects
Q_n	=	nominal load
q	=	distributed loading
	=	shear flow
q_e	=	equivalent static earthquake loading per unit length
q_{cr}	=	buckling loading per unit length
q_{wv}	=	vertical wind pressure applied to super-structure
R	=	radius of curvature of tendon, curved girder, or arch
$\{R_e\}$	=	equivalent nodal loading matrix of finite element
$\{R\}$	=	global nodal loading vector
R_d	=	absolute maximum dynamic response due to vehicles
R_n	=	nominal resistance
R_r	=	factored resistance
R_s	=	absolute maximum static response due to vehicles
R_w	=	nominal railing resistance to transverse load
r	=	radius of gyration of gross cross section
S	=	spacing of beams or webs
	=	shape factor
SM	=	single-mode elastic method
SC	=	mass-damping parameter = Scruton number
S_c	=	section modulus for extreme fiber of section
SE	=	force effect due to settlement
SH	=	shrinkage
	=	force effects due to shrinkage
S_i	=	shape factor of layer of rectangular elastomeric bearing pad
S_{nc}	=	section modulus for extreme fiber of monolithic or noncomposite section where tensile stress is caused by externally applied loads
S_s	=	short-period spectral acceleration coefficients
S_1	=	long-period spectral acceleration coefficients
s	=	spacing of transverse reinforcement
	=	spacing of stirrups
	=	cement rate of hardening coefficient
T	=	thermal
	=	period of vibration
	=	period of motion
	=	total tensile force

T_{burst}	=	tensile force in anchorage zone acting ahead of anchorage device
T_{cr}	=	torsional cracking moment
TG	=	force effect due to temperature gradient
	=	temperature gradient ($\Delta°F$)
TH	=	time history method
T_m	=	period of mth mode of vibration
T_i	=	torsion resulting from a force effect or superimposed deformation
T_n	=	nominal torsional resistance
T_r	=	factored torsional resistance
T_s	=	free torsion moment
TU	=	force effect due to uniform temperature
T_u	=	applied factored torsional moment
T_ω	=	warping torsion moment
$T_{\tilde{\omega}}$	=	distortional moments
t	=	maturity of concrete (day)
	=	thickness of wall
	=	member thickness
t_i	=	age of concrete at time of initial load application (day)
	=	age of concrete at time of initial deck placement (day)
t_s	=	depth of concrete slab
	=	age of concrete (days) at beginning of shrinkage
t_t	=	top flange thickness
t_w	=	web thickness
t_0	=	age of concrete at time of load application
U	=	segment unbalance force
$u(x)$	=	longitudinal displacement along bridge longitudinal direction
V	=	shear force
	=	velocity
	=	coefficient of variance
	=	vertical reaction
V_B	=	base wind velocity of 100 mph at 30.0-ft height
V_{DZ}	=	design wind velocity at design elevation, Z
V_c	=	nominal shear resistance of concrete
V_{cr}	=	critical wind speed for flutter
V_{cr}^g	=	critical wind speed for stay cable
V_{ci}	=	flexure shear strength
V_{cw}	=	nominal shear resistance of concrete
V_s	=	shear strength provided by transverse reinforcing
V_i	=	shear resulting from force effect or superimposed deformation
V_n	=	nominal shear resistance
V_{ni}	=	nominal interface shear resistance
V_p	=	shear resistance provided by component of effective prestressing force in direction of applied shear
V_r	=	factored shear resistance
V_{ri}	=	factored interface shear resistance
V/S	=	volume-to-surface ratio
V_s	=	shear resistance provided by transverse reinforcement
V_u	=	factored shear force
V_{ui}	=	factored interface shear force
V_0	=	friction velocity
V_{30}	=	wind speed at 30 ft above low ground or water level

$v_{sr}(x)$	=	simulated road vertical profile
v	=	highway design speed
	=	wind speed
	=	concrete efficiency factor
v_u	=	shear stress
$v(x)$	=	transverse displacement along bridge longitudinal direction
W	=	width of bearing plate or pad
W_0	=	weight of vehicle
WA	=	water load and stream pressure
WE	=	horizontal wind load on equipment
WL	=	wind on live load
WS	=	wind load on structure
WUP	=	wind uplift on cantilever
w	=	width of clear roadway
w_c	=	unit weight of concrete
$w(x)$	=	weight distribution of super-structure and tributary substructure in the bridge longitudinal direction
X_i	=	unknown redundant force
x_n	=	nominal value
$\bar{x}$	=	average value = mean
y^c	=	distance from cgc to application point of compression force
y_s	=	distance between arch crown and elastic center
y_t	=	distance from neutral axis to extreme tension fiber
Z	=	structure height above low ground or water level
	=	vertical distance from point of load application to ground or water level
	=	location of center of gravity of strands
Z_0	=	friction length of upstream fetch
z	=	vertical distance from center of gravity of cross section
θ	=	torsion angle about girder longitudinal axis
	=	angle of inclination of diagonal compression stresses
	=	cable angle between horizontal and chord
$\tilde{\theta}$	=	distortional angle
α	=	angle of inclination of transverse reinforcement relative to longitudinal axis (degrees)
	=	angle of inclination of tendon force with respect to centerline of member
	=	total angular change in prestressing steel path from jacking end to point under investigation (rad)
	=	angle between cable and horizontal
	=	longitudinal slope of deck at sliding plane
	=	coefficient of thermal expansion
	=	ratio of concrete girder cantilever length to span length
β	=	reliability index
	=	factor indicating ability of diagonally cracked concrete to transmit tension and shear
	=	generalized effective mass
	=	ratio of length of launching nose to that of main girder
	=	rotation angle
	=	panel buckling coefficient
$\beta_E(t)$	=	ratio of modulus of elasticity at time t to 28 days modulus
β_c	=	coefficient describing development of creep with time after loading
$\beta_{cc}(t)$	=	ratio of concrete strength at time t to 28-day concrete strength

β_d	=	ratio of maximum factored permanent moments to maximum factored total load moment
β_1	=	stress block factor taken as ratio of depth of equivalent uniformly stressed compression zone to depth of actual compression zone
γ	=	load factor
	=	generalized mass
	=	ratio of weight per unit length of launching nose to that of main girder
γ_e	=	exposure factor
γ_h	=	correction factor for relative humidity of ambient air
γ_{EQ}	=	load factor for live load applied simultaneously with seismic loads
γ_i	=	load factor
γ_p	=	load factor for permanent loading
γ_{SE}	=	load factor for settlement
γ_{zs}	=	shear deformation
γ_{st}	=	correction factor for specified concrete strength at time of prestress transfer to concrete member
γ_{TG}	=	load factor for temperature gradient
γ_1	=	flexural cracking factor
γ_2	=	prestress factor
γ_3	=	ratio of specified minimum yield strength to ultimate tensile strength of reinforcement
δ	=	small deflection
$\{\delta\}$	=	global displacement vector
δ_b	=	moment or stress magnifier for braced mode deflection
$\delta_g(t)$	=	horizontal ground motion due to earthquake
$\{\delta_e\}$	=	node parameters of finite element
δ_s	=	moment or stress magnifier for unbraced mode deflection
δ_{ij}	=	displacement due to unit force
Δ	=	displacement of column or pier relative to point of fixity for foundation
	=	elongation of tendons
	=	displacement
ΔF	=	force effect, live load stress range due to passage of fatigue load
$(\Delta F)_{TH}$	=	constant-amplitude fatigue threshold
Δ_{fcdp}	=	change in concrete stress at center of gravity of prestressing steel due to all dead loads
Δ_{fpA}	=	loss due to anchorage set
Δ_{fpC}	=	prestress loss due to creep of girder concrete
Δ_{fpES}	=	sum of all losses or gains due to elastic shortening
Δ_{fpF}	=	loss due to friction
Δ_{fpLT}	=	losses due to long-term shrinkage and creep of concrete and relaxation of steel
Δ_{fpR}	=	prestress loss due to relaxation of prestressing strands
Δ_{fpS}	=	prestress loss due to shrinkage of girder concrete
ε	=	normal strain
ε_{cl}	=	compression-controlled strain limit in extreme tension steel
ε_{cu}	=	failure strain of concrete in compression
ε_c	=	concrete ultimate compression strain at extreme fiber = 0.003 in./in.
$\varepsilon_c(t)$	=	creep strain at t days after casting
$\varepsilon_e(t_0)$	=	elastic strain caused when load is applied at t_0 days after casting
$\varepsilon_c(t,\tau_0)$	=	strain caused by creep at t when load is applied at τ_0
$\varepsilon_{cs} = \varepsilon_{sh}$	=	concrete shrinkage strain at given time
ε_{cs0}	=	notational shrinkage coefficient

ε_t	=	net tensile strain in extreme tension steel at nominal resistance
ε_{tl}	=	tension-controlled strain limit in extreme tension steel
ε_x	=	longitudinal strain at mid-depth of member
λ	=	bias factor
	=	slenderness ratio
	=	span ratio
λ_s	=	stability coefficient
λ_{cr}	=	safety factor of bridge buckling
λ_w	=	wall slenderness ratio for hollow columns
μ	=	friction factor
	=	moment magnification factor
	=	reduction factor of modulus of elasticity of cable
μ_0	=	theoretical moment magnifier
ρ	=	density of air
	=	efficiency ratio of a cross section
τ	=	shear stress
ϕ	=	resistance factor
ϕ_a	=	axial resistance factor
ϕ_b	=	bearing resistance factor
ϕ_0	=	notional creep coefficient
ϕ_f	=	flexure resistance factor
ϕ_v	=	shear resistance factor
ϕ_t	=	torsion resistance factor
ϕ_w	=	hollow column reduction factor
η_D	=	ductility factor
η_I	=	importance factor
η_i	=	load modifier
η_R	=	redundancy factor
σ	=	standard deviation of normal distribution
	=	normal stress
ν	=	Poisson's ratio
$\phi(t, t_i)$	=	creep coefficient at time t for loading applied at t_i
ω	=	circular frequency of structure
ω_B	=	vertical bending frequency of bridge
χ	=	radius of curvature in cable
$\chi(t,\tau)$	=	aging coefficient
ψ	=	damping ratio

1 Introduction to Concrete Segmental Bridges

- Development of concrete segmental bridges
- Construction materials and their properties
- Types of concrete segmental bridges and typical construction procedures
- AASHTO stress limits
- Post-tensioning systems
- General design procedures
- Aesthetics

1.1 BRIEF HISTORY AND DEVELOPMENT OF CONCRETE SEGMENTAL BRIDGES

Concrete bridges that are built segment by segment or piece by piece are often called concrete segmental bridges. The basic principles of segmental construction have been in existence for thousands of years and were used by engineers to build arch structures. The Zhao-Zhou Bridge (see Fig. 1-1) built in 615 A.D. in China is the first shallow segmental stone arch bridge in the world. The individual wedge-shaped stones are pushed together by the self-weight of the stones, creating horizontal and vertical reactions at the ends of the arched supports. The reactions are essentially post-tensioning forces applied to the wedged-shaped stones. However, modern segmental bridges are comprised of concrete segments that are connected via post-tensioning forces applied by high-strength steel wires, bars, and/or strands. The first generation of prestressed concrete segmental bridges was constructed in the 1950s as part of the overall reconstruction and expansion of the European post–World War II infrastructure, particularly in Germany and France. As many manufacturing facilities had been damaged during the war, the need for reconstruction created a demand for large highway and railway bridges that could be constructed with minimal factory-manufactured items, such as steel members. The concepts and theory of prestressed concrete segmental bridges were forcibly advanced under these stringent conditions. In 1975, the PCI Committee on Segmental Construction officially defined the segmental construction as a method of construction in which primary load carrying members are composed of individual segments post-tensioned together[1-1].

The first precast concrete segmental bridge was the Luzancy Bridge[1-3] (see Fig. 1-2) over the Marne River in Paris. This bridge was designed by Eugène Freyssinet and was the first bridge application of post-tensioning using high-strength steel. The bridge construction was completed in 1948. The early contributions of Eugène Freyssinet and other pioneers' have led to the development of segment bridge theory and construction practices as we know them today.

The Balduinstein Bridge (see Fig. 1-3) built in 1951[1-4] was designed by Ulrich Finsterwalder and is the first cast-in-place concrete segmental bridge. The bridge was constructed using balanced cantilever construction and a metal form traveler. The form traveler was supported at the end of each successive segment and was used to match cast adjacent segments as construction of the cantilever progressed. Since then, balanced cantilever construction has been widely adopted and has spread quickly throughout the world.

One of the most important innovations in segmental bridge construction is match-casting. Match-casting was first used in a single-span county bridge near Shelton, New York. The precast concrete segmental bridge was built in 1952 and designed by Jean Muller from the Freyssinet Company. The bridge girders were divided into three segments that were cast end to end. The center segment with

FIGURE 1-1 Zhao-Zhou Bridge[1-2], completed in 615 A.D., China (Courtesy of Prof. Bao-Chun Chen).

FIGURE 1-2 Luzancy Bridge, designed by Eugène Freyssinet, 1948.

FIGURE 1-3 Balduinstein Bridge, completed in 1951.

FIGURE 1-4   Choisy-le-Roi Bridge, Paris, completed in 1964.

shear keys on its ends was cast first, and the end segments were cast directly against it. After curing, the segments were transported to the job site and reassembled in the same position in which they were cast in the precasting yard. The units were then post-tensioned with cold joints[1-3]. Although this was a small bridge, it was the first bridge to successfully use match-casting, in which the previous precast segment is used as a template for the adjacent segment. This method reduces imperfections and allows for better fit-up between units at the construction site. This technique has been refined and has become a commonly used method of construction.

The first major precast concrete segmental bridge is the Choisy-le-Roi Bridge designed by Jean Muller (see Fig. 1-4) over the Seine River in France. This bridge was built in 1962 and was opened to traffic in 1964[1-5]. It was the first bridge to use a thin epoxy joint between segmental units. The epoxy is applied to the joints between units and provides a watertight seal that increases the durability and service life of the segmental bridge. The epoxy also serves as a lubricant when placing segments and as a means of achieving perfect alignment between units.

With the increasing need for longer bridge spans, minimizing traffic disruption, eliminating expensive false work, and reducing environmental impacts, segmental bridge construction has gained rapid acceptance and widespread use throughout the world. By 1952, the maximum span length for concrete segmental girder bridges was 113 m (371 ft). This span length was used in the Worms Bridge (see Fig. 1-5), which crosses the Rhine River, in Germany[1-5]. This bridge was constructed using the balance cantilever method and comprises three spans of 100, 113, and 104 meters (m) [(330, 371, and 340 feet (ft)]. In 1965, the Bendorf Bridge (see Fig. 1-6)[1-5], designed by Ulrich Finsterwalder, was constructed over the Rhine at Koblenz, Germany, with a world record span length of 202 m (673 ft). Ulrich Finsterwalder designed the Bendorf Bridge.

FIGURE 1-5   Worms Bridge, Germany.

FIGURE 1-6   Bendorf Bridge, Germany, completed in 1965.

In 1978, the Koror-Babelthuap Bridge (see Fig. 1-7) was constructed with a maximum span length of 241 m (790.68 ft)[1-6, 1-7]. The bridge was designed by Dyckerhoff & Widmann AG and Alfred A. Yee and Associates and had span lengths of 72.20 m–241 m–72.20 m (236.88 ft–790.68 ft–236.88 ft), with girder heights varying from 3.65 to 14.00 m (11.98 to 45.93 ft). Unfortunately, the bridge collapsed in 1996 as a result of excessive deflection[1-6]. Another milestone bridge is the Gateway Bridge[1-8, 1-50] (see Fig. 1-8) over the Brisbane River, Brisbane, Australia. The bridge was built in 1986 and has a total bridge length of 1627 m (5338 ft) with a maximum span length of 260 m (850 ft).

By using lightweight concrete, the maximum span length of concrete segmental girder bridges had reached 301 m (987.53 ft) in 1998. The Stolma Bridge[1-50] (see Fig. 1-9), which crosses the Stolmasundet Strait in Norway, has spans of 94 m–301 m–72 m (308.40 ft–987.53 ft–308.40 ft). Lightweight concrete was used in the middle portion of the main span to improve balance between the main and side spans. Currently, the longest segmental concrete girder bridge is the Shibanpo Yangtze River Bridge (see Fig. 1-10) over the Yangtze River in Chogqing, China. T.Y. Lin International designed this bridge, and construction was completed in 2006. The bridge's main

FIGURE 1-7   Koror-Babelthuap Bridge[1-6], completed in 1978 (Courtesy of Dr. Man-Chung Tang).

FIGURE 1-8 Gateway Bridge, Brisbane, Australia, completed in 1986.

FIGURE 1-9 Stolma Bridge, Hordaland, Norway, completed in 1998.

FIGURE 1-10 Shibanpo Yangtze River Bridge, completed in 2006 (Courtesy of Dr. Man-Chung Tang).

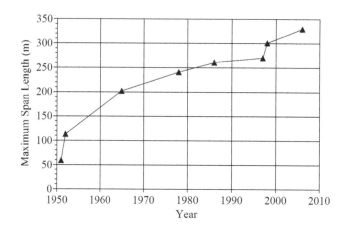

FIGURE 1-11   Development of the World's Longest Concrete Segmental Girder Bridges.

span is 330 m (1080 ft) in length with the central 108 m (350.65 ft) being a steel box girder[1-9]. The progression of the world record spans for concrete segmental girder bridges is shown Fig. 1-11.

Concrete segmental girder bridges are not only widely used for long-span bridges but are also used extensively for elevated viaducts in urban settings. It has been demonstrated on numerous projects that precast concrete segmental bridges can not only significantly reduce construction time and traffic disruptions, but also can provide the most cost-effective bridge solution. One such project is the Lee Roy Selmon Expressway Bridge (Fig. 1-12), Tampa, Florida, which was completed in 2006. The total length of the bridge is 8000 m [5 miles (mi)] with typical span lengths of 42.7 m (140 ft). The cost of the bridge was $700/m^2 ($65/ft^2), which was approximately 30% less than other proposed bridge alternatives[1-10].

Another advantage of the concrete segmental bridges is that with minimal changes in the precasting operation, complex horizontal and vertical alignments, as well as super elevations, can be achieved (see Fig. 1-13).

Segmental construction is not only applied to girder-type bridges but is also widely applied to a variety of complex bridge structures, such as cable-stayed, arch, and rigid frame bridges. Fig. 1-14 shows Skarnsund Bridge across the Trondheim Fjord, north of Trondheim in Norway. It is a concrete girder cable-stayed bridge built in 1991, with a world record span length of 530 m (1739 ft)[1-11].

Figure 1-15 shows a picture for the Hoover Dam Arch Bridge (Mike O'Callaghan -Pat Tillman Memorial Bridge), which has a span length of 323 m (1060 ft) and a vertical clearance of 270 m (900 ft). This bridge consists of concrete arches and steel-concrete composite deck system. It was completed in 2010.

FIGURE 1-12   Lee Roy Selmon Expressway Bridge, Tampa, Florida, completed in 2006.

FIGURE 1-13 I-4 and the Lee Roy Selman Expressway Bridges, Tampa, Florida, completed in 2012.

FIGURE 1-14 Skarnsund Bridge, Norway, completed in 1991 (Used with permission of IABSE[1-11]).

FIGURE 1-15 Hoover Dam Arch Bridge (Mike O'Callaghan-Pat Tillman Memorial Bridge), completed in 2010.

FIGURE 1-16   Pont du Bonhomme Bridge[1-2] in Brittany, France (CRC).

Figure 1-16 shows a picture of the Pont du Bonhomme Bridge, located in Brittany (Bretagne), France. The bridge has span lengths of 67.95 m (223 ft)–146.70 m (418 ft)–67.95 m (223 ft) and was built in 1974. This is a rigid framed bridge with inclined lags.

In the United States, the segmental construction method for prestressed concrete bridges dates back to the early 1950s[1-7]. However, the first segmental box girder bridge was not built until 1973 in Texas. Pine Valley Creek Bridge designed by Dr. Man-Chung Tang, located in San Diego, California, was the first bridge in the United States to be built using the cast-in-place segmental balanced cantilever method (see Fig. 1-17). The bridge has a middle span length of 137.2 m (450 ft)

FIGURE 1-17   Pine Valley Creek Bridge, completed in 1974 (Courtesy of Dr. Man-Chung Tang).

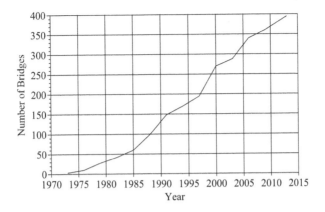

FIGURE 1-18   Segmental Bridges in the United States.

and is about 137.2 m (450 ft) above the valley floor, which made it the highest bridge in the United States at that time. This bridge was completed in 1974. During more recent decades, because of a prevailing desire for expedited construction with minimal traffic disruption, lower life cycle costs, appealing aesthetics, and the need for super-elevated to curved roadway alignment, segmental concrete bridges have developed rapidly and become a primary choice for major transportation projects throughout the United States (see Fig. 1-18).

1.2 MATERIALS

1.2.1 INTRODUCTION

A segmental bridge is typically made of concrete, reinforcing steel, or fiber-reinforced polymer (FRP) bars, high-strength steel, and/or FRP tendons. Though it may not be necessary for bridge engineers to know the detailed molecular constituents of the materials used in segmental bridges, it is important that they understand the mechanical properties of the material used in bridge design. As the reader may already be familiar with the basic properties of the materials, the following discussion focuses on the mechanical properties and behaviors that are directly related to the segmental bridge design.

1.2.2 CONCRETE

1.2.2.1 Introduction

Concrete is a major constituent of all concrete segmental elements and is a composite material normally consisting of aggregates, sand, and gravel bound together by hydrated portland cement, and other chemical admixtures added to improve the durability and overall quality of the concrete. Its mechanical properties may be classified into two categories, short-term properties and long-term properties. The important short-term properties are compression strength, tensile strength, shear strength, and modulus of elasticity. The important long-term properties are creep and shrinkage.

1.2.2.2 Compression Strength

The concrete compression strength, normally termed as concrete strength, refers to the uniaxial compressive strength of a standard test cylinder [6-inch (in.) diameter by 12-in. height] according to ASTM Standards C31 and C39. Concrete gains strength with age. Under normal curing conditions, the strength gain with age can be assumed to vary as given in Table 1-1 proposed by ACI Committee 209[1-12, 1-18, 1-22]. The compression strength f'_c usually means the strength at 28 days of

TABLE 1-1

Strength of Concrete as a Fraction of the 28-Day Strength

Compressive Strength	Age of Concrete (days)				
	3	7	28	60	90
Normal cement (Type I)	0.46	0.70	1.00	1.08	1.12
High early cement (Type III)	0.59	0.80	1.00	1.04	1.06

age. CEB-FIP model code[1-13] provides an approximate relationship between the concrete compressive strength and the age of concrete as follows[1-14]:

$$f'_{ci}(t) = \beta_{cc}(t) f'_c \tag{1-1}$$

where

$$\beta_{cc}(t) = e^{s\left(1 - \sqrt{\frac{28}{t/t_1}}\right)} \tag{1-2}$$

$f'_{ci}(t)$ = concrete compressive strength at time t (day)
f'_c = 28-day concrete compressive strength
$\beta_{cc}(t)$ = ratio of concrete strength at time t to 28-day concrete strength (see Fig. 1-19)
t = age of concrete (days)
t_1 = 1 day
s = cement rate of hardening coefficient
   = 0.20 for rapid-hardening, high-strength concretes
   = 0.25 for normal and rapid-hardening cement
   = 0.38 for slow-hardening cements

For concrete segmental bridges, high compressive strength is necessary to reduce the effect of concrete creep and shrinkage as well as to reduce the cross-sectional area of the concrete section. Depending on the types of admixtures, aggregates, and the time and quality of curing, the compression strength of concrete can be obtained up to 20,000 pounds per square inch (psi) or more.

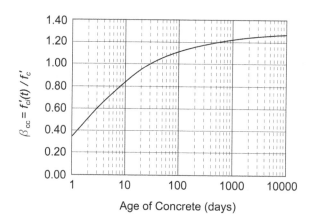

FIGURE 1-19   Variation of Concrete Strengths with Time for Normal and Rapid-harding Cement[1-14].

Currently, the concrete with a compression strength ranging from 5500 to 8500 psi is widely used in segmental bridges.

1.2.2.3 Tensile Strength

Concrete has relatively low tensile strength varying from 10% to 20% of the compression strength. As it is difficult to grip the test sample with test machines, a number of methods are available for testing the tensile strength. The most commonly used method is the cylinder split tensile strength method in accordance with AASHTO T 198 (ASTM C496). For normal-weight concrete with specified compressive strengths up to 10 kilopound per square inch (ksi), the direct tensile strength may be estimated as $0.23\sqrt{f_c'}$ (AASHTO Article 5.4.2.7)[1-15]. For members subjected to bending, the value of the modulus of rupture f_r is used in bridge design, instead of the direct tension strength. The modulus of rupture is determined by testing a plain concrete beam of 6 in. × 6 in. × 30 in. long and loaded at its third point until it fails due to cracking on its tension face (ASTM C78).

The modulus of rupture is sensitive to curing methods. Most modulus of rupture test data on normal-weight concrete is between $0.24\sqrt{f_c'}$ and $0.37\sqrt{f_c'}$ (ksi)[1-16 to 1-19]. AASHTO LRFD Specifications Article 5.4.2.6 recommends the concrete modulus of ruptures f_r for normal-weight concrete for specified concrete strengths up to 15.0 ksi may be taken as

- $f_r = 0.24\sqrt{f_c'}$
- $f_r = 0.20\sqrt{f_c'}$, for calculating the cracking moment of a member

1.2.2.4 Stress-Strain Curve of Concrete and Modulus of Elasticity

It is essential for the reader to keep in mind the stress-strain relationship of concrete in developing all the analysis and design terms and procedures in concrete structures. Figure 1-20 shows some typical stress-strain curves of concrete with different concrete strength[1-20]. From this figure, we can observe (1) the curve is nonlinear, but the gradient of the approximate straight-line portion of the plot increases with concrete compression strength and (2) the lower the concrete strength, the higher the failure strain. Because of these behaviors of the concrete, the

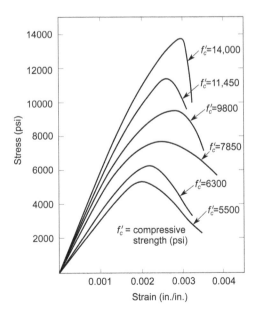

FIGURE 1-20 Typical Stress-Strain Curves of Concrete.

modulus of elasticity is defined as the slope of the straight line that connects the origin to the point corresponding to a given stress of about $0.4f_c'$. For more accurate structural analysis, the tangent modulus is normally used in the nonlinear analysis after the effect of stress-strain curvature becomes significant.

There are many factors that can affect the concrete modulus of elasticity. Test results show that the most important factors are the strength and density of concrete. AASHTO LRFD Specifications Article 5.4.2.4 recommends that the 28-day modulus of elasticity E_c may be taken as:

$$E_c = 33,000 \ K_1 w_c^{1.5} \sqrt{f_c'} \tag{1-3}$$

where
  K_1 = correction factor for source of aggregate to be taken as 1.0, if no physical test results available
  w_c = unit weight of concrete [kilo cubic feet (kcf)]
  f_c' = specified compressive strength of concrete (ksi)

Equation 1-3 is suitable for concrete with unit weights between 0.090 and 0.155 kcf and specified compressive strengths up to 15.0 ksi.

The concrete modulus of elasticity varies with time. CEB-FIP Code[1-13] has provided the following approximate equation describing the variation:

$$E_{ci}(t) = \beta_E(t) E_c \tag{1-4}$$

where
  $\beta_E(t)$ = $\sqrt{\beta_{cc}(t)}$ $\qquad\qquad\qquad\qquad\qquad\qquad\qquad\qquad\qquad$ (1-5)
       = ratio of modulus of elasticity at time t to 28 days modulus (see Fig. 1-21)
  E_c = modulus of elasticity of concrete at an age of 28 days
  $E_{ci}(t)$ = modulus of elasticity at time t (days)
  $\beta_{cc}(t)$ is defined by Eq. 1-2

Under an axial loading, concrete will not only undergo axial deformation as discussed above, but also will experience lateral deformation. The lateral strain is calculated by Poisson's ratio, which is

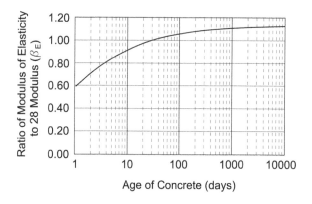

FIGURE 1-21   Variation of Concrete Modulus of Elasticity with Time[1-14].

defined as the ratio of transverse contraction strain to longitudinal extension strain in the direction of applied force. The Poisson's ratio of concrete varies from 0.15 to 0.22. AASHTO specifications recommend the following value can be used if there are no physical tests:

Poisson's ratio: $\mu = 0.2$

1.2.2.5 Creep

1.2.2.5.1 Behaviors of Concrete Creep and Effective Modulus of Elasticity

When concrete is subjected to a sustained load, it not only undergoes an immediate deformation, but its deformation will continue for a long period. The immediate strain is defined as elastic strain, and the strain after due to the sustained loading is called creep strain. Test results on specimens with diameters from 4 to 10 in. under the condition of 50% relative humidity show that the total amount of creep strain at 20 years ranged from 0.5 to 5 times with an average of 3 times the initial strain[1-19]. Concrete creep behaviors are related to many factors, such as the relative humidity for the concrete curing, temperature, water-cement ratio, age at the time of the concrete loading, and type and size of sand aggregates. Concrete continues to creep over its entire service life, but the rate of change becomes very small at the later age. Of the total creep in 20 years, the average values of 25%, 55%, and 75% will occur in the first 2 weeks, 3 months, and 1 year, respectively[1-19, 1-20]. Test results show that the creep can be treated as linear variation with stress when the concrete stress is less than $0.4 f_c'$. A sketch of strain-time curve is shown in Fig. 1-22. Based on the definition of modulus of elasticity, we may define the effective modulus of elasticity as:

$$E_c' = \frac{stress}{elastic\ strain + creep\ strain} \tag{1-6}$$

1.2.2.5.2 Creep Coefficient

1.2.2.5.2.1 General Definition   The behavior of creep is typically described in terms of a creep coefficient

$$\phi(t, t_0) = \frac{\varepsilon_c(t)}{\varepsilon_e(t_0)} \tag{1-7}$$

where
  $\varepsilon_c(t)$ = creep strain at t days after casting
  $\varepsilon_e(t_0)$ = elastic strain caused when load is applied at t_0 days after casting

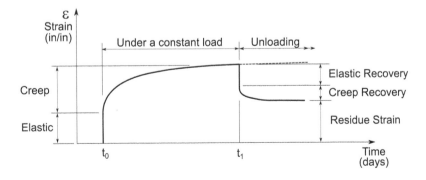

FIGURE 1-22   Variation of Strain with Time[1-14].

Currently, there are two methods for calculating the elastic strain:

- ACI 209[1-12]:

$$\varepsilon_e(t_0) = \frac{\sigma(t_0)}{E(t_0)} \tag{1-8a}$$

where
$E(t_0)$ = modulus of elasticity when load is applied at t_0 days
$\sigma(t_0)$ = stress when load is applied at t_0 days

- CEB-FIP 1990[1-13]:

$$\varepsilon_e(t_0) = \frac{\sigma(t_0)}{E_{28}} \tag{1-8b}$$

where E_{28} = modulus of elasticity at 28 days.

1.2.2.5.2.2   Approximate Methods for Determining Creep Coefficients
1.2.2.5.2.2.1   ACI 209 Method

It is difficult to accurately determine concrete creep coefficient. Several approximate methods have been developed based on numerous tests. AASHTO specifications recommend using the following equation to estimate the creep coefficient for specified concrete strengths up to 15.0 ksi, based on ACI Committee 209:

$$\phi(t, t_0) = 1.9 k_s k_{hc} k_f k_{td} t_0^{-0.118} \tag{1-9a}$$

where
k_s = volume-to-surface ratio factor

$$k_s = 1.45 - 0.13(V/S) \geq 1.0 \tag{1-10}$$

where
V/S = volume-to-surface ratio (in.)
k_{hc} = humidity factor for creep

$$k_{hc} = 1.56 - 0.008H \tag{1-11}$$

where
H = relative humidity (%), may be taken from Fig. 1-23
k_f = strength factor for the effect of concrete strength

$$k_f = \frac{5}{1 + f'_{ci}} \tag{1-12}$$

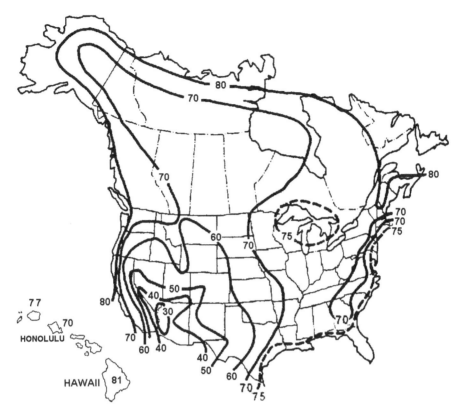

FIGURE 1-23   Annual Average Ambient Relative Humidity (%)[1-15] (Used with permission by AASHTO).

where

f'_{ci} = compressive strength of concrete at time of prestressing for pretensioned members or at time of initial loading for non-prestressed members; f'_{ci} may be taken as 0.80 f'_c (ksi) in design stage

k_{td} = time development factor

$$k_{td} = \frac{t}{61 - 4f'_{ci} + t} \tag{1-13}$$

where

t = maturity of concrete (day), defined as age of concrete between time of loading for creep calculations or end of curing for shrinkage calculations and time being considered for analysis of creep or shrinkage effects

t_0 = age of concrete at time of load application (day)

1.2.2.5.2.2.2   CEB-FIP Code[1-13]

AASHTO specifications also recommend the method for estimating the concrete creep coefficient contained in CEB-FIP Code for the design of the segmental bridges. The concrete creep coefficient equation specified in CEB-FIP Code is:

$$\phi(t, t_0) = \phi_0 \beta_c(t - t_0) \tag{1-9b}$$

where

ϕ_0 = notional creep coefficient
β_c = coefficient describing the development of creep with time after loading
t = age of concrete (days) at the time considered
t_0 = loading age of concrete (days), may be adjusted according to CEB-FIP[1-13], equation 2.1-72

$$\phi_0 = \phi_{RH}\beta(f_c')\beta(t_0)$$ (1-14a)

$$\phi_{RH} = 1 + \frac{1-H}{0.46\left(\frac{h}{100}\right)^{1/3}}$$ (1-14b)

$$\beta(f_c') = \frac{5.3}{(0.1f_c')^{0.5}}$$ (1-14c)

$$\beta(t_0) = \frac{1}{0.1 + t_0^{0.2}}$$ (1-14d)

h = notational size of member [millimeter (mm)] = $2A_c / u$ (mm)
f_c' = 28 days concrete strength [megapascal (MPa)]
H = relative humidity (%)
A_c = area of cross section (mm²)
u = perimeter of member in contact with the atmosphere (mm)

$$\beta_c(t-t_0) = \left[\frac{t-t_0}{\beta_H + (t-t_0)}\right]^{0.3}$$ (1-14e)

$$\beta_H = 1.5h\left(1 + (1.2H)^{18}\right) + 250 \leq 1500$$ (1-14f)

Equation 1-9b is valid for normal concrete strength ranging from 12 MPa (1.74 ksi) to 80 MPa (11.60 ksi) subjected to a compression strength less than $0.4 f_{ci}$ at an age of loading t_i and exposed to mean relative humidity in range of 40% to 100% at temperature from 5° to 30°C.

1.2.2.5.2.2.3 Typical Variations of Creep Coefficients in Concrete Segmental Bridges

In the United States, the creep coefficients for concrete segmental bridges are typically determined using CEB-FIP Code. Some variations of the creep coefficients for typical cast-in-place and precast segmental bridges, calculated using Eq. 1-9b, are shown in Fig. 1-24. The variations shown in Fig. 1-24 are calculated assuming a relative humidity of 70% as well as the loading ages of 3 days and 28 days for cast-in-place and precast segmental bridges, respectively. From this figure, we can see that using higher concrete strength can reduce the creep coefficients and that increasing the loading age of concrete can significantly reduce the creep coefficient.

1.2.2.6 Shrinkage

1.2.2.6.1 General

Shrinkage in concrete is its contraction due to drying and chemical change that is dependent on time and on moisture conditions, not on stresses. The total shrinkage comprises two portions of plastic shrinkage and drying shrinkage. Plastic shrinkage occurs during the first few hours after placing

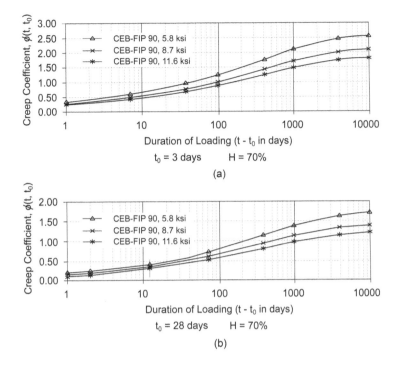

FIGURE 1-24 Variations of Creep Coefficients with Loading Time (Days), (a) Typical Cast-in-Place Segmental, (b) Typical Precast Segmental.

fresh concrete in the forms. Drying shrinkage occurs after the concrete has already obtained its final set and a large portion of the chemical hydration process in the cement gel has been completed.

Like creep, many factors affect the magnitude of shrinkage, such as the relative humidity of the medium, aggregate, water-to-cement ratio, size of the concrete element, and amount of reinforcement. Shrinkage strain can reach 0.0008 or more. For concrete mixtures used typically in segmental bridges, average values for shrinkage range from about 0.0002 to 0.0006. If the concrete is left dry, most of the shrinkage will take place during the first 2 or 3 months. Subject to 50% humidity and temperature of 70°F, the rate of occurrence of shrinkage is comparable to that of creep, and the magnitude of shrinkage is often similar to that of creep produced by a sustained stress of about 600 psi [4.14 [newtons/millimeter squared (N/mm²)]. There is no theoretical method for determining the concrete shrinkage strains. However, two approximate methods have been developed based on numerous experimental researches. The two methods are recommended by AASHTO LRFD specifications and are presented in Section 1.2.2.6.2.

1.2.2.6.2   Estimation of Concrete Shrinkage

1.2.2.6.2.1   AASHTO LRFD Method   For concrete elements that are devoid of shrinkage-prone aggregates, the AASHTO specifications recommend that the shrinkage strain ε_{sh} at time t can be determined as:

$$\varepsilon_{sh} = k_s k_{hs} k_f k_{td} 0.48 \times 10^{-3} \tag{1-15a}$$

where
k_{hs} = humidity factor for shrinkage = $(2.00 - 0.014H)$
H = relative humidity (%)
k_s, k_f, and k_{td} are defined in Eqs. 1-10, 1-12, and 1-13, respectively

Equation 1-15a is applicable for concrete strengths up to 15.0 ksi. The AASHTO specifications also allow that in the absence of more accurate data, the shrinkage coefficients may be assumed to be 0.0002 after 28 days and 0.0005 after 1 year of drying.

1.2.2.6.2.2 CEB-FIP Code Method The equation for determining the shrinkage strain adopted by CEB-FIP Code[1-13], allowed by AASHTO LRFD, is as follows:

$$\varepsilon_{cs}(t,t_s) = \varepsilon_{cs0}\beta_s(t-t_s)$$ (1-15b)

in which $\beta_s(t-t_s)$ is the development coefficient of shrinkage:

$$\beta_s(t-t_s) = \left[\frac{t-t_s}{350(h/100)^2 + t - t_s}\right]^{0.5}$$ (1-16)

where
  h = same as shown in Eq. 1-9b
  t_s = age of concrete (days) at the beginning of shrinkage
  t = age of concrete (days)
  ε_{cs0} = notational shrinkage coefficient

$$\varepsilon_{cs0} = \varepsilon_s\beta_{RH}$$ (1-17)

$$\varepsilon_s = \left[16 + \beta_{sc}(9 - 0.1f'_c)\right] \times 10^{-5}$$ (1-18)

where
  f'_c = 28 days concrete strength (MPa)

$$\beta_{RH} = -1.55\left(1 - H^3\right) \text{ for } 40\% \le H < 99\%$$ (1-19a)

$$\beta_{RH} = 0.25 \text{ for } H \ge 90\%$$ (1-19b)

  β_{sc} = 4, for slowly hardening cements
       = 5, for normal or rapid hardening cements
       = 8, for rapid hardening high-strength cements

Assuming relative humidity of 75% and nominal thickness of 319 mm, the variation of the development coefficient with time can be developed as shown in Fig. 1-25.

1.2.2.7 Thermal Coefficient of Expansion

The concrete thermal coefficient is defined as the change in strain resulting from a one degree change in temperature and should be used based on the test results in the design. For normal concrete, the thermal coefficient ranges from 3.0 to 8.0 × 10.0^{-6}/°F. For lightweight concrete, it varies between 4.0 to 6.0 × 10^{-6}/°F. AASHTO specifications recommend the following value be used for normal-weight concrete: 6.0 × 10^{-6}/°F.

1.2.2.8 Lightweight Concrete

Lightweight concrete has been gaining more applications in segmental bridge construction[1-21], especially for long-span concrete segmental bridges. The unit weight of normal concrete ranges

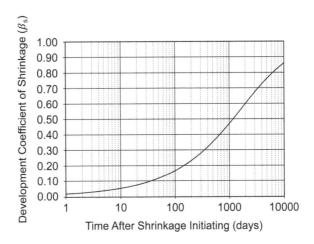

FIGURE 1-25 Variation of Development Coefficient of Shrinkage with Time[1-14].

from 135 to 155 lb/ft³. AASHTO specifications define lightweight concrete as the concrete containing lightweight aggregate and having an air-dry unit weight not exceeding 0.120 kcf. Normally, the density of the lightweight concrete is between 90 to 120 lb/ft³. Lightweight concrete without natural sand is termed "all-lightweight concrete" and lightweight concrete with normal weight sand is termed as "sand lightweight concrete."

Though the mechanical behaviors for lightweight and normal concrete are different, they are generally comparable. For lightweight concrete, the AASHTO specifications recommend

Modulus of rupture:

- For sand: lightweight concrete $= 0.20\sqrt{f_c'}$
- For all: lightweight concrete $= 0.17\sqrt{f_c'}$

Thermal coefficient: $5.0 \times 10^{-6}/°F$

1.2.3 STEEL

1.2.3.1 Reinforcing Steel

Most often, used reinforcement steel in concrete segmental bridge is ASTM A615-85 hot-rolled deformed bars Grade 60, which are available in the 11 sizes listed in Table 1-2. Figure 1-26 shows

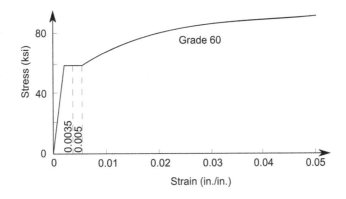

FIGURE 1-26 Typical Stress-Strain Curve for Grade 60 Reinforcing Bars.

TABLE 1-2
Area and Weights of Reinforcing Bars

Bar Designation Number	Nominal Dimensions		
	Diameter [in. (mm)]	Area (in.²)	Unit Weight (lb/ft)
3	0.375 (9)	0.11	0.376
4	0.500 (13)	0.20	0.668
5	0.625 (16)	0.31	1.043
6	0.750 (19)	0.44	1.502
7	0.875 (22)	0.60	2.044
8	1.000 (25)	0.79	2.670
9	1.128 (28)	1.00	3.400
10	1.270 (31)	1.27	4.303
11	1.410 (33)	1.56	5.313
14	1.693 (43)	2.25	7.650
18	2.257 (56)	4.00	13.600

a typical stress-strain curve for Grade 60 reinforcing bars. ASTM specifications define the yield strength of reinforcing bars as the stress at a strain of 0.005. AASHTO and ACI specifications define the yield strength as the stress at a strain of 0.0035, because the concrete crushes at a strain of about 0.0035 and a strain of 0.005 may never be reached in compression members[1-18, 1-22].

AASHTO Article 5.4.3.2 specifies that for steel reinforcement having a minimum yield strength up to 100 ksi, the modulus of elasticity can be taken as

$$E_s = 29,000 \text{ ksi}$$

1.2.3.2 Prestessing Steel

Very high strength steels have to be used in segmental bridges to counterbalance the high creep and shrinkage losses in concrete. There are typically three types of prestessing steels used in post-tension systems: stress-relieved or low-relaxation wires, stress-relieved and low-relaxation strands, and high-strength steel bars. However, to reduce the number of units in the post-tensioning operation, the strands and bars are commonly used in current segmental bridges construction.

1.2.3.2.1 Stress-Relieved and Low-Relaxation Steel Wires and Strands

Stress-relieved wires are processed by cold-drawing high-strength steel bars through a series of dyes conforming to ASTM Standards A421. Strands in the United States are exclusively made of seven wires (see Fig. 1-27) and confirm to AASHTO M 203 (ASTM A416). The seven-wire strands have a straight center wire that is slightly larger than its outside six wires, which are twisted around it tightly on a pitch of 12 to 16 times the normal diameter of the strand. There are two grades available, 250 and 270, where the grade represents the minimum guaranteed breaking stress. The principle properties of the strains are given in Table 1-3.

The typical stress-strain curves of the strands are shown in Fig. 1-28. From this figure, we can see that there is no clear yield point. Various arbitrary methods have been proposed for defining the yield point. The most commonly accepted method for strains is the 1.0% strain criteria (see Fig. 1-28). AASHTO Article 5.4.4 recommends that the yield strengths for the strains be taken as 90% of the tensile strength for low-relaxation strands and 85% for other stress-relieved strands. The modulus of elasticity for the strand may be taken as:

$$E_p = 28,500 \text{ ksi}$$

TABLE 1-3
Properties of Standard Seven-Wire Stress-Relieved Strands

Grade	Nominal Diameter [in. (mm)]	Nominal Area (in.²)	Nominal Weight (lb/1000 ft)	Tensile Strength, f_{pu} (ksi)	Yield Strength, f_{py} (ksi)
250	0.25 (6.35)	0.036	122	250	85% of f_{pu}, except
	0.313 (7.94)	0.058	197		90% of f_{pu} for
	0.375 (9.53)	0.080	272		low-relaxation
	0.438 (11.13)	0.108	367		strand
	0.500 (12.70)	0.144	490		
	0.600 (15.24)	0.216	737		
270	0.375 (9.53)	0.085	290	270	
	0.438 (11.13)	0.115	390		
	0.500 (12.70)	0.153	520		
	0.600 (15.24)	0.217	740		

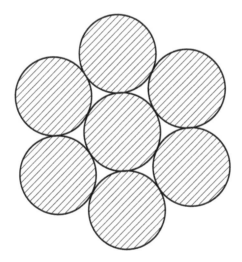

FIGURE 1-27   Cross-Sections of Standard Seven-Wire Strands.

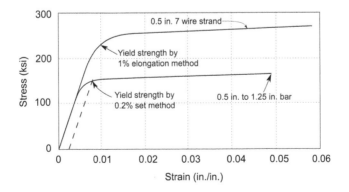

FIGURE 1-28   Typical Stress-Strain Curves for Prestressing Steel.

1.2.3.2.2 Relaxation of Prestressing Strands

The stress in a prestressing strand will decrease with time when the strand is subjected to constant strain[1-20, 1-25, 1-29]. This phenomenon is called steel relaxation and is measured by the loss of steel stress. There are several factors affecting steel relaxation. Test results show that the type of steel, time, and the ratio of the initial prestessing stress to yield stress are three of the most important factors. Based on many test results, the loss of prestressing stress due to steel relaxation over a time interval from t_i to t can be estimated by the following equation[1-15]:

$$f_{pR} = f_{pi} \left(\frac{log24t - log24t_i)}{K'_L} \right) \left(\frac{f_{pi}}{f_{py}} - 0.55 \right) \tag{1-20}$$

where

f_{pi} = stress in prestressing strands at time t_i, taken not less than $0.55f_{py}$
K'_L = 45 for low relaxation strands and 10 for stress-relieved steel
t_i = beginning of time interval (days)
t = end of time interval (days)

Using Eq. 1-20, the variation of prestressing stress with time for stress-relieved strands is illustrated in Fig. 1-29. In this figure, the coordinate is the ratio of remaining prestessing stress f_{pR} to initial prestressing stress f_{pi}, and the abscissa is the time measured in days.

1.2.3.2.3 Post-Tensioning Tendons

To effectively apply post-tensioning forces to segmental bridges and facilitate structural detailing, a number of strands are grouped together inside a plastic or metal duct and then stressed and anchored simultaneously. The set of the strands is typically called a tendon. The properties of typical tendon sizes using 0.6-in.-diameter 7-wire strands for the DYWIDAG Multistrand Post-Tensioning System are given in Table 1-4[1-28].

1.2.3.2.4 High-Strength Steel Bars

The high-strength bars can be grouped into two categories, threaded bars and plain bars. The threaded bars have continuous hot-rolled ridges along the entire bar length. The threads make it easy to couple bars cut to any required length. The plain bars are first cut to the length specified in the project and then the threads at both ends are cold-rolled in the shop according to the specifications of the project. Both plain and deformed bars confirm AASHTO M 275 (ASTM A722). The dimensions and

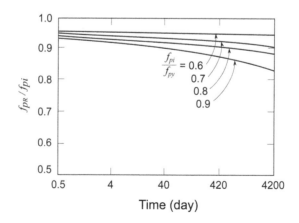

FIGURE 1-29 Variation of Prestressing Stress with Time for Stress-Relieved Strands due to Steel Relaxation[1-18].

TABLE 1-4

Properties of Typical Tendons with 0.6-in.-Diameter 7-Wire Grade 270 Strands

Tendon Size	Nominal Area		Nominal Weight		Guaranteed Ultimate Tensile Strength	
	in.²	mm²	lb/ft	kg/m	kips	kN
3-0.6″	0.651	420	2.22	3.30	175.8	782
4-0.6″	0.868	560	2.96	4.41	234.4	1043
5-0.6″	1.085	700	3.70	5.51	293.0	1303
6-0.6″	1.302	840	4.44	6.61	351.6	1564
7-0.6″	1.519	980	5.18	7.71	410.2	1825
8-0.6″	1.736	1120	5.92	8.81	468.8	2085
9-0.6″	1.953	1260	6.66	9.91	527.4	2346
12-0.6″	2.604	1680	8.88	13.21	703.2	3128
15-0.6″	3.255	2100	11.10	16.52	879.0	3910
19-0.6″	4.123	2660	14.06	20.92	1113.4	4953
27-0.6″	5.859	3780	19.98	29.73	1582.2	7038
37-0.6″	8.029	5180	27.38	40.75	2168.2	9646

properties for some typical used bars are shown in Table 1-5. Though there are several grades of high-strength bars available, Grade 150 is usually used in segmental bridge construction. A typical stress-strain curve for the high-strength bars is shown in Fig. 1-28. The yield strength of high-strength bars is most often determined by the 0.2% offset method, in which the intersection of a line parallel to the initial tangent and drawing from the 0.002 strain with the stress-strain curve is defined as the yield strength (see Fig. 1-28). The yield strengths specified by AASHTO Article 5.4.4.1 are presented in Table 1-5. The modulus of elasticity for the high-strength bars can be taken as:

$$E_p = 30,000 \text{ ksi}$$

TABLE 1-5

Dimensions and Properties of High-Strength Bars

Types	Nominal Diameter [in. (mm)]	Nominal Area (in.²)	Nominal Weight (lb/ft)	Tensile Strength, f_{pu} (ksi)	Yield Strength, f_{py} (ksi)
I (Plain)	0.750 (19)	0.44	1.50	150	85% of f_{pu}
	0.875 (22)	0.60	2.04		
	1.000 (25)	0.78	2.64		
	1.125 (29)	0.99	3.38		
	1.250 (32)	1.23	4.17		
	1.375 (35)	1.48	5.05		
II (Deformed)	0.625 (15)	0.28	0.98	150	80% of f_{pu}
	0.750 (19)	0.42	1.49		
	1.000 (25)	0.85	3.01		
	1.250 (32)	1.25	4.39		
	1.375 (35)	1.58	5.56		
	1.750 (46)	2.62	9.23		
	2.500 (65)	5.20	17.71		

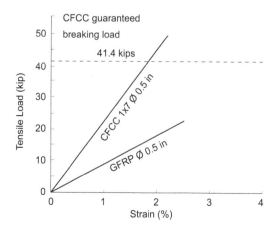

FIGURE 1-30   Typical Stress-Strain Curves for 0.5-in.-diameter FRP.

1.2.4   FRP

Fiber-reinforced polymer (FRP) composites are high-strength, noncorrosive, lightweight material[1-26, 1-27, 1-29 and 1-30]. There are three types of FRP: aramid fiber-reinforced polymer (AFRP), carbon fiber-reinforced polymer (CFRP), and glass fiber-reinforced polymer (GFRP). Figure 1-30 shows some typical stress-strain relationships for GFRP and CFRP–CFCC reinforcing bars. From this figure, it can be seen that FRP does not have a clear yield point and is linear up to failure.

Of the three types of reinforcement bars, GFRP reinforcing bars are the least expensive and the most widely used polymer in bridge structures. Some commercially available GFRP bars and their mechanical properties, which conform to ASTM D7025/ACI 440.3R, are shown in Table 1-6[1-26]. The moduli of elasticity, tensile strengths, and rupture strains shown in the table are the minimum required by AASHTO and vary with different manufacturers.

There are mainly two types of FRP prestressing tendons used in concrete structures: AFRP and CFRP. From the consideration of cost and material properties, CFRP prestressing tendon may be more suitable for bridge structures. Currently, two types of CRFP tendons are available. One is Leadline, developed by Mitsubishi Kasei Corporation of Japan. Another is carbon fiber composite cable (CFCC), developed by Tokyo Rope Mfg. Co., Ltd. and Toho Rayon Co., Ltd., both of Japan. Some typical sections of CFCC tendons are shows in Fig. 1-31, and their properties are presented in Table 1-7.

The application of FRP tendons in concrete segmental bridges is still in the research stage. Research shows that it is feasible to use FRP tendons for post-tensioned concrete segmental bridges.

TABLE 1-6
Properties of GFRP Rebars

Size Number	Nominal Diameter (in.)	Area (in.²)	Weight (lb/ft)	Minimum Tensile Modulus of Elasticity (ksi)	Minimum Guaranteed Tensile Strength (ksi)	Minimum Rupture Strain (%)
3	0.375	0.11	0.122	5700	105	1.2
4	0.500	0.20	0.200	5700	100	1.2
5	0.625	0.31	0.328	5700	95	1.2
6	0.750	0.44	0.443	5700	90	1.2
7	0.875	0.60	0.596	5700	85	1.2
8	1.000	0.79	0.761	5700	80	1.2

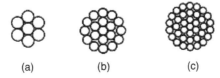

FIGURE 1-31   Typical Sections of CFCC Tendons, (a) CFCC 1 × 7, (b) CFCC 1 × 19, (c) CFCC 1 × 37.

However, the end anchorages are factory-made together with strands that should be ordered in predetermined lengths. These detailed requirements may cause abandonment of the entire cable or other construction issues. Further research is needed on the application of FRP tendons for concrete segmental bridges[1-31].

1.2.5   AASHTO STRESS LIMITS FOR CONCRETE SEGMENTAL BRIDGES

1.2.5.1   Stress Limits for Concrete

The concrete stress limits specified in the AASHTO specifications vary at different loading stages:

- Before losses
- Service limit state after losses
- Construction stage

1.2.5.1.1   Temporary Stresses before Losses

1.2.5.1.1.1   Allowable Compression Stress f_{ca}   The compressive stress is $f_{ca} \leq 0.60 f_c'$ (ksi).

1.2.5.1.1.2   Allowable Tension Stress Limits f_{ta}
 a. Longitudinal stresses through joints in the tensile zone
    - Joints with minimum bonded auxiliary reinforcement through the joints, if the maximum reinforcement tensile stress is $0.5 f_y$: $f_{ta} \leq 0.0948 \sqrt{f_c'}$ (ksi)
    - Joints without the minimum bonded auxiliary reinforcement through the joints: $f_{ta} \leq 0$
 b. Transverse stresses through joints: $f_{ta} \leq 0.0948 \sqrt{f_c'}$ (ksi)

TABLE 1-7
Properties of CFCC Tendons

			Standard Specification of CFCC			
Number of Strands	Designation	Strand Diameter (in.)	Area [mm² (in.²)]	Guaranteed Capacity [kN (kip)]	Mass (g/m)	Elastic Modulus [kN/mm² (ksi)]
1	CFCC U 5.0 ϕ	0.2	15.2 (0.02)	38 (8.5)	30	167 (24,221)
7	CFCC 1 × 7 75.0 ϕ	0.3	31.1 (0.05)	76 (17.1)	60	155 (22,481)
	CFCC 1 × 7 10.5 ϕ	0.4	57.8 (0.09)	141 (31.7)	111	155 (22,481)
	CFCC 1 × 7 12.5 ϕ	0.5	76.0 (0.12)	184 (41.4)	145	155 (22,481)
	CFCC 1 × 7 15.2 ϕ	0.6	115.6 (0.18)	270 (60.7)	221	155 (22,481)
	CFCC 1 × 7 17.2 ϕ	0.7	151.1 (0.23)	350 (78.7)	289	155 (22,481)
19	CFCC 1 × 19 20.5 ϕ	0.8	206.2 (0.32)	316 (71.0)	410	137 (19,870)
	CFCC 1 × 19 25.5 ϕ	1.0	304.7 (0.47)	467 (105.0)	606	137 (19,870)
	CFCC 1 × 19 28.5 ϕ	1.1	401.0 (0.62)	594 (133.5)	777	137 (19,870)
37	CFCC 1 × 37 35.5 ϕ	1.4	591.2 (0.92)	841 (189.1)	1185	127 (18,420)
	CFCC 1 × 37 40.0 ϕ	1.6	798.7 (1.24)	1200 (269.8)	1529	145 (21,030)

c. Stresses in other areas
 – For areas without bonded non-prestressed reinforcement: $f_{ta} \leq 0$
 – In areas with bonded reinforcement sufficient to resist the tensile force in the concrete computed assuming an uncracked section, if the maximum reinforcement tensile stress is $0.5f_y$ and is not to exceed 30 ksi: $f_{ta} \leq 0.19\sqrt{f_c'}$ (ksi)
d. Principal tensile stress at neutral axis of the web: $f_{ta} \leq 0.110\sqrt{f_c'}$ (ksi)

1.2.5.1.2   Permanent Stress Limits at Service Limit State after Losses

1.2.5.1.2.1   Allowable Compression Stress f_{ca}

• Because of the sum of effective prestress and permanent loads: $f_{ca} \leq 0.45\ f_c'$ (ksi)
• Because of the sum of effective prestress, permanent loads, and transient loads as well as during shipping and handling: $f_{ca} \leq 0.60\phi_w f_c'$ (ksi)

where $\phi_w = 1.0$ when the web and flange slenderness ratios are not greater than 15. Otherwise, it shall be calculated according to Section 3.10.2.2.1.

1.2.5.1.2.2   Allowable Tensile Stress f_{ta}

a. Longitudinal stresses through joints in the tensile zone
 – Joints with minimum bonded auxiliary reinforcement through the joints with a maximum tensile stress of $0.5f_y$: $f_{ta} \leq 0.0948\sqrt{f_c'}$ (ksi)
 – Joints without the minimum bonded auxiliary reinforcement through joints: $f_{ta} \leq 0$
b. Transverse stresses through joints: $f_{ta} \leq 0.0948\sqrt{f_c'}$ (ksi)
c. Stresses in other areas
 – For areas without bonded reinforcement: $f_{ta} \leq 0$
 – In areas with bonded reinforcement with a maximum stress of $0.5f_y$, not to exceed 30 ksi: $f_{ta} \leq 0.19\sqrt{f_c'}$ (ksi)
d. Principal tensile stress at neutral axis of the web: $f_{ta} \leq 0.110\sqrt{f_c'}$ (ksi)

1.2.5.1.3   Stress Limits for Construction Loads

1.2.5.1.3.1   Allowable Compression Stress f_{ca}   The compressive stress during construction is $f_{ca} \leq 0.5f_c'$.

1.2.5.1.3.2   Allowable Tensile Stress f_{ta}

• Flexural tension
  Excluding "other loads": $f_{ta} \leq 0.190\sqrt{f_c'}$
  Including "other loads": $f_{ta} \leq 0.220\sqrt{f_c'}$
• Principal tension
  Excluding "other loads": $f_{ta} \leq 0.110\sqrt{f_c'}$
  Including "other loads": $f_{ta} \leq 0.126\sqrt{f_c'}$

The "other loads" include the loads induced by concrete creep, shrinkage, temperature, and water. The requirements for the concrete tensile stress limits should apply to vertically post-tensioned substructures and should not be applied to construction of cast-in-place substructures supporting segmental superstructures.

1.2.5.2   Prestressing Steel

AASHTO Article 5.9.3 specifies that the applied prestressing steel stress during initial prestressing operation or at the service limit state should not exceed the values shown in Table 1-8.

TABLE 1-8

Stress Limits for Prestressing Steel[1-15]

Types of Tensioning	Condition	Stress-Relieved Strand and Plain High-Strength Bars	Low-Relaxation Strand	Deformed High-Strength Bars
Pretensioning	Immediately prior to transfer	$0.70f_{pu}$	$0.75f_{pu}$	
	At service limit state after all losses	$0.80f_{py}$	$0.80f_{py}$	$0.80f_{py}$
Post-tensioning	Prior to seating—short-term	$0.90f_{py}$	$0.90f_{py}$	$0.90f_{py}$
	At anchorages and couplers immediately after anchor set	$0.70f_{pu}$	$0.70f_{pu}$	$0.70f_{pu}$
	Elsewhere along length of member away from anchorages and couplers immediately after anchor set	$0.70f_{pu}$	$0.74f_{pu}$	$0.70f_{pu}$
	At service limit state after losses	$0.80f_{py}$	$0.80f_{py}$	$0.80f_{py}$

1.3 BASIC CONCEPT OF SEGMENTAL CONSTRUCTION

The basic idea for segmental construction is to push individual segments possessing high-compression and low-tensile capacities together by external compression forces. A typical example using this concept is in an ancient stone arch bridge construction (see Fig. 1-32a). First, individual wedge-shaped segments are made from large pieces of stone. Then the segments are rested on the falsework with an arched-shape top face. The units are perfectly matched to each other by a thin layer of lime mortar. Finally, the falsework is removed and the high reaction at the arch ends induced by the self-weight of the segments pushes the individual segments together to form an arch structure to support the anticipated loads. In modern concrete segmental bridges, the individual segments that are made of high-strength concrete are compressed using high-strength steel tendons to apply high compression forces to form a continuous structure (see Fig. 1-32b). The high-strength steel is housed in a hole formed by a duct. The main tasks for bridge engineers in designing for such bridges are to determine how to (a) design the segments, (b) design the steel tendons, and (c) make the segments interact together to effectively carry anticipated loads.

1.4 TYPICAL SEGMENTS

1.4.1 TYPICAL SECTIONS

Theoretically, individual segments can be made in both the longitudinal and transverse directions. However, in current practice, most concrete segmental bridges have an integrated segment in the transverse direction. The typical segmental length in the bridge longitudinal direction is normally determined by transportation and erection requirements. If the segments are transported by trailers over existing roads, the lengths normally range from 8 to 12 ft and the segmental weights range approximately from 60 to 90 tons. It is important to keep the length of the segments equal and keep the segments straight even for curved bridges. Generally, fewer segments will result in more efficient precasting and erection. The cross sections of segmental bridges can have many shapes depending on the required bridge widths. Box shape sections are often used in segmental bridges

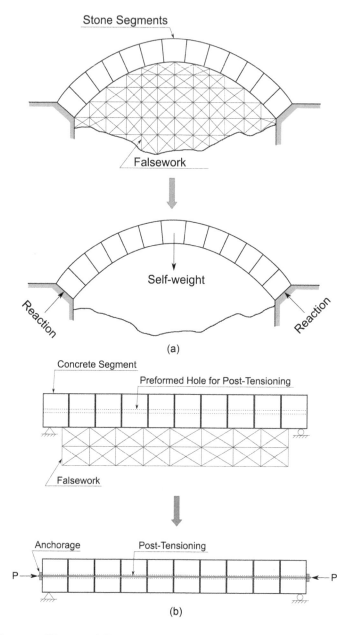

FIGURE 1-32   Concept of Segmental Construction, (a) Store Arch Bridges, (b) Concrete Beam Bridges.

due to their high torsion capacity, and one of the most used sections is the single box cross section due to its comparatively easy construction and aesthetical consideration. Some typical sections are shown in Fig. 1-33[1-32].

1.4.2   PRELIMINARY DIMENSIONS

1.4.2.1   General

A typical section of a single box girder is shown in Fig. 1-34. The dimensions in the cross section vary with different bridges and are determined based on many factors, such as the anticipated

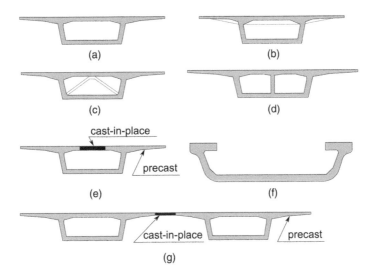

FIGURE 1-33 Typical Sections for Segmental Bridges, (a) Single Box, (b) Single Box with Ribbed Top Slab, (c) Single Box with Struts, (d) Multi-Cell Box, (e) Composite Box, (f) U-Shape, (g) Multi-Box Girder.

design loadings and construction methods. Normally, it is difficult to determine the required bridge cross section through completely theoretical analysis. Fortunately, previous design experiences provide a good reference for bridge designers to assume a preliminary cross section meeting the anticipated requirements with a few tries. Based on the current segmental bridge design and construction practice, AASHTO specifications provide some minimum dimension requirements[1-15] to ensure the structural serviceability, durability, constructability, and maintainability.

1.4.2.2 Girder Height h

From construction and maintenance points of view, the minimum girder depth is normally not less than 6 ft. From a serviceability standpoint, AASHTO Article 5.14.2.3.10d requires that the minimum girder depth should preferably not be less than that limiting the maximum bridge deflection to 1/1000 of the span length due to the design live load plus impact.

If the girder depth is within the following dimensional ranges, generally the deflection criterion specified by AASHTO specification will be satisfied:

a. Constant depth girder h_0
   $1/15 > h_0/L > 1/30$, optimum range 1/18 to 1/20

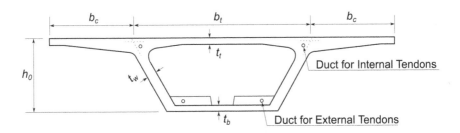

FIGURE 1-34 Dimension Designations of Typical Single Box Girder.

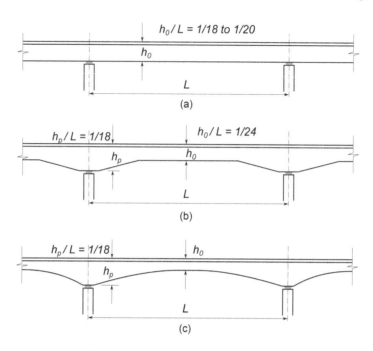

FIGURE 1-35 Recommended Girder Depths, (a) Constant Girder Depth, (b) Variable Girder Depth with Straight Haunches, (c) Variable with Circular or Parabolic Haunches.

where
h_0 = girder depth (ft)
L = span length between supports (ft) (see Fig. 1-35)
b. Constant-depth girder for incrementally launched construction
   The girder depth should preferably be between the following limits[1-15]:
   $1/15 < h_0 / L < 1/12$, for $L = 100$ ft
   $1/13.5 < h_0 / L < 1/11.5$, for $L = 200$ ft
   $1/12 < h_0 / L < 1/11$, for $L = 300$ ft
c. Variable-depth girder with straight haunches (see Fig. 1-35b)
   At pier: $1/16 > h_p / L > 1/20$, optimum: $1/18$
   At center of span: $1/22 < h_0 / L > 1/28$ optimum: $1/24$

where
h_p = girder depth over support
h_0 = girder depth at mid-span
d. Variable-depth girder with circular or parabolic haunches (see Fig. 1-35c)
   At pier: $1/16 > h_p / L > 1/20$, optimum $1/18$
   At center of span: $1/30 > h_0 / L > 1/50$

1.4.2.3   Flange Thickness t_t

The minimum top flange thickness is normally determined by considering the reinforcement layout, post-tensioning duct sizes, and required concrete cover for the steel and ducts. It is preferably 9 in. and not less than 8 in. In anchorage zones where transverse post-tensioning is used, the top flange thickness should not be less than 9 in. The minimum bottom slab thickness is normally recommended to be 7 in. AASHTO specifications also require that the top and bottom flange thickness not be less than 1/30 of the clear span between webs or haunches.

1.4.2.4   Web Thickness t_w

The web thickness is normally determined based on the following three factors:

- Enough capacity to resist vertical and torsional shear forces
- Ability for concrete to be properly placed
- Enough room for properly distributing high prestressing forces at the anchorages when tendon anchors are located in the web

AASHTO Article 5.14.2.3.10b[1-15] recommends the minimum web thickness be taken as follows:

a. $t_{w\_min} = 8.0$ in., for webs with no longitudinal or vertical post-tensioning tendons
b. $t_{w\_min} = 12.0$ in., for webs with only longitudinal (or vertical) post-tensioning tendons
c. $t_{w\_min} = 15.0$ in., for webs with both longitudinal and vertical tendons

1.4.2.5   Web Spacing b_t

Reducing the web number when selecting box cross sections can simplify construction and normally yields a more economical bridge alternative. Currently, the maximum web spacing in the United States has reached 40 ft for a single box girder without struts[1-33]. Normally, the web spacing ranges between 16 to 30 ft. The Florida Department of Transportation Structural Design Guidelines[1-35] specify the maximum clear web spacing measuring between web inside faces should be 32 ft.

1.4.2.6   Length of Top Flange Cantilever b_c

From the consideration of achieving more preferable structural transverse balance state and more uniform moment distribution along bridge transverse direction, AASHTO Article 5.14.2.3.10c[1-15] recommends that the cantilever length of the top flange, measured from the centerline of the web, should normally not exceed 0.45 times the interior span b_t of the top flange measured between the centerline of the webs.

1.4.3   Ducts

1.4.3.1   General

Ducts are used for forming continuous holes in the concrete element to accommodate future internal or external post-tensioning tendons later (see Fig. 1-34). Ducts also act as a barrier to prevent the tendon steel from corrosion. Currently, there are two types of ducts, corrugated metal duct and corrugated plastic duct. Each type typically has two different cross-sectional shapes of circular and oval flat (see Fig. 1-36). Circular ducts are normally used for placing bridge longitudinal

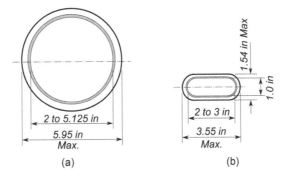

FIGURE 1-36   Shapes of Typical Ducts, (a) Circular Duct, (b) Flat Duct.

TABLE 1-9

Maximum Outside Diameters or Dimensions of Typical Ducts[1-30]

Size and Type	Corrugated Duct (without coupler/with coupler)	Smooth Wall Duct (without coupler/with coupler)
4-0.6" strands	1.54" × 3.55" (flat duct)	3.50"
7-0.6" strands	2.87"	3.50"
12-0.6" strands	3.63"	3.50"
15-0.6" strands	3.95"	4.50"
19-0.6" strands	4.57"	4.50"
27-0.6" strands	5.30"	5.563"
31-0.6" strands	5.95"	5.563"
1" diameter bar	2.87"/4.09"	3.50"/3.50"
1-1/4" diameter bar	2.87"/4.09"	3.50"/3.50"
1-3/8" diameter bar	2.87"/4.09"	3.50"/3.50"
1-3/4" diameter bar	3.63"/ 4.57"	3.50"/4.50"
1-1/2" diameter bar	3.95"/5.95"	4.50"/5.563"
3" diameter bar	4.57"/7.00"	4.50"/6.625"

post-tensioning tendons, and the flat ducts are normally used for placing bridge deck post-tensioning tendons. The internal diameters for the commercially available circular ducts typically range from 2 to 5.125 in., and the internal dimensions for flat ducts are normally 1 in. thick and 2 to 3 in. wide. More information can be found in the manufacture's specifications and Appendix B.

1.4.3.2 Duct Sizes

For the tendons placed by the pull-through method where tendons exceed 400 ft in length, AASHTO Article 5.4.6.2[1-15] calls for a minimum internal cross-sectional area of at least 2.5 times of the net area of the strand tendon. To properly lay out tendon geometries as well as check clearances and concrete covers, the designers also need to know the maximum external dimensions of the ducts, which are provided in Table 1-9[1-30].

1.4.3.3 Locations of Tendons in the Ducts

For a tendon curved in the vertical plane, the tendon will not be in the center of the duct. In bridge design, the tendon should be assumed at the top of the duct in concave upward shape or at the bottom of the duct in convex downward shape (see Fig. 1-37). The distance Z between the tendon center of gravity and the centerline of the duct can be taken as[1-15]

- Ducts with outside diameter not greater than 3 in.: $Z = 0.5$ in
- Ducts with outside diameter greater than 4 in.: $Z = 1.0$ in
- Ducts with remaining sizes: $Z = 0.75$ in

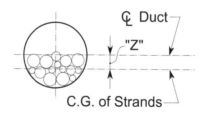

FIGURE 1-37   Location of Tendon in Duct[1-15].

TABLE 1-10

Minimum Center-to-Center Longitudinal Duct Spacing for Post-Tensioning Concrete Bridges

Superstructure Type	Minimum Center-to-Center Vertical Spacing	Minimum Center-to-Center Horizontal Spacing
Precast balanced cantilever segmental bridges	2 × outer duct diameter + 1 in., or outer segmental coupler diameter + 2 in., whichever is greater	2 × outer duct diameter + 1 in., or outer segmental coupler diameter + 2 in., whichever is greater
C.I.P. balanced cantilever segmental bridges	Outer duct diameter + 1.5 × maximum aggregate size, or outer duct diameter + 2 in., whichever is greater	Outer duct diameter + 2½ in.
Post-tensioned I-girder and U-girder bridges	Outer duct diameter + 1.5 × maximum aggregate size, or outer duct diameter + 2 in., whichever is greater (measured along the slope of webs or flanges)	Outer duct diameter + 2½ in.

1.4.3.4 Duct Spacing

To ensure the strands can properly transfer their prestressing force to the surrounding concrete, the strands should be separated sufficiently. AASHTO Article 5.10.3.3.2 requires the clear distance between straight ducts shall not be less than the greater of the duct internal diameter or 4.0 in. for post-tensioning ducts. The minimum center-to-center duct spacings required by the Florida Department of Transportation Structural Design Guidelines[1-30] are provided in Table 1-10.

1.4.3.5 Duct Radius and Tangent Length

To reduce the friction between the tendon and duct and prevent possible abrasion during pulling-through and stressing tendons, the AASHTO specifications[1-15] require that the radius of curvature of tendon ducts shall not be less than 20.0 ft for metal ducts and 30 ft for polyethylene ducts, except in the anchorage areas. In the anchorage areas, the AASHTO specifications allow the use of a minimum radius of 12.0 ft. The Florida Department of Transportation provides more detailed requirements for the minimum duct radii and tangent lengths (see Fig. 1-38) adjacent to anchorages, which are shown in Table 1-11.

1.4.3.6 Bonded and Unbounded Tendons

A post-tensioning tendon placed within the body of concrete is called an internal tendon, and its duct is called an internal duct (see Fig. 1-34). Internal ducts should be completely filled with cement

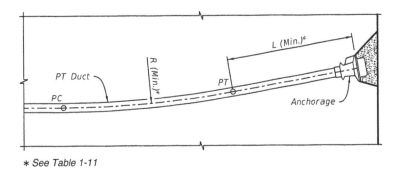

* See Table 1-11

FIGURE 1-38 Radius and Tangent Length Adjacent to Anchorages (Used with permission of FDOT[1-30]).

TABLE 1-11

Minimum Duct Radius and Tangent Lengths of Post-Tensioning Strands

Tendon Size (diameter)	Radius between Two Tangents or Points of Inflection (ft)	Radius and Tangent Length Adjacent to Anchorage (ft)	
		Radius R	Tangent Length
4-0.6"	6	9	3
7-0.6"	6	9	3
12-0.6"	8	11	3
15-0.6"	9	12	3
19-0.6"	10	13	3
27-0.6"	13	16	3.5
31-0.6"	13	16	3.5

grout or flexible filler for corrosion protection after completing post-tensioning. The cement grout within the duct bonds the tendon to the duct and the concrete. The tendon is called a bonded tendon.

A post-tensioning tendon placed outside of the body of concrete, usually inside a box girder, is called an external tendon. The geometry of an external tendon is normally controlled by deviation saddles, which are the concrete block build-out of a web, flange, or web-flange junction (see Fig. 1-34). The tendons in the duct with flexible filler and exterior tendons are called unbounded tendons. Currently, the Florida Department of Transportation requires most of the internal ducts to be filled with flexible filler due to corrosion consideration.

1.4.4   JOINTS BETWEEN SEGMENTS

1.4.4.1   Match-Cast Joint

A match-cast joint is a matching joint produced when a new segment is cast against an existing segment. It should have enough shear keys to transfer shear force from one segment to another. Figure 1-39 shows a typical shear keys configuration. The multiple shear keys in the web are provided to resist design shear forces, and the larger shear keys in the top and bottom slabs are provided to resist slab shear and also help with the alignment of the match-cast segments during construction.

To prevent post-tendons from corrosion due to severe climate conditions, AASHTO specifications require all the match-cast joints to be epoxied before post-tensioning. This type of joint is called an epoxied match-cast joint or a Type A joint. In some counties, the match-cast segments are allowed to be directly post-tensioned together without epoxy in the regions where the tendon corrosion issue due to the environmental conditions is not a concern. This type of match-cast joint is called a dry joint or Type B joint.

1.4.4.2   Cast-in-Place Closure Joint

Cast-in-place closure joints are usually used to join girder segments before post-tensioning the tendons. There are two types of closure joints: reinforced and non-reinforced. For non-reinforced closure joints, the width of a closure joint normally ranges from 4 to 10 in. The width of the closure joints located within a diaphragm shall not be less than 4.0 in.

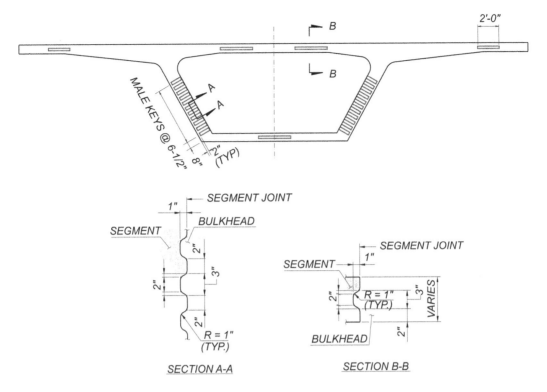

FIGURE 1-39   Typical Arrangement of Segment Shear Keys.

1.5   TYPICAL CONSTRUCTION METHODS

1.5.1   Introduction

The selection of construction methods for concrete segmental bridges is based on the bridge site condition, function, and economic considerations. The design and analytical methods are heavily dependent on the construction method selected. It is essential for the bridge designers to know how the proposed bridge should be built, though the contractors may propose another more efficient method of construction, based on site conditions and their experience. The basic procedures used in current segmental bridge construction practice are discussed in the following sections. More detailed construction methods will be discussed in related chapters.

1.5.2   Span-By-Span Construction Method

The span-by-span construction method is normally used for a relatively short span bridge with span lengths ranging from 100 to 150 ft. In the span-by-span construction, the spans are constructed sequentially. After the first span is installed, the span immediately adjacent to it is constructed, and the process repeats until the completion of the final span. Span-by-span bridges are often built using a movable erection truss under the segments in each span (see Fig. 1-40a). Overhead gantries can also be used in span-by-span construction

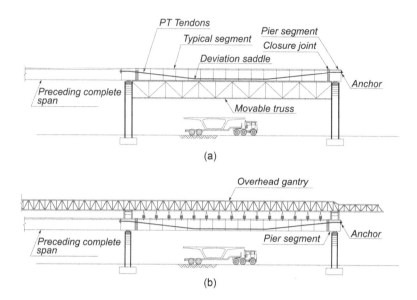

FIGURE 1-40   Span-by-Span Erection Sketch, (a) Movable under Truss, (b) Overhead Gantry.

(see Fig. 1-40b). A typical procedure for erecting one span by the movable erection truss is as follows:

Step 1: Place the erection truss.
Step 2: Erect the pier segments and stabilize with shim packs, jacks, post-tensioned (PT) bars or other means.
Step 3: Erect the typical segments between the preceding span and the pier segment.
Step 4. Apply epoxy between segments and stress together using temporary PT bars until all typical segments in the span are erected and stressed. The required force of the external PT bar may be limited to 50% of the guaranteed ultimate tensile strength (GUTS) for reuse[1-32].
Step 5. Place concrete blocks in the gap between the pier segment and the typical segment, and stress the tendons to approximately 10% GUTS and install the cast-in-place closure joint.
Step 6. After the cast-in-place closure joint reaches the anticipated strength, stress the tendons to their designed forces.
Step 7. Check alignment, grout bearings, remove shim packs, and complete the installation for the span.
Step 8. Move the erection truss to the next span.

1.5.3   BALANCED CANTILEVER SEGMENTAL CONSTRUCTION METHOD

1.5.3.1   Precast Segments

The balanced cantilever method of construction generally involves the progressive placement of segments that are cantilevered on alternate sides of a bridge pier. Post-tensioning tendons hold together a pair of segments on each side (see Fig. 1-41a). Segments can be erected by deck-mounted lifting equipment (see Fig. 1-41a), by an overhead gantry (see Fig. 1-41b), or by cranes. This type of bridge is most suitable for span lengths ranging from 150 to 450 ft[1-32]. The typical construction procedure (see Fig. 1-41a) is described as follows:

Step 1: Construct falsework in end span 1.
Step 2: Erect the end segments in the end span.

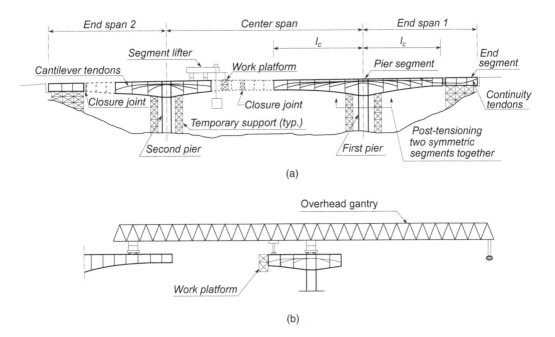

FIGURE 1-41 Precast Segmental Balanced Cantilever Construction, (a) by Lifter, (b) by Overhead Gantry.

Step 3: Erect temporary supports at the first pier location.

Step 4: Erect the pier segment.

Step 5: Place the first pair segments on each side of the pier segment.

Step 6: Erect the post-tendons and apply post-tensioning forces to connect the segments to the pier segment.

Step 7: Erect the next pair of segments on each side of the preceding erected pair of the segments until all segments for the first pier have been erected.

Step 8: Cast closure joint in end span 1.

Step 9: Stress the continuity post-tendons, and finish the erection for end span 1.

Step 10: Move the falsework to the second pier and end span 2.

Step 11: Follow steps 4 to 10, and erect all the remaining precast segments.

Step 12: Cast closure joint in the middle span.

Step 13: Install the continuity tendons in end span 2 and the center span, and complete the entire bridge erection.

1.5.3.2 Cast-in-Place Segments

The basic procedure for cast-in-place segmental bridge construction is the same as that for precast segmental bridge construction. In the construction of a cast-in-place segmental bridge, form travelers are placed at the ends of each balanced cantilever and are used to support the wet concrete as each cast-in-place segment is constructed. As bridge construction progresses, the form traveler advances and is supported by the recently constructed unit (see Fig. 1-42a). This construction method is normally used for longer span lengths ranging from 230 to 850 ft. The cast-in-place balanced cantilever construction method is also used for cable-stayed bridges and arch bridges. In cast-in-place balanced cantilever arch bridge construction, temporary towers and cables are often used for balancing the bending moment during the arch construction (see Fig. 1-42b).

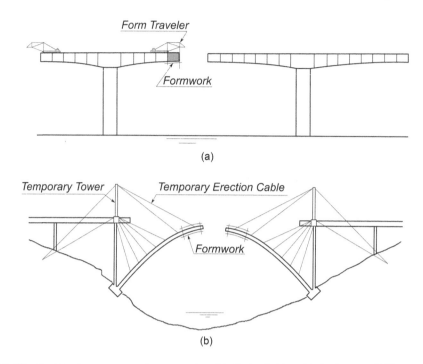

FIGURE 1-42   Cast-in-place segmental construction, (a) Girder Bridge, (b) Arch Bridge[1-32].

1.5.4   PROGRESSIVE PLACEMENT CONSTRUCTION

The bridge construction by the progressive placement method begins at one end of the structure and proceeds to the other end (see Fig. 1-43). The precast segments are erected from one end of the bridge to the other end with successive cantilevers on the same side of the related piers. To reduce the negative bending moment due to the cantilever construction, a temporary support is often used (see Fig. 1-43)[1-32]. This construction method is normally suitable for bridges with span lengths ranging from 100 to 300 ft.

1.5.5   INCREMENTALLY LAUNCHED CONSTRUCTION

The basic concept for the launching construction method is assembling individual segments together by post-tendons behind one or two abutments and then pushing the segments across to the piers using launching equipment (see Fig. 1-44). Though the segments can be precast, they

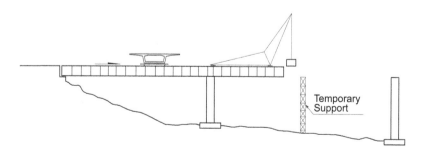

FIGURE 1-43   Progressive Placement Construction[1-32].

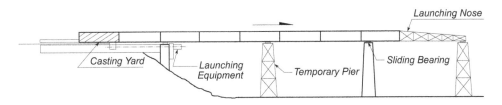

FIGURE 1-44   Incrementally Launched Construction.

are normally cast in place. The incrementally launched method is suitable for both straight and curved bridges with constant radius. With this construction method, the successive cantilever structures will become a continuous structure in the final stage. To reduce the bending moment during construction, a launching nose with lighter self-weight is attached to the first segment. For long-span bridges, temporary supports may be necessary (see Fig. 1-44). This construction method is normally economical for concrete segmental bridges with span lengths ranging from 100 to 200 ft, though a longer span is possible if temporary piers are used. The basic construction procedures are described as follows:

Step 1: Cast the first segment in place.

Step 2: After the concrete reaches a sufficient strength, push the segment forward and attach the launching nose to the lead segment.

Step 3: Cast the second segment directly against the first segment.

Step 4: After the required concrete strength of the second segment has been reached, post-tension the first and second segments together.

Step 5: Push the assembly of the segments forward, and allow casting of the succeeding segments.

Step 6: Repeat steps 1 to 5 until the first segment reaches the opposite abutment as designed.

Step 7: Install additional post-tendons to accommodate the moments in the final bridge structure.

1.5.6   SPLICED PRECAST GIRDER CONSTRUCTION

Because of transportation and erection considerations, currently, the maximum length of the precast, prestressed concrete girder bridges normally will not exceed 200 ft. To increase the girder bridge span length, the girder can be divided into several pieces and then post-tensioned together[1-34] (see Fig. 1-45). This type of bridge is called a spliced girder bridge. The cross section of the prestressed girders can be either an I-section or U-section (see Fig. 1-46). The maximum span length of these bridges normally ranges from 200 to 300 ft.

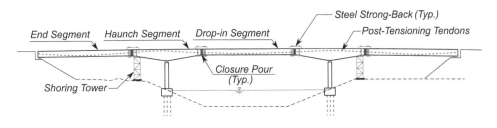

FIGURE 1-45   Spliced Girder Bridge Construction.

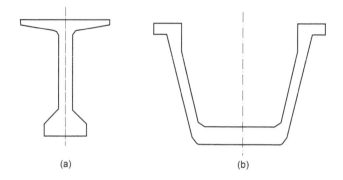

(a)   (b)

FIGURE 1-46   Typical Sections for Spliced Girder Bridges, (a) I-Section, (b) U-Section.

The typical construction method for spliced I-girder bridges is as follows (see Fig. 1-45):

Step 1: Erect temporary shoring towers.
Step 2: Erect haunch segments, and secure to the temporary support towers.
Step 3: Attach the strong backs to the ends of the end segments. The strong backs serve as temporary supports resting on top of the haunch segment ends.
Step 4: Erect end segments, and secure to the haunch segments with the strong backs.
Step 5: Attach the strong backs to the ends of the drop-in segment.
Step 6: Erect the drop-in girder, and connect the strong backs to the haunch segments.
Step 7: Cast closure joints between the haunch segment and the end segment as well as the drop-in segment.
Step 8: Install the continuity post-tensioning tendons, and complete the erection of one girder.

1.6   ECONOMIC SPAN RANGES

One of most important issues in bridge design is properly selecting the type of bridge that is best suited for a specific site condition based on safety, economics, constructability, and aesthetics. This is the most complex issue in bridge design as there is no absolute best alternative in many situations. Based on previously constructed bridges, Table 1-12 gives suitable span ranges for each type of concrete segmental bridges. The information in this table can be used to select feasible bridge alternatives.

TABLE 1-12
Typical Span Ranges of Various Concrete Segmental Bridges

Type of Bridge	Span Range (ft)
Span-by-span box girder	100 to 150
Precast balanced cantilever box girder	150 to 450
Cast-in-plan balanced cantilever box girder	230 to 850
Progressive placement box girder	100 to 300
Incrementally launching box girder	100 to 200
Spliced I-girder	175 to 300
Rigid frame	300 to 850
Arch	200 to 1300
Cable-stay	500 to 3000
Extradosed cable-stay	200 to 900

1.7 POST-TENSIONING SYSTEMS AND OPERATION

1.7.1 INTRODUCTION

Another important issue in segmental bridge construction is how to effectively apply post-tensioning forces to assemble the individual segments together. Currently, there are many manufacturers that have developed different post-tensioning systems available for this purpose, such as the Anderson, AVAR, DYWIDAG, Freyssinet, MEXPRESA, SDI, VSL, and Williams post-tensioning systems. These systems normally include anchorages, steel strands and bars, ducts, jacks, and grout equipment. Based on current practice, the bridge designers may not be required to know the details of the post-tensioning operations. However, it is important for the designer to obtain the general knowledge of post-tensioning methods and systems available to size a constructible bridge structure. Though there are many post-tensioning systems, their basic operational concepts are similar.

For current concrete segmental bridge construction, the multi-strand and bar post-tensioning systems are most commonly used. The anchorages for post-tensioning strands typically use the principle of wedge action producing a friction grip on the strands. Hydraulic jacks are normally used for stressing tendons. The jacks are used to pull the tendons with the reaction acting against the anchorage embedded in the hardened concrete. First the jacks stretch the strands. After the required stressing force is reached, anchor wedges are pushed into a wedge plate to secure the strands. Each company may develop different types of jacks and hydraulic pumps to efficiently and economically stress its tendons. In this section, DYWIDAG[1-28] anchorages and jacks are briefly introduced. The information on the post-tension systems developed by other manufactories is provided in Appendix B, in the FHWA Post-Tensioning Tendon Installation and Grouting Manual[1-23], and in ASBI Construction Practices Handbook[1-32] or on the company websites.

1.7.2 STRESSING EQUIPMENT

DYWIDAG has developed a series of jacks and hydraulic pumps to efficiently and economically stress its tendons. Versatility is provided by changing devices that make one unit adaptable for many different tendon sizes. DYWIDAG jacks have capacities ranging from 56 kips (250 kN) up to 2191 kips (9750 kN). Fig. 1-47 and Table 1-13 show the jacks and the technical data for strand tendons. The jacks can pull up to 37 strands at a time. The technical data of the jacks for THREADBARS® tendons for sizes 1″ to 1-3/4″ diameter are given in Fig. 1-48 and Table 1-14.

It is important for designers to know the jack envelop dimensions when they design a segmental bridge for properly sizing and locating the anchorages. Table 1-15 provides the dimensions for commonly used jacking systems and for post-tensioning tendons and bars. The definitions of the variables in Table 1-15 are shown in Fig. 1-49.

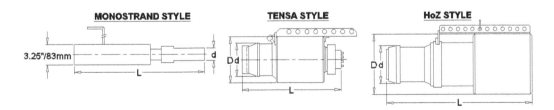

FIGURE 1-47 Types of Jacks and Hydraulic Pump for Strand Tendons (Courtesy of DYWIDAG Systems International).

TABLE 1-13
Technical Data for DYWIDAG Strand Tendons Jacks

Jack type*	Length l‡		Diameter D		Diameter d		Stroke		Piston area		Capacity†		Weight	
	(in)	(mm)	(in)	(mm)	(in)	(mm)	(in)	(mm)	(in²)	(mm²)	(kip)	(kN)	(lb)	(kg)
Mono 0.6	21.500	546	8.500 × 3.250	216 × 83	2.030	52	8.500	216	7.950	5,129	60	267	52	24
HoZ 950	24.450	621	8.000	203	6.300	160	3.940	100	25.100	16,194	218	970	144	65
HoZ 1,700	31.600	803	11.000	279	5.950	151	5.900	150	46.260	29,845	392	1744	354	161
HoZ 3,000	44.760	1137	15.160	385	7.700	196	9.840	250	78.900	50,903	687	3056	884	401
HoZ 4,000	50.000	1270	18.980	482	10.830	275	9.840	250	138.700	89,484	945	4204	1,326	601
Tensa 2,600	30.900‡	785‡	14.570	370	10.630	270	9.840	250	85.200	54,968	572	2544	729	331
Tensa 3,000	30.900‡	785‡	14.570	370	10.630	270	9.840	250	85.200	54,968	680	3025	782	355
Tensa 4,800	39.600‡	1,006‡	18.500	470	10.830	275	11.810	300	135.860	87,651	1,083	4817	1,432	650
Tensa 6,800	45.300‡	1,151‡	22.000	559	15.550	395	11.810	300	191.700	123,677	1,529	6801	2,619	1188
Tensa 8,600	46.000‡	1,168‡	29.500	749	25.600	650	11.810	300	274.700	177,225	2,191	9748	3,912	1774

* Power seating included
† Without friction
‡ Retracted position

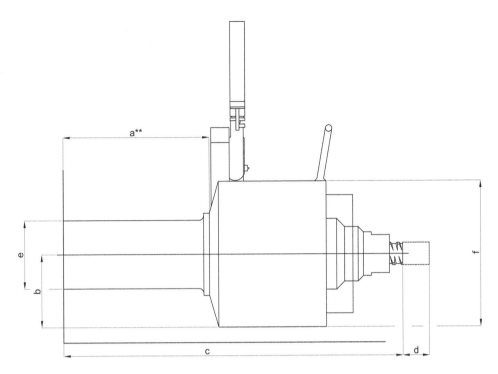

FIGURE 1-48 Types of Jacks for THREADBARS Tendons (Courtesy of DYWIDAG Systems International).

TABLE 1-14
DYWIDAG Jack Technical Data for Bar Sizes 1 in. (26 mm)
to 1-3/4 in. (46 mm)

Jack Capacity	kips/kN	160/720	220/979	330/1500
Bar	in.	1, 1-1/4	1-1/4, 1-3/8	1-3/4
Size	mm	26, 32	32, 36	46
a*	in.	8.85	10.83	18.70
	mm	225	275	475
b	in.	4.00	6.00	6.00
	mm	102	152	152
c	in.	21.00	24.60	41.60
	mm	533	625	1057
e	in.	3.40	4.90	8.50
	mm	86	124	216
f	in.	7.48	10.50	12.00
	mm	190	267	305
Weight	lb	80	110	334
	kg	36	50	152

* Special nose extensions for deep stressing pockets are available upon request.

TABLE 1-15
Jack Envelop Dimensions for Design (in.)[1-35]

Type	Sizes	A	B*	C, D	E	F
Tendon (strands)	4-0.6"	50	86	15	17	11
	7-0.6"	51	92	15	17	15
	12-0.6"	51	92	15	17	15
	15-0.6"	60	120	15	21	19
	19-0.6"	60	120	15	17	19
	27-0.6"	60	120	15	19	24
	31-0.6"	60	120	18	19	25
Bar (diameter)	1-0"	42	72	15	10	11
	1-1/4"	43	72	15	10	11
	1-3/8"	43	72	15	10	11
	1-3/4"	51	92	15	12	15
	2-1/2"	56	92	15	13	16
	3-0"	60	120	15	17	19

* For replaceable tendons and bars with flexible filler.

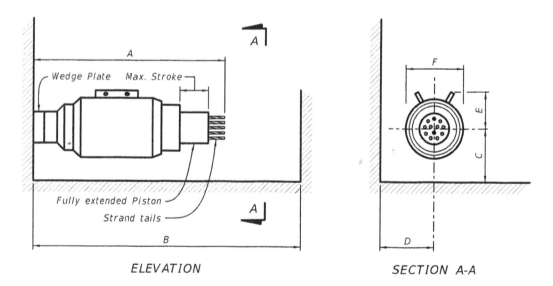

ELEVATION   SECTION A-A

FIGURE 1-49   Definitions for Jack Envelop Dimensions (Used with permission of FDOT[1-35]).

1.7.3 ANCHORAGES

DYWIDAG multiplane and flat anchorages are shown in Figs. 1-50 and 1-51, respectively. For the bridge designers, it is important to obtain the dimensions that will affect the sizes of the structure based on the related manufacturers' specifications. The principal dimensions for DYWIDAG multiplane and flat anchorages are given in Figs. 1-50 and 1-51 and Table 1-16, respectively.

Post-tensioned bars are commonly anchored to the hardened concrete by nuts, washer, and anchor plates. Figure 1-52 shows a typical section for the DYWIDAG bonded bar anchorages. The related technical data are provided in Table 1-17.

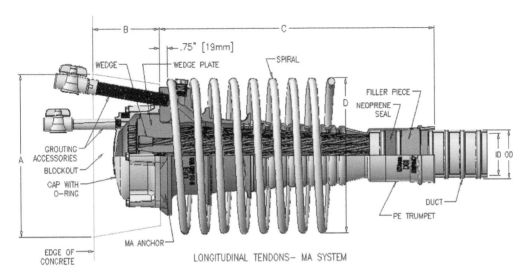

FIGURE 1-50 Typical Longitudinal Post-Tensioning Multiplane Anchorages System (Courtesy of DYWIDAG Systems International).

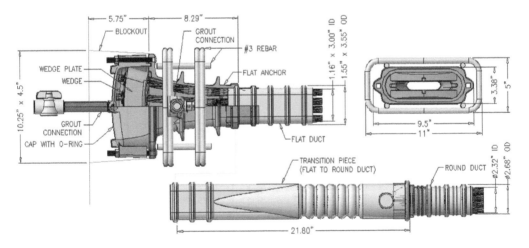

FIGURE 1-51 Typical Transverse Post-Tensioning Anchorages FA System (3-0.6″ & 4-0.6″ Strands) (Courtesy of DYWIDAG Systems International).

1.8   POST-TENSIONING STEEL AND ANCHORAGE PROTECTION

As previous discussed, post-tensioning tendons are key components in concrete segmental bridges. Any damage to the post-tensioning tendons may significantly affect bridge safety and may require complete replacement of the bridge. Previous experiences show some post-tensioning tendons were exposed to salt spray and deck runoff through deteriorated deck joints and some anchorages were penetrated by water due to lack of protection. These cause serious corrosion of post-tensioning steel and significantly affect the durability of concrete segmental bridges. The Florida Department

TABLE 1-16
Principal Dimensions of DYWIDAG Multiplane Anchorages (in.)

Multiplane Anchorage MA Tendon Size using 0.6" and 0.5" Strands		5-0.6" 7-0.5"	7-0.6" 9-0.5"	9-0.6" 12-0.5"	12-0.6" 15-0.5"	15-0.6" 20-0.5"	19-0.6" 27-0.5"	27-0.6" 37-0.5"	31-0.6"	37-0.6"
Block-out (minimum)	A	7.50	12.00	12.50	13.50	15.00	15.00	17.00	17.00	21.00
	B	5.00	6.00	6.00	6.00	6.50	6.50	7.00	8.50	9.00
Transition length	C	15.75	18.01	18.10	24.15	28.15	26.18	28.97	18.00	38.31
Rebar spiral	D	8.25	11.25	9.50	12.88	12.50	14.50	17.00	19.00	22.00
Plastic corrugated duct	ID	1.89	2.32	2.99	3.35	3.94	3.94	4.53	5.04	5.04
	OD	2.32	2.87	3.58	3.95	4.57	4.57	5.30	5.96	5.96
Anchor location	c/c spacing	9.50	12.50	10.50	14.00	13.50	15.50	18.00	20.00	23.00
	Min. edge dist.	6.25	7.75	7.00	8.50	8.50	9.50	11.50	13.25	14.50

Note: Metal spiro ducts are available for all sizes. ID = inner diameter. OD = outer diameter.

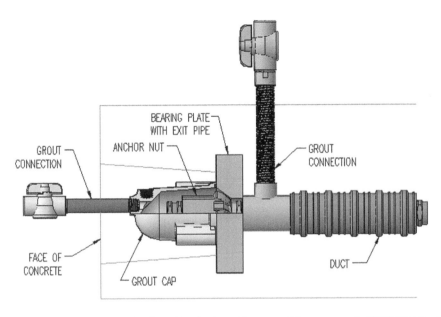

FIGURE 1-52 THREADBARS Post-Tensioning System (Courtesy of DYWIDAG Systems International).

TABLE 1-17
Principal Technical Data for DYWIDAG THREADBARS Anchorages

List of 150 ksi THREADBARS	1″	1-1/4″	1-3/8″	1-3/4″	2-1/2″	3″
Hot rolled = HR; cold rolled = CR	HR	HR	HR	HR	CR	CR
Square bearing plate (in.)	$5 \times 5 \times 1.25$	$7 \times 6 \times 1.5$	$7.5 \times 7 \times 1.75$	$9 \times 9 \times 2$	$14 \times 12 \times 2.5$	$15 \times 15 \times 3$
Rectangular bearing plate (in.)	$6.5 \times 4 \times 1.25$	$8 \times 5 \times 1.5$	$9.5 \times 5 \times 1.75$			
Plastic corrugated duct ID (in.)	2.32	2.32	2.32	2.99	3.35	3.94
Plastic corrugated duct OD (in.)	2.87	2.87	2.87	3.58	3.95	4.57
Bar diameter, max. (in.)	1.20	1.44	1.63	2.01	2.79	3.15
Ultimate load (kips)	128	188	237	400	774	1027

Note: Metal spiro ducts are available for all sizes. ID = inner diameter. OD = outer diameter.

of Transportation (FDOT) has developed the following corrosion protection strategies in the design and detailing of post-tensioned structures[1-30]:

1. Use enhanced post-tensioning systems, such as completely sealed ducts and permanent anchorage caps.
2. Completely fill the ducts and anchorage caps with approved filler, such as concrete grout or flexible filler.
3. Use multi-level anchorage protection as shown in Figs. 1-53 and 1-54[1-36].
4. Provide watertight bridges, such as by using epoxy-sealed joints; avoiding placing holes where leaks would drip onto anchor heads; filling temporary holes with an approved no-shrink, high-bond, high-strength, air-cured concrete or epoxy grout; providing bottom slab drains; and providing drip flanges on the underside of the transverse seat for expansion joint devices. Some of the details will be discussed in Chapter 5.
5. Provide multiple tendon path redundancy to ensure that the loss of any one of the tendons due to corrosion does not critically diminish the overall bridge performance.

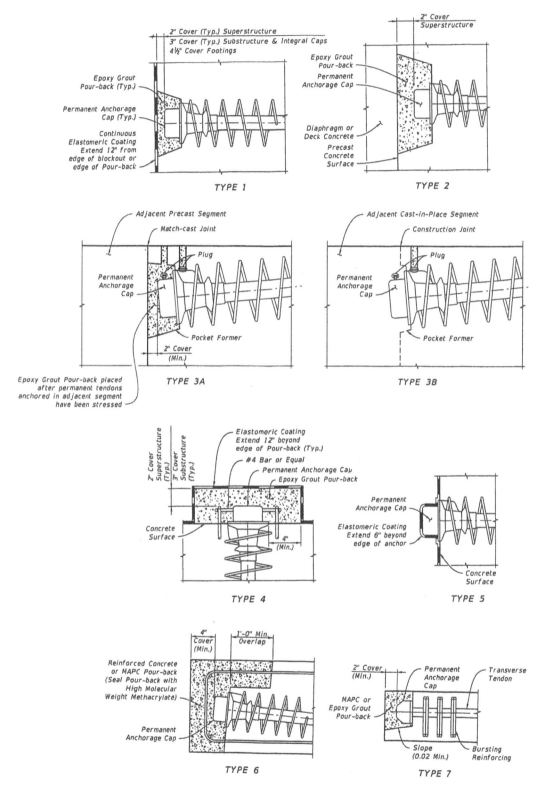

FIGURE 1-53 Typical Anchorage Protection for Post-tensioning Strand Tendons[1-35] (Used with permission by FDOT).

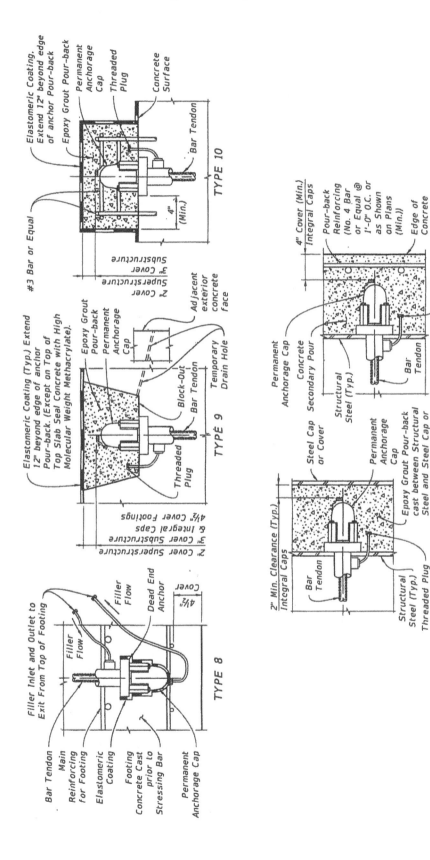

FIGURE 1-54 Typical Anchorage Protection for Post-Tensioning Bar Tendons[1-36] (Used with permission by FDOT).

1.9 GENERAL DESIGN PROCEDURES AND BRIDGE AESTHETICS

1.9.1 GENERAL DESIGN PROCEDURES

The general objectives of bridge design are to achieve the structural safety, constructibility, serviceability, inspectability, economy, and aesthetics. The general design procedures for concrete segmental bridges are similar to those for other types of bridges. There are five basic steps:

    Step 1. Determine bridge span lengths and vertical clearances
    Step 2. Select suitable bridge types
    Step 3. Perform preliminary cost estimation
    Step 4. Evaluate alternatives
    Step 5. Select final bridge alternative and start design and detailing

Step 1. Determine Bridge Span Lengths and Vertical Clearances
    The determination of the bridge length and vertical clearance is one of the most import steps. It is mainly related to bridge function, economic consideration, aesthetics, and site conditions. Generally, the minimum required span length and vertical clearance are analyzed first.
    For bridges crossing roadways, the bridge clearances should meet the requirements of the AASHTO publication "A Policy on Geometric Design of Highways and Streets"[1-37] and the local State Department of Transportation. The typical considerations are

    a. Numbers and widths of travel lanes, shoulders, sidewalks under the bridge, and present and future requirements
    b. Minimum distances from edges of roadway to the piers
    c. Other considerations for the safety of structures and users

For bridges crossing railroads, the minimum bridge span length and vertical clearance should meet the requirements specified in the AREMA Manual for Railway Engineering[1-38]. For bridges over navigable waterways, the minimum vertical and horizontal clearances should be determined based on the needs of navigation together with the U.S. Coast Guard permit requirements and the related local governments.
    After the minimum span length and vertical clearance are determined, the bridge span lengths are refined based on the consideration of economics, aesthetics, and site conditions.
    Step 2: Select Suitable Bridge Types
    Based on the preliminary span arrangements and previous design experiences, such as shown in Table 1-12, select about three most suitable bridge types, including suitable bridge types other than segmental bridges. Typically, for a total deck area of over 110,000 ft^2, concrete segmental bridges are a competitive and desirable alternative.
    Step 3: Perform Preliminary Cost Estimations
    Each of the State Departments of Transportation in the United States typically provides its bridge cost estimation guidelines with the unit prices for bridge deck areas or different materials for different types of bridges based on their previously built bridges. Bridge designers can quickly generate the estimated construction costs for each of the selected bridge alternatives based on preliminary bridge dimensions.
    Step 4: Evaluate Bridge Alternatives
    Evaluate the advantages and disadvantages for each bridge alternative based on the following aspects:

    a. Cost effectiveness
    b. Constructability, including the complexity and period of construction
    c. Durability

d. Maintainability

e. Aesthetics

Step 5: Select Final Bridge Alternative and Start Design and Detailing

After selecting the final bridge alternative, refine the previously assumed bridge dimensions, perform refined design calculations, and start detailing.

1.9.2 SEGMENTAL BRIDGE AESTHETICS

All bridges make an aesthetic impact on the people seeing them and the people using them. Bridges may become the symbols of their communities. Designers should not only design bridges for their functions, cost effectiveness, safety, durability, and maintainability, but also to be aesthetically pleasing to make the best impression on the thousands of people seeing them every day. Concrete segmental bridges are typically used for long-span bridges and major bridge projects and may be more desirable when considering aesthetics. Aesthetics is a matter of people's feelings about perceived objects and belongs to the subjects of philosophy, physiology, and psychology. It is difficult to provide simple aesthetic criteria for structural engineers to follow because there is no universally accepted theory of aesthetics. On signature bridge projects, bridge engineers work closely with architects to create a unique, elegant, practical, and aesthetically pleasing bridge design. For other less high profile segmental bridges, fortunately, some excellent guidelines for bridge aesthetics have been developed based on successful completed projects[1-39 to 1-44]. Some of the practical guidelines are briefly summarized in the following.

1.9.2.1 Harmony with Surroundings

It is important to ensure that bridges integrate congruously into their surrounding terrains and environments by selecting proper member sizes, horizontal alignment, and vertical profile. Alignment with smooth curves, for instance, promotes a feeling of dynamics and increases the play of light and shadow (see Figs. 1-55 and 1-56)[1-41, 1-45].

1.9.2.2 Proportion

Appropriate proportions are fundamental in achieving superior bridge aesthetics. A bridge's proportions encompass the relationships between super- and substructures; between bridge depth and span length; between the height, length, and width of opening; and so forth. These relationships should convey a sense of balance. Using tall, tapered piers with equal proportions and a slender continuous girder superstructure for a bridge crossing a deep valley can create an appealing result (see Fig. 1-57)[1-45].

FIGURE 1-55 Taihu Bridge (Courtesy of Prof. Shi-Jin Yang[1-44]).

FIGURE 1-56   Alignment with Smooth Curves, Montabliz Viaduct, Spain (2008)[1-45] [Used with permission by *SEI Journal* (Taylor & Francis Group), IABSE].

FIGURE 1-57   Slender Continuous Girder Superstructure over a Deep Valley, Montabliz Viaduct, Spain (2008)[1-45] [Used with permission by *SEI Journal* (Taylor & Francis Group), IABSE].

FIGURE 1-58   Pitan Bridge[1-46], 1994 [Used with permission by *SEI Journal* (Taylor & Francis Group), IABSE].

1.9.2.3   Simplicity of Details

Keep a bridge's design as simple as possible. An uncomplicated bridge structure with fewer super-fluous components not only reduces maintenance work but also results in a more elegant composition (see Figs. 1-58[1-46] and 1-59).

1.9.2.4   Artistic Shaping

An aesthetically pleasing bridge structure can often be achieved by slightly changing the shape of the bridge elements.

1. Use varying girder depth to produce an approximate constant web shear for a huanched girder bridge (see Fig. 1-60)[1-47].
2. Use tapering that follows the moment diagram shape for a tall pier (see Fig. 1-61).
3. Use ribbing, striation, or rustication to treat the piers (see Fig. 1-62).
4. Use sculptured pier (see Fig. 1-63).
5. Use ornamental pieces to hide supports (see Fig. 1-64).
6. Use architectural pylons (see Figs. 1-65 and 1-66).

FIGURE 1-59   Natchez Trace Parkway Arch Bridge [Courtesy of Mr. John Corven (designer)].

FIGURE 1-60   Shin Chon Bridge, Korea, 2007[1-47] [Used with permission by *SEI Journal* (Taylor & Francis Group), IABSE].

FIGURE 1-61   Tall Pier with Tapering Shape (Courtesy of Prof. Shi-Jin Yang).

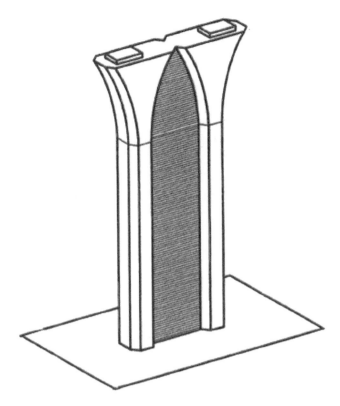

FIGURE 1-62   Pier with Architectural Treatments, after FDOT[1-35].

FIGURE 1-63   Sculptured Pier, Clearwater Memorial Causeway Bridge, 2005.

FIGURE 1-64   Hiding Supports, (a) I-35W Saint Anthony Falls Bridge, 2008, (b) Lee Roy Selmon Crosstown Expressway, 2004 (both designed by Figg Engineering).

FIGURE 1-65   Yumekake Bridge, Japan, 2010[1-48] [Used with permission by *SEI Journal* (Taylor & Francis Group), IABSE].

FIGURE 1-66 Ganter Bridge by Christian Menn[1-49] [Used with permission by *SEI Journal* (Taylor & Francis Group), IABSE].

REFERENCES

1-1. PCI Committee on Segmental Concrete Construction, "Recommended Practice for Segmental Construction in Prestressed Concrete," *PCI Journal*, Vol. 20, No. 2, March–April 1975, pp. 22–41.

1-2. Chen, W. F., and Duan, L., *Bridge Engineering Handbook*, CRC Press, Boca Raton, London, New York, Washington, D.C., 2014.

1-3. Podolny, W., and Muller J., *Construction and Design of Prestressed Concrete Segmental Bridges*, John Wiley & Sons, New York, 1982.

1-4. Finsterwalder Ulrich, "Prestressed Concrete Construction," *Journal of ACI*, Vol. 62, No. 9, 1965, pp. 1037–1046.

1-5. Podolny, W., "An Overview of Precast Prestressed Segmental Bridges," *PCI Journal*, January–February 1979, pp. 56–87.

1-6. Tang, M. C., "Koror-Babelthuap Bridge—A World Record Span," ASCE Convention, Chicago, October 16–20, 1978.

1-7. Muller, J., "Ten Years of Experience in Precast Segmental Construction," *PCI Journal*, Vol. 20, No. 1, January–February 1975, pp. 28–61.

1-8. Newson, N. R., *Prestressed Concrete Bridges: Design and Construction*, Thomas, Telford, London, New York, 2003.

1-9. Tang, M. C., "Segmental Bridges in Chongqing, China," *Journal of Bridge Engineering*, ASCE, Vol. 20, Issue 8, 2015, pp. B4015001–1 to B4015001–10.

1-10. Figg, L., and Patt, W. D., "Precast Concrete Segmental Bridges—America's Beautiful and Affordable Icons, *PCI Journal*, September–October 2004, pp. 26–39.

1-11. Hansvold, C., "Skarnsundet Bridge," *Proceedings of IABSE/FIP International Conference on Cables-Stayed and Suspension Bridges*, Deauville, October 12–15, 1994, pp. 247–256.

1-12. ACI Committee 209, "Prediction of Creep, Shrinkage and Temperature Effects in Concrete Structures: Designing for Creep and Shrinkage in Concrete Structures," ACI Publication SP-76, ACI, Detroit, 1982, pp. 193–300.

1-13. Comite Euro-International du Beton, *CEB-FIP MODEL CODE 1990 - Design Code*, Thomas Telford, London, 1993.

1-14. Corven, J., *Post-Tensioned Box Girder Design Manual*, Federal Highway Administration, U.S. Department of Transportation, Washington D.C., 2015.

1-15. AASHTO, *LRFD Bridge Design Specifications*, 7th ed., 2014, Washington D.C.

1-16. Walker, S., and Bloem, D. L., "Effect of Aggregate Size on Properties of Concrete," *Journal of the American Concrete Institute*, American Concrete Institute, Farmington Hills, MI, Vol. 57, No. 3, September 1960, pp. 283–298.

1-17. Khan, A. A., Cook, W. D., and Mitchell, D., "Tensile Strength of Low, Medium, and High-Strength Concretes at Early Ages," *ACI Materials Journal*, American Concrete Institute, Farmington Hills, MI, Vol. 93, No. 5, September–October 1996, pp. 487–493.

1-18. Lin, T. Y., and Burns, N. H., *Design of Prestressed Concrete Structures*, John Wiley & Sons, Inc., New York, 1981.

1-19. Troxell, G. E., Raphael, J. M., and Davis, R. E., "Long-time Creep and Shrinkage Tests of Plain and Reinforced Concrete," *Proceedings American Society for Testing and Materials*, Vol. 58, 1958, pp. 1101–1120.

1-20. Nawy, E. G., *Prestressed Concrete—A Fundamental Approach*, 3rd edition, Prentice Hall, Upper Saddle River, NJ, 1999.

1-21. Castrodale, R. W., "Lightweight Concrete for Long Span Bridges," ASBI Convention, Dallas, November 3, 2015.

1-22. MacGregor, J. G., *Reinforced Concrete—Mechanics and Design*, Prentice Hall, Upper Saddle River, NJ, 1988.

1-23. FHWA, *Post-Tensioning Tendon Installation and Grouting Manual*.

1-24. Magura, D. D., Sozen, M. A., and Siess, C. C., "A Study of Stress Relaxation in Prestressing Reinforcement," *Journal of Prestressed Concrete Institute*, Vol. 9, No. 2, 1964, pp. 13–17.

1-25. PTI, *Post-Tensioning Manual*, 6th edition, Phoenix, 2006.

1-26. AASHTO, *AASHTO LRFD Bridge Design Guide Specifications for GFRP-Reinforced Concrete Bridge Decks and Railings*, Washington, D.C., 2009.

1-27. ACI 440.1R-15, *Guide for the Design and Construction of Structural Concrete Reinforced with Fiber-Reinforced Polymer (FRP) Bars*, ACI Committee 440, Farmington Hills, 2015.

1-28. DYWIDAG, *DYWIDAG Post-Tensioning Systems, Multi-strand System, Bar System, Repair and Strengthening*, DYWIDAG-Systems International, April 2006.

1-29. ACI 440.4R-04, *Prestressing Concrete Structures with FRP Tendons*, ACI Committee 440, Farmington Hills, 2004.

1-30. Florida Department of Transportation, *Structural Design Guidelines*, January 2016, Tallahassee, FL.

1-31. Yang X., Zhrevand, P., Mirmiran, A., Arockiasamy, M., and Potter, W., "Post-tensioning of Segmental Bridges Using Carbon-Fiber-Composite Cable," *PCI Journal*, May–June 2015.

1-32. ASBI, *Construction Practices Handbook for Concrete Segmental and Cable-Supported Bridges*, American Segmental Bridge Institute, Buda, Transportation Research Board, National Research Council, Washington, D.C., 2008.

1-33. Huang, D. Z., and Hu, B., "Evaluation of Cracks of a Large Single Cell Precast Concrete Segmental Box Girder Bridge without Internal Struts," *Journal of Bridge Engineering*, ASCE, Vol. 20, No. 5, 2015.

1-34. Huang, D. Z., Arnold, S, and Hu, B. (2012) "Evaluation of Cracks in a Spliced Prestressed Concrete I-Girder Bridge," *Journal of the Transportation Research Board*, No. 2313, Transportation Research Board of the National Academies, Washington, D.C.

1-35. Florida Department of Transportation, *Structures Detailing Manual*, Tallahassee, FL, 2018.

1-36. Florida Department of Transportation, *Design Standards*, 2017 edition, Tallahassee, FL.

1-37. American Association of State Highway and Transportation Officials (AASHTO), *A Policy on Geometric Design of Highways and Streets*, 4th edition, Washington D.C., 2001.

1-38. American Railway Engineering and Maintenance-of-Way Association (AREMA), *Manual for Railway Engineering*, Washington, D.C., 2003.

1-39. Transportation of Research Board (TRB), *Bridge Aesthetics around the World*, National Research Council, Washington, D.C., 1991.

1-40. Maryland Department of Transportation (MDOT), *Aesthetic Bridges Users' Guide*, Baltimore, MD, 1993.

1-41. Menn, C., "Aesthetics in Bridge Design," *Bulletin of the IASS*, Vol. 26, No. 88, April 1986, pp. 53–62.

1-42. Barker, C. M., and Puckett, J. A., *Design of Highway Bridges Based on AASHTO LRFD Bridge Design Specifications*, John Wiley & Sons, New York, 1997.

1-43. Freyermuth, C. L., "Ten Years of Segmental Achievements and Projections for the Next Century," *PCI Journal*, May–June 1999, pp. 36–52.

1-44. Yang, S. J., "Aesthetic Consideration of Tai Hu Bridge Design," *Journal of Chinese Highway*, No. 5, 1994, pp. 14–18.

1-45. Pantaleon, M. J., Revilla, R., and Olazabat, "Montabliz Viaduct, Cantabria (Spain)," *Journal of Structural Engineering, International*, IABSE, Vol. 18, No. 3, 2008, pp. 222–226.

1-46. Cheng, K. M., "The Pitan Bridge, Taiwan," *Journal of Structural Engineering, International*, IABSE, Vol. 4, No. 4, 1994, pp. 231–234.

1-47. Starossek, U., "Shin Chon Bridge, Korea," *Journal of Structural Engineering, International*, IABSE, Vol. 19, No. 1, 2009, pp. 79–84.

1-48. Akiyama, H., "Extradosed Prestressed Concrete Bridge with High-Strength Concrete, Japan—Yumekake Bridge," *Journal of Structural Engineering, International*, IABSE, Vol. 21, No. 3, 2009, pp. 366–371.

1-49. Voget, T., and Schellenberg, K., "The Impact of the Sunniberg Bridge on Structural Engineering, Switzerland," *Journal of Structural Engineering*, International, IABSE, Vol. 25, No. 4, 2015, pp. 381–388.

1-50. Tang, M. C., and Ho, T., "Evolution of Bridge Technology," ASCE/SEI Workshop, Washington D.C., 2008.

2 Loads on Bridges and General Design Methods

- Types of loads on bridges
- Basic theoretical background of AASHTO LRFD specifications
- General design equation
- Load combinations
- Load factors and resistance factors

2.1 INTRODUCTION

A highway bridge should be designed to safely support all anticipated loads on it during its service life. Thus, there are two basic issues:

- What types of loads should be considered in bridge design?
- How is the "safety" in the bridge design measured?

Though it may not be necessary for bridge designers to fully grasp the detailed design theory, it is important that they understand the basic concepts of the design methods.

2.2 TYPES OF LOADS ON CONCRETE SEGMENTAL BRIDGES

2.2.1 GENERAL DESCRIPTION

The loads on concrete segmental bridges can be grouped into two categories[2-1]:

I. Permanent Loads
   - Dead loads
   - Forces due to earth
   - Prestressing forces
   - Forces due to concrete creep and shrinkage
   - Locked-in force effects resulting from the construction process
   - Force effect due to settlement
II. Transient Loads
   - Live loads due to vehicles and pedestrians
   - Force effects due to moving vehicles
   - Wind loads
   - Force effect due to temperature
   - Water load
   - Ice load
   - Blast loading
   - Earthquake load

It may be impossible to accurately determine all possible loads on a bridge when it is being designed. However, a bridge designed based on the loads specified in the AASHTO specifications[2-1] is deemed to have enough capacity to carry anticipated loads that it will experience during its service life.

2.2.2 Dead Loads

The dead load of a bridge includes the weight of the structure and any equipment attached to it, such as appurtenances, utilities, earth cover, wearing surface, and future overlays. The unit weights shown in Table 2-1 are recommended by the AASHTO LRFD specifications and can be used for bridge design when more precise information is not available. The unit weight of reinforced concrete is generally taken as 0.005 kcf greater than the unit weight of plain concrete.

2.2.3 Live Loads

2.2.3.1 Vehicle Live Load

2.2.3.1.1 Introduction

There are many types of vehicles with different dimensions and weights that might pass over highway bridges during their service life. However, it is not possible for bridge designers to know what types of vehicles will use the bridges, and it is not economical to consider some excessively heavy vehicles in the design of normal bridges. To ensure the safety, based on the results of truck weight studies, the AASHTO Highway Bridge Design Specifications provide some hypothetical vehicular loading models representing the majority of vehicles traveling on the U.S. highway system. The AASHTO loads simulate most vehicular traffic patterns and prohibit overweight or excessively heavy vehicles from using the bridges. Vehicle loading is a moving mass that induces larger loading than its equivalent static load. The effect of a moving vehicle is separately considered as two parts of static and dynamic loadings in the AASHTO specifications. The static effect is designated as a design vehicle live load (LL) and the dynamic effect is designated as a dynamic load allowance or impact factor (IM).

2.2.3.1.2 Design Vehicle Live Load (LL)

In the AASHTO specifications, the vehicular live loading on bridges is designated as HL-93 and includes two load models:

- Model I: Design truck plus design lane load (see Fig. 2-1a)
- Model II: Design tandem plus design lane load (see Fig. 2-1b)

TABLE 2-1
Unit Weights of Typical Materials

Material	Unit Weight (kcf)
Aluminum	0.175
Bituminous wearing surfaces	0.140
Cast iron	0.450
Compacted sand, silt, or clay	0.120
Lightweight concrete	0.110
Sand-lightweight concrete	0.120
Normal weight concrete with $f_c' \leq 5.0$ ksi	0.145
Normal weight concrete with $5.0 < f_c' \leq 15.0$ ksi	$0.140 + 0.001 f_c'$
Steel	0.490
Fresh water	0.0624
Stone masonry	0.170
Hard wood	0.060
Soft wood	0.050

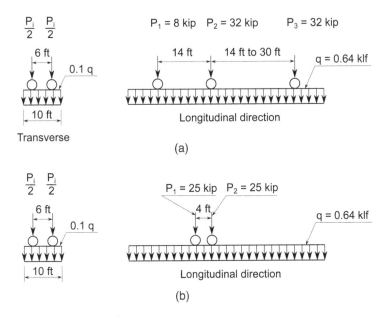

FIGURE 2-1 Design Vehicle Live Load Models, (a) Truck and Lane Model, (b) Tandem and Lane Model.

Each design lane under consideration should be occupied by either the design truck or tandem, coincident with the lane load. The loads are assumed to occupy a width of 10.0 ft placed transversely within a design lane.

The wheel loads may be modeled as concentrated loads or as patch loads. The tire contact area of a wheel is assumed to be a single rectangle, whose width is 20.0 in. and length is 10.0 in.

2.2.3.1.3   Application of Design Vehicular Live Loads

2.2.3.1.3.1   General Loading Application

2.2.3.1.3.1.1   Bridge Transverse Direction

The design live loadings shown in Fig. 2-1 can be applied to one loading lane to the maximum loading lane numbers accommodated on the bridge. The number of loading lanes is different from the design lane number designated in AASHTO's *A Policy on Geometric Design of Highways and Streets*[2-2]. The loading lane number (N_L) should be determined as

$$N_L = integal \ part \left(\frac{w}{12} \right) \tag{2-1}$$

where w is the clear roadway width in feet between curbs and/or barriers (see Fig. 2-2).

The extreme live load force effect should be determined by considering each possible combination of number of loaded lanes multiplied by a corresponding multiple presence factor presented in Table 2-2. The multiple presence factors are used to account for the probability of simultaneous lane occupation by the full HL93 design live load. The center of any wheel load of the design truck or tandem should not be positioned transversely closer to the following values (see Fig. 2-2):

- 1.0 ft from the face of the curb or railing, for the design of the deck overhang
- 2.0 ft from the edge of the design lane or the face of the curb or railing, for the design of all other components

TABLE 2-2
Multiple Presence Factors

Number of Loaded Lanes	Multiple Presence Factors
1	1.20
2	1.00
3	0.85
>3	0.65

2.2.3.1.3.1.2   Bridge Longitudinal Direction

In bridge longitudinal direction, the design live loads should be applied based on the types of bridge responses required.

1. *Positive moment, shear, and torsion.* Two design live load models shown in Fig. 2-1 should be loaded to analyze the bridge individually, and the extreme effect should be taken as the larger forces generated by either models and should be used for the design.
2. *Negative moments between points of contraflexure under a uniform load on all spans and reaction at interior piers.* The extreme force effect should be taken as the larger of the load-ings generated by the models shown in Fig. 2-3 for the design. The extreme force effect should be determined based on 90% of the effect of two design trucks spaced a minimum of 50.0 ft between the lead axle of one truck and the rear axle of the other truck, combined with 90% of the effect of the design lane load (see Fig. 2-3). The distance between the 32.0-kip axles of each truck should be taken as 14.0 ft. The two design trucks should be placed in adjacent spans to produce maximum force effects. Axles that do not contribute to the extreme force effect under consideration should be neglected.
3. *Bridge deflection.* The limits of bridge vertical deflections specified in the AASHTO LRFD specifications are developed based on previous AASHTO standard specifications[2-3]. Thus, the heavier live loads required in the AASHTO LRFD specifications are reduced when calculating the bridge deflections. The AASHTO LRFD specifications recommend using the larger of the deflections determined by the following two types of live load applications if the owner has not provided any optional live load deflection criteria.
   - Apply the design truck alone.
   - Apply 25% of the design truck and design lane load.

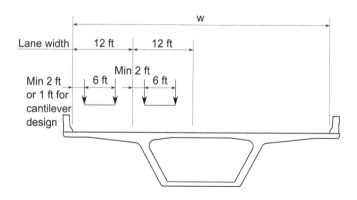

FIGURE 2-2   Design Live Load Application in Transverse Direction.

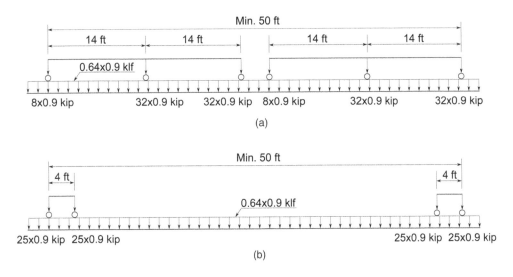

FIGURE 2-3 Design Live Load Application for Negative Moment and Reaction, (a) Truck and Lane Model, (b) Tandem and Lane Model.

2.2.3.1.3.2 Fatigue Load Application   For determining the fatigue loading, use one design truck or axles with a constant spacing of 30 ft between the 32 kip axles discussed in Section 2.2.3.1.2 without lane loading. The fatigue loading calculated should be multiplied by the dynamic allowance of 1.33. The fatigue loading should be applied by changing both transverse and longitudinal locations, ignoring the striped lanes, to produce maximum stress or deflection.

The single-lane average daily truck traffic (ADTT) should be used for determining the frequency of the fatigue load and can be calculated using the following equation in the absence of more accurate data:

$$ADTT_S = ADDT \times p \tag{2-2}$$

where
    $ADDT$ = number of trucks per day in one direction averaged over bridge design life
    p = fraction of traffic in a single lane taken as follows:
        = 1.00, for single lane
        = 0.85, for two lanes
        = 0.80, for three or more lanes

2.2.3.2   Pedestrian Live Load
AASHTO Specification Article 3.6.1.6 requires that a pedestrian load of 0.075 ksf be applied to all sidewalks wider than 2.0 ft and that it be considered simultaneously with the vehicular design live load in the vehicle lane.

2.2.4   DYNAMIC LOADING DUE TO MOVING VEHICLES

2.2.4.1   Dynamic Loading Analysis
Moving vehicles will cause larger loading effect on bridges than their equivalent static loads (see Fig. 2-4). Currently, most of the highway bridge design specifications adopted in various countries use a specific "dynamic load allowance" or "impact factor" to approximately account for the induced effects caused by moving vehicles. However, many bridge specifications also allow designers to

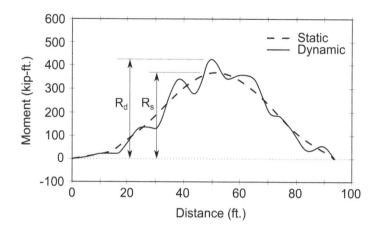

FIGURE 2-4 Typical Time Histories in a Simply Supported Girder Bridge due to Moving Vehicle.

determine the dynamic load allowances based on the analysis of vehicle and bridge interaction and/ or by test results. The dynamic load allowance or impact factor is defined as

$$IM = \frac{R_d}{R_s} \tag{2-3}$$

in which R_d and R_s are the absolute maximum response for dynamic and static studies, respectively.

The interaction between moving vehicles and a bridge is complex and is related to many factors, such as the type of bridge, the weight and stiffness of the vehicle, road surface roughness, and vehicle speed. In recent years, numerous research has been performed on vehicle-induced dynamic loading of highway bridges[2-4 to 2-7]. In this research, the vehicle and bridge were considered as an interactive system. The AASHTO LRFD specifications design truck HL-93 was modeled as a three-dimensional model that consisted of five rigid masses that represent a tractor, semitrailer, steer-wheel axle set, tractor wheel axle set, and trailer wheel axle set (see Fig. 2-5). Each of the masses is connected by springs. In the model, the tractor and semitrailer are each assigned three degrees of freedom (DOFs), corresponding to the vertical displacement (y_{ti}) rotation about the transverse axis (pitch θ_{ti}) and rotation about the longitudinal axis (roll ϕ_{ti}). Each wheel axle set is provided with two DOFs in the vertical and roll directions. The tractor and trailer are interconnected at the pivot point (the so-called fifth wheel point).

Test and theoretical results indicate that the deck surface is one of the most important factors that affect the bridge dynamic loading. The bridge road surface roughness can be simulated as a stationary gaussian random process:

$$v_{sr}(x) = \sum_{i=1}^{N} \sqrt{4S(\omega_i)\Delta\omega} \cos(\omega_i x + \theta_i) \tag{2-4}$$

where
$\quad v_{sr}(x)$ = simulated road vertical profile
$\quad x$ = longitudinal location of generated point
$\quad S(\omega_i)$ = power spectral density (PSD) function, which indicates the conditions of deck vertical profiles[2-4]
$\quad \omega_i$ = circular frequency
$\quad \theta_i$ = random number uniformly distributed from 0 to 2π, and $N = 200$

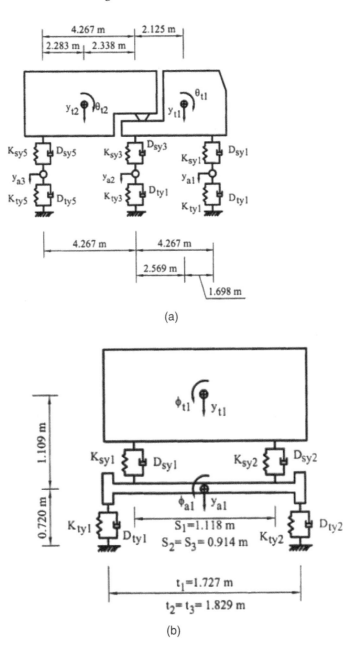

FIGURE 2-5 Analytical Vehicle Model, (a) Front View, (b) Side View.

Deck profiles can be classified into four types of very good, good, poor, and very poor. Existing survey data show that most bridge deck profiles in the United States can normally be classified as "good." Figure 2-6 shows two simulated good profiles.

Research results indicate: (1) Generally, maximum impact factor of moment, shear, and deflection for concrete segmental bridges will not exceed 0.25. (2) Normally, the heavier the static loading, the smaller the dynamic loading will be. (3) If a bridge has a longitudinal hinge at the midspan, the bridge dynamic loading will increase with vehicular speed.

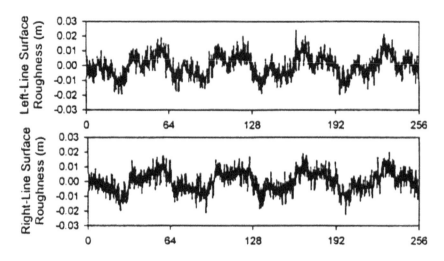

FIGURE 2-6   Typical Good Road Profiles.

2.2.4.2   Dynamic Loading Allowance: IM

It may be difficult or unnecessary to design normal bridges based on the tested or theoretically analyzed dynamic loading. To simplify the dynamic load analysis, AASHTO Specifications Article 3.6.2 specifies the dynamic loading be considered through increasing the static effects of the design truck or tandem by a percentage. The percentage increase is called a dynamic load allowance or impact factor as shown here:

IM = 0.75, for deck joints
    = 0.15 for fatigue and fracture limit state
    = 0.33 for all other limit states

If IM is determined based on an analysis for dynamic interaction between a bridge and the live load as approved by the owner, the dynamic load allowance used in the design should be not less than 50% of the above dynamic load allowance given by the AASHTO specifications.

2.2.4.3   Centrifugal Forces: CE

When a vehicle travels on a curvilinear path, it produces a force perpendicular to the tangent of the path. This force is called a centrifugal force, and it causes the vehicle to change its driving direction. The centrifugal force has not been included in the dynamic load allowance discussed above. If the vehicle moves at a constant speed and takes time Δt to travel from Point A to Point B (see Fig. 2-7), the vehicle speed toward the center of the circle O can be obtained as

$$\Delta v = \Delta s \frac{v}{R} \tag{2-5a}$$

and the vehicle acceleration toward to the circle's center O as

$$a = \frac{\Delta v}{\Delta t} = \frac{v}{R}\frac{\Delta s}{\Delta t} = \frac{v^2}{R} \tag{2-5b}$$

According to Newton's second law of motion, the centrifugal force can be determined by

$$F = \frac{W_0}{g}a = \frac{W_0}{g}\frac{v^2}{R} \tag{2-6}$$

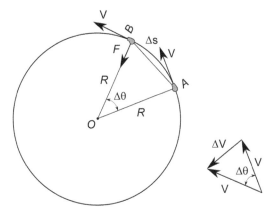

FIGURE 2-7 Centrifugal force.

where

W_0 = weight of the vehicle (kips)
v = highway design speed (ft/s)
g = gravitational acceleration: 32.2 (ft/s²)
R = radius of curvature of traffic lane (ft)

AASHTO HL-93 design live load consists of design lane and design truck or tandem loads, and the specifications stipulate that the centrifugal force is only applied to the design truck or tandem. The specified live load combination of the design truck and lane load represents a group of exclusion vehicles that produce force effects of at least 4/3 of those caused by the design truck alone on short- and medium-span bridges. To account for the lane loading effect on the centrifugal force, the AASHTO specifications increase the weight of the design truck and tandem by 4/3 for calculating the centrifugal force, i.e.,

$$CE = \frac{4}{3}\frac{W}{g}\frac{v^2}{R}$$ (2-7)

where W is the total axle weights of the design truck or design tandem. Centrifugal forces shall be applied horizontally at a distance 6.0 ft above the roadway surface.

2.2.4.4 Braking Force: BR

When a vehicle brakes, the braking force will be transmitted to the bridge deck through its tires. The magnitude of the braking force is related to the mass of the vehicle and its deceleration. Based on Newton's second law of motion, the braking force can be determined as follows:

$$F = \frac{W_0}{g}\frac{v}{t}$$ (2-8)

where

W_0 = weight of vehicle (kips)
v = vehicle speed at time of braking (ft/s)
g = gravitational acceleration: 32.2 (ft/s²)
t = time taken for truck to coming to a complete stop (s)

If the brake is suddenly applied to a truck with a weight of W_0 traveling at 55 mph (80.65 m/s) and it is assumed that the truck takes 10 s to stop completely, the bracing force is

$$F = \frac{W_0}{32.2} \frac{80.65}{10} = 0.25W_0$$

Based on the aforementioned and AASHTO standard specifications[2-3], the AASHTO LRFD specifications stipulate the design braking force BR shall be taken as the greater of:

- 25% of the axle weights of the design truck or design tandem
- 5% of the design truck plus lane load or 5% of the design tandem plus lane load

The design braking force should be placed in all design lanes with the multiple presence factors shown in Table 2-2 and be assumed to act horizontally at a distance of 6.0 ft above the roadway surface.

2.2.4.5   Vehicular Collision Force: CT

During the past two decades, several severe collisions have occurred between large heavy vehicles and bridge piers, some of which were catastrophic. In 1998, the AASHTO specifications introduced the requirement that vehicle collision forces shall be considered in bridge design. When a vehicle hits a bridge pier or an abutment, the interaction forces between the vehicle and the structure are related to the masses of the vehicle and pier, vehicle speed, vehicle and pier mechanical properties, angle of collision, etc. To accurately determine the structural response as a result of vehicular collision is a complex analysis problem. Theoretical analysis and test results have shown that the peak value of the interaction force between the vehicle and pier is very high[2-8] and cannot be directly used for bridge design. Currently, an equivalent static force is employed for bridge design. The equivalent static force is that force required to produce the same deflection at a given point that would be produced by a dynamic event.

The AASHTO specifications require that abutments and piers located within a distance of 30.0 ft to the edge of the roadway shall be designed for a collision force of 600 kips if there are no facilities to redirect or absorb the collision load. The equivalent static force of 600 kips is assumed to act in a direction of 0 to 15 degrees, with the edge of the pavement in a horizontal plane, at a distance of 5.0 ft above ground. The equivalent static force of 600 kips is based on the results from full-scale crash tests of rigid columns impacted by 80.0-kip tractor trailers at 50 mph[2-1].

2.2.5   Wind Loads

2.2.5.1   Introduction

Wind loads are considered to be dynamic loads. Their magnitude depends on many factors, such as the size and shape of the bridge, angle of attack of the wind, shielding effect of the terrain, and wind speed. The dynamic behavior as a result of wind–structure interaction is complicated, computationally intensive, and requires extensive knowledge to correctly interpret results, particularly when the structure experiences phenomenon such as buffeting, vortex shedding, and galloping. Generally, the wind force due to a constant wind speed on a solid object can be expressed by the equation

$$F_w = C_D C_A \rho A \frac{v^2}{2} \tag{2-9}$$

where
C_D = drag coefficient determined by test
C_A = modification factor for object area, wind direction, etc.

ρ = density of air
A = area of object facing wind
v = wind speed

However, the natural wind speed is rarely constant and varies along the height of structures due to the surrounding terrain. A peak wind velocity may be reached in a short period; it may remain for a while or it may decrease rapidly. If the time it takes to reach the peak pressure is equal to or greater than the inverse of the natural frequency of the structure, the wind loading may be treated as a static load equal to the peak pressure. Research results show that the afore-mentioned is true for most bridges. To simplify the wind load calculations in bridge design, the AASHTO specifications provide some equations for estimating the wind pressures for bridges with a span-to-depth ratio and structural components with a length-to-width ratio not exceed-ing 30.0. For the remaining bridges, aeroelastic force effects should be taken into account in the bridge design.

2.2.5.2   Wind Pressure on Structures: WS

2.2.5.2.1   Horizontal Wind Pressure Applied to Superstructure

Based on the AASHTO LRFD specifications[2-1], the design horizontal wind pressure on superstruc-tures can be determined by

$$p_D = P_B \left(\frac{V_{DZ}}{V_B} \right)^2 = P_B \frac{V_{DZ}^2}{10,000} \tag{2-10}$$

where
P_B = base wind pressure given in Table 2-3 (ksf)
V_{DZ} = design wind velocity at design elevation, Z (mph)
 = $2.5V_0 \left(\frac{V_{30}}{V_B} \right) \ln \left(\frac{Z}{Z_0} \right)$
V_B = base wind velocity of 100 mph at 30.0-ft height
V_{30} = wind velocity at 30.0 ft above design water level or above low ground (mph)
Z = height of structure at which wind loads are being calculated, measured from low ground or from water level (>30.0 ft)
V_0 = friction velocity provided in Table 2-4 (mph)
Z_0 = friction length of upstream fetch, taken as values shown in Table 2-4 (ft)

The wind force on the structure should be calculated by multiplying the design wind pressure p_D by the exposed area.

TABLE 2-3
Base Wind Pressures P_B (ksf)

Skew Angle of Wind	Arches, Trusses, Columns		Girders	
(degree)	Lateral	Longitudinal	Lateral	Longitudinal
0	0.075	0.000	0.050	0.000
15	0.070	0.012	0.044	0.006
30	0.065	0.028	0.041	0.012
45	0.047	0.041	0.033	0.016
60	0.024	0.050	0.017	0.019

TABLE 2-4
Friction Velocity V_0 and Friction Length Z_0

Structural Location	Open County	Suburban	City
V_0 (mph)	8.20	10.90	12.00
Z_0 (ft)	0.23	3.28	8.20

The parameters of friction velocity and length are used to consider the effect of upstream surface conditions on the wind speed. In Table 2-4, "open area" is defined as open terrain with scattered obstructions having heights generally less than 30.0 ft; "suburban" means urban and suburban areas or the terrain with many closely spaced single-family or larger dwellings; "city" means the large city centers with at least 50% of the buildings having a height exceeding 70.0 ft. More detailed definitions for the types of bridge locations can be found in ASCE 7-10[2-9]. The skew angle of wind in Table 2-3 is measured from a perpendicular to the longitudinal axis. Both lateral and longitudinal loads shall be applied simultaneously.

If the wind is taken as normal to the structure, the base wind pressures of 0.075 ksf shown in Table 2-3 should be divided into 0.050 ksf and 0.025 ksf and distributed to the related windward and leeward members, respectively, for truss, arch, and column structures. The total wind loading shall not be taken to be less than 0.30 kips per linear foot (klf) in the plane of a windward chord and 0.15 klf in the plane of a leeward chord on arch components, and not less than 0.30 klf on beam or girder spans.

2.2.5.2.2 Vertical Wind Pressure Applied to the Superstructure

AASHTO LRFD Article 3.8.2 requires that a vertical line wind load q_{wv} be considered in the design of bridges. The vertical line load should be applied at the windward quarter-point of the deck together with the horizontal wind loads.

$$q_{wv} = 0.02 \times total\ deck\ width\ (\text{klf}) \tag{2-11}$$

2.2.5.2.3 Horizontal Wind Pressure Applied to the Substructure

The AASHTO LRFD specifications require that a base wind pressure of 0.040 ksf be applied directly to the substructure in the transverse or longitudinal direction. The wind force on the substructure should be calculated by multiplying the base wind pressure by the exposed area. If the wind direction is taken skewed to the substructure, this wind pressure should be resolved into components perpendicular to the end and front elevations of the substructure. The wind pressure should be applied simultaneously with the wind loads from the superstructure.

2.2.5.3 Wind Pressure Applied to Vehicles: WL

The AASHTO LRFD specifications also require applying a wind load of 0.10 klf acting normal to and 6.0 ft above the roadway on vehicles. The 0.10-klf wind load is based on a long row of randomly sequenced passenger cars, commercial vans, and trucks exposed to a 55-mph design wind speed.

2.2.6 Earthquake Loads

2.2.6.1 Introduction

Bridges should be designed to have a low probability of collapse but may suffer significant damage and disruption to service when subject to earthquake ground motions that have a 7% probability of exceedance in 75 years. The effect of an earthquake on a bridge depends on the magnitude of the

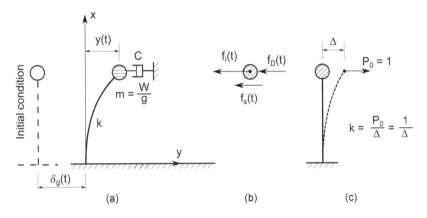

FIGURE 2-8 Lumped-Mass Single-Degree-of-Freedom Model, (a) Analytical Model, (b) Equilibrium Forces, (c) Lateral Stiffness of the Pier.

earthquake, the soil properties, the bridge mechanical characteristics, the distribution of weight, etc. It is difficult to accurately determine the effect of an earthquake on a bridge. The typical approach is to simplify the bridge as a single degree of freedom (SDOF) vibration system and to consider the earthquake-induced lateral forces acting in any direction at the center of gravity of the structure with a magnitude equal to a percentage of the weight of the structure. The lateral loads are then statically applied to the structures.

Figure 2-8 shows a lumped-mass SDOF bridge vibration model. The entire bridge superstructure is assumed to be rigid, and its total weight is lumped together at the top of the pier with a lateral stiffness of k; the effect of the pier mass is neglected. The lateral stiffness is defined as the horizontal force required to produce a unit horizontal displacement at the top of the pier and can be determined by using the following equation:

$$k = \frac{1}{\Delta}$$
(2-12)

where Δ is the horizontal displacement at the top of the pier produced by the horizontal force $P_0 = 1$ (see Fig. 2-8c). When an earthquake occurs, the ground around a bridge moves and the motion is transferred to the bridge's substructure and then to the superstructure. Assuming that the horizontal ground motion caused by the earthquake is denoted by δ_g and that the horizontal displacement of the superstructure is represented by $y(t)$, based on d'Alembert's principle, the variation of the inertial force of the superstructure with time t is

$$f_I(t) = \left(\ddot{\delta}_g(t) + \ddot{y}(t)\right)m$$
(2-13)

where
$\ddot{\delta}_g(t)$ = acceleration of substructure at ground elevation
$\ddot{y}(t)$ = acceleration of superstructure

$m = \frac{W}{g}$ = mass of superstructure

W = total weight of superstructure
g = gravitational acceleration

The idealized superstructure is subjected to three forces (see Fig. 2-8b): the inertial force, the elastic force applied by the pier, and the damping force due to the energy loss during the vibration.

The elastic force at the top of the pier caused by the displacement $y(t)$ is

$$f_S(t) = ky(t) \qquad (2\text{-}14)$$

The damping force can be assumed to be proportional to velocity $\dot{y}(t)$:

$$f_D(t) = c\dot{y}(t) \qquad (2\text{-}15)$$

where c = damping ratio.

All the forces acting on the should be in equilibrium; therefore,

$$f_I(t) + f_S(t) + f_D(t) = 0 \qquad (2\text{-}16)$$

Substituting Eqs. 2-12 to 2-15 into Eq. 2-16, the equation of motion for the SDOF model can be written as

$$m\ddot{y}(t) + c\dot{y}(t) + ky(t) = -m\ddot{\delta}_g(t) \qquad (2\text{-}17\text{a})$$

or

$$\ddot{y}(t) + \frac{c}{m}\dot{y}(t) + \omega^2 y(t) = -\ddot{\delta}_g(t) \qquad (2\text{-}17\text{b})$$

where $\omega = \sqrt{\frac{k}{m}}$ = circular frequency of the structure that represents the number of cycles of motion for a period of 2π seconds.

The frequency of motion of the structure is

$$f = \frac{\omega}{2\pi} = \text{cycles per second (Hz)}$$

The time required to complete one cycle is called the structural period of motion and is

$$T = \frac{1}{f} \qquad (2\text{-}18\text{a})$$

The right term of Eq. 2-17a represents the support excitation loading due to an earthquake. If the earthquake acceleration and the structural mechanical properties are known, Eqs. 2-17a and 2-17b can be solved using the Duhamel integral[2-10] to obtain the displacement response of the structure and the earthquake loading $f_S(t)$ applied to the superstructure.

The lumped-mass method is based on the assumptions that the entire superstructure is rigid and its mass is uniformly distributed in the bridge longitudinal and transverse directions. This method may overestimate the bridge earthquake responses. In this case, the single-mode method described below may provide more accurate results.

Any structure of arbitrary form can be treated as an SDOF system if it is assumed that its displacements are restricted to a single shape. Figure 2-9 shows a bridge pier having distributed mass $m(x)$ and stiffness EI(x). By analogy with the forgoing analysis for the lumped-mass SDOF system, the equilibrium of this pier at any location x can be written as

$$f_I(x, t) + f_S(x, t) + f_D(x, t) = 0 \qquad (2\text{-}19)$$

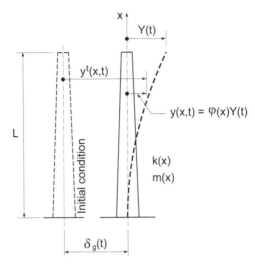

FIGURE 2-9   Generalized-Coordinate Single-Degree-of-Freedom Model.

where

$f_I(x, t)$ = distributed inertial force at x
$f_S(x, t)$ = distributed elastic forces at x
$f_D(x, t)$ = distributed damping force at x

Assuming the lateral displacements of the pier are given by the product of a single shape function $\varphi(x)$ and a generalized-coordinate amplitude $Y(t)$, then

$$y(x, t) = \varphi(x)Y(t)$$

The equation of motion of the generalized SDOF system can be developed conveniently by the principle of virtual work, which states that the external virtual work performed by the external loadings acting through their corresponding virtual displacements is equal to the internal virtual work, i.e.,

$$\delta W_E = \delta W_I \tag{2-20a}$$

where

δW_E = external virtual work = $-\int_0^L [f_I(x, t) + f_S(x, t) + f_D(x, t)]\delta y(x, t)dx$

δW_I = internal virtual work = $\int_0^L M(x, t)\delta y(x, t)'' dx$

$M(x, t)$ = moment in pier
$\delta y(x, t)$ = virtual displacement
$\delta y(x, t)''$ = virtual curvature

Substituting δW_E and δW_I into Eq. 2-20a yields the equation of motion for the pier:

$$\bar{m}\ddot{Y}(t) + \bar{c}\dot{Y}(t) + \bar{k}Y(t) = \beta\ddot{\delta}_g \tag{2-20b}$$

or

$$\ddot{Y}(t) + \frac{\overline{c}}{\overline{m}} \dot{Y}(t) + \frac{\overline{k}}{\overline{m}} Y(t) = \frac{\beta}{\overline{m}} \ddot{\delta}_g \qquad (2\text{-}20c)$$

where

$\overline{m} = \int_0^L m(x)\varphi(x)^2 dx$ = generalized mass

$\overline{c} = c \int_0^L EI(x)\ddot{\varphi}(x)^2 dx$ = generalized damping

$\overline{k} = \int_0^L EI(x)\ddot{\varphi}(x)^2 dx$ = generalized stiffness

$\beta = \int_0^L m(x)\varphi(x)dx$ = generalized effective mass

By comparison of Eqs. 2-17b and 2-20c, we can notice that both of the equations have only one variable and that factor $\frac{\beta}{\overline{m}}$ characterizes the difference between the lumped and the generalized SDOF responses. The dynamic response to earthquake loads applied to the pier can be obtained by the method similar to that of the lumped-mass SDOF system described above.

2.2.6.2   Earthquake Loads Determined by AASHTO LRFD Specifications

2.2.6.2.1   General

The AASHTO LRFD specifications require analysis of seismic effects only for multispan bridges and recommend that four different analytical methods be used based on the bridge configurations and operation importance (see Table 2-5): uniform load (UL) elastic method, single-mode (SM) elastic method, multimode (MM) elastic method, and time history (TH) method.

The "regular bridges" shown in Table 2-5 mainly include the straight bridge or curved bridges with a maximum subtended angle not greater than 90°, which meet the following conditions:

- maximum span number less than 7
- maximum span length ratio from span to span not greater than 1.5

The detailed requirements for regular bridges can be found in the AASHTO specifications[2-1].

The UL, SM, and MM methods are based on the design response spectrum adopted by the specifications and indicating the relationship between the bridge responses and bridge vibration periods (see Section 2.2.6.2.2). The TH method is used to develop time histories of the bridge responses typically though inputting the time histories of accelerations using step-by-step numerical methods. In this section, the UL and SM methods will be briefly discussed.

TABLE 2-5
Minimum Analysis Requirements for Seismic Effects for Multispan Bridges

Seismic Zone	Other Bridges		Essential Bridges		Critical Bridges	
	Regular	Irregular	Regular	Irregular	Regular	Irregular
1	N/A	N/A	N/A	N/A	N/A	N/A
2	SM/UL	SM	SM/UL	MM	MM	MM
3	SM/UL	MM	MM	MM	MM	TH
4	SM/UL	MM	MM	MM	TH	TH

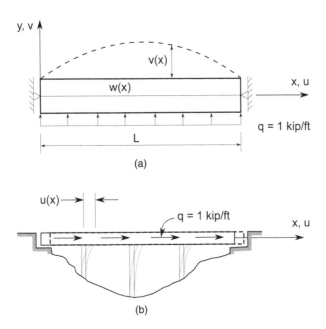

FIGURE 2-10   Bridge Model for Single-Mode Methods, (a) Plan View and Transverse Loading, (b) Elevation View and Longitudinal Loading.

2.2.6.2.2   *Uniform Load (UL) Elastic Method*

The uniform load elastic method is essentially a lumped-mass method and is based on the assumption that the inertial forces acting on the superstructure are uniform. In this method, the vibration period of the bridge is first determined based on bridge average stiffness. Then, the earthquake effects on the bridge can be obtained based on the response spectrum recommended by the related specifications (see Fig. 2-11).

Figure 2-10 shows the transverse displacement $v(x)$ and the longitudinal displacements $u(x)$ for a typical bridge under unit loading $q = 1$ kip/ft. In the UL method, the maximum displacements v_{max} and u_{max} are used to determine the average equivalent lateral stiffness K_v and longitudinal stiffness K_u of the bridge, i.e.,

$$K_v = \frac{L}{v_{max}} \qquad (2\text{-}21a)$$

$$K_u = \frac{L}{u_{max}} \qquad (2\text{-}21b)$$

where
  L = total length of bridge (ft)
  v_{max} = maximum displacement in bridge transverse direction due to unit load $q = 1$ kip/ft
  u_{max} = maximum displacement in bridge longitudinal direction due to unit load $q = 1$ kip/ft

Assuming the distributed dead load of the bridge superstructure and tributary substructure is $w(x)$, then the total bridge weight (kips) can be expressed as

$$W = \int_0^L w(x)dx$$

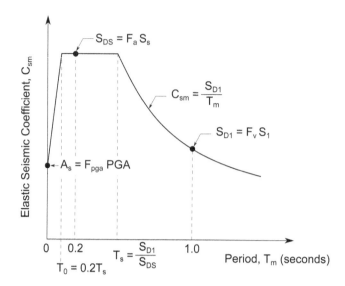

FIGURE 2-11 AASHTO Design Response Spectrum[2-1].

where $w(x)$ = dead load distribution of superstructure and tributary substructures.

The period of the bridge is

$$T_m = 2\pi\sqrt{\frac{W}{gK}} \tag{2-18b}$$

where

K = bridge equivalent stiffness = K_v or K_u
g = acceleration of gravity (ft/s²)

The equivalent static earthquake loading per unit length along the bridge can be determined as

$$q_e = \frac{C_{sm}W}{L} \tag{2-22}$$

where C_{sm} = elastic seismic response coefficient = function of period T_m (see Fig. 2-11).

If the period T_m is known for the mth vibration mode, then the elastic seismic response coefficient can be obtained using the following equations (see Fig. 2-11):

$$C_{sm} = A_s + (S_{DS} - A_S)\left(\frac{T_m}{T_0}\right), \text{ if } T_m \le T_0 \tag{2-23a}$$

$$C_{sm} = S_{DS}, \text{ if } T_0 \le T_m \le T_S \tag{2-23b}$$

$$C_{sm} = \frac{S_{D1}}{T_m}, \text{ if } T_m > T_S \tag{2-23c}$$

where

A_S = $F_{pga}PGA$
S_{DS} = F_aS_S
S_{D1} = F_VS_1 = acceleration coefficient
T_0 = $0.2T_S(\text{s})$
T_S = $\frac{S_{D1}}{S_{DS}}(\text{s})$

PGA = peak ground acceleration coefficient on rock (Site Class B) (see Fig. 2-12)
S_S = horizontal response spectral acceleration coefficients at 0.2-s period on rock (see Fig. 2-13)
S_1 = horizontal response spectral acceleration coefficients at 1.0-s period on rock (see Fig. 2-14)
F_{pga} = site factors for zero period (see Table 2-6)
F_a = site factors in short-period range (see Table 2-6)
F_V = site factors in long-period range (see Table 2-6)

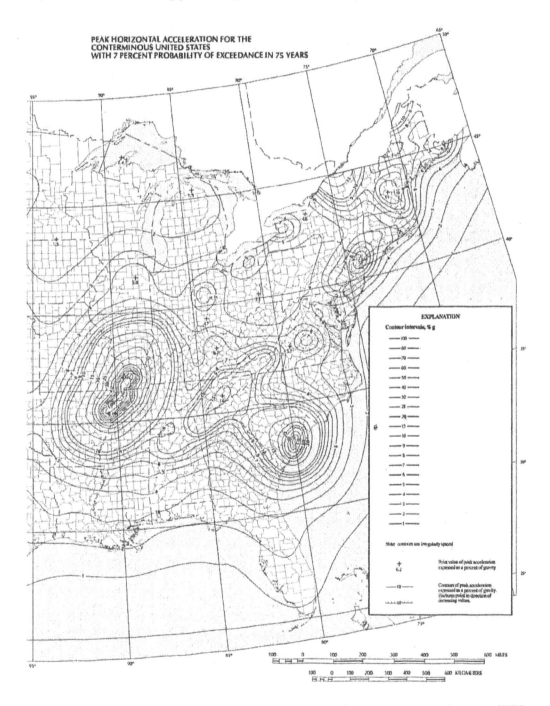

FIGURE 2-12   PGA with 7% Probability of Exceedance in 75 years[2-1] (Used with permission by AASHTO).

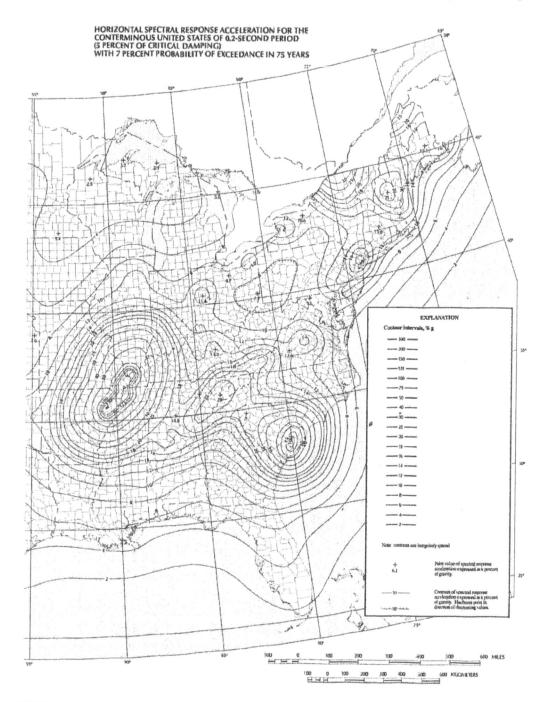

FIGURE 2-13 S_S with 7% Probability of Exceedance in 75 years[2-1] (Used with permission by AASHTO).

The values of PGA, S_S, and S_1 can be found in the AASHTO LRFD specifications or from state ground motion maps approved by the owner. Some of the ground motion maps contained in AASHTO[2-1] are shown in Figs. 2-12 to 2-14. The values of site factors, F_{pga}, F_a, and F_v, can be found in Table 2-6; the definitions for the site classes are presented in Table 2-7. The values of shear wave velocity $\overline{v}_s$ contained in Table 2-7 can be determined by the procedures recommended by the AASHTO specifications[2-1].

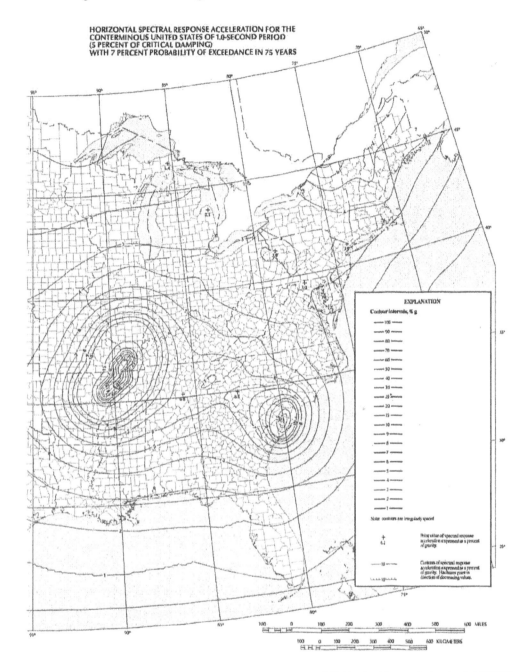

FIGURE 2-14 S_1 with 7% Probability of Exceedance in 75 years[2-1] (Used with permission by AASHTO).

To reflect the variation in seismic risk in different areas and specify different design requirements for analytical methods, minimum support lengths, column design details, etc., each bridge should be assigned to one of the four seismic zones based on the acceleration coefficient S_{D1}:

Seismic Zone 1: $S_{D1} \leq 0.15$
Seismic Zone 2: $0.15 \leq S_{D1} \leq 0.30$
Seismic Zone 3: $0.30 \leq S_{D1} \leq 0.50$
Seismic Zone 4: $0.50 < S_{D1}$

TABLE 2-6

Values of Site Factors, F_{pga}, F_a, and F_v

Site Factors		Site Class				
		A	B	C	D	E
F_{pga}	PGA < 0.10	0.8	1.0	1.2	1.6	2.5
	PGA < 0.20	0.8	1.0	1.2	1.4	1.7
	PGA < 0.30	0.8	1.0	1.1	1.2	1.2
	PGA < 0.40	0.8	1.0	1.0	1.1	0.9
	PGA < 0.50	0.8	1.0	1.0	1.0	0.9
F_a	$S_S < 0.25$	0.8	1.0	1.2	1.6	2.5
	$S_S < 0.50$	0.8	1.0	1.2	1.4	1.7
	$S_S < 0.75$	0.8	1.0	1.1	1.2	1.2
	$S_S < 1.00$	0.8	1.0	1.0	1.1	0.9
	$S_S < 1.25$	0.8	1.0	1.0	1.0	0.9
F_v	$S_1 < 0.1$	0.8	1.0	1.7	2.4	3.5
	$S_1 < 0.2$	0.8	1.0	1.6	2.0	3.2
	$S_1 < 0.3$	0.8	1.0	1.5	1.8	2.8
	$S_1 < 0.4$	0.8	1.0	1.4	1.6	2.4
	$S_1 < 0.5$	0.8	1.0	1.3	1.5	2.4

TABLE 2-7

Site Class Definitions

Site Class	Soil Type and Profile
A	Hard rock with measured shear wave velocity, $\bar{v}_s > 5000$ ft/s
B	Rock with 2500 ft/s $< \bar{v}_s < 5000$ ft/s
C	Very dense soil and soil rock with 1200 ft/s $< \bar{v}_s < 2500$ ft/s
D	Stiff soil with 600 ft/s $< \bar{v}_s < 1200$ ft/s
E	Soil profile with $\bar{v}_s < 600$ ft/s

2.2.6.2.3   Single-Mode (SM) Spectral Method

The SM spectral method is essentially the generalized-coordinate SDOF method. From Eqs. 2-17b, 2-20c and 2-22 as well as recognizing that the difference between the UL and SM models is the factor $\frac{\beta}{m}$, the equivalent earthquake loading in bridge longitudinal and transverse directions can be written as

$$\text{Longitudinal direction: } q_{ex} = \frac{C_{sm}\beta_x}{\bar{m}_x}w(x)u(x) \tag{2-24a}$$

$$\text{Transverse direction: } q_{ey} = \frac{C_{sm}\beta_y}{\bar{m}_y}w(x)v(x) \tag{2-24b}$$

where

$$\bar{m}_x = \int_0^L w(x)u(x)^2 dx$$
$$\bar{m}_y = \int_0^L w(x)v(x)^2 dx$$
$$\beta_x = \int_0^L w(x)u(x)dx$$
$$\beta_y = \int_0^L w(x)v(x)dx$$

and

$v(x)$ = displacement function of superstructure in transverse direction due to unit load (see Fig. 2-10a)

$u(x)$ = displacement function of superstructure in longitudinal direction due to unit load (see Fig. 2-10b)

The elastic seismic response coefficient C_{sm} can be determined from Eq. 2-23 or Fig. 2-11 based on the bridge longitudinal and transverse vibration periods T_{mx} and T_{my}:

$$T_{mx} = 2\pi \sqrt{\frac{\bar{m}_x}{g\bar{k}_x}} \qquad (2\text{-}25a)$$

$$T_{my} = 2\pi \sqrt{\frac{\bar{m}_y}{g\bar{k}_y}} \qquad (2\text{-}25b)$$

where
$$\bar{k}_x = \int_0^L u(x)dx$$
$$\bar{k}_y = \int_0^L v(x)dx$$

2.2.7   WATER LOADS (WA)

Substructures subject to flowing water should be designed to withstand water pressure, and the effects of buoyancy should be considered in the design. From the foregoing discussion for wind pressure, the longitudinal stream pressure can be written as follows:

$$p_{wx} = \frac{C_D V^2}{1,000} \qquad (2\text{-}26a)$$

where
p_{wx} = longitudinal pressure of flowing water (ksf)
V = design velocity of water determined by the owner (ft/s)
C_D = drag coefficient for piers specified by AASHTO[2-1]
C_D = 0.7 for semicircular-nosed pier
C_D = 1.4 for square-ended pier and debris lodged against the pier
C_D = 0.8 for wedged-nosed pier with nose angle 90° or less

The lateral stream pressure can be written as:

$$p_{wy} = \frac{C_L V^2}{1,000} \qquad (2\text{-}26b)$$

where
p_{wy} = lateral pressure of flowing water (ksf)
C_L = lateral drag coefficient related to the angle θ, between the direction of flow and the longitudinal axis of the pier, specified by AASHTO[2-1]
      = 0.0 for $\theta = 0°$
      = 0.5 for $\theta = 5°$
      = 0.7 for $\theta = 10°$
      = 0.9 for $\theta = 20°$
      = 1.0 for $\theta \geq 30°$

The total drag force is the product of the stream pressure and the projected surface exposed.

2.2.8   Vessel Collision Loads (CV)

Some catastrophic failures of bridges due to collisions by aberrant vessels have occurred in the past in the United States and the world. Currently, the AASHTO specifications require that all bridge components in a navigable waterway crossing, located in design water depths not less than 2.0 ft, shall be designed for vessel impact. The determination of the impact load on a bridge structure during a ship collision is complex and depends on many factors, such as the structural type and shape of the ship's bow, degree of water ballast carried in the forepeak of the bow, size and velocity of the ship, the pier shape, and mechanical characteristics. Based on the research results obtained by Woisin[2-11], the AASHTO specifications recommend that the head-on ship collision impact force on a pier can be calculated as

$$P_s = 8.15 \, V\sqrt{DWT} \tag{2-27}$$

where
   P_s = equivalent static vessel impact force (kip)
   DWT = deadweight tonnage of vessel (tonne)
   V = vessel impact velocity (ft/s)

More information regarding the determination of vessel collision forces can be found in AASHTO[2-1] and IABSE, 1983[2-12]. It is a common practice to use physical protection systems, such as fenders, pile-supported structures, dolphins, and islands, to reduce or to eliminate the exposure of bridge substructures to vessel collision. The owner shall establish and/or approve the bridge operational classification, the design velocity of vessels, and protective systems for the bridge.

2.2.9   Ice Loads (IC)

In cold climates, floating ice can cause high forces to bridge piers and may even cause bridge collapse. The ice loads on the substructures can be classified as dynamic and static loadings. The dynamic forces caused by a moving ice floe hitting the bridge pier are related to the shape, size, strength, and speed of the moving floe. The static forces may be caused by thermal expansion of ice. Ice forces on piers should be determined based on the site conditions, such as the expected thickness of ice, the direction of its movement, and the height of its action determined by field investigations, review of public records, etc. More information regarding ice loads can be found in AASHTO LRFD specifications[2-1].

2.2.10   Temperature Loads

Temperature change will cause bridge deformations and thus cause additional internal forces if the deformations have been restrained. The effects of temperature on bridges should be evaluated in bridge design.

2.2.10.1   Uniform Temperature

The uniform temperature change will cause bridge horizontal movements. The AASHTO specifications[2-1] provide Procedure A or Procedure B for determining the temperature ranges. Both of the procedures can be used for bridge design. The temperature ranges for Procedure A are shown in Table 2-8. The designer should note that each state in the United States may have different temperature ranges that supersede the temperature ranges specified in the AASHTO specifications.

   The design thermal movement range ΔL can be determined as:

$$\Delta L = \alpha \, L \left(T_{max} - T_{min} \right) \tag{2-28}$$

TABLE 2-8

Design Temperature Ranges for Procedure A for Concrete Bridges

Climate	T_{min}	T_{max}
Moderate	10°F	80°F
Cold	0°F	80°F

where

L = expansion length (in.)
α = coefficient of thermal expansion (in./in./°F)
T_{min} = minimum design temperature
T_{max} = maximum design temperature

2.2.10.2   Temperature Gradient

The superstructure temperature distribution is a complex phenomenon, and it varies along the depth of the structure. It depends on many factors, such as structure type, locations, and geometry. To simplify the design, the AASHTO specifications provide an idealized positive vertical temperature gradient as shown in Fig. 2-15 for concrete bridges. The temperature magnitudes T_1 and T_2 shown in Fig. 2-15 are provided in Table 2-9. The solar radiation zones for the United States are presented in Fig. 2-16. Temperature value T_3 shall be taken as 0.0°F, unless a site-specific study is made to determine an appropriate value, but it shall not exceed 5°F. The negative vertical temperature gradient can be obtained by multiplying the values specified in Table 2-9 by −0.30 for plain concrete decks and −0.20 for decks with an asphalt overlay.

The analysis of the stresses and structure deformations due to both positive and negative temperature gradients will be discussed in Chapter 4.

2.2.11   MISCELLANEOUS LOADS

Other forces, such as blast loading, earth pressures, and forces caused by foundation settlements, should also be considered in the design of the bridges under particular conditions. The designer should consult the design specifications or perform an evaluation using sound engineering judgment

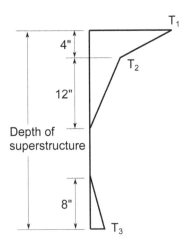

FIGURE 2-15   Positive Vertical Temperature Gradient in Concrete.

TABLE 2-9
Basic Temperature Gradients[2-1]

Solar Radiation Zone	T_1 (°F)	T_2 (°F)
1	54	14
2	46	12
3	41	11
4	38	9

for the particular condition. The forces caused by elastic shortening, shrinkage, and creep will be discussed in Chapter 4.

2.2.12 CONSTRUCTION LOADS

Most concrete segmental bridges have different structural systems during construction than their final structural systems. It is essential that bridge engineers adequately estimate the construction loads and stresses that the structure will be subject to during construction. In additional to the applicable loads discussed in previous sections, the following construction loads should be considered for capacity checks during bridge construction.

   a. Differential load (DIFF)
      DIFF = 2% of dead load applied to one cantilever for balanced cantilever construction (kip)
   b. Distributed construction live load (CLL)
      CLL = allowance for miscellaneous construction equipment, apart from major specialized erection equipment
      = 0.010 ksf of deck area
      = 0.010 ksf on one cantilever and 0.005 ksf on the other in cantilever construction
   c. Specialized construction equipment load (CEQ)
      CEQ = dead weight from material delivery trucks, form traveler, launching gantry, or major auxiliary structure and maximum loads applied to the structure by equipment during the lifting of segments (kip)
   d. Dynamic load from equipment (IE)
      IE is determined according to the type of machinery anticipated (kip).

FIGURE 2-16   Solar Radiation Zones for the United States[2-1] (Used with permission by AASHTO).

  e. Horizontal wind load on equipment (WE)
     WE = 0.1 ksf of exposed surface (ksf)
  f. Wind uplift load on cantilever (WUP)
     WUP = 0.005 ksf of deck area for balanced cantilever construction applied to one side only
  g. Dynamic load due to accidental release (AI)
     AI = 100% of segment weight (kip)

2.3   GENERAL DESIGN METHODS

2.3.1   INTRODUCTION

As previously mentioned, one of the main purposes for bridge design is to provide a structure that can safely carry all anticipated loads during its service life. The questions are how to measure the safety and how safe is safe enough. As in the foregoing discussions, there are many variable loads that may be applied to the highway bridges. It would be very expensive and unreasonable to design all bridges with 100% safety, particularly because types of risk other than load are acceptable matter, such as the risk of severe earthquakes, hurricane, volcanic eruption, and fire.

Currently, the AASHTO Standard Specifications for Highway Bridges[2-3] contains two design methods; the allowable stress design (ASD) method and the load factor design (LFD) method. In the ASD method, the stress of the material is limited in the material elastic range under the service loads. Therefore, the ASD method is also named the service load design (SLD) method. The ASD method is the earliest design methodology, and it was primarily used to design metal structures at the beginning. In this method, the effects of the design loads are limited to a fraction of the yield strength F_y or other properties of the material. For example, based on the AASHTO Standard Specifications for Highway Bridges, the design equation for an axial tension steel member in a superstructure, subjected to Load Combination Group I, can be written as

$$f_{tD} + f_{tL} \leq 0.55 \, F_y = \frac{F_y}{1.82} \qquad (2\text{-}29a)$$

or

$$1.82 \, f_{tD} + 1.82 \, f_{tL} \leq F_y \qquad (2\text{-}29b)$$

where
  f_{tD} = tensile stress caused by dead loads
  f_{tL} = tensile stress caused by live loads
  F_y = yield stress of steel

From Eq. 2-29a, it can be observed there is a safety factor of 1.82 for both dead load and live load. The safety factor is used to account for the unpredictability of the various loads based on the engineers' confidence. The authors of the code subjectively determined the safety factor. However, it is apparent that the dead load can more accurately be determined than the live load. The use of the same safety factor for both dead and live loads becomes questionable and has been addressed by the LFD method. The design equation 2-29b is revised as

$$1.30 \, f_{tD} + 2.17 f_{tL} \leq F_y \qquad (2\text{-}30)$$

From Eq. 2-30, we can observe that the safety factor discussed above has been moved to the left side of Eq. 2-30 and becomes two load factors: 1.30 for the dead loads and 2.17 for the live load, respectively. A comparison between Eqs. 2-29b and 2-30 indicates that the LFD method reduces the dead load safety factor to 1.30 and increases the life load safety factor to 2.17. Though the LFD method provides more uniform safety for the structures, the load factors were still determined subjectively by the authors of the code based on previous experiences. Both the ASD and LFD methods only account for the variability of loads.

The main goal of the AASHTO LRFD specifications[2-1] is to ensure that the bridge structures designed according to the code provisions will have more uniform safety levels for different limit states by specifying both load and resistance factors determined based on probability-based reliability theory. As the design method contained involves both load factors and resistance factors, the design method is called load and resistance factor design (LRFD).

2.3.2 BASIC THEORY OF THE **LRFD** METHOD

2.3.2.1 Mathematical Models of Load and Resistance Variations

Strictly speaking, everything in the world will not be exactly the same and will vary with time, locations, and many other factors; the same is true for the bridge loads and structural resistances. Figure 2-17a illustrates a variation of concrete strengths based on the test results from 238 samples. The abscissa is the concrete compression strength, and the coordinate represents the number of times for a particular concrete strength (interval). Figure 2-17b shows a histogram of vehicle gross weights obtained from weigh-in-motion (WIM) vehicle records collected in 2007. The abscissa is

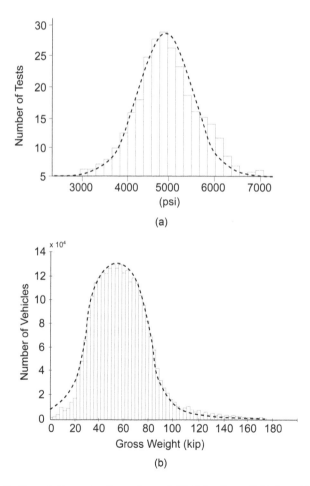

FIGURE 2-17 Distribution of Concrete Strengths and Vehicle Gross Weights, (a) Concrete Strengths, (b) Vehicle Gross Weights.

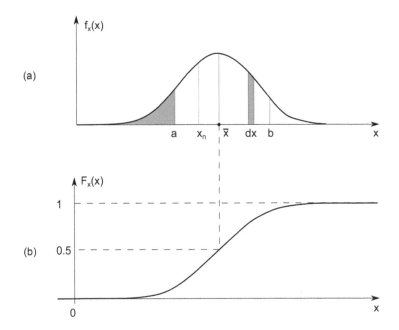

FIGURE 2-18 Idealization of the Distributions of Load and Resistance, (a) Normal Density Function, (b) Cumulative Distribution Function.

the vehicle gross weight, and the coordinate represents the number of times for a particular gross weight of vehicle (interval).

From Fig. 2-17, it can be observed that both the distribution of concrete strengths and vehicle gross weights can be approximately described by a bell-shape curve as shown by the dotted lines. In general, the natural phenomenon of load and residence distributions can be described by a continuous function shown in Fig. 2-18a. In Fig. 2-18a, the abscissa x is mathematically called the normal random variable that represents the events (i.e., tested or measured results) in a sample space mapped on the axis of the real number, such as the load or the resistance. The sample space is defined as all possible results of the random variable. The curve is called a normal density function $f_x(x)$ representing the distribution of the frequency of occurrence (possibility) in range dx within the sample space and can be written as:

$$f_x(x) = \frac{1}{\sigma\sqrt{2\pi}} e^{-\frac{(x-\bar{x})^2}{2\sigma^2}} \qquad (2\text{-}31)$$

where $\bar{x}$ = average value = mean.

For a continuous random variable

$$\bar{x} = \int_{-\infty}^{\infty} x f_x(x)\,dx \qquad (2\text{-}32a)$$

For a discrete random variable, such as tested concrete strengths and measured vehicle weights,

$$\bar{x} = \frac{\sum_{i=1}^{n} x_i}{n} \qquad (2\text{-}32b)$$

where
  x_i = tested or measured values
  n = number of samples
  σ = standard deviation used for measuring dispersion of tested or measured data

For a continuous random variable,

$$\sigma = \sqrt{\int_{-\infty}^{\infty} (x - \bar{x})^2 f_x(x)dx} \qquad (2\text{-}33a)$$

For a discrete random variable, such as tested concrete strengths and measured vehicle weights,

$$\sigma = \sqrt{\frac{\sum_{i=1}^{n} (x_i - \bar{x})^2}{(n-1)}} \qquad (2\text{-}33b)$$

It is common practice not to use the mean values in actual design. A value different from the mean value is usually used in the design and called nominal value x_n (see Fig. 2-18a). The nominal value for the resistance will usually be smaller than the mean value, and the normal value for the load will be normally greater than the mean value. There is a difference between the nominal value and the mean value. This difference is referred to as "bias" and defined as the ratio of the mean value to nominal value:

  Bias factor:

$$\lambda = \frac{\bar{x}}{x_n} \qquad (2\text{-}34)$$

To further quantify a measurement of dispersion for tested data, it is common to use the coefficient of variance expressed as a ratio of the standard deviation to the mean value:

  Coefficient of variance:

$$V = \frac{\sigma}{\bar{x}} \qquad (2\text{-}35)$$

The total area under the bell-shaped curve represents the possibility of the entire events (such as all test results) in the sample space and is equal to 1. From Fig. 2-18a, the probability of random variable $x \leq a$ can be written as:

$$P(x \leq a) = \int_{-\infty}^{a} f_x(x)dx \qquad (2\text{-}36a)$$

where $P(x \leq a)$ denotes the probability that random variable $x \leq a$. Similarly,
  $P(a \leq x \leq b) = \int_a^b f_x(x)dx$ = probability that random variable x is within $a \leq x \leq b$
  From probability theory, we can obtain

$$P(\bar{x} - \sigma \leq x \geq \bar{x} + \sigma) = 0.6826 \qquad (2\text{-}36b)$$
$$P(\bar{x} - 2\sigma \leq x \geq \bar{x} + 2\sigma) = 0.9770 \qquad (2\text{-}36c)$$
$$P(\bar{x} - 3\sigma \leq x \geq \bar{x} + 3\sigma) = 0.9987 \qquad (2\text{-}36d)$$

Mathematically, $F_x(x) = \int_{-\infty}^{x} f_x(x)dx$ is referred to as the cumulative distribution function and is shown in Fig. 2-18b.

From Fig. 2-18b, it can be observed that:

$$0 \leq F_x(x) \leq 1 \tag{2-37}$$

$$F_x(-\infty) = 0 \tag{2-38}$$

$$F_x(\infty) = 1 \tag{2-39}$$

2.3.2.2 Probability of Failure and Safety Index

Structural failure can be defined as the realization of the failure of a member of predefined limit states; i.e., its resistance is smaller than its response due to the loads. As with the previous discussion, it may be impossible to design all bridges to resist all possible loads during its service life. The AASHTO LRFD specifications provide the load and resistance factors to ensure that each possible limit states is reached only with an acceptable small possibility of failure.

If both the load and resistance are normal random variables, the normal distribution functions for each variable can be plotted in the same coordinate system (see Fig. 2-19)[2-13]. In Fig. 2-19, the abscissa represents the value of load Q or resistance R, and the coordinate is their distributions $f_{(R,Q)}$. The mean value of load $\bar{Q}$ and the mean value of resistance $\bar{R}$ are also shown in Fig. 2-19. The "nominal" values, or the numbers that are used by designers to calculate the load and the resistance, are denoted as Q_n and R_n individually. In bridge design, the nominal load and the nominal resistance are factored, not the mean values. From Eq. 2-34, we can obtain:

$$\bar{R} = \lambda_R R_n \tag{2-40a}$$

$$\bar{Q} = \lambda_Q Q_n \tag{2-40b}$$

where λ_R and λ_Q are bias factors for resistance and load, respectively.

The overlapped area (shaded area) in Fig. 2-19 represents the structural failure area in which resistance $R <$ load Q. The objective of the bridge design is to separate the distribution of resistance from the distribution of load, by carefully selecting the load factor γ and resistance factor ϕ such that the area where the load is greater than the resistance is tolerably small.

To simplify the discussion, a new random variable $RQ = R - Q =$ the difference between the resistance and load can be defined. As both R and Q are normal random variables, based on the probability theory, RQ is also a normal random variable. Its distribution function f_{RQ} is shown in Fig. 2-20. From Fig. 2-20, it can be seen that structural failure occurs when $f_{RQ} \leq 0$, i.e., $R \leq Q$. The shaded area on the left side of the vertical axis shown in Fig. 2-20 represents the probability of failure. The

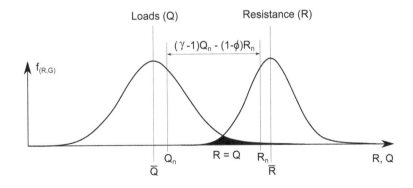

FIGURE 2-19   Separation of Load and Resistance (CRC).

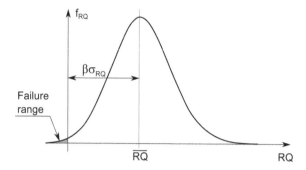

FIGURE 2-20 Definition of Reliability Index, β (CRC).

larger the mean $\overline{RQ}$, the smaller the probability of failure. From Eqs. 2-35 and 2-36, we can observe that it is easier to use the number of standard deviation to measure the probability of failure:

$$\overline{RQ} = \beta\sigma_{RQ} \tag{2-41}$$

where β is defined as the safety index or reliability index and σ_{RQ} denotes the standard deviation of the difference RQ. The larger the β value, the less likely is the probability of failure.

From probability theory, the mean and standard deviation of the difference RQ are:
Mean:

$$\overline{RQ} = \overline{R} - \overline{Q} \tag{2-42}$$

Standard deviation:

$$\sigma_{RQ} = \sqrt{\sigma_R^2 + \sigma_Q^2} \tag{2-43}$$

Then,

$$\beta = \frac{\overline{R} - \overline{Q}}{\sqrt{\sigma_R^2 + \sigma_Q^2}} \tag{2-44}$$

From Eq. 2-44, it can be observed that the reliability index β can be obtained if the distributions of the load and resistance are known. From Eqs. 2-36b to 2-36d, it can be seen that if β = 2, approximately 97.7% of the values under the bell-shaped curve are on the right side of the vertical axis; i.e., the possibility of the failure is 2.3%. If β = 3.0, 99.87% of the values will be under the bell-shaped curve on the right side of the vertical axis; i.e., the probability of failure is only 1.3/1000. Based on the analysis of existing bridges, the AASHTO specifications use target reliability index β = 3.5 for stength limit state I, which corresponds to the probability of failure of 2/10,000. The relationship between the probability of failure and the reliability index is illustrated in Fig. 2-21.

2.3.2.3 Determining Load and Resistance Factors

The design objective is that the factored resistance must be greater than or equal to the sum of factored loads:

$$\phi R_n = \gamma Q_n = \sum \gamma_i x_i \tag{2-45}$$

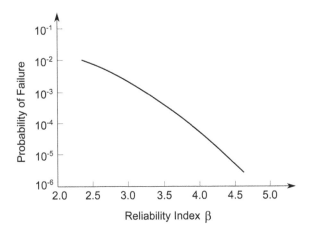

FIGURE 2-21 Variation of Probability Failure with the Reliability Index.

where

ϕ = resistance factor
γ = generalized load factor
γ_i = load factor corresponding to load x_i

From Eqs. 2-40 (a) and 2-44, we can obtain

$$\bar{R} = \lambda_R R_n = \bar{Q} + \beta\sqrt{\sigma_R^2 + \sigma_Q^2}$$

or

$$R_n = \frac{\bar{Q} + \sqrt{\sigma_R^2 + \sigma_Q^2}}{\lambda_R} \tag{2-46}$$

Substituting Eq. 2-46 into Eq. 2-45 yields

$$\phi = \frac{\lambda_R \sum \gamma_i x_i}{\bar{Q} + \beta\sqrt{\sigma_R^2 + \sigma_Q^2}} \tag{2-47}$$

From Eq. 2-47, it can be seen that there are still two variables after the code-writer body chooses the reliability index. One way to resolve this issue is by first selecting the load factor and then calculating the resistance factors.

Factored loads can be assumed as an average value of load plus some number of standard deviation of the load, i.e.,

$$\gamma_i x_i = \bar{x}_i + n\sigma_i \tag{2-48}$$

Using Eq. 2-35, Eq. 2-48 can be written as

$$\gamma_i x_i = \bar{x}_i + nV_i\bar{x}_i$$

or

$$\gamma_i = \lambda_R\left(1 + nV_i\right) \tag{2-49}$$

From Eq. 2-49, the load factor γ_i can be calculated if the distribution of load x_i is known. The general procedures for determining load and resistance factor are:

Step 1: Calculate load factors by Eq. 2-49.

Step 2. Using Eq. 2-47 for a given set of load factors and assumed reliability index, calculate the resistance factors for various types of structural members and for various load components, e.g., shear and moment, on the various structural components. A large number of values for the resistance factor can be obtained with computer simulations of a representative body of structural members.

Step 3. Group the resistance factors by structural member and load component to determine if they cluster around convenient values. If the results are closely clustered, a suitable combination for the load and resistance factors has been obtained.

Step 4. If the results are not closely clustered, a new trial set of load factors can be used and the process is repeated until the resistance factors do cluster around a workable number of narrowly defined values.

More detailed information regarding the calibration of the LFRD code can be found in References 2-13 to 2-18.

2.3.3 GENERAL PROVISIONS OF THE AASHTO LRFD METHOD

2.3.3.1 General Design Equation

The issue of safety is usually codified by an application of the general statement that the design resistances must be greater than or equal to the design load effects. The basic design equation in the AASHTO LRFD Bridge Specifications that should be satisfied for all limit states and for global structures and local members[2-1] is

$$\sum \eta_i \gamma_i Q_i \leq \phi R_n = R_r \tag{2-50}$$

where

η_i = load modifier

    = $\eta_R \eta_D \eta_I \geq 0.95$, if a maximum load factor of γ_i specified in Table 2-10 is used

    = $\frac{1}{\eta_R \eta_D \eta_I} \leq 1.0$, if a minimum load factor of γ_i specified in Table 2-10 is used

γ_i = load factor, taken as in Table 2-10

ϕ = resistance factor (see Section 2.3.3.5)

η_R = redundancy factor

For the strength limit state:

$\eta_R \geq 1.05$, for nonredundant members

    = 1.00, for conventional levels of redundancy and foundation elements

    ≥ 0.95, for exceptional levels of redundancy

For all other limit states, $\eta_R = 1.00$.

η_D = ductility factor

For the strength limit state:

$\eta_D \geq 1.05$, for nonductile components and connections

    = 1.00, for conventional designs and details

    ≥ 0.95, for components and connections with additional ductility requirements beyond those required by AASHTO LRFD specifications

TABLE 2-10
Load Combinations and Load Factors

Load Combination Limit State	DC DD DW EH EV ES EL PS CR SH	LL IM CE BR PL LS	WA	WS	WL	FR	TU	TG	SE	Use One of These at a Time				
										EQ	BL	IC	CT	CV
Strength I (unless noted)	γ_p	1.75	1.00	-	-	1.00	0.50/1.20	γ_{TG}	γ_{SE}	-	-	-	-	-
Strength II	γ_p	1.35	1.00	-	-	1.00	0.50/1.20	γ_{TG}	γ_{SE}	-	-	-	-	-
Strength III	γ_p	-	1.00	1.40	-	1.00	0.50/1.20	γ_{TG}	γ_{SE}	-	-	-	-	-
Strength IV	γ_p	-	1.00	-	-	1.00	0.50/1.20	-	-	-	-	-	-	-
Strength V	γ_p	1.35	1.00	0.40	1.00	1.00	0.50/1.20	γ_{TG}	γ_{SE}	-	-	-	-	-
Extreme Event I	γ_p	γ_{EQ}	1.00	-	-	1.00	-	-	-	1.00				
Extreme Event II	γ_p	0.50	1.00	—	—	1.00	—	—	—	—	1.00	1.00	1.00	1.00
Service I	1.00	1.00	1.00	0.30	1.00	1.00	1.00/1.20	γ_{TG}	γ_{SE}					
Service II	1.00	1.30	1.00	—	—	1.00	1.00/1.20							
Service III	1.00	0.80	1.00	—	—	1.00	1.00/1.20	γ_{TG}	γ_{SE}					
Service IV	1.00	-	1.00	0.70	—	1.00	1.00/1.20	-	1.00					
Fatigue I— LL, IM, and CE only	—	1.50												
Fatigue II— LL, IM, and CE only	—	0.75												

Source: Used with permission by AASHTO.[2-1]

For all other limit states, $\eta_D = 1.00$.
η_I = importance factor

For the strength limit state:
$\eta_I \geq 1.05$ for critical or essential bridges
　　$= 1.00$ for typical bridges
　　≥ 0.95 for relatively less important bridges

For all other limit states, $\eta_I = 1.00$.
Q_i = nominal force effect, including deformation, stress, or stress resultant
R_n = nominal resistance, determined based on the dimensions as shown on the plans and on permissible stresses, deformations, or specified strength of materials
R_r = factored resistance $= \phi R_n$

2.3.3.2   Design Limit States

2.3.3.2.1   General

A design limit state is defined as a condition beyond which the bridge or component ceases to satisfy the design requirements. We have previously mentioned the design limit state that mostly mains the strength limit state. Except for the strength requirement, the designed bridges should meet the service, fatigue, and extreme loading condition requirements. The strength limit state is mostly related to bending capacity, shear capacity, and stability. The service limit state is related to gradual deterioration, such as cracking, deflection, or vibration; user comfort; or maintenance costs. The fatigue limit state is related to loss of strength under repeated loads and is intended to limit crack growth under repetitive loads to prevent fracture during the design life of the bridge. The extreme event limit state is intended to ensure the structural survival of a bridge during a major earthquake, flood, vessel collision, etc., whose return period may be significantly greater than the design life of the bridge.

2.3.3.2.2   Strength Limit States

The strength limit state is related to the loss of the load-carrying capacity and is intended to provide sufficient strength or resistance to satisfy Eq. 2-50 for the statistically significant load combinations specified in Table 2-10 that a bridge is expected to experience in its design life. Strength limit states include the evaluation of resistance to bending, shear, torsion, and axial load. The AASHTO specifications[2-1] specify five strength limit state load combinations to include different design considerations.

Strength I: This strength limit state is the basic load combination for the normal vehicular use of the bridge without wind.

Strength II: This strength limit state is the load combination for use of the bridge by owner-specified special design vehicles, permit vehicles, or both without wind. If a permit vehicle is traveling unescorted, or if control is not provided by the escorts, the other lanes may be assumed to be occupied by the vehicular live load specified in AASHTO specifications. For bridges, longer than the permit vehicle, addition of the lane load, preceding and following the permit load in its lane, should be considered.

Strength III: This strength limit state is the load combination for the bridge exposed to wind velocities exceeding 55 mph.

Strength IV: This strength limit state is for a very high ratio of dead load to live load force effect in bridge superstructures. The calibration process for the load and resistance factors has been carried out for a large number of bridges with spans not exceeding 200 ft. Some limited research of a few bridges with spans up to 600 ft show that the ratio of dead and live load force effects is rather high and could result in a set of resistance factors different from those found acceptable for small- and medium-span bridges for the primary components of large bridges. If the ratio of dead load to live load force effect exceeds about 7.0, this load combination may govern the design.

Strength V: This limit state is the load combination for normal vehicular use of the bridge with a wind velocity of 55 mph.

2.3.3.2.3   Service Limit States

The service limit state is to control the stresses, deflections, and crack widths of bridge components that occur under regular service conditions. Controlling the stresses ensures the bridge and its components will normally deform in the elastic range under service loads. Accumulation of permanent deformations may cause serviceability problems. Bridges with visible deflections may not be acceptable to the public or may cause excessive vibration and discomfort to drivers. Though concrete cracks by themselves may not affect a bridge's performance, they lead to corrosion of steel, spalling of concrete, penetration of salt, etc. There are four service limit state load combinations contained in the AASHTO specifications.

Service I: This service limit state is the load combination for the normal operational use of the bridge with a wind speed of 55 mph. All loads are taken as their actual values, and extreme load conditions are excluded. This combination is used for checking compression stress of pre-stressed concrete, for controlling crack width in reinforced concrete structures and for trans-verse analysis relating to tension in concrete segmental girders. This combination is also used for checking deflection of certain buried structures and for the investigation of slope stability.

Service II: This service limit state is the load combination for controlling yielding of steel structures and slip of slip-critical connections due to vehicular live load.

Service III: This service limit state is the load combination for longitudinal analysis to control cracks due to the tension in prestressed concrete superstructures and due to the principal tension in the webs of segmental concrete girders.

Service IV: This service limit load combination is for only controlling cracks due to the ten-sion in prestressed concrete columns.

2.3.3.2.4   Fatigue and Fracture Limit States

The fatigue and fracture limit state is related to the accumulation of damage and eventually failure under repeated loads. It includes a set of restrictions on stress range caused by the design truck. The restrictions are related to the number of stress range excursions expected to occur during the design life of the bridge. There are two fatigue and fracture limit states specified in AASHTO specifications.

Fatigue I: This fatigue and fracture load combination is for infinite load-induced fatigue life.

Fatigue II: This fatigue and fracture load combination is for finite load-induced fatigue life.

2.3.3.2.5   Extreme Event Limit States

The extreme event limit state is to ensure the structural survival of a bridge that may undergo a sig-nificant inelastic deformation during a major earthquake or flood, or collision by a vessel, vehicle, or ice floe. The probability of these events occurring simultaneously is very low; therefore, they are specified to be applied separately in the AASHTO specifications.

Extreme Event I: This load combination is used for considering earthquake effect.

Extreme Event II: This load combination is used for ice load, collision by vessels and vehicles, check floods, and certain hydraulic events.

2.3.3.3   Load Combinations and Load Factors for Design

2.3.3.3.1   General

As previously discussed, a bridge may be subjected to many types of loads in its service life. However, only a limited number of them will likely act on a bridge simultaneously. Because it is difficult for engineers to decide what combination of loads and magnitudes will be applied simul-taneously, the AASHTO LRFD specifications have predefined load combinations with load factors based on the reliability theory. These load factors and load combinations vary with each limit state. The load factors for the different load combinations are shown in Table 2-10. The descriptions of each load designation are defined as follows:

CR = force effects due to creep
DD = downdrag force
DC = dead load of structural components and attachments

DW = dead load of wearing surfaces and utilities
EH = load of horizontal earth pressure
EL = miscellaneous locked-in force effects resulting from construction process
ES = earth surcharge load
EV = vertical pressure due to dead load of earth fill
PS = secondary forces from post-tensioning for strength limit states
     = total prestress forces for service limit states
BL = blast loading
SH = force effects due to shrinkage
BR = vehicle braking force
CE = vehicle centrifugal force
CT = vehicle collision force
CV = vessel collision force
EQ = earthquake load
FR = friction load
IC = ice load
IM = vehicle dynamic load allowance
LL = live load due to moving vehicles
LS = surcharge due to live load
PL = pedestrian live load
SE = forces due to settlement
TG = forces due to temperature gradient
TU = forces due to uniform temperature
WA = water load and stream pressure
WS = wind load on structure
WL = wind on live load

2.3.3.3.2   Load Combinations

In addition to the load combinations given in Table 2-10, for segmentally constructed bridges, the following combination should be investigated at the service limit state:

$$\sum \gamma Q = DC + DW + EH + EV + ES + WA + CR + SH + TG + EL + PS \qquad (2\text{-}51)$$

Load combinations that include settlement should also be considered for the cases without settlement.

2.3.3.3.3   Load Factors

Most of the load factors have been shown in Table 2-10. However, it may be necessary in some instances to determine the factors based on the types of loads and their effects on the structures.

2.3.3.3.3.1   Load factors γ_p of Permanent Loads   The load factors γ_p for permanent loads shown in Tables 2-11 and 2-12 vary with the types of loads being considered. The factors should be selected to produce the worst-case extreme factored force effects. For each load combination, both positive and negative extreme effects should be investigated. In load combinations where one force effect decreases another, the minimum load factor should be applied to the load to reduce the force and in turn maximize the combined load effects on the structure. For permanent force effects, the load factor that produces the more critical combination should be chosen from Table 2-11. If the permanent load increases the stability or load-carrying capacity of a structure, the minimum value of the load factor for that permanent load should also be investigated.

2.3.3.3.3.2   Load Factor for Temperature   There are two factors provided for the force effect due to uniform temperature TU: γ_{TU} (larger value) is for deformations and γ_{TU} (small value) is for all

TABLE 2-11

Load Factors for Permanent Loads, γ_p[2-1]

Type of Load	Note of Cases	Load Factor Maximum	Load Factor Minimum
DC	Component and attachments	1.25	0.90
	Strength IV only	1.50	0.90
DD	Piles, α Tomlinson method	1.40	0.25
	Piles, λ method	1.05	0.30
	Drilled shafts, O'Neill and Reese (1999) method	1.25	0.35
DW	Wearing surfaces and utilities	1.50	0.65
EH	Active	1.50	0.90
	At-rest	1.35	0.90
	AEP for anchored walls	1.35	N/A
EL	Locked-in construction stresses	1.00	1.00
EV	Overall stability	1.00	N/A
	Retaining walls and abutments	1.35	1.00
	Rigid buried structure	1.30	0.90
	Rigid frames	1.35	0.90
ES	Earth surcharge	1.50	0.75

other effects. The load factor for temperature gradient γ_{TG} should be considered on a project-specific basis and may be taken as follows:

γ_{TG} = 0.0, for the strength and extreme event limit states,
    = 1.0, for the service limit state if live load is not considered
    = 0.50, for the service limit state if live load is considered.

2.3.3.3.3.3 Load Factors for Settlement and for Earthquake The load factor for settlement, γ_{SE}, should be considered on a project-specific basis and may be taken as

$$\gamma_{SE} = 1.0$$

The load factor for live load in Extreme Event I, γ_{EQ}, should be determined on a project specific basis.

2.3.3.3.3.4 Load Factors for Jacking and Post-Tensioning Forces The AASHTO specifications require that the design jacking forces for jacking in service should not be less than 1.3 times the

TABLE 2-12

Load Factors for Permanent Loads due to Superimposed Deformations, γ_p[2-1]

Bridge Component		PS	CR, SH
Concrete segmental bridges, superstructures, and concrete substructures		1.0	See γ_p for DC, Table 2-11
Concrete superstructures—nonsegmental		1.0	1.0
Substructures for nonsegmental bridges	using Ig	0.5	0.5
	using $I_{effective}$	1.0	1.0
Steel substructures		1.0	1.0

TABLE 2-13

Load Factors for Construction Load Combinations at Service Limit State

Load Factors			Load Combination					
			a	b	c	d*	e†	f‡
Loads	Dead load	DC	1.0	1.0	1.0	1.0	1.0	1.0
		DIFF	1.0	0.0	1.0	1.0	0.0	0.0
		U	0.0	1.0	0.0	0.0	1.0	0.0
	Live load	CEQ, CLL	1.0	1.0	0.0	1.0	1.0	1.0
		IE	1.0	1.0	0.0	0.0	1.0	1.0
		CLE	0.0	0.0	0.0	0.0	0.0	1.0
	Wind load	WS	0.0	0.0	0.7	0.7	0.3	0.3
		WUP	0.0	0.0	0.7	1.0	0.0	0.0
		WE	0.0	0.0	0.0	0.7	0.3	0.3
	Other loads	CR	1.0	1.0	1.0	1.0	1.0	1.0
		SH	1.0	1.0	1.0	1.0	1.0	1.0
		TU	1.0	1.0	1.0	1.0	1.0	1.0
		TG	γ_{TG}	γ_{TG}	γ_{TG}	γ_{TG}	γ_{TG}	γ_{TG}
		WA	1.0	1.0	1.0	1.0	1.0	1.0
	Earth loads	EH, EV, ES	1.0	1.0	1.0	1.0	1.0	1.0

* Equipment not working.
† Normal erection.
‡ Moving equipment.

permanent load reaction at the bearing. If the bridge is not closed to traffic during the jacking operation, the jacking load should also include a live load reaction determined based on the maintenance of traffic plan, multiplied by the load factor for live load.

The design force for post-tensioning anchorage zones should be 1.2 times the maximum jacking force.

2.3.3.4 Load Combinations and Load Factors for Segmental Bridge Construction

The construction loads are discussed in Section 2.2.12. In segmentally constructed bridges, the construction loads and conditions often control the bridge sections and detailing. It is important for bridge engineers to check bridge capacities for both service and strength limit states during construction with the assumed conditions shown in the contract documents.

2.3.3.4.1 Load Combinations for Construction at the Service Limit State

The construction load combinations at the service limit states are given in Table 2-13. The designations for the various construction loads shown in this table are given below.

DC = weight of supported structure
DIFF = differential load for balanced cantilever construction
DW = superimposed dead load
CLL = distributed construction live load: in balanced cantilever bridge construction, this load may be taken as 0.005 ksf on one cantilever and 0.01 ksf on another as AASHTO specifications recommended
CEQ = specialized construction equipment load
IE = dynamic load from equipment, may be taken as 10% of the lift weight for very gradual lift segments with small dynamic effects

CLE = longitudinal construction equipment load
U = segment unbalance for primarily to balanced cantilever construction.
WS = horizontal wind load on structures
WE = horizontal wind load on equipment
WUP = wind uplift on cantilever
A = static weight of precast segment
AI = dynamic response due to accidental release or application of a precast segment load
CR = creep effects
SH = shrinkage
TU = effect due to uniform temperature variation
TG = effect due to temperature gradients

2.3.3.4.2   Load Combinations for Construction at Strength Limit States
2.3.3.4.2.1   Load Combination for Superstructures
- For maximum force effects:

$$\sum \gamma Q = 1.1(DC + DIFF) + 1.3(CEQ + CLL) + A + AI \tag{2-52a}$$

- For minimum force effects:

$$\sum \gamma Q = DC + CEQ + A + AI \tag{2-52b}$$

2.3.3.4.2.2   Load Combination for Substructures
- For strength limit states I, III, and V:

The load combinations are the same as specified in Table 2-10, in which load factors for the specified loads are as follows:

$$DIFF \text{ and } CEQ: \gamma_{DC}$$
$$WUP: \gamma_{WS}$$
$$WE: \gamma_{WL}$$
$$CLL: \gamma_{LL}$$

- Construction strength load combinations are the same as those specified in Eqs. 2-52a and 2-52b.

2.3.3.5   Resistance Factors
AASHTO LRFD specifications specify the resistance factors based on different limit states.

I. Service limit state and fatigue and fracture limit state, generally:

$$\phi = 1.00$$

II. Extreme event limit
- For compression-controlled sections with spirals or ties in Seismic Zones 2, 3, and 4:

$$\phi = 0.90$$

- For others:

$$\phi = 1.00$$

TABLE 2-14

Resistance Factor for Joints in Segmental Construction[2-1]

Type of Concrete	Tendon Bonding Condition	ϕ_f Flexure	ϕ_v Shear
Normal weight	Fully bonded	0.95	0.90
	Unbonded or partially bonded	0.90	0.85
Sand-lightweight	Fully bonded	0.90	0.70
	Unbonded or partially bonded	0.85	0.65

III. Strength limit states
   - For tension-controlled reinforced concrete sections: $\phi = 0.90$
   - For tension-controlled prestressed concrete sections: $\phi = 1.00$
   - For shear and torsion:
         Normal weight concrete: $\phi = 0.90$
         Lightweight concrete: $\phi = 0.80$
   - For compression-controlled sections with spirals or ties: $\phi = 0.75$
   - For bearing on concrete: $\phi = 0.70$
   - For compression in strut-and-tie models: $\phi = 0.70$
   - For compression in anchorage zones:
         Normal weight concrete: $\phi = 0.80$
         Lightweight concrete: $\phi = 0.65$
   - For tension in steel in anchorage zones: $\phi = 1.00$
   - For resistance during pile driving: $\phi = 1.00$
   - For joints in segmental construction, the resistance factors are taken from Table 2-14.

For sections in which the net tensile strain in the extreme tension steel at nominal resistance is between the compression-controlled strain limit and tension-controlled strain limit, the load factor ϕ should be determined by a linear interpolation from 0.75 to that for tension-controlled sections:
   For prestressed members:

$$0.75 \le \phi = 0.75 + \frac{0.25(\varepsilon_t - \varepsilon_{cl})}{(\varepsilon_{tl} - \varepsilon_{cl})} \le 1.0 \qquad (2\text{-}53a)$$

For non-prestressed members:

$$0.75 \le \phi = 0.75 + \frac{0.15(\varepsilon_t - \varepsilon_{cl})}{(\varepsilon_{tl} - \varepsilon_{cl})} \le 0.9 \qquad (2\text{-}53b)$$

TABLE 2-15

Strain Limits

Steel Yield Strength f_y (ksi)	Non-prestressed		Prestressed	
	Compression Control ε_{cl}	Tension Control ε_{tl}	Compression Control ε_{cl}	Tension Control ε_{tl}
60	0.0020	0.0050	0.0020	0.005
75	0.0028	0.0050		
80	0.0030	0.0056		
100	0.0040	0.0080		

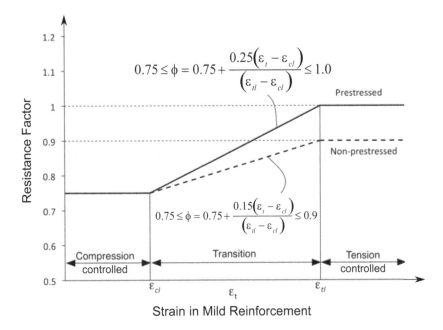

FIGURE 2-22 Variation of Resistance Factors ϕ with the Net Tensile Strain ε_t.[2-1].

where

ε_t = net tensile strain in extreme tension steel at nominal resistance (in./in.)

ε_{cl} = compression-controlled strain limit in extreme tension steel, taken as Table 2-15 (in./in.)

ε_{tl} = tension-controlled strain limit in extreme tension steel taken as Table 2-15 (in./in.)

The variation of the resistance factors ϕ with the net tensile strain ε_t is illustrated in Fig. 2-22.

REFERENCES

2-1. American Association of State Highway and Transportation Officials (AASHTO), *LRFD Bridge Design Specifications*, 7th edition, Washington D.C., 2014.

2-2. AASHTO, *A Policy on Geometric Design of Highways and Streets*, 4th edition, Washington D.C., 2001.

2-3. AASHTO, *Standard Specifications for Highway Bridges*, 17th edition, Washington D.C., 2002.

2-4. Huang, D. Z., Wang, T. L., and Shahawy, M., "Vibration of Thin Walled Box Girder Bridges Excited by Vehicles," *Journal of Structural Engineering*, Vol. 121, No. 9, 1995, pp. 1330–1337.

2-5. Huang, D. Z., and Wang, T. L., "Impact Analysis of Cable-Stayed Bridges," *Journal of Computers and Structures*, Vol. 43, No. 5, 1992, pp. 897–908.

2-6. Huang, D. Z., T. L. Wang, and M. Shahawy. "Vibration of Horizontally Curved Box Girder Bridges due to Vehicles," *Journal of Computer and Structures*, Vol. 68, 1998, pp. 513–528.

2-7. Wang, T. L., Huang, D. Z., and Shahawy, M., "Dynamic Behavior of Continuous and Cantilever Thin-Walled Box Girder Bridges," *Journal of Bridge Engineering*, ASCE, Vol. 1, No. 2, 1996, pp. 67–75.

2-8. El-Tawil, S., Severino, E., and Fonseca P., "Vehicle Collision with Bridge Piers," *Journal of Bridge Engineering*, ASCE, Vol. 10, No. 3, 2005, pp. 345–353.

2-9. American Society of Civil Engineers (ASCE), *Minimum Design Loads for Buildings and Other Structures*, ASCE-10, New York, 2010.

2-10. Clough, R. W., and Penzien, J., *Dynamics of Structures*, 2nd ed., McGraw-Hill Inc., New York, 1993.

2-11. Woisin, G., "The Collision Tests of the GKSS." In *Jahrbuch der Schiffbautechnischen Gesellschaft*, Vol. 70. Berlin, Germany, 1976, pp. 465–487.

2-12. International Association of Bridge and Structural Engineers, "Ship Collision with Bridges and Offshore Structures." In *International Association of Bridge and Structural Engineers Colloquium*, Copenhagen, Denmark. 3 vols., 1983.

2-13. Chen, W. F., and Duan, L., *Bridge Engineering Handbook*, CRC Press, Boca Raton, London, New York, Washington, D.C., 2000.

2-14. Barker, R. M., and Puckett, J. A., *Design of Highway Bridges Based on AASHTO LRFD Bridge Design Specifications*, John Wiley & Sons, New York, 1997.

2-15. Kulicki, J. M., Mertz, D. R., and Wassef, W. G., "LRFD Design of Highway Bridges," NHI Course 13061, Federal Highway Administration, Washington, D.C., 1994.

2-16. Nowak, A. S., "Calibration of LRFD Bridge Design Code," Department of Civil and Environmental Engineering Report, UMCE 92-25, University of Michigan, Ann Arbor, 1993.

2-17. Nowak, A. S., and Lind, N. C., "Practical Bridge Code Calibration," *Journal of the Structural Division*, ASCE, Vol. 105, No. ST12, 1979, pp. 2497–2510.

2-18. Nowak, A. S., "Reliability of Structures," Class Notes, Department of Civil and Environmental Engineering, University of Michigan, Ann Arbor, MI, 1995.

3 Fundamentals of Segmental Bridge Analysis and Design

- Basic theory of prestressed concrete
- Estimation of required concrete section and prestressing steel
- Losses of prestressing
- Bending and torsion of straight and curved girders
- Theory and AASHTO methods for determining strength resistances for flexural and torsion members
- Theory and AASHTO methods for determining strength resistances for axial and biaxial members
- Theory and AASHTO methods for determining strength resistances for anchorage zones
- Service checks and reinforcement details required by AASHTO specifications

3.1 BASIC CONCEPTS OF PRESTRESSED CONCRETE STRUCTURES

3.1.1 GENERAL BEHAVIORS OF PRESTRESSED CONCRETE STRUCTURES

Prestressed concrete can be generally defined as concrete in which internal stresses have been introduced with a magnitude and have been distributed such that external loadings are effectively resisted by the internal stresses[3-1]. In bridge structures, prestress is commonly introduced by tensioning high-strength steel reinforcements, which are generally called tendons as discussed in Chapter 1. There are two methods of prestressing concrete: pretensioning and post-tensioning.

Pretensioning is a method of prestressing in which the tendon is tensioned before the concrete is placed. First, the tendons are prestressed and temporarily anchored using some retaining bulkheads; then, the concrete is placed and allowed to harden. Finally, the tendons are released after the concrete has gained sufficient strength. The tendon force is transferred to the concrete by the bond between the concrete and tendons. In the post-tensioning method, the tendons are tensioned after the concrete has hardened. Ducts are placed within the concrete section with the designed geometry and adequately secured prior to concrete being poured. After the concrete has developed the designed strength, the tendons are tensioned using jacks and are anchored against the hardened concrete section using anchorage devices as previously discussed in Chapter 1. It is evident that the post-tensioning method is almost always used in segmental bridge construction.

It is important for the designer to understand the basic concept of prestressing. The general concept can be easily explained by investigating the behavior of a simply supported concrete I-beam subject to an external load when prestressing forces are introduced (Fig. 3-1). From Fig. 3-1a, it can be observed that the beam will be bent in a concave-upward shape under the external load, including its self-weight. The top fiber segment $\overline{ab}$ is shortened to segment $\overline{a'b'}$, which indicates that the top fiber is in compression. The bottom fiber segment $\overline{cd}$ is elongated to segment $\overline{c'd'}$, which indicates that the bottom fiber is in tension. As concrete is a brittle material that is weak in tension and the joint between the segments in the concrete segmental construction may have zero tensile capacity, introducing prestressing forces is the only reasonable choice to make the structure effectively carry the design loadings in segmental construction.

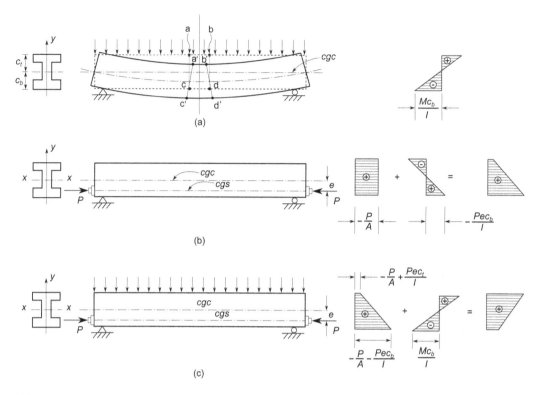

FIGURE 3-1 Stress Distribution in a Concrete Beam with Straight Tendon, (a) Dead Load, (b) Post-Tensioning Force, (c) Dead Load and Post-Tensioning Force.

Assuming the moment at the midspan induced by the external load as M, the normal stress at any point across the section due to the external load can be written as

$$f = \frac{My}{I} \tag{3-1a}$$

where
   I = moment of inertia for section
   y = distance from centroidal axis or neutral axis

The top and bottom stresses caused by the external load are

$$\text{Top fiber stress due to dead load: } f^t = \frac{Mc_t}{I} \tag{3-1b}$$

$$\text{Bottom fiber stress due to dead load: } f^b = -\frac{Mc_b}{I} \tag{3-1c}$$

In Eqs. 3-1b and 3-1c, c_t and c_b are the distances from the centroidal axis to the top and bottom fibers of the beam, respectively (see Fig. 3-1a).

The normal stress distribution along the girder depth caused by the external load and self-weight is shown in Fig. 3-1a. A plus sign is used for compressive stress and a minus sign is used for tensile stress throughout the text unless noted otherwise. Also, a moment causing top fiber in compression and bottom fiber in tension is defined as a positive moment.

Assuming a straight tendon is placed at a distance e below the center of gravity of the concrete (cgc), the tensile force P in the tendon produces an equal compressive force P in the concrete beam, which can be resolved as a concentric force P and a moment Pe. The top and bottom normal stress due to the post-tensioning force P can be written as

$$\text{Top fiber stress due to prestressing: } f^t = \frac{P}{A_c} - \frac{Pec_t}{I} \qquad (3\text{-}2a)$$

$$\text{Bottom fiber stress due to prestressing: } f^b = \frac{P}{A_c} + \frac{Pec_b}{I} \qquad (3\text{-}2b)$$

where A_c = area of concrete section.

From Eq. 3-2, we can see that the first term on the right side is the section average compression stress induced by the force P loaded along the cgc line and that the second term is the bending stress induced by the negative bending moment $-Pe$. The distribution of the normal stress caused by the prestressing force is illustrated in Fig. 3-1b. Comparing Eq. 3-1c with Eq. 3-2b, it can be observed that the tensile stress induced by the external load is counteracted by the compressive stress due to the pre-stessing force. The total top and bottom stresses due to both external loads and prestressing force are

$$\text{Top fiber stress due to external loads and prestressing: } f^t = \frac{Mc_t}{I} + \frac{P}{A_c} - \frac{Pec_t}{I} \qquad (3\text{-}3a)$$

$$\text{Bottom fiber stress due to external loads and prestressing: } f^b = -\frac{Mc_b}{I} + \frac{P}{A_c} + \frac{Pec_b}{I} \qquad (3\text{-}3b)$$

The distributions of the normal stress at the midspan due to both external load and prestressing are shown in Fig. 3-1c. From Fig. 3-1c, we can see that the tensile stress caused by external loads can be completely eliminated by properly selecting the location of the tendon and magnitude of the pre-stressing force. Therefore, the concrete will have no tensile stress and will not crack. Thus, a brittle concrete member can be designed to behave as a pseudo elastic material by the introduction of a prestressing force. From this point of view, if the tensile stress induced by the service design loads is entirely compensated for, smaller than, or equal to the concrete modulus of rupture, the prestressed structure can be analyzed as an elastic structure.

If the total tensile stress f^b determined by Eq. 3-3b is equal to or greater than zero, i.e., if the tensile stress caused by service load is completely eliminated by prestressing force, the concrete is considered to be fully prestressed. Generally, full prestressing is designed to eliminate concrete ten-sile stress in the direction of prestressing under service loads and may cause significant camber. Full prestressing may not be economical in some cases. If some tensile stresses will be induced under service load, the concrete is named as partial prestressed. However, actually, the segmental bridges designed based on AASHTO specifications may be still classified as full prestressed structures, though some tensile stresses are allowed under service load. A more common definition for partial prestressing concretes may be that the partial concrete is the concrete that is designed to allow sig-nificant tensile stresses to occur at service load and that the tensile regions are typically reinforced with non-prestressed reinforcement[3-1, 3-2].

To further understand the basic behavior of a prestressed concrete beam, Fig. 3-2 shows a schematic plot of the variation of deflection at midspan with loading for the simply supported post-tensioned beam shown in Fig. 3-1. In Fig. 3-2, point A represents the loading stage with effec-tive post-tensioning force and full beam self-weight. Point B indicates the decompression loading stage with zero tensile stress as the bottom fiber, and point C represents the concrete cracking loading stage. From Section 1.2.5, the AASHTO LRFD specifications[3-3] normally specify the service limit loads between the decompression load and the concrete cracking load. After exceed-ing concrete cracking loading, the beam deflects more quickly and the starts to behave similar to a conventionally reinforced concrete beam. Finally, the beam at its midsection reaches its ultimate

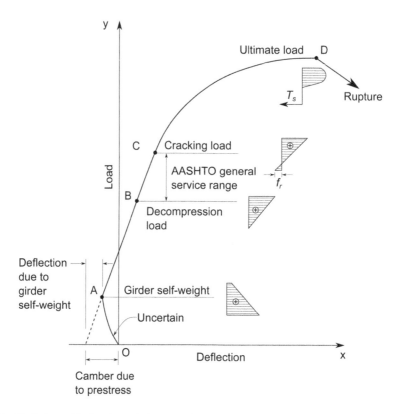

FIGURE 3-2   Variation of Deflection at Midspan with Loads for a Post-Tensioned Concrete Beam.

strength as depicted at point D in the figure. Figure 3-3 illustrates the variation of tendon stress with loading. It can be observed from this figure that the tendon stress has minimal variation with load before reaching the cracking load. After reaching the cracking load, the tensile stresses being resisted by the concrete are released and transferred to the tendon. When this occurs, there is a sudden stress increase in the bonded tendon, from point B to point C as shown in Fig. 3-3. Many test results[3-1, 3-4 to 3-8] show that the tendon stress approaches very closely to its ultimate strength at the rupture of the beam if compression failure does not occur in the concrete and there are no shear and/or bonding failures. The variation of the tendon stress with load can be approximately drawn as from point C to point D. The variation of unbonded tendon stress with load is also shown as a dotted line in Fig. 3-3. Unbonded tendons are frequently used in segmental bridges, such as external tendons, tendons enclosed in ducts injected with grease and wax. From Fig. 3-3, it can be seen that the unbonded tendon stress increases with the load more slowly than the bonded tendon stress does. This is because any strain in the unbonded tendon will be distributed throughout its entire length, while the strain in the bonded tendon is the same as the concrete at the level of the tendon and varies with the section bending moment[3-5 to 3-9]. There is a tendency for beams with unbonded tendons to develop large cracks before rupture. The large cracks concentrate strains at some local sections in the concrete, which lower the beam's ultimate strength. Thus, generally, the ultimate load of the beam with unbonded tendons is less than that for a corresponding beam with bonded tendons.

In the foregoing discussion, it has not been mentioned whether the post-tensioned beam is monolithic or consists of precast segments. Research shows that the behavior of a post-tensioned concrete segmental beam with epoxied joints is essentially the same as that of a monolithic beam.

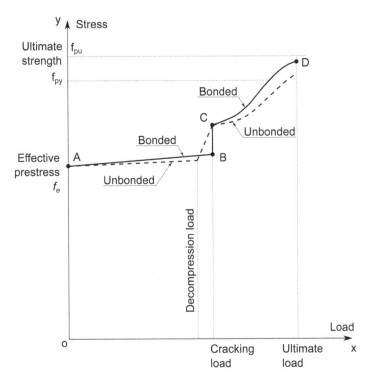

FIGURE 3-3 Variation of Tendon Stresses at Midspan with Loads for a Post-Tensioned Concrete Beam.

Test results show the epoxied joints constructed according to the AASHTO specifications normally have higher strength than the surrounding concrete section and any cracks that develop usually occur away from the epoxied joint[3-5 to 3-8, 3-10, 3-11].

3.1.2 Bending Analysis for Prestressed Girders

A bridge with external loading may undergo both vertical bending and torsion deformations. Normally, the cross section of a bridge is symmetrical about its vertical axis or can be treated as a symmetrical section about its vertical axis. Thus, the bending and the torsion analysis can be separated. The post-tensioning tendons in the majority of segmental bridges are used to counteract in-plane bending and shear due to external loadings.

3.1.2.1 C-Line Method

From the discussion in Section 3.1.1, we can see that in prestressed concrete structures, the tensile stress caused by external load is entirely resisted by the prestressing force and that the compression force is completely resisted by concrete under service load if the small tensile capacity of concrete is neglected. Thus, we can consider the post-tensioned concrete structure as a composite structure with the concrete taking the compression force and the tendon taking the tensile force. These two types of materials form a couple to resist external moment. Based on this concept, the determination of the concrete stress becomes a determination of the magnitude and location of the resulting concrete force. Figure 3-4b shows a free-body diagram of a segment of a prestressed beam under external vertical load as shown in Fig. 3-4a. In this figure, T represents the tendon tensile force and C represents the resulting concrete compression force. The distance between T and C is defined as the internal level arm length and is denoted as d. In Fig. 3-4, y_c represents

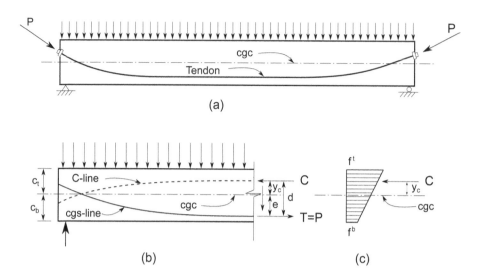

FIGURE 3-4 C-Line Method, (a) Post-Tensioned Concrete Beam, (b) Free-Body, (c) Concrete Normal Stress Distribution Determined by the C-Line Method.

the distance from the center of gravity of the concrete section (cgc) to the application point of the compression force C and e denotes the eccentricity from the center of gravity of the tendon (cgs) to the cgc. As from the discussion in Section 3.1.1, the prestessing force P varies very little under service load and can be assumed to be a constant, so the concrete compression force C can be obtained based on the equilibrium condition because the sum of the horizontal forces is equal to zero, i.e.,

$$\sum H = 0: C = T = P \tag{3-4}$$

The couple C-T forms an internal moment resisting external moment M. Thus, the location of the concrete compression force C can be determined based on the equilibrium condition that the sum of the moment is equal to zero as follows:

$$\sum M = 0: M = Cd = Td = Pd \tag{3-5}$$

From Fig. 3-4b, we can see

$$d = (e + y_c) \tag{3-6}$$

Substituting Eq. 3-6 into Eq. 3-5 yields

$$y_c = \frac{M}{P} - e \tag{3-7}$$

where
    M = moment due to external loadings
    y_c = distance from cgc to application point of compression force C

From Fig. 3-4b, the concrete compression stresses at the top and bottom fibers can be written as

$$f^t = \frac{P}{A_c} + \frac{Py_c c_t}{I} \tag{3-8a}$$

$$f^b = \frac{P}{A_c} - \frac{Py_c c_b}{I} \tag{3-8b}$$

where
c_t = distance from cgc to top fiber
c_b = distance from cgc to bottom fiber

Equations 3-8a and 3-8b can be rewritten as

$$f^t = \frac{P}{A_c}\left(1 + \frac{y_c c_t}{\frac{I}{A_c}}\right) = \frac{P}{A_c}\left(1 + \frac{y_c c_t}{r^2}\right) \tag{3-9a}$$

$$f^b = \frac{P}{A_c}\left(1 - \frac{y_c c_b}{\frac{I}{A_c}}\right) = \frac{P}{A_c}\left(1 - \frac{y_c c_b}{r^2}\right) \tag{3-9b}$$

where $r = \sqrt{\frac{I}{A_c}}$ = radius of gyration of the section.

From Eq. 3-7, it can be seen that y_c varies with the external moment M, i.e., varies along with the beam longitudinal location. At the end supports, $M = 0$ and the level arm length $d = 0$; at the midspan of the beam with a uniform external loading as shown in Fig. 3-4a, y_c becomes its maximum value. Function y_c represents the trajectory of the concrete resulting compression force and is called the center of compression line (C-line). From Eqs. 3-8 and 3-9, it can be observed that the concrete stress can be determined if the C-line is known. The method for determining the post-tensioned concrete stress based on the C-line function is often called the C-line method. The reader can verify that Eqs. 3-3 and 3-9 will give the same results.

3.1.2.2 Equivalent Load Method

In the equivalent load method, the tendon forces are treated as an external load acting against the concrete elements. This method can be classified as the load balancing method developed by T. Y. Li[3-11]. However, the authors believe that the equivalent load concept may more easily or more directly be applied to complex segmental bridges analyzed by the finite-element method[3-12 to 3-14].

3.1.2.2.1 Equivalent Loads at Anchorage Locations

The magnitude of the equivalent load at the tendon anchorage location is equal to the tendon force, and its direction of application is in the tangent direction at the anchorage. The equivalent force can be resolved into any concentrated forces and moments applied at the center of gravity of the concrete section. In the cases shown in Figs. 3-5 and 3-6, the equivalent end concentrated forces and the moment for the tendon force P at point A can be written as

$$\text{Equivalent horizontal force: } H_A = P\cos\alpha_A \tag{3-10}$$
$$\text{Equivalent vertical force: } V_A = P\sin\alpha_A \tag{3-11}$$
$$\text{Equivalent moment: } M_A = Pe_A\cos\alpha_A \tag{3-12}$$

where α_A = tangential angle of tendon at anchorage A (see Figs. 3-5a and 3-6a).

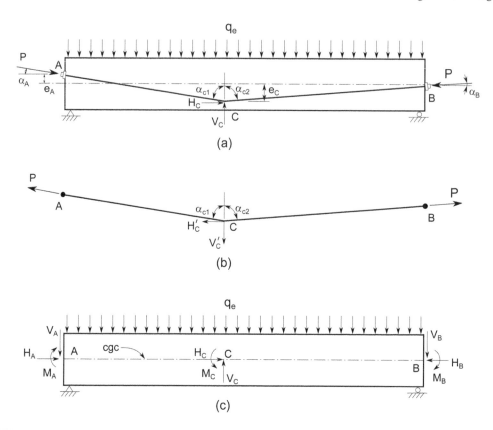

FIGURE 3-5 Equivalent Loads for Harped Tendon, (a) Post-Tensioned Beam, (b) Free-Body Diagram for Tendon, (c) Elastic Analytic Model.

3.1.2.2.2   Equivalent Loads for Broken Line Tendons

Figure 3-5a shows a post-tensioned concrete beam by a harped tendon under external loading q_e. A free-body diagram of the tendon is shown in Fig. 3-5b. Some concentrated forces at point C, such as H_C' and V_C', are necessary to keep the tendon in equilibrium because the tendon can only resist tensile force. The concentrated forces H_C' and V_C' are balanced by the reactions H_C and V_C at point C in the concrete. Then the equivalent forces at point C can be written as

$$\text{Equivalent horizontal force: } H_C = P(\sin\alpha_{C2} - \sin\alpha_{C1}) \tag{3-13}$$

$$\text{Equivalent vertical force: } V_C = P(\cos\alpha_{C1} + \cos\alpha_{C2}) \tag{3-14}$$

$$\text{Equivalent moment: } M_C = Pe_C(\sin\alpha_{C2} - \sin\alpha_{C1}) \tag{3-15}$$

where
α_{C1} and α_{C2} = angles of tendon segments AC and CB in vertical direction at point C, respectively;
e_C = eccentricity of tendon at point C from cgc line (see Fig. 3-5)

Thus, the post-tensioned concrete beam can be treated as an elastic beam (see Fig. 3-5c) in the analysis.

3.1.2.2.3   Equivalent Loads for Curved Tendons

Figure 3-6a shows a post-tensioned concrete girder with a parabolic tendon under an external loading q_e. Assume the eccentricities from both anchorages A and B to the cgc line are the same and

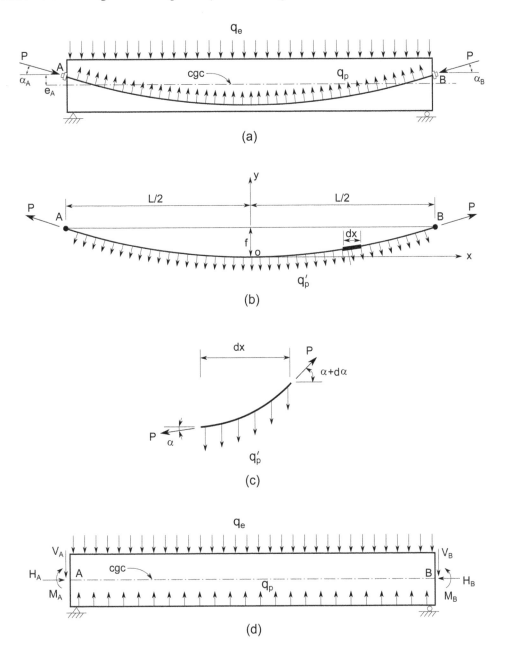

FIGURE 3-6 Equivalent Loads for Curved Tendon, (a) Post-Tensioned Beam, (b) Free-Body Diagram for Tendon, (c) Free-Body Diagram of Infinitesimal Length dx, (d) Elastic Analytic Model.

denote as e_o. A free-body diagram of the tendon is shown in Fig. 3-6b. If the coordinate system is taken as shown in Fig. 3-6b, the shape of the parabolic tendon can be written as

$$y = \frac{4f}{l^2}x^2$$ (3-16)

where
    f = distance from AB line to tendon at midspan (see Fig. 3-6b)
    l = distance between anchorages

From Fig. 3-6b, it can be observed that there must be a continuous distributed loading q_p' acting along the tendon to maintain the curved tendon shape as the tendon can only resist tensile force. The distributed loading q_p' is provided by the concrete reaction q_p due to the post-tensioning force P. Take an infinitesimal element dx from the tendon as a free body as shown in Fig. 3-6c. From Fig. 3-6c and using the force equilibrium condition in the y-direction, we can obtain:

$$\sum Y = 0:\ P\sin\alpha + q_p'dx = P\sin(\alpha + d\alpha) \tag{3-17}$$

where α = tangential angle = $\frac{dy}{dx}$ (see Fig. 3-6c).

Assuming that α = small, we have $\sin\alpha \approx \alpha$ and $\sin(\alpha + d\alpha) \approx \alpha + d\alpha$. Then, from Eq. 3-17 we can get

$$q_p' = \frac{d\alpha}{dx}P = \frac{d^2 y}{dx^2}P \tag{3-18}$$

Substituting Eq. 3-16 into Eq. 3-18, we can obtain the equivalent loading that has equal magnitude in opposite direction as q_p':

$$q_p = q_p' = \frac{8fP}{l^2} \tag{3-19}$$

The elastic analytic model by the equivalent load method for the post-tensioned concrete beam is shown in Fig. 3-6d. Note that the small horizontally distributed loading along the tendon location has been neglected by assuming the curvature of the tendon is small. From Fig. 3-6d, it can be observed that the vertical external loading q_e has been partially balanced by q_p and could be completely balanced by the post-tensioned tendon if the magnitude and location of the tendon are properly selected.

3.1.3 Basic Design Concepts

The design of a bridge cross section is essentially a trial and adjustment process as the self-weight of the bridge is unknown when the design begins. Normally, the initial geometric dimensions of the bridge's cross section is selected to satisfy the bridge flexural requirements of concrete stress and tendon stress limitations under service loading. The adequacy of other design requirements, such as the strength, shear and torsion capacity, deflection, cracks, and fatigues are subsequently checked. In this section, some simple methods for estimating the preliminary concrete and tendon areas as well as the concept of tendon layout are discussed. It should be mentioned that it is always a good idea to reference the cross-sectional properties of similar existing concrete segmental bridges.

3.1.3.1 Estimation of Concrete and Tendon Areas

3.1.3.1.1 Estimation Based on Service Moment Requirement

From the viewpoint of girder flexural behavior, a box girder can be treated as an I-girder shown in Fig. 3-7. In practice, normally, the girder depth h is assumed based on the minimum girder depths requirements discussed in Section 1.4.2.2 and by referencing the dimensions of similar existing bridges. As discussed in Section 3.1.2.1, a prestressed beam section can be treated as a section acted on by the internal couple C-T. For a preliminary section design, the concrete section can be assumed to have no tension as shown in Fig. 3-7b. Under service loading, the level arm d for the internal couple ranges from 60% to 80% of the girder height h_0. Assuming that the total bending moment $M_t = DL + LL$, the required effective tendon force P can be determined as

$$P = T = \frac{M_t}{\alpha h_0} \tag{3-20}$$

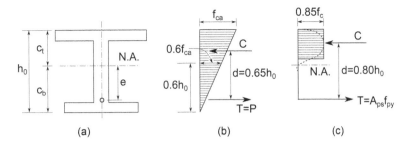

FIGURE 3-7 Estimation of Concrete and Tendon Areas, (a) Girder Section, (b) Stress Distribution at Service Loading, (c) Stress Distribution at Ultimate Loading.

where
M_t = section total service bending moment
h_0 = girder height
α = 0.65 for I-girder section
= 0.70 for positive moment at midspan for box-girder section
= 0.75 for negative moment over pier for box-girder section

The required tendon area A_{ps} is

$$A_{ps} = \frac{P}{f_{pe}} = \frac{M_t}{\alpha h_0 f_{pe}} \tag{3-21}$$

where f_{pe} = tendon effective prestress.

Assuming the concrete allowable compression stress is f_{ca} and the average concrete compression stress is $0.6 f_{ca}$, the total concrete compression force $0.6 f_{ca} A_c$ is equal to the total tensile force of tendon. Thus, the required concrete area A_c is

$$A_c = \frac{A_{ps} f_{pe}}{0.6 f_{ca}} \tag{3-22}$$

where f_{ca} = concrete allowable compression stress.

For permanent loading, the AASHTO specifications require that $f_{ca} = 0.45 f_c'$ (refer to Section 1.2.5.1.2). From Eq. 3-21 and Eq. 3-22, it can be seen that the required concrete and tendon areas can be determined if we know the girder total moment and girder height.

3.1.3.1.2 Estimation Based on Ultimate Moment

If the preliminary design begins at the ultimate-load level, the required design moment should at least be equal to the factored moment M_u. As discussed above, at this loading stage, the behavior of a prestressing girder is the same as that of a reinforced girder and the concrete stress distribution can be assumed as the equivalent rectangular block with an average concrete compression stress of $0.85 f_c'$ (see Fig. 3-7c). If the distance between the center of gravity of the tendon and the mid-depth of the flange is approximately $0.85 h_0$, then the level arm of the moment couple C-T, d, can be assumed to be $0.80 h_0$ (see Fig. 3-7c). The required tendon area can be written as:

$$A_{ps} = \frac{M_u}{0.80 h_0 f_{py}} \tag{3-23}$$

where f_{py} = tendon yield strength = $0.9 f_{pu}$ for low relaxation strands (see Section 1.2.3.2).

Assuming the concrete area of the equivalent compressive block as A_c, then total concrete compressive force C is

$$C = 0.85 f_c' A_c \qquad (3\text{-}24)$$

From the equilibrium of couple C-T, the required concrete area can be obtained as

$$A_c = \frac{M_u}{0.85 f_c' 0.8 h_0} = \frac{M_u}{0.68 f_c' h_0} \qquad (3\text{-}25)$$

3.1.3.2 Determination of Post-Tensioning Force and Efficiency Ratio of Cross Section

Assuming the permissible concrete tensile stress as f_{ta} and multiplying I/c_b on both side of Eq. 3-3b yield

$$\frac{f_{ta} I}{c_b} = -M + \frac{PI}{A c_b} + Pe \qquad (3\text{-}3c)$$

The left term of Eq. 3-3c represents the moment required for inducing the permissible concrete stress at the bottom fiber and denotes M_{ba}, i.e.,

$$M_{ba} = \frac{f_{ta} I}{c_b} \qquad (3\text{-}26)$$

Rearranging Eq. 3-3c, the required prestressing force P can be written as

$$P = \frac{M + M_{ba}}{e + \frac{I}{A c_b}} = \frac{M + M_{ta}}{e + \frac{I}{A c_b c_t} c_t} = \frac{M + M_{ba}}{e + \rho c_t} \qquad (3\text{-}27)$$

where

$$\rho = \frac{I}{A c_b c_t} = \frac{r^2}{c_b c_t} \qquad (3\text{-}28)$$

From Eq. 3-28, it can be seen that ρ is a constant without unit for a given cross section. From Eq. 3-27, we can observe that the denominator of the term on the right side of the equation is the level arm length of the couple moment C-T. The larger the value of ρ, the smaller the required prestressing force P. Therefore, the ρ value represents the efficiency in using the materials. Thus ρ is called the efficiency ratio of a cross section. If the concrete is assumed to be concentrated in thin flanges with negligible web concrete area, $\rho = 1$. A rectangular section has $\rho = 0.33$. For normal box girder sections, $\rho = 0.60$.

3.1.3.3 Determination of Tendon Placement Limiting Zone

The tensile stress at extreme fibers under service loading cannot exceed the allowable tensile stresses specified by AASHTO LRFD (see Section 1.2.5). It is useful to know the limiting zone in the concrete section where the tendon can be placed without causing a tensile stress larger than its allowable value at extreme fibers. The limiting zone is often called the tendon placement envelope. For simplicity, we discuss the tendon limiting zone based on a simply supported beam. However, the basic principle can also be applied to cantilever and continuous bridges, which will be discussed in Section 4.3.2.4.

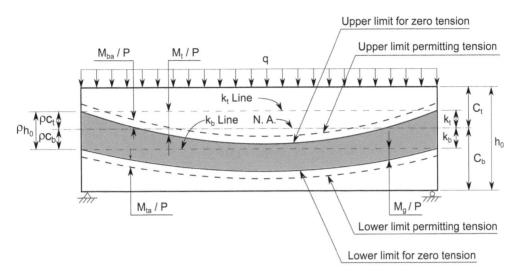

FIGURE 3-8   Limiting Zone for Tendon with Permissible Tensile Stress at Concrete Extreme Fibers.

Figure 3-8 shows a simply supported girder with external uniform loading q, and it is assumed that the concrete section and the area of the tendon have been already determined. First, let us examine the girder ends where there is no bending moment. Form Eq. 3-9a and Eq. 3-7, the concrete normal stress at top fiber can be written as

$$f^t = \frac{P}{A}\left(1 - \frac{ec_t}{r^2}\right)$$

Setting the top fiber tensile stress equal to zero yields the allowable eccentricity below the neutral axis as

$$e_b = k_b = \frac{r^2}{c_t} \tag{3-29a}$$

where k_b = lower kern point, at which any compression force applied will cause zero tensile stress at the top fiber. To ensure the top fiber tensile stress is not greater than zero, the eccentricity of the tendon below the neutral axis should not be greater than k_b at the girder ends. Similarly, from Eq. 3-9b and Eq. 3-7 we can obtain the top limiting eccentricity above the neutral axis e_t and up kern point k_t as

$$e_t = k_t = \frac{r^2}{c_b} \tag{3-29b}$$

To ensure that the bottom fiber tensile stresses are not greater than zero, the eccentricity of the tendon above the neutral axis should not be greater than k_t at the girder ends. For a girder with a constant cross section, the k_b line and k_t line are straight (see Fig. 3-8).

From Eq. 3-28, we have

$$c_t = \frac{r^2}{\rho c_b} \tag{3-30a}$$

$$c_b = \frac{r^2}{\rho c_t} \tag{3-30b}$$

Substituting Eq. 3-30a and Eq. 3-30b into Eq. 3-29a and Eq. 3-29b, respectively, yields:

$$k_b = \rho c_b \tag{3-31a}$$

$$k_t = \rho c_t \tag{3-31b}$$

The range in which the tendons can be arranged without introducing any tensile stresses in the end sections is

$$k_b + k_t = \rho(c_b + c_t) = \rho h_0 \tag{3-32}$$

From Eq. 3-32, it can be observed that the larger the efficiency ratio ρ, the larger the kern area.

Within the girder span, each section is subjected to some bending moment caused by the girder self-weight and external loading. Because any positive moment will cause compression in the top fiber and tension in the bottom fiber, the girder's minimum positive moment should be considered when determining bottom eccentricity limit and the maximum positive moment should be considered when determining the top eccentricity limit. The minimum moment in the simply supported girder is caused by girder self-weight and is denoted as M_g, and the maximum moment includes both girder self-weight and external loading and is denoted as M_t. Assuming the permissible concrete tensile stress is f_{ta}, from Eq. 3-3a, it is not difficult to obtain the bottom eccentricity limit e_b as

$$e_b = \frac{I}{Ac_t} + \frac{M_g}{P} - \frac{If_{ta}}{Pc_t} \tag{3-33a}$$

Considering the radius of gyration of the section $r = \sqrt{\frac{I}{A_c}}$ and $\frac{If_{ta}}{c_t}$ represents the moment M_{ta} required for causing the concrete permissible tensile stress f_{ta} at the top fiber, Eq. 3-33a can be written as

$$e_b = \frac{r^2}{c_t} + \frac{M}{P} + \frac{M_{ta}}{P} = k_b + \frac{M_g}{P} + \frac{M_{ta}}{P} \tag{3-33b}$$

From Eq. 3-33b, it can be observed that the lower tendon eccentricity will be moved down from the k_b line due to the moments. The permissible eccentricity will move down from the k_b line by $\frac{M_g}{P}$ if there is no tensile stress allowed and by $\frac{M_g}{P} + \frac{M_{ta}}{P}$ if the allowable tensile stress in the concrete is considered.

Similarly, from Eq. 3-3b, we can obtain the top tendon eccentricity limit e_t within the span as

$$e_t = -\frac{r^2}{c_t} + \frac{M_t}{P} + \frac{M_{ba}}{P} = -k_t + \frac{M_t}{P} + \frac{M_{ba}}{P} \tag{3-33c}$$

where $M_{ba} = \frac{If_{ta}}{c_b}$ = moment required to cause the allowable tensile stress f_{ta} at bottom fiber. The minus sign on the right side of the equation indicates the eccentricity is located above the cgc line or neutral axis.

The tendon placement range for the simply supported girder is schematically illustrated in Fig. 3-8. For a more complex bridge, the basic theory for plotting the tendon limiting zone is the same as for the simply supported girders, though it is more complex (see Section 4.3.2.4).

3.2   LOSSES OF PRESTRESSING

The prestressing force in concrete after the tendon is finally seated at its anchorage is always less than the force applied at the jack during post-tensioning. The prestressing force will be reduced in a progressive process over a long period of about 5 years. The bridge designer provides the tendon force requirements in the contract plans by either specifying the required jacking force at the ends or the final effective prestressing force at some location along the length of the tendon. The difference between the effective prestressing force and the jacking force is called losses of prestress. The prestress losses can generally be grouped into two categories:

I. Instantaneous losses
- Elastic shortening of concrete
- Duct friction
- Anchorage set

II. Time-dependent losses
- Shrinkage of concrete
- Creep of concrete
- Relaxation of prestressing steel

3.2.1 INSTANTANEOUS LOSSES

3.2.1.1 Losses Due to Elastic Shortening of Concrete in Post-Tensioning Members

In post-tensioned segmental bridges, it is almost always necessary to use multiple tendons to stress the superstructure. The tendons are stressed one at a time. The prestressing force is gradually applied to the concrete, and the magnitude of the concrete shortening increases with the post-tensioning of each tendon. The tendon stressed first would suffer the maximum amount of prestressing loss due to the elastic shortening of the concrete by the subsequent applications of prestressing from all the other tendons. The last tendon to be stressed will not suffer any losses due to elastic concrete shortening as all the shortening will have occurred when the prestress in the last tendon is measured. Figure 3-9 shows a simply supported girder with two straight tendons. The eccentricities of tendon 1 and tendon 2 are denoted as e_1 and e_2, respectively. Assuming tendon 1 is post-tensioned first (see Fig. 3-9b) and then tendon 2 is tensioned, the girder will be shortened due to the post-tensioning force F_2 in tendon 2 as shown in Fig. 3-9c.

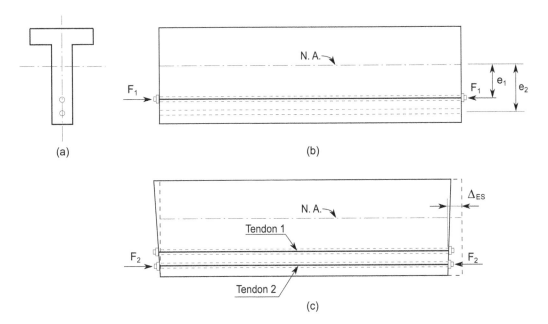

FIGURE 3-9 Effect of Elastic Shortening of Concrete Girder, (a) Cross Section, (b) Beam Elevation View with Tendon 1, (c) Effect of Tendon 2 on Tendon 1.

The increased unit shortening of the concrete at the level of tendon 1 due to tendon 2 is

$$\varepsilon_{Es12} = \frac{f_{cs12}}{E_{ci}}$$ (3-34a)

f_{cs12} = concrete stress at level of tendon 1 due to tendon 2

$$f_{cs12} = \frac{F_2}{A_c}\left(1 + \frac{e_1 e_2}{r^2}\right)$$ (3-34b)

E_{ci} = concrete modulus of elasticity at time prestressing is applied
F_2 = prestressing force in tendon 2
A_c = girder concrete section area
e_1 and e_2 = eccentricities for tendons 1 and 2, respectively
r = radius of gyration

Then, the prestressing loss in tendon 1 due to the unit shortening ε_{12} can be written as:

$$\Delta f_{pEs12} = E_p \varepsilon_{Es12} = n f_{cs12}$$ (3-35)

where
ε_{Es12} = reduced strain in tendon 1 due to tendon 2
n = $\frac{E_p}{E_{ci}}$
E_p = modulus of elasticity of prestressing steel

Assuming the cross-sectional area of tendon 1 is A_{ps1}, and then the total prestressing loss in tendon 1 is

$$\Delta F_{ps1} = A_{PS1}\Delta f_{pEs12}$$ (3-36)

The reduced prestressing force ΔF_{ps1} in turn recovers part of the concrete elastic shortening that occurred during the post-tensioning of tendon 1. The unit elongation at the level of tendon 1 due to the reduced prestressing force ΔF_{ps1} is

$$\varepsilon_{11} = -\frac{\Delta F_{ps1}}{A_c E_{ci}}\left(1 + \frac{e_1^2}{r^2}\right) = -\frac{n A_{ps1} f_{cs12}}{A_c E_{ci}}\left(1 + \frac{e_1^2}{r^2}\right)$$ (3-37)

The actual shortening of the concrete at the level of tendon 1

$$\varepsilon_1 = \varepsilon_{12} + \varepsilon_{11} = \frac{f_{cs12}}{E_{ci}} - \frac{\Delta F_{ps1}}{A_c E_{ci}}\left(1 + \frac{e_1^2}{r^2}\right)$$ (3-38)

From Eq. 3-35, the total prestressing loss in tendon 1 due to tendon 2 is

$$\Delta f_{pEs} = n f_{cs12}\left(1 - \frac{n A_{ps1}}{A_c}\left(1 + \frac{e_1^2}{r^2}\right)\right)$$ (3-39)

Normally, $\frac{n A_{ps1}}{A_c} \approx 0$; that is, the effect of the concrete shortening due to the reduced prestressing force of tendon 1 itself can be neglected. Thus, Eq. 3-39 and Eq. 3-35 are identical.

If there are many tendons, it is apparent that the calculation of the prestressing losses due to multiple tendons can be quite complicated. As previously discussed, the first tendon stressed is shortened by the subsequent stressing of all other tendons and the last tendon is not shortened by any other tendons. In practice, it is accurate enough to use an average value of stress change and apply it to all tendons equally. The AASHTO specifications propose that the following equation be used to determine the loss due to elastic shortening in post-tensioned members,

$$\Delta f_{pES} = \frac{N-1}{2N} \frac{E_p}{E_{ci}} f_{cgp} \tag{3-40}$$

where
N = number of identical prestressing tendons
f_{cgp} = sum of concrete stresses at center of gravity of prestressing tendons due to prestressing
force after jacking and self-weight of member at sections of maximum moment (ksi).

For post-tensioned structures with bonded tendons, f_{cgp} may be taken at the center section of the span or, for continuous construction, at the section of maximum moment. For post-tensioned structures with unbonded tendons, the f_{cgp} value may be calculated as the stress at the center of gravity of the prestressing steel averaged along the length of the member. For slab systems, the value of Δf_{pES} may be taken as 25% of that obtained from Eq. 3-40.

3.2.1.2 Losses Due to Duct Friction

The friction between the tendon and its surrounding materials causes the loss of post-tensioning force along the tendon. The frictional losses are generally comprised of two parts; curvature effect and wobble effect. Curvature effect is the prestressing loss that is caused by the intentional curvature of the tendon and is related to the coefficient of friction μ between the contact materials and the normal pressure N on the contact area applied by the tendon due to the curvature. The wobble effect indicates the prestressing loss caused by the misalignment of the post-tensioning duct and the deviation of the tendon duct from a specified profile. Thus, the wobble effect is proportional to the tendon length, regardless of whether the tendon is straight or curved. The wobble effect is accounted for by the wobble friction coefficient K, which is defined as the friction force per foot of tendon as a result of a unit tendon force. The values of μ and K are related to many factors, such as the types of steels, types of ducts, construction methods, and should be based on experimental data. In the absence of such data, the AASHTO specifications recommend that the values shown in Table 3-1 can be used. The values of μ and K used in the design shall be shown in the contract plan and be verified during the construction.

3.2.1.2.1 Curvature Effect

Figure 3-10a shows a post-tensioned girder with a curved tendon. Assuming the post-tensioning force to be P, the curved tendon exerts a distributed radial force to the duct as previously discussed.

TABLE 3-1

Friction Coefficients for Post-Tensioning Tendons[3-3]

Type of Steel	Type of Duct	K (per foot)	μ
Wire or strand	Rigid and semi-rigid galvanized metal sheathing	0.0002	0.15–0.25
	Polyethylene	0.0002	0.23
	Rigid steel pipe deviators for external tendons	0.0002	0.25
High-strength bars	Galvanized metal sheathing	0.0002	0.30

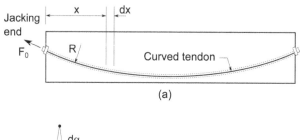

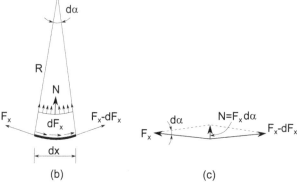

FIGURE 3-10 Theoretical Model for Analyzing Curvature Effect, (a) Post-Tensioned Girder with Curved Tendon, (b) Infinitesimal Length, (c) Force Equilibrium.

Taking an infinitesimal length of the tendon dx as shown in Fig. 3-10b, the tendon force changes from F_x at the left end of dx to F_x, dF_x at the right end due to the friction force, dF_x, between the tendon and duct. If the tendon radius is R, from Fig. 3-10b, we have

$$d\alpha \approx \tan d\alpha = \frac{dx}{R} \tag{3-41}$$

where $d\alpha$ = change in angle of dx from its left end to right end.

From Fig. 3-10c, the resultant of the distributed radial force N is

$$N = F_x \sin\frac{d\alpha}{2} + (F_x - dF_x)\sin\frac{d\alpha}{2} \tag{3-42a}$$

Considering $\sin\frac{d\alpha}{2} \approx \frac{d\alpha}{2}$ and $dF_x \sin\frac{d\alpha}{2} \approx 0$, Eq. 3-42a can be written as

$$N = F_x d\alpha \tag{3-42b}$$

Then the friction force acting on the infinitesimal length dx is

$$dF_x = -\mu N = -\mu F_x d\alpha \tag{3-43a}$$

Rearranging Eq. 3-43a, we have

$$\frac{dF_x}{F_x} = -\mu d\alpha \tag{3-43b}$$

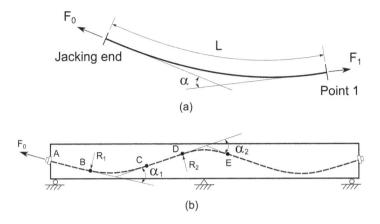

FIGURE 3-11 Determination of Friction Losses, (a) Curved Tendon Segment, (b) Curved Tendon with both Straight and Curved Segments.

Assuming that the tendon force is F_0 at the jacking end and is F_1 at point 1 in the tendon and the angle of the tendon changing by α from the jacking end to point 1 (see Fig. 3-11a), by integrating both sides of Eq. 3-43b, we have

$$\int_{F_0}^{F_1} \frac{dF_x}{F_x} = \int_0^\alpha -\mu d\alpha \tag{3-44}$$

$$\ln\left(\frac{F_1}{F_0}\right) = -\mu\alpha \tag{3-45a}$$

$$F_1 = F_0 e^{-\mu\alpha} = F_0 e^{-\mu L/R} \tag{3-45b}$$

where
  e = base of Napierian logarithms,
  $\alpha = L/R$ for circular curve tendon

3.2.1.2.2 Wobble Effect

As discussed, the only difference between the curvature effect and wobble effect is that the wobble effect is proportional to tendon length, while the curvature effect is proportional to the angle change of the tendon. Thus, substituting the loss KL for $\mu\alpha$ in Eq. 3-45b, the tendon force at any point 1 can be written as

$$F_1 = F_0 e^{-KL} \tag{3-46}$$

3.2.1.2.3 Total Loss Due to Friction

Combining Eq. 3-45b and Eq. 3-46, we have

$$F_1 = F_0 e^{-(KL+\mu\alpha)} \tag{3-47a}$$

or in terms of stresses

$$f_1 = f_0 e^{-(KL+\mu\alpha)} \tag{3-47b}$$

From Eq. 3-47b, the prestressing loss due to friction from jacking end to point 1 can be written as

$$\Delta f_{PF} = f_0 - f_1 = f_0 \left(1 - e^{-(KL+\mu\alpha)}\right) \tag{3-48}$$

where
f_0 = tendon stress at jacking point
f_1 = tendon stress at point 1

The total friction loss of a tendon can be calculated from section to section, such as AB, BC, CD, DE, as shown in Fig. 3-11b. The reduced stress at the end of a segment can be used to determine the frictional loss of the next segment. In calculating the friction loss from the jacking end A to point E for the tendon shown in Fig. 3-11b, $\alpha = \alpha_1 + \alpha_2$ and L = total tendon length from A to E.

Based on Eq. 3-48, AASHTO LRFD specifications recommend:
The prestressing losses due to friction between the internal tendons and the duct wall are taken as:

$$\Delta f_{pF} = f_{pj} \left(1 - e^{-(Kx+\mu\alpha)}\right) \tag{3-49a}$$

The prestressing losses due to friction between the external tendons across a single deviator pipe are taken as:

$$\Delta f_{pF} = f_{pj} \left(1 - e^{-\mu(\alpha+0.04)}\right) \tag{3-49b}$$

where
f_{pj} = stress in prestressing steel at jacking (ksi)
x = length of prestressing tendon from jacking end to any point under consideration (ft)
K = wobble friction coefficient (per foot of tendon, see Table 3-1)
μ = coefficient of friction (see Table 3-1)
α = sum of absolute values of angular change of prestressing steel path from jacking end, or from nearest jacking end, to point under consideration (radians)
e = base of Napierian logarithms

Value 0.04 radians in Eq. 3-49b is for an inadvertent angle change. This angle may be taken as zero in cases where the deviation angle is strictly controlled or precisely known.

3.2.1.2.4 Approximate Method for Determining Friction Losses
As we know, e^{-x} can be expanded as

$$e^{-x} = 1 - x + \frac{x^2}{2!} - \cdots + (-1)^n \frac{x^n}{n!} \tag{3-50a}$$

If x is small, Eq. 3-50a can be written as

$$e^{-x} \approx 1 - x \tag{3-50b}$$

If $x = 0.2$, the error caused by Eq. 3-50b will be less than 2%, and if $x = 0.3$, the error caused by Eq. 3-50b will be less than 5%. Using Eq. 3-50b, then Eq. 3-48 can be written as:

$$\Delta f_{PF} = f_0 \left(KL + \mu\alpha\right) \tag{3-51}$$

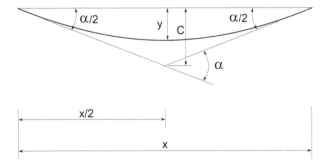

FIGURE 3-12   Approximation for Determining the Center Angle of a Tendon.

For any shapes of tendons as shown in Fig. 3-12, if tendon vertical draping y is small in comparison with its span x, the projection length of the tendon along the axis of the girder can be used in calculating the prestressing loss. The angular change of the tendon can be determined as follows:

$$y = \frac{c}{2}$$

$$\tan\frac{\alpha}{2} = \frac{c}{x/2} \approx \frac{\alpha}{2}$$

$$\alpha = \frac{8y}{x} \tag{3-52}$$

3.2.1.3   Losses Due to Anchor Set

As discussed before, when a tendon is tensioned to its design value at the jacking end, the wedges are made snug and the jack is released. The tendon pulls the wedges and seats them into the wedge plate (see Fig. 3-13) [3-15]. The friction wedges used to hold the strand will slip for a short distance before the strand can be firmly gripped. This movement is called anchorage set, and it varies with different types of post-tensioning systems. For wedge-type strand anchors, the set may vary between 0.125 in.

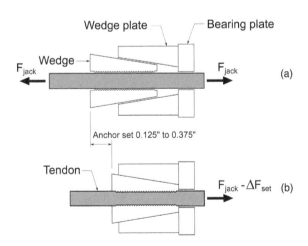

FIGURE 3-13   Anchorage Set, (a) Before Releasing the Jack, (b) After Releasing the Jack.

and 0.375 in. Florida Department of Transportation Structures Design Guidelines[3-9] recommend the following values be used for the anchor set in design:

Δ_{set} = 0.375 in. for strands
Δ_{set} = 0.500 in. for parallel wires
Δ_{set} = 0.0625 in. for bars

The magnitude of the anchor set assumed for the design and used to calculate anchor set loss should be shown in the contract documents and verified during construction.

Because of friction, the loss due to anchorage set may affect only part of the prestressed member. Figure 3-14b shows the variation of tendon force before and after anchor set for one end stressing. The point of zero movement (point O) due to the anchor set can be determined by trial and adjustment method as follows:

a. Using Eq. 3-47a, determine the variation of tendon prestressing force before releasing the jack, as shown in a line ABCD in Fig. 3-14b.
b. Assume the location of zero movement at point O, $x = l_{AO}$.
c. Calculate the friction losses from point O to anchorage end A due to anchorage set by Eq. 3-47a or using the mirror image method about the horizontal line O′O (see Fig. 3-14b) from the losses determined before releasing the jack, i.e.,

$$F_{pFA'} = F_{pFA} - 2\left(F_{pFA} - F_{pFO}\right) = 2F_{pFO} - F_{pFA} \qquad (3\text{-}53a)$$

$$F_{pFB'} = F_{pFB} - 2\left(F_{pFB} - F_{pFO}\right) = 2F_{pFO} - F_{pFB} \qquad (3\text{-}53b)$$

where
$F_{pFA}, F_{pFB}, F_{pFO}$ = tendon prestressing forces at points A, B, O, respectively, before releasing jack, considering friction losses
$F_{pFA'}, F_{pFB'}$ = tendon prestressing forces at points A′ and B′ respectively, after releasing the prestressing jack, and considering friction losses.

d. Use the average change of the tendon force within segments AB and BO as shown in the shaded area in Fig. 3-14b to calculate the change of tendon length, i.e.,

$$\Delta = \frac{\Delta F_{average}^{AB} l_{AB} + \Delta F_{average}^{BO} l_{BO}}{A_s E_S}$$

where

$$\Delta F_{average}^{AB} = \frac{F_{PFA} - F_{PFA'} + F_{PFB} - F_{PFB'}}{2},$$

$$\Delta F_{average}^{BO} = \frac{F_{PFB} - F_{PFB'}}{2}$$

l_{AB}, l_{BO} = lengths for segments AB, BO, respectively (see Fig. 3-14a).
A_s, E_S = area and modulus of elasticity of tendon, respectively.

e. If Δ is equal or close to the assumed anchor set, such as 0.375 in., then the assumed location of zero movement is correct. Otherwise, revise the x until the calculated anchor set is close to the assumed one.

If tendons are continuous over multiple spans, their friction losses could be high. The tendon can be stressed on both ends. Their friction losses can be significantly reduced (see Fig. 3-14c).

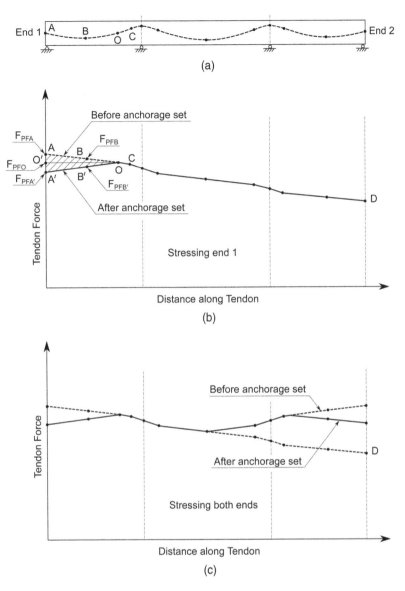

FIGURE 3-14 Variation of Tendon Prestessing Force after Anchorage Set, (a) Post-Tensioned Girder, (b) One-End Stressing, (c) Two-End Stressing.

3.2.1.4 Elongation of Tendons

It is often necessary to document the tendon elongation caused by post-tensioning force in the contract plans so that tendon forces and measured elongation can be verified in the field. It also assists the contractor to select the proper stressing equipment. The elongation of a tendon is normally calculated based on an average force over a length of tendon segment and then summed together, i.e.,

$$\Delta = \sum_i \frac{\bar{F}_i l_i}{A_s E_s} \tag{3-54}$$

where

$\bar{F_i}$ = average force of tendon segment
l_i = length of tendon segment

3.2.2   TIME DEPENDENT LOSSES

3.2.2.1   Time Dependent Losses for Segmental Bridges

As discussed in Chapter 1, both tendon and concrete have properties that change with time after a sustained stress is imposed on them. Thus, prestress losses due to creep and shrinkage of concrete and steel relaxation are both time dependent and interdependent. To account for these changes in successive time intervals, a step-by-step method should be used for concrete segmental bridge design[3-3] by dividing the time into short intervals over which constant strain is assumed. As the analysis is time consuming, computer programs are normally used for determining the long-term effect of concrete creep and shrinkage and steel relaxation in segmental bridge design. The changes in strain due to concrete creep and shrinkage as well as the loss of prestressing stress due to steel relaxation were discussed in Sections 1.2.2. and 1.2.3. The AASHTO LRFD specifications permit using three different models for analyzing the effects of creep and shrinkage in concrete segmental bridge design. The three models are contained in the AASHTO LRFD, CEP-FIP, and ACI codes. The CEP-FIP Model Code has been used extensively in the United States in the last 30 years. In this section, only some approximate methods will be discussed for preliminary design use. More detailed analytical procedures for evaluating the long-term effect of concrete creep and shrinkage will be discussed in Sections 4.3.4 and 4.3.5.

3.2.2.2   Estimations of Time Dependent Losses for Preliminary Design of Segmental Bridges

The AASHTO specifications provide two approximate methods for estimating time dependent losses; the refined estimation method and lump-sum method.

3.2.2.2.1   Refined Estimation Method

The time dependent losses in a concrete segmental bridge include three components of creep of concrete, shrinkage of concrete, and relaxation of steel. The total long-term time-dependent prestress loss Δf_{pLT} can be written as:

$$\Delta f_{pLT} = \Delta f_{pS} + \Delta f_{pC} + \Delta f_{pR}$$ (3-55)

where

Δf_{pS} = prestress loss due to shrinkage (ksi)
Δf_{pC} = prestress loss due to creep (ksi)
Δf_{pR} = prestress loss due to relaxation of prestressing strands (ksi)

3.2.2.2.1.1   Estimation of Loss Due to Shrinkage   As tendons and the surrounding concrete interact, it is evident that the prestressing stress losses as a result of concrete shrinkage and creep is smaller in bonded tendons than in deboned tendons. The AASHTO LRFD specifications use a transformed section coefficient K_f to account for this effect [3-1, 3-16]. Based on Eq. 1-15a, the long-term prestressing loss Δf_{pS} due to shrinkage can be estimated as

$$\Delta f_{pS} = \varepsilon_{sf} E_p K_f$$ (3-56)

where

K_f = 1.0, for unbonded tendon
K_f = $\dfrac{1}{1+\frac{E_p}{E_{ci}}\frac{A_{ps}}{A_c}\left(1+\frac{A_c e_{pc}^2}{IC}\right)\left(1+0.7\phi_b(t_f,t_i)\right)}$, for bonded tendon,

ε_{sf}	= shrinkage strain from time of prestressing to final time calculated by Eq. 1-15a
e_{pc}	= distance between prestressing force and centroid of composite section (in.), positive when prestressing force is below centroid of section
A_c	= area of gross section of concrete girder (in.²)
I_c	= moment of inertia of concrete girder gross section (in.⁴)
$\phi_b(t_f,t_i)$	= girder creep coefficient at final time calculated by Eq. 1-9a
t_f	= final age (days)
t_i	= age at prestressing (days)

3.2.2.2.1.2 Estimation of Loss Due to Creep   Based on Eq. 1-9a, the prestressing stress losses due to creep can be estimated as

$$\Delta f_{pC} = \frac{E_p}{E_{ci}} f_{cgp} \phi(t_f,t_i) K_f \tag{3-57}$$

where

f_{cgp}	= sum of concrete stresses at center of gravity of prestressing tendons due to the prestressing force after jacking and self-weight of member at section with maximum moment (ksi)
$\phi(t_f,t_i)$	= girder creep coefficient at final time calculated by Eq. 1-9a

3.2.2.2.1.3 Estimation of Loss Due to Relaxation   The long-term prestress loss due to the relaxation of prestressing strands between the time of transfer and the deck placement, Δf_{pR}, can be estimated determined as:

$$\Delta f_{pR} = \frac{f_{pt}}{K_L}\left(\frac{f_{pt}}{f_{py}} - 0.55\right) \tag{3-58}$$

where

f_{pt}	= stress in the prestressing strands immediately after transfer, not less than $0.55f_{py}$ (ksi)
K_L	= 30 for low relaxation strands and 7 for other prestressing steel

Δf_{pR} can be assumed to be equal to 2.4 ksi for low-relaxation strands.

For more accurate estimation of prestressing loss due to steel relaxation, Eq. 1-20 can be used.

3.2.2.2.2 Lump-Sum Method
For a quick check of the results obtained by computer program or to perform preliminary design of segmental concrete bridges, the long-term prestress loss Δf_{pLT} due to creep of concrete, shrinkage of concrete, and relaxation of steel can be estimated using the following simple equation[3-3], though it may provide less accurate results:

$$\Delta f_{pLT} = 10.0 \frac{f_{pi} A_{pi}}{A_g} r_h r_{st} + 12.0 r_h r_{st} + \Delta f_{pR} \tag{3-59}$$

where

$$r_h = 1.7 - 0.01H \tag{3-60}$$

$$r_{st} = \frac{5}{1 + f'_{ci}} \tag{3-61}$$

f_{pi}	= prestressing steel stress immediately prior to transfer (ksi)
A_g	= gross area of section (in.²)

A_{pi} = area of prestressing steel (in.^{2})
H = average annual ambient relative humidity (%)
Δf_{pR} = estimate of relaxation loss
    = 2.4 ksi for low relaxation strands
    = 10.0 ksi for stress-relieved strands

3.3 BENDING AND TORSION OF I-GIRDER

3.3.1 INTRODUCTION

I-sections are the simplest type of sections used in segmental bridges and are often used in spliced girder bridges. For easy understanding, the bending and torsion of I-girders are discussed first and then those of box girders will be discussed in Section 3.4. An I-girder bridge is normally subjected to different types of loadings. The normal stress in the girder mainly consists of two portions, i.e.,

$$f_N = f_b + f_\omega \tag{3-62}$$

where
   f_b = normal stress due to bending
   f_ω = normal stress due to warping torsion

The shear stress in the girder mainly includes three portions, i.e.,

$$f_s = f_{bv} + f_v + f_{\omega v} \tag{3-63}$$

where
   f_{bv} = shear stress due to bending
   f_v = shear stress due to shear force
   $f_{\omega v}$ = shear stress due to warping torsion

3.3.2 BENDING

The normal stress due to vertical bending as shown in Fig. 3-15a is well known and can be written as

$$f_b = \frac{My}{I} \tag{3-64}$$

The shear stress caused by bending can be derived as follows:
   Taking an infinitive element dz as shown in Fig. 3-15b and considering the horizontal force equilibrium (in z-direction) of a portion of the element as shown, we have

$$f_v t dz = \int f_2 \, dA' - \int f_1 \, dA' = \int \frac{(M + dM)y}{I} dA' - \int \frac{My}{I} dA' \tag{3-65a}$$

Rearrange the above equation. We have

$$f_v = \frac{1}{It} \frac{dM}{dz} \int y \, dA' \tag{3-65b}$$

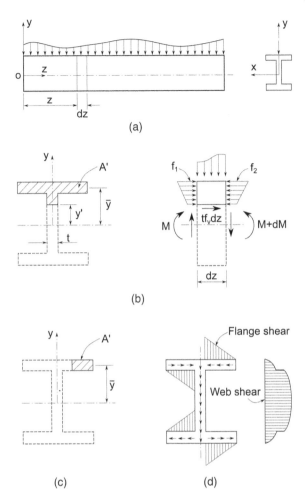

FIGURE 3-15   Shear Stress Distribution Due to Bending, (a) I-Girder in Bending, (b) Free Body for Determining Web Shear, (c) Free Body for Determining Flange Shear, (d) Shear Stress Distribution.

where
f_v = shear stress, which is assumed to be constant through thickness
t = thickness of member's cross-sectional area, measured at point where f_v is to be determined
I = moment of inertia of entire cross-sectional area about neutral axis
A' = member's cross-sectional area in consideration (see Figs. 3-15b and 3-15c)

In Eq. 3-65b, $\frac{dM}{dz}$ represents the section shear force V and the integral represents the moment of the area A' about the neutral axis. Then, Eq. 3-65b can be simplified as

$$f_v = \frac{VQ}{It}$$ (3-65c)

where
V = internal shear force
Q = $\bar{y}A'$ = moment of area A' about neutral axis
$\bar{y}$ = distance from neutral axis to centroid of A'

The shear stress distribution determined from Eq. 3-65c is shown in Fig. 3-15d.

3.3.3 TORSION

When a load is applied away from a girder shear center, such as an eccentrically loaded vertical load or wind load, the girder will twist and bend. The shear center is defined as the point in the cross section through which the lateral (or transverse) loads must pass to produce bending without twisting. For the I-girders with two axes of symmetry, the shear center is also the center of rotation. There are two types of torsion: free torsion and warping torsion. Free torsion means that the member is allowed to freely deform in its longitudinal direction after a torque is applied. Free torsion is also called uniform torsion or St. Venant's torsion, which assumes plane cross sections normal to the axis of the member remains plane after twisting (see Fig. 3-16). Warping torsion means that the longitudinal deformation of a member is restrained and sectional normal stress is developed after the applied torque (see Fig. 3-17). The longitudinal deformation is referred to as warping due to torsion. The warping torsion is also called nonuniform torsion. In an I-girder, the external torque is always resisted by two types of internal torsions: free-torsion moment and warping torsion moment, i.e.,

$$T = T_s + T_\omega \tag{3-66}$$

where

T_s = free torsion
T_ω = warping torsion

If the angle of the twist is denoted as θ, the free torsion can be written as

$$T_s = JG\frac{d\theta}{dz} \tag{3-67}$$

$$J = \frac{1}{3}\sum b_i t_i^3 \tag{3-68}$$

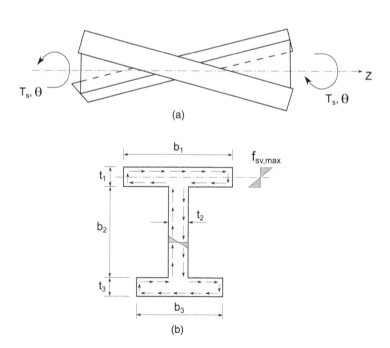

FIGURE 3-16   Free Torsion of I-Girder, (a) Free Torsion, (b) Distribution of Free Torsion Shear Stress.

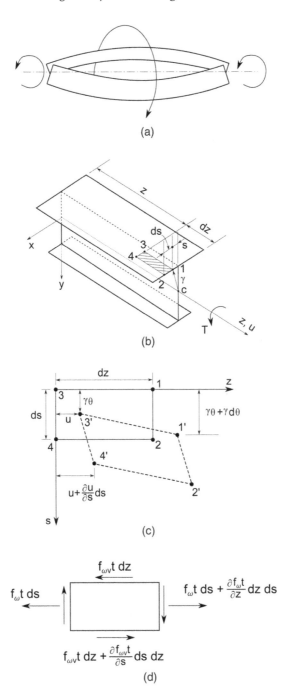

FIGURE 3-17 Warping Torsion of an I-Girder, (a) Nonuniform Torsion Deformation, (b) Typical I-Girder under Torsion, (c) Small Element Deformation, (d) Equilibrium of Stress.

where
J = torsional constant = free torsional moment of inertia
G = shear modulus of inertia
t = slab thickness
b = slab width

The shear stress caused by free torsion is called free torsional stress or St. Venant's torsional shear stress. The distribution of free torsional stress is shown in Fig. 3-16. The maximum shear stress is located at the exterior faces of the slab and can be written as

$$f_{sv,\,max} = \frac{T_s t}{J} \tag{3-69}$$

Theoretically, free torsion can only occur in a member with a constant circular cross section. The deformation in the longitudinal direction due to external torque will be restrained in a member with other types of sections. When warping deformation is constrained, the member undergoes nonuniform torsion. An example of nonuniform torsion is illustrated in Fig. 3-17a. From this figure, we can see that the warping restraint causes in-plane bending deformation of the flanges in addition to twisting the flanges. The bending deformation is accompanied by a shear force in each flange. Now let us discuss the warping torsion of an I-girder as shown in Fig. 3-17b. The curvilinear coordinate axis s is taken along the perimeter of the cross section. Assuming the sections at z and $z + dz$ rotate about the torsion center C by θ and $\theta + d\theta$, respectively, due to the action of the torsional moment T, a small element 1 2 3 4 taken from the segment $dz - ds$, as shown in the figure, will translate to new positions 1′ 2′ 3′ 4′ (see Fig. 3-17c). The shear deformation of the small element can be written as

$$\gamma_{zs} = r\frac{d\theta}{dz} + \frac{\partial u}{\partial s} \tag{3-70}$$

where r = the perpendicular distance from the torsion center to the flange of the I-section in consideration (see Fig. 3-17b).

The shear deformation in the z-s plane at mid-depth of the flange and web is small and can be treated as zero. Thus, integrating with respect to the s coordinate axis gives the warping deformation as

$$u(z,s) = -\int_0^s \frac{r d\theta}{dz} ds + u_0(z) = -\omega(s)\frac{d\theta}{dz} + \omega_0 \tag{3-71}$$

where ω_0 = torsional warping function at the origin $s = 0$.

Torsional warping function:

$$\omega(s) = \int_0^s r ds \tag{3-72}$$

If the origin of s is selected at the cross points between the web and the flanges as shown in Fig. 3-17b, $\omega_0 = 0$.

Warping normal stress can be written as

$$f_\omega = E\frac{\partial u}{\partial z} = -E\omega\theta'' + E\omega_0' \tag{3-73}$$

As there is only torque acting on the section; from the conditions of equilibrium, the axial force in the z-direction N_z, moment about the x-axis M_x, and moment about the y-axis M_y should be equal to zero, i.e.,

$$N_z = \int f_\omega\, dA = 0, M_x = \int f_\omega y\, dA = 0, M_y = \int f_\omega x\, dA = 0$$

Substituting Eq. 3-73 into the above equations yields

$$\theta'' \int \omega \, dA - \omega_0' = 0 \tag{3-74}$$

$$\theta'' \int \omega y \, dA - \omega_0' \int y \, dA = 0 \tag{3-75}$$

$$\theta'' \int \omega x \, dA - \omega_0' \int x \, dA = 0 \tag{3-76}$$

where the prime denotes the derivative with respect to the z-axis. From Eq. 3-74, $\omega_0' = 0$ only if

$$\int \omega \, dA = 0 \tag{3-77}$$

From Eq. 3-77, we can determine the location where $\omega_0 = 0$. In general, if the origin of s is selected at the shear center, $\omega_0 = 0$.

From Eqs. 3-75 and 3-76, we can determine the location of the torsion center.

Similar to the normal bending stress in a beam, the warping normal stress also induces warping shear stress, which is also called secondary shear stress and is uniformly distributed along the thickness. Figure 3-17d shows a small element in equilibrium cut from the I-girder. From the figure, it can be deduced that

$$\frac{\partial f_{\omega v} t}{\partial s} = -\frac{\partial f_\omega t}{\partial z} \tag{3-78}$$

where $f_{\omega v}$ = secondary shear stress due to warping.

Substituting Eq. 3-73 into Eq. 3-78 and performing integration on both sides of the above equation with respect to the s coordinate axis, we can obtain the second shear flow as

$$q_\omega = f_{\omega v} t = -E\theta''' \int_0^s \omega t \, ds = -E\theta''' S_\omega \tag{3-79}$$

where S_ω is referred to as the static moment with respect to the warping function, in an analogy to the definition of the static moment in bending.

$$S_\omega = \int_0^s \omega t \, ds \tag{3-80}$$

The products of the secondary shear flow q_ω and the perpendicular distance r from the torsional center, as shown in Fig. 3-17, is referred to as the secondary torsional moment or warping torsional moment T_ω and can be written as

$$T_\omega = \int_0^s q_\omega r \, ds = -E\theta''' \int_0^s \omega^2 t \, ds = -EI_\omega \theta''' \tag{3-81}$$

where $I_\omega = \int_0^s \omega^2 t \, ds$ is called the torsional warping constant and represents the geometric moment of inertia with respect to the warping function ω.

From Eqs. 3-79 and 3-81, the warping stress also can be written as

$$f_{\omega v} = \frac{q_\omega}{t} = \frac{T_\omega S_\omega}{I_\omega t} \tag{3-82}$$

Comparing Eq. 3-82 with Eq. 3-65c, it can be seen that the bending shear stress and warping shear stress have a similar expression. Moreover, by selecting the origin of the s-coordinate at the shear center and $\omega_0 = 0$, from Eq. 3-73, the warping normal stress can be written as

$$f_\omega = \frac{M_\omega \omega}{I_\omega} \tag{3-83}$$

where

$$M_\omega = EI_\omega \theta'' \tag{3-84}$$

M_ω is referred to as the bi-moment. Figure 3-18 shows the physical meaning of the bi-moment.

Comparing Eq. 3-83 with Eq. 3-64, we can find that the flexural normal stress and warping normal stress have a similar expression.

From Eqs. 3-66, 3-67, and 3-81, we can obtain the torsion differential equation as follows:

$$GJ\theta' - EI_\omega\theta''' = T \tag{3-85a}$$

If a girder is subjected to a uniformly distributed torque m_{zt}, Eq. 3-85a can be rewritten as

$$\frac{d^4\theta}{dz^4} - \alpha^2 \frac{d^2\theta}{dz^2} = \frac{m_{zt}}{EI_\omega} \tag{3-85b}$$

where

$$\alpha = \sqrt{\frac{GJ}{EI_\omega}} \tag{3-86}$$

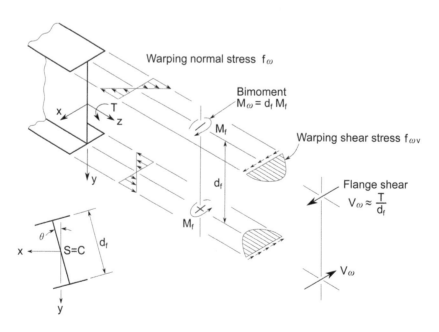

FIGURE 3-18 Distributions of Warping Normal and Shear Stresses in I-Girder and Bi-Moment.

The solution of Eq. 3-85b is

$$\theta = A\sinh\alpha z + B\cosh\alpha z + Cz + D + \frac{m_{zt}}{2GJ}z(l-z) \qquad (3\text{-}87)$$

The integration constants A, B, C, and D can be determined based on the boundary conditions of the girder in consideration. Once the integration constants have been determined, the pure torsion moment, warping torsion moment, and bi-moment can be obtained from Eqs. 3-67, 3-81, and 3-84, respectively.

If the I-girders with wide top and bottom flanges, such as $\alpha l \leq 0.4$ (l = span length), the pure torsion T_s is very small. The flange shear can be estimated as

$$V_\omega \approx \frac{T}{d_f} \qquad (3\text{-}88)$$

where T = external torque.

The warping shear and normal stresses can be estimated by treating the top and bottom flanges as a laterally supported beam loaded by V_ω.

3.4 BENDING AND TORSION OF BOX GIRDERS

3.4.1 INTRODUCTION

When a box girder with symmetric cross section as shown in Fig. 3-19 is subjected to a vertical eccentric load P with eccentricity e from its centerline, the applied load can be resolved as a flexural force P (Fig. 3-19b) and a torsion moment Pe (Fig. 3-19c). The torsion moment can be further resolved as pure torsion and distortion as shown in Fig. 3-19d and e. In Fig. 3-19, O represents the center of gravity, C represents the torsion center, and D represents the distortion center. Under flexure loading, the box girder will bend about its neutral axis (x-axis); under pure torsion, the box girder will rotate about its torsion center C as a rigid box; under distortion loading, the shape of the

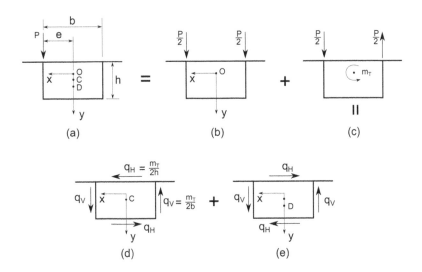

FIGURE 3-19 Box Girder Subjected to Eccentric Loading, (a) Eccentric Loading, (b) Flexure Loading, (c) Torsion, (d) Pure Torsion, (e) Distortion[3-17].

box section will deform and the webs and top and bottom slabs will rotate about its distortion center D[3-17 to 3-20]. The longitudinal normal stress f_N in the cross section includes three portions:

Longitudinal normal stress:

$$f_N = f_b + f_\omega + f_{d\omega} \tag{3-89}$$

where

f_b, f_ω, $f_{d\omega}$ are the normal stresses due to bending, warping torsion, and distortion, respectively. The shear stress f_{Tv} in the cross section consists of four portions:

$$f_{Tv} = f_v + f_{sv} + f_{\omega v} + f_{dv} \tag{3-90}$$

where

f_v, f_{sv}, $f_{\omega v}$, f_{dv} are the shear stresses due to bending, free torsion, warping torsion, and distortion, respectively.

If a vertical loading, as shown in Fig. 3-19, is located between the webs, it also will induce local transverse bending in addition to the aforementioned deformation. The transverse bending moment M_t includes two portions:

$$M_t = M_{dt} + M_{lt} \tag{3-91}$$

where M_{dt}, M_{lt} are the transverse bending moment due to distortion and local bending, respectively.

The sketches for the longitudinal normal stress, shear stress, and transverse moment distributions are illustrated in Fig. 3-20.

3.4.2   BENDING

3.4.2.1   Shear Stress

The basic methods for determining the normal and shear stresses of a box girder due to bending are similar to those for an I-girder. However, for a box section, we normally do not know the locations

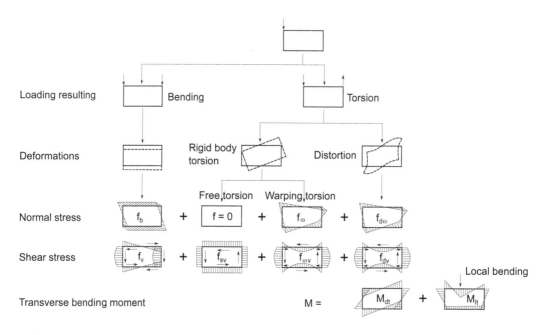

FIGURE 3-20   Types of Longitudinal Stresses and Transverse Moment in a Box Girder.

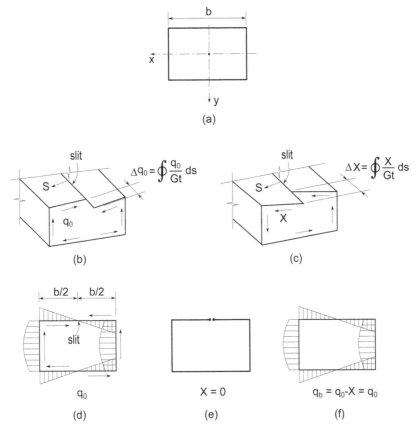

FIGURE 3-21 Determination of Shear Flow Due to Bending in a Box Section, (a) Box Section, (b) Longitudinal Displacement at Slit Due to Shear Flow q_0, (c) Longitudinal Displacement at Slit Due to Unknown Shear Flow X, (d) Distribution of Shear Flow q_0, (e) Distribution of Shear Flow X, (f) Distribution of Actual Shear Flow.

where the shear stress is equal to zero and we cannot directly use Eq. 3-65c to determine the shear stress. We can first transfer the closed box section into an open section by inserting a slit with an unknown shear flow X as shown in Fig. 3-21c. Then using the compatibility condition of deformation, we can determine the unknown shear flow X. The actual shear flow f_b is the sum of shear flow q_0 in the imaged open section (Fig. 3-21b) and X (Fig. 3-21c).

The shear flow q_0 acting on the open section can be determined from Eq. 3-65c. The total relative longitudinal displacement at the slit due to q_0 can be obtained by integrating the shear deformation along the curvilinear coordinate axis s which is taken along the perimeter of the cross section, i.e.,

$$\Delta_{q0} = \oint \frac{q_0}{Gt} ds \tag{3-92}$$

where

$$q_0 = \frac{VQ}{I}$$

Similarly, the displacement at the slit due to the unknown shear flow X can be written as

$$\Delta_X = \oint \frac{X}{Gt} ds \tag{3-93}$$

The relative displacement at the slit should be equal to zero, i.e.,

$$\oint \frac{q_0}{Gt} ds - \oint \frac{X}{Gt} ds = 0 \tag{3-94}$$

Thus

$$X = \frac{\oint \frac{q_0}{t} ds}{\oint \frac{ds}{t}} \tag{3-95}$$

The actual shear flow is

$$q_b = q_0 - X \tag{3-96}$$

Most box girder sections in a segmental bridge are symmetrical about their vertical centerlines as shown in Fig. 3-21a. If we insert a slit at its centerline as shown in Fig. 3-21b, we will find the actual shear flow q_b at the location is equal to zero, i.e., $X = 0$ (see Fig. 3-21e).

For a multi-cell box girder section, more than one slit should be inserted so that the box section becomes an open section. For example, for a two-cell box girder as shown in Fig. 3-22, two slits are needed to make the box section an open section. There are two unknown shear flows X_1 and X_2 which can be determined based on two compatibility conditions of deformation at their slit locations, i.e.,

$$\oint q_0 \frac{ds_1}{t} - X_1 \oint \frac{ds_1}{t} + X_2 \int_a^b \frac{ds_2}{t} = 0 \tag{3-97a}$$

$$\oint q_0 \frac{ds_2}{t} + X_1 \int_a^b \frac{ds_1}{t} - X_2 \oint \frac{ds_2}{t} = 0 \tag{3-97b}$$

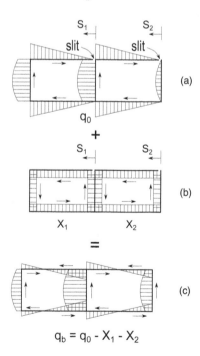

FIGURE 3-22 Shear Flows in Multi-cell Box Girder Section, (a) Shear Distribution Due to Shear Flow q_0 in Open Section, (b) Shear Distribution Due to Unknowns X_1, and X_2, (c) Actual Shear Distribution.

where s_1 and s_2 represent the curvilinear coordinates taken along the perimeter of the cell numbers 1 and 2, respectively.

By solving the simultaneous Eqs. 3-97, we can obtain X_1 and X_2. Then the actual shear flow can be calculated as

$$q_b = q_0 - X_1 - X_2 \tag{3-98}$$

The sketches for q_b, q_0, X_1, and X_2 are illustrated in Fig. 3-22.

3.4.2.2 Flexural Normal Stress Distribution in the Flanges and Effective Width

The flexural normal stress distribution in the top and bottom flanges of a box girder section varies along its x-direction as shown in Fig. 3-23 and does not have a constant value determined based on elementary beam theory. The maximum flexural normal stress occurs at the junctions between the flange and the web and may be significantly different than that obtained from the elementary beam theory. This phenomenon is caused by the lag of shear strain in the flanges and is called shear lag phenomenon. The shear lag phenomenon can be analyzed using the theory of elasticity by treating the flange plate as a plane-stress problem [3-21]. Theoretical and tested results show that the shear lag phenomenon varies with different types of loadings and the location along the girder length. For bridge engineers, it may not be necessary to accurately determine and explicitly consider the shear lag phenomenon in bridge design. It is more convenient to define the effective width of the flange plate for practical use. The effective width b_{0m} is defined as the width by which the normal stress determined by the elementary beam theory is identical to the maximum stress at the junction point of the web and flange plate. The maximum normal stress at the junction point is determined based on shear lag theory on actual flange width, i.e. (see Fig. 3-23),

$$b_{0m} = \frac{\int_0^b f_b(x)\,dx}{f_{b,\,max}} \tag{3-99}$$

where
$f_b(x)$ = actual normal stress distribution along flange plate
$f_{b,\,max}$ = maximum stress at junction point of web and flange plate

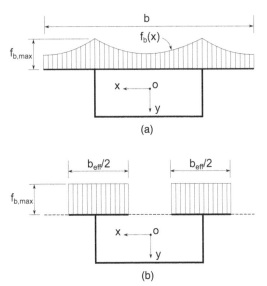

(a)

(b)

FIGURE 3-23   Shear Lag Phenomenon and Effective Width, (a) Distribution of Normal Stress in Top Flange, (b) Effective Flange Width.

The effective width varies with different loading conditions and locations along the girder length. The distributions of the normal and shear stresses at the midspan are more uniform than those at the supports. For simplicity, based on many test and theoretical analysis, AASHTO specifications provide a simple method for determining the effective widths.

Figure 3-24a shows a typical box section with the designations for the physical width b and effective width b_e for each of the flanges. In this figure,

b_i = physical flange width of flange i
b_{ei} = effective flange width of flange i

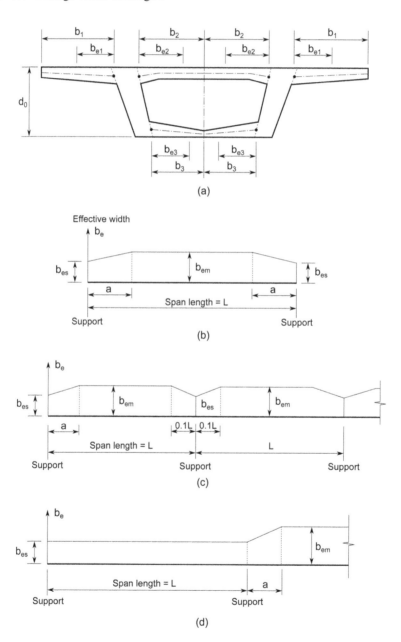

(a)

(b)

(c)

(d)

FIGURE 3-24 Designations of Effective Flange Widths, (a) Typical Section, (b) Simply Supported Bridges, (c) Continuous Bridges, (d) Cantilever Bridges.

Figure 3-24b to d illustrate the variations of the effective flange widths with the span length for simple span, continuous span and cantilever bridges, respectively. In these figures,

b_{em} = effective flange width for interior portions of span (in.)
b_{es} = effective flange width at supports
a = portion of span subject to a transition in effective flange width taken as the lesser of physical flange width on each side of the web shown in Fig. 3-24 or one-quarter of the span length (in.)

The effective flange widths for segmental concrete box beams and single-cell, cast-in-place box beams can be determined as follows.
The effective flange width

$$b_{em} = b_{es} = b, \text{ If } b \le 0.1 \, l_i \text{ or } b \le 0.3 \, d_o \qquad (3\text{-}100)$$

where
b = physical flange width
d_o = girder depth
l_i = notional span length
l_i = 1.0L for single span
     = 0.8L and 0.6L for end span and interior span of continuous span
     = 1.5L for cantilever

For the other cases, the effective widths can be determined based on the ratio of b/l_i from Fig. 3-25.

3.4.3 TORSION

3.4.3.1 Pure Torsion
Similar to the I-girders, the pure torsion is resisted by both free torsion and warping torsion. In practical segmental box girder bridges, the warping torsion is comparatively small and can be

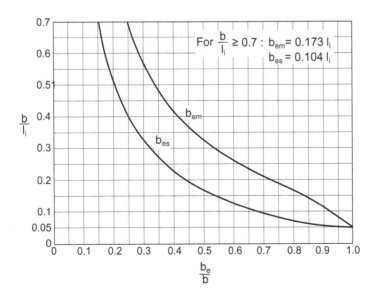

FIGURE 3-25 Variation of the Effective Flange Width Coefficients with Ratios of b/l_i.[3-3].

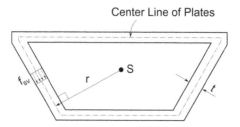

FIGURE 3-26   Free Torsion of a Single-Cell Box Section.

neglected and will not be discussed in this section as its basic analytical theory is the same as that for I-girders.

Figure 3-26 shows a single-cell box girder subjected to a torque T_s. Assuming the shear stress f_{sv} is uniformly distributed across its thin wall thickness t, there is a constant shear flow $q_s = tf_{sv}$ along the girder plates. Letting r denote the distance from the mid-depth of the thin wall plate thickness to the shear center or torsion center C, the moment produced by the shear flow over the entire cross section should be equal to the torque T_s:

$$\oint rq_s\,ds = T_s \tag{3-101}$$

Thus

$$q_s = tf_{sv} = \frac{T_s}{2A} \tag{3-102}$$

where $A = \frac{1}{2}\oint r\,ds$ = area enclosed by centerlines of the box girder plates as shown in the dotted lines in Fig. 3-26.

Equating the strain energy of torsion to the work done by the torque within a small member element dz yields:

$$\frac{1}{2}\oint tf_{sv}\gamma_s\,ds = \frac{1}{2}T_s\frac{d\theta}{dz}\,dz \tag{3-103}$$

With Hooke's law $\gamma_s = f_{sv}/G$ and f_{sv} from Eq. 3-102, we have

$$T_s = GJ\frac{d\theta}{dz} = GJ\theta' \tag{3-104}$$

where J = free torsion moment of inertia of a box girder section:

$$J = \frac{4A^2}{\oint(ds/t)} \tag{3-105}$$

For a three-cell single box girder subjected to a pure torsion T_s, as shown in Fig. 3-27, the shear flows q_1, q_2, and q_3 in cells 1, 2, and 3, respectively, can be determined as follows.

Assuming that A_1, A_2, A_3 are the areas of cells 1, 2, and 3, respectively, and based on Eq. 3-102 and the torsion equilibrium condition, we have

$$2A_1q_1 + 2A_2q_2 + 2A_2q_3 = T_s \tag{3-106}$$

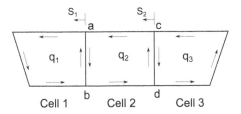

FIGURE 3-27   Free Torsion in a Multi-Cell Single Box Section.

As the pure torsion angle θ is the same for each of the cells, from Eq. 3-101, three equations corresponding to each of the cells can be obtained:

$$q_1 \oint \frac{ds_1}{t} - q_2 \int_a^b \frac{ds_2}{t} = GJ_1\theta' \tag{3-107a}$$

$$q_2 \oint \frac{ds_{21}}{t} - q_1 \int_a^b \frac{ds_1}{t} - q_3 \int_c^d \frac{ds_3}{t} = GJ_2\theta' \tag{3-107b}$$

$$q_3 \oint \frac{ds_1}{t} - q_2 \int_c^d \frac{ds_2}{t} = GJ_3\theta' \tag{3-107c}$$

where
  s_1, s_2, and $s_3 =$ curvilinear coordinates taken along perimeter of cells 1, 2, and 3, respectively.
  J_1, J_2, and $J_3 =$ free torsion moments of inertia for cells 1, 2, and 3, respectively.
  Solving the simultaneous Eqs. 3-106 to 3-107c, we can obtain the shear flows q_1, q_2, and q_3. The shear flows in the interior webs ab and cd can be written as

$$q_{ab} = q_1 - q_2$$

$$q_{cd} = q_2 - q_3$$

It is apparent that $q_1 = q_2 = q_3$ if the areas of the three cells are the same. In this case, the shear flow is zero in the interior webs and the three-cell box section can be simply treated as a single cell by neglecting the interior webs.

3.4.3.2   Distortion
A box girder subjected to a distortion loading m_T as shown in Fig. 3-28a will be deformed as shown in Fig. 3-28b. Each of the four angles will be deformed by a distortional angle $\tilde{\theta}$. Let the rectangular coordinate axes (x, y, z) be taken at its distortion center D. Analogously to the I-girder torsion warping, the distortional warping can be derived from Eq. 3-70 with $\gamma_{zs} = 0$. The result is similar to Eq. 3-71:

$$u(z,s) = \frac{d\tilde{\theta}}{dz}\left[-\int_0^s r_D ds + C_1 \right] \tag{3-108a}$$

where
  $C_1 =$ integration constant = distortional warping function at origin $s = 0$. For a symmetrical box girder in the vertical direction as shown in Fig. 3-28, if the origin $s = 0$ is selected at the symmetrical y-axis, $C_1 = 0$.

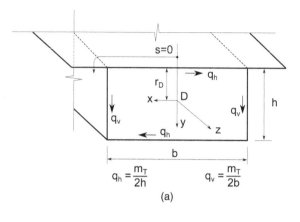

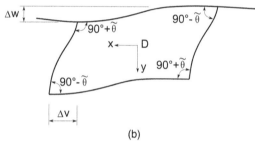

FIGURE 3-28 Distortion of a Box Section, (a) Box Girder Subjected Distortional Loads, (b) Distortional Deformation.

r_D = perpendicular distance from distortional center to box sides, i.e., along the curvilinear coordinate s-axis as shown in Fig. 3-28.

Equation 3-108a can be rewritten as

$$u(z,s) = \frac{d\tilde{\theta}}{dz}\tilde{\omega}(s) \tag{3-108b}$$

where

$\tilde{\omega}(s)$ is defined as distortional warping function:

$$\tilde{\omega}(s) = -\int_0^s r_D ds + C_1 \tag{3-109}$$

By Hooke's law, the distortional normal warping stress can be given as

$$f_{\tilde{\omega}} = E\varepsilon = E\frac{\partial u}{\partial z} = E\frac{\partial^2 \theta}{\partial z^2}\tilde{\omega} \tag{3-110}$$

where E = modulus of elasticity.

As the distortion will not produce any additional axial force and bending moment, we have

$$N_z = \int f_{\tilde{\omega}} \, dA = 0 \tag{3-111}$$

$$M_x = \int f_{\tilde{\omega}} y \, dA = 0 \tag{3-112}$$

$$M_y = \int f_{\tilde{\omega}} x \, dA = 0 \tag{3-113}$$

Substituting Eq. 3-110 into Eq. 3-111, we can determine C_1. Substituting Eq. 3-110 into Eqs. 3-112 and 3-113, we can determine the location of the distortional center.

Similar to Eq. 3-78 for the warping torsion, we have

$$\frac{\partial f_{\tilde{\omega}v}t}{\partial s} = -\frac{\partial f_{\tilde{\omega}}t}{\partial z} \tag{3-114}$$

where $f_{\tilde{\omega}v}$ = distortional warping shear stress.

Substituting Eq. 3-110 into Eq. 3-114 and performing integration on both sides of the equation above with respect to the s coordinate axis, we can obtain the distortional shear flow as

$$q_{\tilde{\omega}} = f_{\tilde{\omega}v}t = -E\tilde{\theta}''' \int_0^s \tilde{\omega}t ds = -E\tilde{\theta}''' S_{\tilde{\omega}} \tag{3-115}$$

where $S_{\tilde{\omega}}$ is referred to as the static moment with respect to the distortional warping function, in an analogy to the definition of the static moment in bending.

$$S_{\tilde{\omega}} = \int_0^s \tilde{\omega}t ds \tag{3-116}$$

The products of the distortional shear flow $q_{\tilde{\omega}}$ and the perpendicular distance r_D from the distortional center, as shown in Fig. 3-28, are referred to as the distortional moments $T_{\tilde{\omega}}$ and can be written as

$$T_{\tilde{\omega}} = \int_0^s q_{\tilde{\omega}} r_D ds = -E\tilde{\theta}''' \int_0^s \tilde{\omega}^2 t ds = -EI_{\tilde{\omega}}\tilde{\theta}''' \tag{3-117}$$

where $I_{\tilde{\omega}} = \int_0^s \tilde{\omega}^2 t ds$ is called the distortional warping constant and represents the geometric moment of inertia with respect to the distortional warping function $\tilde{\omega}$.

From Eqs. 3-115 and 3-117, the distortional warping stress can also be written as

$$f_{\tilde{\omega}v} = \frac{q_{\tilde{\omega}}}{t} = \frac{T_{\tilde{\omega}}S_{\tilde{\omega}}}{I_{\tilde{\omega}}t} \tag{3-118}$$

Comparing Eq. 3-118 with Eq. 3-65c, it can be seen that the bending shear stress and warping shear stress have a similar expression. Moreover, Eq. 3-110 can be written as

$$f_{\tilde{\omega}} = \frac{M_{\tilde{\omega}}\tilde{\omega}}{I_{\tilde{\omega}}} \tag{3-119}$$

where

$$M_{\tilde{\omega}} = EI_{\tilde{\omega}}\tilde{\theta}'' \tag{3-120}$$

$M_{\tilde{\omega}}$ is referred to as distortional bi-moment.

From Eqs. 3-118 and 3-119, to determine the distortional warping and shear stresses requires the distortional angle function $\tilde{\theta}(z)$, which can be determined by solving its differential equation. The differential equation for distortion can be developed by the principle of energy as follows.

The total potential energy due to box girder distortion includes three portions, i.e.,

$$U = U_e + U_s + U_m \tag{3-121}$$

where

U_e = strain energy due to distortional warping stress
U_s = strain energy due to distortional moment, i.e., transverse deformation of box section
U_m = work done by distortional forces

$$U_e = \frac{1}{2E} \int\int_0^l f_{\tilde{\omega}}^2 \, dA \, dz = \frac{E}{2} \int \tilde{\omega}^2 \, dA \int_0^l \left(\frac{\partial^2 \tilde{\theta}}{\partial z^2}\right)^2 dz = \frac{EI_{\tilde{\omega}}}{2} \int_0^l \left(\frac{\partial^2 \tilde{\theta}}{\partial z^2}\right)^2 dz \tag{3-122}$$

$$U_s = \frac{K_D}{2} \int_0^l \tilde{\theta}^2 dz \tag{3-123}$$

where K_D = the stiffness of the box section against the distortion, i.e. the distortional moment required for producing unit distortional angle $\tilde{\theta}$ which can be determined as follow:

$$K_D = \frac{h^2}{\Delta} \tag{3-124}$$

where

h = box height (see Fig. 3-29)
Δ = horizontal displacement at top flange due to unit horizontal force P along top flange, which can be determined by taking a unit-length box segment and treating it as a simply supported frame structure (see Fig. 3-29).

From Fig. 3-28, we have

$$U_m = -\int_0^l [q_V \Delta w + q_H \Delta v] dz = -\int_0^l \frac{m_T}{2}\left(\frac{\Delta w}{b} + \frac{\Delta v}{h}\right) dz \tag{3-125a}$$

$$U_m = -\int_0^l \frac{m_T}{2} \tilde{\theta} dz \tag{3-125b}$$

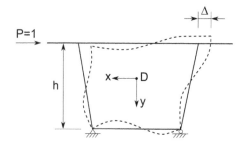

FIGURE 3-29   Analytical Model for Determining Distortional Rigidity of Box Girders.

where Δw and Δv are the relative displacements of the box corners (see Fig. 3-28b).

An elastic system is in equilibrium when the total potential energy is a minimum, which means that the variation of the total potential energy equals zero:

$$\delta U = EI_{\tilde{\omega}} \int_0^l \frac{d^2\tilde{\theta}}{dz^2} \delta\left(\frac{d^2\tilde{\theta}}{dz^2}\right) dz + K_D \int_0^l \tilde{\theta}\delta\tilde{\theta}dz - \int_0^l \frac{m_T}{2}\delta\tilde{\theta}dz$$

$$= \left[M_{\tilde{\omega}}\delta\left(\frac{d\tilde{\theta}}{dz}\right)\right]_0^l - \int_0^l EI_{\tilde{\omega}}\frac{d^3\tilde{\theta}}{dz^3}\delta\left(\frac{d\tilde{\theta}}{dz}\right)dz + \int_0^l \left(K_D\tilde{\theta} - \frac{m_T}{2}\right)\delta\tilde{\theta}dz \qquad (3\text{-}126)$$

$$= [M_{\tilde{\omega}}\delta\tilde{\theta}]_0^l - [T_{\tilde{\omega}}\delta\tilde{\theta}]_0^l + \int_0^l \left(EI_{\tilde{\omega}}\frac{d^4\tilde{\theta}}{dz^4} + K_D\tilde{\theta} - \frac{m_T}{2}\right)\delta\tilde{\theta}dz = 0$$

As Eq. 3-126 is valid for any boundary condition, the following equation must be satisfied:

$$EI_{\tilde{\omega}}\frac{d^4\tilde{\theta}}{dz^4} + K_D\tilde{\theta} = \frac{m_T}{2} \qquad (3\text{-}127)$$

The above differential equation for distortion can be written as

$$\frac{d^4\tilde{\theta}}{dz^4} + 4\beta^4\tilde{\theta} = \frac{m_T}{2EI_{\tilde{\omega}}} \qquad (3\text{-}128)$$

where $\beta = \sqrt[4]{\frac{K_D}{4EI_{\tilde{\omega}}}}$

If m_T is uniformly distributed, a solution for the distortional angle $\tilde{\theta}$ can be obtained as

$$\tilde{\theta} = A\ \sin\beta z\ \sinh\beta z + B\ \sin\beta z\ \cosh\beta z + C\ \cos\beta z\ \sinh\beta z + D\ \cos\beta z\ \cosh\beta z + \frac{m_T}{2K_D} \qquad (3\text{-}129)$$

After obtaining the distortional angle $\tilde{\theta}$, we can calculate the distortional warping normal and shear stresses from Eqs. 3-110 and 3-115, respectively. The distributions of the typical distortional warping normal and shear stresses are sketched in Fig. 3-30. The transverse bending stress due to the distortion can be obtained by treating the unit length of the box girder as a frame structure as shown in Fig. 3-29.

A method often used for distortional analysis of box girders in practice is the beam-on-elastic-foundation (BEF) analogy method[3-22], which is discussed next.

As we know, the differential equation for a beam with flexural rigidity EI on an elastic foundation with a spring constant k and subjected to a uniformly distributed load q can be written as

$$EI\frac{d^4y}{dz^4} + ky = q \qquad (3\text{-}130)$$

Comparing Eq. 3-127 with Eq. 3-130, we can see that both equations are in the same form. If we assume a beam with flexural rigidity $EI_{\tilde{\omega}}$ on an elastic foundation with a spring constant K_D and subjected to a uniformly distributed load $\frac{m_T}{2}$ as shown in Fig. 3-31, the method for analyzing a BEF can be used for box girder distortional analysis. The determination of the distortional angle is transferred to the determination of the deflection of the BEF. Figure 3-31 shows a typical analogy model

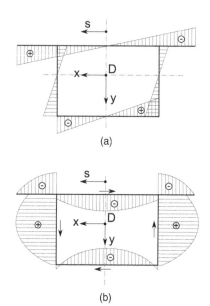

(a)

(b)

FIGURE 3-30 Distributions of Distortional Warping Normal and Shear Stresses, (a) Warping Normal Stress, (b) Warping Shear Stress.

where

$EI_{\bar{\omega}}$ = flexural rigidity of analogy beam
K_D = spring constant of analogy beam
K_{D0} = stiffness of internal struts against distortional deformation, which can be determined as a truss structure as shown in Fig. 3-29
$\frac{m_T}{2}$ = uniform loading of analogy beam

As the diaphragms in concrete segmental bridges are normally very strong, the distortions at the bridge supports can be treated as zero; i.e., there is no deflection at the supports.

Based on the BEF analogous method, Wright et al.,[3-22] developed some simple equations for calculating the distortional warping normal and the transverse bending stresses due to uniformly distributed torques and concentrated torques for single cell box girders as shown in Fig. 3-32. The designers may use these equations to determine the distortion-related stresses and estimate what internal struts spacing and stiffness may be best for large box girders.

Distortional transverse bending stress of a box section can be determined as:

$$f_{\bar{\omega}t} = C_t F_d \beta \frac{1}{2a} (ml \ or \ T) \tag{3-131}$$

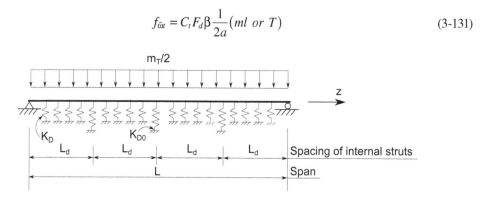

FIGURE 3-31 BEF Analogy for Distortional Analysis.

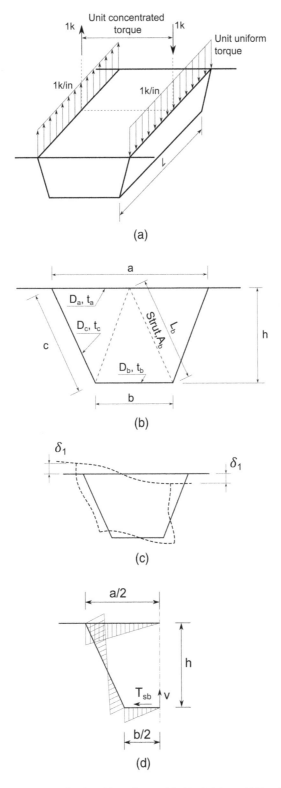

FIGURE 3-32   A Typical Box under Torsional Loading and Its Definitions, (a) Loading Cases, (b) Definitions of Dimensions, (c) Deformation Due to Unit Torque, (d) Transverse Moment Distribution Due to Unit Torque.

where

m = uniform torque per unit length
l = spacing between diaphragms or internal struts
T = concentrated torque

At the bottom corner of box: $F_d = \frac{bv}{2S}$,
At the top corner of box: $F_d = \frac{a}{2S}\left(\frac{b}{a+b} - v\right)$
For any locations in the box slabs: $F_d = \frac{M_t}{S}$
M_t = transverse moment in slabs under consideration, which is directly calculated from Fig. 3-32(d) based on the values of v and T_{sb}

$$T_{sb} = \frac{ab}{(a+b)h} \tag{3-132}$$

S = section modulus of web at location under consideration
v = compatibility shear at center of bottom flange = shear in bottom flange per unit torsional load (see Fig. 3-32d) and can be written as

$$v = \frac{\dfrac{1}{D_c}\left[(2a+b)abc\right] + \dfrac{1}{D_a}ba^3}{(a+b)\left[\dfrac{a^3}{D_a} + \dfrac{2c\left(a^2+ab+b^2\right)}{D_c} + \dfrac{b^3}{D_b}\right]} \tag{3-133}$$

a, b, c = top flange, bottom flange, and web lengths

$$D_a = Et_a^3 / 12\left(1-\mu^2\right)$$
$$D_b = Et_b^3 / 12\left(1-\mu^2\right)$$
$$D_c = Et_c^3 / 12\left(1-\mu^2\right)$$

t_a, t_b, t_c = top flange, bottom flange, and web thickness (in.)
μ = Poisson's ratio
$\beta = \left(\frac{1}{EI\delta_1}\right)^{0.25}$ = stiffness parameter
δ_1 = deflection as shown in Fig. 3-32c and can be written as:

$$\delta_1 = \frac{ab}{24(a+b)}\left\{\frac{c}{D_c}\left[\frac{2ab}{a+b} - v(2a+b)\right] + \frac{a^2}{D_a}\left(\frac{b}{a+b} - v\right)\right\} \tag{3-134}$$

I = moment of inertia of box section
C_t = coefficient found from Fig. 3-33b and d.

In Fig. 3-33, q is the dimensionless ratio of diaphragm or bracing stiffness to box stiffness per unit length. It is defined as

$$q = \frac{EA_b}{L_b l \delta_1}\delta_b^2 \tag{3-135}$$

where

E = modulus of elasticity
A_b = area of one diaphragm bracing member (see Fig. 3-32b)
L_b = length of bracing member

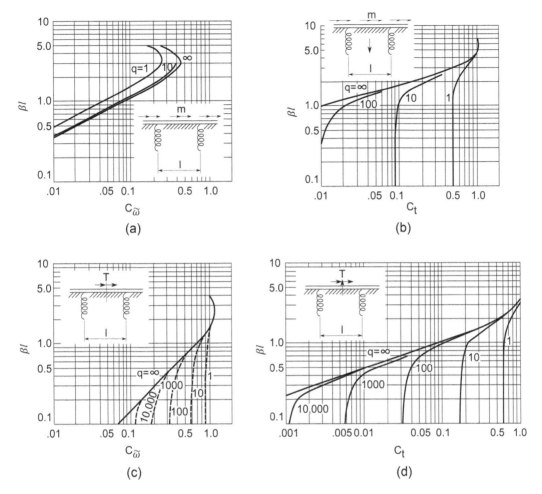

FIGURE 3-33 Coefficients $C_{\tilde{\omega}}$ and C_t for Determining Normal Distortional Warping and Transverse Bending Stresses, (a) $C_{\tilde{\omega}}$ for Mid-panel or Midspan with Uniform Torque, (b) C_t for Mid-panel or Midspan with Uniform Torque, (c) $C_{\tilde{\omega}}$ for Mid-panel or Midspan with Concentrated Torque, (d) C_t for Mid-panel or Midspan with Concentrated Torque.

$$\delta_b = \frac{2\left(1+\dfrac{a}{b}\right)}{\sqrt{1+\left(\dfrac{a+b}{2h}\right)^2}}\,\delta_1 \tag{3-136}$$

Normal distortional warping stress at any point in the section can be determined as:

$$f_{\tilde{\omega}} = \frac{C_{\tilde{\omega}}y}{I\beta a}\left(ml \text{ or } T\right) \tag{3-137}$$

where

$\quad y \quad$ = distance along vertical axis of box from neutral axis to point under consideration.

$\quad C_{\tilde{\omega}} \quad$ = coefficient found from Fig. 3-33a and c.

In Fig. 3-33, l is the spacing of diaphragms or lateral strut bracings. First, calculate the value of βl and q then coefficients C_t and $C_{\bar{\omega}}$ can be found in this figure. Using Eqs. 3-131 and 3-137, we can determine the normal distortional warping and transverse bending stresses.

3.5 BENDING AND PURE TORSION OF CURVED GIRDERS

Curved concrete segmental bridges are often used in bridge engineering, and their bending and torsion are coupled with each other. It is normally complex to accurately analyze such structures, and there may be few practicing engineers using hand calculation to analyze curved segmental bridges, especially for continuous curved bridges. Though there are many computer programs available, it is still useful for engineers to understand the basic theory for analyzing curved bridges.

3.5.1 EQUILIBRIUM OF FORCES

Let us consider a curved girder subjected to a distributed vertical load p and distributed torque about the z-axis as shown in Fig. 3-34 with a moving right-hand rectangular coordinate system taken at the centroidal axis (x, y, z). Take a small segment $dz = Rd\phi$ with its external loading and internal loading shown in Fig. 3-35. In Figs. 3-34 and 3-35,

Q_y = shear force in y-direction
M_x = bending moment about x-axis

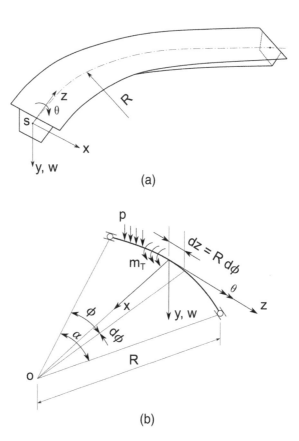

(a)

(b)

FIGURE 3-34   Coordinate System and Analytical Model of a Typical Curved Girder, (a) Typical Curved Girder, (b) Analytical Model and Applied Loadings.

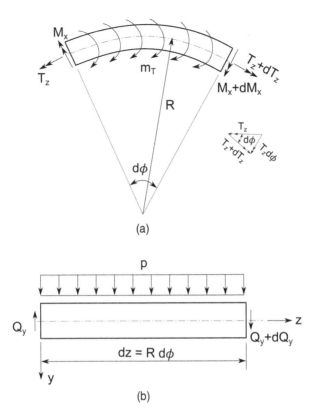

FIGURE 3-35 Infinitive Curved Segment in Equilibrium, (a) Plan View, (b) Elevation View.

T_z = torsion moment about z-axis
w = deflection in y-direction
θ = rotation angle about z-axis
α = center angle of curved girder
R = radius of curvature on centroidal beam axis
ϕ = angular coordinate axis

From the equilibrium condition $\sum F_y = 0$, we have

$$Q_y + dQ_y - Q_y + pdz = 0$$

Thus,

$$\frac{dQ_y}{dz} + p = 0 \tag{3-138}$$

The equilibrium of moment about the x-axis $\sum M_x = 0$ yields:

$$M_x + dM_x - M_x - Q_y Rd\phi + T_z d\phi + \frac{p(Rd\phi)^2}{2} = 0$$

Neglecting the term $(Rd\phi)^2$ and rearranging the above equation, we have

$$\frac{dM_x}{dz} + \frac{T_z}{R} - Q_y = 0 \tag{3-139}$$

From the equilibrium condition for the torque $\sum T_z = 0$, we have

$$T_z + dT_z - T_z - M_x d\phi + m_T dz = 0$$

Rearranging the above equation yields

$$\frac{dT_z}{dz} - \frac{M_x}{R} + m_T = 0 \tag{3-140}$$

Differentiating Eq. 3-139 with respect to z and substituting Eq. 3-140 into the resulting equation, we have

$$\frac{d^2 M_x}{dz^2} + \frac{M_x}{R^2} + p - \frac{m_T}{R} = 0 \tag{3-141}$$

Solving the above equation,

$$M_x = A \sin\phi + B \cos\phi - pR^2 + m_T R \tag{3-142}$$

For a simply supported girder, we have

$$\phi = 0: M_x = 0$$
$$\phi = \alpha: M_x = 0$$

Using the above boundary conditions, we can determine the integration constants A and B in Eq. 3-142.

From Eq. 3-140, the torsional moment T_z can be written as

$$T_z = \int \left(\frac{M_x}{R} - m_T \right) dz + C \tag{3-143}$$

Using the compatibility conditions of displacements, the integration constant C can be determined.

3.5.2   RELATION BETWEEN INTERNAL FORCE AND DISPLACEMENT

Figure 3-36 shows the change of the deflection angle about the x-axis and the rotation angle about the z-axis from section ϕ to section $\phi + d\phi$. It is obvious from this figure that we can find the deflection angle about the x-axis of the section $\eta = \frac{dw}{dz} - \theta d\phi$.

Similar to the relationship between the change of the curvature and the internal bending moment in straight girder, we have

$$\frac{d^2 w}{dz^2} - \frac{\theta}{R} = -\frac{M_x}{EI_x} \tag{3-144}$$

From Fig. 3-36c, it can be seen that the section torsional angle β of a curved girder consists of two portions: One is the section rotation angle in reference to the x-axis θ and the other is caused by vertical deflection w/R (see Fig. 3-36), that is

$$\beta = \theta + \frac{w}{R} \tag{3-145}$$

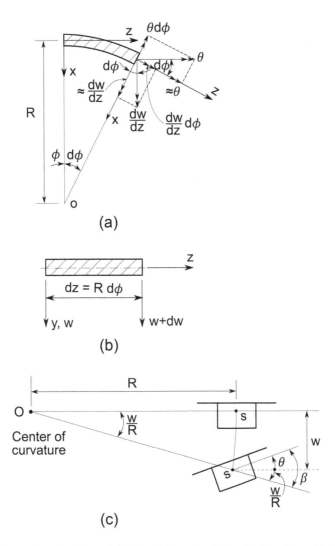

FIGURE 3-36 Deformations in a Curved Girder, (a) Elevation View, (b) Plan View, (c) Side View.

Similar to Eq. 3-104, we have

$$\frac{d\beta}{dz} = \frac{T_s}{GJ} \tag{3-146a}$$

or

$$\frac{d\theta}{dz} + \frac{1}{R}\frac{dw}{dz} = \frac{T_s}{GJ} \tag{3-146b}$$

Differentiating both sides of Eq. 3-144 and substituting the term $\frac{d\theta}{dz}$ from Eq. 3-146b into the differentiated equation, we have

$$\frac{d^3w}{dx} + \frac{1}{R^2}\frac{dw}{dz} + \frac{1}{EI_x}\frac{dM_x}{dz} - \frac{T_z}{GJR} = 0 \tag{3-147}$$

From Eqs. 3-142 and 3-143, we can obtain the bending moment M_x and torsional moment T_z. Then, we can easily solve the vertical deflection w from Eq. 3-147 and rotation angle θ from Eq. 3-144.

3.6 REQUIREMENTS AND DETERMINATION OF STRENGTH RESISTANCES FOR FLEXURAL AND TORSION MEMBERS

3.6.1 GENERAL REQUIREMENTS

The requirements of the strength limit states must be satisfied at each location along the length of the designed member. The strength of a cross section calculated using the current established methods is called nominal strength or nominal resistance. Each designed cross section must be satisfied for the following equations:

For flexural and torsion members:

$$M_r = \phi_f M_n \geq \eta_i \gamma_i M_i \tag{3-148a}$$

$$V_r = \phi_v V_n \geq \eta_i \gamma_i V_i \tag{3-148b}$$

$$T_r = \phi_t T_n \geq \eta_i \gamma_i T_i \tag{3-148c}$$

where

M_r	=	factored flexure resistance
M_n	=	nominal flexure resistance
V_r	=	factored shear resistance
V_n	=	nominal shear resistance
T_r	=	factored torsion resistance
T_n	=	nominal torsion resistance
ϕ_f, ϕ_v, ϕ_t	=	resistance factors for flexure, shear and torsion, respectively
η_i	=	load modifier as specified in Section 2.3.3.1
γ_i	=	load factor as specified in Section 2.3.3.3
M_i	=	bending moment resulting from force effect or superimposed deformation
V_i	=	shear resulting from force effect or superimposed deformation
T_i	=	torsion resulting from force effect or superimposed deformation

3.6.2 DETERMINATION OF FLEXURAL STRENGTH

3.6.2.1 Girders with Bonded Tendons

As discussed before, after the concrete stress at the level of the prestressing steel reaches the concrete fracture strength f_r, the prestressed girder behaves essentially the same as a reinforced concrete girder. The ultimate theory in flexure and the basic design concept of reinforced concrete are applicable to the prestessing concrete. The only difference is that the prestessing tendon undergoes two different stages of deformations: before decompression (elastic beam) and after decompression (normal reinforced beam, cracked beam).

In determining the section's ultimate flexural strength, we assume:

a. The strain distribution is linear; i.e., plane sections remain planes after bending and the maximum allowable concrete strain $\varepsilon_{cu} = 0.003$ in./in. (see Fig. 3-37b).
b. As concrete is weak in tension, the concrete in the tension zone of the section is neglected and the tension steel is assumed to take all tensile stresses (see Fig. 3-37c).
c. An equivalent rectangular stress block, as shown in Fig. 3-37c, can be used to calculate the total concrete compression force. The equivalent stress block has a depth a and average stress $0.85 f_c'$.

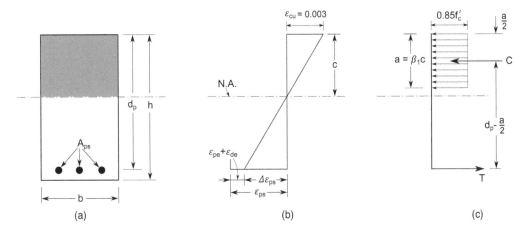

(a)   (b)   (c)

FIGURE 3-37 Idealization of Stress and Strain Distribution of a Prestressing Beam in Ultimate Flexure Design, (a) Typical Section, (b) Strain Distribution, (c) Idealization of Stress Distribution.

From the equilibrium condition of tension force T = compression force C, we have

$$A_{ps}f_{ps} = 0.85f_c'ba \tag{3-149a}$$

$$a = \beta_1 c \tag{3-149b}$$

$$c = \frac{A_{ps}f_{ps}}{0.85f_c'\beta_1 b} \tag{3-149c}$$

$$\beta_1 = 0.85 \text{ for } f_c' \leq 4.0 \ ksi \tag{3-149d}$$

$$\beta_1 = 0.85 - 0.05\left(f_c' - 4.0\right) \geq 0.65 \text{ for } f_c' > 4.0 \ ksi \tag{3-149e}$$

The nominal flexure strength is equal to the tension force multiplied by the moment arm:

$$\left(d_p - \frac{a}{2}\right)$$

that is,

$$M_n = A_{ps}f_{ps}\left(d_p - \frac{a}{2}\right) \tag{3-150}$$

where f_{ps} = average stress in prestressing steel at the nominal bending resistance.

The magnitude of the tendon prestressing stress f_{ps} at failure cannot be easily determined. However, it can be determined using the strain compatibility condition shown in Fig. 3-37b through different loading stages until the failure limit state is reached. The strains for bonded prestressing steel can be expressed as follows.

$$\varepsilon_{ps} = \varepsilon_{pe} + \varepsilon_{de} + \Delta\varepsilon_{ps} \tag{3-151a}$$

The strain due to the effective prestressing force:

$$\varepsilon_{pe} = \frac{f_{pe}}{E_{PS}} \tag{3-151b}$$

where f_{pe} = effective stress.

The decompression strain, when the surrounding concrete stress at the prestressing steel level is zero:

$$\varepsilon_{de} = \frac{P_e}{A_c E_c}\left(1 + \frac{e^2}{r^2}\right) \tag{3-151c}$$

where P_e is effective prestressing force. For the definitions of the other variables refer to Section 3.1.1. ε_{de} is normally small and can be taken as zero.

The strain due to overload above the decompression load and corresponding to the maximum compression strain at the extreme fiber 0.003 in./in. (see Fig. 3-37b) is:

$$\Delta\varepsilon_{ps} = \varepsilon_{cu}\left(\frac{d_p - c}{c}\right) \tag{3-151d}$$

There are several approximate methods[3-23 to 3-28] for predicting f_{ps} based on tested results. If $f_{pe} = \frac{P_e}{A_{ps}} \geq 0.50 f_{pu}$, the AASHTO LRFD specifications recommend that f_{ps} for bonded prestressing steel be determined as

$$f_{ps} = f_{pu}\left(1 - k\frac{c}{d_p}\right) \tag{3-152a}$$

where

$$k = 2\left(1.04 - \frac{f_{py}}{f_{pu}}\right) \tag{3-152b}$$

Substituting Eq. 3-152a into 3-149a, we can rewrite Eq. 3-149c as:

$$c = \frac{A_{PS}f_{pu}}{0.85 f_c'\beta_1 b + kA_{ps}\frac{f_{pu}}{d_p}} \tag{3-153}$$

If $f_{pe} < 0.50 f_{pu}$ or when the arrangement of prestressing steel cannot be lumped into a single layer, the following procedure by using the strain compatibility method can be used for determining f_{ps} and the nominal flexure strength.

Step 1: Assume the location of the neutral axis c.
Step 2: Calculate the tendon strain $\Delta\varepsilon_{ps}$ by Eq. 3-151d.
Step 3: Calculate total tendon strain: $\varepsilon_{ps} \approx \varepsilon_{pe} + \Delta\varepsilon_{ps}$.
Step 4: Use the tendon strain-stress curve or following approximate equations[3-28] to calculate the tendon stress:

   For 250 ksi strand:
   $f_{ps} = 28,500\varepsilon_{ps}$ (ksi) for $\varepsilon_{ps} \leq 0.0076$
   $f_{ps} = 250 - \frac{0.04}{\varepsilon_{ps}-0.0064}$ (ksi) for $\varepsilon_{ps} > 0.0076$

   For 270 ksi strand:
   $f_{ps} = 28,500\varepsilon_{ps}$ (ksi) for $\varepsilon_{ps} \leq 0.0086$
   $f_{ps} = 270 - \frac{0.04}{\varepsilon_{ps}-0.007}$ (ksi) for $\varepsilon_{ps} > 0.0086$

Step 5: Check if the concrete compression force C is close enough to the resulting steel tension force T. If not, revise c until $C \approx T$.

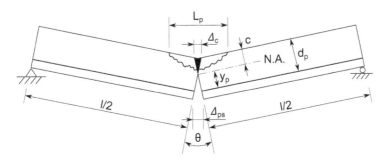

FIGURE 3-38   Failure Mechanism of a Simply Supported Girder.

3.6.2.2   Girders with Unbonded Tendons

Unlike the bonded tendon, unbonded tendon strains in a concrete girder are not compatible with surrounding concrete and are averaged over its entire unbonded length as previously discussed. Thus, it is difficult to predict the tendon stress corresponding to the section ultimate flexural strength. Many tested results[3-29 to 3-32] indicate that the stresses in unbounded tendons increase only slightly before concrete cracking or joint opening and that the girder's ultimate strength is achieved after the formation of a collapse mechanism. Based on these observations, a simplified failure mechanism for a post-tensioned girder with unbonded tendons is proposed[3-29 to 3-32] and shown in Figs. 3-38 and 3-39 for a simply supported and a continuous girder, respectively.

If the distance from the neutral axis to the tendon is y_p and assuming the angle of rotation of the midsection is θ, the tendon elongation can be written as

$$\Delta_{ps} = y_p\theta = \left(d_p - c\right)\theta \tag{3-154a}$$

For continuous spans, more plastic hinges are required for forming a flexural failure mechanism. We assume that the ultimate capacity is reached when a mechanism forms in one critical span with one midspan hinge and one or two support hinges, as shown in Fig. 3-39. Tests results show that the rotation at a support hinge is only one-half of the rotation at a midspan hinge. Thus, based on Eq. 3-154a, the total elongation of the tendon can be written as

$$\Delta_{ps} = \left(d_p - c\right)\left(1 + \frac{N_s}{2}\right)\theta \tag{3-154b}$$

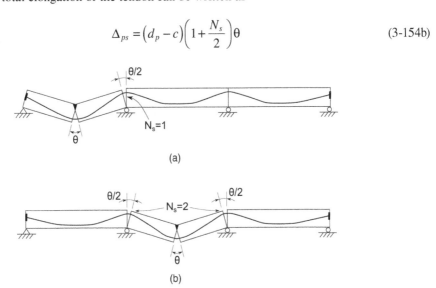

FIGURE 3-39   Failure Mechanism for a Continuous Girder, (a) One Support Hinge, (b) Two Support Hinges.

where N_s = number of support hinges required to form a flexure mechanism. If the critical span is the end span, $N_s = 1$; if the critical span is interior span, $N_s = 2$; for a simple span bridge, $N_s = 0$.

The tendon strain increased due to the section rotation is

$$\Delta\varepsilon_{ps} = \frac{\Delta_{ps}}{l_{ps}} = \frac{(d_p - c)\theta}{l_e} \tag{3-155a}$$

where
l_{ps} = tendon length
l_e = effective tendon length

$$l_e = \frac{l_{ps}}{1 + N_s/2} = \frac{2l_{ps}}{2 + N_s} \tag{3-155b}$$

c = distance from neutral axis to compression fiber (see Fig. 3-38)

Assuming the concrete shortening of the extreme compression fiber Δ_c is entirely induced by the deformation over the length of plastic hinge L_p, then we have

$$\Delta_c = \varepsilon_c L_P \tag{3-156}$$

where
ε_c = concrete ultimate compression strain at extreme fiber = 0.003 in./in.
L_P = length of plastic hinge

Using the assumption that plane sections remain planes after bending, the angle of rotation of the midsection can be written as

$$\theta = \frac{\Delta_c}{c} = \frac{\varepsilon_c L_P}{c} \tag{3-157}$$

Substituting Eq. 3-157 into Eq. 3-155a, we have

$$\Delta\varepsilon_{ps} = \frac{\Delta_{ps}}{l_e} = \frac{\varepsilon_c L_P}{c}\frac{(d_p - c)}{l_e} \tag{3-158}$$

Assuming the tendon remaining in the elastic range for simplifying the calculation, the increased tendon stress due to the plastic hinge can be written as

$$\Delta f_{ps} = E_{ps}\Delta\varepsilon_{ps} = E_{ps}\frac{\varepsilon_c L_P}{c}\frac{(d_p - c)}{l_e} \tag{3-159}$$

Test results show $\frac{L_P}{c}$ can be approximately assumed as 10.0. Substituting $E_{ps} = 30{,}000$ ksi and $\varepsilon_c = 0.003$, we have

$$\Delta f_{ps} = 900\frac{(d_p - c)}{l_e} \tag{3-160}$$

The total tendon stress includes effective prestressing stress f_{pe} and Δf_{ps}, i.e.,

$$f_{ps} = f_{pe} + 900\frac{(d_p - c)}{l_e} \le f_{py} \tag{3-161}$$

3.6.2.3 Girders with Both Bonded and Unbonded Tendons

3.6.2.3.1 Introduction

The bending behaviors of the members with both bond and unbond tendons are still not fully understood and need further research[3-11]. The Current AASHTO specifications[3-3] recommend two methods for estimating the bending capacity. One is the detailed analytical method in which the strain compatibility conditions as previous discussed should be used. Another one is a simplified analytical method, which is discussed below.

3.6.2.3.2 Simplified Method

The flexural capacity of the members with both bond and unbond tendons may be estimated using Eq. 3-150 or 3-164 for bonded tendons. In Eq. 3-153 or 3-165a, the tendon yield force $A_{ps}f_{pu}$ can be approximated as

$$A_{ps}f_{pu} = A_{psb}f_{pu} + A_{psu}f_{pe} \tag{3-162}$$

where
A_{psb} = area of bonded prestressing steel (in.^{2})
A_{psu} = area of unbonded prestressing steel (in.^{2})

The nominal bending resistance M_n can be determined using the following equivalent bonded tendon area:

$$A_{ps} = \frac{\left(A_{psb}f_{pu} + A_{psu}f_{pe}\right)}{f_{pu}} \tag{3-163}$$

3.6.2.4 Nominal Flexure Residence for General Cross Sections Recommended by AASHTO Specifications

Based on the theory discussed in previous sections, it is not difficult to derive the following nominal flexure resistance equation for a more typical section as shown in Fig. 3-40, which represents most of box sections.

$$M_n = A_{ps}f_{ps}\left(d_p - \frac{a}{2}\right) + A_s f_s\left(d_s - \frac{a}{2}\right) - A'_s f'_c\left(d'_s - \frac{a}{2}\right) + 0.85f'_c(b - b_w)h_f\left(\frac{a}{2} - \frac{h_f}{2}\right) \tag{3-164}$$

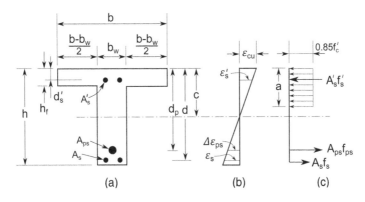

FIGURE 3-40 Typical Section and Design Assumptions of a Prestressing Concrete Girder, (a) Typical Section, (b) Strain Assumption, (c) Force Equilibrium.

where

A_{ps} = area of prestressing steel (in.^{2})

d_p = distance from extreme compression fiber to the centroid of prestressing tendons (in.)

A_s = area of nonprestressed tension reinforcement (in.^{2})

f_s = stress in mild steel tension reinforcement at nominal flexural resistence (ksi)

d_s = distance from extreme compression fiber to centroid of nonprestressed tensile reinforcement (in.)

A'_s = area of compression reinforcement (in.^{2})

f'_s = stress in mild steel compression reinforcement at nominal flexural resistance (ksi)

d'_s = distance from extreme compression fiber to centroid of compression reinforcement (in.)

f'_c = compressive strength of concrete at 28 days (ksi)

b = width of compression face of member or effective width of flange (in.)

b_w = web width (in.)

a = $c\beta_1$ (in.)

β_1 = stress block factor specified in Eqs. 3-149d and 3-149e

h_f = compression flange depth (in.)

c = distance from extreme compression fiber to neutral axis, given by Eqs. 3-165a and 3-165b

f_{ps} = average stress in prestressing steel at nominal bending resistance determined as follows:

- For bonded tendons:
  Use Eq. 3-152a in which

$$c = \frac{A_{ps}f_{pu} + A_s f_s - A'_s f'_s - 0.85 f'_c (b - b_w) h_f}{0.85 f'_c \beta_1 b_w + k A_{ps} \frac{f_{pu}}{d_p}} \tag{3-165a}$$

For a rectangular section, set $b_w = b$ in Eq. 3-165a.
- For unbonded tendons:
  Use Eq. 3-161 in which

$$c = \frac{A_{ps}f_{ps} + A_s f_s - A'_s f'_x - 0.85 f'_c (b - b_w) h_f}{0.85 f'_c \beta_1 b_w} \tag{3-165b}$$

For unbonded tendons, the distance from the extreme compression fiber to the neutral axis c can be determined by assuming that the tendon prestressing steel has yielded, i.e., $f_{ps} = f_{py}$ or assuming $f_{ps} = f_{pe} + 15$ (ksi) as a first estimation of the average stress.

The distance from the centroid of the tensile force to the extreme compression fiber is defined as effective depth as (seen Fig. 3-40)

$$d_e = \frac{A_{ps}f_{ps}d_p + A_s f_s d_s}{A_{ps}f_{ps} + A_s f_s} \tag{3-166}$$

The distance measured perpendicular to the girder neutral axis between the resultant of tensile force and compression force is defined as effective shear depth d_v and can be calculated as

$$d_v = \frac{M_n}{P_{ps} + P_s} = \frac{M_n}{P_c} \tag{3-167}$$

The AASHTO specifications require that the effective shear depth d_v is not smaller than the greater of $0.9d_e$ or 0.72 times the girder depth h.

3.6.2.5 AASHTO Specifications on Minimum Flexural Reinforcement and Control of Cracking

3.6.2.5.1 Minimum Flexural Reinforcement

To reduce the probability of brittle failure, the factored flexural resistance M_r at any section of a noncompression-controlled flexural component should be at least the lesser of 1.33 times the factored moment required M_u and 1.33 times M_{cr}, i.e.,

$$M_r \geq 1.33 M_u$$

or

$$M_r \geq 1.33 M_{cr} = 1.33\gamma_3 \left[\left(\gamma_1 f_r + \gamma_2 f_{cpe} \right) S_c - M_{dnc} \left(\frac{S_c}{S_{nc}} - 1 \right) \right] \tag{3-168}$$

where

f_r = modulus of rupture of concrete

f_{cpe} = compressive stress in concrete due to effective prestress forces at extreme tensile fiber induced by externally applied loads (ksi)

M_{dnc} = total unfactored dead load moment (kip-in.)

S_c = section modulus for extreme tensile fiber in composite section induced by externally applied loads (in.^{3})

S_{nc} = section modulus for extreme tensile fiber in monolithic or noncomposite section induced by externally applied loads (in.^{3})

γ_1 = flexural cracking factor
    = 1.2 for precast segmental structures
    = 1.6 for all other concrete structures

γ_2 = prestress factor
    = 1.1 for bonded tendons
    = 1.0 for unbonded tendons

γ_3 = ratio of specified minimum yield strength to ultimate tensile strength of reinforcement
    = 0.67 for A615, Grade 60 reinforcement
    = 0.75 for A706, Grade 60 reinforcement
    = 1.00 for prestressed concrete structures

3.6.2.5.2 Flexural Cracking Control and Maximum Reinforcement Spacing

Tested results show that several bars at moderate spacing are more effective in controlling cracking than one or two larger bars. AASHTO LRFD specifications require that the spacing of mild steel reinforcement in the layer closest to the tension face shall satisfy the following:

$$s \leq \frac{700\gamma_e}{\beta_s f_{ss}} - 2d_c \tag{3-169}$$

where

$$\beta_s = 1 + \frac{d_c}{0.7\left(h - d_c\right)} \tag{3-170}$$

γ_e = exposure factor
    = 1.00 for Class 1 exposure condition
    = 0.75 for Class 2 exposure condition

d_c = thickness of concrete cover measured from extreme tension fiber to center of flexural reinforcement (in.)

f_{ss} = calculated tensile stress in mild steel reinforcement at service limit state, not to exceed $0.60f_y$ (ksi)

h = overall depth of component (in.)

Exposure Condition Class 1 normally applies where the appearance or corrosion are less concern. Exposure Condition Class 2 typically applies to transverse design of segmental concrete box girders, the decks, and substructures exposed to water.

3.6.2.6 Summary of Flexural Strength Checking

For an easier reference, the flow chart for flexural strength checking is shown in Fig. 3-41.

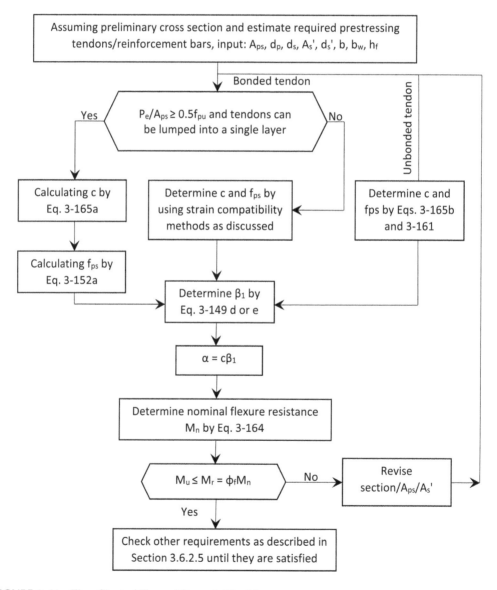

FIGURE 3-41   Flow Chart of Flexural Strength Checking.

3.6.3   Determination and Check of Longitudinal Shear and Torsion Strengths

3.6.3.1   Shear Strength

3.6.3.1.1   General

The determination of the shear strength for prestressed-concrete sections is much more complex than its flexure strength. Generally, the nominal shear strength can be written as

$$V_n = V_c + V_s + V_p \tag{3-171}$$

where

V_c = shear strength provided by concrete
V_s = shear strength provided by transverse reinforcing
V_p = shear resistance provided by component of effective prestressing force in direction of applied shear

For segmental concrete bridges, the post-tensioned force is treated as an external load and $V_p = 0$. If the direction of the post-tensioned tendon is perpendicular to the cross section in question, $V_p = 0$ and Eq. 3-171 is the same as that for reinforced concrete section. Although much theoretical and experimental research has been done on how to determine the shear strength, the difference of the shear strengths predicted by different design codes worldwide can be as much as doubled for some particular sections [3-33], which indicates that further effort in seeking a simple and more accurate method for predicating the shear strength is necessary. Currently, there are two main methods for determining the shear strength that are allowed in post-tensioned concrete bridge design: semi-empirical method and simplified modified compression field theory (MCFT). For an easy understanding, a brief discussion for stress transformation and principal stresses is given before discussing these methods.

3.6.3.1.2   Plane Stress Transformation and Principal Stresses

Figure 3-42a shows a typical state of stress for an infinitesimal element taken from any location in the girder web. The outward normal stresses from the side of the element are defined as positive when they act in the positive x- or y-directions. The angle defined by the orientation of the plane is positive provided it follows clockwise as shown in the figure. If the normal-stress components, f_x, f_y and one shear stress τ_{xy} as shown are known, then the state of stress in an element for any other orientation, θ as shown in Fig. 3-42 can be determined. Figure 3-42b shows a free-body diagram of segment ABC. Simply using the equilibrium conditions along the x'-axis and the y'-axis, and using the trigonometric identities $\sin 2\theta = 2\sin\theta\cos\theta$, $\sin^2\theta = (1 - \cos 2\theta)/2$ and $\cos^2\theta = (1 + \cos 2\theta)/2$, we can obtain:

$$f_{x'} = \frac{f_x + f_y}{2} + \frac{f_x - f_y}{2}\cos 2\theta + v_{xy}\sin 2\theta \tag{3-172a}$$

$$v_{x'y'} = \frac{f_x - f_y}{2}\sin 2\theta + v_{xy}\cos 2\theta \tag{3-173}$$

Substituting $\theta = \theta + 90°$ into Eq. 3-172a yields

$$f_{y'} = \frac{f_x + f_y}{2} - \frac{f_x - f_y}{2}\cos 2\theta - v_{xy}\sin 2\theta \tag{3-172b}$$

From Eqs. 3-172a and 3-172b, it can be seen that the normal stresses vary with the section angle θ. The orientation of the plane with maximum normal stress can be determined by satisfying

$$\frac{df_{x'}}{d\theta} = -\frac{f_x - f_y}{2}2\sin 2\theta + 2v_{xy}\cos 2\theta = 0 \tag{3-174}$$

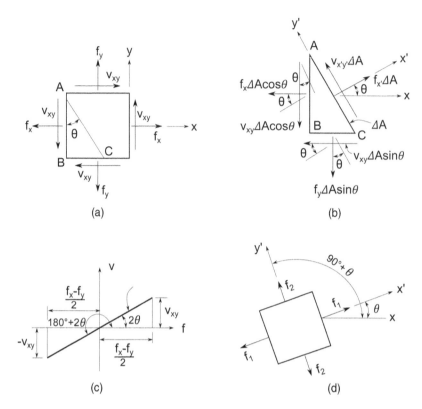

FIGURE 3-42   Stress Transformation and Principal Stress, (a) Infinitesimal Stress Element, (b), Free-Body Diagram of Stress Element, (c) Geometrical Conditions for Principal Planes, (d) Principal Stresses.

Solving Eq. 3-174, we obtain the orientation of the planes of maximum and minimum normal stress.

$$\tan 2\theta = \frac{v_{xy}}{\left(f_x - f_y\right)/2} \tag{3-175}$$

From Eq. 3-175 and the trigonometrical relations as shown in Fig. 3-42c, it is easy to obtain $\sin 2\theta$ and $\cos 2\theta$. Substituting $\sin 2\theta$ and $\cos 2\theta$ into Eqs. 3-172a and 3-172b, we have

$$f_{1,2} = \frac{f_x + f_y}{2} \pm \sqrt{\left(\frac{f_x - f_y}{2}\right)^2 + v_{xy}^{\ 2}} \tag{3-176}$$

$$v_{12} = 0$$

where
$f_{1,2}$ = maximum and minimum normal stress acting at a point = principal stresses in principal planes (see Fig. 3-42d)
v_{12} = shear stress in principal planes. That is there are no shear stress acting on the principal planes.

The state of stress at a point can be expressed as an equation of circle called Mohr's circle. Moving the first term on the right side to the left side of Eq. 3-172a, eleminating the parameter θ by squaring both sides of Eqs. 3-172a and 3-173, and adding the equations together, we have

$$\left[f_{x'} - \left(\frac{f_x + f_y}{2}\right)\right]^2 + v_{x'y'}^2 = \sqrt{\left(\frac{f_x - f_y}{2}\right)^2 + v_{xy}^{\ 2}} = R \tag{3-177}$$

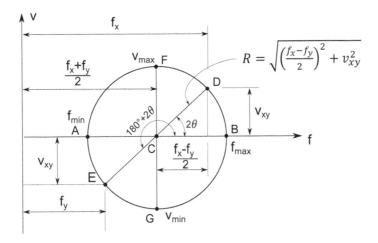

FIGURE 3-43 Mohr's Circle.

Equation 3-177 represents a circle with radius $R = \sqrt{\left(\frac{f_x - f_y}{2}\right)^2 + v_{xy}{}^2}$ and center on the y-axis at point $C\left(\frac{f_x + f_y}{2}, 0\right)$ as shown in Fig. 3-43. In this figure, the abscissa represents the normal stress and positive to the right and the coordinate represents shear stress and positive as upward. Each point on Mohr's circle represents the normal stress and shear stress acting on the plane defined by a specific direction θ. From Mohr's circle as shown in this figure, we can easily obtain the maximum and minimum normal stresses (points B and A), the maximum and minimum shear stresses (points F and G), and their orientations θ, which is equal to half of the angle as shown in the figure.

If we know the normal stress f_x, f_y, and shear stress v_{xy} at a point, the Mohr's circle can be easily constructed as follows:

1. Establish a coordinate system as shown in Fig. 3-43.
2. Plot two reference points $D(f_y, v_{xy})$ and $E(f_x, -v_{xy})$.
3. Determining the center of circle C by connecting points D and E.
4. Draw the circle with the center of circle C and radius of CD or CE.

3.6.3.1.3   Semi-empirical Method for Determining V_c and V_s

3.6.3.1.3.1   Shear Strength Provided by Concrete V_c
As discussed before, concrete is weak in tension strength and is easily cracked by external loadings. Many test results show that there are essentially two types of shear failures for a presstressed concrete girder except for the girder end zones: inclined web cracking caused by high principal stress and flexural shear cracking, normally starts at the bottom of the girder as a result of bending and then it gradually propagates into an inclined shear cracking (see Fig. 3-44).

Let V_{cw} denote the nominal shear resistance provided by concrete when inclined cracking results from excessive principal tension in the web (kip) and V_{ci} denote the nominal shear resistance provided by concrete when inclined cracking results from combined shear and moment (kip). The AASHTO LRFD specifications require that the concrete shear strength V_c be taken as the lesser of V_{cw} and V_{ci}.

3.6.3.1.3.1.1   Web Shear Strength V_{cw}
The web shear strength V_{cw} in AASHTO LRFD is determined based on the magnitude of the maximum tensile stress at the center of gravity of the concrete cross section. The shear strength

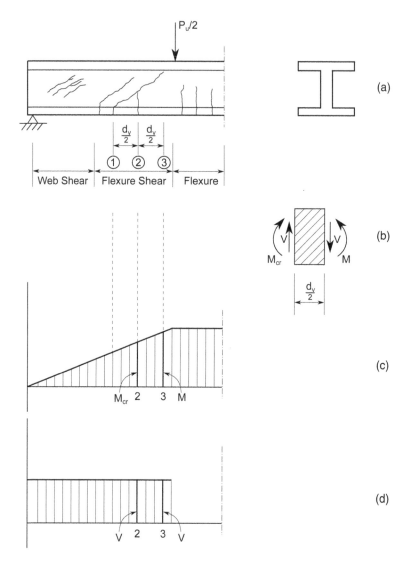

FIGURE 3-44 Crack Types and Flexural Shear Failure Development, (a) Crack Patterns, (b) Small Element in Equilibrium, (c) Distribution of Moment, (d) Distribution of Shear.

is equal to an average concrete tensile strength f_{ta} which is determined based on tested results. Thus, we need to know the stress state and the principal stress at the center of gravity of the section.

The concrete normal stress at the neutral axis is only caused by the post-tensioned tendon and is equal to

$$f_{pc} = \frac{P_{ps}}{A_c} \qquad (3\text{-}178)$$

where
    P_{ps} = prestressing force
    A_c = area of concrete section

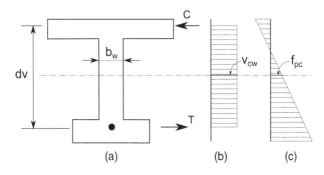

FIGURE 3-45 Assumption of Web Shear Stress Distribution, (a) Typical Section, (b) Shear Distribution, (c) Normal Stress Distribution.

Assuming the shear stress is uniformly distributed within the effective shear depth d_v (see Fig. 3-45), the web shear stress can be written as

$$v_{cw} = \frac{V_{cw}}{d_v b_w}$$

(3-179)

where
V_{cw} = nominal shear resistance
d_v = effective shear depth as defined in Eq. 3-167

The principal tensile stress can be obtained based on Eq. 3-176 and should be equal to or less than the specified average concrete tensile strength f_{tav}, that is

$$f_{tav} = \sqrt{v_{cw}^2 + \left(\frac{f_{pc}}{2}\right)^2} - \frac{f_{pc}}{2}$$

(3-180)

Solving Eq. 3-180 yields

$$v_{cw} = f_{tav}\sqrt{1 + \frac{f_{pc}}{f_{ta}}}$$

(3-181)

Based on test results, $f_{tav} \approx 0.06\sqrt{f_c'}$; then, Eq. 3-181 can be further simplified as

$$v_{cw} = 0.06\sqrt{f_c'} + 0.3 f_{pc}$$

(3-182a)

Thus, the concrete shear strength can be written as

$$V_{cw} = \left(0.06\sqrt{f_c'} + 0.3 f_{pc}\right) d_v b_w$$

(3-182b)

3.6.3.1.3.1.2 Flexure Shear Strength V_{ci}

As discussed before, flexure shear cracking is initiated due to bending moment and then the shear force causes the inclined crack from the initial vertical crack (see Fig. 3-44a). From the free body shown in Fig. 3-44b, the change in moment between sections 2 and 3 is

$$M - M_{cr} = \frac{V d_v}{2}$$

(3-183a)

Rearranging Eq. 3-183a, we have

$$V = \frac{M_{cr}}{M/V - d_v/2} \tag{3-183b}$$

where M_{cr} = moment causing flexural cracking at section due to externally applied loads (kip-in).

$$M_{cr} = S_t\left(f_r + f_{cpe} - \frac{M_d}{S_t}\right) \tag{3-184}$$

where
S_t = section modulus for extreme tensile fiber (in.^{3})
f_{cpe} = compressive stress in concrete due to effective prestress forces at extreme tensile fiber (ksi)
M_d = total unfactored dead load moment (kip-in.)
f_r = concrete modulus of ruptures (ksi, see Section 1.2.2.3)

Extensive test results show an additional vertical shear of about $0.02\sqrt{f_c'}b_w d_v$ plus the girder self-weight V_d are needed to fully develop the inclined crack as shown in Fig. 3-44a. Thus, the total vertical shear at section 2 is

$$V_{ci} = 0.02\sqrt{f_c'}b_w d_v + V_d + \frac{M_{cr}}{M/V - d_v/2} \tag{3-185a}$$

Equation 3-185a shows a good agreement with experimental data. The value of M in Eq. 3-185a is the maximum factored moment at the section under consideration and V is the factored shear force occurring simultaneously with M. Thus, AASHTO LFRD specifications simply modify Eq. 3-185a by eliminating term $d_v/2$ as

$$V_{ci} = 0.02\sqrt{f_c'}b_w d_v + V_d + \frac{V_i M_{cr}}{M_{max}} \geq 0.06\sqrt{f_c'}b_w d_v \tag{3-185b}$$

where
V_d = shear force at section due to unfactored dead load included by both DC and DW (kip)
M_{max} = maximum factored moment at section induced by externally applied loads (kip-in)
V_i = factored shear force induced by externally applied loads occurring simultaneously with M_{max} (kip)

3.6.3.1.3.2 Shear Strength Provided by Transverse Reinforcing V_s
To prevent diagonal shear cracks from developing in post-tensioned segmental bridges, reinforcement is typically provided as shown in Fig. 3-46. To resist shear, it is most effective to place the reinforcement on an incline in the direction of the maximum tensile stresses. However, to simplify the construction and accommodate current practice, the shear reinforcement is placed perpendicular to the girder's longitudinal axis. Based on the results of numerous experimental researches, the shear failure model for evaluating the shear strength provided by the transverse reinforcement can be simulated as an arched section with compression in the top concrete and tied at the bottom by the longitudinal tension bars as shown in Fig. 3-46. From Fig. 3-46, the shear strength provided by the transverse reinforcement can be written as (AASHTO Eq. 5.8.3.3-4)

$$V_s = \frac{A_v f_y d_v (\cot\theta + \cot\alpha)\sin\alpha}{s} \tag{3-186a}$$

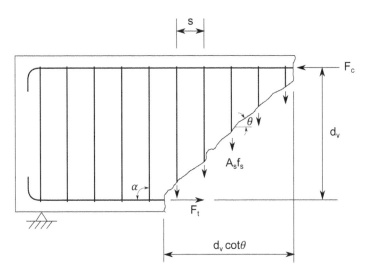

FIGURE 3-46   Classic Model for Determining Shear Strength Provided by Transverse Reinforcement.

where

d_v = effective shear depth (in.)

s = spacing of shear reinforcement (in.)

α = angle of inclination of shear reinforcement to longitudinal axis (degrees)

A_v = area of shear reinforcement within a distance s (in.^{2})

θ = angle of inclination of diagonal compressive stresses as determined as follows:

$\cot\theta = 1.0$, if $V_{ci} \leq V_{cw}$

$\cot\theta = 1.0 + 3\left(\frac{f_{pc}}{\sqrt{f_c'}}\right) \leq 1.8$, if $V_{ci} > V_{cw}$

If $\alpha = 90°$ as shown in Fig. 3-46, Eq. 3-186a reduces to

$$V_s = \frac{A_v f_y d_v \cot\theta}{s} \tag{3-186b}$$

3.6.3.1.4   Simplified Modified Compression Field Theory (MCFT) for Determining V_c and V_s

3.6.3.1.4.1   Introduction
Figure 3-47a shows some typical principal stress trajectories in a girder support area. The solid lines represent the principal tensile stress trajectories, and the dotted lines indicate principal compression stress trajectories. If the principal tensile stress exceeds the concrete fracture strength, the web will crack along the direction of the principal compression stress trajectories and form a series of compression struts. In the modified compression field theory, the equilibrium conditions in the cracked web are idealized as Fig. 3-47b. In Fig. 3-47b, f_1, f_2, and N_v denote the principal tensile, principal compression stresses, and axial tensile force, respectively. The stirrup spacing is denoted as s.

After the inclined web cracks are developed, the compression diagonal is the dominant support in the web. For this reason, the method for determining the web shear strength based on the analytical model as shown in Fig. 3-47b is called compression field theory. Originally, the theory assumed that the principal tensile stress is equal to zero once web cracking occurred. However, tested results indicate that the principal tensile stress will not vanish if crack width is small. The theory is latter modified to include the principal tensile stress as shown in Fig. 3-47 and is called modified compression field theory (MCFT).

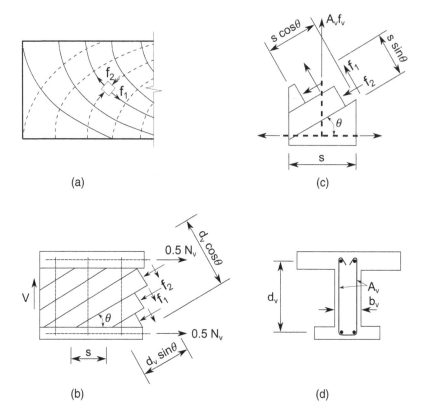

FIGURE 3-47 Modified Compression Field Model, (a) Principal Stress Trajectories, (b) Idealized Free-Body Diagram for Cracked Reinforced Concrete Web, (c) Tensile Force in Web Reinforcement, (d) Cross Section.

3.6.3.1.4.2 Shear Strength Based on MCFT Using the equilibrium condition of vertical forces, from Fig. 3-47b, we have

$$V = f_2 b_v d_v \cos\theta\sin\theta + f_1 b_v d_v \sin\theta\cos\theta = b_v d_v \sin\theta\cos\theta(f_2 + f_1)$$

Rearranging the above equation, we can obtain the principal compression stress as

$$f_2 = \frac{V}{b_v d_v \sin\theta\cos\theta} - f_1 = \frac{v}{\sin\theta\cos\theta} - f_1 \qquad (3\text{-}187)$$

where $v = \frac{V}{b_v d_v}$ = the average shear stress.

From Fig. 3-47c and using the equilibrium condition of vertical forces, we have

$$A_v f_v = f_2 s b_v \sin^2\theta - f_1 s b_v \cos^2\theta$$

where f_v = stress in vertical reinforcement

Assuming that the vertical steel is yielded when the limit state is reached and substituting Eq. 3-187 into the above equation, we can obtain the shear resistance of the reinforced concrete web:

$$V = f_1 b_v d_v \cot\theta + \frac{A_v f_y d_v}{s} \cot\theta \qquad (3\text{-}188a)$$

The first term on the right side of Eq. 3-188a represents the contribution of concrete, and the second term represents the contribution of reinforcement, i.e.,

$$V_c = f_1 b_v d_v \cot \theta \qquad (3\text{-}188b)$$

$$V_s = \frac{A_v f_y d_v}{s} \cot \theta \qquad (3\text{-}188c)$$

However, the principal tensile stress f_1 varies in both principal tensile and compression stress directions and is difficult to accurately determine. Based on extensive experimental results, including Walraven[3-34], Collins and Mitchell[3-35, 3-36], and others[3-37 to 3-39], the following simplified method to determine f_1 has been proposed.

Figure 3-48a shows an idealized web crack pattern by shear force V. When the cracks are initially developed, the principal stress will not vanish along the cracks that are represented by section A-A in Fig. 3-48b. When the cracks are further developed, the concrete tensile stress is reduced to

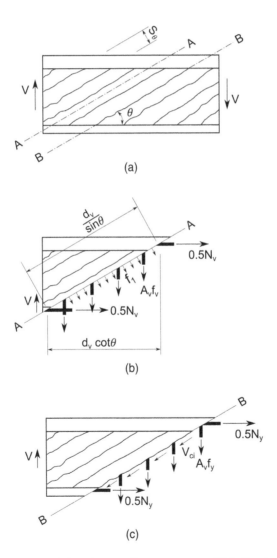

(a)

(b)

(c)

FIGURE 3-48   Models for Determining Average Stresses across a Web Crack, (a) Idealized Web Crack Pattern, (b) Average Stresses between Cracks, (c) Local Stresses at a Crack [3-36].

zero and the concrete aggregate mechanism is active along the cracks, which are denoted as section B-B in Fig. 3-48c. The local stresses along sections A-A and B-B are shown in Fig. 3-48b and c, respectively. Both types of stress distributions shown in Fig. 3-48b and c must balance the same shear force V. Thus, we have

$$A_v f_v \frac{d_v \cot \theta}{s} + f_1 \frac{b_v d_v}{\sin \theta} \cos \theta = A_v f_y \frac{d_v \cot \theta}{s} + v_{ci} \frac{b_v d_v}{\sin \theta} \sin \theta \qquad (3\text{-}189a)$$

Rearranging Eq. 3-189a we have

$$f_1 = v_{ci} \tan \theta + \frac{A_v}{b_v s} \left(f_y - f_v \right) \qquad (3\text{-}189b)$$

Based on extensive test results[3-38], the limiting value of v_{ci} is recommended to be taken as

$$v_{ci} \leq \frac{0.069 \sqrt{f_c'}}{0.3 + \frac{24w}{a_{max} + 0.63}} (\text{ksi}) \qquad (3\text{-}190)$$

where

a_{max} = the maximum aggregate size (in.)
w = $\varepsilon_1 s_\theta$ = the crack width (in.)
ε_1 = concrete principal tensile strain
s_θ = average crack spacing (in.)

Assuming $f_v = f_y$ when the limit state is reached and substituting Eq. 3-190 into Eq. 3-189b, then substituting into Eq. 3-188a, we have

$$V_c = \beta b_v d_v \sqrt{f_c'} \ (\text{ksi}) \qquad (3\text{-}191)$$

where

$$\beta \leq \frac{0.069}{0.3 + 24w/(a_{max} + 0.63)} \qquad (3\text{-}192a)$$

For simplicity, Collins and Mitchell[3-35] assumed $a_{max} = 0.787$ in. (20 mm) and $s_\theta = 11.811$ in. (300 mm) and obtain an upper bound value of β as

$$\beta \leq \frac{0.069}{0.3 + 200\varepsilon_1} \qquad (3\text{-}192b)$$

From Eqs. 3-191 and 3-192b, it can be seen that the larger the principal tensile strain ε_1 is, the smaller the concrete shear capacity will be. The normal tensile strain ε_1 can be obtained based on the compatibility of Mohr's strain circle instead of Mohr's stress circle discussed in Section 3.6.3.1.2 and some trigonometric identities as follows:

$$\varepsilon_1 = \varepsilon_x + \left(\varepsilon_x - \varepsilon_2 \right) cot^2 \theta \qquad (3\text{-}193)$$

As the principal tensile stress f_1 is much smaller than the principal compression stress f_2, we can write f_2 as

$$f_2 = \frac{v}{\sin \theta \cos \theta}$$

Assuming maximum concrete shear strain as –0.002 at peak compression stress, ε_2 can be obtained from the constitutive relationship established based on experimental results [3-38] as

$$\varepsilon_2 = -0.002\left(1 - \sqrt{1 - \frac{v(0.8 + 170\varepsilon_1)}{f_c'\sin\theta\cos\theta}}\right)$$

Substituting the equation above into Eq. 3-193, we have

$$\varepsilon_1 = \varepsilon_x + \left(\varepsilon_x + 0.002(1 - \sqrt{1 - \frac{v(0.8 + 170\varepsilon_1)}{f_c'\sin\theta\cos\theta}}\right)\cot^2\theta \qquad (3\text{-}194)$$

where ε_x = the average horizontal strain in the web. In the AASHTO LRFD specifications, ε_x is assumed to be the strain at mid-depth of the girder and conservatively estimated to be half of the strain ε_s at the level of the centroid of the flexural tensile reinforcement (see Fig. 3-49), i.e.,

$$\varepsilon_x = \frac{\varepsilon_s}{2} = \frac{\dfrac{|M_u|}{d_v} + 0.5N_u + |V_u - V_p| - A_{ps}f_{po}}{2(E_sA_s + E_pA_{ps})} \qquad (3\text{-}195)$$

where
A_{ps} = area of prestressing steel on flexural tension side (in.²)
A_s = area of non-prestressed steel on flexural tension side (in.²)
f_{po} = $0.7f_{pu}$ for normal cases (ksi)
N_u = factored axial force, taken as positive for tensile and negative for compressive (kip)
M_u = absolute value of the factored moment $\geq |V_u - V_p|d_v$ (kip-in.)
V_u = factored shear force (kip)
V_p = component of prestressing force in direction of sheer force (kip)

If the value of ε_s calculated from Eq. 3-195 is negative, it should be taken as zero or be calculated according to the AASHTO LRFD specifications.

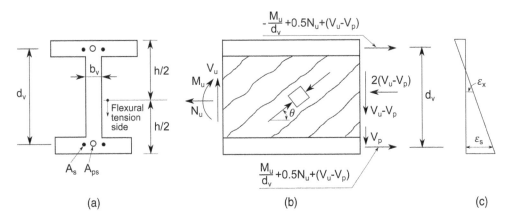

(a)   (b)   (c)

FIGURE 3-49 Typical Forces and Determination of Stain ε_s in MCFT Model, (a) Idealized Section, (b) Sectional Forces, (c) Calculated Strain.

From Eq. 3-194, we can see that the principal tensile strain ε_1 cannot be directly solved as the angle θ is unknown and an iterative method should be used to solve Eq. 3-194.

3.6.3.1.4.3 Simplified MCFT method   From Eq. 3-194, we can see that an iterative method is necessary for determining ε_1 and θ. It is certainly inconvenient for practicing engineers. Based on the numerical analysis performed by Bentz et al.[3-33], the AASHTO specifications recommend the following simplified MCFT method for determining the section shear resistance.

$$V_c = 0.0316\beta b_v d_v \sqrt{f_c'} \tag{3-196a}$$

$$V_s = \frac{A_v f_y d_v}{s} \cot\theta \tag{3-196b}$$

where

$$\beta = \frac{4.8}{1+750\varepsilon_s}$$

$$\theta = 29 + 3500\varepsilon_s$$

$$\varepsilon_s = \frac{\frac{|M_u|}{d_v} + 0.5N_u + |V_u - V_p| - A_{ps}f_{po}}{\left(E_s A_s + E_p A_{ps}\right)}$$

For the description of the remaining notations and restrictions refer to Eq. 3-195.

3.6.3.1.5   Shear Strength Provided by Vertical Component of Effective Prestressing Force V_p
The contribution of tendon force to shear strength can be simply written as

$$V_p = A_{ps} f_{ps} \sin\alpha \tag{3-197}$$

where α = the angle of the tendon at the section under consideration relative to the horizontal axis.

3.6.3.1.6   Shear Resistance in Haunched Girders
In a continuous concrete segmental bridge, the girder depth often varies along its longitudinal length (see Fig. 3-50). When an external moment M is applied, the bottom concrete chord in the truss model is in compression. The inclined compression force will increase the shear resistance. If

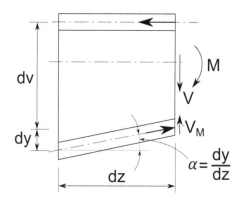

FIGURE 3-50   Shear Resistance in Haunched Girders.

α as shown in Fig. 3-50 is small in comparison with θ as shown in Fig. 3-49, the amount of the shear resistance added can be written as

$$V_M \approx \frac{M}{d_v} \tan \alpha \qquad (3\text{-}198)$$

where
- α = angle of inclination of bottom slab
- M = moment corresponding to maximum shear under consideration
- d_v = effective shear depth

3.6.3.1.7 Critical Shear Section Near Supports

Vertical loads on beams close to the support are transferred directly to the support through compression arching action without inducing additional stresses in the stirrups. The critical shear section can be taken as the effective shear depth dv from the internal face of the support as shown in Fig. 3-51[3-3, 3-40].

3.6.3.1.8 Minimum Longitudinal Reinforcement

From Fig. 3-47b and considering that there is not any external axial loading, we can obtain

$$N_v = f_2 b_v d_v cos^2\theta - f_1 b_v d_v sin^2\theta$$

Using Eq. 3-187, the above equation yields

$$N_v = (v \cot\theta - f_1) b_v d_v = V \cot\theta - f_1 b_v d_v \qquad (3\text{-}199a)$$

From Fig. 3-48, the two groups of stresses shown in Fig. 3-48b and c should give the same horizontal force, that is.

$$N_v + \frac{f_1 b_v d_v}{\sin\theta} \sin\theta = N_y + \frac{v_{ci} b_v d_v}{\sin\theta} \cos\theta \qquad (3\text{-}199b)$$

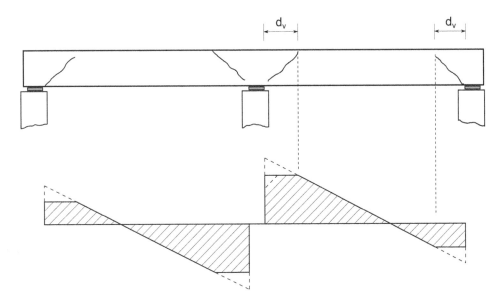

FIGURE 3-51 Critical Shear Location Near Supports[3-40].

where N_y is the total longitudinal force and is assumed to be equally distributed in the top and bottom steel as shown in Fig. 3-48c.

Substituting v_{ci} obtained from Eq. 3-189a into Eq. 3-199b and rearranging the equation, we have

$$N_y = N_v + f_1 b_v d_v + \left[f_1 - \frac{A_v}{b_v s}(f_y - f_v) \right] b_v d_v \cot^2 \theta$$

Assuming $f_v = f_y$ when the limit state is reached yields

$$N_y \geq N_v + f_1 b_v d_v + f_1 \, b_v d_v \cot^2 \theta \tag{3-200}$$

N_y should be taken by all the longitudinal reinforcement on the entire cross section, including both reinforcing and prestressing steel. Substituting Eq. 3-199a into Eq. 3-200 and considering that N_y is equally distributed in the top and bottom steel as shown in Fig. 3-48, we have

$$N_y = 2\left(A_s f_y + A_{ps} f_{ps} \right) \geq V \cot \theta + f_1 b_v d_v \cot^2 \theta$$

where A_s and A_{ps} are the areas of longitudinal reinforcing steel and post-tensioning tendon on the flexure tensile side, respectively. Substituting Eq. 3-188a into the equation above, we can obtain

$$A_s f_y + A_p f_{ps} \geq \left(V_c + 0.5 V_s \right) \cot \theta \tag{3-201a}$$

From Eq. 3-171, we can get $V_c = V_n - V_s - V_P$. Then the longitudinal reinforcement required by the shear can be rewritten as

$$A_s f_y + A_p f_{ps} \geq \left(\frac{V_u}{\phi_v} - 0.5 V_s - V_p \right) \cot \theta \tag{3-201b}$$

For a more general section as shown in Fig. 3-49, the longitudinal tensile steel should also resist the tensile force due to the applied bending moment M_u and axial force N_u, except for the shear V_u. Thus Eq. 3-201b can be revised as (AASHTO)

$$A_s f_y + A_p f_{ps} \geq \frac{M_u}{d_v \phi_f} + 0.5 \frac{N_u}{\phi_a} + \left(\frac{V_u}{\phi_v} - 0.5 V_s - V_p \right) \cot \theta \tag{3-201c}$$

where

V_s = shear resistance $\leq V_u / \phi_v$ (kip)

ϕ_f, ϕ_a, ϕ_v = resistance factors for moment, axial, and shear resistances, respectively

3.6.3.1.9 AASHTO Specifications on Minimum Transverse Reinforcement and Spacing

3.6.3.1.9.1 Minimum Transverse Reinforcement Based on the experimental results, the AASHTO specifications require that transverse reinforcement should be provided if Eq. 3-214 specified in Section 3.6.3.2, or Eq. 3-216a specified in Section 3.6.3.4, or the following Eq. 3-202 is met:

$$V_u > 0.5 \phi_v \left(V_c + V_p \right) \tag{3-202}$$

To restrain the growth of diagonal cracks and increase the ductility of the section, the following minimum transverse reinforcement for normal-weight concrete members should be provided:

$$A_v \geq 0.0316\sqrt{f_c'}\,\frac{b_v s}{f_y} \qquad (3\text{-}203)$$

where

A_v = area of transverse reinforcement within distance s (in.^{2})
b_v = effective width of web (in.)
s = spacing of transverse reinforcement (in.)
f_y = yield strength of transverse reinforcement (ksi) ≤ 100 ksi

For post-tensioned segmental box girder bridges, if transverse reinforcement is not required, the minimum area of transverse shear reinforcement per web shall not be less than the equivalent of two No. 4 Grade 60 reinforcement bars per foot.

3.6.3.1.9.2 Maximum Spacing The spacing of the transverse reinforcement should not exceed the following:

$$s \leq 0.8dv \leq 24.0 \ in, \text{ for } v_u < 0.125 \ f_c',$$
$$s \leq 0.4dv \leq 12.0 \ in, \text{ for } v_u \geq 0.125 \ f_c',$$

where

$$v_u = \frac{|V_u - \phi V_p|}{\phi b_v d_v} = \text{effective factored shear stress (ksi)}$$

The spacing of closed stirrups required to resist torsion shear in post-tensioned segmental concrete box girder bridges should not exceed one-half of the shortest dimension of the cross section, nor 12.0 in.

3.6.3.2 Torsional Resistance

When a box girder is subjected to a torque T as shown in Fig. 3-52, the torque T is mainly resisted by the sectional shear stress τ in the slabs within the closed box section. The shear stresses are caused by pure torsion. The effect of the shear stress in the cantilever slab is very small and can be neglected. Although the warping stress in the section will be induced by the torque, it can be generally neglected in the determination of the section torsional resistance.

Assuming the wall thickness is t, the shear flow q in the wall can be written as

$$q = t\tau \qquad (3\text{-}204)$$

The shear flow in the thin walls caused by girder free torsion can be considered to be constant along the enclosed box girder slabs. Taking an infinitesimal distance dv along the shear flow path, the torsion moment about the torsion center O due to the shear flow can be written as

$$T = q\oint r\,dv \qquad (3\text{-}205a)$$

where r = the distance from the torsional center to the centerline of the web in consideration.

From Fig. 3-52a, rdv = twice area of the shaded triangle. Thus

$$\oint r\,dv = 2A_0 \qquad (3\text{-}205b)$$

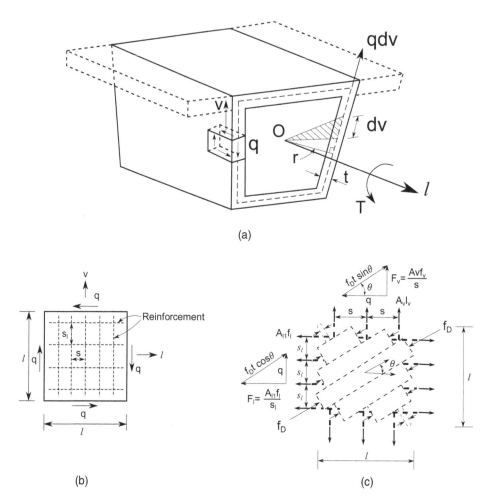

FIGURE 3-52 Analytical Model for Torsional Resistance, (a) Shear Flow q in Box Section Subject to Torsion T, (b) Shell Element in Pure Shear, (c) Truss Model.

where A_0 = enclosed area by the dotted lines shown in Fig. 3-52a.

From Eqs. 3-205a and 3-205b, the shear flow can be written as

$$q = \frac{T}{2A_0} \tag{3-206}$$

Currently, most of the design codes use truss theory[3-39] in the determination of nominal torsional resistance. Taking a unit square element from the box girder web, its free-body diagram and truss model are shown in Fig. 3-52b and c, individually. Assuming that the unit square has transverse and longitudinal reinforcement with spacing of s and s_l, respectively (see Fig. 3-52), and denoting the stresses of the transverse and longitudinal reinforcement as f_v and f_l, respectively, we have the unit forces acting in the transverse (v) and longitudinal directions (l) as

$$F_v = \frac{A_v f_v}{s} \tag{3-207a}$$

$$F_l = \frac{A_{l1} f_l}{s_l} \tag{3-207b}$$

where
   A_v = area of one transverse rebar
   A_{l1} = area of one longitudinal rebar

From Fig. 3-52c and the trigonometrical relations, we have

$$q = F_v \cot\theta = \frac{A_v f_v}{s}\cot\theta \qquad\qquad (3\text{-}208a)$$

$$q = F_l \tan\theta = \frac{A_{l1} f_l}{s_l}\tan\theta \qquad\qquad (3\text{-}208b)$$

$$q = f_{Dt}\sin\theta\cos\theta \qquad\qquad (3\text{-}208c)$$

If the reinforcements in both the transverse and the longitudinal directions have yielded, from Eqs. 3-208a, 3-208b, and 3-208c, we have

$$\tan\theta = \sqrt{\frac{A_v s_l}{A_{l1} s}} \qquad\qquad (3\text{-}209)$$

Substituting Eq. 3-208a into Eq. 3-205a and assuming the steel yields at the limit state, we can obtain the nominal torsion strength

$$T_n = \frac{2A_0 A_v f_y}{s}\cot\theta \qquad\qquad (3\text{-}210)$$

where θ is the angle of crack determined by Eq. 3-209.
   The required torsional reinforcement in the transverse direction is

$$A_v = \frac{T_n s}{2A_0 f_y \cot\theta} \qquad\qquad (3\text{-}211)$$

Because the right terms in Eqs. 3-208a and 3-208b are equal and assuming that both the transverse and longitudinal reinforcements have yielded, we can obtain the required torsional reinforcement in the longitudinal direction as

$$A_{l1} = \frac{A_v}{s}\left(s_l \cot^2\theta\right) \qquad\qquad (3\text{-}212a)$$

The required total longitudinal reinforcement in the box section can be written as

$$A_l = A_{l1}\frac{p_h}{s_1} = \frac{A_v p_h}{s}\left(\cot^2\theta\right) \qquad\qquad (3\text{-}212b)$$

where
   A_l = total area of longitudinal reinforcement in entire box section
   p_h = perimeter of centerline of transverse reinforcement located in outmost webs and top and bottom slabs of box girder

Substituting A_v obtained from Eq. 3-211 into Eq. 3-212b, we have

$$A_l = \frac{T_n p_h}{2A_0 f_y}\cot\theta \qquad\qquad (3\text{-}212c)$$

AASHTO LRFD specifications simplify Eq. 3-212c by taking $\cot\theta \approx 1$ as

$$A_l = \frac{T_n p_h}{2 A_0 f_y} \tag{3-213}$$

The minimum shear and torsion resistance is ensured by the requirement for minimum transverse reinforcing. Except for segmental box girder bridges, the AASHTO LRFD specifications require that the effects of torsion be considered in the web design if the factored torsional moment T_u is greater than one-fourth of the torsional cracking moment, i.e.,

$$T_u > 0.25\phi_t T_{cr} \tag{3-214}$$

$$T_{cr} = 0.125\sqrt{f_c'}\,\frac{A_{cp}^2}{p_c}\sqrt{1+\frac{f_{pc}}{0.125\sqrt{f_c'}}} \tag{3-215a}$$

where

T_u = factored torsional moment (kip-in.)
T_{cr} = torsional cracking moment (kip-in.)
A_{cp} = total area of the longitudinal reinforcement enclosed by outside perimeter of the cross section (in.²)
p_c = length of outside perimeter of the section (in.)
f_{pc} = compressive stress in concrete after prestress losses at centroid of cross section (ksi)
ϕ_t = torsional resistance factor

For axial box members with or without bending moments:

$$\frac{A_{cp}^2}{p_c} \leq 2 A_0 b_v \tag{3-215b}$$

3.6.3.3 Transverse Reinforcement for Sections Subjected to Combined Shear and Torsion

For sections subjected to combined shear and torsion, the shear due to torsion will add on one side of the section and offset on the other side (see Fig. 3-53). The transverse reinforcement is designed for the side where the vertical shear and the shear caused by the torsion are in the same direction.

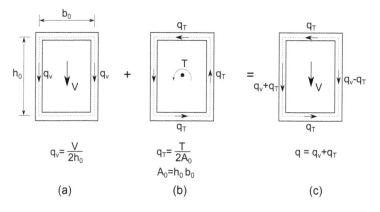

FIGURE 3-53 Shear Flow q in Box Section Subjected to Combined Shear and Torsion, (a) Shear Stress Due to Vertical Shear, (b) Shear Stress Due to Torsion, (c) Total Shear Stress.

The web should be designed for the highest shear and its concurrent torsion and/or the highest torsion and its concurrent shear.

3.6.3.4 AASHTO Simplified Methods for Determining Shear and Torsion Resistances for Segmental Box Bridges

3.6.3.4.1 Normal Torsional Residence

For concrete segmental box girder bridges with normal-weight concrete, the condition for considering the torsional effects is different from Eq. 3-214 and is shown as follows:

$$T_u > \frac{1}{3}\, \phi_v T_{cr} \qquad (3\text{-}216a)$$

in which

$$T_{cr} = 0.0632 K \sqrt{f_c'}\, 2 A_0 b_e \qquad (3\text{-}216b)$$

$$K = \sqrt{1 + \frac{f_{pc}}{0.632\sqrt{f_c'}}} \le 2.0 \qquad (3\text{-}216c)$$

where
T_u = factored torsional moment (kip-in.)
T_{cr} = torsional cracking moment (kip-in.)
K ≤ 1.0, when the tensile stress in extreme fiber exceeds $0.19\sqrt{f_c'}$, calculated on basis of gross section properties, due to factored load and effective prestress force after losses
f_{pc} = unfactored compressive stress in concrete after prestress losses at centroid (ksi)
ϕ_v = shear resistance factor

From Eqs. 3-210 and 3-212c and assuming the concrete cracking angle $\theta = 45°$, the nominal torsional resistance T_n and the minimum additional longitudinal reinforcement for torsion, A_ℓ can be written as

$$T_n = \frac{2 A_0 A_v f_y}{s} \qquad (3\text{-}217a)$$

$$T_u \le \phi_v T_n \qquad (3\text{-}217b)$$

$$A_l \ge \frac{T_u p_h}{2\phi_v A_0 f_y} \qquad (3\text{-}218)$$

where
T_u = factored torsional moment (kip-in.)
f_y = yield strength of additional longitudinal reinforcement (ksi)

3.6.3.4.2 Nominal Shear Resistance

In the design of segmental box girder bridges, the shear component of the primary effective longitudinal prestress force acting in the direction of the applied shear being considered V_p and its secondary shear effects PS are considered as external loads with load factor equal to 1.0.

If factored load $V_u > 0.5\phi_v V_c$, transverse shear reinforcement should be provided. The nominal shear resistance V_n should be taken as the lesser of those determined by Eqs. 3-219a and 3-220, i.e.,

$$V_n = V_c + V_s \qquad (3\text{-}219a)$$

or

$$V_n = 0.379\sqrt{f_c'}b_v d_0 \tag{3-220}$$

where

$$V_c = 0.0632\ K\sqrt{f_c'}b_v d_0 \tag{3-219b}$$

$$V_s = \frac{A_v f_y d_0}{s} \tag{3-219c}$$

b_v = effective web width

d_0 = greater of $0.8h$ or distance between extreme compression fiber and centroid of pre-stressing reinforcement

If the effects of torsion are required to be considered by Eq. 3-216a, the cross-sectional dimensions should also satisfy:

$$\frac{V_u}{b_v d_0} + \frac{T_u}{2A_0 b_e} \le 0.474\sqrt{f_c'} \tag{3-221}$$

where V_u = factored shear including any normal component from the primary prestressing force (kip).

The definitions of the remaining notations are the same as those defined before.

3.6.3.5 Summary of Shear Strength Checking

For clarity, the flow chart for shear strength checking is shown in Fig. 3-54.

3.6.4 INTERFACE SHEAR STRENGTH

Interface shear strength should be checked at interfaces between dissimilar materials, between two concretes cast at different times, or between different elements of the cross section, such as web and flange interfaces in box girders (see Fig. 3-55a) and blister and flange (see Fig. 3-55b).

The factored interface shear resistance V_{ri} should meet the following equation:

$$V_{ri} = \phi_v V_{ni} \ge V_{ui} \tag{3-222}$$

where

V_{ni} = nominal interface shear resistance (kip)

V_{ui} = factored interface shear force (kip)

ϕ_v = shear resistance factor (see Section 2.3.3.5)

It is assumed that a crack develops along the interface when there is an interface shear failure. The reinforcement across the interface is assumed to have yielded and induces a normal compression force (see Fig. 3-55b) to the crack. The normal compression forces develop a friction force to prevent the displacement of the crack. Because the reinforcement stress cannot exceed its yield stress, the nominal shear resistance of the interface plane can be easily obtained from Fig. 3-55b as follows:

$$V_{ni} = cA_{cv} + \mu\left(A_{vf} f_y \sin\alpha + P_c\right) + A_{vf} f_y \cos\alpha \tag{3-223a}$$

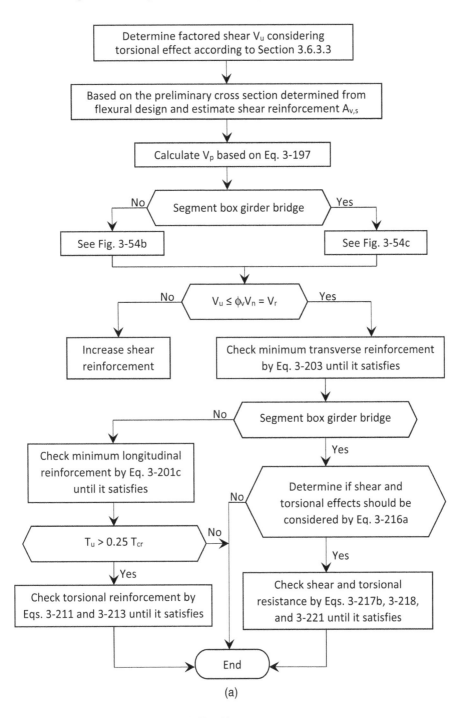

FIGURE 3-54 Flow Chart of Shear Strength Checking.

(Continued)

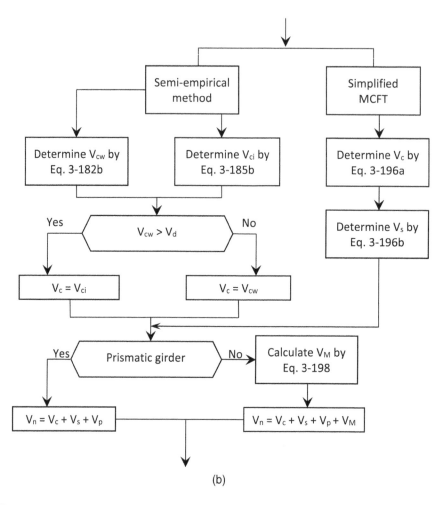

(b)

FIGURE 3-54   (Continued)

(Continued)

where

A_{cv} = $b_{vi}L_{vi}$ = interface area of concrete (see Fig. 3-55b) (in.²)

A_{vf} = area of interface shear reinforcement crossing shear plane within area A_{cv} (in.²)

b_{vi} = interface width (see Fig. 3-55b) (in.²)

L_{vi} = interface length (see Fig. 3-55b) (in.²)

c = cohesion factor as shown in Table 3-2, taken as 0 for brackets, corbels, and ledges (ksi)

μ = friction factor as shown in Table 3-2

f_y = yield stress of reinforcement ≤ 60 (ksi)

P_c = permanent compressive force perpendicular to shear plane (kip)

α = inclined angle of reinforcement cross shear plane (see Fig. 3-55b)

The first, second, and third terms on the right side of Eq. 3-223a represent the contributions due to concrete cohesive force, shear friction force, and the inclined reinforcement, respectively. If $\alpha = 90°$, Eq. 3-223a is the same as Eq. 5.8.4.1-3 contained in the AASHTO LRFD specifications, which has been based on the experimental results[3-40 to 3-44] and does not include the effect of the rebar incline angle.

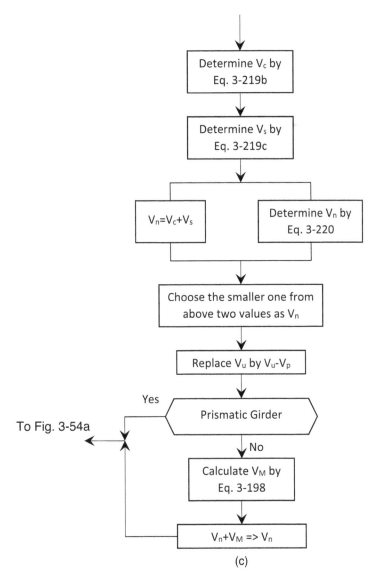

FIGURE 3-54   (Continued)

Based on many test results available, the nominal interface shear resistance should meet the following equation:

$$V_{ni} \le K_1 \ f'_c A_{cv} \tag{3-223b}$$
$$\le K_2 \ A_{cv}$$

where
  K_1 = fraction of concrete strength available to resist interface shear (see Table 3-2) (ksi)
  K_2 = limiting interface shear resistance as shown in Table 3-2 (ksi)

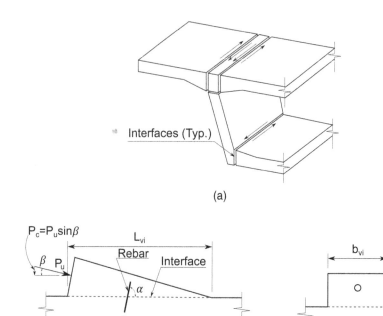

FIGURE 3-55 Interface Shear, (a) Box Girder, (b) Blister.

TABLE 3-2
Values of Coefficients c, μ, K_1, and, K_2[3-3]

Conditions	c (ksi)	μ	K_1 (ksi)	K_2 (ksi)
A cast-in-place concrete slab on clean concrete girder surfaces, free of laitance with surface roughened to an amplitude of 0.25 in.	0.28	1.0	0.3	Normal-weight: 1.8 Lightweight: 1.3
Normal-weight concrete placed monolithically	0.40	1.4	0.25	1.5
Lightweight concrete placed monolithically, or non-monolithically against a clean concrete surface, free of laitance with surface intentionally roughened to an amplitude of 0.25 in.	0.24	1.0	0.25	1.0
Normal-weight concrete placed against a clean concrete surface, free of laitance, with surface intentionally roughened to an amplitude of 0.25 in.	0.24	1.0	0.25	1.5
Concrete placed against a clean concrete surface, free of laitance, but not intentionally roughened	0.075	0.6	0.2	0.8
Concrete anchored to as-rolled structural steel by headed studs or by reinforcing bars where all steel in contact with concrete is clean and free of paint	0.025	0.7	0.2	0.8

Note: $c = 0$ for brackets, corbels, and ledges

3.7 REQUIREMENTS AND DETERMINATION OF STRENGTH RESISTANCES FOR AXIAL AND BIAXIAL BENDING MEMBERS

3.7.1 INTRODUCTION

A member under axial compression with and without bending is often called a column. Strictly speaking, all columns are subjected to bending, because external axial loads are rarely concentric and the columns are rarely perfectly straight. Prestressing steel is often used in piles and tall piers to increase their bending capacity. As the prestressing steel in a compression member will be deflected with its surrounding bonded concrete, there is no change in the eccentricity of the prestressing steel in regard to the concrete section. Thus, no matter how the compression member is deflected, there is no P-Δ effect due to the prestressing force. In addition, under external compressive load, the column will shorten and the prestressing force will be decreased. The majority of the prestress will be lost at the ultimate compression strength of the concrete. The effect of the prestressing on the ultimate compression strength of concrete is comparatively small. The theory, analysis, and design of pre-stressed compression members are similar to those of reinforced concrete members.

3.7.2 LOAD-MOMENT INTERACTION

In determining the resistance of a column, the previous assumptions for beam analysis are still used, i.e.:

1. A plane cross section remains plane after axial and bending deformations.
2. Compatibility of strain between the concrete and the prestressing steel is valid.
3. The section is considered to have failed if the strain in the concrete at the extreme compression fiber reaches 0.003 in./in., or if the strain at mid-depth reaches 0.002 in./in.
4. The actual stress distribution in the concrete compressive zone is replaced by an equivalent rectangular block in analysis and design.

Figure 3-56a shows a typical column loaded with an eccentrically placed vertical load P_n with an eccentricity e. This loading case can be replaced by a column subjected to a concentric axial load P_n and a bending moment $M_n = P_n e$. Similar to the reinforced concrete columns, the failures of a pre-stressing concrete column can also be classified as three models. If the tensile steel strain is greater or equal to its yield strain when the concrete crushes ($\varepsilon_u = 0.003$) at ultimate load, the column is called tension failure. If the tensile steel is in the elastic range when the concrete crushes, the failure of the column is called compression failure. If the mild tensile steel yields and the tensile strain of the prestressing steel is equal to a strain increment of 0.0012 to 0.0020 in./in. beyond the service load level when the concrete crushes, this condition is referred to balanced condition.

The nominal axial load capacity P_n and the nominal bending capacity M_n of a column are related to each other. The variation trajectory of the M_n with P_n is called the load-moment interaction diagram of the column (see Fig. 3-57). The distributions of strain and stress at the limit state for three main control points in the interaction diagram are shown in Fig. 3-56b, c, and d for three loading cases: $M_n = 0$, no tension with the neutral axis location at extreme tension fiber, and $P_n = 0$, respectively. In these figures, the actual parabolic distributions of concrete compression stress are replaced by the equivalent rectangular block depth $a = \beta_1 c$, as done in the flexural beams. From Fig. 3-56b to d, it can be seen that the distributions of strain for the three specific cases can be determined based on the column failure assumptions. Thus, P_n and M_n can be determined based on column equilibrium conditions. For example, consider the concentrically loaded column as shown in Fig. 3-56b. Using the equilibrium condition in axial direction, its nominal axial resistance can be written as

$$P_n = 0.85 f_c'\left(A_g - A_{st} - A_{ps}\right) + f_y A_{st} - A_{ps}\left(f_{ps} - E_p \varepsilon_{cu}\right) \tag{3-224}$$

$$M_n = 0$$

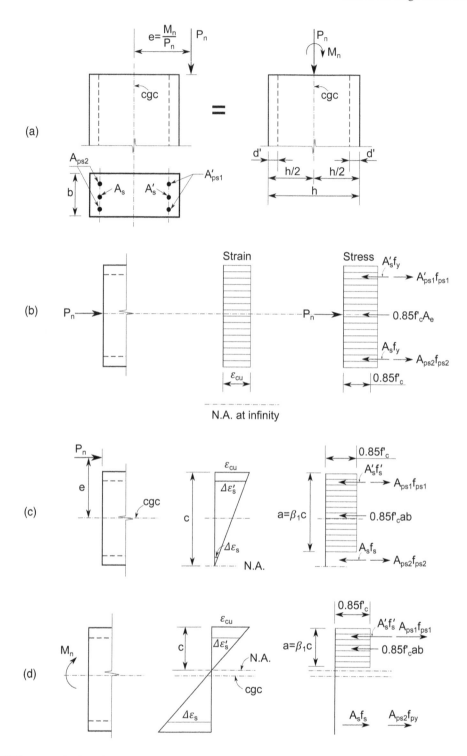

FIGURE 3-56 Typical Column Loading Cases and Analytical Assumptions, (a) Eccentric Loading and Combined Bending and Axial Load, (b) Axial Loading Case, (c) Zero Tension Case, (d) Pure Bending Case[3-4].

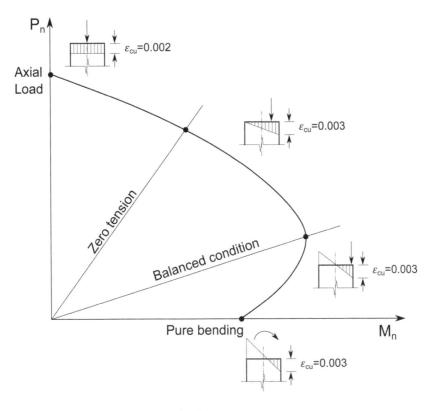

FIGURE 3-57   Axial Load–Moment Interaction Diagram.

where
f_c' = strength of concrete (ksi)
A_g = gross area of section (in.²)
A_{st} = total area of longitudinal reinforcement (in.²)
f_y = yield strength of reinforcement (ksi)
A_{ps} = area of prestressing steel (in.²)
E_p = modulus of elasticity of prestressing tendons (ksi)
f_{ps} = stress in prestressing steel after losses (ksi)
ε_{cu} = failure strain of concrete in compression (in./in.)

For any other loading cases, the location of the neutral axis cannot be directly determined and the distributions of the strain and stress are unknown. It is normal to first assume the location of the neutral axis and then use the trial and adjustment method to determine the column axial and flexural resistances. If we know the neutral axis location at the column limit state, it is not difficult to determine the column capacity by using three types of conditions: equilibrium conditions, geometrical conditions (compatibility), and physical conditions (constitutive law). Figure 3-58 shows the strain and stress distributions for a typical nonslender prestressed column by assuming the distance between the neutral axis and the extreme compression fiber to be c. For simplicity, only prestressing steel is considered. In Fig. 3-58b, ε_{cp} represents the uniform strain in the concrete due to effective prestressing force P_e. The nominal axial load capacity P_n and bending capacity M_n can be derived as follows.

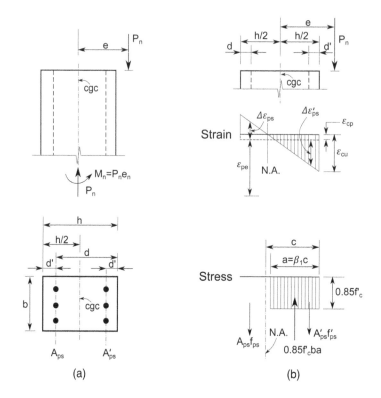

FIGURE 3-58 Distributions of Strain and Stress in Typical Eccentrically Loaded Nonslender Column, (a) Cross Section, (b) Strain and Stress Distributions.

Based on the strain geometric condition as shown in Fig. 3-58b, the strain increments of the prestessing tendons on compression and tension sides can be written as

Strain increment of the tendons on the compression side:

$$\Delta \varepsilon'_{ps} = \varepsilon_{cu} \left(\frac{c - d'}{c} \right) - \varepsilon_{cp} \tag{3-225a}$$

Strain increment of the tendons on the tensile side:

$$\Delta \varepsilon_{ps} = \varepsilon_{cu} \left(\frac{d - c}{c} \right) + \varepsilon_{cp} \tag{3-225b}$$

The tendon stress on the compression side can be written as

$$f'_{ps} = E_{ps} \left(\varepsilon_{pe} - \Delta \varepsilon'_{ps} \right) \tag{3-226a}$$

The tendon stress on the tensile side can be written as

$$f_{ps} = E_{ps} \left(\varepsilon_{pe} + \Delta \varepsilon_{ps} \right) \tag{3-226b}$$

From the equilibrium conditions and referring to Fig. 3-58b, we can obtain

$$P_n = 0.85 f'_c ba - A_{ps} f_{ps} - A'_{ps} f'_{ps} \tag{3-227}$$

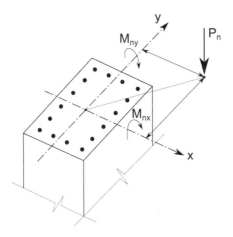

FIGURE 3-59   Typical Column Subjected to Biaxial Bending.

$$M_n = 0.85 f'_c ba \left(\frac{h}{2} - \frac{a}{2} \right) + A_{ps} f_{ps} \left(d - \frac{h}{2} \right) - A'_{ps} f'_{ps} \left(\frac{h}{2} - d' \right) \qquad (3\text{-}228)$$

The following general procedures can be used for determining the nominal axial loading and flexure resistances of a prestressed column for a specific loading case:

Step 1: Assume the location of the neutral axis c.
Step 2: Determine the strain increments of the tendons $\Delta \varepsilon'_{ps}$ and $\Delta \varepsilon_{ps}$, using Eq. 3-225a and 3-225b, respectively.
Step 3: Determine the stresses of the tendons f'_{ps} and f_{ps}, using Eq. 3-226a and 3-226b.
Step 4: Calculate the normal axial resistance P_n using Eq. 3-227 and the normal flexure resistance M_n using Eq. 3-228.
Step 5: Check if $M_n / P_n \approx$ actual eccentricity e. If not, revise c until $M_n / P_n \approx e$.

For the compression members subject to biaxial bending as shown in Fig. 3-59, using a similar method, a failure interaction surface for the biaxial bending column as shown in Fig. 3-60 can be developed.

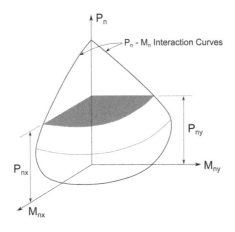

FIGURE 3-60   Typical Failure Surface Interaction Diagram for Biaxial Bending Column.

3.7.3 DETERMINATION OF AXIAL LOAD AND BIAXIAL FLEXURE RESISTANCES BY AASHTO LRFD SPECIFICATIONS

3.7.3.1 Axial Resistance

As there are always unintended eccentricities in actual columns, the AASHTO specifications reduce the axial normal resistance determined using Eq. 3-224 by 0.85 and 0.80 for members with spiral reinforcement and for members with tie reinforcement, respectively[3-3, 3-44] i.e.

$$P_n = k_\beta \left[k_c f_c' \left(A_g - A_{st} - A_{ps} \right) + f_y A_{st} - A_{ps} \left(f_{pe} - E_p \varepsilon_{cu} \right) \right] \tag{3-229}$$

where

k_β = 0.85, for members with spiral reinforcement
　　= 0.80, for members with tie reinforcement
k_c = 0.85, for $f_c' \le 10$ (ksi)
　　= $0.85 - \left(f_c' - 10 \right) 0.02$
P_n = nominal axial resistance (kip)
ε_{cu} = failure strain of concrete in compression (in./in.), assuming 0.003.

The definitions of the remaining notations in Eq. 3-229 are the same as those shown in Eq. 3-224.

3.7.3.2 Biaxial Flexure Resistance

It may be too time consuming for practicing engineers to determine the resistances of a biaxial flexure noncircular member based on equilibrium and strain compatibility conditions. The AASHTO specifications provide two simple approximate equations for evaluating the resistance of biaxial flexure members, based on the reciprocal load method proposed by Bressler[3-6] and Bressler-Parme contour method[3-4].

$$\frac{1}{P_{rxy}} = \frac{1}{P_{rx}} + \frac{1}{P_{ry}} - \frac{1}{\phi_a P_{n0}}, \text{ if the factored axial load} \ge 0.10 \phi_a f_c' A_g \tag{3-230a}$$

$$\frac{M_{ux}}{M_{rx}} + \frac{M_{uy}}{M_{ry}} \le 1.0, \text{ if the factored axial load} < 0.10 \phi_a f_c' A_g \tag{3-230b}$$

where

P_{n0} = nominal axial resistance = $k_c f_c' \left(A_g - A_{st} - A_{ps} \right) + f_y A_{st} - A_{ps} \left(f_{pe} - E_p \varepsilon_{cu} \right)$
ϕ_a = resistance factor for axial compression members
P_{rxy} = ϕP_{nxy}, factored axial resistance in biaxial flexure (kip)
P_{rx} = ϕP_{nx}, factored axial resistance, with eccentricity $e_x = 0$ (kip)
P_{ry} = ϕP_{ny}, factored axial resistance, with eccentricity $e_y = 0$ (kip)
ϕ = residence factor determined based on Eqs. 2-53a and 2-53b
M_{rx} = ϕM_{nx} = uniaxial factored flexural resistance about x-axis (kip-in.)
M_{ry} = ϕM_{ny} = uniaxial factored flexural resistance about y-axis (kip-in.)
e_x = eccentricity of the applied factored axial force in x-direction, i.e., = M_{uy}/P_u (in.)
e_y = eccentricity of applied factored axial force in y-direction, i.e., = M_{ux}/P_u (in.)
M_{ux} = factored moment about x-axis (kip-in.)
M_{uy} = factored moment about y-axis (kip-in.)
P_u = factored axial force (kip)

The factored axial resistances P_{rx}, P_{ry}, M_{rx}, and M_{ry} can be determined by using a method similar to the one discussed in Section 3.7.2. For example, P_{rx} and M_{rx} can be determined as follows:

Step 1: Set $e_x = 0$, and assume the distance from x-axis to the neutral axis as c.
Step 2: Use the geometric relationships similar to Eqs. 3-225a and 3-225b, and determine the strain increments of the reinforcement and tendons: $\Delta \varepsilon_s'$, $\Delta \varepsilon_s$, $\Delta \varepsilon_{ps}'$, and $\Delta \varepsilon_{ps}$.

Step 3: Use the material physical relationships similar to Eqs. 3-226a and 3-226b, and determine the stresses of the reinforcement and tendons f_s', f_s, f_{ps}', f_{ps}.

Step 4: Calculate normal axial resistance P_{nx} using Eq. 3-229 and normal flexure resistance M_{nx} using Eq. 3-164.

Step 5: Check if $M_{nx}/P_{nx} \approx$ actual eccentricity $e_y = \frac{M_{ux}}{P_u}$. If not, revise c and repeat steps 1 to 4 until $M_{nx}/P_{nx} \approx e_y$.

Step 6: Calculate residence factor ϕ using Eq. 2-53a or 2-53b.

Step 7: Determine factored resistances $P_{rx} = \phi P_{nx}$.

The factored axial resistance P_{rx} and P_{ry} should not be greater than the product of the resistance factor ϕ_a and the maximum nominal compressive resistance given by Eq. 3-229.

3.7.4   LIMITATIONS OF REINFORCEMENT FOR AXIAL COMPRESSION MEMBERS

3.7.4.1   Maximum Longitudinal Reinforcement for Axial Members

For noncomposite compression components, the AASHTO specifications require that the maximum area of prestressed and non-prestressed longitudinal reinforcement should satisfy the following:

$$\frac{A_s}{A_g} + \frac{A_{ps}f_{pu}}{A_g f_y} \leq 0.08 \qquad (3\text{-}231)$$

and

$$\frac{A_{ps}f_{pe}}{A_g f_c'} \leq 0.30 \qquad (3\text{-}232)$$

where

A_s = area of non-prestressed tension steel (in.^{2})
A_g = gross area of section (in.^{2})
A_{ps} = area of prestressing steel (in.^{2})
f_{pu} = tensile strength of prestressing steel (ksi)
f_y = yield strength of reinforcing bars (ksi)
f_{pe} = effective prestress of prestressing steel (ksi)

3.7.4.2   Minimum Longitudinal Reinforcement for Axial Members

For noncomposite compression components, the AASHTO specifications require that the minimum area of prestressed and non-prestressed longitudinal reinforcement should satisfy the following equation:

$$\frac{A_s}{A_g} + \frac{A_{ps}f_{pu}}{A_g f_c'} \geq 0.135 \frac{f_c'}{f_y} \qquad (3\text{-}233\text{a})$$

If the design normal concrete strength is not greater than 15 ksi and the unfactored permanent loads do not exceed $0.4 A_g f_c'$, the reinforcement ratio determined by Eq. 3-233a should not be greater than the following:

$$\frac{A_s}{A_g} + \frac{A_{ps}f_{pu}}{A_g f_c'} < 0.0150 \qquad (3\text{-}233\text{b})$$

Additional limits on reinforcement for compression members in Seismic Zones 2 to 4 can be found in the AASHTO LRFD specifications[3-3].

3.7.5 General Consideration of Slenderness Effects

Research results indicate that the geometrical nonlinear effects are relatively small if the slenderness ratio of a compression member is small. Slenderness ratio is defined as

$$\lambda = \frac{kl_u}{r} \tag{3-234a}$$

where
  k = effective length factor (see Table 4-3)
  l_u = column unsupported length
  r = radius of gyration

The AASHTO specifications allow the effects of slenderness to be neglected if members not braced against side sway have a slenderness ratio less than 22. The effects of slenderness can also be neglected if the members braced against side sway have a slenderness ratio that meets the following equation:

$$\lambda < 34 - 12\frac{M_1}{M_2} \tag{3-234b}$$

where M_1 and M_2 = smaller and larger end moments, respectively.

If the slenderness ratio of a member is smaller than 100, the approximate method for determining the P-Δ effect discussed in Section 4.3.7.3 can be used.

3.7.6 Summary of Design Procedures of Compression Members

For clarity, the design procedures for compression members are summarized in Fig. 3-61.

3.8 ANALYSIS AND STRENGTH VALIDATION OF ANCHORAGE ZONE

3.8.1 Pretensioning Anchorage Zone

Pretensioning strands in I-girders and U-girders are mostly located in their bottom flanges. These pretensioning forces cause high vertical splitting forces at each end of the girder in their thin webs for a distance of about one-quarter of the girder depth. These vertical splitting forces may be especially high for larger-depth girders. Figure 3-62 shows an example with typical stress distributions in the vertical and longitudinal directions of a beam. Research results[3-12 to 3-14, 3-53] show that the magnitude of vertical tensile force varies with girder depths. The relationship between the girder depths and the ratios of the total splitting force to the pretensioning force is shown in Table 3-3[3-13]. In this table, V_T and P_{Ts} represent the total vertical splitting force and total pretensioning force, respectively. Adequate reinforcement should be provided to resist the splitting force. The AASHTO specifications stipulate that the vertical reinforcement required

---

TABLE 3-3

Effect of Girder Depth

Girder Depth (in.)	58	63	70	78
V_T/P_{Ts}	2.53%	3.44%	4.85%	7.8%

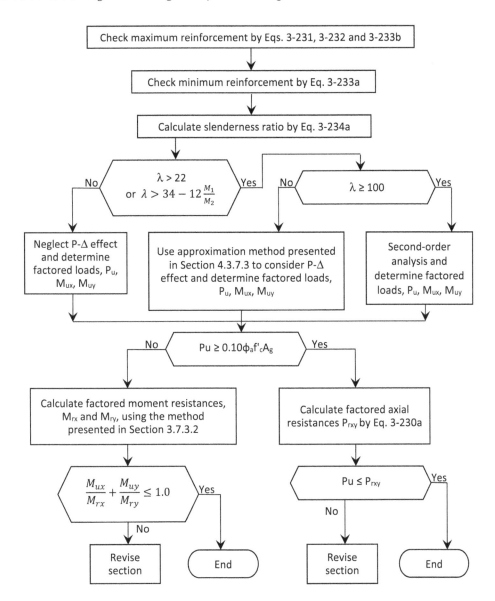

FIGURE 3-61 Flow Chart for Design of Compression Members.

to resist vertical splitting forces that develop in the end anchorage zones of pretensioned beams shall be calculated as:

$$P_r = f_s A_s \tag{3-235}$$

where
  f_s = stress in steel not to exceed 20 ksi which is to control crack
  A_s = total area of reinforcement located within distance $h/4$ from end of beam (in.²)
  h = total girder height (in.)

The current AASHTO specifications require the resistance calculated from Eq. 3-235 should not be less than 4% of the total prestressing force at transfer. Research[3-13] shows that the 4% specified

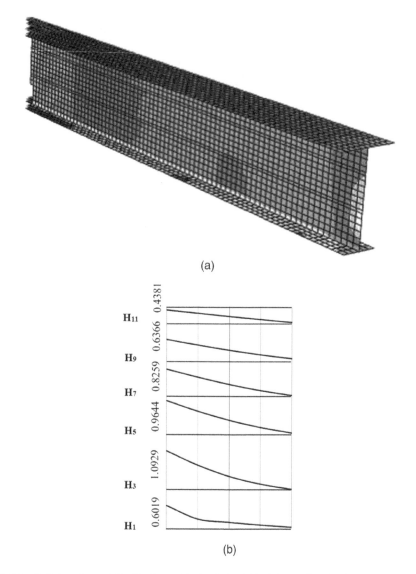

FIGURE 3-62 Splitting Stress Distributions, (a) Analytical Model, (b) Tensile Stress (ksi) along Girder Longitudinal Direction, (c) Tensile Stress (ksi) along Girder Vertical Direction.

(Continued)

may be unsafe for girder depths greater than 70 in. It is recommended that 8% be used for girder depths greater or equal to 78 in.

3.8.2 POST-TENSIONED ANCHORAGE ZONE

3.8.2.1 General

As previously discussed, the prestressing forces in a concrete segmental bridge are normally imposed by post-tensioning tendons that are anchored at the girder ends or other locations via anchorage devices. The highly concentrated post-tensioning force spreads out gradually to a more evenly distributed stress over the entire cross section of the girder at some distance from the anchorage. Figure 3-63

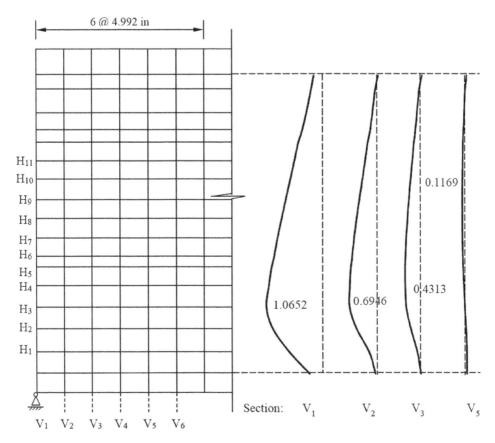

H_i – Horizontal Section Number
V_i – Vertical Section Number

(c)

FIGURE 3-62 (Continued)

shows some typical contour plots of the principal tensile (positive) and compression (negative) stresses for the case of a concentric end anchor[3-45 to 3-47]. In this figure

h = girder height
P = post-tensioning force
f_0 = P/A_0 = average compression stress in the section
f_{burst} = bursting stress

From this figure, we can observe (1) the compressive stresses immediately ahead of the bearing plate are very high and become nearly uniform about one girder depth from the bearing plate and (2) the horizontal post-tensioning force not only causes compressive stresses along the girder's longitudinal axis, but it also causes high compression and tensile forces along the girder's vertical axis. Along the loaded edge, there are high tensile stress concentrations, which are normally named spalling stresses. The area between the anchorage and the section about one girder height from the anchorage is subjected to transverse tensile stresses, which are called bursting stresses. The zone from the anchorage to the section where the compressive stress becomes uniform is called the anchorage zone. The AASHTO specifications require that the length of the anchorage zone in the direction of

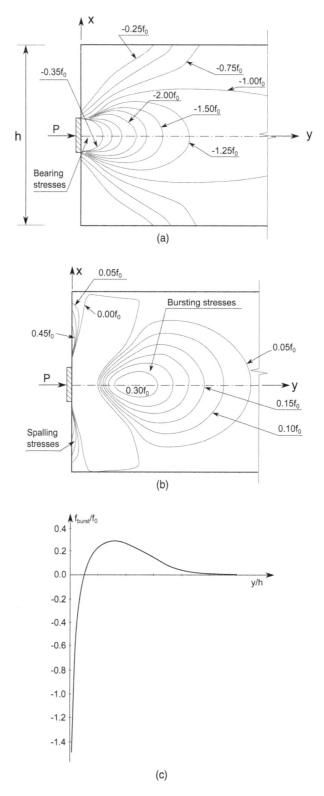

FIGURE 3-63   Typical Stress Distributions in Anchorage Zone, (a) Distribution of Principal Compression Stress, (b) Distribution of Principal Tensile Stress, (c) Variation of Bursting Stress along y-Axis.

the tendon should not be less than the greater of the transverse dimensions of the anchorage zone and should not be taken as more than one and one-half times that dimension.

The main concerns in the anchorage zone design are the high compressive stresses immediately ahead of the anchor plate and the tensile stresses in the girder's transverse direction. For this reason, in the AASHTO specifications, the anchorage zone is defined as two typical zones, namely, the local zone and the general zone. The local zone is the highly stressed area immediately ahead of the anchor plate, and the general zone mainly defines the lateral tensile stress area. In cases where the manufacturer of the anchorage devices have not provided edge distance recommendations, or the recommended edge distances have not been independently verified, the width of the local zone should be taken as the greater of (1) the corresponding bearing plate size, plus twice the minimum concrete cover required for the particular application and environment, and (2) the sum of the outer dimension of any required confining reinforcement and the required concrete cover over the confining reinforcing steel for the particular application and environment (see Fig. 3-64b). The length of the local zone shall not be taken as greater than 1.5 times the width of the local zone. The ASSHTO Article 5.10.9.7.1 provides more information regarding the location zone definition. For the responsibilities of the anchorage device suppliers and the designers, AASHTO specifications define the entire anchorage zone as the general zone (refer to Fig. 3-64a). The anchorage device suppliers are responsible for furnishing anchorage devices that satisfy the local zone anchor efficiency requirements established by the AASHTO LRFD Bridge Construction Specifications[3-48] and should provide the information for the auxiliary and confining

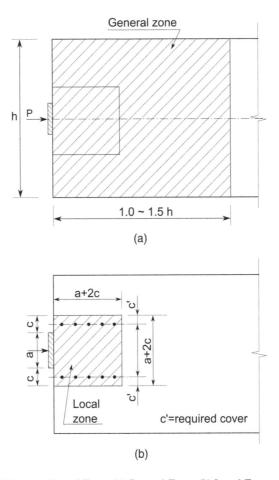

FIGURE 3-64   General Zone and Local Zone, (a) General Zone, (b) Local Zone.

reinforcement, minimum anchor spacing, minimum edge distance, and minimum concrete strength at the time of stressing. The Engineer of Record is responsible for the entire design for the general zone, including the reinforcement, the location of the tendons, and anchorage devices.

3.8.2.2 Design of Local Zone

The AASHTO specifications specify that anchorages be designed at the strength limit states for the factored jacking forces. The factored jacking force cannot be greater than the factored bearing resistance, i.e.,

$$P_u \le P_r = \phi_b f_n A_b \tag{3-236a}$$

where

$$f_n = 0.7 f'_{ci} \sqrt{\frac{A}{A_g}} \le 2.25 f'_{ci} \tag{3-236b}$$

ϕ_b = bearing resistance factor (see Section 2.3.3.5)
A = maximum area of portion in bearing plate mounting surface that is proportional to sizes of bearing plate and concentric with it and does not overlap with proportioned areas for adjacent anchorage devices (see Fig. 3-65c) (in.^{2})
A_g = gross area of bearing plate (see Fig. 3-65c) (in.^{2})
A_b = effective net area of bearing plate, minus area of openings in the bearing plate (see Fig. 3-65b) (in.^{2})
f'_{ci} = nominal concrete strength at time of application of tendon force (ksi)

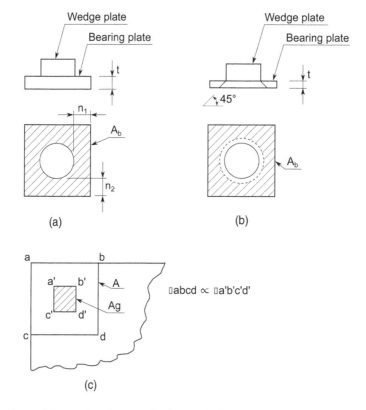

FIGURE 3-65 Area of Supporting Concrete Surface and Effective Area of Bearing Plate, (a) Typical Stiff Bearing Plate, (b) Typical Flexible Bearing Plate, (c) Supporting Area and Gross Area.

A_b can be taken as the full bearing plate area A_g minus the wedge hole area for stiff bearing plates (see Fig. 3-65b). The stiff bearing plate material should not yield at the factored tendon force, and its slenderness meets following equation:

$$\frac{n_i}{t} \le 0.08\left(\frac{E_b}{f_b}\right)^{0.33}$$

(3-237)

where

t	=	thickness of bearing plate (in.)
E_b and f_b	=	modulus of elasticity and stress, respectively, at edge of wedge hole or holes of bearing plate (ksi)
n_i	=	width of base plate beyond wedge hole (see Fig. 3-65a)

For a flexible bearing plate, the wedge hole area should be calculated by increasing hole dimensions by a distance equal to the thickness of the bearing plate (see Fig. 3-65b). The determination of A_b for other types of anchorage devices can be found in Article 5.10.9.7.2 of the AASHTO specifications. It should be mentioned that Eqs. 3-236 and 3-237 were developed based on tested results.

3.8.2.3   Design of General Zone

3.8.2.3.1   General

Except the requirement by Eq. 3-236a, the concrete compressive strength and reinforcement tensile strength in the general zone should be investigated. The factored concrete compressive stress f'_{cu} and factored reinforcement steel tensile stress f_u normally should meet the following equations:

Factored concrete compressive stress:

$$f'_{cu} \le 0.7\phi f'_{ci}$$

(3-238a)

Factored bonded reinforcement tensile stress:

$$f_u \le f_y$$

(3-238b)

Factored bonded prestressed reinforcement:

$$f_{pu} \le f_{pe} + 15000 \text{ psi}$$

(3-238c)

AASHTO specifications allow three main types of analytical methods for analyzing the concrete compressive stresses and the bursting stresses:

- Refined elastic stress analytical models
- Strut-and-tie models
- Approximate methods

3.8.2.3.2   Elastic Stress Analytical Method

Elastic analysis of anchorage zone is often used and has been found acceptable, though it may provide a conservative design because the cracks in the anchorage zone may cause stress redistributions[3-12, 3-45].

Currently, the finite-element method is mostly used in the elastic analysis of the anchorage zone. The following procedures can be used for determining the required tensile reinforcement.

a. Properly develop the finite-element meshes based on the girder shape and tendon locations (see Fig. 3-66a). For an I-girder or box web, shell-plate finite elements can be used to modeling the structure. For diaphragms and other thick members, three-dimensional solid elements should be used.

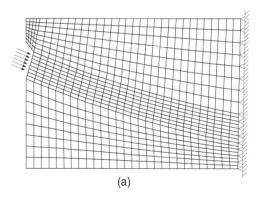

(a)

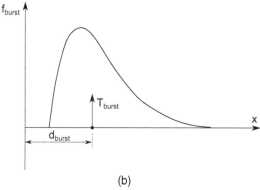

(b)

FIGURE 3-66   Elastic Stress Analysis by Finite-Element Method[3-45], (a) Finite Element Meshes, (b) Variation of Vertical Tensile Stress along Girder Longitudinal Direction.

b. Develop the variation of tensile stresses along the girder tendon path (see Fig. 3-66b) or along another direction of concern.

c. Integrate the tensile bursting stresses in both the longitudinal and the transverse directions, and obtain the magnitude T_{burst} and location d_{burst} of the resulting bursting force.

d. The required bursting reinforcement area is determined as

$$A_{burst} = \frac{T_{burst}}{f_y} \tag{3-239}$$

where T_{burst} = total bursting force due to factored post-tensioning forces.

e. The required A_{burst} is distributed in the lesser of the following distances, with the centroid of the bursting reinforcement coinciding with the location d_{burst} in the design.
   – 2.5 d_{burst}
   – 1.5 times the corresponding lateral dimension of the section

f. The spacing of the burst reinforcement should be not greater than either 24.0 bar diameters or 12.0 in.

To control the cracks of the anchorage zones, it may be a good practice to keep the maximum reinforcement stress not exceeding 20 ksi with service loadings.

Under factored tendon force, the concrete compressive stress calculated at the location ahead of the anchorage devices cannot be greater than that calculated by Eq. 3-238a.

3.8.2.3.3 *Strut-and-Tie Models*

The analytical method for the anchorage zones based on the strut-and-tie models is a simplified method to consider the state of equilibrium of a structure[3-45, 3-46, 3-49 to 3-51]. In a strut-and-tie model, the trajectories of principal stresses in concrete are assumed to be straight lines between points called nodes of the model and the distributed compressive stresses are carried by compression members named struts. The tensile stresses are assumed to be carried entirely by the reinforcing bars. The effect of several reinforcing bars is normally lumped into one single tension member that is called a tie (see Fig. 3-67). Based on many of experimental results, the strut-and-tie model yields more realistic results than elastic stress analysis. When significant cracks have developed,

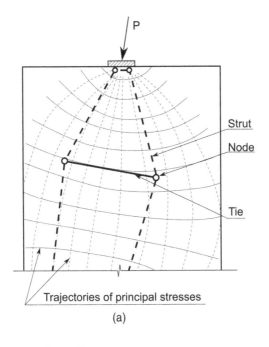

(a)

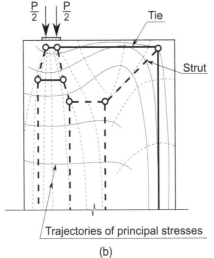

(b)

FIGURE 3-67   Trajectories of Principal Stresses and Strut-and-Tie Model, (a) Inclined Concentric Loading, (b) Vertical Eccentric Loading.

the compressive trajectory stresses in the concrete tend to congregate into straight lines that can be treated as straight compressive struts and balanced by tension ties at the nodes. The strut-and-tie model is based on the static theorem, which is the basis for the theory of plasticity of structures. It may state that the load computed based on the assumed stress distribution that satisfies the equilibrium condition and in no point exceeds the strength of materials is the lower bound of actual ultimate load. In other words, it is possible to obtain a conservative estimate of the capacity of a structure by assuming an internal stress distribution with respect to the material properties.

Generally, in plastic design, there is no unique solution to a given problem. The designer has the freedom to develop the geometry of a strut-and-tie model. A proper strut-and-tie model can be obtained through tests or comparison and evaluation of various strut-and-tie model configurations with the results obtained from finite-element method. Figure 3-68 provides some typical strut-and-tie models for analyzing anchorage zones.

In concrete segmental bridges, external prestressing tendons are often anchored at the end diaphragms which are supported on the box girder slabs. The strut-and-tie models of the diaphragms are different from those for a girder with a uniform rectangular cross section. Two typical strut-and-tie models for the flanged sections are shown in Fig. 3-69a and b. From these figures, we can see the required bursting reinforcement may be significantly larger than for the beams with a continuous rectangular section.

In a strut-and-tie model, nodes are critical elements. The adequacy of the nodes is ensured by limiting the bearing pressure under the anchorage device. Local zones that satisfy Eq. 3-236a may be considered as properly detailed and are adequate nodes. The dimensions of the strut ahead of the bearing plate normally are determined based on the sizes of the bearing plate and the selected strut-and-tie model. The local zone node for developing the strut-and-tie model can be selected at a depth of $a/4$ ahead of the anchorage plate (see Fig. 3-70b). Then, the width of the strut can be determined based on the direction of the strut α. For special anchorage devices based on the acceptance test Article 10.3.2.3 of AASHTO LRFD Bridge Construction Specifications[3-48], the dimensions of the strut can be determined at a larger distance from the node, assuming that the width of the strut increases with the distance from the local zone (see Fig. 3-70c and d[3-3]). The thickness of the strut for ordinary anchorage devices is equal to the length of the bearing plate b. The thickness of the strut for the special anchorage devices is taken as

$$t' = b\left[1 + a\left(\frac{1}{b} - \frac{1}{t}\right)\right] \tag{3-240}$$

3.8.2.3.4 Approximate Methods

Based on parameter analysis of members with rectangular cross sections, Burdet[3-46, 3-47] proposed some approximate equations to estimate the anchorage zone concrete strength, the bursting force, and the location of the bursting force. These equations have been adopted by the AASHTO LRFD specifications and can be used for analyzing the anchorage zones with rectangular cross section or close to rectangular cross sections, such as the post-tensioned I-girders.

The factored concrete compressive stress ahead of the anchorage devices can be determined as

$$f_u = \frac{0.6P_u k}{A_b\left[1 + l_c\left(\frac{1}{b_e} - \frac{1}{t}\right)\right]} \tag{3-241}$$

where
  P_u = factored tendon force (kip)
  t = member thickness (in.)
  l_c = longitudinal length of confining reinforcement of local zone ≤ larger of $1.15a_e$ or $1.15b_e$ (in.)

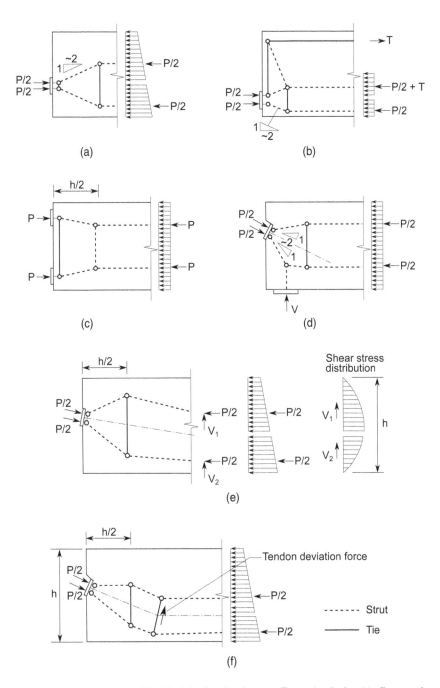

FIGURE 3-68 Typical Strut-and Tie Models for Anchorage Zone Analysis, (a) Concentric or Small Eccentricity, (b) Large Eccentricity, (c) Multiple Anchors, (d) Eccentric Anchor and Support Reaction, (e) Inclined and Straight Tendon, (f) Inclined and Curved Tendon[3-4].

k = modification factor for closely spaced anchorages

      = 1, when $s \geq 2a_e$

      = $1 + \left(2 - \frac{s}{a_e}\right)\left(0.3 + \frac{n}{15}\right)$

a_e = lateral dimension of effective bearing area measured along larger dimension of cross section as shown in Fig. 3-71 (in.)

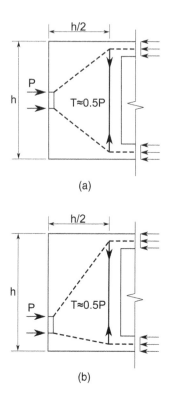

(a)

(b)

FIGURE 3-69 Typical Strut-and-Tie Models for Solid Diaphragms Subjected to External Post-tensioning Forces, (a) Concentric Loading, (b) Eccentric Loading.

b_e = lateral dimension of effective bearing area measured along smaller dimension of cross section as shown in Fig. 3-71 (in.)

s = center-to-center spacing of anchorages (in.)

n = number of anchorages in a row

A_b = effective bearing area = larger of A_{plate} and A_{conf} (in.²)

A_{plate} = anchor bearing plate area ≤ $1.273A_{conf}$

A_{conf} = bearing area of confined concrete in local zone
     = A_{plate}, if $a_e > 2a$ or $b_e > 3b$ (assuming $a_e > b_e$ and $a > b$) (see Fig. 3-71)

The bursting force and its location measured from the anchorage face in the anchorage zones can be estimated by the following equations:

$$T_{burst} = 0.25 \sum P_u \left(1 - \frac{a}{h}\right) + 0.5 \left|\sum P_u \sin \alpha\right| \tag{3-242}$$

$$d_{burst} = 0.5(h - 2e) + 5e \sin \alpha \tag{3-243}$$

where

T_{burst} = tensile force in anchorage zone (see Fig. 3-72b) (kip)

d_{burst} = distance from anchorage device to centroid of bursting force (in.)

P_u = factored tendon force (kip)

a = dimension of anchorage device or group of devices in direction considered (in.)

e = eccentricity (see Fig. 3-72a); taken as positive (in.)

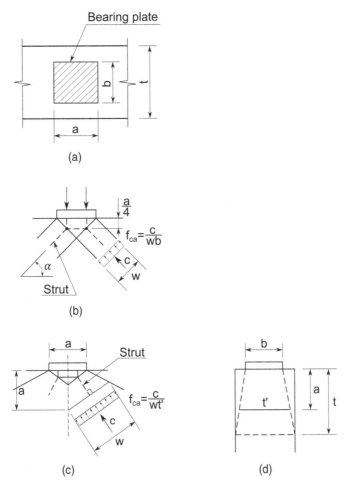

FIGURE 3-70 Critical Sections for Compressive Struts in Anchorage Zone, (a) Plan View, (b) Ordinary Anchorage Devises, (c) Special Anchorage Devices, (d) Determination of t'[3-3].

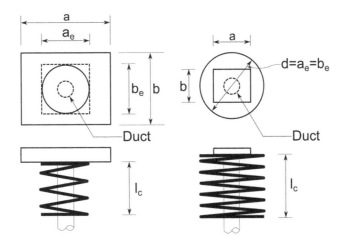

FIGURE 3-71 Effective Bearing Area[3-3].

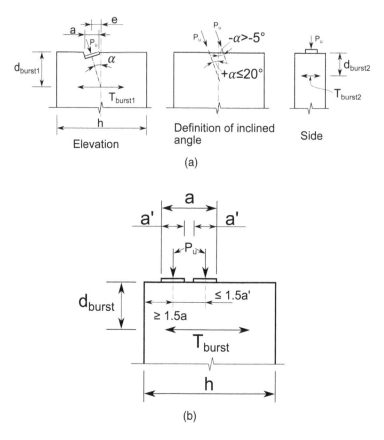

FIGURE 3-72 Burst Force and Its Location, (a) Inclined Tendons, (b) Closely Spaced Anchorage Devices.

h = dimension of cross section in direction considered (in.)

α = angle of inclination of a tendon force with respect to centerline of member; sign of angle is defined as shown in Fig. 3-72a and angle should be within $-5°$ to $20°$.

The AASHTO specifications also provide the following approximate equations for estimating the edge tension force and busting force for multiple slab anchorages:

Edge tension force between anchorages (kips):

$T_1 = 0.10P_u\left(1 - \dfrac{a}{s}\right)$, if the edge distance of a slab anchorage $\geq$ two plate width or one slab thickness

$= 0.25P_u$, for remaining cases   (3-244)

where

P_u = factored tendon load of individual anchor (kip)

a = anchor plate width as shown in Fig. 3-73 (in.)

s = anchorage spacing as shown in Fig. 3-73 (in.)

The bursting force along the slab thickness direction (kip) is

$$T_2 = 0.20P_u\left(1 - \frac{a}{s}\right)$$ (3-245)

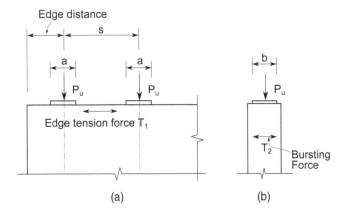

FIGURE 3-73   Edge Force and Burst Force of Multiple Slab Anchorages, (a) Elevation, (b) Side View.

3.9   GENERAL STRUCTURAL DESIGN BY STRUT-AND-TIE MODEL

3.9.1   INTRODUCTION

In Section 3.8.2.3.3, we have discussed the strut-and-tie model and its application to the design of post-tensioning anchorage zones. The AASHTO LRFD specifications also require that the strut-and-tie model be used for designing deep footing, pier caps, pile caps, or other cases in which the distance between the centers of applied load and the supporting reactions is not greater than twice the member thickness. The basic theory for the strut-and-tie model was discussed in Section 3.8.2.3. In this section, we will briefly discuss how to check the strengths of struts, ties, and nodes. The methods presented below are developed based on a large number of experimental results.

3.9.2   DETERMINATION OF COMPRESSIVE STRUT STRENGTH

3.9.2.1   Strength of Unreinforced Struts

The compression strength of a strut is the same as that for axially loaded members and can be determined as

$$P_n = f_{cu} A_{cs} \tag{3-246a}$$

where
   P_n = strut nominal compressive resistance (kip)
   f_{cu} = limiting strut compressive stress (ksi)

$$f_{cu} = \frac{f_c'}{0.8 + 170\varepsilon_l} \le 0.85 f_c' \tag{3-246b}$$

$$\varepsilon_l = \varepsilon_s + (\varepsilon_s + 0.002)\cot^2 \theta_s \tag{3-246c}$$

   θ_s = smallest angle between strut and adjoining ties (degrees)
   ε_s = tensile strain in concrete in direction of tension tie (in./in.); taken as tensile strain due to factored loads in reinforcing bars; taken as zero for a tie consisting of prestressing
   A_{cs} = effective cross-sectional area of strut determined as in Fig. 3-74 (in.²)

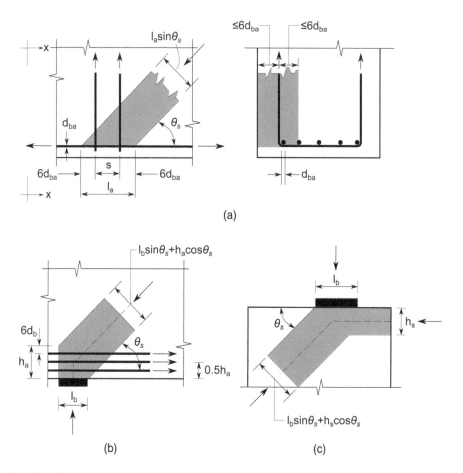

FIGURE 3-74 Effective Cross-Sectional Area of Struts, (a) Strut Anchored by Reinforcement, (b) Strut Anchored by Reinforcement at Bearing, (c) Strut Anchored by Strut and Bearing[3-3].

3.9.2.2 Strength of Reinforced Strut

If a compressive strut has reinforcing steel detailed to develop its yield stress in compression and parallel to the strut, the strength of the strut can be calculated as

$$P_n = f_{cu}A_{cs} + f_y A_s \tag{3-247}$$

where A_s = area of reinforcement in strut (in.^{2}).

3.9.3 DETERMINATION OF TENSION TIE STRENGTH

The nominal tension strength of a tie can be determined as

$$P_n = A_s f_y + A_{ps}(f_{pe} + f_y) \tag{3-248}$$

where
A_s and A_{ps} = areas of mild steel reinforcement and prestressing steel in tie, respectively (in.^{2})
f_y and f_{pe} = yield stresses of mild steel and stress in prestressing steel after losses, respectively (ksi)

3.9.4 Stress Limits in Node Regions

The compression stress limits in nodal zones of a strut-and-tie model are related to the degree of the confinement in these zones by the concrete. Generally, if there is no confining reinforcement provided, AASHTO specifications require that the concrete compressive stress f_{ca} in the node regions should meet the following:

$f_{ca} \leq 0.85\phi_b f_c'$, for node regions bounded by compressive struts
$\leq 0.75\phi_b f_c'$, for node regions anchoring a one-direction tension tie
$\leq 0.65\phi_b f_c'$, for node regions anchoring tension ties in more than one direction
ϕ_b = resistance factor of bearing on concrete (see Section 2.3.3.5)

3.10 SERVICE STRESS CHECK REQUIRED BY AASHTO LRFD SPECIFICATIONS

3.10.1 Stress Validation for Prestressing Tendons

The tendon stress due to prestressing force, at the service limit state, at the strength and extreme event limit states should be checked and cannot exceed the values specified in Section 1.2.5.2.

3.10.2 Concrete Stress Validations

3.10.2.1 Temporary Stresses before Losses

The temporary compressive stress cannot exceed $0.60f_c'$ as discussed in Section 1.2.5.1.1.1, and the temporary tensile stress cannot exceed the values specified in Section 1.2.5.1.1.2.

3.10.2.2 Stresses at Service Limit State after Losses

3.10.2.2.1 Compression Stresses

The concrete compression stress should be checked for the Service Limit I Load Combination. The sum of the concrete compressive stresses due to permanent loadings should not exceed $0.45f_c'$, and the sum of the compressive stresses due to all the permanent and transient loadings should not be greater than $0.6\phi_w f_c'$ as discussed in Section 1.2.5.1.2.1. The reduction factor ϕ_w should be determined as:

$$\phi_w = 1.0, \text{ if } \lambda_w \leq 15$$
$$= 1.0 - 0.025(\lambda_w - 15), \text{ if } 15 < \lambda_w \leq 25 \qquad (3\text{-}249)$$
$$= 0.75, \text{ if } 25 < \lambda_w \leq 35$$

where
λ_w = $\frac{X_u}{t}$ = wall slenderness ratio for hollow rectangular columns or box girders
X_u = clear length of constant thickness portion of a wall between walls or fillets between walls (in.)
t = thickness of wall (in.)

3.10.2.2.2 Tension Stresses

The tensile stresses in the bridge longitudinal direction should be checked for load combination Service III. The tensile stresses in the transverse analysis of box girder bridges should be checked for load combination Service I. The maximum tensile stresses should not exceed the value specified in Section 1.2.5.1.

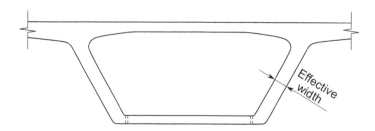

FIGURE 3-75   Effective Width of Web.

3.10.2.2.3   Principal Stresses in Webs of Segmental Concrete Bridges

The principal tensile stress at the neutral axis of the critical web should be checked for both the Service III limit state and during construction. The maximum tensile stress should not exceed the values specified in Section 1.2.5.1.

The principal tensile stress is determined from the maximum shear and/or maximum shear combined with the shear from torsion stress and axial stress at the natural axis using classical beam theory. The vertical force component of draped tendons is considered as to be an external force in the reduction of the shear for segmental bridges. The effective width of the web is measured perpendicular to the plane of the web (see Fig. 3-75).

REFERENCES

3-1. Lin, T. Y., and Burns, N. H., *Design of Prestressed Concrete Structures*, John Wiley & Sons, Inc., New York, 1981.

3-2. Menn, C., *Prestressed Concrete Bridges*, Birkhauser Verlag, Basel, Boston, Berlin, 1990.

3-3. AASHTO, *LRFD Bridge Design Specifications*, 7th edition, American Association of State Highway and Transportation Officials, Washington DC, 2014.

3-4. Nawy, E. G., *Prestressed Concrete—A Fundamental Approach*, 3rd edition, Prentice Hall, New Jersey, 1999.

3-5. Yamazaki, J., Kattula, B. T., and Mattock, A. H., "A Comparison of the Behavior of Post-Tensioned Concrete Beam with and without Bond," Report SM69-3, University of Washington, College of Engineering, Structures and Mechanics, December 1969.

3-6. Mattock, A. H., Yamazaki, J., and Kattula, B. T., "Comparative Study of Prestressed Concrete Beam, with and without Bond," *Journal of the American Concrete Institute*, Vol. 68, No. 68, February 1971, pp. 116–125.

3-7. Kashima, S., and Breen, J. E., "Construction and Load Tests of a Segmental Precast Box Girder Bridge Model," Research Report No. 121-5, Center for Highway Research, University of Texas at Austin, February 1975.

3-8. MacGregor, R. J. G., Kreger, M. E., and Breen, J. E., "Strength and Ductility of a Three-Span Externally Post-Tensioned Segmental Box Girder Bridge Model," Research Report No. 365-3F, Center for Transportation Research Bureau of Engineering Research, University of Texas at Austin, January 1989.

3-9. FDOT, "Structural Design Guidelines," Florida Department of Transportation, Tallahassee, FL, 2018.

3-10. Moreton, A. J., "Test of Epoxy-Glued Joint for a Segmental Precast Bridge Deck," *International Journal of Adhesion and Adhesives*, April 1982, Vol. 2, No. 2, April 1982, pp. 97–101.

3-11. Yuan, A. M., He, Y., Dai, H., and Cheng, L., "Experimental Study of Precast Segmental Bridge Box Girders with External Unbounded and Internal Bonded Post-tensioning under Monotonic Vertical Loading," ASCE, *Journal of Bridge Engineering*, Vol. 20, Issue 4, April 2014.

3-12. Huang, D. Z., and Hu, B., "Evaluation of Cracks a Large Single Cell Precast Concrete Segmental Box Girder Bridge without Internal Struts," *Journal of Bridge Engineering*, ASCE,

3-13. Huang, D. Z., Arnold, S., and Hu, B., "Cracking Analysis of Florida Barge Canal Spliced I-girder Bridge," TRB, *Journal of the Transportation Research Board*, Transportation Research Board of the National Academies, Washington. D.C., No. 2313, 2012, pp. 83–91.

3-14. Huang, D. Z., and Shahawy, M., "Analysis of Tensile Stresses in Transfer Zone of Prestressed Concrete U-Beams," TRB, *Journal of the Transportation Research Board*, Transportation Research Board of the National Academies. Washington, DC, No. 1928, 2005, pp. 134–141.

3-15. Coven, J., "Post-Tensioned Box Girder Design Manual," Federal Highway Administration, U.S. Department of Transportation, Report No. FHWA-HIF-15-016, 2016.

3-16. Tadros, M. K., Al-Omaishi, N., Seguirant, S. P., and Gallt, J. G., "Prestress Losses in Pretensioned High-Strength Concrete Bridge Girders," NCHRP Report 496. Transportation Research Board, National Research Council, Washington, D.C., 2003.

3-17. Fan, L. C., *Bridge Engineering*, China Communication Publishing House, Peking, China, 1980.

3-18. Li, M. Z., and Wan, G. H., *Analysis of Bridge Structures*, China Communication Publishing House, Peking, China, 1990.

3-19. Li, G. H., *Theory of Bridge and Structures*, Shanghai Science and Technology Publishing House, Shanghai, China, 1983.

3-20. Huang, D. Z., and Li, G. H., "Elastic and Inelastic Stability Analysis of Truss Bridges," *Journal of China Civil Engineering*, Vol. 16, No. 4, 1988.

3-21. Nakai, H., and Yoo, C. H., *Analysis and Design of Curved Steel Bridges*, McGraw-Hill Book Company, New York, 1998.

3-22. Wright, R. N., Abdel-Samad, S. R., and Robinson, A. R., "BEF Analogy for Analysis of Box Girder," *Proceedings of American Society of Civil Engineers*, Vol. 97, No. 7, 1968, pp. 1719–1743.

3-23. Loov, R. E., "A General Equation for the Steel Stress for Bonded Prestressed Tendons," *PCI Journal*, Prestressed Concrete Institute, Chicago, IL, Vol. 33, No. 6, November–December 1988, pp. 108–137.

3-24. Naaman, A. E., "Discussion of Loov 1988," *PCI Journal*, Prestressed Concrete Institute, Chicago, IL, Vol. 34, No. 6, November–December 1989, pp. 144–147.

3-25. Naaman, A. E., "A New Methodology for the Analysis of Beams Prestressed with Unbonded Tendons." In *External Prestressing in Bridges*, ACI SP-120. A. E. Naaman and J. Breen, eds. American Concrete Institute, Farmington Hills, MI, pp. 339–354.

3-26. Naaman, A. E., "Unified Design Recommendations for Reinforced Prestressed and Partially Prestressed Concrete Bending and Compression Members," *ACI Structural Journal*, American Concrete Institute, Farmington Hills, MI, Vol. 89, No. 2, March–April 1992, pp. 200–210.

3-27. Naaman, A. E., and Alkhairi, F. M., "Stress at Ultimate in Unbonded Prestressing Tendons—Part I: Evaluation of the State-of-the-Art; Part II: Proposed Methodology," *ACI Structural Journal*, American Concrete Institute, Farmington Hills, MI, September–October 1991, November–December 1991, respectively.

3-28. Precast Prestressed Concrete Institute, *PCI Design Handbook*, 6th edition, PCI, Chicago, 2004.

3-29. Roberts-Wollmann, C., Kreger, M. E., Rogowsky, D. M., and Breen, J. E., "Stresses in External Tendons at Ultimate," *ACI Structural Journal*, Vol. 102, No. 2, 2005.

3-30. MacGregor, R. J. G., Kerger, M. E., and Breen, J. E., "Strength and Ductility of a Three-Span Externally Post-Tensioned Segmental Box Girder Bridge Model," Research Report No. 365–3F, Center for Transportation Research, The University of Texas at Austin, 1989.

3-31. MacGregor, R. J. G., "Strength and Ductility of Externally Post-Tensioned Segmental Box Girders," PhD. Dissertation, University of Texas at Austin, Austin, 1989.

3-32. MacGregor, R. J. G., Kerger, M. E., and Breen, J. E., "Strength and Ductility of a Three-Span Externally Post-Tensioned Segmental Box Girder Bridge Model." In *External Prestressing in Bridges*, ACI SP-120, A. E. Naaman and J. Breen, eds., ACI, Farmington Hills, MI, 1990, pp. 315–338.

3-33. Bentz, E. C., Vecchio, F. J., and Collins, M. P., "The Simplified MCFT for Calculating the Shear Strength of Reinforced Concrete Elements," *ACI Structural Journal*, American Concrete Institute, Farmington Hills, MI, Vol. 103, No. 4, July–August 2006, pp. 614–624.

3-34. Walraven, J. C., "Fundamental Analysis of Aggregate Interlock," *Journal of the Structural Division, ASCE*, Vol. 107, No. ST11, November 1981, pp. 2245–2270.

3-35. Collins, M. P., Mitchell, D., Adebar, P. E., and Vecchio, F. J., "A General Shear Design Method," *ACI Structural Journal*, American Concrete Institute, Farmington Hills, MI, Vol. 93, No. 1, 1996, pp. 36–45.

3-36. Collins, M. P., and Mitchell, D., *Prestressed Concrete Structures*, Prentice-Hall, Englewood Cliffs, NJ, 1991.

3-37. Hawkins, N. M., and Kuchma, D. A., *Simplified Shear Design of Structural Concrete Members*, NCHRP Report 549. Transportation Research Board, National Research Council, Washington, DC, 2006.

3-38. Barker, R. M., and Puckett, J. A., Design of Highway Bridges Based on AASHTO LRFD Bridge Design Specifications, John Wiley & Sons, New York, 1997.

3-39. Hsu, T. T. C., *Unified Theory of Reinforced Concrete*, CRC Press, Boca Raton, 1993.

3-40. MacGregor, J. G., Reinforced Concrete, Mechanics and Design, Prentice, Hall, Englewood Cliffs, NJ, 1988.

3-41. Loov, R. E., "A General Equation for the Steel Stress for Bonded Prestressed Tendons," *PCI Journal*, Prestressed Concrete Institute, Chicago, IL, Vol. 33, No. 6, November–December 1988, pp. 108–137.

3-42. Loov, R. E., and Patnaik, A. K., "Horizontal Shear Strength of Composite Concrete Beams with a Rough Interface," *PCI Journal*, Vol. 39, No. 1, Prestressed Concrete Institute, Chicago, IL, January–February 1994, pp. 48–69. See also "Reader Comments," *PCI Journal*, Prestressed Concrete Institute, Chicago, IL, Vol. 39, No. 5, September–October 1994, pp. 106–109.

3-43. Mattock, A. H., "Shear Transfer in Concrete Having Reinforcement at an Angle to the Shear Plane," In Vol. 1, *Shear in Reinforced Concrete*, SP-42. American Concrete Institute, Farmington Hills, MI, 1974, pp. 17–42.

3-44. Shahrooz, B. M., Miller, R. A., Harries, K. A., and Russell, H. G., "Design of Concrete Structures Using High-Strength Steel Reinforcement," NCHRP Report 679. Transportation Research Board, National Research Council, Washington, DC, 2011.

3-45. Sanders, D. H., "Design and Behavior of Post-Tensioned Concrete Anchorage Zones," Ph.D. dissertation, University of Texas, Austin, TX, August 1990.

3-46. Wollmann, G. P., *Anchorage Zones in Post-Tensioned Concrete*, University of Texas, Austin, TX, May 1992.

3-47. Burdet, O. L., "Analysis and Design of Anchorage Zones in Post-Tensioned Concrete Bridges," Ph.D. dissertation, University of Texas, Austin, TX, May 1990.

3-48. *AASHTO LRFD Bridge Construction Specifications*, 3rd edition, American Association of State Highway and Transportation Officials, Washington D.C., 2010.

3-49. Schlaich, J., Schäfer, K., and Jennewein, M., "Towards a Consistent Design of Structural Concrete," *PCI Journal*, Prestressed Concrete Institute, Chicago, IL, Vol. 32, No. 3, May–June 1987, pp. 74–151.

3-50. Morsch, E., *Under die Berechnung der Glenkquqder*, Beton und Eisen, No. 12, 1924, pp. 156–161.

3-51. Beaupre, R. J., Powell, L. C., Breen, J. E., and Kreger, M. E., "Deviation Saddle Behavior and Design for Externally Post-Tensioned Bridges," Research Report 365-2. Center for Transportation Research, July 1988.

3-52. Marshall, W. T., and Mattock, A. H., "Control of Horizontal Cracking in the Ends of Prestressed Concrete Girders," *Journal of the Prestressed Concrete Institute*, Vol. 7, 1962, pp. 56–74.

4 General Analytical Theory of Super-Structures

- Determination of girder displacement and the integral by diagram multiplication method
- Analysis of continuous girder bridges by force method, displacement method, and moment distribution method
- Theory and procedures of finite-element method
- Analysis of secondary forces due to post-tendons and design considerations
- Analysis of secondary forces due to temperature
- Analysis of secondary forces due to concrete creep and shrinkages
- Geometrical and material non-linear analysis
- Stability analysis by finite-element method
- Bridge transverse analysis

4.1 INTRODUCTION

Modern concrete segmental bridges are almost all statically indeterminate structures. Though there are many powerful commercial computer programs available for analyzing these types of bridges, it is still very helpful for the bridge designers to clearly know the fundamental analytical methods for analyzing statically indeterminate structures. This chapter includes three sections: fundamental analytical methods of statically indeterminate structures, bridge longitudinal analysis, and bridge transverse analysis.

4.2 FUNDAMENTALS OF ANALYSIS OF INDETERMINATE BRIDGE STRUCTURES

4.2.1 DETERMINATION OF GIRDER DISPLACEMENTS

In analyzing a statically indeterminate structure, it is critical to know the relationship between internal forces and displacements of a member, including its slopes. In concrete segmental bridges, the displacements of the super-structures are mainly caused by bending and axial deformations. Two methods that are often used in bridge engineering for determining girder deformations are the conjugate-beam method and the method of virtual work.

4.2.1.1 Conjugate-beam method

Figure 4-1a shows a beam with arbitrary vertical loadings and the corresponding deformation. The sign convention is taken as a positive deflection for upward deflection and positive when the slope angle is measured counterclockwise from the x-axis; a positive moment is one that causes the beam to bend concave upward. The deformation of a small segment dx is shown in Fig. 4-1b. Based on Saint-Venant's principle, each cross section can be assumed to remain plane after deformation and the angle between them is denoted as $d\theta$. The arc dx represents a portion of the elastic curve along the neutral axis. Letting R denote the radius of curvature of the arc dx, which is measured from the center of curvature to dx, the strain of the arc ds located at a distance y away from dx can be written as

$$\varepsilon = \frac{(R-y)d\theta - Rd\theta}{Rd\theta} \tag{4-1a}$$

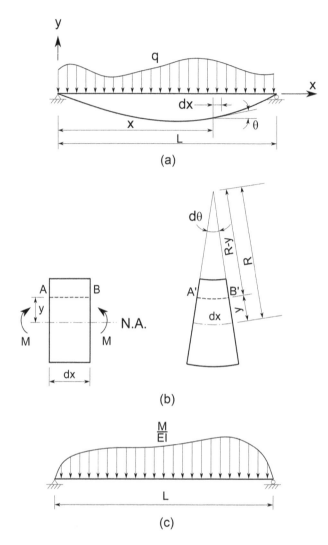

FIGURE 4-1 Beam Deformation and Conjugate Beam, (a) Real Beam with External Loading, (b) Bending Deformation of a Small Element, (c) Conjugate Beam.

Based on Hooke's law, Eq. 4-1a can be written as

$$\varepsilon = \frac{f}{E} = -\frac{y}{ER} \tag{4-1b}$$

where
    E = modulus of elasticity
    f = normal stress at location y, negative for compression

$$f = -\frac{My}{I} \tag{4-1c}$$

Substituting Eq. 4-1c into Eq. 4-1b yields

$$\frac{1}{R} = \frac{M}{EI} \tag{4-2}$$

Substituting $dx = Rd\theta$ into Eq. 4-2 yields

$$d\theta = \frac{M}{EI}dx \text{ or } \frac{d\theta}{dx} = \frac{M}{EI} \qquad (4\text{-}3)$$

Because the slope of elastic curve is small for actual bridge structures, we can assume $\theta \approx dy / dx$; thus Eq. 4-3 becomes

$$\frac{d^2y}{dx^2} = \frac{M}{EI} \qquad (4\text{-}4)$$

Based on mechanics of materials, we have

$$\frac{dV}{dx} = q \qquad (4\text{-}5)$$

$$\frac{d^2M}{dx^2} = q \qquad (4\text{-}6)$$

Comparing Eq. 4-3 with Eq. 4-5 and then Eq. 4-4 with Eq. 4-6, it can be seen that the relationship between displacements and $\frac{M}{EI}$ is similar to the relationship between the internal force and external loading. If we compare the shear V with the slope, the moment M with the deflection, and the external load q with the M / EI which applies to an imaginary beam as shown in Fig. 4-1c, the real beam deflection and slope can be calculated by determining the bending moment and shear of the imaginary beam. The imaginary beam is called a conjugate beam. The method for determining girder deflections and slopes is called the conjugate-beam method. The boundary conditions at the conjugate-beam supports should be determined based on the corresponding slope and deflection of the real beam at its supports. Table 4-1[4-1] shows the conjugate beam boundary conditions for different support conditions.

The following steps can be used for determining the deflections and slopes of a girder through the conjugate-beam method:

1. Determine the bending moment of the real beam.
2. Develop the conjugate beam with the same beam length as the real beam.
3. Impose boundary conditions on the conjugate beam based on Table 4-1.
4. Apply the imaginary distributive load M / EI to the conjugate beam. If the real beam undergoes an angle of displacement, apply a concentrated shear force equal to the angle at the corresponding point in the conjugate beam. If the real beam undergoes a settlement at a support, apply a concentrated moment equal to the settlement at the conjugate beam support.
5. The slope at a point in the real beam is equal to the shear at the corresponding point in the conjugate beam.
6. The displacement of a point in the real beam is equal to the moment at the corresponding point in the conjugate beam.

TABLE 4-1

Support Conditions in Real Beam and Conjugate Beam

Real Beam	Pin	Roller	Fixed	Free	Internal Roller	Hinge
Conjugate Beam	Pin	Roller	Free	Fixed	Hinge	Internal Roller

4.2.1.2 Method of Virtual Work

The method of virtual work is based on the principle of virtual work and is an energy method that is based on the principles of conservation of energy, which states that the work done by all the external forces acting on a structure is transformed into internal work or strain energy that is developed when the structure deforms. For more complicated loadings or for more complicated structures, it is often convenient to use energy methods in determining structural displacements. In the principle of virtual work, the total virtual work done by external virtual loads through the displacements due to the real loads is equal to the total strain energy developed by the internal virtual forces through the internal deformations caused by the real loads.

Figure 4-2a shows a beam under an arbitrary loading q. As discussed before, the deflections of the beam are primarily caused by its bending moment $M(x)$. To determine the displacement or the slope at an arbitrary point C using the principle of virtual work, the following procedure can be used:

1. To determine the vertical displacement at point C due to external load q, place a unit virtual load $P_0 = 1$ on the beam at point C in the direction of the desired displacement (see Fig. 4-2c). If the slope at point C is to be determined, place a unit virtual bending moment $M_0 = 1$ at the point (see Fig. 4-2d).
2. Calculate the internal virtual moment $m(x)$ due to a unit virtual force P_0 (see Fig. 4-2c) or virtual moment $m_\theta(x)$ due to a unit virtual moment $M_0 = 1$ (see Fig. 4-2d).
3. Calculate the actual bending moment $M(x)$ due to external loading q (see Fig. 4-2b).

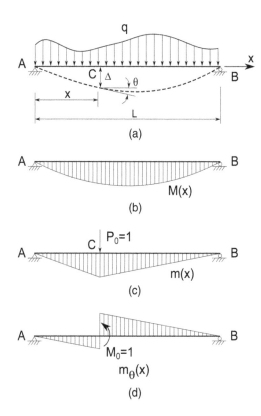

FIGURE 4-2 Deflections and Slopes Determined by the Method of Virtual Work, (a) Beam under an Arbitrary Loading q, (b) Moment $M(x)$ Due to External Load q, (c) Moment $m(x)$ Due to a Unit Vertical Load $P_0 = 1$, (d) Moment $m_\theta(x)$ Due to Unit Moment $M_0 = 1$.

4. Calculate the internal virtual strain energy developed by the actual moment $M(x)$ through the virtual rotation angle $\theta(x)$ in the entire beam.

5. Apply the principle of virtual work to determine the desired displacement or rotation, i.e.,

$$\textit{External Virtual Work = Internal Virtual Strain Energy} \tag{4-7}$$

For example, to determine the vertical displacement Δ at point C, we can use Eq. 4-7 and Eq. 4-3 as

$$1 \times \Delta = \int_0^L m(x)\,d\theta = \int_0^L \frac{m(x)M(x)}{EI}\,dx \tag{4-8a}$$

$$\Delta = \int_0^L \frac{m(x)M(x)}{EI}\,dx \tag{4-8b}$$

Similarly, to determine the slope θ at point C, using Eq. 4-7 we have:

$$1 \times \theta = \int_0^L m_\theta(x)\,d\theta = \int_0^L \frac{m_\theta(x)M(x)}{EI}\,dx \tag{4-9a}$$

$$= \int_0^L \frac{m_\theta(x)M(x)}{EI}\,dx \tag{4-9b}$$

Equations 4-8b and 4-9b are often called Maxwell-Mohr's integral and can be easily integrated using a diagram multiplication method, which is a graphical method that was developed by a Russian engineer named Vereshchagin in 1925[4-2]. The diagram multiplication method is described as follows.

As discussed above, moment $M(x)$, in Eqs. 4-8 and 4-9, represents the bending moment due to an arbitrary external load and may have an arbitrary shape as shown in Fig. 4-3. Moments m(x) and $m_\theta(x)$ in Eqs. 4-8 and 4-9 indicate the moments due to a unit load that are always a linear function or that can be divided into several linear functions as shown in Figs. 4-2c, 4-2d, and 4-3.

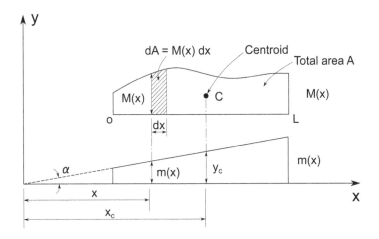

FIGURE 4-3   Integral by Diagram Multiplication Method.

The typical functions $M(x)$ and $m(x)$ are illustrated in Fig. 4-3. Assuming the slope of the function $m(x)$ as α, we have

$$m(x) = x \tan \alpha \tag{4-10}$$

If EI is constant, the Maxwell-Mohr integral can be written as

$$\int_o^L \frac{m(x)M(x)}{EI}dx = \frac{\tan \alpha}{EI}\int_o^L xM(x)dx = \frac{\tan \alpha}{EI}\int_o^L xdA \tag{4-11}$$

The integral $\int_o^L xdA$ represents the static moment of area of the bending moment diagram $M(x)$ and can be expressed by the total area A and the coordinate of the centroid x_c.

That is,

$$\int_o^L xdA = Ax_c \tag{4-12}$$

Because $x_c \tan \alpha = y_c$, we can rewrite Eq. 4-11 as

$$\int_o^L m(x)M(x)dx = Ax_c \tan \alpha = Ay_c \tag{4-13}$$

where

    A = area of bending moment diagram $M(x)$
    y_c = ordinate of $m(x)$ at x_c where centroid of A is located (see Fig. 4-3)

Equation 4-13 indicates that Maxwell-Mohr's integral can be replaced with elementary algebraic expression, i.e., the area of the bending moment diagram $M(x)$ multiplied by the coordinate of $m(x)$ at its centroid location.

4.2.2 Analysis of Indeterminate Bridge Structures by Force Method

The force method has sometimes been referred to as the compatibility method or the method of consistent displacements. This method consists of three steps: First, transfer the statically indeterminate structure into a statically determinate structure by inserting redundant forces. Then determine the redundant forces by the conditions of displacement compatibility. Finally, use the principle of superposition to solve the indeterminate structure.

Figure 4-4a shows a three-span continuous girder with five unknowns of four vertical reactions and one horizontal reaction. Because only three equilibrium equations are available for solving this structure, the girder is indeterminate to the second degree. Consequently, two additional equations are necessary to find a solution. To obtain these equations, we can use the principle of superposition and consider the compatibility of displacement at the two interior supports 1 and 2. This is done by temporary removing "redundant" supports 1 and 2 and replacing them with two unknown reactions X_1 and X_2. Then, the girder becomes a statically determinate structure and is referred to as the primary structure and can be resolved as three statically determined structures (see Figs. 4-3b, c, and d). Based on the condition of displacement compatibility, the displacements at supports 1 and 2 should be equal to zero, i.e.,

$$\Delta_{11} + \Delta_{12} + \Delta_{1P} = 0 \tag{4-14a}$$

$$\Delta_{22} + \Delta_{21} + \Delta_{2P} = 0 \tag{4-14b}$$

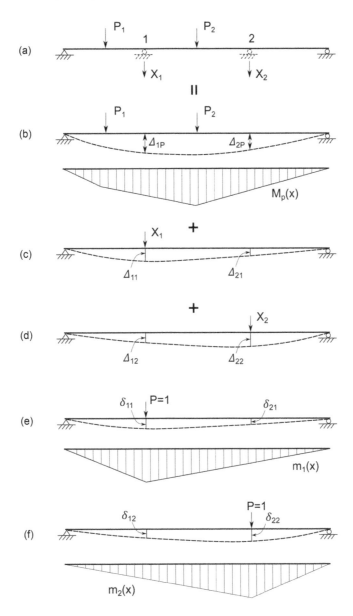

FIGURE 4-4 Analysis of Indeterminate Structures by Force Method, (a) Three-Span Continuous Beam and Primary Structure, (b) Displacement and Moment in the Primary Structure due to External Loads, (c) Displacement of the Primary Structure due to an Unknown Force X_1, (d) Displacement of the Primary Structure Due to an Unknown Force X_2, (e) Displacement and Moment in the Primary Structure Due to a Unit Force at Support 1, (f) Displacement and Moment in the Primary Structure Due to a Unit Force at Support 2.

where

$\Delta_{11}, \Delta_{12}, \Delta_{1P}$ = displacements at support 1 due to X_1, X_2, and external loadings P_1

$\Delta_{22}, \Delta_{21}, \Delta_{2P}$ = displacements at support 2 due to X_2, X_1, and external loadings P_2

The first letter in this double-subscript notation refers to the location where the deflection is specified, and the second letter refers to the force which induces the deflection.

From Eq. 4-8, the displacements due to the unknown forces can be written as

$$\Delta_{11} = X_1 \delta_{11} = X_1 \int \frac{m_1(x)m_1(x)}{EI} dx \tag{4-15a}$$

$$\Delta_{12} = X_2 \delta_{12} = X_2 \int \frac{m_1(x)m_2(x)}{EI} dx \tag{4-15b}$$

$$\Delta_{22} = X_2 \delta_{22} = X_2 \int \frac{m_2(x)m_2(x)}{EI} dx \tag{4-15c}$$

$$\Delta_{21} = X_1 \delta_{21} = X_1 \int \frac{m_1(x)m_2(x)}{EI} dx \tag{4-15d}$$

$$\Delta_{1P} = \int \frac{m_1(x)M_P(x)}{EI} dx \tag{4-15e}$$

$$\Delta_{2P} = \int \frac{m_2(x)M_P(x)}{EI} dx \tag{4-15f}$$

where
   δ_{11}, δ_{21} = displacements at supports 1 and 2 due to unit force applied at support 1 (see Fig. 4-4e)
   δ_{22}, δ_{12} = displacements at supports 2 and 1 due to unit force applied at support 2 (see Fig. 4-4f)
   $M_P(x), m_1(x), m_2(x)$ = moments due to external loadings, unit force at support 1, and unit force at support 2, respectively (see Figs. 4-4b, e, and f)

From Eqs. 4-15b and 4-15d, we can observe that $\delta_{12} = \delta_{21}$. This is true for the general case as

$$\delta_{ij} = \delta_{ji} \tag{4-16}$$

δ_{ij} is also called the flexibility coefficient, which represents the displacement due to a unit force. Substituting Eqs. 4-15a through 4-15f into Eqs. 4-14a and 4-14b yields

$$\delta_{11}X_1 + \delta_{12}X_2 + \Delta_{1P} = 0 \tag{4-17a}$$
$$\delta_{22}X_2 + \delta_{12}X_1 + \Delta_{2P} = 0 \tag{4-17b}$$

Solving the above equations, we can obtain X_1 and X_2. The moment of the continuous girder can be written as

$$M(x) = M_P(x) + X_1 m_1(x) + X_2 m_2(x) \tag{4-18}$$

4.2.3   ANALYSIS OF INDETERMINATE BRIDGE STRUCTURES BY DISPLACEMENT METHOD

The force method described previously uses the forces as unknowns and is easily understood. However, since all the unknowns will involve the equilibrium equations, the force method is often limited to structures that are not highly indeterminate. The displacement method works the opposite way by using displacements as its unknowns. It generally requires less computational work and can be easily programmed on a computer and used to analyze complicated indeterminate structures. Similar to the force method, the displacement method also consists of three steps: First, take displacements at the joints as unknowns. Then, determine the fixed-end forces due to external loads and write the beam end forces in terms of the displacements by assuming that the joints are fixed. Finally, release the fixed ends and use the force equilibrium conditions at the ends to solve the displacements. Once the displacements are obtained, the joint forces can be determined by using the load-displacement relations.

Figure 4-5a shows a three-span continuous beam subjected to two concentrated loads P_1 and P_2. As there are no vertical and horizontal displacements at supports A, B, C, D and there are no bending moments at the ends of the beam, the slopes θ_B and θ_C at joints B and C can be considered as unknowns. The angles are measured in radians, and are positive in the clockwise. Taking a free body at joints B and C separately as shown in Fig. 4-5b, we have

$$M_{1BB} + M_{2BB} + M_{2BC} + F_{MB} = 0 \qquad\qquad (4\text{-}19a)$$

$$M_{2CB} + M_{2CC} + M_{3CC} + F_{MC} = 0 \qquad\qquad (4\text{-}19b)$$

where

F_{MB}, F_{MC} = fixed-end moments at B and C due to external loads by assuming joints B and C
    as fixed, positive clockwise when acting on beam (see Fig. 4-5c and d)

M_{1BB}, M_{2BB}, M_{2BC}, M_{2CB}, M_{2CC}, M_{3CC} = beam end moments at supports due to slopes θ_A and
    θ_B (see Fig. 4-5e to h) by assuming joints B and C as fixed.

The beam end moments are positive when acting in a clockwise direction on the beam. Each of the notations has three subscripts which refer to the span number, the location of the moment, and the location of the slope that induces the moment, respectively.

The fixed-end moments and beam end moments can be easily obtained using the conjugate beam method as described in Section 4.2.1.1. Figure 4-5i shows the conjugate beam corresponding to the real beam shown in Fig. 4-5c. The sum of the moments at end A' of the conjugate beam must be equal to zero, i.e.,

$$\frac{P_1 L}{4EI}\left(\frac{L}{4}\frac{L}{3} + \frac{L}{4}\frac{2L}{3}\right) - \frac{F_{MB}L}{2EI}\frac{2L}{3} = 0$$

Rearranging the above equation yields

$$F_{MB} = \frac{3P_1 L}{16} \qquad\qquad (4\text{-}20)$$

Figure 4-5j illustrates the conjugate beam corresponding to the real beam shown in Fig. 4-5f. From $\sum M_{D'} = 0$, we have

$$M_{3CC} = \frac{3EI}{L}\theta_C \qquad\qquad (4\text{-}21)$$

Figure 4-5k shows the conjugate beam corresponding to the real beam shown in Fig. 4-5g. Taking $\sum M_{B'} = 0$ and $\sum M_{C'} = 0$ individually, we have

$$\frac{M_{2BB}}{EI}\frac{L}{2}\frac{L}{3} - \frac{M_{2CB}}{EI}\frac{L}{2}\frac{2L}{3} = 0$$

$$\frac{M_{2BB}}{EI}\frac{L}{2}\frac{2L}{3} - \frac{M_{2CB}}{EI}\frac{L}{2}\frac{L}{3} - \theta_B L = 0$$

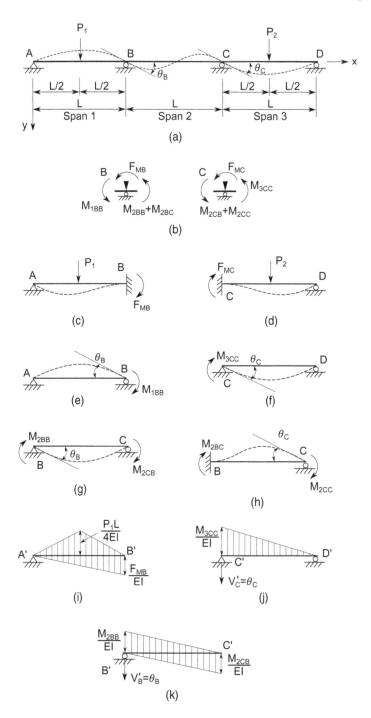

FIGURE 4-5 Illustration of Analyzing a Three-Span Continuous Girder by Displacement Method, (a) Unknown Displacements, (b) Equilibrium at Nodes B and C, (c) Fixed-End Moment at Node B, (d) Fixed-End Moment at Node C, (e) End Moment in Span 1 due to Unknown Slope θ_B, (f) End Moments of Span 3 due to Unknown Slope θ_C, (g) End Moments in Span 2 due to Unknown Slope θ_B, (h) End Moment in Span 2 due to Unknown Slope θ_C, (i) Conjugate Beam for Determining Fixed-End Moment at Node B, F_{MB}, (j) Conjugate Beam for Determining End Moment M_{3CC} at Node C in Span 3, (k) Conjugate-Beam for Determining End Moments M_{2BB} and M_{2CB} in Span 2.

Solving the above equations, we can obtain

$$M_{2BB} = \frac{4EI}{L}\theta_B \tag{4-22}$$

$$M_{2CB} = \frac{2EI}{L}\theta_B \tag{4-23}$$

In a similar manner, we obtain

$$F_{MC} = \frac{-3P_2L}{16} \tag{4-24}$$

$$M_{1BB} = \frac{3EI}{L}\theta_B \tag{4-25}$$

$$M_{2BC} = \frac{2EI}{L}\theta_C \tag{4-26}$$

$$M_{2CC} = \frac{4EI}{L}\theta_C \tag{4-27}$$

Substituting Eqs. 4-20 to 4-27 into Eq. 4-19, we obtain:

$$7i\theta_B + 2i\theta_c = -\frac{3P_1L}{16} \tag{4-28a}$$

$$2i\theta_B + 7i\theta_c = \frac{3P_2L}{16} \tag{4-28b}$$

where $i = EI / L$ = span stiffness coefficient.
    Writing Eqs. 4-28a and 4-28b in matrix form yields

$$[K]\{\delta\} = \{P\} \tag{4-28c}$$

where

$$[K] = \begin{bmatrix} 7i & 2i \\ 2i & 7i \end{bmatrix} = \text{structural stiffness matrix}$$

$$\{\delta\} = \begin{Bmatrix} \theta_B \\ \theta_C \end{Bmatrix} = \text{displacement matrix}$$

$$\{P\} = \begin{Bmatrix} -\dfrac{3P_1L}{16} \\ \dfrac{3P_2L}{16} \end{Bmatrix} = \text{load matrix}$$

Solving Eq. 4-28c yields

$$\theta_B = \frac{-L^2(7P_1 + 2P_2)}{240EI}$$

$$\theta_C = \frac{L^2(2P_1 + 7P_2)}{240EI}$$

After the unknown angular displacements are obtained, the internal forces can be easily determined by taking a free body for each of the spans, which are simply supported beams, and applying the end moments calculated by Eqs. 4-21 to 4-27 and the external loadings.

From Eq. 4-28c, it can be seen that the structural stiffness matrix is independent of the external loadings. For any other loading cases, it is only necessary to change the load matrix based on the actual loading conditions. For bridge engineers, it is not necessary to develop the equations of fixed-end moments each time. Table 4-2 provides the fixed-end moments for some frequently used loading cases for the bridge engineers' use.

4.2.4   Analysis of Indeterminate Bridge Structures by Moment Distribution Method

4.2.4.1   Introduction

The moment distribution method is a displacement method of structural analysis that is essentially an iteration method that can be carried out to any desired degree of accuracy. This method is easy to follow and to remember. Many engineers still use this method to analyze some relatively simple structures, such as segmental transverse analysis, though there are many computer programs

---

TABLE 4-2
Typical Fixed-End Moments[4-3]

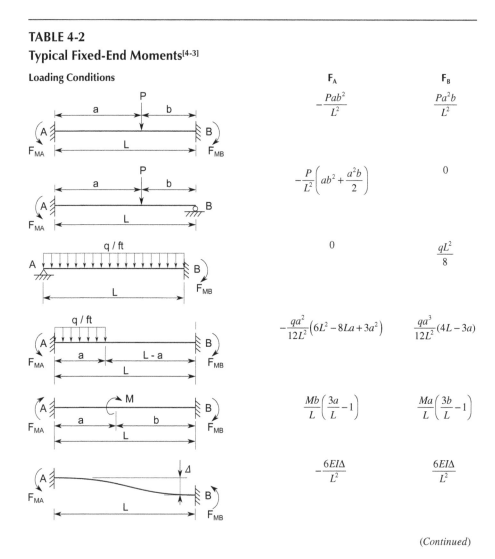

Loading Conditions	F_A	F_B
	$-\dfrac{Pab^2}{L^2}$	$\dfrac{Pa^2b}{L^2}$
	$-\dfrac{P}{L^2}\left(ab^2+\dfrac{a^2b}{2}\right)$	0
	0	$\dfrac{qL^2}{8}$
	$-\dfrac{qa^2}{12L^2}\left(6L^2-8La+3a^2\right)$	$\dfrac{qa^3}{12L^2}(4L-3a)$
	$\dfrac{Mb}{L}\left(\dfrac{3a}{L}-1\right)$	$\dfrac{Ma}{L}\left(\dfrac{3b}{L}-1\right)$
	$-\dfrac{6EI\Delta}{L^2}$	$\dfrac{6EI\Delta}{L^2}$

(Continued)

TABLE 4-2 *(Continued)*
Typical Fixed-End Moments[4-3]

Loading Conditions	F_A	F_B

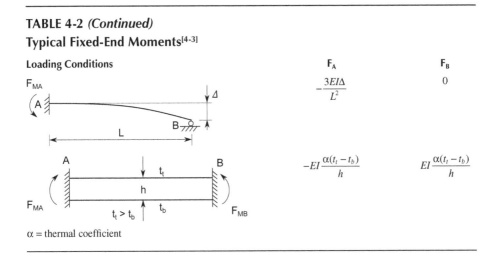

| | $-\dfrac{3EI\Delta}{L^2}$ | 0 |
| | $-EI\dfrac{\alpha(t_t - t_b)}{h}$ | $EI\dfrac{\alpha(t_t - t_b)}{h}$ |

α = thermal coefficient

available. Before we explain the moment distribution method, two important concepts of distribution factor and carryover factor should be discussed.

4.2.4.2 Distribution Factor (DF)

Figure 4-6 shows a frame structure with an external loading P. Using the displacement method to analyze this structure, we take the angular displacement θ_B at joint B as an unknown. First add an additional restraint at joint B and determine the fixed-end moment F_{MB} (Fig. 4-6a). Then, release the added restraint and apply the fixed-end moment to the actual structure. The fixed-end moment causes joint B to rotate by an angle θ_B (see Fig. 4-6b). The fixed-end moment F_{MB} should be balanced by the end moments M_{BA}, M_{BC}, and M_{BD} (see Fig. 4-6b). From Eqs. 4-24, 4-22, and 4-21, we have

$$F_{MB} = \frac{-3PL_2}{16} \tag{4-29}$$

$$M_{BA} = \frac{4EI}{L_1}\theta_B = k_1\theta_B \tag{4-30a}$$

$$M_{BC} = \frac{3EI}{L_3}\theta_B = k_2\theta_B \tag{4-30b}$$

$$M_{BD} = \frac{4EI}{L_2}\theta_B = k_3\theta_B \tag{4-30c}$$

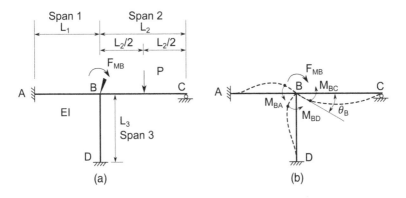

FIGURE 4-6 Concept of Moment Distribution, (a) Fixed-End Moment, (b) Distribution of Fixed-End Moment.

where k_1, k_2, k_3 = span stiffness coefficients corresponding to spans AB, BC, and BD, respectively. They are only related to the beam stiffness EI, span length, and beam end support conditions. They are not related to external loadings and the type of the structures and can be determined from the equations given in Table 4-2.

As joint B is in equilibrium, we have

$$M_{BA} + M_{BD} + M_{BC} + F_{MB} = 0 \tag{4-31}$$

Substituting Eq. 4-29 and Eqs. 4-30a to 4-30c into Eq. 4-31 yields

$$\theta_B = \frac{1}{k_1 + k_2 + k_3}\left(\frac{3PL_2}{16}\right)$$

Substituting the above equation into Eqs. 4-30a to 4-30c yields

$$M_{BA} = \frac{k_1}{k_1 + k_2 + k_3}\left(\frac{3PL_2}{16}\right) \tag{4-30e}$$

$$M_{BC} = \frac{k_2}{k_1 + k_2 + k_3}\left(\frac{3PL_2}{16}\right) \tag{4-30f}$$

$$M_{BD} = \frac{k_3}{k_1 + k_2 + k_3}\left(\frac{3PL_2}{16}\right) \tag{4-30g}$$

Equations 4-30e to 4-30g can be written as a general form

$$M_i = \frac{k_i}{\Sigma k_i}\, F_M = DF_i F_M \tag{4-32}$$

where

M_i = span i end moment, positive clockwise
F_M = fixed-end moment, positive clockwise
DF_i = $\frac{k_i}{\Sigma k_i}$ = distribution factor for span i. The distribution factor indicates the fraction of the fixed moment taken by each of the spans connected with the joint concerned.
k_i = span stiffness coefficient for span i.
Σk_i = sum of span stiffness coefficients for all spans connected to joint concerned

4.2.4.3 Carryover Factor

Consider the beam in Fig. 4-7a. From Eqs. 4-22 and 4-23, we can obtain

$$M_B = C_{AB}M_A = \frac{1}{2}M_A \tag{4-33}$$

where

M_A, M_B = beam end moment at ends A and B, respectively
$C_{AB} = \frac{1}{2}$ = carryover factor, which represents the fraction of M_A that is "carried over" from the pin end A to the fixed end B

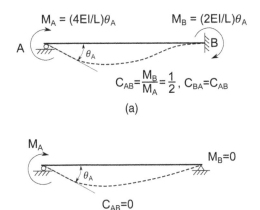

FIGURE 4-7   Carryover Factors, (a) Fixed at Support B, (b) Hinge at Support B.

Apparently, if end B is a hinge (see Fig. 4-6b), $C_{AB} = 0$. If end A is fixed, the applied M_A will not be transferred to anywhere and $C_{AB} = 0$.

4.2.4.4   Procedure for Moment Distribution

Moment distribution is based on the principle of successively locking (fixing) and unlocking (unfixing) the joints of a structure to allow the moments at the joints to be distributed and balanced. For an easy understanding, let's consider a three-span continuous beam with a constant moment of inertia I and modulus of elasticity E and having the same span length L with loadings as shown in Fig. 4-8.

The following procedures, as shown in Fig. 4-8, can be used to solve the internal forces of the continuous beam:

Step 1: Determine the distribution factors based on Eq. 4-32 (see Fig. 4-8):

$$DF_{BC} = DF_{CB} = \frac{4}{3+4} = \frac{4}{7}$$

$$DF_{BA} = DF_{CD} = \frac{3}{3+4} = \frac{3}{7}$$

For hinge supports A and D: $DF_{AB} = DF_{DC} = 0$.

Step 2: Determine the carryover factors based on Section 4.2.4.3:

$$C_{BC} = C_{CB} = \frac{1}{2}$$

$$C_{AB} = C_{BA} = C_{CD} = C_{DC} = 0$$

Step 3: Locking joints B and C, determine fixed-end moments F_{MB} and F_{MC}, based on the equations shown in Table 4-2:

$$F_{MB} = \frac{qL^2}{8}$$

$$F_{MC} = -\frac{3PL}{16}$$

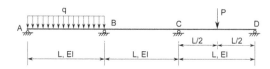

Analytical Steps	Joint	A	B		C		D
	Member	AB	BA	BC	CB	CD	DC
1	DF	0	3/7	4/7	4/7	3/7	0
2	C	0	0	1/2	1/2	0	0
3	FM	0	$\dfrac{qL^2}{8}$	0	0	$\dfrac{-3PL}{16}$	0
4	Dist	0	$\dfrac{-3qL^2}{56}$	$\dfrac{-qL^2}{14}$	$\dfrac{3PL}{28}$	$\dfrac{9PL}{112}$	0
5	Carry	0	0	$\dfrac{3PL}{56}$	$\dfrac{-qL^2}{28}$	0	0
6	Dist		$\dfrac{-9PL}{392}$	$\dfrac{-3PL}{98}$	$\dfrac{qL^2}{49}$	$\dfrac{3qL^2}{196}$	
7	ΣM	0	$\dfrac{qL^2}{14} - \dfrac{9PL}{392}$	$\dfrac{-qL^2}{14} + \dfrac{9PL}{392}$	$\dfrac{3PL}{28} - \dfrac{3qL^2}{196}$	$\dfrac{-3PL}{28} + \dfrac{3qL^2}{196}$	0
8	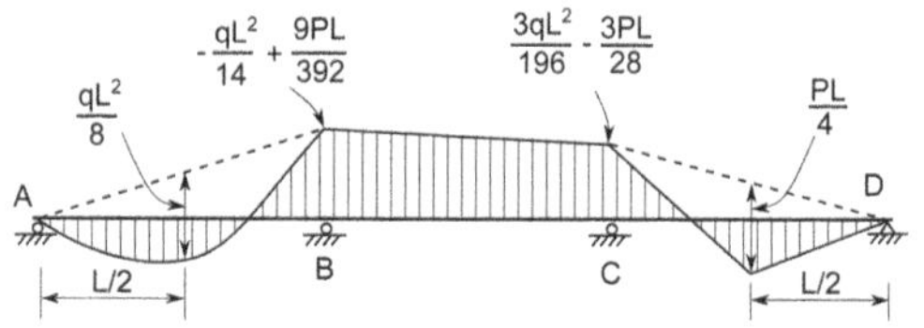						

FIGURE 4-8   Analysis of Continuous Girders by Moment Distribution Method.

Step 4: Unlocking joints B and C, distribute the fixed-end moment to the related members:

$$M_{BA} = DF_{BA} \times (-F_{MB}) = -\frac{3qL^2}{56}$$

$$M_{BC} = DF_{BC} \times (-F_{MB}) = -\frac{qL^2}{14}$$

$$M_{CB} = DF_{CB} \times (-F_{MC}) = \frac{3PL}{28}$$

$$M_{CD} = DF_{CD} \times (-F_{MC}) = \frac{9PL}{112}$$

Step 5: Locking joints B and C again, calculate the carryover moments that are the new fixed moments at joints B and C:

$$F_{MBN} = C_{CB} \times M_{CB} = \frac{3PL}{56}$$

$$F_{MCN} = C_{BC} \times M_{BC} = -\frac{qL^2}{28}$$

Step 6: Unlocking joints B and C again, distribute the new fixed-end moment from the carryover moments to the related members, similar to step 4.

Step 7: Sum the member end moments, including the fixed-end moments:

$$M_{AB}^B = \frac{qL^2}{8} - \frac{3qL^2}{56} - \frac{9PL}{392} = \frac{qL^2}{14} - \frac{9PL}{392}$$

$$M_{BC}^B = -\frac{qL^2}{14} + \frac{3PL}{56} - \frac{3PL}{98} = -\frac{qL^2}{14} + \frac{9PL}{392}$$

$$M_{CB}^C = \frac{3PL}{28} - \frac{qL^2}{28} + \frac{qL^2}{49} = \frac{3PL}{28} - \frac{3qL^2}{196}$$

$$M_{DC}^C = -\frac{3PL}{16} + \frac{9PL}{112} + \frac{3qL^2}{196} = -\frac{3PL}{28} + \frac{3qL^2}{196}$$

If $M_{AB}^B + M_{BC}^B$ and $M_{CB}^C + M_{DC}^C$ are equal to or close to zero, finish the calculations of the member end moments. Otherwise continue steps 5 and 6 until the results meet the required accuracy.

Step 8: Draw the moment distribution diagram as shown in Fig. 4-8.

To provide a more efficient design, girders used for long-span bridges and box girder slabs are designed to be nonprismatic, that is, to have a variable moment of inertia. The equations shown in Table 4-2 may not be used for nonprismatic elements. New equations for determining the beam end moments, fixed-end moments, distribution factors, and carryover factors can be developed based on the conjugate method or the principle of virtual work as discussed in Sections 4.2.1.1 and 4.2.1.2, respectively. Figure 4-9 shows the beam end moments, carryover factor, and fixed-end moments for a special tapered beam with uniform thickness in the transverse direction.

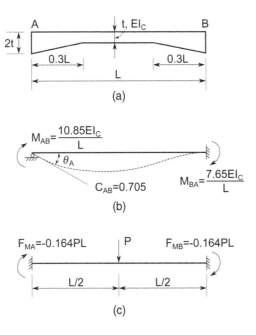

FIGURE 4-9   Beam End Moments, Carryover Factor, and Fixed-End Moment of a Tapered Girder, (a) Typical Tapered Girder, (b) End Moments and Carryover Factor, (c) Fixed End Moments.

4.2.5 Bridge Analysis by Finite-Element Methods

4.2.5.1 Introduction

The finite-element method for analyzing bridge structures is essentially an approximate numerical method. This method involves first modeling the structure using small interconnected elements called finite elements. Every interconnected element is linked to every other element through common interfaces, often called nodes. The unknowns at each of the nodes can be either taken as forces or displacements; however, to simplify programming and computation, the finite-element method is commonly based on the displacement method, i.e., using displacements as the unknowns. Then develop the relations between the nodal forces and nodal displacements by assuming simple approximate displacement functions to replace the actual structural displacements within each of the finite elements. Finally, the force equilibrium conditions at the nodes are used to set up force equilibrium equations to find the displacements and internal force of the structure. For practicing engineers, it may not be necessary to know the detailed theory of the finite-element method and how the computer programs used were developed based on its theory. It is definitely beneficial for the engineers to know the basic procedures of the finite-element method to ensure their finite-element model of the structure is properly developed.

4.2.5.2 Procedure for Finite-Element Method (FEM)

Take a continuous beam as shown in Fig. 4-10a, for example, to describe the basic theory and procedure of the finite-element method.

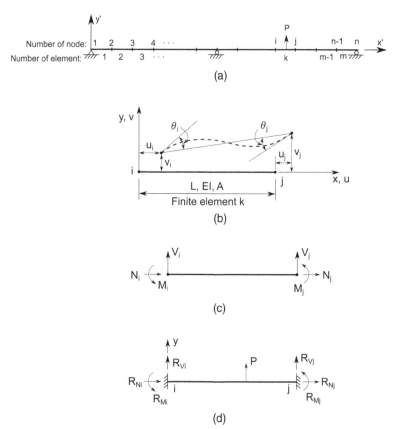

FIGURE 4-10 Bridge Analysis by Two-Dimensional Finite-Element Method, (a) Numbering of Finite Elements and Nodes, (b) Nodal Displacements of Finite-Element, (c) Nodal Forces of Element, (d) Equivalent Nodal Loadings due to External Loading.

Step 1: Discretize the continuous girder into m finite elements. To ensure the accuracy of the analytical results, normally the number of finite elements in each span is not less than 10, depending on the degree of variation of the girder cross section and loading conditions, as the cross section within an element is generally assumed to be uniform (see Fig. 4-10b). The node numbers on the left and right sides of element k are denoted as i and j, respectively.

Step 2: Select displacement functions within the element, and express the functions in terms of the nodal parameters.

Polynomials are frequently used as functions because they are convenient to work within the finite-element formulation. Assume the strain along the beam axial direction is constant; then the displacement of the beam along the x-direction can be assumed to be a linear function, i.e.,

$$u(x) = a_1 + a_2 x \tag{4-34}$$

From the strength of material, we have

$$\frac{d^3 v}{dx^3} = \frac{V}{EI} \tag{4-35}$$

Thus, to ensure the required shear force V can be determined by the finite-element method, the beam displacement function in the y-direction should be at least a cubic polynomial, i.e.,

$$v(x) = b_1 + b_2 x + b_3 x^2 + b_4 x^3 \tag{4-36}$$

There are a total of six unknowns in the displacement functions $u(x)$ and $v(x)$. Therefore, we need three boundary conditions at each end of the finite-element. Thus, we choose three nodal parameters as linear displacements in the x- and y-directions: u and v, respectively, and the angular displacement θ at each node. The node parameters of the finite-element can be written in matrix form as

$$\{\delta_e\} = \left\{ \begin{array}{c} \{\delta_{ei}\} \\ \{\delta_{ej}\} \end{array} \right\} \tag{4-37}$$

where
$\{\delta_e\}$ = displacement parameter vector of finite-element

$\{\delta_{ei}\} = \left\{ \begin{array}{c} u_i \\ v_i \\ \theta_i \end{array} \right\}$ = displacement parameter vector at node i (see Fig. 4-10b)

$\{\delta_{ej}\} = \left\{ \begin{array}{c} u_j \\ v_j \\ \theta_j \end{array} \right\}$ = displacement parameter vector at node j (see Fig. 4-10b)

Substituting the following boundary conditions of the finite-element into Eqs. 4-34 and 4-36, respectively,

$$x = 0: u = u_i, v = v_i, \frac{dv}{dx} = \theta_i$$

$$x = L: u = u_j, v = v_j, \frac{dv}{dx} = \theta_j$$

we can obtain:

$$\left\{ \begin{matrix} u \\ v \end{matrix} \right\} = [N]\{\delta_e\}$$

(4-38)

where

$$[N] = \text{shape function of element} = \begin{bmatrix} N_1 & 0 & 0 & N_2 & 0 & 0 \\ 0 & N_3 & N_4 & 0 & N_5 & N_6 \end{bmatrix}$$

$$[N_1] = 1 - \bar{x}$$

$$[N_2] = \bar{x}$$

$$[N_3] = 1 - 3(\bar{x})^2 + 2(\bar{x})^3$$

$$[N_4] = -l(1 - \bar{x})^2 \bar{x}$$

$$[N_5] = 3(\bar{x})^2 - 2(\bar{x})^3$$

$$[N_6] = l(1 - \bar{x})\bar{x}^2$$

$$\bar{x} = \frac{x}{L}$$

Step 3: Define the relationships between generalized strain and displacements.

As we treat the continuous beam as a two-dimensional structure, there are three generalized strains necessary for determining the required internal forces, i.e.,

$$\{\varepsilon\} = \left\{ \begin{matrix} \varepsilon_x \\ k_z' \\ k_z \end{matrix} \right\} = \left[\frac{du}{dx} \ \frac{d^3v}{dx^3} \ \frac{d^2v}{dx^2} \right]^T = [B]\{\delta_e\}$$

(4-39)

where

$\varepsilon_x \ = \ \frac{du}{dx} = $ uniform longitudinal strain

$k_z \ = \ \frac{d^2v}{dx^2} = $ curvature of beam deflection

$k_z' \ = \ \frac{d^3v}{dx^3} = $ first derivative of curvature

$$[B] = \begin{bmatrix} \dfrac{dN_1}{dx} & 0 & 0 & \dfrac{dN_2}{dx} & 0 & 0 \\ 0 & \dfrac{d^3N_3}{dx^3} & \dfrac{d^3N_4}{dx^3} & 0 & \dfrac{d^3N_5}{dx^3} & \dfrac{d^3N_6}{dx^3} \\ 0 & \dfrac{d^2N_3}{dx^2} & \dfrac{d^2N_4}{dx^2} & 0 & \dfrac{d^2N_5}{dx^2} & \dfrac{d^2N_6}{dx^2} \end{bmatrix} = \text{generalized strain matrix}$$

Step 4: Establish the relationships between the generalized forces of the finite-element and the generalized strains.

From Hooke's law and Eqs. 4-4 and 4-35, it is easy to write the generalized forces as

$$\{F\} = \left\{ \begin{matrix} N \\ V \\ M \end{matrix} \right\} = [D]\{\varepsilon\}$$

(4-40)

where

$$[D] \quad = \begin{bmatrix} EA & 0 & 0 \\ 0 & EI & 0 \\ 0 & 0 & EI \end{bmatrix} = \text{elastic matrix}$$

N, V, M = axial force, shear force, and moment, respectively.

Step 5: Derive the element stiffness matrix.

The nodal forces (see Fig. 4-10c) at the ends of the finite-element can be written as

$$\{F_e\} = \begin{bmatrix} N_i & V_i & M_i & N_j & V_j & M_j \end{bmatrix}^T \tag{4-41}$$

Assuming the finite-element has undergone a virtual displacement δuv^*; then the nodes of the finite-element have a virtual displacement $\delta\{\delta_e\}^*$ and the generalized strain in the finite-element has a virtual strain $\delta\{\varepsilon_e\}^*$. The virtual energy within the element due to the virtual generalized strain can be written as

$$\delta U^* = \int \delta\{\varepsilon\}^{T*}\{F\}dx = \delta\{\delta_e\}^{T*}\int [B]^T[D][B]dx\{\delta_e\} \tag{4-42}$$

The virtual work done by the nodal forces can be written as

$$\delta W^* = \delta\{\delta_e\}^{T*}\{F_e\} \tag{4-43}$$

Because the finite-element is in equilibrium, $\delta U^* = \delta W^*$, thus we have

$$\{F_e\} = \int [B]^T[D][B]dx\{\delta_e\} = [k]\{\delta_e\} \tag{4-44}$$

where

$$[k] = \int [B]^T[D][B]dx = \text{stiffness matrix of finite element} \tag{4-45}$$

Substituting matrixes $[D]$ and $[B]$ into Eq. 4-45 yields

$$[k] = \begin{bmatrix} k_{ii} & k_{ij} \\ k_{ji} & k_{jj} \end{bmatrix} \tag{4-46}$$

$$[k_{ii}] = \begin{bmatrix} \dfrac{EA}{L} & 0 & 0 \\ 0 & \dfrac{12EI}{L^3} & \dfrac{6EI}{L^2} \\ 0 & \dfrac{6EI}{L^2} & \dfrac{4EI}{L} \end{bmatrix}$$

$$[k_{jj}] = \begin{bmatrix} \dfrac{EA}{L} & 0 & 0 \\ 0 & \dfrac{12EI}{L^3} & -\dfrac{6EI}{L^2} \\ 0 & -\dfrac{6EI}{L^2} & \dfrac{4EI}{L} \end{bmatrix}$$

$$\left[k_{ij}\right]=\left[k_{ji}\right]=\begin{bmatrix} -\dfrac{EA}{L} & 0 & 0 \\[2mm] 0 & -\dfrac{12EI}{L^3} & -\dfrac{6EI}{L^2} \\[2mm] 0 & \dfrac{6EI}{L^2} & \dfrac{2EI}{L} \end{bmatrix}$$

The stiffness matrix of a finite-element represents the forces induced by unit displacements at its ends and is not related to external loadings.

Step 6: Determine equivalent nodal loadings due to external loadings within the finite-element.

The equivalent nodal loadings of the finite-element (see Fig. 4-10d) can be easily obtained using the conjugate-beam method and written as

$$\{R_e\}=\begin{Bmatrix} R_{Ni} \\ R_{Vi} \\ R_{Mi} \\ R_{Nj} \\ R_{Vj} \\ R_{Mj} \end{Bmatrix}=\begin{Bmatrix} 0 \\[2mm] P\left(1+\dfrac{2a}{L}\right)\left(1-\dfrac{a}{L}\right)^2 \\[3mm] Pa\left(1-\dfrac{a}{L}\right)^2 \\[3mm] 0 \\[1mm] P-R_{Ni} \\[2mm] -Pa^2\dfrac{(L-a)}{L^2} \end{Bmatrix} \tag{4-47}$$

where a = distance from left end of element to the loading P (see Table 4-2).

Step 7: Coordinate system transformation.

Normally the local coordinate systems of the finite elements in arch bridges, curved girder bridges, and cable-stayed bridges are different from the global coordinate system of the structure. The stiffness matrix, the nodal forces, and loadings of each of the finite elements should be transformed to the global coordinate system before assembling the global equations of equilibrium. Assuming the angle between the global coordinate system $\bar{x}-\bar{o}-\bar{y}$ and local coordinate system $x-o-y$ is denoted as α (see Fig. 4-11); then, we have

$$\{\delta_e\}=[L]\{\bar{\delta}_e\} \tag{4-48}$$

$\{\bar{\delta}_e\}$ = nodal displacements in global coordinate system

$$[L]=\begin{bmatrix} L_1 & 0 \\ 0 & L_1 \end{bmatrix} \tag{4-49}$$

$$[L_1]=\begin{bmatrix} \cos\alpha & \sin\alpha & 0 \\ -\sin\alpha & \cos\alpha & 0 \\ 0 & 0 & 1 \end{bmatrix}$$

$[L]$ is called the coordinate transformation matrix.

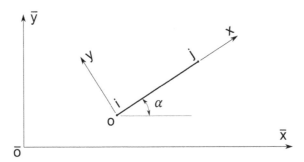

FIGURE 4-11   Coordinate Transformation.

Substituting Eq. 4-48 into Eqs. 4-42 and 4-43, we can obtain the stiffness matrix of the finite-element in the global coordinate system as

$$\left[\bar{k}\right]=[L]^{T}[k][L] \tag{4-50}$$

Step 8: Assemble global equations of equilibrium,
The forces at each of the finite-element nodes should be in equilibrium. Equating the total nodal forces at each of the nodes to the total equivalent nodal loadings yields the structural global or total equations of equilibrium:

$$[K]\{\delta\}+\{R\}=0 \tag{4-51}$$

where
$\{\delta\}$ = global displacement vector with 3n unknown nodal degrees of freedom
$[K]$ = global stiffness matrix
$\{R\}$ = global nodal loading vector

Step 9: Introduce boundary conditions.
A simple way to impose the structural boundary conditions on Eq. 4-51 is to eliminate the equations corresponding to the displacements that are equal to zero.
Step 10: Solve Eq. 4-51 and obtain the displacements.
Step 11: From Eq. 4-44, we can obtain the end reactions of each of the finite elements. Imposing the end reactions and the related external loadings on the elements that are simply supported, we can obtain the internal forces of the elements at any locations.

4.2.5.3   Common Finite Elements Used in Segmental Bridge Analysis
The two-dimensional beam element described in Section 4.2.5.2 is simple and can often yield enough accuracy for the design of straight bridges. However, the two-dimensional finite-element model cannot account for the effect of the structural torsion and other finite elements should be used to predicate more accurate structural behaviors.

4.2.5.3.1   Plane Grid Element
The plane grid element is widely used in bridge structural modeling as it can account for the effect of torsion with the same number of nodal parameters as the plane beam element (see Fig. 4-12). Its nodal parameter vector can be written as

$$\{\delta_{e}\}=\left\{\begin{array}{c}\{\delta_{ei}\}\\\{\delta_{ej}\}\end{array}\right\} \tag{4-52}$$

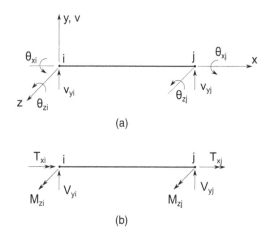

FIGURE 4-12 Plane Grid Element, (a) Element Displacements, (b) Element Nodal Forces.

$$\{\delta_{ei}\} = \begin{bmatrix} v_{yi} & \theta_{xi} & \theta_{zi} \end{bmatrix}^T$$

$$\{\delta_{ej}\} = \begin{bmatrix} v_{yj} & \theta_{xj} & \theta_{zj} \end{bmatrix}^T$$

i, j = left and right ends of finite element
v = linear displacements in y-directions (see Fig. 4-12)
θ_x, θ_z = angular displacements about x-axis, z-axis, respectively

Following the same procedure described in Section 4.2.5.2 and assuming the $v(x)$ and $\theta(x)$ as a three-order parabola, it is not difficult to derive the stiffness matrix of plane grid beam element[4-4,4-5].

The nodal forces (see Fig. 4-12b) at the ends of the finite-element can be written as

$$\{F_e\} = \begin{Bmatrix} \{F_{ei}\} \\ \{F_{ej}\} \end{Bmatrix} \tag{4-53}$$

where

$$\{F_{ei}\} = \begin{bmatrix} V_{yi} & T_{xi} & M_{zi} \end{bmatrix}^T$$

$$\{F_{ej}\} = \begin{bmatrix} V_{yj} & T_{xj} & M_{zj} \end{bmatrix}^T$$

V_y = shear forces along y-direction
T_x = pure torsion moment
M_z = bending moments about z-axis

4.2.5.3.2 Three-Dimensional Beam Element

Figure 4-13 shows a three-dimensional beam element. There are six nodal parameters at each of the ends of the finite-element. They can be written in matrix form as

$$\{\delta_e\} = \begin{Bmatrix} \{\delta_{ei}\} \\ \{\delta_{ej}\} \end{Bmatrix} \tag{4-54}$$

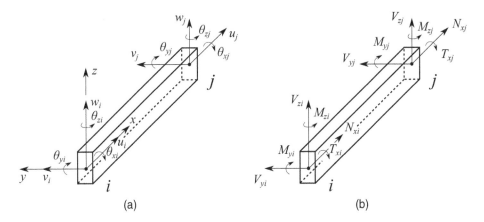

FIGURE 4-13   Three-Dimensional Beam Element, (a) Nodal Displacements, (b) Nodal Forces.

where

$$\{\delta_{ei}\} = \left[\begin{array}{cccccc} u_i & v_i & w_i & \theta_{xi} & \theta_{yi} & \theta_{zi} \end{array}\right]^T$$

$$\{\delta_{ej}\} = \left[\begin{array}{cccccc} u_j & v_j & w_j & \theta_{xj} & \theta_{yj} & \theta_{zj} \end{array}\right]^T$$

i, j = left and right ends of finite-element
u, v, w = linear displacements in x-, y-, z-directions respectively (see Fig. 4-13a)
$\theta_x, \theta_y, \theta_z$ = angular displacements about x-axis, y-axis, z-axis, respectively

Following the same procedure described in Section 4.2.5.2 and assuming the $u(x)$ and $\theta_x(x)$ as linear functions and $v(x)$ and $w(x)$ as third-order parabolas, it is not difficult to derive the stiffness matrix of three-dimensional beam element [4-3].

The nodal forces (see Fig. 4-13b) at the ends of the finite-element can be written as

$$\{F_e\} = \left\{\begin{array}{c} \{F_{ei}\} \\ \{F_{ej}\} \end{array}\right\} \tag{4-55}$$

where

$$\{F_{ei}\} = \left[\begin{array}{cccccc} N_{xi} & V_{yi} & V_{zi} & T_{xi} & M_{yi} & M_{zi} \end{array}\right]^T$$

$$\{F_{ej}\} = \left[\begin{array}{cccccc} N_{xj} & V_{yj} & V_{zj} & T_{xj} & M_{yj} & M_{zj} \end{array}\right]^T$$

N_x = axial force along x-axis
V_y, V_z = shear forces along y- and z-directions, respectively
T_x = pure torsion moment
M_y, M_z = bending moments about y-axis and z-axis, respectively

It should be mentioned that the cross section of the beam element is treated as solid. As most concrete segmental bridges have closed sections and relatively strong distortional rigidity, the three-dimensional beam model can normally provide enough accuracy for most practical bridge designs, though the warping and distortional effects have been neglected in the three-dimensional beam model.

4.2.5.3.3 Three-Dimensional Thin-Wall Element

For large box girders or open sections, especially for curved girder bridges, the warping and distortional torsions may have a significant effect on the bridge internal force distribution and should be considered in bridge design. As discussed in Chapter 3, the warping torsion moment is related to the second derivative of the torsion angle. Thus, comparing the nodal parameters in the three-dimensional beam element, we need at least nine nodal parameters at each of the nodes of the three-dimensional thin-wall finite-element to consider the warping and distortional effect, i.e.,

$$\{\delta_e\} = \begin{Bmatrix} \{\delta_{ei}\} \\ \{\delta_{ej}\} \end{Bmatrix} \tag{4-56}$$

where

$$\{\delta_{ei}\} = \begin{bmatrix} u_i & v_i & w_i & \theta_{xi} & \theta_{yi} & \theta_{zi} & \theta'_{xi} & \tilde{\theta}_{xi} & \tilde{\theta}'_{xi} \end{bmatrix}^T$$

$$\{\delta_{ej}\} = \begin{bmatrix} u_j & v_j & w_j & \theta_{xj} & \theta_{yj} & \theta_{zj} & \theta'_{xj} & \tilde{\theta}_{xj} & \tilde{\theta}'_{xj} \end{bmatrix}^T$$

i, j = left and right ends of finite-element
u, v, w = linear displacements in x-, y-, z-directions, respectively (see Fig 4-14a)
θ_x = torsion angle about torsion center
θ_y, θ_z = angular displacements about y-axis, z-axis, respectively
$\tilde{\theta}_x$ = distortional angle
$\theta'_x, \tilde{\theta}'_x$ = the first derivatives of torsion and distortional angles

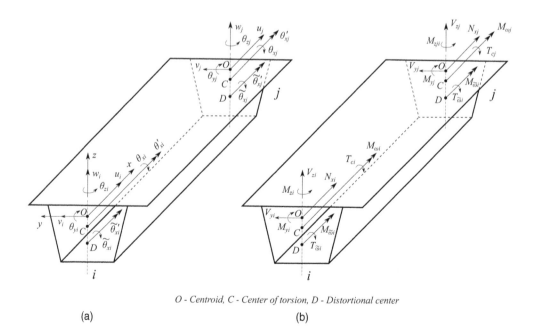

O - Centroid, C - Center of torsion, D - Distortional center

(a)   (b)

FIGURE 4-14 Three-Dimensional Thin-Wall Finite-Element, (a) Nodal Displacements, (b) Nodal Forces.

Using the same procedures described in Section 4.2.5.2 and assuming the $u(x)$ as a linear function and $v(x)$, $w(x)$, $\theta(x)$, $\tilde{\theta}(x)$ as third-order parabolas, we can obtain the stiffness matrix of the three-dimensional thin-wall finite-element[4-6 to 4-8].

The nodal forces (see Fig. 4-14b) at the ends of the finite-element can be written as

$$\{F_e\} = \left\{ \begin{matrix} \{F_{ei}\} \\ \{F_{ej}\} \end{matrix} \right\}$$ (4-57)

where

$$\{F_{ei}\} = \begin{bmatrix} N_{xi} & V_{yi} & V_{zi} & T_{ci} & M_{yi} & M_{zi} & M_{\omega i} & T_{\tilde{\omega} i} & M_{\tilde{\omega} i} \end{bmatrix}^T$$

$$\{F_{ej}\} = \begin{bmatrix} N_{xj} & V_{yj} & V_{zj} & T_{cj} & M_{yj} & M_{zj} & M_{\omega j} & T_{\tilde{\omega} j} & M_{\tilde{\omega} j} \end{bmatrix}^T$$

N_x = axial force along x-axis
V_y, V_z = shear forces along y- and z-directions, respectively
$T_c, T_{\tilde{\omega}}$ = pure torsion and distortional warping moments
M_y, M_z = bending moments about y-axis and z-axis, individually
$M_\omega, M_{\tilde{\omega}}$ = torsion and distortional warping bi-moments

4.2.5.3.4   Shell-Plate Finite-Element

Most of the current commercially available computer programs have a choice to select shell-plate finite elements to model segmental bridges. Normally, the bridge model using shell-plate finite elements can provide more accurate analytical results. However, this type of bridge model involves a large number of finite elements and huge output data, which are often more difficult to be interpreted and process for design.

Figure 4-15a shows a four-node quadrilateral shell-plate finite-element that has four nodes numbered as i, j, k, l. There are six nodal parameters at each of the nodes with a total of 24 parameters. The nodal parameter vector of the finite-element can be written as

$$\{\delta_e\} = \left\{ \begin{matrix} \{\delta_{ei}\} \\ \{\delta_{ej}\} \\ \{\delta_{ek}\} \\ \{\delta_{el}\} \end{matrix} \right\}$$ (4-58)

where

$$\{\delta_{ei}\} = \begin{bmatrix} u_i & v_i & w_i & \theta_{xi} & \theta_{yi} & \theta_{zi} \end{bmatrix}^T$$

$$\{\delta_{ej}\} = \begin{bmatrix} u_j & v_j & w_j & \theta_{xj} & \theta_{yj} & \theta_{zj} \end{bmatrix}^T$$

$$\{\delta_{ek}\} = \begin{bmatrix} u_k & v_k & w_k & \theta_{xk} & \theta_{yk} & \theta_{zk} \end{bmatrix}^T$$

$$\{\delta_{el}\} = \begin{bmatrix} u_l & v_l & w_l & \theta_{xl} & \theta_{yl} & \theta_{zl} \end{bmatrix}^T$$

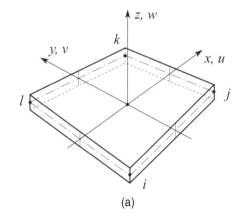

(a)

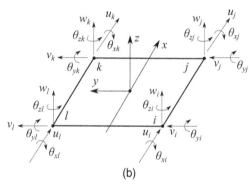

(b)

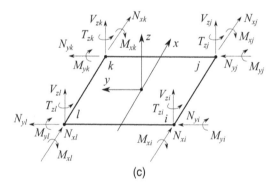

(c)

FIGURE 4-15   Shell-Plate Finite-Element, (a) Typical Shell-Plate Finite-Element, (b) Nodal Displacements, (c) Nodal Forces.

u, v, w = linear displacements in x-, y-, z-directions, respectively (see Fig. 4-15b)
$\theta_x, \theta_y, \theta_z$ = angular displacements about x-axis, y-axis, z-axis, respectively

Assume the displacement functions as polynomials as follows:

$$v(x,y) = b_1 + b_2 x + b_3 y + b_4 xy + b_5 x^2 + b_6 x^3 \tag{4-59}$$

$$u(x,y) = b_7 + b_8 x + b_9 y + b_{10} xy + b_{11} y^2 + b_{12} y^3 \tag{4-60}$$

$$w(x,y) = b_{13} + b_{14} x + b_{15} y + b_{16} xy + b_{17} x^2 + b_{18} y^2 + b_{19} x^2 y + b_{20} xy^2$$
$$+ b_{21} x^3 + b_{22} y^3 + b_{23} x^3 y + b_{24} xy^3 \tag{4-61}$$

Using 24 boundary conditions at the four nodes, we can determine the above unknowns b_i and determine the displacement functions. Following the procedures described in Section 4.2.5.2, we can obtain the stiffness matrix of the shell-plate finite-element[4-9].

The nodal forces of the shell-plate finite-element can be written as

$$\{F_e\} = \begin{Bmatrix} \{F_{ei}\} \\ \{F_{ej}\} \\ \{F_{ek}\} \\ \{F_{el}\} \end{Bmatrix} \tag{4-62}$$

where

$$\{F_{ei}\} = \begin{bmatrix} N_{xi} & N_{yi} & V_{zi} & M_{xi} & M_{yi} & T_{zi} \end{bmatrix}^T$$

$$\{F_{ei}\} = \begin{bmatrix} N_{xj} & N_{yj} & V_{zj} & M_{xj} & M_{yj} & T_{zj} \end{bmatrix}^T$$

N_x, N_y = axial force along x-axis and y-axis
V_z = shear force in z-direction
M_x, M_y = moments about x-axis and y-axis, respectively
T_z = torsion about z-axis

4.3 LONGITUDINAL ANALYSIS OF SEGMENTAL BRIDGES

4.3.1 INTRODUCTION

Most concrete segmental bridges are indeterminate structures. Any deformations experienced by the bridge as a result of post-tensioning, concrete shrinkage, steel creep, temperatures, and support settlements will induce additional structural internal forces, which should be considered in bridge design. As segmental bridges have very limited flexural cracking under service loads, the elastic theory for indeterminate bridge structures can be applied with sufficient accuracy at the service load limit state. The AASHTO LRFD specifications require that the factored load of a post-tensioned concrete box girder bridge of super-structure sections shall not be smaller than the strength at all sections for all limit states. The loads being factored to achieve a reliable design are determined based on elastic analytical methods. Thus, the following discussion is mainly based on the assumptions of elastic materials.

4.3.2 ANALYSIS OF SECONDARY FORCES DUE TO POST-TENSIONING TENDONS AND DESIGN CONSIDERATIONS

4.3.2.1 Analysis of Post-Tensioned Continuous Girders by Force Method

For a better understanding of the secondary forces due to post-tensioning forces, let's examine a two-span continuous beam with a curved tendon as shown in Fig. 4-16. The shapes of the tendon in both spans are assumed to be parabola with rises f_1 and f_2 for spans 1 and 2, respectively. The tendon at the beam ends is located at the centroid of the girder and at the middle support C has an eccentricity e (see Fig. 4-16a). Assuming an effective tendon prestressing force of P after loss, the moment induced by the tendon can be simply found by multiplying the tendon force with the related eccentricity as shown in Fig. 4-16c. Under this bending moment, the beam will tend to deflect at the

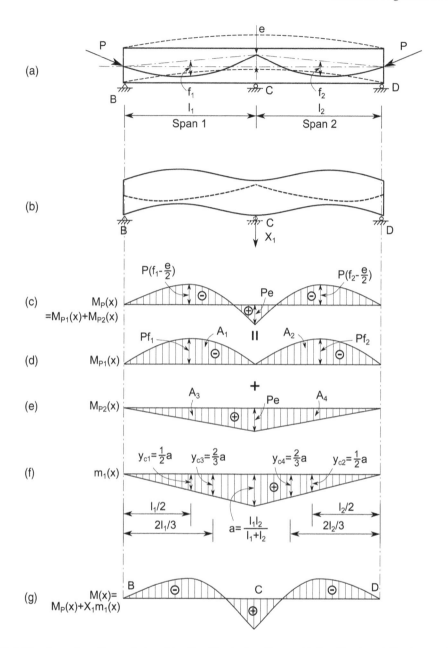

FIGURE 4-16 Secondary Forces Induced by Post-Tensioning Forces, (a) Post-Tensioned Continuous Girder, (b) Deflection of Girder Due to Secondary Support Reaction, (c) Moment Due to Post-Tensioning Force in Primary Structure, (d) Primary Moment Due to Tendon without Eccentricity at Support C, (e) Primary Moment due to Tendon with Eccentricity e at Support C, (f) Moment Due to $X_1 = 1$, (g) Sum of Primary and Second Moments.

support C as shown in the dotted lines in Fig. 4-16a. If the beam is restrained at support C, a reaction X_1 must be applied on the beam to hold it there (see Fig. 4-16b). The reaction X_1 is often called a secondary force and can be determined by the force method discussed in Section 4.2.2, by treating the moment induced by post-tensioning forces as the moment induced by external loads.

To determine the second force X_1, we can temporarily remove support C and replace it with an unknown reaction X_1. Then, the continuous beam becomes a simple beam (see Fig. 4-16b). Using the condition that the displacement as a result of both the post-tensioning force p and the unknown force X_1 at support C must be equal to zero, we can obtain the following equation similar to Eq. 4-14:

$$\delta_{11}X_1 + \Delta_{1P} = 0 \tag{4-63}$$

where
Δ_{1P} = displacement at support C in primary structure due to post-tensioning force P
$\quad = \int \frac{m_1(x)M_P(x)}{EI} dx$ (see Eq. 4-15e)
δ_{11} = displacement at support C in the primary structure due to a unit force
$\quad = \int \frac{m_1(x)m_1(x)}{EI} dx$ (see Eq. 4-15a)

Resolving $M_P(x)$, the moment in the primary structure induced by the post-tensioning force, into two portions as shown in Figs. 4-16c and d and using Eq. 4-13, we can obtain

$$\Delta_{1P} = \int \frac{m_1(x)M_P(x)}{EI} dx = A_1 y_{c1} + A_2 y_{c2} + A_3 y_{c3} + A_4 y_{c4}$$

$$= -\frac{Pl_1 l_2}{3EI(l_1 + l_2)}(f_1 l_1 + f_2 l_2 - e l_1 - e l_2) \tag{4-64}$$

where
A_i = subareas of the moment $M_p(x)$ (see Figs. 4-16d and e)
y_{ci} = value of $m_1(x)$ at location where centroid of A_i is located (see Fig. 4-16f)

Similarly, we can obtain

$$\delta_{11} = \int \frac{m_1(x)m_1(x)}{EI} dx = \frac{l_1^2 l_2^2}{3EI(l_1 + l_2)} \tag{4-65}$$

Substituting Eqs. 4-64 and 4-65 into Eq. 4-63 yields the reaction at support C due to post-tensioning force P as

$$X_1 = -\frac{\Delta_{1P}}{\delta_{11}} = \frac{P(f_1 l_1 + f_2 l_2 - e l_1 - e l_2)}{l_1 l_2} \tag{4-66a}$$

If $f_1 = f_2 = f$ and $l_1 = l_2 = l$, Eq. 4-66a can be written as

$$X_1 = \frac{2P(f - e)}{l} \tag{4-66b}$$

The additional moment at support C due to X_1 can be obtained as

$$M'_C = X_1 m_1(l_1) = P\left(\frac{f_1 l_1 + f_2 l_2}{l_1 + l_2} - e\right) \tag{4-67a}$$

If $f_1 = f_2 = f$ and $l_1 = l_2 = l$, Eq. 4-67a can be written as

$$M'_C = P(f - e) \tag{4-67b}$$

The total moment of the beam over the support C is

$$M_C = M_P + M'_C = Pe + P\left(\frac{f_1 l_1 + f_2 l_2}{l_1 + l_2} - e\right) = P\left(\frac{f_1 l_1 + f_2 l_2}{l_1 + l_2}\right) \tag{4-68a}$$

If $f_1 = f_2 = f$ and $l_1 = l_2 = l$, Eq. 4-68a can be written as

$$M_C = Pf \qquad (4\text{-}68b)$$

The moment caused by the post-tensioning force in the primary structure is often called the primary moment and the moment induced by the support reaction is called the secondary moment. The total moment (see Fig. 4-16g) in the beam can be obtained from Eq. 4-18, i.e.,

$$M(x) = M_P(x) + X_1 m_1(x)$$

4.3.2.2 General Procedures for Analyzing Post-Tensioned Continuous Bridge Structures by the Equivalent Load Method

The equivalent load method described in Section 3.1.2.2 can be applied to analyze any of the post-tensioned continuous bridges. The method treats the post-tensioning forces as external loads, and the entire analytical procedures are the same as those for analyzing any statically intermediate structures.

Figure 4-17 shows a two-span continuous post-tensioned girder. Assuming the curvatures and bent angles of the tendon along the span are small, the procedures for elastic analysis of the girder due to the post-tensioning force can be described as follows:

Step 1: Determining the equivalent loads at the girder ends. From Eqs. 3-10 to 3-12, we have
Axial loads: $H_A = P\cos\alpha_1 \approx P$, $H_C = P\cos\alpha_5 \approx P$
Vertical loads: $V_A = P\sin\alpha_1 \approx P\alpha_1$, $V_C = P\sin\alpha_5 \approx P\alpha_5$
End moments: $M_A = e_1 P\cos\alpha_1 \approx Pe$, $M_C = e_2 P\cos\alpha_5 \approx Pe_2$

Step 2: Determine the equivalent loads within the spans. As flat curvatures are assumed, both parabolic and circular curves can be considered to have the same effect in producing the equivalent loads and a uniform distributed load can be assumed along the length of curve. From Eq. 3-17, the uniform load can be taken as the change in the slope between two end tangents times the post-tensioning force, i.e.,
In span AB: $q_{AB} = \frac{P\sin\alpha_2}{l_1} \approx \frac{P\alpha_2}{l_1}$
In span BC: $V_{BC} = P\sin\alpha_4 \approx P\alpha_4$
Step 3: Determine equivalent loads over support B.

$$q_B = \frac{P\sin\alpha_3}{l_2} \approx \frac{P\alpha_3}{l_2}$$

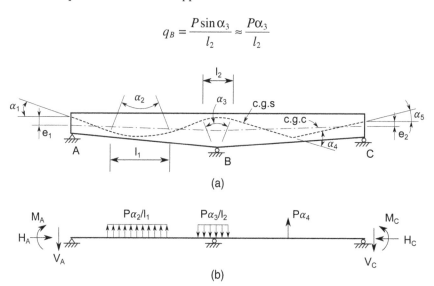

(a)

(b)

FIGURE 4-17 Equivalent Loads in a Continuous Beam, (a) Post-Tensioned Girder and Tendon Geometry, (b) Analytical Model.

Step 4: Apply the equivalent loads on the continuous girder (see Fig 4-17b), and analyze the girder by either of the methods described in Section 4.2.

4.3.2.3   Useful Concepts and Properties of Post-Tensioned Continuous Bridge Structures

As actual bridge design is essentially a trial-and-adjustment process in an effort to obtain the best alternative, it is important for the designers to be well acquainted with the basic mechanical behaviors of prestressed continuous structures to ensure an efficient bridge design.

1. **Linear change of C-line between two consecutive supports**

Using the equivalent loading method described above, it is easy to verify that the post-tensioning force in simply supported beam is self-balanced and will not cause any support reaction (Fig. 4-18a). From Eq. 3-5, we can also see that the C-line (compression line) of a simply supported prestressed girder without external loading must coincide with the T-line; i.e., the C-line coincides with the center of gravity of post-tensioning steel (cgs) (see Fig. 4-18a). However, the post-tensioning force in a continuous girder can cause secondary moment M' as discussed in Section 4.3.2.1. Thus, the C-line in a continuous girder will be shifted away from the cgs line by a deviation (see Fig. 4-18b).

$$\Delta = \frac{M'}{P} \tag{4-69}$$

Because the secondary moment M' is solely produced by the secondary reaction at the support, the secondary moment varies linearly between two consecutive supports. If the post-tensioning force P is assumed to be constant between the supports, the deviation Δ also varies linearly. Thus, we can conclude that the post-tensioning force in a continuous girder may shift the C-line from the cgs line. However, the intrinsic shape (covertures and bends) of the new C-line between the two consecutive supports remains unchanged.

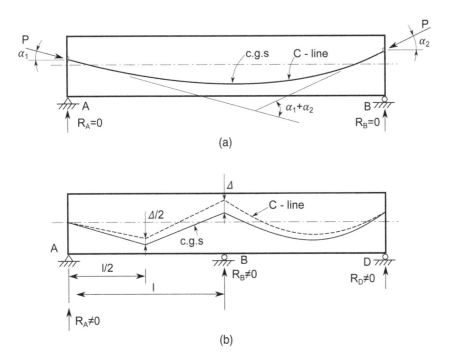

FIGURE 4-18   Linear Change of C-Line, (a) Simply Supported Beam, (b) Continuous Beam.

2. Linear Transformation

From Eqs. 4-68a and 4-68b, it can be observed that the eccentricity e over the interior support will not affect the sum of the primary and secondary moments. That means the total moment in continuous girder due to post-tensioning force remain unchanged if the cgs line is moved up and down over the interior support without changing the intrinsic shape of the line within each of the individual spans. The type of tendon relocation is often called the linear transformation of cables. The behavior of a post-tensioned continuous girder by linear transformation is true for any other continuous girders and can be generally stated as[4-10]: In a continuous girder, the cgs line can be linearly transformed without changing the location of the resulting C-line without external loadings. In other words, linear transformation of cgs line will not change the concrete stress. From Eqs. 4-66a and 4-66b, it can be observed that adjusting the eccentricity e over the support can adjust the secondary support reaction induced by post-tensioning force. The property of linear transformation can be used to adjust the secondary support reaction and moment. Increasing the eccentricity at the interior support will reduce the support reaction.

3. Concordant cable

From Eq. 4-67a, if $e = \frac{f_1 l_1 + f_2 l_2}{l_1 + l_2}$ by linear transformation, the secondary moment M' is equal to zero and support reaction $X_1 = 0$. That means the C-line is coincident with the cgs line. The cgs line is called concordant cable in the continuous girder. A concordant cable induces no support reaction in the supports, and the analytic method for the continuous beam is the same as a simply supported beam. If a cable induces secondary moment, the cable is termed nonconcordant. In design, it is convenient to first select a best concordant cable in resisting external loadings and then linearly transform it to give a more practical location without changing its C-line. However, there is no important reason to choose one or the other of concordant and nonconcordant cables. Both concordant and nonconcordant cables may produce a desirable C-line and satisfy other practical requirements.

The condition for a concordant cable is that no secondary moment will be induced by its prestressing force. From Eqs. 4-66a and 4-67a, the following equation must meet:

$$\Delta_{1P} = \int \frac{m_1(x) M_P(x)}{EI} dx = 0 \tag{4-70}$$

where $M_P(x)$ is the total moment induced by a concordant cable. Let's select a concordant cable proportional to moment diagram $M(x)$ caused by any combination of external loadings, i.e.,

$$M_P(x) = kM(x) \tag{4-71}$$

where k is an arbitrary constant.

Substituting Eq. 4-71 into Eq. 4-70 yields

$$\Delta_{1P} = \int \frac{m_1(x) kM(x)}{EI} dx = k \int \frac{m_1(x) M(x)}{EI} dx = 0 \tag{4-72}$$

Equation 4-72 indicates that the shape of the moment diagram due to external loadings in a continuous girder is one location of a concordant cable.

4.3.2.4   General Procedures for Tendon Layout in Continuous Prestressed Structures

As discussed, designing a continuous structure is essentially a procedure of trial and adjustment. Though there are many previously designed existing bridges that can serve as good examples when determining the cable geometry, the following procedures are useful in reaching a desirable result for the structural design of any continuous bridges.

Step 1: Based on the assumed sections, determine the moment distribution M_D due to dead load, the maximum moment distribution M_{max}, and the minimum moment distribution M_{min} at control

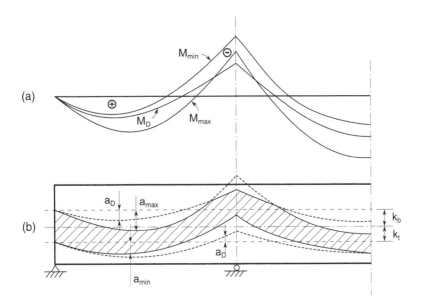

FIGURE 4-19   Tendon Placement Limits in Continuous Girder, (a) Moment Envelop, (b) Limits of Tendon Placement.

sections. The maximum and minimum moments shall include all effects of dead loads, live loads, and other external loads (see Fig. 4-19a).

Step 2: Determine the tendon placement limits by the method described in Section 3.1.3.3. Assuming that no tensile stresses are allowed in the concrete section, the limits can be drawn as follows: (a) determine the top and bottom kern lines for the girder, k_t and k_b, respectively, using Eqs. 3-31a and 3-31b (see Fig. 4-19b); (b) from the bottom kern line k_b, taking positive moment downward and negative moment upward, draw $a_{min} = \frac{M_{min}}{P}$ and $a_D = \frac{M_D}{P}$ (see Fig. 4-19b); (c) from the top kern line k_t, taking positive moment downward and negative moment upward, draw $a_{max} = \frac{M_{max}}{P}$ and $a_D = \frac{M_D}{P}$ (see Fig. 4-19b). The shaded area enclosed by the four lines determined by a_{max}, a_{min}, and a_D is the tendon placement limits without tensile stress in the concrete. If some tensile stress in the concrete is allowed, the limits will be enlarged by an amount discussed in Section 3.1.3.3.

Step 3: Select a cgs line approximately proportional to the moment diagrams (approximate to concordant cable) within the limits, considering: (a) set the eccentricity as large as possible at the control section where the largest moment developed at the state limit of failure; (b) whenever possible, the prestressing force at any sections is sufficient to resist the average service moment at that section; (c) choose the tendon profile that has less friction loss and that has a tendon curvature that meets the requirement by related specifications.

Step 4: Use the linear transformation method to adjust the cgs line to a more practical location without changing the C-line.

Step 5: Perform refined structural analysis, check the capacity at control sections, and revise sections and adjust cable locations as necessary. If needed, adjust the secondary support reaction by either linear transformation or changing the tendon locations over the interior support within 0.25 span range on each side of the support.

4.3.3   ANALYSIS OF SECONDARY FORCES DUE TO TEMPERATURE

4.3.3.1   Introduction

Temperature change may cause significant stresses in a concrete segmental bridge and damage the bridge. As discussed in Section 2.2.10, there are basically two temperature models of uniform and

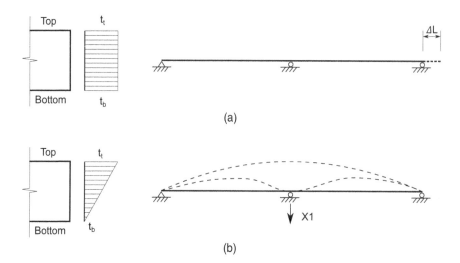

FIGURE 4-20 Effect of Temperature on Structures, (a) Uniform Distributed Temperature, (b) Gradient Temperature.

gradient that should be considered in the bridge design. If the temperature is uniformly distributed across the cross section as shown in Fig. 4-20a, there is only axial displacement in the bridge, which is normally accommodated by the bridge expansion joints, and there are no stresses induced by temperature. If the temperature varies linearly from the top to the bottom faces of a girder as shown in Fig. 4-20b, the continuous girder tends to bend upward and a support reaction will be induced in the girder, which is commonly referred to as a secondary force. Based on experimental results, the distribution of temperatures along the bridge cross-section is nonlinear as discussed in Section 2.2.10.2. The nonlinear distribution of temperature will not only cause the axial displacement and bending deformation, but it will also introduce restrained internal stresses. When analyzing statically indeterminate structures, the temperature can be treated as an external loading. The key is to establish the relationship between the temperature and the deformation. Then, using the displacement compatibility or force equilibrium condition solves the internal forces as discussed in Sections 4.2.2 and 4.2.3.

4.3.3.2 Deformation Due to Linear Distributed Temperature Gradient

Figure 4-21 shows a simply supported girder subjected to a linearly distributed temperature gradient with top and bottom temperatures t_t and t_b, respectively (see Fig. 4-21). Assuming (a) the temperature is constant along the girder longitudinal direction; (b) the material is homogeneous; and (c) the plane section remains plane after deformation and taking an infinitive element dx as shown in Fig. 4-21c, we can obtain the elongation of dx at the neutral axis due to the temperature (see Fig. 4-21c) as

$$d\Delta = \alpha t_N dx \qquad (4\text{-}73a)$$

where

α = coefficient of thermal expansion (in./in./°F)

t_N = $\frac{t_t y_b + t_b y_t}{h}$ = temperature at neutral axis

The corresponding strain is

$$\varepsilon_N = \frac{d\Delta}{dx} = \alpha t_N \qquad (4\text{-}73b)$$

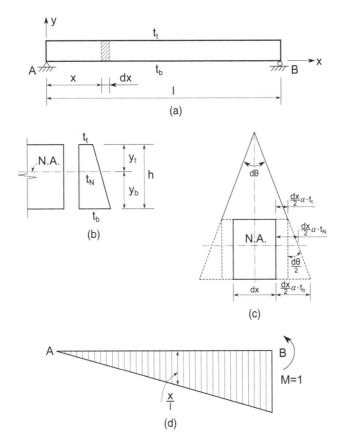

FIGURE 4-21 Deformation Due to Linear Distributed Temperature, (a) Simply Supported Girder, (b) Distribution of Temperature, (c) Deformation of an Infinitive Element, (d) Virtual Moment $m_\theta(x)$.

The rotation angle of the dx induced by the temperature (see Fig. 4-21c):

$$d\theta = \frac{\alpha(t_b - t_t)dx}{h} \tag{4-74a}$$

The corresponding curvature is

$$k_b = \frac{d\theta}{dx} = \frac{d^2 y}{dx^2} = \frac{\alpha(t_b - t_t)}{h} \tag{4-74b}$$

From Eq. 4-73a, we can obtain the total elongation of the girder

$$\Delta = \int_0^l \alpha t_N dx = \alpha t_N l \tag{4-73c}$$

Using the principle of virtual work, we can determine the deflections and rotation angles at any locations. For example, to obtain the angle at support B, we can apply a virtual moment $M = 1$ at support B. Then, the virtual moment (see Fig. 4-21d) in the beam can be written as

$$m_\theta(x) = \frac{x}{l} \tag{4-75}$$

Substituting Eqs. 4-74a and 4-75 into Eq. 4-9b yields the rotation angle due to the temperature at support B as

$$\theta_B = \int_0^l \frac{x}{l} \frac{\alpha(t_b - t_t)}{h} dx = \frac{l\alpha(t_b - t_t)}{2h} \tag{4-76}$$

4.3.3.3 Deformation Due to Nonlinear Distributed Temperature Gradient and Internal Restrained Stress

Assume that the simply supported girder shown in Fig. 4-21a has an arbitrary cross section as shown in Fig. 4-22a and is subjected an arbitrary non linear distributed temperature (see Fig. 4-22b). Assuming that there is no restraint between the longitudinal fibers of the cross section, the free longitudinal strain deformation of any fiber at location y due to the temperature can be written as

$$\varepsilon_t(y) = \alpha T(y) \tag{4-77}$$

The longitudinal fibers actually restrain each other, and the plane section can be assumed to remain plane after deformation. Because the temperature is assumed to be constant along the girder length, the curvatures of the girder are constant / unchanged along the length of the girder (see Eq. 4-74b). From Fig. 4-22d, the actual strain distribution can be written as

$$\varepsilon_a(y) = \varepsilon_0 + k_b y \tag{4-78}$$

where
ε_0 = strain at bottom face of girder, i.e., $y = 0$
k_b = curvature of section

The internal restrained strain can be written as

$$\varepsilon_r(y) = \varepsilon_t(y) - \varepsilon_a(y) = \alpha T(y) - (\varepsilon_0 + k_b y) \tag{4-79}$$

The corresponding restrained stress is

$$f_r(y) = E\varepsilon_r(y) = E\left[\alpha T(y) - (\varepsilon_0 + k_b y)\right] \tag{4-80}$$

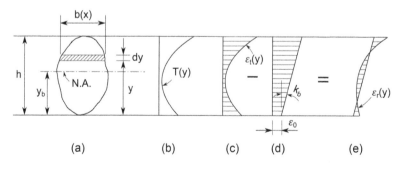

FIGURE 4-22 Variation of Strain due to Nonlinear Distributed Temperature, (a) Arbitrary Section, (b) Distribution of Temperature, (c) Free Strain, (d) Linear Strain, (e) Restrained Strain.

As there are no other external loads acting on the girder, the total axial force N in the girder longitudinal direction and the total moment M_{NA} about the neutral axis should be equal to zero, i.e.,

$$\sum N = 0:$$

$$\int_0^h f_r(y)b(x)dy = E\left[\alpha\int_0^h T(y)b(y)dy - \varepsilon_0\int_0^h b(y)dy - k_b\int_0^h yb(y)dy\right] = 0 \qquad (4\text{-}81)$$

Noticing the following equations:

Section area: $A = \int_0^h b(y)dy$
Static moment about $y = 0$: $S = \int_0^h yb(y)dy = Ay_b$

Equation 4-81 can be rewritten as

$$\varepsilon_0 = \frac{\alpha}{A}\int_0^h T(y)b(y)dy - y_b k_b \qquad (4\text{-}82\text{a})$$

From $\sum M_{NA} = 0$, we have

$$\int_0^h f_r(y)b(x)(y-y_b)dy = E\left[\alpha\int_0^h T(y)b(y)(y-y_b)dy - \varepsilon_0\int_0^h b(y)(y-y_b)dy - \right.$$
$$\left. k_b\int_0^h yb(y)(y-y_b)dy\right] = 0 \qquad (4\text{-}83)$$

Noticing that the static moment about the neutral axis is equal to zero, i.e.,

$$\int_0^h b(y)(y-y_b)dy = 0$$

Moment of inertia: $I = \int_0^h yb(y)(y-y_b)dy$

Equation 4-83 can be written as

$$k_b = \frac{\alpha}{I}\int_0^h T(y)b(y)(y-y_b)dy \qquad (4\text{-}84\text{a})$$

The restrained stress in the section due to temperature can be obtained by substituting Eqs. 4-82a and 4-84a into Eq. 4-80.

For the temperature distribution as shown in Fig. 2-15, the integrals in Eqs. 4-82a and 4-84a can be numerically performed by dividing the girder section into n strips along the girder vertical

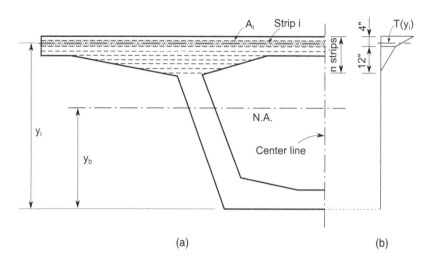

FIGURE 4-23 Determination of Generalized Strains Due to Gradient Temperature, (a) Strip Divisions, (b) Variation of Temperature.

direction within the temperature variation area (see Fig. 4-23), assuming the temperature in each strip is constant. Thus, Eqs. 4-82a and 4-84a can be written as

$$\varepsilon_0 = \frac{\alpha}{A}\sum_{i=1}^{n}T(y_i)A_i - y_b k_b \tag{4-82b}$$

$$k_b = \frac{\alpha}{I}\sum_{i=1}^{n}T(y_i)A_i(y_i - y_b) \tag{4-84b}$$

where
   n = number of strips, normally taking $n = 8$ to 10 can give an accurate enough result.
   A_i = area of strip i.

4.3.3.4   Determination of Secondary Forces Due to Temperature Gradient

The secondary forces in a continuous girder due to temperature can be determined based on the discussion in Sections 4.2.2 to 4.2.5, by treating the temperature as a type of external loadings.

4.3.3.4.1   Using the Force Method

For simplicity, let's examine the two-span continuous girder shown in Fig. 4-24a with a constant cross section as shown in Fig 4-23a and determine its secondary force due to the temperature gradient shown in Fig. 4-23b, using the force method. As there is no restraint in the girder longitudinal direction, there is only one unknown. Taking the moment $X_1 = M_t$ at support B as an unknown, its primary structure is shown in Fig. 4-24b. Using the condition that the relative rotation angle at B caused by both temperature and the unknown moment X_1 should be equal to zero, we have (refer to Section 4.2.2).

$$\delta_{11}X_1 + \Delta_{1T} = 0 \tag{4-85}$$

where
   δ_{11} = relative angle due to $X_1 = 1$

$$\delta_{11} = \int \frac{m_1(x)m_1(x)}{EI}dx \tag{4-86}$$

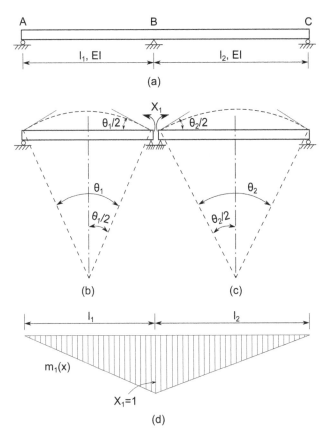

FIGURE 4-24 Secondary Forces Due to Gradient Temperature, (a) Two-Span Continuous Girder, (b) Primary Structure, (c) Moment $m(x)$ due to Unit Moment.

$m_1(x)$ = moment due to $X_1 = 1$ (see Fig. 4-24c)
Δ_{1T} = relative angle at support B due to temperature

$$= -\left(\frac{\theta_1 + \theta_2}{2}\right) \tag{4-87}$$

θ_1 = change of angle between sections A and B due to temperature in primary structure (see Fig. 4-24b)

$$\theta_1 = \int_A^B k_b dx = k_b l_1 \tag{4-88}$$

θ_2 = change of angle between sections B and C due to temperature in primary structure (see Fig. 4-24c)

$$\theta_2 = \int_B^C k_b dx = k_b l_2 \tag{4-89}$$

Substituting Eqs. 4-86 and 4-87 into Eq. 4-85, we can obtain the secondary moment at support B, i.e., $X_1 = M_t$.

4.3.3.4.2 Using the Displacement Method

If the displacement method is used to determine the secondary forces due to the temperature gradient, the only difference from the determination of the secondary forces due to external loadings is the determination of the fixed-end forces induced by the temperature. For a general case, the fixed-end axial forces and moments can be developed from Eqs. 4-82a and 4-84a or Eqs. 4-82b and 4-84b as

$$N_i = EA(\varepsilon_0 + k_b y_b) \tag{4-90a}$$

$$N_j = -EA(\varepsilon_0 + k_b y_b) \tag{4-90b}$$

$$M_i = EIk_b \tag{4-90c}$$

$$M_j = -EIk_b \tag{4-90d}$$

where

N_i, N_j = fixed-end axial forces at i and j ends, respectively
M_i, M_j = fixed-end moments at i and j ends, respectively

See Fig. 4-22 for the remaining notations.

4.3.3.4.3 Using the Finite-Element Method

If the finite-element method is used to analyze the effects of the temperature gradient on a bridge structure, the vector of equivalent nodal loads of the finite-element due to the temperature should be established and can be written as (see Fig. 4-25)

$$\{F_e\} = \begin{bmatrix} -N_i & 0 & -M_i & -N_j & 0 & -M_j \end{bmatrix}^T \tag{4-91}$$

where N_i, N_j, M_i, M_j can be determined from Eqs. 4-90a to 4-90d, respectively.

4.3.3.4.4 Normal Stress Due to the Temperature Gradient

After the secondary axial force N_t and moment M_t are obtained, the normal stress at any location of the section due to the temperature gradient can calculated as

$$f_t = \frac{N_t}{A} - \frac{M_t(y - y_b)}{I} + E\left[\alpha T(y) - \varepsilon_0 - k_b y\right] \tag{4-92}$$

Note that the normal stress is positive in tension in Eq. 4-92, which is a common sign convention in the finite-element method.

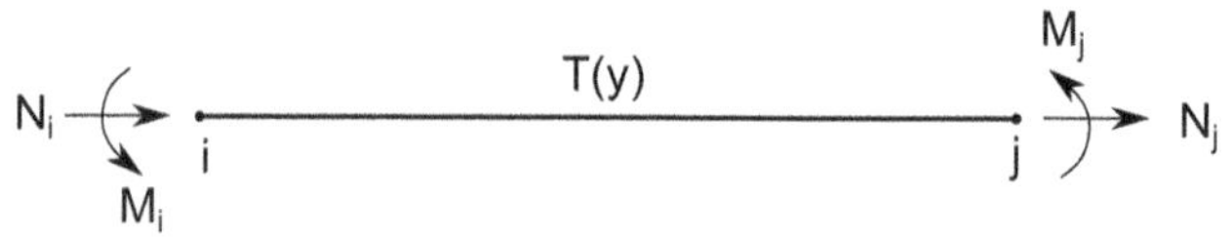

FIGURE 4-25 Equivalent Nodal Loads due to Temperature.

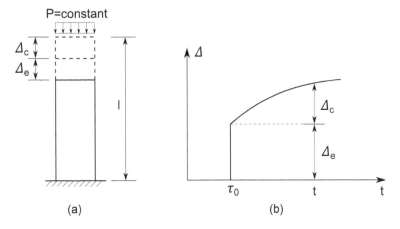

FIGURE 4-26 Deformation of a Concrete Column under Constant Loading, (a) Column under Constant Axial Loading P, (b) Variation of Column Shortening with Time.

4.3.4 ANALYSIS OF SECONDARY FORCES DUE TO CONCRETE CREEP

4.3.4.1 Relationships between Creep Strain and Stress

4.3.4.1.1 Relationships between Creep Strain and Stress with Constant Concrete Stress

As discussed in Section 1.2.2.5, a concrete structure under a sustained load will undergo two types of deformation, i.e., an immediate elastic deformation Δ_e and a time-depended deformation caused by concrete creep Δ_c (see Fig. 4-26). The total strain can be written as

$$\varepsilon(t,\,\tau_0) = \frac{\Delta_e}{l} + \frac{\Delta_c}{l} = \varepsilon_e(\tau_0) + \varepsilon_c(t,\tau_0) \tag{4-93a}$$

where

$\varepsilon(t,\,\tau_0)$ = total strain

$\varepsilon_e(\tau_0) = \dfrac{\Delta_e}{l}$ = immediate elastic strain when load is applied at τ_0

$\varepsilon_c(t,\tau_0) = \dfrac{\Delta_c}{l}$ = strain caused by creep at t when load is applied at τ_0

Experimental results have shown that the relationship between the creep strain and the external loading can be treated as linear if the concrete stress does not exceeded 50% of the concrete strength, i.e.,

$$\varepsilon_c(t,\tau_0) = \frac{f(\tau_0)}{E(\tau_0)}\phi(t,\tau_0) = \varepsilon_e(\tau_0)\phi(t,\tau_0) \tag{4-94}$$

where

$\varepsilon_c(t,\tau_0)$ = creep strain

$\phi(t,\tau_0)$ = creep coefficient equal to ratio of creep strain to elastic strain (see Section 1.2.2.5)

$f(\tau_0)$ = concrete stress at τ_0 when load is applied

$E(\tau_0)$ = concrete modulus of elasticity at time $\tau_0 = E_{28}$ in CEB-FIP code

The linear relationship between the stress and creep strain is often called linear creep theory which allows that the superposition principle can be used for analyzing the effects of creep due to different loading stages.

Substituting Eq. 4-94 into 4-93a yields

$$\varepsilon(t, \tau_0) = \varepsilon_e(\tau_0)(1 + \phi(t, \tau_0)) = \frac{f(\tau_0)}{E(\tau_0)}(1 + \phi(t, \tau_0)) \qquad (4\text{-}93\text{b})$$

4.3.4.1.2 Relationships between Creep Strain and Stress with Time-dependent Concrete Stress

For concrete segmental bridges, the concrete stress varies with time due to restrained creep deformation, time-dependent creep coefficient, etc. The relationship between strain and stress becomes complex. Accurate analysis of the effect of concrete creep may be very difficult, though vast research has been done on this topic. With the development of computers, the complicity of the stress-strain relationship due to creep and its effect on the structures can be satisfactorily resolved through numerical analysis. In this section, three commonly used expressions for the strain-stress relationship will be discussed for both hand and computer analysis.

a. General Relationship between Strain and Stress Due to Creep

Figure 4-27 shows an arbitrary stress-time curve due to restrained creep strain. We can divide the range between τ_0 and t into n small time intervals. Based on the principle of supposition and Eq. 4-93b, the variation of the strain with stress can be written as:

$$\varepsilon(t, \tau_0) = \frac{f(\tau_0)}{E(\tau_0)}(1 + \phi(t, \tau_0)) + \sum_{i=1}^{n} \frac{\Delta f(\tau_i)}{E(\tau_i)}[1 + \phi(t, \tau_i)] \qquad (4\text{-}93\text{c})$$

The first term on the right side in Eq. 4-93c represents the strain caused by initial stress, and the second term represents the strain increment due to the stress changing after τ_0. If the stress varies with time as a continuous function, Eq. 4-93c can be written in the form

$$\varepsilon(t, \tau_0) = \frac{f(\tau_0)}{E(\tau_0)}(1 + \phi(t, \tau_0)) + \int_{\tau_0}^{t} \frac{df(\tau)}{d\tau} \frac{1}{E(\tau)}[1 + \phi(t, \tau)]d\tau \qquad (4\text{-}93\text{d})$$

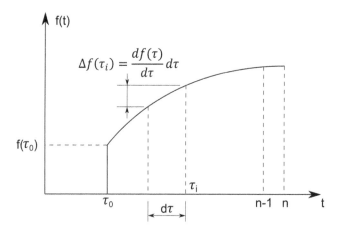

FIGURE 4-27   Variation of Stress due to Restrained Creep Strain with Time.

The relationship between creep strain and stress can be expressed as

$$\varepsilon_c(t, \tau_0) = \frac{f(\tau_0)}{E(\tau_0)}\phi(t, \tau_0) + \int_{t_0}^{t} \frac{df(\tau)}{d\tau}\frac{1}{E(\tau)}\left[1+\phi(t,\tau)\right]d\tau \tag{4-93e}$$

b. Differential Relationships between Strain and Stress Increments

Equations 4-93a to 4-93e developed based on the assumption that the initial loading age on an entire structure is the same. In the segmental construction, different elements of a segmental bridge may have different loading times. We need to establish the differential relationship between the strain and stress increments.

Differentiating both sides of Eq. 4-93b yields

$$d\varepsilon(t, \tau) = d\left[\frac{f(\tau)}{E(\tau)}\left(1+\phi(t,\tau)\right)\right] = \frac{df(\tau)}{E(\tau)} + \frac{f(\tau)}{E(\tau)}d\phi(t,\tau) \tag{4-95}$$

The first right term on the right side in Eq. 4-95 represents the elastic strain caused by the incremental stress within a small time frame $d\tau$. The second term represents the creep strain due to initial stress at τ. Note that the time of initial stressing is designated as τ, which also varies.

As discussed in Section 1.2.2.5, the concrete creep coefficient is related to many factors and each country may use different creep coefficient functions, which may complicate solving the creep differential equations of equilibrium, Eq. 4-95. For this reason, in 1937, Dischinger[4-11 and 4-12] proposed a simple mathematical creep model as follows:

$$\phi(t, \tau) = \phi_o\left(e^{-\beta\tau} - e^{-\beta t}\right) = \phi_o e^{-\beta\tau}\left(1-e^{-\beta(t-\tau)}\right) \tag{4-96}$$

where
ϕ_o = $\phi(\infty, 0)$ = ultimate creep coefficient, at $\tau_0 = 0$, $t = \infty$
β = creep incremental ratio ranging from 1 to 4, which is related to humidity, concrete properties, etc.

From Eq. 4-96, it can be observed that the derivative of the above creep model with time t is the same as for any initial loading at time τ. A creep coefficient function having this characteristic is often called aging theory (see Fig. 4-28), i.e.,

$$\frac{d\phi(t, \tau_2)}{dt} = \frac{d\phi(t, \tau_1)}{dt} = \phi_o\beta e^{-\beta t} \tag{4-97}$$

The analytical method of the structures based on Dischinger's creep model is often called Dischinger's method[4-11 to 4-14]. Though Eq. 4-96 may cause comparatively large errors under some conditions and is not currently often used in concrete segmental bridge design, the differential equation of equilibrium can be easily established and solved. Therefore, it still has theoretical value and can be used to develop some approximate equations for estimating the effects of creep. From Eq. 4-95, the total strain using Dischinger's creep model can be written as

$$\varepsilon(t, \tau_0) = \varepsilon(\tau_0) + \int_{\tau_0}^{t} \frac{df(\tau)}{E(\tau)} + \int_{\tau_0}^{t} \frac{f(\tau)}{E(\tau)}d\phi(t,\tau) \tag{4-93f}$$

where $\varepsilon(\tau_0)$ = instantaneous elastic strain due to loadings applied at τ_0.

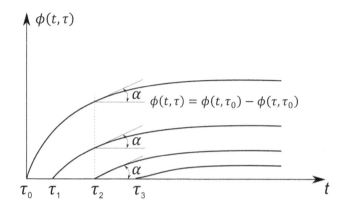

FIGURE 4-28 Dischinger's Creep Model Using Aging Theory.

c. Creep Strain-Stress Relationship by Effective Modulus

Normally, it is difficult to obtain closed form solutions for Eqs. 4-93d and 4-95. Moreover, in concrete segmental bridges, each of the segments may have different loading times and undergo different structural behaviors. It is more convenient to write the creep strain-stress relationship in algebraic form[4-15 to 4-17]. The following well-known creep strain-stress relationship was first proposed by Trost[4-15, 4-16] and revised by Bezant[4-17]:

$$\varepsilon_c(t,\tau_0)=\frac{f(\tau_0)}{E(\tau_0)}\phi(t,\tau_0)+\frac{f_c(t,\tau_0)}{E(\tau_0)}\left(1+\chi(t,\tau_0)\phi(t,\tau_0)\right) \tag{4-93g}$$

where $\chi(t,\tau)$ is called the aging coefficient[4-14 to 4-16], which can be written as

$$\chi(t,\tau_0)=\frac{\int_{\tau_0}^t \frac{df(\tau)}{d\tau}\phi(t,\tau)d\tau}{f_c(t,\tau_0)\phi(t,\tau_0)}=\frac{1}{1-\frac{E_R(t,\tau_0)}{E(\tau_0)}}-\frac{1}{\phi(t,\tau_0)} \tag{4-98a}$$

where $E_R(t,\tau_0)=$ relaxation function = stress at time t caused by a unit strain introduced at time τ_0. The relaxation function is related to many factors and can be written as

$$E_R(t,\tau_0)=\frac{E(\tau_0)}{1+\phi(t,\tau_0)}$$

τ_0 = time at first load application (in days)
t = time in days from casting of concrete (in days)

Using Dischinger's creep model Eq. 4-96, the aging coefficient can be approximately written as[4-14, 4-15]

$$\chi(t,\tau_0)=\frac{1}{1-e^{-\phi(t,\tau_0)}}-\frac{1}{\phi(t,\tau_0)} \tag{4-98b}$$

Equation 4-93g is often called the Trost[4-15] creep strain-stress algebraic expression and can be written as

$$\varepsilon_c(t,\tau_0)=\frac{f(\tau_0)}{E(\tau_0)}\phi(t,\tau_0)+\frac{f_c(t,\tau_0)}{E_\phi(t,\tau_0)} \tag{4-93h}$$

where $E_\phi(t,\tau_0)$ is called the effective modulus or age-adjusted effective modulus[4-14]:

$$E_\phi(t,\tau_0) = \frac{E(\tau_0)}{1+\chi(t,\tau_0)\phi(t,\tau_0)} = \beta(t,\tau_0)E(\tau_0) \tag{4-99}$$

$\beta(t,\tau_0)$ = $\frac{1}{1+\chi(t,\tau_0)\phi(t,\tau_0)}$ = effective modulus coefficient
$f(\tau_0)$ = initial concrete stress at τ_0
$f_c(t,\tau_0)$ = concrete stress change from τ_0 to t due to concrete creep

The first term on the right side of Eq. 4-93h represents the strain increment caused by the initial stress due to creep, while the second term represents the total strain increment due to the stress increment caused by creep. The total strain from loading time τ_0 to t can be written as

$$\varepsilon(t,\tau_0) = \frac{f(\tau_0)}{E(\tau_0)}(1+\phi(t,\tau_0)) + \frac{f_c(t,\tau_0)}{E_\phi(t,\tau_0)} \tag{4-93i}$$

From Eq. 4-93i, it can be observed that the strain-stress relationship including the effect of concrete creep is similar to that for conventional elastic structures by using the concept of effective modulus. Therefore, basic elastic analytic methods of structures can be applied to creep analysis of structures with some minor modifications.

4.3.4.2 Determination of Displacements Due to Creep under Constant Loadings

The key issue for analyzing a statically indeterminate structure due to the effect of creep is how to determine the displacements caused by the creep strains. For simplicity, first, we discuss the creep-induced displacements under constant loads.

Figure 4-29 shows a cantilever girder with a constant loading q. To determine the displacement Δ_{jp} at point j, we can use the principle of virtual work described in Section 4.2.1.2 and apply a unit virtual force P = 1 at point j. Then Δ_{jp} can be written as

$$1 \cdot \Delta_{jp} = \int_0^l \int \varepsilon_{(x,y)}(t,\tau_0) f_{1(x,y)} dA dx \tag{4-100a}$$

where
$f_{1(x,y)}$ = $\frac{m(x)y}{I_x}$ = stress in location (x, y) due to unit virtual load $P = 1$
$m(x)$ = moment due to unit virtual load $P = 1$ (see Fig. 4-29f)
I_x = moment of inertia at section x
$\varepsilon_{(x,y)}(t,\tau_0)$ = strain in location (x, y) at time t, due to loading q applied at time τ_0 (see Eq. 4-93c)

$$\varepsilon_{(x,y)}(t,\tau_0) = \frac{f_{p(x,y)}}{E}(1+\phi(t,\tau_0))$$

$f_{p(x,y)}$ = stress at location (x, y) due to loading q

Thus, Eq. 4-100a can be written as

$$\Delta_{jp} = \int_0^l \frac{m(x)}{EI_x}\left[\int f_{p(x,y)} y dA (1+\phi(t,\tau_0))\right]dx = \int_0^l \frac{m(x)M_p(x)}{EI_x}dx(1+\phi(t,\tau_0)) \tag{4-100b}$$

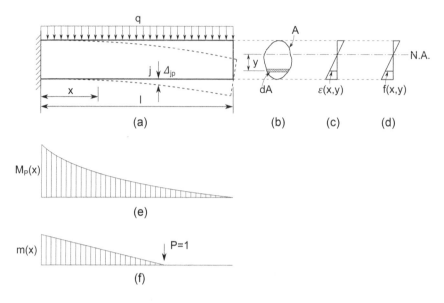

FIGURE 4-29 Determination of Displacement with Constant Loading Including Creep Effect, (a) Cantilever Beam, (b) Typical Section, (c) Strain Distribution, (d) Stress Distribution, (e) Moment due to External Load, (f) Moment due to Unit Force.

where $M_p(x)$ = moment due to loading q (see Fig. 4-29e). The first term on the right side of Eq. 4-100b represents the instantaneous elastic displacement due to the loading, and the second term is creep displacement. From Eq. 4-100b, we can see that the displacement under a constant loading including the effect of concrete creep varies linearly with loading. It can also be directly derived from Eq. 4-93b.

4.3.4.3 Determination of Displacements Due to Time-Dependent Loading

4.3.4.3.1 Dischinger's Method

Equation 4-95 indicates that the total strain increment during dt consists of two portions: the elastic strain induced by the stress increment $df(t)$ and the creep strain caused by stress $f(t)$, which includes initial stress f_0 and creep stress $f_c(t)$, i.e.,

$$f(t) = f_0 + f_c(t) \tag{4-101}$$

From Eqs. 4-95 and 4-100b, the displacement increment $d\Delta_{jp}$ induced by time-dependent loadings during dt can be written as

$$d\Delta_{jp} = \int_0^l \frac{dM(t)m_j}{EI}dx + \int_0^l \frac{M_0 m_j}{EI}dxd\phi(t,\,\tau) + \int_0^l \frac{M(t)m_j}{EI}dxd\phi(t,\,\tau) \tag{4-102}$$

where M_0, m_j, $M(t)$ = initial moment induced by different loads, moment produced by unit virtual force, and moment caused by creep redundant forces, respectively. They are all functions of x.

4.3.4.3.2 Effective Modulus Method

Substituting Eq. 4-93h into Eq. 4-100a yields the displacement increment due to creep at point j as

$$d\Delta_{jp} = \int_0^l \int \left[\frac{f(\tau_0)}{E(\tau_0)}\phi(t,\tau_0) + \int_{\tau_0}^t \frac{f_c(t,\tau_0)}{E_\phi(t,\tau_0)} \right] \frac{m(x)y}{I}dAdx$$

$$= \int_0^l \frac{M_0(x)m(x)}{E(\tau_0)I}\phi(t,\tau_0)dx + \int_0^l \frac{M_t(x)m(x)}{E_\phi(t,\tau_0)I}dx \qquad (4\text{-}103a)$$

where $M_t(x)$ = redundant moment caused by concrete creep. The first term on the right side of Eq. 4-103a represents the creep displacement caused by the initial forces in the primary structure. The second term on the right side of the equation indicates the displacements due to redundant forces induced by concrete creep form time τ_0 to t.

Substituting Eq. 4-93i into 4-100a yields the total displacement at point j as

$$\Delta_{jp} = \int_0^l \int \left[\frac{f(\tau_0)}{E(\tau_0)}(1+\phi(t,\tau_0)) + \int_{\tau_0}^t \frac{f_c(t,\tau_0)}{E_\phi(t,\tau_0)} \right] \frac{m(x)y}{I} dA dx$$

$$= \int_0^l \frac{M_0(x)m(x)}{E(\tau_0)I}(1+\phi(t,\tau_0))dx + \int_0^l \frac{M_t(x)m(x)}{E_\phi(t,\tau_0)I}dx \qquad (4\text{-}103b)$$

The first term on the right side of Eq. 4-103b represents the elastic and creep displacements caused by the initial forces in the primary structure, and the second term on the right side indicates the displacement due to the redundant forces caused by creep.

4.3.4.4 Determination of Secondary Force Due to Creep

In a statically indeterminate structure, concrete creep will induce secondary internal forces which can be analyzed by any of the methods discussed in Section 4.2. Based on the discussions in Sections 4.3.4.1 and 4.3.4.3, three methods for determining the secondary forces due to concrete creep are discussed in this section.

4.3.4.4.1 Dischinger's Method

Consider the span-by-span bridge shown in Fig. 4-30 as an example to describe the analytical procedures. For simplicity, assume the bridge has a constant moment of inertia I and modulus of elasticity E and is built in two stages. The first stage is to build segment 0-2, which is simply supported and the initial loading time of the first segment is τ_1 (see Fig. 4-30a). The second stage is to build segment 2-3, which is connected with segment 0-2 and the initial loading time for the second segment is τ_2 as shown in Fig. 4-30b.

The internal forces at time t due to the effect of concrete creep can be determined as follows.

Step 1: Determine the initial moment at loading time τ_2.

In loading stage I, the structure is statically determined and the concrete creep will not cause secondary forces. Its moment can be easily determined as shown in Fig. 4-30c and is denoted as $M_{01}(x)$. In loading stage II, the structure becomes a continuous girder and the second stage loading (see Fig. 4-30b) will induce secondary forces. Assume the unknown moment at support 1 as X_1 at the time τ_2 and the primary structure becomes two simply supported girders (see Fig. 4-30d). Using the principle of virtual work, the relative angle at support 1 can be written as

$$\Delta_{1P}^0 = \int_0^{2l} \frac{M_{0P}(x)m_1(x)}{EI}dx \qquad (4\text{-}104)$$

where
$M_{0P}(x)$ = moment due to Stage II loading in primary structure (see Fig. 4-30e)
$m_1(x)$ = moment due to unit moment at support 1 as shown in Fig. 4-30f

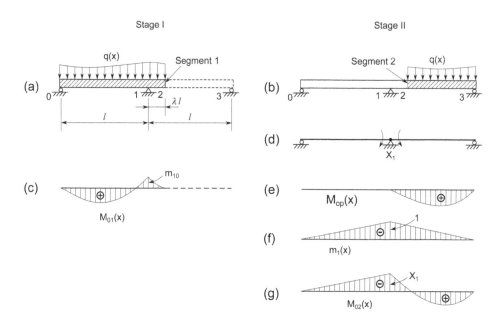

FIGURE 4-30 Determination of Creep Effect by Dischinger's Method, (a) Construction Stage I, (b) Construction Stage II, (c) Moment Caused by Stage I Loading, (d) Primary Structure for Analyzing Initial Redundant Force, (e) Moment Caused by Stage II Loading in Primary Structure, (f) Moment Caused by Unit Moment in Primary Structure, (g) Initial Moment Caused by Stage II Loading at Time τ_2 in Continuous Beam[4-11].

The relative angle due to the unknown moment X_1 can be written as

$$\Delta_{11} = X_1 \int_0^{2l} \frac{m_1(x)m_1(x)}{EI} dx = X_1 \delta_{11} \tag{4-105}$$

Using the compatibility of the relative angle at support 1, we have

$$X_1 \delta_{11} + \Delta_{1P}^0 = 0 \tag{4-106}$$

Solving Eq. 4-106 yields:

$$X_1 = -\frac{\Delta_{1P}^0}{\delta_{11}} \tag{4-107}$$

The bending moment due to second stage loading (see Fig. 4-30g) can be written as

$$M_{02}(x) = X_1 m_1(x) + M_{0P}(x) \tag{4-108}$$

The total initial moment in the girder is

$$M_0(x) = M_{01}(x) + M_{02}(x) \tag{4-109}$$

Step 2: Determine the displacement due to the effect of concrete creep
   Under the initial moment $M_0(x)$, the concrete creep will cause secondary forces in the continuous girder. Again, take the moment X_{1t} at support 1 as an unknown and the primary structure is shown in Fig. 4-31a. Based on the principle of superposition, the initial moment Eq. 4-109 can be rewritten as

$$M_0(x) = X_{10} m_1(x) + M_P(x) \tag{4-110}$$

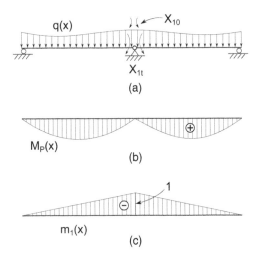

FIGURE 4-31 Determination of the Secondary Force Increment due to Creep, (a) Initial Forces and Unknown Redundant Force, (b) Moment Caused by External Loads in Primary Structure, (c) Moment Caused by Unit Moment.

where

X_{10} = initial force at support 1

X_{10} = $m_{10} + X_1$, for this example

m_{10} = moment at support 1 due to loading in stage I (see Fig. 4-30c)

$M_P(x)$ = moment in primary structure caused by both stage I and II loadings (see Fig. 4-31b)

From Eq. 4-102, the displacement caused by the initial moment $M_0(x)$ due to concrete creep can be written as

$$d\Delta_{1p} = \int_0^{2l} \frac{dM(t)m_1(x)}{EI}\,dx + \int_0^{2l} \frac{M_0 m_1(x)}{EI}\,dxd\phi(t,\tau) + \int_0^{2l} \frac{M(t)m_1(x)}{EI}\,dxd\phi(t,\tau) \qquad (4\text{-}111)$$

where

$M(t)$ is the moment increment due to concrete creep and can be written as

$$M(t) = X_{1t}m_1(x) \qquad (4\text{-}112a)$$

$$dM(t) = dX_{1t}m_1(x) \qquad (4\text{-}112b)$$

Substituting Eqs. 4-110, 4-112a, and 4-112b into Eq. 4-111 and considering different loading stages, we have

$$d\Delta_{1p} = dX_{1t}\int_0^{2l} \frac{m_1(x)m_1(x)dx}{EI} + X_{10}\left[\int_0^{(1+\lambda)l} \frac{m_1(x)m_1(x)dx}{EI}\,d\phi(t,\tau_1) \right.$$

$$\left. + \int_{(1+\lambda)l}^{l} \frac{m_1(x)m_1(x)dx}{EI}\,d\phi(t,\tau_2) \right] + \int_0^{(1+\lambda)l} \frac{M_P(x)m_1(x)dx}{EI}\,d\phi(t,\tau_1)$$

$$+ \int_{(1+\lambda)l}^{l} \frac{M_P(x)m_1(x)dx}{EI} d\phi(t, \tau_2) + X_{1t}\left[\int_{0}^{(1+\lambda)l} \frac{m_1(x)m_1(x)dx}{EI} d\phi(t, \tau_1) \right.$$

$$\left. + \int_{(1+\lambda)l}^{l} \frac{m_1(x)m_1(x)dx}{EI} d\phi(t, \tau_2) \right]$$

$$= dX_{1t}\delta_{11} + X_{10}d\phi(t, \tau_2)\left[\delta_{11}^I \frac{d\phi(t, \tau_1)}{d\phi(t, \tau_2)} + \delta_{11}^{II} \right] + d\phi(t, \tau_2)\left[\delta_{1P}^I \frac{d\phi(t, \tau_1)}{d\phi(t, \tau_2)} + \delta_{1P}^{II} \right]$$

$$+ X_{1t}d\phi(t, \tau_2)\left[\delta_{11}^I \frac{d\phi(t, \tau_1)}{d\phi(t, \tau_2)} + \delta_{11}^{II} \right] \tag{4-113}$$

where $\delta_{11} = \int \frac{m_1(x)m_1(x)dx}{EI}$, $\delta_{1P} = \int \frac{M_P(x)m_1(x)dx}{EI}$, and the superscripts I and II represent the integral ranges for construction stages I and II, respectively.

Assuming that we need to determine the displacement increment $d\Delta_{1p}$ due to the concrete creep at time t_n days from the casting time of the concrete segment 2 (taking the casting time of segment 2 as a reference (see Fig. 4-32a and b), then, for segment 1, the days from the casting time equal $t_n + \tau_1$ (see Fig. 4-32c). From Eq. 4-97, we have

$$\frac{d\phi(t, \tau_1)}{d\phi(t, \tau_2)} = \frac{\phi_o \beta e^{-\beta(t_n + \tau_1)}}{\phi_o \beta e^{-\beta(t_n)}} = e^{-\beta \tau_1} \tag{4-114}$$

Substituting Eq. 4-114 into Eq. 4-113 yields

$$d\Delta_{1p} = dX_{1t}\delta_{11} + X_{10}d\phi(t, \tau_2)\left[\delta_{11}^I e^{-\beta \tau_1} + \delta_{11}^{II} \right] + d\phi(t, \tau_2)\left[\delta_{1P}^I e^{-\beta \tau_1} + \delta_{1P}^{II} \right]$$
$$+ X_{1t}d\phi(t, \tau_2)\left[\delta_{11}^I e^{-\beta \tau_1} + \delta_{11}^{II} \right] \tag{4-115}$$

Step 3: Establish the incremental differential equation by imposing the displacement compatibility condition. As the relative angle at support 1 is equal to zero, i.e., $d\Delta_{1p} = 0$, Eq. 4-115 can be written as

$$\left[\delta_{11}^*(X_{1t} + X_{10}) + \delta_{1P}^* \right]d\phi(t, \tau_2) + \delta_{11}dX_{1t} = 0 \tag{4-116}$$

where

$$\delta_{11}^* = \delta_{11}^I e^{-\beta \tau_1} + \delta_{11}^{II} \tag{4-117a}$$

= displacement at support 1 caused by a unit moment at support 1 in primary structure of stage II

$$\delta_{1P}^* = \delta_{1P}^I e^{-\beta \tau_1} + \delta_{1P}^{II} \tag{4-117b}$$

= displacement at support 1 caused by initial external loadings in primary structure of stage II at time τ_2

Using the boundary condition: $t = \tau_1$, $X_{1t} = 0$ and solving Eq. 4-116 yields the moment increment at support 1 due to concrete creep:

$$X_{1t} = \left(X_B^* - X_{10}\right)\left[1 - e^{-\frac{\delta_{11}^*}{\delta_{11}}[\phi(t,\tau_1) - \phi(\tau_2,\tau_1)]} \right] = \left(X_B^* - X_{10}\right)\left[1 - e^{-\frac{\delta_{11}^*}{\delta_{11}}\phi(t,\tau_2)} \right] \tag{4-118a}$$

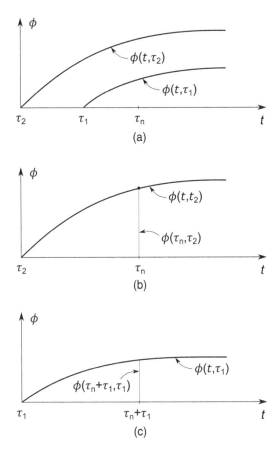

FIGURE 4-32 Variation of Creep Coefficients with the Loading Time t in Reference to Segment 2, (τ_2) (a) Creep Functions for Segments 1 and 2, (b) Creep Coefficient at Time t_n for Segment 2, (c) Creep Coefficient at Time t_n for Segment 1.

where $X_B^* = -\frac{\delta_{1P}^*}{\delta_{11}^*}$ = stability force of structural creep system. If the first and second segments have the same casting time, $\delta_{11}^* = \delta_{11}$ and $\delta_{1P}^* = \delta_{1P}$. Equation 4-118a can be written as

$$X_{1t} = \left(X_B - X_{10}\right)\left[1 - e^{-[\phi(t,\tau_1) - \phi(\tau_2,\tau_1)]}\right] = \left(X_B - X_{10}\right)\left[1 - e^{-\phi(t,\tau_2)}\right] \qquad (4\text{-}118b)$$

From Eq. 4-118a or Eq. 4-118b, it can be observed that one of the ways to reduce the forces caused by concrete creep is to increase the initial loading times. If the bridge is built in one stage and all concrete segments are cast at the same time, then $X_B^* = X_{10}$. Thus

$$X_{1t} = 0 \qquad (4\text{-}118c)$$

Therefore, we can conclude that the concrete creep will not induce any redundant forces if the bridge is built in one stage and all the concrete is cast at the same time.

After obtaining the redundant force due to creep X_{1t}, the moment in the girder due to creep can be written as

$$M_{1t} = X_{1t}m_1\left(x\right) \qquad (4\text{-}119)$$

The total moment in the girder can be obtained by adding the initial moment (see Eq. 4-110) and the creep moment (see Eq. 4-118) as

$$M(x) = (X_{10} + X_{1t})m_1(x) + M_P(x) \qquad (4\text{-}120)$$

4.3.4.4.2 Effective Modulus Method

Take a simplified segmental bridge built by the balanced cantilever method as an example to discuss the procedure for determining the secondary forces in statically indeterminate structures due to concrete creep using the effective modulus method, as shown in Figs. 4-33a and b[4-14, 4-16, 4-17]. To simplify the discussion, it is assumed that the bridge is built in three phases. In phase I, cantilever girder 0-1 is built, and in phase II cantilever girder 1-2 is built. Finally, during phase III, the two cantilever girders become a fixed girder by connecting them at midpoint 1 at time τ_0. In this example, we will find the moment and shear distribution 1000 days after the completion of the girder. For simplicity, assuming (1) the girder cross-section and modulus are constant, i.e., EI is constant; (2) the creep coefficients for the cantilever girders I and II are $\phi_I(1000, \tau_0) = 1$ and $\phi_{II}(1000, \tau_0) = 2$, respectively; and (3) the aging coefficients for the cantilever girders I and II are $\chi_I(1000, \tau_0) = 0.55$ and $\chi_{II}(1000, \tau_0) = 0.65$, individually.

The following procedures can be used for determining the secondary forces due to concrete creep by effective modulus method.

Step 1: Select a primary structure as shown in Fig 4-33c, and insert two unknown forces of ΔX_1 (moment) and ΔX_2 (shear force).

Step 2: Determine initial moment due to external loading $q = 2$ kips/ft (see Fig. 4-33d)

Step 3: Draw moment distributions due to unit moments and shear forces as shown in Fig. 4-33e and f.

Step 4: Determine the effective modulus

From Eq. 4-99, we have

Cantilever 0-1:

$$E_{I\phi} = \frac{E}{1 + 0.55 \times 1} = 0.645E$$

Cantilever 1-2:

$$E_{II\phi} = \frac{E}{1 + 0.65 \times 2} = 0.435E$$

Step 5: Determine the displacement increments due to creep at location 1 due to external loading in primary structure

Relative angle increment (see the first term on the right side of Eq. 4-103a):

$$\Delta_{1P} = \int_0^{50} \frac{M_P(x)m_1(x)\phi_I(1000, \tau_0)}{EI} dx + \int_{50}^{100} \frac{M_P(x)m_1(x)\phi_{II}(1000, \tau_0)}{EI} dx$$

Integrating the above equation using the diagram multiplication method (Eq. 4-13) yields:

$$\Delta_{1P} = \frac{-50 \times 2500}{3EI} \times 1 \times 1 + \frac{-50 \times 2500}{3EI} \times 1 \times 2 = \frac{-125000}{EI}$$

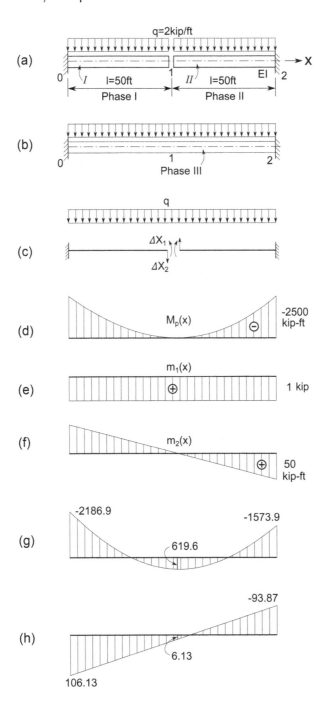

FIGURE 4-33 Determination of Final Internal Forces due to Creep in a Girder Built by the Cantilever Method, (a) Construction Phase I and Phase II, (b) Construction Phase III, (c) Primary Structure, (d) Moment Distribution due to the Dead Load in Phases I and II in the Primary Structure, (e) Moment Distribution due to $\Delta X_1 = 1$, (f) Moment Distribution due to $\Delta X_2 = 1$, (g) Final Moment Distribution, (h) Final Shear Distribution.

Relative displacement increment:

$$\Delta_{2P} = \int_0^{50} \frac{M_P(x)m_2(x)\phi_I(1000,\tau_0)}{EI}dx + \int_{50}^{100} \frac{M_P(x)m_2(x)\phi_{II}(1000,\tau_0)}{EI}dx$$

$$= \frac{50 \times 2500}{3EI} \times \frac{50 \times 3}{4} \times 1 + \frac{-50 \times 2500}{3EI} \times \frac{50 \times 3}{4} \times 2 = \frac{-1562500}{EI}$$

Step 6: Determine the displacement increments due to creep at location 1 due to a unit force couple in the primary structure.

Relative angle increment due to $\Delta X_1 = 1$:

$$\delta_{11} = \int_0^{50} \frac{m_1(x)m_1(x)}{E_{I\phi}I}dx + \int_{50}^{100} \frac{m_1(x)m_1(x)}{E_{II\phi}I}dx$$

$$= \frac{1 \times 50 \times 1}{0.645EI} + \frac{1 \times 50 \times 1}{0.435EI} = \frac{192.5}{EI}$$

Relative angle increment due to $\Delta X_2 = 1$:

$$\delta_{12} = \int_0^{50} \frac{m_1(x)m_2(x)}{E_{I\phi}I}dx + \int_{50}^{100} \frac{m_1(x)m_2(x)}{E_{II\phi}I}dx$$

$$= \frac{-1 \times 50 \times 25}{0.645EI} + \frac{1 \times 50 \times 25}{0.435EI} = \frac{935.6}{EI}$$

Relative displacement increment due to $\Delta X_1 = 1$:

$$\delta_{21} = \delta_{12} = \frac{935.6}{EI}$$

Relative displacement increment due to $\Delta X_2 = 1$:

$$\delta_{22} = \int_0^{50} \frac{m_2(x)m_2(x)}{E_{I\phi}I}dx + \int_{50}^{100} \frac{m_2(x)m_2(x)}{E_{II\phi}I}dx$$

$$= \frac{50 \times 50 \times 2 \times 50}{0.645EI \times 2 \times 3} + \frac{50 \times 50 \times 50 \times 2}{0.435EI \times 2 \times 3} = \frac{160384.9}{EI}$$

Step 7: Determine the redundant forces caused by creep, using the displacement compatibility conditions at joint 1.

The relative angle and displacement at joint 1 should be equal to zero. Thus, we have

$$\delta_{11}\Delta X_1 + \delta_{12}\Delta X_2 + \Delta_{1P} = 0$$
$$\delta_{21}\Delta X_1 + \delta_{22}\Delta X_2 + \Delta_{2P} = 0$$

Substituting δ_{11}, δ_{12}, Δ_{1P}, δ_{21}, δ_{22}, Δ_{2P} into the above equations yields

$$192.5\Delta X_1 + 935.6\Delta X_2 - 125000 = 0$$
$$935.6\Delta X_1 + 160384.9\Delta X_2 - 1562500 = 0$$

Solving the above equations yields

$$\Delta X_1 = 619.6 \text{ kips-ft}$$

$$\Delta X_2 = 6.13 \text{ kips}$$

Total moments:

$$\text{Joint 0: } M_0 = M_P + \Delta X_1 + \Delta X_2 l$$

$$= -2500 + 619.6 - 6.13 \times 50 = -2186.9 \text{ kips-ft}$$

Joint 1: $M_1 = \Delta X_1 = 619.6$ kips-ft
Joint 2: $M_2 = -2500 + 619.6 + 6.13 \times 50 = -1573.9$ kips-ft
Total shear:
Joint 0: $V_0 = 100 + 6.13 = 106.13$ kips
Joint 1: $V_1 = 6.13$ kips
Joint 2: $V_2 = -100 + 6.13 = -93.87$ kips

The moment and shear distributions of the girder at 1000 days after it was closed at midpoint are shown in Fig. 4-33g and h. It can be observed from this example that concrete creep reduces the negative moment and increases the positive moment.

4.3.4.4.3   Finite-Element Method Based on Effective Modulus

To simplify the explanation, the analytical procedures for determining the effect of concrete creep using the finite-element method are first discussed for bridges built in two stages and then for the bridges built in multiple stages.

4.3.4.4.3.1   Typical Analytical Procedures for Bridges Constructed in Two Stages   From Eq. 4-93h, the stress increment due to concrete creep can be written as

$$f_c(t,\tau_0) = E_\phi(t,\tau_0)\left[\varepsilon_c(t,\tau_0) - \frac{f(\tau_0)}{E(\tau_0)}\phi(t,\tau_0)\right]$$

$$= \beta(t,\tau_0)E(\tau_0)\varepsilon_c(t,\tau_0) - \beta(t,\tau_0)\phi(t,\tau_0)E(\tau_0)\varepsilon(\tau_0) \qquad (4\text{-}121)$$

Equation 4-121 shows the linear relationship between creep stress and strains. Thus, the relationship between the nodal forces and displacements of a finite-element due to concrete creep is similar to Eq. 4-44 and can be directly written as:

$$\{F_{ce}\} = \left(\beta(t,\tau_0)[k]\{\delta_{ce}\} - \phi(t,\tau_0)\beta(t,\tau_0)[k]\{\delta_{e0}\}\right.$$

$$= [k_\phi]\{\delta_{ce}\} - \phi(t,\tau_0)[k_\phi]\{\delta_{e0}\} = [k_\phi]\left[\{\delta_{ce}\} - \{\delta_{c0}^*\}\right] \qquad (4\text{-}122a)$$

where
$[k]$ = elastic stiffness matrix of finite-element

$$[k_\phi] = \beta(t,\tau_0)[k] \qquad (4\text{-}122b)$$

$$= \text{creep stiffness matrix of finite element}$$

$\{\delta_{ce}\}$ = element displacement vector due to concrete creep
$\{\delta_{e0}\}$ = element displacement vector due to initial forces
$\{\delta_{c0}^*\}$ = $\phi(t,\tau_0)\{\delta_{e0}\}$ = element displacement vector caused by initial forces due to creep

As $\{\delta_{e0}\}$ is known before any creep analysis, the nodal forces caused by $\{\delta_{e0}\}$ can be written as

$$\{F_{C0}'\} = \phi(t,\tau_0)[k_\phi]\{\delta_{e0}\} \tag{4-123}$$

where $\{F_{C0}'\}$ is a vector of the fixed-end forces caused by $\{\delta_{e0}\}$ due to creep. Thus, Eq. 4-122a can be re-written as

$$\{F_{ce}\} = [k_\phi]\{\delta_{ce}\} + \{F_{e0}\} \tag{4-122c}$$

where

$$\{F_{e0}\} = -\{F_{C0}'\} \tag{4-122d}$$

The first term on the right side of Eqs. 4-122a and 4-122c represents the element nodal forces due to displacements caused by concrete creep, and the second term $\{F_{e0}\}$ indicates the element nodal loads caused by the initial nodal forces of the element due to concrete creep and is often called equivalent creep nodal loads.

$$\{F_{e0}\} = \phi(t,\tau_0)\beta(t,\tau_0)\{F_e\} \tag{4-122e}$$

where $\{F_e\}$ = vector of initial node forces of the element.

As each node in a structure is in equilibrium during concrete creep provided there is no new additional external loads, we have

$$\sum \{F_{ce}\} = 0: [K_\phi]\{\delta_c\} + \{F_0\} = 0 \tag{4-124}$$

where
$\quad [K_\phi]$ = global creep stiffness matrix
$\quad \{\delta_c\}$ = global displacement vector due to creep
$\quad \{F_0\}$ = global vector of equivalent creep nodal loads

Solving Eq. 4-124 yields the displacements due to creep. Then, the nodal forces due to creep can be obtained.

Take the same example presented in Section 4.3.4.4.2 for describing the procedure for analyzing creep effect for bridges constructed in two stages by the finite-element method[4-14] (stage II and stage III are considered as one stage).

Step 1: For simplicity, we divide the fixed beam into two finite elements I and II as shown in Fig. 4-34a.

Step 2: Calculate the effective modulus coefficients based on Eq. 4-99.

Element I:

$$\beta(1000,\tau_0)_I = \frac{1}{1+\chi_I(1000,\tau_0)\phi_I(1000,\tau_0)} = \frac{1}{1+0.55\times1} = 0.645$$

Element II:

$$\beta(1000,\tau_0)_{II} = \frac{1}{1+\chi_{II}(1000,\tau_0)\phi_{II}(1000,\tau_0)} = \frac{1}{1+0.65\times2} = 0.435$$

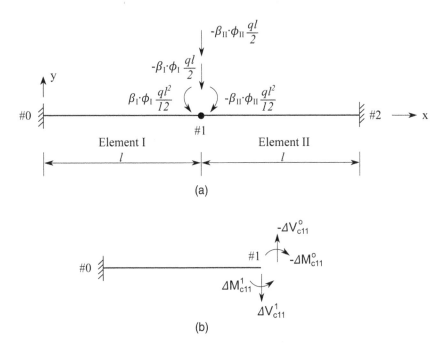

FIGURE 4-34 Determination of Secondary Forces due to Concrete Creep Using the Finite-Element Method, (a) Equivalent Nodal Loadings due to Creep, (b) Internal Force due to Creep.

Step 3: Determine the creep stiffness matrix of elements. As joints 0 and 2 are fixed and there is no axial deformation, there are only two unknown displacements; the rotation angle and the vertical deflection, the displacement vector of the finite-element can be therefore reduced to two, i.e.,

$$\{\delta_c\} = \left\{ \begin{array}{c} v_{c1} \\ \theta_{c1} \end{array} \right\}$$

The sizes of the related element stiffness matrix as shown in Eq. 4-46 are reduced from 6×6 to 2×2. From Eq. 4-122b, the creep stiffness matrix of the finite-element can be written as
Element I:

$$\left[k_\phi \right]_I = \left(\beta_I (t, \tau_0) \right) [k] = 0.645E \left[\begin{array}{cc} \dfrac{12I}{l^3} & -\dfrac{6I}{l^2} \\ -\dfrac{6I}{l^2} & \dfrac{4I}{l} \end{array} \right]$$

Element II:

$$\left[k_\phi \right]_{II} = \left(\beta_{II} (t, \tau_0) \right) [k] = 0.435E \left[\begin{array}{cc} \dfrac{12I}{l^3} & \dfrac{6I}{l^2} \\ \dfrac{6I}{l^2} & \dfrac{4I}{l} \end{array} \right]$$

Step 4: Determine the nodal loadings caused by the initial forces due to creep.

The equivalent nodal loadings of the elements due to creep are equal to the fixed-end moment and shear with opposite sign (see Fig. 4-34a) and can be calculated using Eq. 4-122e:

Element I:

$$\left\{ \begin{array}{c} \Delta V^0_{c11} \\ \Delta M^0_{c11} \end{array} \right\} = \beta_I(1000,\tau_0)\phi_I(1000,\tau_0)\{F_e\} = 0.645 \times 1 \times \left\{ \begin{array}{c} -\dfrac{ql}{2} \\ \dfrac{ql^2}{12} \end{array} \right\} = \left\{ \begin{array}{c} -32.25 \\ 268.75 \end{array} \right\}$$

Element II:

$$\left\{ \begin{array}{c} \Delta V^0_{c12} \\ \Delta M^0_{12} \end{array} \right\} = \beta_{II}(1000,\tau_0)\phi_{II}(1000,\tau_0)\{F_e\} = 0.435 \times 2 \times \left\{ \begin{array}{c} -\dfrac{ql}{2} \\ -\dfrac{ql^2}{12} \end{array} \right\} = \left\{ \begin{array}{c} -43.5 \\ -362.5 \end{array} \right\}$$

Step 5: Determine the global creep stiffness matrix and the global load vector.

As there is no transformation between the local and the global coordinate systems, the global stiffness matrix and load vector can be written as

Global creep stiffness matrix

$$[K_\phi] = [k_\phi]_I + [k_\phi]_I = \begin{bmatrix} 1.08\dfrac{12EI}{l^3} & -0.21\dfrac{6EI}{l^2} \\ -0.21\dfrac{6EI}{l^2} & 1.080\dfrac{4EI}{l} \end{bmatrix}$$

Global load matrix

$$\left\{ \begin{array}{c} \Delta V^0_{c1} \\ \Delta M^0_{c1} \end{array} \right\} = \left\{ \begin{array}{c} \Delta V^0_{c11} \\ \Delta M^0_{c11} \end{array} \right\} + \left\{ \begin{array}{c} \Delta V^0_{c12} \\ \Delta M^0_{c12} \end{array} \right\} = \left\{ \begin{array}{c} -75.75 \\ -93.75 \end{array} \right\}$$

Step 6: Establish global creep equations of equilibrium, and determine the nodal displacements

Similar to Eq. 4-124, the global creep equations of equilibrium are

$$\begin{bmatrix} 1.08\dfrac{12EI}{l^3} & -0.21\dfrac{6EI}{l^2} \\ -0.21\dfrac{6EI}{l^2} & 1.08\dfrac{4EI}{l} \end{bmatrix} \left\{ \begin{array}{c} v_{c1} \\ \theta_{c1} \end{array} \right\} = \left\{ \begin{array}{c} -75.75 \\ -93.75 \end{array} \right\}$$

Solving the equations above yields

$$\left\{ \begin{array}{c} v_{c1} \\ \theta_{c1} \end{array} \right\} = \left\{ \begin{array}{c} \dfrac{-1817.65l^2}{6EI} \\ \dfrac{-110.06l}{EI} \end{array} \right\}$$

Step 7: Determine the internal force increments at joint 1 due to creep, which are the sum of the nodal forces and the fixed-end loads due to creep (see Fig. 4-34b), i.e.,

Element I

The nodal forces induced by v_{c1} and θ_{c1}:

$$\left\{ \begin{array}{c} \Delta V_{c11}^1 \\ \Delta M_{c11}^1 \end{array} \right\} = 0.645E \begin{bmatrix} \dfrac{12I}{l^3} & -\dfrac{6I}{l^2} \\ -\dfrac{6I}{l^2} & \dfrac{4I}{l} \end{bmatrix} \left\{ \begin{array}{c} v_{c1} \\ \theta_{c1} \end{array} \right\} = \left\{ \begin{array}{c} -38.38 \\ 888.45 \end{array} \right\}$$

Total section forces:

$$\left\{ \begin{array}{c} \Delta V_{c11} \\ \Delta M_{C11} \end{array} \right\} = \left\{ \begin{array}{c} \Delta V_{c11}^1 \\ \Delta M_{c11}^1 \end{array} \right\} - \left\{ \begin{array}{c} \Delta V_{c11}^0 \\ \Delta M_{c11}^0 \end{array} \right\} = \left\{ \begin{array}{c} -6.13\ (kips) \\ 619.7\ (kips-ft) \end{array} \right\}$$

Element II:

The nodal forces induced by v_{c1} and θ_{c1}:

$$\left\{ \begin{array}{c} \Delta V_{c12}^1 \\ \Delta M_{c12}^1 \end{array} \right\} = 0.435E \begin{bmatrix} \dfrac{12I}{l^3} & \dfrac{6I}{l^2} \\ \dfrac{6I}{l^2} & \dfrac{4I}{l} \end{bmatrix} \left\{ \begin{array}{c} v_{c1} \\ \theta_{c1} \end{array} \right\} = \left\{ \begin{array}{c} -37.37 \\ -982.18 \end{array} \right\}$$

Total section forces:

$$\left\{ \begin{array}{c} \Delta V_{c12} \\ \Delta M_{C12} \end{array} \right\} = \left\{ \begin{array}{c} \Delta V_{c12}^1 \\ \Delta M_{c12}^1 \end{array} \right\} - \left\{ \begin{array}{c} \Delta V_{c12}^0 \\ \Delta M_{c12}^0 \end{array} \right\} = \left\{ \begin{array}{c} 6.13\ (kips) \\ -619.7\ (kips-ft) \end{array} \right\}$$

Step 8: Calculate the final internal forces at time 1000 days, which are equal to the sum of the internal forces in stage I and the increments due to creep:

At joint 0:

$$M_0 = -2500 + 619.7 - 6.13 \times 50 = -2186.8 \text{ kips-ft}$$
$$V_0 = 100 + 6.13 = 106.13 \text{ kips}$$

At joint 1:

$$M_1 = 619.7 \text{ kips-ft}$$
$$V_1 = 6.13 \text{ kips}$$

At joint 2:

$$M_2 = -2500 + 619.7 + 6.13 \times 50 = -1573.82 \text{ kips-ft}$$
$$V_2 = -100 + 6.13 = -93.87 \text{ kips}$$

It can be observed that there is little differences between the results obtained by the finite-element method based on the displacement method and those obtained in Section 4.3.4.4.2.

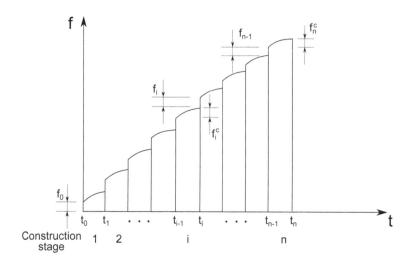

FIGURE 4-35   Variation of Stress with Construction Stages and Definitions of Stress Increments.

4.3.4.4.3.2   Typical Analytical Procedures for Bridges Constructed in Multiple Stages   As discussed in Chapter 1, concrete segmental bridges are built segment by segment and their structural systems often change with the different stages of construction. Because the initial forces vary with different construction stages, it is more convenient to develop the relationship between stress and strain in incremental form for analyzing the effect of concrete creep on structural behaviors.

Assume a segmental bridge is built in n stages. The variation of stress with time is shown in Fig. 4-35. From Eq. 4-93e, the total concrete strain due to creep at time t_m ($1 \le m \le n$) can be written as

$$\varepsilon_m^t = \sum_{i=0}^{m-1} \frac{f_i}{E} \phi(t_m, t_i) + \sum_{i=1}^{m} \int_{t_{i-1}}^{t_i} \frac{1}{E} \frac{df_i^c}{d\tau} \left[1 + \phi(t_m, \tau) \right] d\tau \qquad (4\text{-}125a)$$

where
    f_i = initial elastic stress increment at stage I due to external loads, including post-tensioning force (see Fig. 4-35)
    f_i^c = creep stress increment from time t_{i-1} to t_i (see Fig. 4-35)

At time $m-1$, the total concrete strain due to creep is

$$\varepsilon_{m-1}^t = \sum_{i=0}^{m-2} \frac{f_i}{E} \phi(t_{m-1}, t_i) + \sum_{i=1}^{m-1} \int_{t_{i-1}}^{t_i} \frac{1}{E} \frac{df^c}{d\tau} \left[1 + \phi(t_{m-1}, \tau) \right] d\tau \qquad (4\text{-}125b)$$

At stage m (from t_{m-1} to t_m) creep strain is

$$\varepsilon_m = \varepsilon_m^t - \varepsilon_{m-1}^t$$

$$= \sum_{i=0}^{m-1} \frac{f_i}{E} \left[\phi(t_m, t_i) - \phi(t_{m-1}, t_i) \right] + \frac{f_{m-1}}{E} \phi(t_{m-1}, t_{m-1})$$

$$+ \sum_{i=1}^{m-1} \int_{t_{i-1}}^{t_i} \frac{1}{E} \frac{df_i^c}{d\tau} \left[\phi(t_m, \tau) - \phi(t_{m-1}, \tau) \right] d\tau + \int_{t_{m-1}}^{t_m} \frac{1}{E} \frac{df_i^c}{d\tau} \left[1 + \phi(t_m, \tau) \right] d\tau \qquad (4\text{-}126)$$

The first term on the right side of Eq. 4-126 represents the sum of the strains caused by the initial stresses in stages 1 to $m-1$ from time t_{m-1} to t_m. The second term is equal to zero if the creep function is continuous. The third term indicates the sum of the strains caused by the creep strain increment during the time interval between t_{m-1} and t_m for stages 1 to $m-1$. The fourth term represents the strain caused by the creep stress increment from time t_{m-1} to t_m for stage m.

Using Lagrange's median theory, Eq. 4-126 can be written as

$$\varepsilon_m = \sum_{i=0}^{m-1} \frac{f_i}{E}\left[\phi(t_m, t_i) - \phi(t_{m-1}, t_i)\right] + \frac{f_{m-1}}{E}\phi(t_{m-1}, t_{m-1})$$

$$+ \sum_{i=1}^{m-1} \frac{f_i^c}{E}\left[\phi(t_m,\tau_\zeta) - \phi(t_{m-1},\tau_\zeta)\right] + \frac{f_i^c}{E}\left[1 + \chi(t,\tau_0)\phi(t_m,t_{m-1})\right]$$

$$= \sum_{i=0}^{m-1} \frac{f_i}{E}\Delta\phi_{mi} + \frac{f_{m-1}}{E}\phi(t_{m-1}, t_{m-1}) + \sum_{i=1}^{m-1} \frac{f_i^c}{E}\Delta\phi_{mi}^a + \frac{f_m^c}{E_\phi(t_m,t_{m-1})}$$

$$= \sum_{i=0}^{m-1} \frac{f_i}{E}\Delta\phi_{mi} + \frac{f_{m-1}}{E}\phi(t_{m-1}, t_{m-1}) + \sum_{i=1}^{m-1} \frac{f_i^c}{E_\phi(t_i,t_{i-1})}\beta(t_i,t_{i-1})\Delta\phi_{mi}^a + \frac{f_m^c}{E_\phi(t_m,t_{m-1})} \quad (4\text{-}127)$$

where

$$\tau_\zeta \approx \frac{t_{i-1} + t_i}{2}$$

$$\chi(t,\tau_0) = \frac{\int_{t_{m-1}}^{t_m} \frac{df_i^c}{dt}\phi(t_m,\tau)d\tau}{f_m^c\phi(t_m,t_{m-1})} = \text{aging coefficient (see Eq. 4-98a)}$$

$E_\phi(t_m,t_{m-1}) = $ effective modulus (see Eq. 4-99)

$\Delta\phi_{mi} = \phi(t_m, t_i) - \phi(t_{m-1}, t_i) = $ creep coefficient increment from t_{m-1} to t_m

$$\Delta\phi_{mi}^a = \phi(t_m,\tau_\zeta) - \phi(t_{m-1},\tau_\zeta)$$

Similar to Eq. 4-127, the corresponding total nodal displacements caused by the concrete creep from time t_{m-1} to t_m for stage m can be written as

$$\{\delta_m\} = \sum_{i=0}^{m-1} \{\delta_{0i}\} \Delta\phi_{mi} + \{\delta_{0m-1}\}\phi(t_{m-1}, t_{m-1})$$

$$+ \sum_{i=1}^{m-1}\left[\left(\{\delta_i\} - \{\delta_{0i}^c\}\right)\beta(t_i,t_{i-1})\Delta\phi_{mi}^a + \left(\{\delta_m\} - \{\delta_{0m}^c\}\right)\right] \quad (4\text{-}128)$$

where

First subscript 0 = initial nodal displacements before the stage under consideration

$\{\delta_{0i}\}$ = elastic nodal displacements caused by initial forces in stage i

$\{\delta_{0i}^c\}$ = creep nodal displacements caused by initial forces in stage i before the stage under consideration

$\{\delta_{0m}^c\}$ = creep nodal displacements caused by initial forces in stage m

$\{\delta_i\}$ = nodal displacements due to creep from time t_{i-1} to t_i for stage i

The first and second terms on the right side of Eq. 4-128 represent the change in the initial nodal displacements due to creep from time t_{m-1} to t_m. The third and fourth terms represent the nodal displacements caused by the stress increment due to creep in stage m. Rearranging Eq. 4-128 yields the creep nodal displacements caused by the initial forces in stage m:

$$\{\delta_{0m}^c\} = \sum_{i=0}^{m-1}\{\delta_{0i}\}\Delta\phi_{mi} + \{\delta_{0m-1}\}\phi(t_{m-1}, t_{m-1}) + \sum_{i=1}^{m-1}\left[\{\delta_i\} - \{\delta_{0i}^c\}\right]\beta(t_i, t_{i-1})\Delta\phi_{mi}^a \quad (4\text{-}129)$$

The following procedures can be used to develop a computer program for analyzing the effects of concrete creep:

Step 1: Calculate the initial nodal elastic displacements $\{\delta_{00}\}$ due to external and post-tensioning forces after instantaneous losses at time 0 for stage I of the structure, i.e., by solving the following equations:

$$[K_1]\{\delta_{00}\} + \{R_1\} = 0$$

where
  $[K_1]$ = global stiffness matrix for structure in stage I
  $\{R_1\}$ = global load vector in stage I

The initial nodal elastic forces of element $j\left\{F_{01}^e\right\}_j$ can be calculated as

$$\left\{F_{01}^e\right\}_j = [k]_j\left\{\delta_{00}^e\right\}_j$$

where the superscript e indicates "element" and $[k]_j$ is the elastic stiffness matrix of element j.

Step 2: Calculate the creep nodal displacements $\{\delta_{01}^c\}$ caused by initial internal forces during stage I. In stage I, $\{\delta_0\} = \{\delta_{00}^c\} = 0$; thus, from Eq. 4-129, we have

$$\{\delta_{01}^c\} = \{\delta_{00}\}\Delta\phi_{10} + \{\delta_{00}\}\phi(t_0, t_0)$$

Step 3: Calculate the effective global creep stiffness matrix based on Eq. 4-122b, i.e.,

$$\left[K_\phi^1\right] = \beta(t, \tau_0)[K_1]$$

Step 4: Calculate the equivalent nodal loads, based on Eq. 4-122d, i.e.,

$$\{F_{01}^c\} = -\left[K_\phi^1\right]\{\delta_{01}^c\}$$

Step 5: Solve the following equations and obtain the creep displacements $\{\delta_1\}$ from time t_0 to t_1.

$$\left[K_\phi^1\right]\{\delta_1\} + \{F_{01}^c\} = 0$$

Step 6: Calculate the nodal forces of element $j\left\{F_{01}^{ec}\right\}_j$ due to the creep displacements $\{\delta_1\}$

$$\left\{F_{01}^{ec}\right\}_j = \left[k_\phi^1\right]_j\{\delta_1\}_j$$

where $\left[k_\phi^1\right]_j$ is the effective stiffness matrix of element j in stage I.

Step 7: Impose new loads on the stage II structure, and determine the initial elastic nodal displacements at time t_1 by solving the following equations:

$$[K_2]\{\delta_{01}\} + \{R_2\} = 0$$

where

$[K_2]$ = global stiffness matrix for structure in stage II
$\{R_2\}$ = global load vector in stage II

The initial nodal elastic forces of element j $\{F^e_{02}\}_j$ can be calculated as

$$\{F^e_{02}\}_j = [k]_j \{\delta^e_{01}\}_j$$

Step 8: Calculate the creep nodal displacements $\{\delta^c_{02}\}$ caused by the initial internal forces during stage II based on Eq. 4-129, i.e.,

$$\{\delta^c_{02}\} = \sum_{i=0}^{2-1} \{\delta_{0i}\}\Delta\phi_{2i} + \{\delta_{01}\}\phi(t_1,\ t_1) + \left[\{\delta_1\} - \{\delta^c_{01}\}\right]\beta(t_1,t_0)\Delta\phi^a_{21}$$

Step 9: Follow steps 3 to 6 and obtain
The creep displacements from time t_1 to t_2: $\{\delta_2\}$ and the nodal forces of element j: $\{F^{ec}_{02}\}_j$
Step 10: Repeat steps 6 to 8 until the analysis for stage n is completed.
Step 11: Calculate total nodal displacements and nodal forces.
Total nodal displacements: $\{\delta\} = \sum_{i=0}^{n-1}\{\delta_{0i}\} + \sum_{i=1}^{n}\{\delta^c_{0i}\}$
Total nodal forces in element j: $\{F^e\}_j = \sum_{i=1}^{n}\left[\{F^e_{0i}\}_j + \{F^{ec}_{0i}\}_j\right]$
Step 12: Calculate section forces.
Using equilibrium conditions, the internal forces at any section can be obtained by imposing the nodal forces to the element in which the section under consideration is located and the related external loads within the element.

It should be mentioned that we have neglected the effect of the reinforcement in the concrete on the structural creep behaviors in above discussion.

4.3.4.4.4   Estimation of Moments Due to Concrete Creep

Equation 4-120 developed based on Dicschinger's assumptions can be used to estimate the effect of concrete creep. Considering Eq. 4-118b, we rewrite Eq. 4-120 as

$$M(x) = \left[X_{10}m_1(x) + M_P(x)\right] + \left[(X_B m_1(x) + M_P(x)) - (X_{10}\, m_1(x) + M_P(x))\right] \left[1 - e^{-[\phi(t,\tau_1) - \phi(\tau_2,\tau_1)]}\right] \qquad (4\text{-}130)$$

For the balanced cantilever constructed bridges as shown in Fig. 1-41, $X_{10} = 0$, $X_{10}m_1(x) + M_P(x)$ represents the moment due to the loads in the first structure; $X_B m_1(x) + M_P(x)$ represents the moment due to the loads in the second stage structure. Based on Eq. 4-130, for bridges constructed in two stages similar as shown in Fig. 1-41, the total moment due to dead load and creep at time t can be generally written as

$$M_{Dt}(x) = M_{D1}(x) + \left[M_{D2}(x) - M_{D1}(x)\right]\left[1 - e^{-[\phi(t,\tau_0) - \phi(\tau,\tau_0)]}\right] \qquad (4\text{-}131)$$

where

$M_{Dt}(x)$ = final dead load moment at time t in final structural system
$M_{D1}(x)$ = initial moment due to dead load in first stage structure
$M_{D2}(x)$ = moment due to dead load in second stage structure
τ_0 = initial loading time
τ = time at which first stage structural system is changing to second stage

It should be mentioned that Eq. 4-131 is derived based on the assumption that all the segments have the same casting time. The equation can also be used for estimating the creep moments due to post-tensioning forces. The secondary moments due to post-tensioning forces vary with time, and it can be approximated by treating the post-tensioning force before creep loss as a constant and then the secondary moment due to creep can be calculated using the following average effective coefficient C:

$$C = \frac{P_e}{P_0} \tag{4-132}$$

where

P_e = average final post-tensioning force after creep
P_0 = average post-tensioning force before creep

The final creep moment due to post-tensioning forces can be estimated as

$$M_{Pt}(x) = M_{P1}(x) + CX_{1Pt}m_1(x) \tag{4-133}$$

where

$M_{Pt}(x)$ = final moment due to post-tensioning forces in final structural system
$M_{P1}(x)$ = initial moment due to post-tensioning forces in first stage structure
$m_1(x)$ = moment in primary structure due to $X_{1Pt} = 1$
X_{1Pt} = secondary moment caused by post-tensioning forces before creep loss in second stage structure

The final combined moment due to both dead load and post-tensioning forces is

$$\begin{aligned} M_t(x) &= M_{Dt}(x) + M_{Pt}(x) \\ &= M_{D1}(x) + X_{1Dt}m_1(x) + M_{P1}(x) + CX_{1Pt}m_1(x) \end{aligned} \tag{4-134}$$

where X_{1Dt} = secondary moment caused by dead load due to creep in second stage structure.

Equations 4-131 and 4-134 can be used for estimating the creep effect on the bridges constructed by both span-by-span and balanced cantilever methods.

Based on the results of numerous analytical studies on continuous segmental bridges built using the cantilever method, the following simpler equation can also be used to estimate the final moment caused by the dead load due to the effect of creep in the preliminary bridge design stage[4-18]:

$$M_{Df} = M_{Dca} + 0.8(M_{Dco} - M_{Dca}) \tag{4-135}$$

where

M_{Df} = final moment in continuous girder due to dead loads (see Fig. 4-36c)
M_{Dca} = moment due to dead loads calculated based on cantilever girder system (see Fig. 4-36a)
M_{Dco} = moment due to dead loads calculated based on continuous girder system (see Fig. 4-36b)

4.3.5 ANALYSIS OF SECONDARY FORCES DUE TO SHRINKAGE

The behavior of concrete shrinkage and its variation with time were discussed in Section 1.2.2.6. As discussed, concrete shrinkage is not related to concrete stress and can be treated as an equivalent temperature increment uniformly distributed across the section in structural analysis. Then, the method discussed in Section 4.3.3 can be used to analyze the temperature-induced secondary forces. The equivalent uniform temperature increment can be written as

$$\Delta T_s = \frac{\varepsilon_{cs}(t,t_0)}{\alpha} \tag{4-136}$$

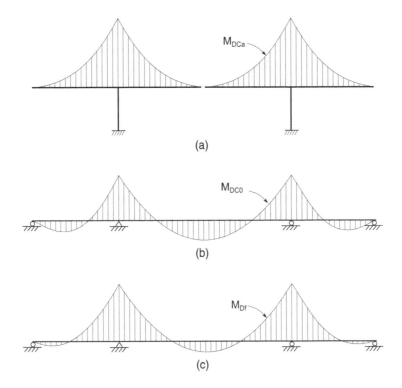

FIGURE 4-36 Estimation of Redistributed Moment Induced by Dead Loads Due to Creep in a Continuous Girder Built by the Cantilever Method, (a) Moment Calculated based on the Cantilever Girder System, (b) Moment Calculated based on the Continuous Girder System, (c) Final Moment Distribution.

where

$\varepsilon_{cs}(t,t_0)$ = concrete shrinkage strain

α = coefficient of thermal expansion (in./in./°F)

Apparently, if there is no restraint in the structure's axial direction, no secondary forces will be induced by concrete shrinkage.

In the finite-element method, shrinkage can be treated as an external load and the equivalent nodal forces of a two-dimensional element can be written as

$$\{R_e\} = \left[-\varepsilon_{cs}(t,t_0)EA \ \ 0 \ \ 0 \ \ \varepsilon_{cs}(t,t_0)EA \ \ 0 \ \ 0 \ \right]^T \tag{4-137}$$

4.3.6 Analysis of Secondary Forces Due to Settlements of Supports

If significant nonuniform settlements of supports are expected in a statically indeterminate bridge structure, the secondary forces induced by the settlements of the supports should be considered in the bridge design. Any of the methods discussed in Section 4.2 can be used for analyzing the secondary forces.

If supports 1 and 2 shown in Fig. 4-4 have settled by Δ_1 and Δ_2, respectively, the compatibility equations by the force method can be expressed based on Eqs. 4-17a and 4-17b:

$$\delta_{11}X_1 + \delta_{12}X_2 + \Delta_1 = 0 \tag{4-138a}$$

$$\delta_{12}X_1 + \delta_{22}X_2 + \Delta_2 = 0 \tag{4-138b}$$

Using the displacement method, we can treat the relative displacement as an external load and use the same procedure discussed in Section 4.2.3 to determine the secondary forces due to the

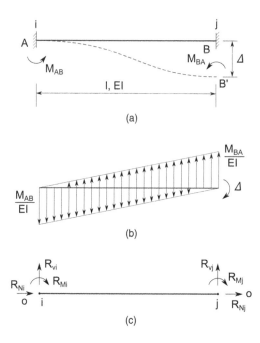

FIGURE 4-37 Fixed-End Moments Due to Support Deflection, (a) Support Deflection, (b) Loads on Conjugate Beam, (c) Equivalent Nodal Forces of Finite-Element.

nonuniform settlements. The fixed-end moments due to relative displacement between ends A and B as shown in Fig. 4-37a can be obtained by the conjugate beam method as follows. From the conjugate beam shown in Fig. 4-37b, summing the moment about B,

$$-\left(\frac{M_{AB}l}{2EI}\right)\left(\frac{2l}{3}\right)+\left(\frac{M_{BA}l}{2EI}\right)\left(\frac{l}{3}\right)+\Delta=0 \tag{4-139}$$

As $|M_{AB}|=|M_{BA}|$, solving Eq. 4-139 yields:

$$M_{BA}=-M_{AB}=\frac{6EI}{l^2}\Delta \tag{4-140}$$

Using the finite-element method, we can treat the relative displacement as an external load and use the same procedure described in Section 4.2.5 to determine the secondary forces due to the nonuniform settlements. The equivalent nodal loads of a two-dimensional element as shown in Fig. 4-37c are the same as the fixed-end forces with opposite sign and the load vector of the finite-element can be written as

$$\{R_e\}=\begin{Bmatrix} R_{Vi} \\ R_{Ni} \\ R_{Mi} \\ R_{Vj} \\ R_{Nj} \\ R_{Mj} \end{Bmatrix}=\begin{Bmatrix} \dfrac{12EI\Delta}{l^3} \\ 0 \\ \dfrac{6EI}{l^2}\Delta \\ \dfrac{-12EI\Delta}{l^3} \\ 0 \\ \dfrac{-6EI}{l^2}\Delta \end{Bmatrix} \tag{4-141}$$

where Δ is positive if the relative position between nodal i and nodal j is as shown in Fig. 4-37a, i.e., node i is in up direction (y-axis direction).

4.3.7   GEOMETRICAL NONLINEAR ANALYSIS

4.3.7.1   Introduction

Some slender members in segmental bridges may be subjected to heavy compression axial forces, such as concrete piles in the foundation, tall piers, arch ribs, towers, and the main girders in cable-stayed bridges. The axial forces in these members may significantly increase their bending moment, which should be considered in the bridge design due to large structural deformation, including the inevitable structural imperfectness and initial bending moments. The relationships between the internal forces and displacements are no longer linear. The structural analysis should be based on the large deflection theory, i.e., the theory of nonlinear analysis. Accurate nonanalysis needs to be carried out through a step-by-step analysis by dividing the external loading into a number of small amounts of loading. For this reason, such analysis is often done by computers. In this section, first the finite-element method is discussed. Then an approximate method the AASHTO specifications recommend will be presented.

4.3.7.2   Geometrical Nonlinear Analysis by FEM

Let's examine a simply supported beam subjected to both vertical and axial loads as shown Fig. 4-38. Consider a small element dx, and analyze its axial deformation, which includes two components. The first is due to the beam shortening in the longitudinal direction (see Fig. 4-38b), and the second is caused by the vertical deflections of the beam (see Fig. 4-38c). From Fig. 4-38b, the elongation of dx due to the longitudinal displacement is

$$\Delta_u = u + du - u = du \tag{4-142a}$$

From Fig. 4-38c, it can be observed:

$$ds = \sqrt{dx^2 + dv^2} = \sqrt{1 + \left(\frac{dv}{dx}\right)^2}\, dx \approx \left[1 + \frac{1}{2}\left(\frac{dv}{dx}\right)^2\right] dx \tag{4-143}$$

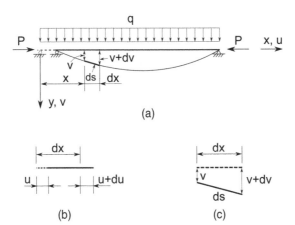

FIGURE 4-38 Large Deflection of Beam, (a) Deflection, (b) Longitudinal Deformation, (c) Vertical Deformation.

The elongation of dx due to the beam vertical deflection is

$$\Delta_v = ds - dx = \frac{1}{2}\left(\frac{dv}{dx}\right)^2 dx \qquad (4\text{-}142\text{b})$$

From Eqs. 4-142a and 4-142b, we can obtain the axial strain as

$$\varepsilon_x = \frac{\Delta_u + \Delta_v}{dx} = \frac{du}{dx} + \frac{1}{2}\left(\frac{dv}{dx}\right)^2 \qquad (4\text{-}144)$$

The first term of the right side of Eq. 4-144 represents the linear relationship between the strain and deformation, and the second term indicates the nonlinear relationship between strain and deformation.

Then the generalized strains considering large deflections can be revised from Eq. 4-39 as

$$\{\varepsilon\} = \left\{\begin{array}{c} \varepsilon_x \\ k_z' \\ k_z \end{array}\right\} = \left[\frac{du}{dx} \ \frac{d^3v}{dx^3} \ \frac{d^2v}{dx^2}\right]^T + \left[\frac{1}{2}\left(\frac{dv}{dx}\right)^2 \ 0 \ 0\right]^T = [B]\{\delta_e\} \qquad (4\text{-}145)$$

where

$$[B] = [B_0] + [B_L] \qquad (4\text{-}146)$$

$$[B_0] = \begin{bmatrix} \dfrac{dN_1}{dx} & 0 & 0 & \dfrac{dN_2}{dx} & 0 & 0 \\[2mm] 0 & \dfrac{d^3N_3}{dx^3} & \dfrac{d^3N_4}{dx^3} & 0 & \dfrac{d^3N_5}{dx^3} & \dfrac{d^3N_6}{dx^3} \\[2mm] 0 & \dfrac{d^2N_3}{dx^2} & \dfrac{d^2N_4}{dx^2} & 0 & \dfrac{d^2N_5}{dx^2} & \dfrac{d^2N_6}{dx^2} \end{bmatrix} \qquad (4\text{-}147)$$

$$[B_L] = \begin{bmatrix} [N_{14}] & [0] \\ [0] & [N_{14}] \end{bmatrix} \begin{bmatrix} \{\delta_e\} & \{0\} & \{0\} \end{bmatrix} \begin{bmatrix} [N_{14}] & [0] \\ [0] & [N_{14}] \end{bmatrix} \qquad (4\text{-}148)$$

$$[N_{14}] = \begin{bmatrix} 0 & 0 & 0 \\ 0 & \dfrac{dN_3}{dx} & 0 \\ 0 & 0 & \dfrac{dN_4}{dx} \end{bmatrix} \qquad (4\text{-}149)$$

$[B_0]$ is called the linear generalized strain matrix and $[B_L]$ is called the nonlinear generalized strain matrix.

Using the same procedures as discussed in step 4 of Section 4.2.5.2, we can obtain the nodal forces of the element as

$$\{F_e\} = \int [B]^T \{F\} dx - \{R_e\} \qquad (4\text{-}150\text{a})$$

$$\{F_e\} = \int [B]^T [D][B]\{\delta_e\} dx - \{R_e\} \qquad (4\text{-}150\text{b})$$

where $\{R_e\}$ = nodal loads of finite-element due to external loading (see Eq. 4-47).

From Eq. 4-148, we can see that the generalized strain matrix includes nodal displacements. Therefore, large deflection nonlinear equations cannot be directly solved. Normally, a step-by-step method should be used to solve this type of equation, such as the Newton-Raphson method[4-4, 4-8]. For this reason, we need to write Eq. 4-150 in differential form.

Differentiating both sides of Eq. 4-150a yields

$$d\{F_e\} = \int d[B]^T \{F\} dx + \int [B]^T d\{F\} dx \tag{4-151}$$

Substituting 4-40 and 4-146 into Eq. 4-151 yields

$$d\{F_e\} = \int d[B_L]^T \{F\} dx + \int [B]^T [D][B] d\{\delta_e\} dx \tag{4-152}$$

The first term on the right side of Eq. 4-152 represents the nodal force increments caused by the initial forces due to large deflection and can be written as

$$\int d[B_L]^T \{F\} dx = [k_g]\{\delta_e\} \tag{4-153}$$

where $[k_g]$ is often called the initial stress matrix of the element or the geometrical stiffness matrix.

The second term on the right side of Eq. 4-152 can be expressed as

$$\int [B]^T [D][B] d\{\delta_e\} dx = ([k_0]+[k_L])\{\delta_e\} \tag{4-154}$$

where

$$[k_0] = \int [B_0]^T [D][B_0] dx \tag{4-155}$$

$$[k_L] = \int \left\{[B_0]^T [D][B_L]+[B_L]^T [D][B_0]+[B_L]^T [D][B_L]\right\} dx \tag{4-156}$$

$[k_0]$ is often referred to as the linear stiffness matrix, which is the same as Eq. 4-46, and $[k_L]$ is called the large deflection matrix.

Substituting Eqs. 4-153 through 4-156 into Eq. 4-152 yields

$$d\{F_e\} = [k_T] d\{\delta_e\} \tag{4-157}$$

where

$$[k_T] = ([k_0]+[k_g]+[k_L]) \tag{4-158}$$

$[k_T]$ is termed the tangential stiffness matrix.

Equation 4-157 can be written in incremental form as

$$\{\Delta F_e\} = [k_T]\{\Delta \delta_e\} \tag{4-159}$$

The summation of nodal forces at each of the nodes plus nodal loads in the structure should be equal to zero. Thus we can obtain global equations of equilibrium as

$$\left[\left[K_0\right]+\left[K_g\right]+\left[K_L\right]\right]\{\Delta\delta\}+\{\Delta R\}=0 \tag{4-160}$$

where

$\left[K_0\right]$ = global linear stiffness matrix
$\left[K_g\right]$ = global initial stress or geometrical matrix
$\left[K_L\right]$ = global large deflection stiffness matrix
$\{\Delta\delta\}$ = vector of global deflection increments
$\{\Delta R\}$ = vector of global nodal load increments

Equation 4-160 can be solved using a step-by-step method in which the external loads are divided into a number of small loads[4-3 to 4-25].

4.3.7.3 Approximate Method for Considering P-Δ Effect

4.3.7.3.1 Background of Theory

From the discussion in Section 4.3.7.2, we see that accurate nonlinear analysis is time-consuming and is normally suitable for computer analysis. As mentioned in Section 3.7.5, if the slenderness ratio of a compression member is less than 100, the AASHTO specifications allow using an approximate method to consider the P-Δ effect (geometric nonlinear effect). In this section, an approximate method for considering the P-Δ effect of compression members subjected to a constant axial loading will be discussed. For simplicity, let's consider a compression-bending member with end moment M_0 subjected to a compression force P as shown in Fig. 4-39a. Taking an isolated body as shown in Fig. 4-39b and using $\Sigma M = 0$, we have

$$M + Py = M_0 \tag{4-161}$$

Differentiating twice on both sides of Eq. 4-161 and considering Eq. 4-4, we have

$$EI\frac{d^4y}{dx^4} + P\frac{d^2y}{dx^2} = 0 \tag{4-162a}$$

Rearranging Eq. 4-162a yields

$$\frac{d^4y}{dx^4} + \alpha^2\frac{d^2y}{dx^2} = 0 \tag{4-162b}$$

where $\alpha^2 = \frac{P}{EI}$
   The general solution of Eq. 4-162b is

$$y = A\sin\alpha x + B\cos\alpha x + Cx + D$$

Using boundary conditions:

$$x = 0,\ y = 0,\ EI\frac{d^2y}{dx^2} = M_0$$

$$x = l,\ y = 0,\ EI\frac{d^2y}{dx^2} = M_0$$

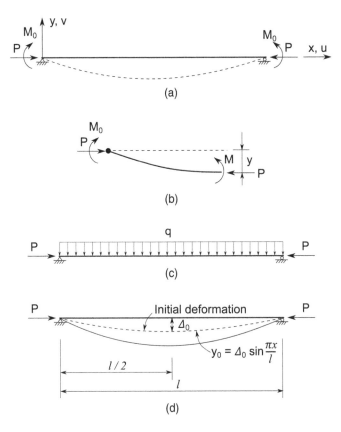

FIGURE 4-39 Compression and Bending Members, (a) Compression-Bending Member with End Moment, (b) Isolated Body, (c) Compression-Bending Member with Vertical Uniform Loading, (d) Compression-Bending Member with Initial Deformation.

we can obtain

$$y = \frac{M_0}{P}\left(\frac{\sin\alpha x}{\sin\alpha l} + \frac{\sin\alpha(l-x)}{\sin\alpha l} - 1\right) \tag{4-163}$$

$$M = -EI\frac{d^2y}{dx^2} = M_0\left(\frac{\sin\alpha x}{\sin\alpha l} + \frac{\sin\alpha(l-x)}{\sin\alpha l}\right) \tag{4-164a}$$

As the maximum moment is at mid-span, substituting $x = \frac{l}{2}$ into Eq. 4-164a, we can obtain:

$$M_{max} = \mu M_0 \tag{4-164b}$$

where μ is called the moment magnification factor.

$$\mu = \frac{1}{\cos\frac{\alpha l}{2}} \approx 1 + \frac{1}{1-1.23\beta} \tag{4-165a}$$

where

$$\beta = \frac{P}{P_e}$$

$$P_e = \frac{\pi^2 EI}{l^2} = \text{Euler buckling load}$$

The maximum moment magnification factor μ varies with different loading conditions. If the beam is subjected to uniform transverse loading (see Fig. 4-39c), the magnification factor for the moment at its mid-span can be derived using a similar method as described above and can be expressed as

$$\mu = 1 + \frac{\pi^2}{9.6}\frac{1}{1-\beta} \tag{4-165b}$$

For a beam with an initial deformation y_0 that can be described as

$$y_0 = \Delta_0 \sin\frac{\pi x}{l} \tag{4-166}$$

its moment magnification factor for the mid-span moment can be expressed as

$$\mu = 1 + \frac{1}{1-\beta} \tag{4-165c}$$

4.3.7.3.2   Method Proposed by AASHTO LRFD Specifications

The AASHTO LRFD specifications[4-19] present a simplified method for estimating the P-Δ effect on a beam-column member, which is based the theory described above. The factored moments or stresses can be increased to account for the effects of deformations as follows:

$$M_c = \mu_b M_b + \mu_s M_s \tag{4-167a}$$

where

$$\mu_b = \frac{C_m}{1-\frac{P_u}{\phi_k P_e}} \geq 1 \tag{4-167b}$$

$$\mu_s = \frac{1}{1-\frac{\Sigma P_u}{\phi_k \Sigma P_e}} \tag{4-167c}$$

where
M_b = moment due to factored gravity loads that induce no appreciable sidesway, calculated by conventional first-order elastic frame analysis, such as the end moments and moment due to transverse loading, always positive
M_s = moment due to factored lateral or gravity loads that induce sidesway Δ greater than $l_u / 1500$, calculated by conventional first-order elastic frame analysis, such as moment due to initial member deformations; always positive
P_u = factored axial load
ϕ_k = stiffness reduction factor; 0.75 for concrete members and 1.0 for steel members
P_e = Euler buckling load

$$P_e = \frac{\pi^2 EI}{(kl_u)^2} \tag{4-167d}$$

k = effective length factor in plane of bending (see Tables 4-3 and 4-4)
l_u = unsupported length, for arch bridge, equal to one-half length of rib
C_m = $0.6 + 0.4\frac{M_1}{M_2}$, if members are braced against sidesway at their ends and there are no transverse loads between supports; for all other cases and arch ribs, C_m should be taken as 1.0

TABLE 4-3
Effective Length Factors for Straight Members[4-19]

Support Conditions

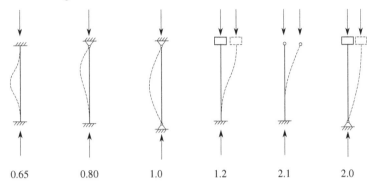

Design k Values	0.65	0.80	1.0	1.2	2.1	2.0

M_1 = smaller end moment
M_2 = larger end moment

The ratio M_1 / M_2 is positive if the member is bent in single curvature (see Fig. 4-39); otherwise the ratio is negative.

As the design is based on a factored applied axial force, the stiffness of the reinforced concrete beams or columns should be reduced in determining the Euler's buckling P_e and slenderness ratio. The AASHTO specifications recommend that the effective stiffness can be estimated as the larger of the following:

$$EI_e = \frac{\frac{E_c I_g}{5} + E_s I_s}{1 + \alpha_d}$$ (4-168a)

$$EI_e = \frac{\frac{E_c I_g}{2.5}}{1 + \alpha_d}$$ (4-168b)

where
E_c = concrete modulus of elasticity
I_g = moment of inertia of concrete gross section about centroidal axis
E_s = longitudinal steel modulus of elasticity
I_s = longitudinal steel moment of inertia about centroidal axis
α_d = ratio of maximum factored permanent load moments to maximum factored total load moment; positive

TABLE 4-4
Effective Length Factors for Arches

Rise-to-Span Ratio	3-Hinged	2-Hinged	Fixed
0.1 to 0.2	1.16	1.04	0.70
0.2 to 0.3	1.13	1.10	0.70
0.3 to 0.4	1.16	1.16	0.72

4.3.8 MATERIAL NON LINEAR ANALYSIS

4.3.8.1 Introduction

As discussed in Chapter 2, the safety of a bridge structure is codified by an application of the general statement that the design resistances must be greater than, or equal to, the design load effects. The design load effects are determined based on elastic theory by assuming that the material remains elastic until structural failures. However, for statically indeterminate structures, this statement is not true. For indeterminate structures, after the material in one member exceeds its elastic limit, the internal forces in the structure will be redistributed and more loads can be added. The AASHTO specifications have addressed this issue by specifying load modifiers. Another design concern is to ensure the structural stability and to check the buckling loads of the bridge. As all structural elements cannot be perfectly fabricated and built, Euler's buckling loading will never be reached and the process for determining the buckling load essentially becomes determining the ultimate capacity, which is often called the second category of buckling. To obtain more rational bridge ultimate capacities, especially for long-span bridges, it may be necessary to consider the material's nonlinear characteristics and perform inelastic structural analysis. The main difference between the elastic analysis and the inelastic analysis is that the modulus of elasticity varies with stresses after the stress exceeds the material's elastic limit for inelastic structures. Thus, the inelastic structural analysis should be carried out by iterative methods. First, the external loads are divided into a number of small amounts that are gradually applied to the structure. In each of the loading stages, the modulus of elasticity of an element is determined based on the initial internal forces calculated in the last loading stage and treated as a constant. Then, the basic analytical procedure for elastic analysis can be used for inelastic analysis with modified modulus of elasticity of the materials. Currently, there are several analytical methods for considering the effects of material non-linearity on the behavior of structures. One such method is the modified stiffness method, which can easily be implemented into computer programs for the elastic analysis of structures. The modified stiffness method is discussed in Section 4.3.8.2.

4.3.8.2 Modified Stiffness Method

Though the stiffness of a structural element varies with external loadings after its stress exceeds the material's elastic limit, the relationship between the internal force and the generalized strain can be assumed to be linear if the load increment is small enough, i.e.,

$$N_i = \xi_i EA\varepsilon_i = A_{ei}\varepsilon_i \tag{4-169a}$$

$$M_i = \eta_i EI\kappa_i = B_{ei}\kappa_i \tag{4-169b}$$

where

N, M = axial force and moment of section of concerned, respectively
i = load stage number
ξ, η = reduction factors of axial stiffness and bending stiffness, respectively
A_e, B_e = effective axial stiffness and effective bending stiffness, respectively
ε, κ = strain and curvature along the sectional centroid

Thus, the inelastic analysis of bridge structures can be performed by using the elastic analysis methods previously discussed by replacing the EI and EA by B_e and A_e, respectively. This is the basic concept for the modified stiffness method. The remaining question is how to determine the effective axial and bending stiffnesses of A_e and B_e.

The following assumptions are made when determining A_e and B_e:

a. The plane section remains plane until its ultimate strength; i.e., the strains are linearly distributed over the girder depth.
b. Tendons are assumed to be bonded to the concrete, and their strain change is the same as the reinforcement.

c. The effects of shear stress and strain are neglected.

d. Stress-strain relationships of the materials are known or can be assumed as shown in Fig. 4-40, such as for concrete, the stress-strain relationship may written as

$$f_c(\varepsilon) = E_c\varepsilon + \left(\frac{3f_c'^2}{\varepsilon_{c0}^2} - \frac{2E_c}{\varepsilon_{c0}}\right)\varepsilon^2 + \left(\frac{E_c}{\varepsilon_{c0}^2} - \frac{2f_c'}{\varepsilon_{c0}^3}\right)\varepsilon^3 \qquad (4\text{-}170)$$

where ε_{c0} is the strain corresponding to concrete strength f_c'.

e. The variation of the element internal forces can be assumed to be linear, and the effective stiffness of the element can be determined by the average internal forces of the element.

For simplicity, consider a rectangular box section as shown in Fig. 4-41, for example, and assume the tendon area $A_{ps1} = A_{ps2} = 0$ to discuss the basic procedures for determining the effective stiffness.

Step 1: Divide the section into m strips along the girder depth, and denote the area of strip j as A_j (see Fig. 4-41).

Step 2: Assume the axial strain and curvature along the girder centroid as

$$\varepsilon_i^{(0)} = \frac{N}{A_{ei}} \qquad (4\text{-}171)$$

$$\kappa_i^{(0)} = \frac{M}{B_{ei}} \qquad (4\text{-}172)$$

where subscript i denotes the load stage and the superscript 0 represents the initial assumed strain or curvature, which can be taken as the values obtained for the previous loading stage.

Step 3: Determine the average concrete stress f_{cj} for each of the area strips.

The average strain in strip j is

$$\varepsilon_j = \varepsilon_i + y_j\kappa_i \qquad (4\text{-}173)$$

where y_j = distance from center of strip j to section centroid line.

From the concrete stress-strain curve, we can obtain the average stress in strip j f_{cj}. If $\varepsilon_j \geq 0$ (concrete is in tension), we can take $f_{cj} = 0$.

Step 4: Determine the average steel stress f_{sj} for each of the top and bottom reinforcements.

The strains at the centers of the top and bottom reinforcements are

$$\varepsilon_{s1} = \varepsilon_i + y_{s1}\kappa_i \qquad (4\text{-}174)$$

$$\varepsilon_{s2} = \varepsilon_i - y_{s2}\kappa_i \qquad (4\text{-}175)$$

where y_{s1} and y_{s2} are the distances from the centers of the top and bottom reinforcements to the section's centroid. From the stress-strain curve of steel as shown in Fig. 4-40c, we can obtain the steel stress. From the assumed stress-strain curve as shown in Fig. 4-40c, if $|\varepsilon_{sj}| \geq \varepsilon_y$, the modulus of elasticity of the steel can be taken as $E_s = 0$.

Step 5: Using the equilibrium conditions, we can obtain the initial estimated axial force $N_e^{(0)}$ and moment $M_e^{(0)}$ as

$$N_e^{(0)} = \sum_{j=1}^{m} f_{cj}A_j + f_{s1}A_{s1} + f_{s2}A_{s2} \qquad (4\text{-}176)$$

$$M_e^{(0)} = \sum_{j=1}^{m} f_{cj}A_jy_j + f_{s1}A_{s1}y_1 - f_{s2}A_{s2}y_2 \qquad (4\text{-}177)$$

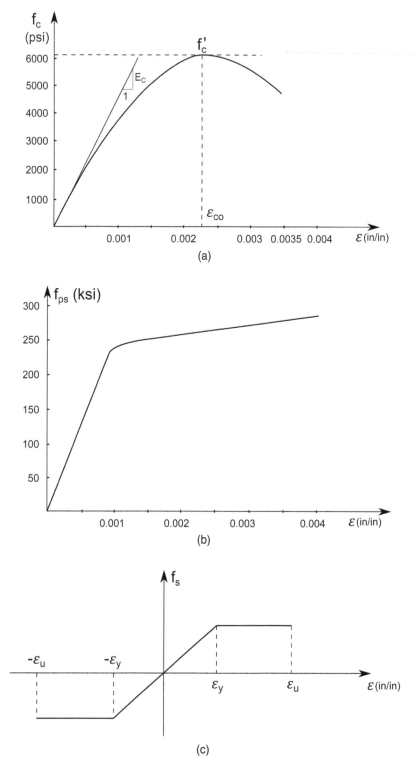

FIGURE 4-40 Stress-Strain Relationships of Materials, (a) Concrete, (b) Prestressing Tendon, (c) Reinforcement.

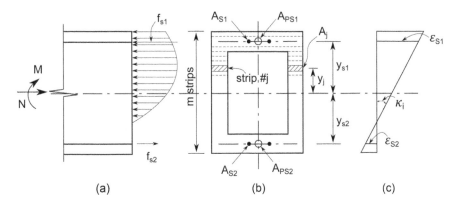

FIGURE 4-41 Analytical Model for Effective Stiffness, (a) Stress Distribution, (b) Cross Section and Area Strips, (c) Strain Distribution.

Step 6: Check if the assumed strain and curvature are sufficiently accurate.

The differences between the actual and the assumed axial forces and moments are

$$\Delta N = N_e^{(0)} - N \tag{4-178}$$

$$\Delta M = M_e^{(0)} - M \tag{4-179}$$

If ΔN and ΔM are both small or equal to the allowable errors, then the assumed $\varepsilon_i^{(0)}$ and $\kappa_i^{(0)}$ are correct and we can proceed to the next load stage of analysis. Otherwise, the assumed stress and curvature should be adjusted as follows:

$$\varepsilon_i^{(1)} = \varepsilon_i^{(0)} + \varepsilon_l \tag{4-180}$$

$$\kappa_i^{(1)} = \kappa_i^{(0)} + \kappa_l \tag{4-181}$$

where ε_l and κ_l are the adjusted amounts of the strain and curvature. To make ΔN and ΔM become zero in the next iteration calculation, ε_l and κ_l must meet the following equations:

$$\begin{cases} \dfrac{\partial(\Delta N)}{\partial \varepsilon} \varepsilon_l + \dfrac{\partial(\Delta N)}{\partial \kappa} \kappa_l + \Delta N = 0 \\[4mm] \dfrac{\partial(\Delta M)}{\partial \varepsilon} \varepsilon_l + \dfrac{\partial(\Delta M)}{\partial \kappa} \kappa_l + \Delta M = 0 \end{cases} \tag{4-182}$$

where

$$\frac{\partial(\Delta N)}{\partial \varepsilon} = \frac{\partial(N_e)}{\partial \varepsilon} = \sum_{j=1}^{m} (E_t) A_j$$

$$\frac{\partial(\Delta N)}{\partial \kappa} = \frac{\partial(N_e)}{\partial \kappa} = \sum_{j=1}^{m} (E_t) A_j y_j$$

$$\frac{\partial(\Delta M)}{\partial \varepsilon} = \frac{\partial(M_e)}{\partial \varepsilon} = \sum_{j=1}^{m} (E_t) A_j y_j$$

$$\frac{\partial(\Delta M)}{\partial \kappa} = \frac{\partial(M_e)}{\partial \kappa} = \sum_{j=1}^{m} (E_t) A_j y_j^2$$

Solving Eq. 4-182, yields ε_l and κ_l.

Step 7: Repeat steps 3 to 6, until ΔN and ΔM meet the required accuracy.

Step 8: Calculate effective stiffness.

Compressive stiffness: $A_i = \frac{N}{\varepsilon_i}$

Bending stiffness: $B_i = \frac{M}{\kappa_i}$

Now that the effective compressive and the effective bending stiffness have been completed, we can analyze the structure using the same procedure for elastic analysis by substituting A_i and B_i for EA and EI, respectively. The applied loading is increased, and steps 2 through 9 are repeated until the structure fails.

4.3.9 STABILITY ANALYSIS BY FINITE-ELEMENT METHOD

When the in-plane loading of a bridge, beam, column, arch, or cable-stayed bridges, reaches a critical loading, even very small deformation may suddenly induce large deformation of the bridge either in-plane or out-plane. The phenomenon is termed buckling or the first category of instability. The critical loading is also called buckling loading. All bridges should be designed with enough capacity to prevent bridge buckling. Actually, the buckling load of a bridge structure can never be reached because the bridge could not be built perfectly according to the design plans as previously mentioned. The ultimate loading capacity corresponding to a certain loading pattern can be obtained through nonlinear analysis, and the ultimate capacity is less than but reasonably close to the theoretical buckling loading, provided the bridge's final geometry well matches the design plans. Thus, the safety factors of bridge elastic buckling are always assigned a high value of over 3 in the design. Although, the stability analysis of most bridges in the final stage of design is generally performed using the finite-element method through computer software programs, it is beneficial for bridge engineers to understand the basic theory and method. Because the basic theory of the finite-element method and nonlinear analysis procedures have been discussed in previous sections, in this section, we will consider two-dimensional structures and only discuss the basic analytical procedures for buckling analysis.

Step 1: Determine the finite-element stiffness matrix for each of the finite elements in the bridge structure by Eq. 4-46.

Step 2: Assemble the global stiffness matrix $[K]$.

Step 3: Select the worst in-plane loading case $q(x)$ (symmetrical in transverse direction and any patterns in longitudinal direction (x). The most unfavorable load case includes both dead and live loads and generally will induce maximum axial loadings in the members. In the analysis, several load cases may be selected for comparison.

Step 4: Determine the global load vector $\{R\}$.

Step 5: Solve Eq. 4-51, and determine the axial loads N for each of the finite elements.

Step 6: Determine the finite-element geometrical matrixes based on Eq. 4-153. For a two-dimensional beam element, its geometrical matrix can be written as

$$[k_g] = \frac{N}{l} \begin{bmatrix} 0 & 0 & 0 & 0 & 0 & 0 \\ 0 & \frac{6}{5} & -\frac{l}{10} & 0 & -\frac{6}{5} & -\frac{l}{10} \\ 0 & -\frac{l}{10} & \frac{2l^2}{15} & 0 & \frac{l}{10} & -\frac{l^2}{30} \\ 0 & 0 & 0 & 0 & 0 & 0 \\ 0 & -\frac{6}{5} & \frac{l}{10} & 0 & \frac{6}{5} & \frac{l}{10} \\ 0 & -\frac{l}{10} & -\frac{l^2}{30} & 0 & \frac{l}{10} & \frac{2l^2}{15} \end{bmatrix} \tag{4-183}$$

where N = axial force of finite-element due to $q(x)$.

Step 7: Assemble the global geometrical matrix $[Kg]$.

Step 8: Assuming that the bridge buckling loading $q_{cr}(x)$ is λ_{cr} times the design loading $q(x)$, i.e.,

$$q_{cr}(x) = \lambda_{cr} q(x) \tag{4-184}$$

where

$q(x)$ = most unfavorable design loading pattern, including dead and live loadings

λ_{cr} = safety factor of bridge buckling

Then, the buckling equations of the bridge can be written as

$$\left([K_0] - \lambda_{cr}[K_g]\right)\{\delta\} = 0 \tag{4-185}$$

where

$\{\delta\}$ = vector of global small buckling displacements

$[K_0]$ = global stiffness matrix and represents elastic nodal force due to small buckling displacements with a unit of one

$[K_g]$ = global geometric matrix and represents additional nodal forces induced by in-plane vertical design loading $q(x)$ due to small buckling displacements

$[K_g]$ is the function of initial axial loadings N_i ($i = 1, 2, ..., m$), where m is the total number of elements.

Step 9: Determine the safety factor of bridge buckling.

Equation 4-185 is valid only if the following denominator determinant is equal to zero:

$$\left\|[K] - \lambda_{cr}[G]\right\| = 0 \tag{4-186}$$

Solving Eq. 4-186, we can obtain the safety factor of bridge buckling λ_{cr}.

The detailed theory and procedures for deriving Eqs. 4-185 and 4-186 can be found in References 4-20 to 4-25.

4.3.10   Bridge Modeling by the Finite-Element Method

4.3.10.1   Straight Bridges

4.3.10.1.1   Three-Dimensional Beam Model

For a straight box girder bridge, a two-dimensional beam model can normally be used for analyzing the girder bending moments and vertical shears accurately enough for bridge design. However, this type of model cannot be used for analyzing bridge torsion. Thus, it is better to use the three-dimensional beam model for analyzing segmental box girder bridges, especially for single box girder bridges subject to large torsion forces. Figure 4-42 shows a typical three-dimensional beam model of a single box girder bridge. This model is often called the three-dimensional spine model. The selection of the number of finite elements in each span is related to the types of analytical loads, girder section variation, etc. Normally, the number of finite elements in one span is not recommended to be less than 10. For most bridge cases, 20 elements in each of the spans should be adequate. For creep analysis, it is reasonable to assume the same casting time for each of the elements. The end diaphragms over the supports along the bridge transverse direction can be treated as a rigid member (see Fig. 4-42).

4.3.10.1.2   Grid Model

For multi-box (see Fig. 4-43a) or multi-cell box (see Fig. 4-43b) girder bridges, plane grid or three-dimensional grid models can be used for analyzing the bridges (see Fig. 4-43)[4-4, 4-5]. For multi-box

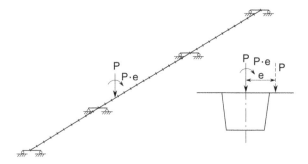

FIGURE 4-42   Three-Dimensional Beam Model.

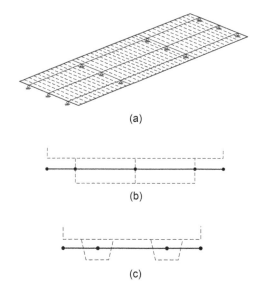

FIGURE 4-43   Grillage Modeling, (a) Typical Grid Model, (b) Multi-Cell Box Girder Modeling, (c) Multi-Box Girder Modeling.

girder bridges, each of the boxes is treated as one longitudinal beam plus two additional longitudinal edge beams along each of the barrier walls (see Fig. 4-43a and c). For multi-cell box girder bridges, each of the webs plus its attributed top and bottom slabs is treated as one longitudinal beam (see Fig. 4-43b).

4.3.10.1.3   Shell-Plate Model

If local stress distributions or the effects of section deformation are to be analyzed, the shell-plate model can be used (see Fig. 4-44)[4-9, 4-26]. The aspect ratio of the shell-plates should be less than two to ensure the accuracy of the results, especially in the areas concerned.

4.3.10.2   Curved Bridge

Based on numerical analysis[4-27 to 4-33], horizontally curved concrete box girder bridges can be analyzed using three approaches:

   I. A horizontally curved concrete box girder bridge with central angles up to 12 degrees within one span may be designed as a straight bridge, and the effect of the curvatures can be neglected.

FIGURE 4-44   Shell-Plate Model.

II. If the center angle is between 12° and 34° the bridge can be modeled as a single spine beam divided into a number of straight elements. The maximum length of the elements should not be greater than 0.06 times the radius R (see Fig. 4-45).

III. If the center angle of any of the spans exceeds 34°, the effect of section deformation should be considered and more accurate analytical models should be developed, such as the shell-plate finite-element model (see Fig. 4-46) and curved thin-wall box girder model.

4.3.11   REMARKS ON DEFLECTION AND CAMBER CALCULATIONS

Deflection and camber calculations are very important issues in segmental bridge constructions. Under service loadings, post-tensioned concrete girders generally do not crack and can be treated as elastic structures. The deflections can be computed by any of the methods described in this chapter. There are two difficulties in accurately calculating the deflections. First, the concrete modulus of elasticity E_c normally varies with different stress level and the age of concrete. The second difficulty is that the value of the creep coefficient as well as the duration and the magnitudes of applied loads may not always be known in advance. Nevertheless, for most segmental bridges, an accuracy of about 10% to 20% is typically sufficient and can be obtained[4-10], based on methods specified in the AASHTO specifications.

Camber is defined as the amount by which the concrete profile at the time of casting and installing must differ from the theoretical geometric profile grade to compensate for all structural dead load, post-tensioning, long and short-term time-dependent deformations (creep and shrinkage), and effects of construction loads and sequence of erection. Camber calculations should be based on the modulus of elasticity and the maturity of the concrete when each increment of load is added

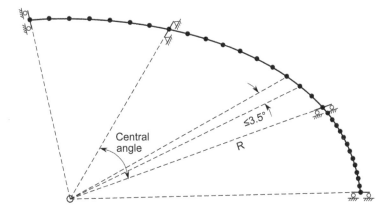

FIGURE 4-45   Curved Girder Modeling[4-33].

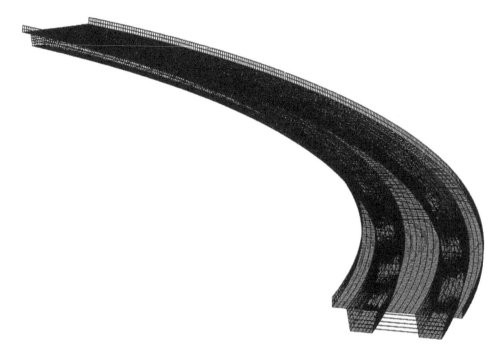

FIGURE 4-46   Shell-Plate Model for Curved Girders[4-31].

or removed. The deflections due to the concrete creep and shrinkage should be determined based on the equations for estimating the creep and shrinkage specified in Sections 1.2.2.5 and 1.2.2.6 or Article 5.4.2.3 in the AASHTO specifications[4-19].

The AASHTO specifications stipulate that the contract documents shall require that the deflections of segmentally constructed bridges be calculated prior to the casting of segments and that the deflections shall be based on the anticipated casting and erection schedules. The computed deflections are used as a guide for checking the actual deflection measurements during segment erection.

The Florida Department of Transportation (FDOT) provides a more specific requirement that for segmental box girders, the specialty engineer shall provide the camber curves, and the engineer of record (EOR) shall check them. For other bridge types, the EOR shall provide and check the camber curves.

4.4   TRANSVERSE ANALYSIS OF SEGMENTAL BRIDGES

4.4.1   INTRODUCTION

Theoretically, longitudinal and transverse behaviors of a bridge structure are related to each other. However, for simplicity in bridge designs, the longitudinal and transverse analyses are typically separated. Thus, some assumptions are always made in the bridge analysis. Though considerable efforts have been made to find a simple way of performing transverse analysis in segmental bridges, such as the effective width method[4-33], most of the simple methods do not provide reliable results that can be used for the design of box girder bridges, especially for single box girder bridges. Currently most bridge engineers use three-dimensional models to perform the transverse analysis because of the complexity and variations in the cross sections and loadings along the length of the segmental bridges. In the following section, a conventional simplified method is first discussed. Then, a more complex three-dimensional method is presented.

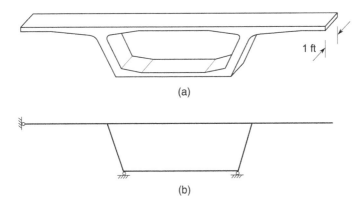

1 ft

(a)

(b)

FIGURE 4-47 Transverse Analytic Model of Box Girder Bridge with Longitudinal Uniform Loads, (a) Analytical Strip, (b) Analytical Model.

4.4.2 TWO-DIMENSIONAL ANALYSIS

The basic concept of the two-dimensional method of the transverse analysis is to first consider the load distribution effect of the loaded top slab of the box section in the bridge longitudinal direction. Then, take a unit length in the bridge's longitudinal direction and treat it as a plane frame structure that can be analyzed for bending moments at critical locations.

4.4.2.1 Transverse Analysis with Uniform Loadings

Longitudinally uniformly distributed loads include the self-weight of barrier walls, box girder, future wearing surface, and lane loadings. Strictly speaking, the section transverse moments within a unit length along a girder's longitudinal direction at mid span will be smaller than that along the support areas. Analytical results show that the transverse response due to the uniformly distributed loads can be conservatively analyzed by modeling the box section as a frame structure with unit width as shown in Fig. 4-47. The plane frame structure can be analyzed using the force method, the displacement method, or finite-element method as discussed in Section 4.2. It should be mentioned that the flanges shall be analyzed as variable-depth sections, considering the fillets between the flanges and webs.

4.4.2.2 Transverse Analysis with Truck Loadings

For concentrated truck loadings, it is important to first identify how the concentrated loads are distributed along the section concerned in the girder longitudinal direction. However, this is not as straight forward as it seems. Most segmental bridge designers currently use influence surfaces developed by Homberg[4-34] and Pucher[4-35] to obtain the deck bending moment per unit longitudinal length for the transverse analysis when using hand-calculations. The general procedures are as follows[4-33]:

1. Calculate the maximum slab fixed-end moments (see Fig. 4-48) using pre-developed influence surfaces (see Fig. 4-49).
2. Apply the fixed-end moments to a two-dimensional model as shown in Fig. 4-50 and determine the redistributed moments in the members' control sections.
3. Calculate the final moments for each section by summing the fixed-end moment and the redistributed moments.

Take the typical section as shown in Fig. 4-48 for example to illustrate the analytical procedures in determining the bending moments at typical control sections in the deck slabs by using influence surfaces.

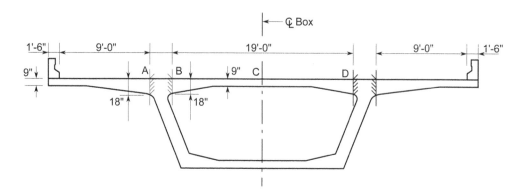

FIGURE 4-48 Typical Locations for Transverse Analysis.

4.4.2.2.1 Determination of Live Load Moments in the Cantilever Slab

Figure 4-49a shows the influence surface for a cantilever slab at its end A. The length of the cantilever slab is denoted as l and is assumed to have an infinite width in its longitudinal direction. Its coordinate system is shown in the figure. Its abscissa (y-axis) is denoted as a multiple of the cantilever length. The influence surface is developed by assuming that the ratio of slab thickness at the tip to that at its fixed end is 2 (see left cross section in Fig. 4-49a) and can be used for analyzing the box girder shown in Fig. 4-48. The most unfavorable truck loading conditions along the cantilever slab are shown in Fig. 4-49b. Based on the dimensions of the design truck (see Section 2.2.3.1.2), the design wheel loads are positioned as shown in Fig. 4-49b to produce the maximum negative moment in the cantilever. The influence coordinates corresponding the locations of the wheels can be determined by interpolating the values

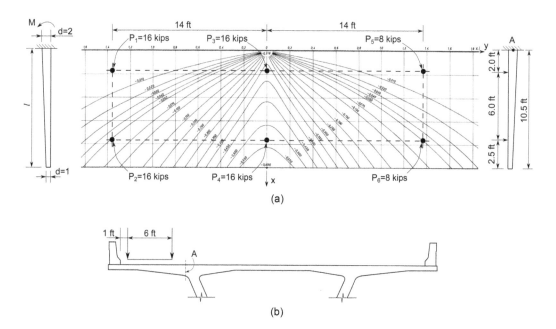

FIGURE 4-49 Wheel Load Positions on Influence Surface for Negative Moment in Cantilever Slab, (a) Wheel Loads on Influence Surface of Section A, (b) Truck Transverse Position on Bridge Deck.

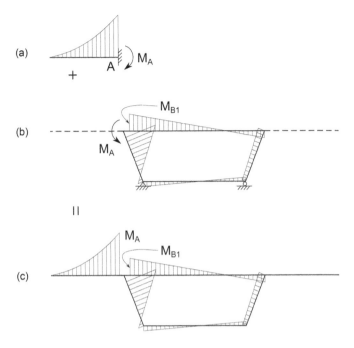

FIGURE 4-50 Transverse Moment Sketches Due to One Truck on Cantilever Slab, (a) Fixed-End Moment in Cantilever, (b) Fixed-End Moment Redistribution on Box Slabs, (c) Total Section Transverse Moment Sketch.

shown in the influence surface. Then, we can calculate the bending moment at the end of cantilever A as

$$M_A = \sum_{i=1}^{6} \eta_i P_i = -(0.0045 + 0.055 + 0.525 + 0.358) \times 16 + (0.0045 + 0.050) \times 8$$

$$= -15.5 \; k - ft \, / \, ft$$

where
  η_i = influence coordinates corresponding to wheel loads P_i
  P_i = wheel loads

The transverse moment distribution sketch of the box section in the cantilever slab is illustrated in Fig. 4-50a.

4.4.2.2.2 Determination of Negative Live Load Moments in the Top Slab of the Box

Figure 4-51a shows the most unfavorable truck transverse loading positions, which can produce maximum negative bending moments over the web at section B. The maximum negative moment caused by the two trucks can be obtained by summing the effects due to the truck on the cantilever and the truck on the box section as follows:

Step 1: Apply M_A determined from Section 4.4.2.2.1 to the two-dimensional frame structure as shown in Fig. 4-50b, and determine M_{B1} at section B due to M_A by any methods discussed in Section 4.2.

Step 2: Determine the fixed moment M_{B2} by using the method described in Section 4.4.2.2.1, based on the wheel load positions as shown Fig. 4-51a and b (see Fig. 4-52a).

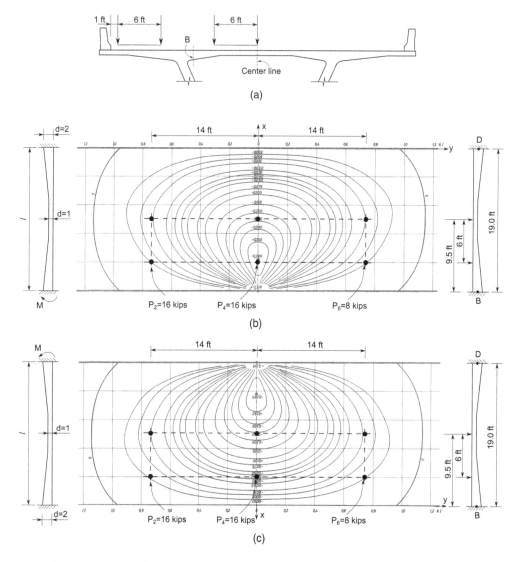

FIGURE 4-51 Wheel Load Positions on Influence Surfaces for Negative Moment in Box Top Slab, (a) Truck Transverse Positions on Bridge Deck, (b) Wheel Loads on Influence Surface of Section B, (c) Wheel Loads on Influence Surface of Section D.

Step 3: Determine the fixed moment M_D by using the method described in Section 4.4.2.2.1, based on the wheel load positions as shown in Fig. 4-51c (see Fig. 4-52a).

Step 4: Apply the fixed-end moments M_{B2} and M_D to the two-dimensional frame structure as shown in Fig. 4-52a with a unit longitudinal length, and determine the moment M_{B3} caused by M_{B2} and M_D by using any methods discussed in Section 4.2.

Step 5: Calculate the total negative moment at section B by summing moments M_{B1}, M_{B2} and M_{B3}, i.e.,

$$M_B = M_{B1} + M_{B2} + M_{B3}$$

4.4.2.2.3 Determination of Positive Live Load Moments in the Top Slab of the Box

The maximum positive moment of the top slab is located at its mid-span. The truck positions that induce the maximum positive live load moment in the top slab are illustrated in Fig. 4-53a. The

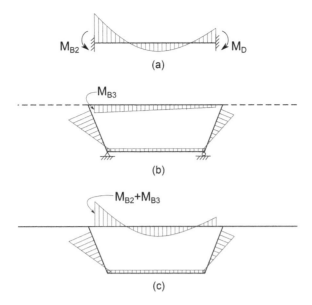

FIGURE 4-52 Determination of Maximum Transverse Negative Moment on the Box Top Slab due to Trucks, (a) Fixed End Moment in the Box Top Slab, (b) Fixed-End Moment Redistribution in the Box Slabs, (c) Total Section Transverse Moment Sketch due to Trucks on Top Slab.

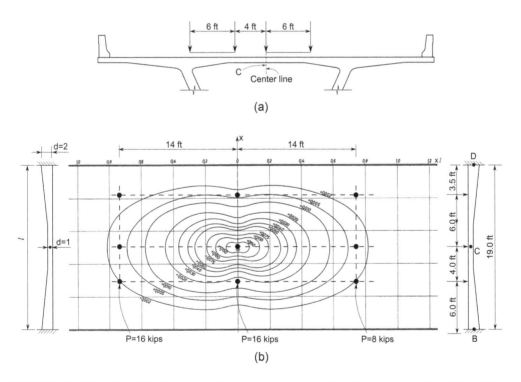

FIGURE 4-53 Wheel Load Positions on Influence Surface for Position Moment in Box Top Slab, (a) Truck Transverse Position on Bridge Deck, (b) Wheel Loads on Influence Surface of Section C.

influence surface at mid-span of a fixed-end slab is shown in Fig. 4-53b. Using the influence surfaces shown in Fig. 4-51b and c and Fig. 4-53b, the maximum positive moment can be obtained using the following procedures:

Step 1: Using the procedures described in Section 4.4.2.2.1, determine the maximum moment M_{C1} (see Fig. 4-54a) at the mid-section C of the box's top slab by locating the wheel loads on the influence surface of section C (see Fig. 4-53b).

Step 2: Determine the fixed-end moments M_B and M_D (see Fig. 4-54a) by applying the truck loadings as shown in Fig. 4-53a on the influence surfaces of sections B and D (see Fig. 4-51b and 4-51c), respectively.

Step 3: Determine the redistributed moment at section C, M_{C2}, by applying the fixed moments M_B and M_D to the two-dimensional frame structure (see Fig. 4-54b) using any methods as described in Section 4.2.

Step 4: Obtain the total positive moment M_C at section C by summing M_{C1} and M_{C2} (see Fig. 4-54c), i.e.,

$$M_C = M_{C1} + M_{C2}$$

More influence surfaces are available for various geometries (relative thicknesses) and support conditions in References 4-34 and 4-35, which can be used for determining the transverse moments of box sections. Test and analytical results[4-36] show that the transverse analysis based on fixed-end moments described above generally provide conservative results for the design. A bridge barrier wall may reduce the cantilever bending moment by over 30%. Though this benefit cannot be used in bridge design, it may be used when determining whether or not to permit trucks to pass over the segmental bridges.

4.4.3 THREE-DIMENSIONAL ANALYSIS

Though the two-dimensional method discussed in the previous section can be performed using hand-calculations, it is still time consuming and may yield overly conservative results. As there are currently numerous commercially available computer programs, the first choice of most bridge designers is to use computers to perform the bridge transverse analysis. A bridge longitudinal segment of about twice the deck width can be taken from the bridge for the transverse

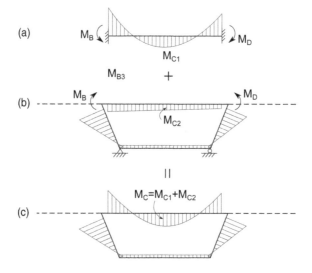

FIGURE 4-54 Transverse Moment Sketches due to Trucks on the Box's Top Slab, (a) Fixed-End Moment in the Box's Top Slab, (b) Fixed-End Moment Redistribution on Box's Slabs, (c) Total Section Transverse Moment Sketch due to the Trucks on the Box's Top Slab.

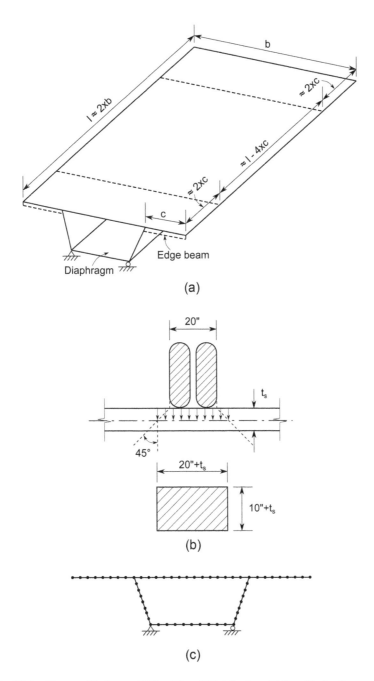

FIGURE 4-55 Finite-Element Meshes and Wheel Load Distribution, (a) Box Girder Segment for Transverse Analysis, (b) Tire Contact Area, (c) Finite-Element Meshes.

analysis (see Fig. 4-55a). However, the transverse moments of the deck vary along the bridge longitudinal directions. Analytical results show that the maximum transverse moments occur in the end portion of the girder within a distance of about 2 times the length of the cantilever slab from the ends of the support (see Fig. 4-55a) if there is no edge beam or thickened cantilever slab provided at the end of the girder. As an edge beam is generally always provided at the end of the box girder to resist a large cantilever moment, the transverse moments of the slab in the end portion of the girder will generally be smaller than those in the center. Thus, the transverse moments at the mid-span determined by the finite-element method can be safely used for the bridge transverse design.

The box girder segment can be modeled as a thin-wall structure and divided into a number of quadrilateral shell-plate elements and analyzed by the finite-element method. Based on Article 3.6.1.2.5 of the AASHTO specifications[4-19], the truck loads can be treated as uniform loading distributed on the tire rectangular contact area plus deck thickness in its width and length (see Fig. 4-55b). To ensure the accurate analytical results can be obtained, the following should be considered in the finite-element meshing (see Fig. 4-55c).

a. Add nodes at the locations with abrupt change in the slab thickness.
b. Each element can be reasonably treated as constant thickness.
c. External loads can be reasonably distributed to the related nodes using level theory.
d. Recommend that the number of elements in the cantilever slabs is not less than 8 and that the number in the top slab of the box is not less than 16. If there are no external loads in the web and bottom slab, the number of finite elements are normally not less than 6, otherwise, not less than 8.
e. The aspect ratio of the finite elements is not greater than 2.

REFERENCES

4-1. Hibbeler, R. C., *Structural Analysis*, 9th edition, Prentice Hall, New York, 2016.

4-2. Karnovsky, I., and Lebed, O., *Advanced Methods of Structural Analysis*, Springer, New York, 2010.

4-3. Shi, T., Shi, Z. Y., and Huang, D. H., *Analysis of Bridge Structures by Computers*, Publishing House of Tongji University, Shanghai, China, 1985.

4-4. Huang, D. Z., Wang, T. L., and Shahawy, M., "Impact Studies of Multi-girder Concrete Bridges," *Journal of Structural Engineering, ASCE*, Vol. 119, No. 8, 1992, pp. 2387–2402.

4-5. Wang, T. L., Huang, D. Z., and Shahawy, M., "Dynamic Responses of Multi-girder Bridges," *Journal of Structural Engineering, ASCE*, Vol. 118, No. 8, 1992, pp. 2222–2238.

4-6. Huang, D. Z., Wang, T. L., and Shahawy, M., "Vibration of Thin-walled Box Girder Bridges Excited by Vehicles," *ASCE, Journal of Structural Engineering*, Vol. 121, No. 9, 1995, pp. 1330–1337.

4-7. Huang, D. Z., Wang, T. L., and Shahawy, M., "Dynamic Behavior of Continuous and Cantilever Thin-Walled Box Girder Bridges," *ASCE, Journal of Bridge Engineering*, Vol. 1, No. 2, May 1996, pp. 67–75.

4-8. Li, G. H., "*Theory and Analysis of Structures and Bridges*," Shanghai Science Publish House, Shanghai, China, 1988.

4-9. Zienkiewicz, O. C., "*The Finite-Element Method*," McGraw-Hill, London, 1977.

4-10. Lin, T. Y., and Burns, N. H., "*Design of Prestressed Concrete Structures*," John Wiley & Sons, New York, 1981.

4-11. Fan, L. C., *Design of Prestessed Continuous Concrete Bridges*, China Communication Press, Beijing, China, 1980.

4-12. Fan, L. C., *Bridge Engineering*, China Communication Press, Beijing, China, 1988.

4-13. Shao, X. D., *Bridge Engineering*, China Communication Press, Beijing, China, 2014.

4-14. Xiang, H. F., et al., *Advanced Theory of Bridge Structural Analysis*, The People's Communication Press, Beijing, China, 2001.

4-15. Trost, H., "Implications of the Principle in Creep and Relation Problem for Concrete and Prestressed Concrete," Beton-und, Staklbetonbau, No. 10, 1967 (in German).

4-16. Trost, H., and Woff, H. J., "*On Realistic Determination of Stresses in Prestressed Concrete Structures Built by Segments*," Der Bauingenieur, Vol. 45, 1970, pp. 155–179 (in German).

4-17. Bezant, Z. P., "Prediction of Concrete Effectives Using Age-Adjusted Effective Modulus Method," *Journal of ACI*, Vol. 69, 1972.

4-18. PTI, *Post-Tensioning Manual*, Post-Tensioning Institute, 6th edition, Phoenix, AZ, 2001.

4-19. AASHTO, *LRFD Bridge Design Specifications*, 7th edition, American Association of State Highway and Transportation Officials, Washington D.C., 2014.

4-20. Huang, D. Z., "Lateral Stability of Arch and Truss Combined System," Master Degree Thesis, Tongji University, Shanghai, China, 1985.

4-21. Huang, D. Z., "Elastic and Inelastic Stability of Truss and Arch-Stiffened Truss Bridges," Ph.D. Degree Thesis, Tongji University, Shanghai, China, 1989.

4-22. Huang, D. Z., and Li, G. H., "Inelastic Lateral Stability of Truss Bridges," *Journal of Tongji University*, Vol. 10, No. 4, 1988.

4-23. Huang, D. Z., Li, G. H., and Xiang, H. F., "Inelastic Lateral Stability of Truss Bridges with Inclined Portals," *Journal of Civil Engineering of China*, Vol. 24, No. 3, 1991.

4-24. Huang, D. Z., and Li, G. H., "Inelastic Lateral Stability of Arch-Stiffened Truss Bridges," *Journal of Tongji University*, Vol. 10, No. 4, 1991.

4-25. Huang, D. Z., "Lateral Bending and Torsion Buckling Analysis of Jiujian Yangtze River Arch-Stiffened Truss Bridge," *Proceedings, IABSE Congress*, Stockholm, 2016.

4-26. Huang, D. Z., and Hu, B., "Evaluation of Cracks in a Large Single-Cell Precast Concrete Segmental Box Girder Bridge without Internal Struts," *ASCE, Journal of Bridge Engineering*, Vol. 20, No. 8, 2014, pp. B4014006-1 to B4014006-8.

4-27. Nutt, Redfield & Valentine in association with David Evans and Associates and Zocon Consulting Engineers, "Development of Design Specifications and Commentary for Horizontally Curved Concrete Box-Girder Bridges," NCHRP Report 620, Transportation Research Board, National Research Council, Washington, D.C., 2008.

4-28. Song, S. T., Chai, Y. H., and Hida, S. E., "Live Load Distribution Factors for Concrete Box-Girder Bridges," *Journal of Bridge Engineering*, American Society of Civil Engineers, Vol. 8, No. 5, 2003, pp. 273–280.

4-29. Huang, D. Z., Wang, T. L., and Shahawy, M., "Dynamic Behavior of Horizontally Curved I-Girder Bridges," *Journal of Computers and Structures*, Vol. 57, No. 4, 1995, pp. 702–714.

4-30. Huang, D. Z., Wang, T. L., and Shahawy, M., "Vibration of Horizontally Curved Box Girder Bridges Due to Vehicles," *Journal of Computers and Structures*, Vol. 68, 1998, pp. 513–528.

4-31. Huang, D. Z., "Dynamic Analysis of Steel Curved Box Girder Bridges," *Journal of Bridge Engineering, ASCE*, Vol. 6, No. 6, 2001, pp. 506–513.

4-32. Huang, D. Z., "Full-Scale Test and Analysis of a Curved Steel-Box Girder Bridge," *Journal of Bridge Engineering, ASCE*, Vol.13, No. 5, 2008, pp. 492–500.

4-33. Coven, J., "Post-Tensioned Box Girder Design Manual," Federal Highway Administration, U.S. Department of Transportation, Report No. FHWA-HIF-15-016, 2016.

4-34. Homberg, H., *Fahrbahnplatten mit Verandlicher Dicke*, Springer-Verlag, New York, NY, 1968.

4-35. Pucher, A., *Influence Surfaces of Elastic Plates*, 4th edition, Springer-Verlag, New York, NY, 1964.

4-36. Rambo-Roddenberry, M., Kuhn, D., and Tindale, G., "Barrier Effect on Transverse Load Distribution for Prestressed Concrete Segmental Box Girder Bridges," *ASCE, Journal of Bridge Engineering*, Vol. 21, No. 6, 2016, pp. 04016019-1 to 04016019-10,

5 Design of Span-by-Span Construction and Common Details of Concrete Segmental Bridges

- Span arrangement and determination of preliminary dimensions
- Types of segments and detailing
- Longitudinal analysis and tendon layout
- Transverse analysis and tendon layout
- Analysis and design of diaphragms and blisters
- Design example - span-by-span bridge

5.1 INTRODUCTION

The basic concepts of span-by-span bridges are discussed in Chapter 1. Generally speaking, span-by-span bridges are the simplest of all the types of segmental bridges and often the most cost effective[5-1]. If the site condition permits, span-by-span design is the first choice as a segmental bridge alternative. Based on the recently constructed bridges in Florida, with a total deck area of approximately 110,000 ft², the span-by-span concrete segmental bridge alternative can compete with cast-in-place concrete bridge and I-girder bridge alternatives.

The ASBI *Construction Practice Handbook for Concrete Segmental and Cable-supported Bridges* defines the span-by-span bridge as a bridge in which an entire span is erected and becomes self-supporting, before erecting the next span. Based on this definition, span-by-span bridges may be grouped into three types:

a. Entire spans of precast segments are constructed on the ground or on barges. Then the segments are lifted on the erection truss and post-tensioned together (see Fig. 1-40).
b. Entire spans are cast on the ground or on a barge and then lifted into position.
c. Entire spans are cast-in-place using a movable scaffolding system (MSS), which consists of superstructure formworks and support structures.

The first method is most often used in practice as it does not require special construction equipment. The first method will be discussed in this chapter.

5.2 LONGITUDINAL DESIGN

5.2.1 Span Arrangement

5.2.1.1 Span Length and Girder Depth

Generally, the first thing in bridge design is to determine the approximate span lengths based on the required bridge vertical and horizontal clearances as well as the site conditions. As previously discussed, the normal economic span range for the span-by-span segmental bridges is from 100 to 150 ft. Though the span length of 150 to 180 ft may be feasible, it may be difficult to arrange all

313

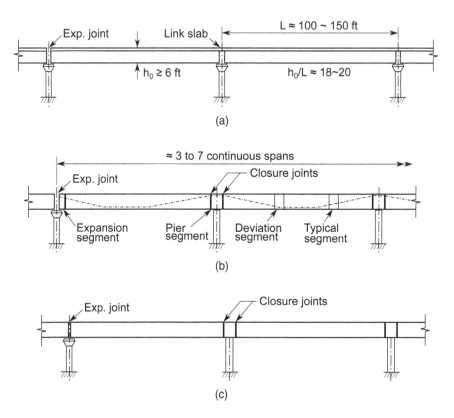

FIGURE 5-1 Span Layout, (a) Simple Span, (b) Typical Continuous Spans, (c) Continuous Spans with Fixed Super-Structure–Pier Connections.

the post-tensioning tendons' anchors in the pier diaphragms and to design the erection trusses[5-2]. The girder depth for span-by-span bridges is typically constant with a minimum girder depth of 6 ft to accommodate both pre-casting and maintenance. The economic ratios of span length to girder depth usually range from 18 to 25 (see Fig. 5-1). The smaller ratios may yield more economic designs due to the simplicity of the tendon arrangement, while larger ratios may be required for vertical clearance and aesthetics.

5.2.1.2 Span Arrangement

There are two ways of arranging the spans:

A. Simple Spans

With this arrangement, the segments in each of the spans are individually post-tensioned together. Then link slabs are used to connect the top slabs of the adjacent spans to limit the number of expansion joints and to increase deck smoothness (see Fig. 5-1a). The simple span arrangement has several advantages, some of which include: (1) No secondary forces will be induced by post-tensioned forces, creep and shrinkage, differential settlements, thermal gradients, etc. (2) Cast-in-place joints can be eliminated and reduce the construction time. However, there are some apparent shortcomings, such as it being a single path structure, which generally needs more post-tendons and reinforcement.

B. Continuous Spans

The continuous span arrangement is the most common scheme in highway bridges. Normally, continuous spans are constructed with the post-tendons in adjacent spans overlapping the pier segments (see Fig. 5-1b). Closure joints are placed on one or both sides of the pier segment to ensure that they have been installed correctly (see Fig. 5-1b). If there are no significant horizontal forces in the superstructure, up to seven continuous spans can be arranged and neoprene or mechanical bearings can be used to accommodate the horizontal movements. If a bridge is expected to experience high horizontal forces, such as a bridge located in high seismic areas, it may be a good choice to connect the super structure with the piers (see Fig. 5-1c). In this case, the number of continuous spans should be reduced to ensure that there are no excessive forces in the piers due to concrete creep, shrinkage, temperature, etc.

5.2.2 Selection of Typical Section

The majority of cross-sections used in span-by-span bridges are box sections, either single box or multi-box sections (see Fig. 1-33). Box sections possess the following properties:

    a. As discussed in Section 3.1.3.2, the box section has a high efficiency ratio of 0.6. The box girder often requires the least amount of post-tensioning tendons. Its larger flanges can resist larger post-tensioning force without changing its C-T arm length (refer to Section 3.1.2).

    b. It can resist larger torsion moments.

Reducing the number of webs can simplify box girder precasting and normally yield a more economic design. The Florida Department of Transportation allows a maximum web spacing of up to 32 ft[5-3]. For straight segmental box girder bridges, the web spacing of up to 40 ft is still feasible if it is properly designed[5-4].

    Generally, the preliminary dimensions of a box section can be determined based on the information contained in Section 1.4.2 and Article 5.14.2.3.10 of the AASHTO specifications[5-5]. The segmental box girder standards for span-by-span construction developed by AASHTO, PCI, and ASBI[5-6] are valuable sources in selecting segmental box girder cross sections. These design standards were developed based on an extensive survey of existing segmental box girder bridges built in the United States and Canada[5-7, 5-8]. These standards should be used as frequently as possible to increase the reuse of casting equipment. These standards are suitable for the bridges with span lengths ranging from 100 to 150 ft, and bridge widths ranging from 28 to 45 ft, and girder heights ranging from 6 to 8 ft. The typical sections with a girder height of 8 ft are shown in Fig. 5-2. Additional information as well as other standards are included in Appendix A.

5.2.3 Layout of Longitudinal Post-Tensioning Tendons and Bars

5.2.3.1 Introduction

Two types of prestressing tendons and bars are typically used in the bridge's longitudinal direction. The longitudinal tendons are the key elements in resisting the self-weight, live loads, and other external loads. Longitudinal post-tensioning bars are normally used as temporary elements during construction, though they can also be used as permanent load resisting elements. To improve construction efficiency, it is preferable to use a number of large-size longitudinal tendons. However, to ensure structural redundancy, a minimum of four tendons per web should be provided[5-3] for span-by-span segmental bridges. The longitudinal tendons used to construct span-by-span segmental bridges can be embedded in the webs or flanges of a box section and are called internal tendons. The longitudinal tendons placed within the box, but outside of the concrete section, are called external tendons. Internal tendons may be used in high seismic zones for high ultimate strength requirement and increasing ductility. This arrangement can more effectively use the tendons at the mid-span. However, with many internal tendons, it is often more difficult to form a continuous span by overlapping the tendons within the length of the pier segment. External tendons are often used to circumvent this restriction in practice. External tendon can reduce friction, streamline grouting and

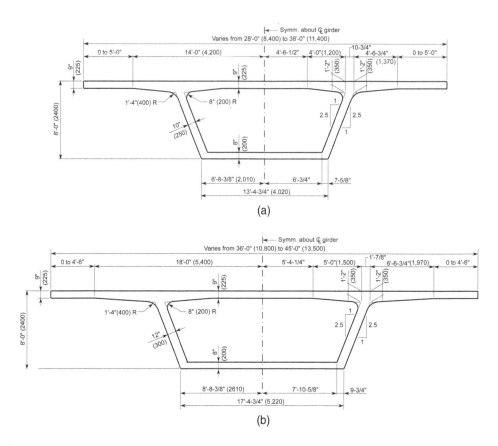

FIGURE 5-2   AASHTO-PCI-ASBI Segmental Box Girder Standards with Girder Height of 8 ft for Span-by-Span Construction, (a) Deck Width from 28 to 38 ft, (b) Deck Width from 36 to 45 ft (Used with permission of ASBI).

inspections, simplify the precasting operation, and are straightforward to replace, etc. The internal and external tendons may be combined together and used in segmental bridges constructed by the span-by-span method. Recent research show that the load capacity of bridges will increase as the ratio of the internal bounded tendons to the external unbounded tendons increase[5-9, 5-10]. It is important to use repeated tendon layout in a given project to streamline construction.

5.2.3.2   Internal Tendons
Figure 5-3 shows a typical internal tendon layout[5-2]. For an easier stressing operation, it is preferable to locate the entire tendon stressing ends at the face of the pier segment (see Fig. 5-3a).

5.2.3.3   External Tendons
External tendons are anchored at the pier segments and deviated typically at 2 to 3 deviation saddles, which are placed at the bottom corners between the web and the bottom slab (see Fig. 5-4). The post-tendons are enclosed in polyethylene or steel ducts.

The tendon's minimum radius and spacing should meet the requirements discussed in Chapter 1. As most recently constructed span-by-span bridges typically have a single box cross-section with external tendons running through the full span length, the redundancy of the post-tensioning load paths is directly related to the number of tendons within the span. To ensure that the loss of one of the tendons due to corrosion does not critically affect the overall performance of the bridge, the selected tendon

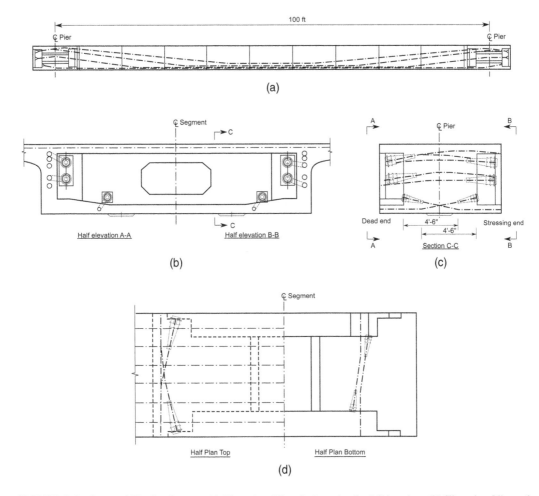

FIGURE 5-3 Internal Tendon Layout, (a) Elevation View in Longitudinal Direction, (b) Elevation View of Pier Segment, (c) Side View of Pier Segment, (d) Plans View of Pier Segment[5-2].

size normally does not exceed 19-0.6″ diameter strands and represents over 12.5% of the total post-tensioning. FDOT's *Structures Design Guidelines* requires that a minimum of four tendons per web be provided[5-3, 5-11, 5-12].

5.2.3.4 Future Tendons

For the purpose of future maintenance and rehabilitation, such as deflection adjustment, accommodation of future additional dead loads, and replacing of damaged tendons, the AASHTO specifications require that some empty ducts, anchorage attachments, and deviation block attachments be provided to accommodate future tendons installation. The future tendons should be located symmetrically about the girder center line and shall provide at least 10% of the original designed capacities for both positive and negative moments.

5.2.3.5 Temporary Post-Tensioning Bars

The temporary post-tensioning bars for erecting the segments should be distributed to provide a minimum compressive stress of 0.030 ksi and an average stress of 0.040 ksi across the joint for epoxy curing. There are three methods to install the PT bars. The first method is to provide a small

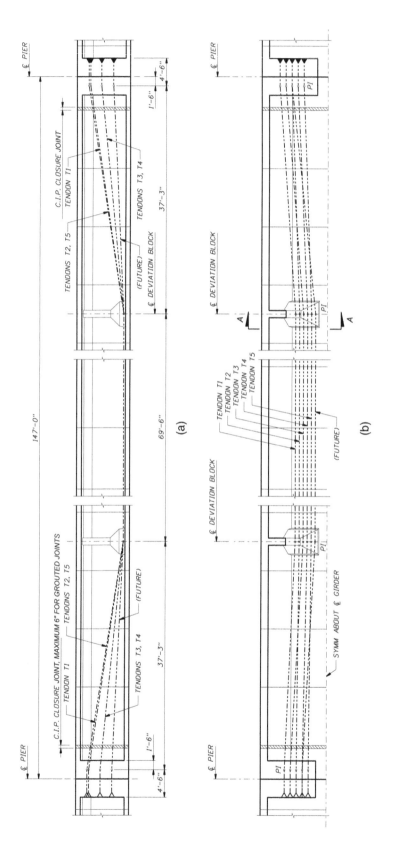

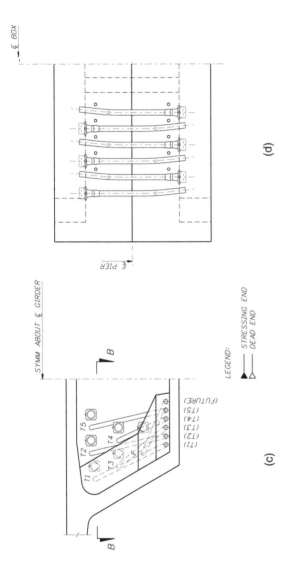

(d)

(c)

LEGEND:
STRESSING END
DEAD END

FIGURE 5-4 External Tendon Layout, (a) Elevation View, (b) Half of Plan View, (c) Section *A-A* View, (d) Section *B-B* View.

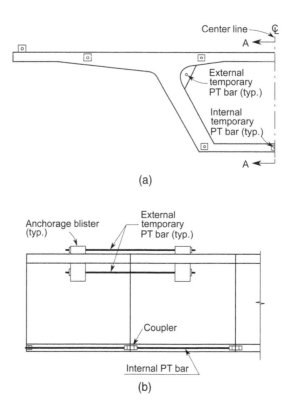

(a)

(b)

FIGURE 5-5   Typical Arrangement of Temporary Post-Tensioning Bars, (a) Elevation View, (b) *A-A* View.

blister at the interior corners of the box or at the tip of the top slab (see Fig. 5-5). This method allows easy access to both ends of the bars, and the bars can be removed for reuse. However, the reused bars are typically stressed to a maximum of 50% of their ultimate strength. This method may have little impact on the forms and casting operations. The second method is to use internal PT bars (see Fig. 5-5). As the subsequent bars are coupled onto the previous set, these temporary bars are normally designed as permanent elements and are stressed to higher levels. The temporary bars cannot be removed and should be properly grouted to prevent corrosion. A typical temporary bars arrangement is shown in Fig. 5-5. The third method is to use a bracket/blister attached to the segment through the preformed openings to anchor the bars (see Fig. 5-5b). This method is similar to the first, but it eliminates the blisters. In the third method, it is imperative that the brackets be carefully detailed and constructed to avoid local spalling.

In designing temporary PT bars, two loading stages are typically considered.

a. Under the self-weight of the segment and construction loads, the minimum joint compression stress produced by the PT bars should not be less than those specified in Section 5.2.5.2 during the curing/open time of the epoxy (about 45 to 60 min).

b. Under the self-weight of the segment, construction loads and permanent cantilever tendons, the minimum joint compression stress should not be less than zero after 1 or 2 hours after the curing/open time of the epoxy.

5.2.3.6 Minimum Clearance Requirements at Anchorages for Replaceable and Typical Used Sizes of Tendons and Bars

For the tendons that utilize flexible filler and are designed as replaceable, the minimum clearances at the anchorage area shown in Table 5-1 should be met for a proper operation[5-11, 5-12].

TABLE 5-1

Minimum Clearances at Anchorages for Replaceable Tendons

Anchorage Type	Location	Minimum Clearance
Stressing end	Near deviator	B (see Table 1-15)
	At intermediate diaphragm	A (see Table 1-15) $+ 1' - 0''$
	Near minor obstruction	
	At other location	$A + \Delta_T{}^*$
Non stressing end	Near abutment or other structure/location	$2' - 6'' + \Delta_T{}^*$

* Maximum design thermal expansion, including adjacent structures.

The commonly used tendon sizes are 4, 7, 12, 15, 19, 27, and $31 - 0.6''$ diameter strands, and the commonly used post-tensioning bars are $1 - 0''$, $1 - 1/4''$, $1 - 3/8''$, $1 - 3/4''$, $2 - 1/2''$, and $3 - 0''$ diameter deformed bars.

5.2.4 SPECIAL SEGMENTS AND DETAILING

5.2.4.1 Introduction

With the exception of the typical segment discussed in Section 1.4, there are usually three other types of special segments that serve the special purpose of anchoring and deviating tendons. They are called deviator segments, expansion pier segments, and interior pier segments.

5.2.4.2 Deviator Segment

A deviator segment is a segment with deviation saddles that are built out of the web, flange, or the junction between the web and flange to provide for change of direction of extenal tendons (see Fig. 5-6). There are several types of deviation saddles that are successfully used in span-by-span bridges, such as the full-depth diaphragm type (Fig. 5-6), the partial-depth diaphragm type (Fig. 5-7a and b), the rib type (Fig. 5-7c and d), and the blister type (Fig. 5-8). The full-depth diaphragm type is very rigid and is a good choice for more than three tendons or where large tendons are used, though it may complicate the internal formwork used in the precasting machines. The Florida Department of Transportation requires that all the deviation saddles be placed in segments with full-depth diaphragms to reduce the force in the webs and flanges. It also stipulates that deviation saddles are not permitted in segments with transverse bottom slab ribs. The deviation blister type is typically used for deviating single tendons and used in segmental bridges built by the balanced cantilever construction method. The formwork for the deviation blisters can normally be combined with anchorage blister formwork.

The tendon ducts pass through smooth round preformed holes with diabolos in the deviator (see Fig. 5-9). The radius of the ducts in the deviator should be designed to be greater or equal to the values specified in Table 1-11, and the minimum clear distance between the top of the bottom slab to the edge of the diabolo should be 1 in. A 2-in.-diameter drain hole/vent hole should be placed on each side of the deviator to prevent water from ponding near post-tensioning components (see Figs. 5-6 and 5-9)[5-11 to 5-13]. The deviation blisters should be detailed to be set back a minimum of 12 in. from the joint.

5.2.4.3 Pier Segments

5.2.4.3.1 Introduction

The pier segments have a thick diaphragm that transmits large concentrated forces from the longitudinal post-tensioning tendons to the webs and large loadings from the super-structure to the

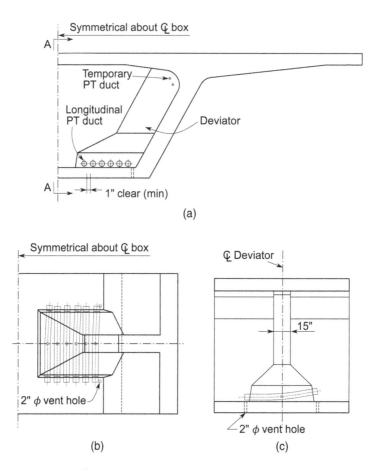

FIGURE 5-6 Typical Deviator Segments with Full-Height Diaphragm, (a) Elevation, (b) Bottom Slab Plan View, (c) Section *A-A* View.

substructure. These are the most complex elements in span-by-span segmental bridges. Currently, there is no simple and accurate design method for these elements. Based on previous experience, the thickness of the diaphragm is generally not less than 6 ft (see Figs. 5-10 and 5-11). The Florida Department of Transportation requires that the minimum thickness of segment pier diaphragms containing external tendons be 4 ft[5-3].

5.2.4.3.2   Interior Pier Segment

The interior pier segment is normally the heaviest segment in a concrete segmental bridge. The longitudinal post-tensioning tendons on the adjacent spans are overlapped at the interior pier segment diaphragm (see Fig. 5-10). The segment can be precasted either in two pieces (see Fig. 5-10c) or in one piece (see Fig. 5-10d) or built using other methods. The ducts are normally embedded in the diaphragm, and the minimum radius *R* and the tangential length *L* of the ducts should meet the requirement shown in Table 1-11 (see Fig. 5-11). For maintenance, the diaphragm should be provided with an access opening. The minimum diaphragm access opening size is normally 32 in. wide × 42 in. tall (see Fig.5-10a).

5.2.4.3.3   Expansion Pier Segment

Except for the functions of the interior pier segment, the expansion pier segment also accommodates the expansion joint devices and is provided with edge beams for larger transverse

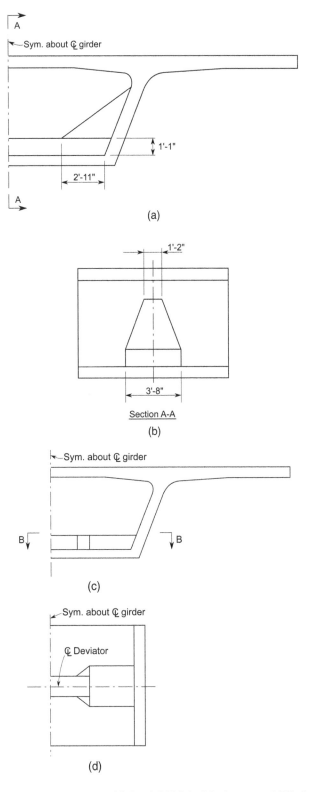

FIGURE 5-7 Typical Deviator Segments with Partial-Height Diaphragm and Rib Saddle, (a) Elevation of Partial-Height Diaphragm, (b) Section *A-A* View, (c) Elevation of Rib Type Saddle (d) Section *B-B* View.

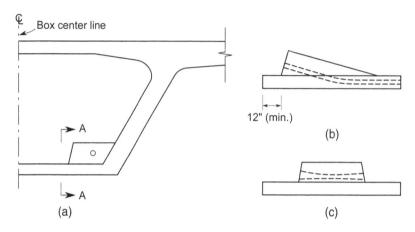

FIGURE 5-8 Typical Deviation Blisters, (a) Elevation View, (b) *A-A* View for Internal Tendons, (c) *A-A* View for External Tendons.

loadings (see Fig. 5-12). The anchorages at this location are vulnerable to damage by water penetration; therefore, designers should pay special attention and take the necessary precautions to protect the anchorages from damage, such as by using Type I anchorage protection (four-level protection) (see Figs. 1-53 and 1-54). At expansion joints, it is necessary to provide a recess and continuous expansion joint device seat to receive the assembly, anchorage bolts, and frames of the expansion joint, i.e., a finger or modular type joint (see Fig. 5-12). The anchorages should be protected from dripping water by means of skirts, baffles, V-grooves, or drip flanges (see Fig. 5-13).

5.2.4.3.4   Minimum Clearance Requirements at Pier Expansion Joint Anchorages

As discussed in Section 1.7, to ensure that there is enough space for jacking, during the sizing of the pier segments, the minimum clearance as shown in Fig. 5-14 should be provided to meet the requirements shown in Tables 1-15 and 5-1.

5.2.5   SEGMENT JOINTS

5.2.5.1   Introduction

The segment joints refer to the connections between precast segments. They can be classified into four types of segment joint: epoxy joint, dry joint, non-reinforced closure joint (grouted joint), and cast-in-place reinforced joint.

5.2.5.2   Epoxy Joint

The epoxy joint refers to a joint with match cast surfaces between adjacent precast segments that are epoxied before being assembled into the final structure. These joints are often called Type A joints. Each of the joints is bonded together with contact pressure of approximately 40 psi applied uniformly to the joint during the epoxy curing period for about 20 hours[5-13, 5-14], and the minimum compression stress in the joint should be 30 psi. The stress limitation is to prevent uneven thickness across the match case joint which can lead to systematic error in geometry control[5-14]. The application of epoxy is typically 1/16-in. thick applied to both faces of the match cast joints. The epoxy on the faces has the following functions:

   a. Lubricate the joint during assembling of the segments to facilitate the alignment between segments.
   b. Prevent water and chlorides intrusion.

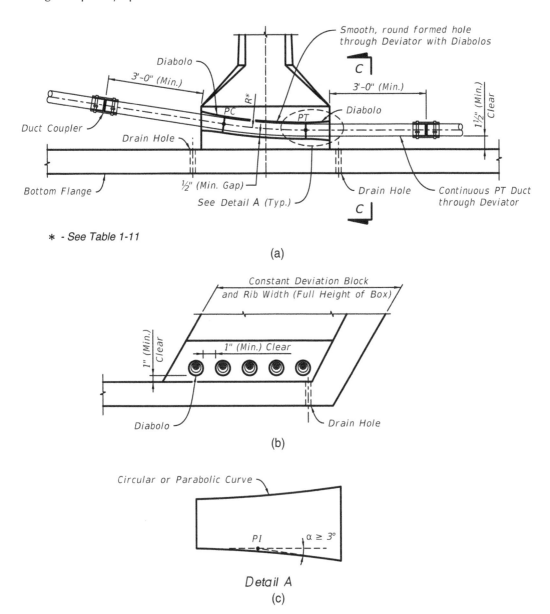

FIGURE 5-9 Deviator Details, (a) Elevation View, (b) Side View, (c) Typical Diabolo Geometry (Used with permission of FDOT).

c. More uniformly distribute compression and shear stresses.

d. Prevent cementitious grout in the tendon duct from leaking out.

5.2.5.3 Dry Joint

The dry joint means that the match cast surfaces of adjacent precast segments are not epoxied during erection. This type of joint is also called a Type B joint and is not currently allowed to be used in the United States due to corrosion concerns. However, the dry joint is still being used in other countries in regions where there are no freeze/thaw cycles and where no de-icing chemicals are used.

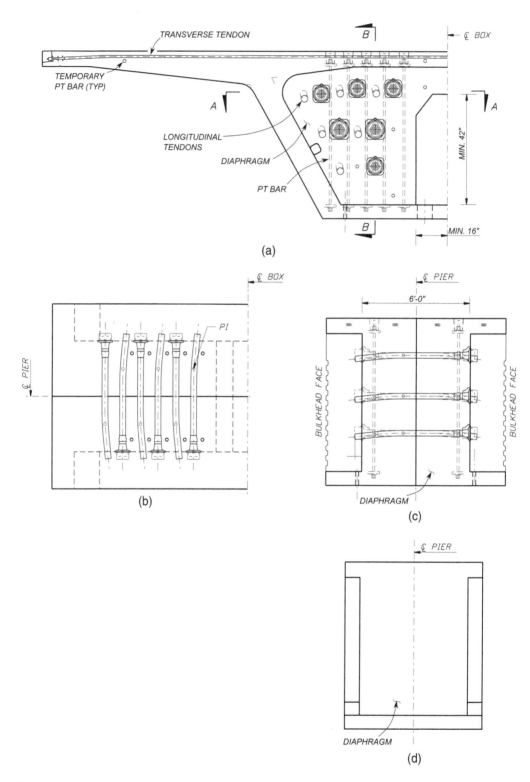

(a)

(b)

(c)

(d)

FIGURE 5-10 Typical Pier Continuous Segment, (a) Elevation View, (b) Plan View, (c) Side View with Two-Pieces, (d) Side View with One Piece.

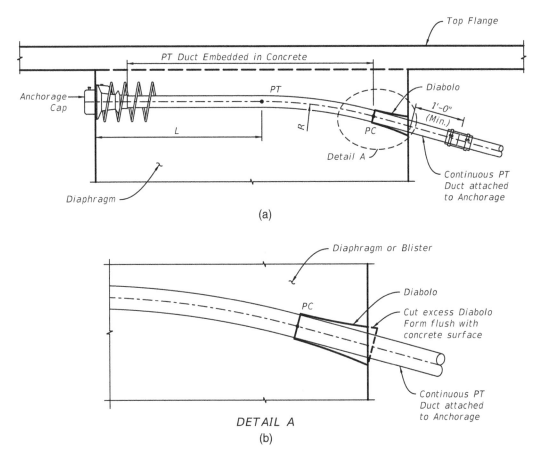

FIGURE 5-11 Tendon Anchorage in Interior Pier Segment, (a) Tendon Geometry, (b) Detail A (Used with permission of FDOT).

5.2.5.4 Non-Reinforced Closure Joint

The non-reinforced closure joint is also called a grouted joint. The lengths of the closure joints used in existing bridges typically range from 6 in to 9 in. The Florida Department of Transportation currently specifies a maximum length of 6 in. In span-by-span constructed bridges with external tendons, the cast-in-place closures are normally provided at each end of each span between a pier segment and the adjacent segment. The function of the closure joint is to provide a space for placing the typical segments between pier segments and to accommodate small adjustments to the precast length and erected geometry of the segments. The same concrete as the precast segments is used in the closures. Temporary concrete spacer blocks are first placed in the closures at suitable locations, and then about 10% of the permanent longitudinal post-tensioning is applied after all the segments have been placed, positioned, and aligned[5-6, 5-14].

5.2.5.5 Cast-in-Place Reinforced Joint

The cast-in-place reinforced closure joint is normally used in precast balanced cantilever construction and is located between the tips of the cantilevers in the middle of a span and between the tip of the cantilever and the adjacent precast segments which are erected on falsework for end spans. Normally, the space between adjacent precast elements should not be smaller than 18 in.

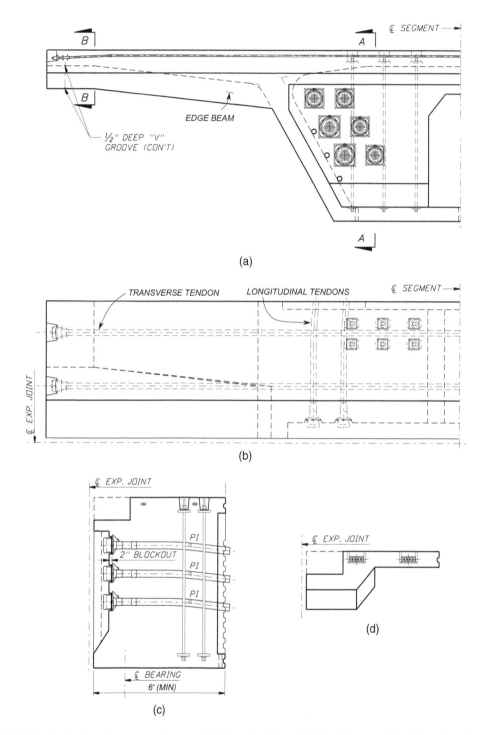

FIGURE 5-12 Typical Expansion Segment, (a) Elevation View, (b) Plan View, (c) Side View—Section *A-A* View, (d) Side View of Edge Beam—Section *B-B* View.

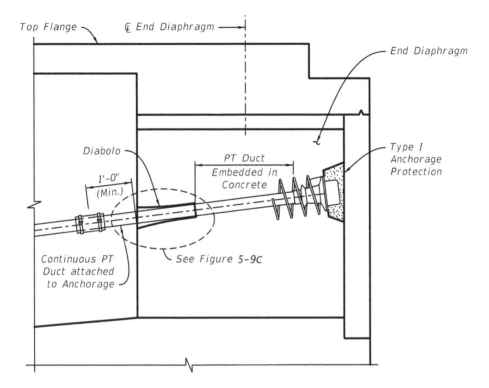

FIGURE 5-13 Details of Tendon Anchorage in Expansion Segment (Used with permission of FDOT).

5.2.6 Design Considerations for Maintenance

5.2.6.1 Clear Height of Box Girder
Whenever possible the clear height of box girders should not be smaller than 6 ft.

5.2.6.2 Access Opening
For the safety of bridge inspectors and the maintenance work crew, designers should give the utmost consideration to accessibility. Access openings should be provided in the bottom flange and the interior diaphragms with the following consideration[5-3]:

1. Provide access openings in the bottom flange, and space the opening so that the distance from any location within the box girder to the nearest opening is not greater than 300 ft. The space between openings is usually less than 600 ft. If feasible, place one opening near each abutment.
2. Provide an access opening through all interior diaphragms.
3. The access opening shall have minimum dimensions of 32 in. wide × 42 in. tall [5-12].

5.2.6.3 Lighting
Provide interior lighting located at approximately equal intervals not exceeding 50 ft (see Fig. 5-15)[5-13].

5.2.7 Longitudinal Analysis and Estimation of Post-Tensioning Tendons

5.2.7.1 Introduction
As discussed in Sections 1.5.2 and 5.2.1.2, the static system of span-by-span bridges usually changes during construction, from simply supported to continuous. Thus, it is necessary to analyze the

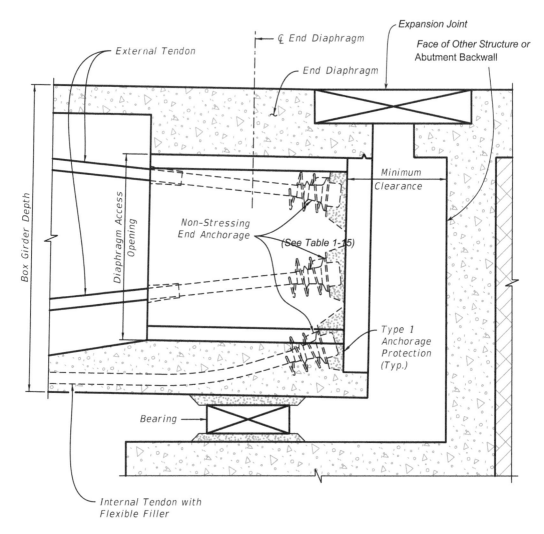

FIGURE 5-14   Minimum Clearance at Pier Expansion Joint Anchorages[5-12].

bridges for time-dependent effects. Design loads become the function of the casing dates of all of precast components. It is standard practice to first assume a feasible construction sequence, reasonably estimate the construction schedule, perform the necessary analysis, and then modify the analysis during construction based on the actual schedule and construction sequence. The casting dates of segments are normally determined by working backwards from the approximate end of construction, and the last segment assembled in the bridge is at least 28 days old. Then the casting dates of other segments can be determined by assuming the time required for casting each segment. The final structural responses may be determined by assuming 10,000 days for developing all the time-dependent effects[5-14].

The basic methods of structural analysis were discussed in Chapter 4. It may be time consuming to perform the time-dependent analysis of the bridges by hand. Fortunately, there are many commercially available computer programs that can be used to perform this analysis. It is only necessary for the designers to input the correct information and construction sequences for the computer program to perform the analyses. The computer programs can perform a stage-by-stage analysis using assumed post-tensioning layout. In this section, we first discuss the general analytical

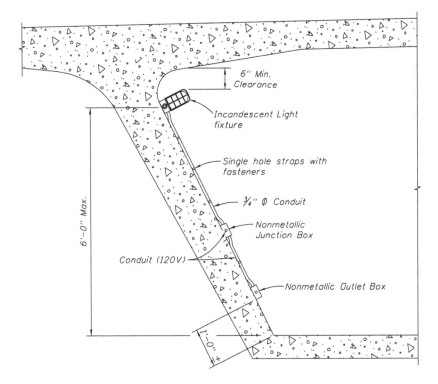

6" Min.
Clearance

Incandescent Light
fixture

Single hole straps with
fasteners

¾" ⌀ Conduit

Nonmetallic
Junction Box

Conduit (120V)

Nonmetallic Outlet Box

6'-0" Max.

1'-0"
+

FIGURE 5-15   Lighting Details (Used with permission of FDOT).

methods of the span-by-span bridges due to dead and live loads. Then, the estimation of required post-tensioning tendons is presented.

5.2.7.2   Analysis of Permanent Load Effects

The effects of permanent loadings, including the self-weight, other dead load, and post-tensioning forces, on the bridge should be performed by time-dependent analysis. Based on the construction sequences and the loading stages described in Section 1.5.2, the structural responses to the permanent loadings can be determined stage-by-stage using the methods presented in Chapter 4. The sketch of bending moment distributions due to the girder's self-weight are illustrated in Fig. 5-16.

5.2.7.3   Analysis of Live Load Effects

The design live loads discussed in Section 2.2.3 should be positioned to produce maximum effects at control sections. This can be done through influence lines as shown in Fig. 5-17. Most commercial computer programs are capable of analyzing the effects of live loads. From Fig. 5-17, it can be seen that all the influence lines have both positive and negative portions. That means the design live load may induce both positive and negative moments and shear at a section depending on where the design loads are positioned. The design live loads should be located in the range of the influence line with the same sign of the influence coordinates to produce maximum absolute values of the effects at the section of concern. Figure 5-18 shows the typical live load positions for the maximum positive moment of the interior span (Fig. 5-18a), maximum negative moment (Fig. 5-18b), and reaction over support (Fig. 5-18c). The moment and shear envelops can be developed by summing the effects due to the permanent loadings and the maximum and minimum effects due to the live loads separately. Some sketches of typical moment and shear envelopes for a four-span continuous bridge due to dead loads and live load are shown in Fig. 5-19 and can be used to estimate the required post-tendons and preliminarily layout of the tendons.

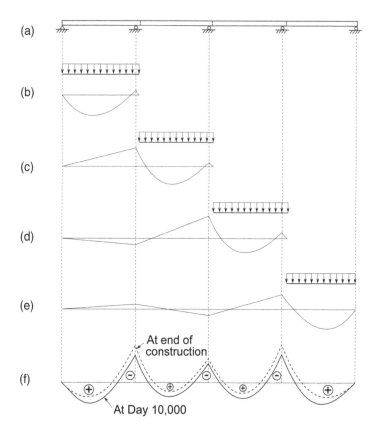

FIGURE 5-16 Moment Distributions in Different Construction Stages, (a) Four Continuous Spans Bridge Built by Span-by-Span Method, (b) Self-weight and Moment Distribution in Stage I, (c) Self-weight and Moment Distribution in Stage II, (d) Self-weight and Moment Distribution in Stage III, (e) Self-weight and Moment Distribution in Stage IV, (f) Moment Distribution in Completed Structure Due to Self-weight.

5.2.7.4 Estimation of Post-Tensioning Tendons

The required post-tensioning tendons can be estimated by meeting the service limit states for positive moment based on the methods discussed in Section 3.1.3.1. As the preliminary bridge cross sections are normally determined based on previous experience, constructability and maintenance requirements, site conditions, etc., Eq. 3-27 can be used for the estimation of the required area of the post-tensioning tendons as:

$$A_{ps} = \frac{M_{max} - M_{ba}}{f_{pe}\left(e + \frac{I}{Ac_b}\right)} \tag{5-1}$$

where

M_{max} = maximum positive moment caused by the dead and live loads

M_{ba} = $\frac{f_{ta}I}{c_b}$ = moment inducing the allowable concrete tensile stress at the bottom fiber

f_{ta} = allowable concrete tensile stress (positive)

f_{pe} = tendon effective stress (may approximately be taken as 0.75 jacking stress)

e = distance between centroids of post-tensioning tendons and section

c_b = distance from bottom to centroid of section

A, I = area and modulus of inertia of section, respectively

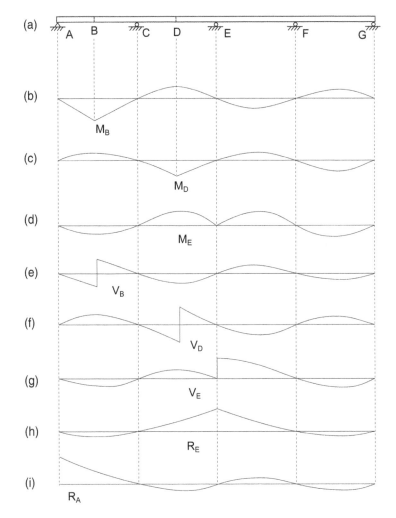

FIGURE 5-17   Typical Influence Lines of Continuous Beams, (a) Four-Span Continuous Beam, (b) Moment at Section B, (c) Moment at Section D, (d) Moment at Support E, (e) Shear at Section B, (f) Shear at Section D, (g) Shear at Section E, (h) Reaction at Support E, (i) Reaction at Support A.

5.2.8   CAPACITY VERIFICATIONS

5.2.8.1   Introduction

After the preliminary bridge dimensions and the layout of the post-tensioning tendons have been determined, the bridge structure can be more accurately analyzed. The capacities of the bridge should be met for all the limit states described in Section 2.3.3.2 for all applicable load combinations presented in Section 2.3.3.3. The following limit states typically have to be satisfied.

5.2.8.2   Service Limit State Stress Check

The service limit states for concrete segmental bridges include service I, service III, and special load combination as discussed in Section 2.3.3. The concrete stresses should satisfy the requirements discussed in Section 1.2.5.1. The stress checks typically include:

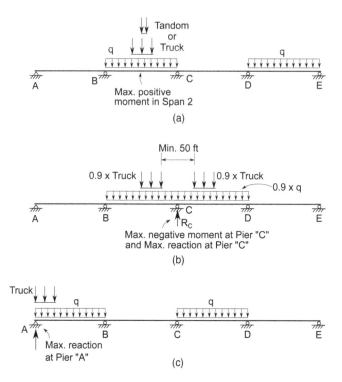

FIGURE 5-18 Typical Most Unfavorable Loading Positions for Live Loads, (a) Loading Positions for Maximum Positive Moment in Span 2, (b) Maximum Negative Moment and Reaction at Support *C*, (c) Maximum Reaction at Support *A*.

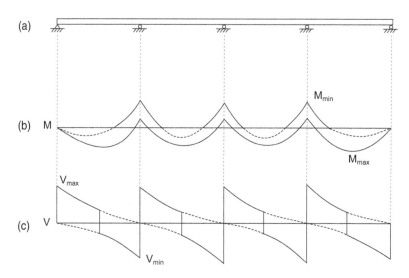

FIGURE 5-19 Typical Moment and Shear Envelops Due to Dead and Live Loads, (a) Four-Span Continuous Beam, (b) Moment Envelop, (c) Shear Envelop.

5.2.8.2.1   Service I

Service I is for checking concrete compression stresses. The stresses are typically verified in the following stages:

- Temporary flexure stresses after the tendons are stressed and before grouting.
- Flexure stresses after the bridge is open to traffic before long-term losses.
- Flexure stresses in the bridge during operation and after long-term losses.

5.2.8.2.1.1   Temporary Compression Stress Check
The normal stresses in prestressed concrete due to dead loads (DC) and post-tensioning tendons (PT) after stressing all tendons and before grouting should meet the following:

$$f_{c\_max} \leq 0.6 \; f_c' \; \text{ksi}$$

5.2.8.2.1.2   Permanent Stresses after Losses
The normal stresses in prestressed concrete due to the load combination $DC + (LL + IM) + 0.3 \, WS + WL + FR + TU + 0.5TG$ after losses shall be satisfied:

Compression: $f_{c\_max} \leq 0.45 f_c'$ ksi (for the sum of effective prestress and permanent load)
$f_{c\_max} \leq \phi_w 0.6 f_c'$ ksi (for the sum of effective prestress, permanent load, and transient loads)

5.2.8.2.2   Service III

Service III is for checking concrete tensile stresses. The stresses are typically verified in the following stages:

- Temporary flexure stresses after the tendons are stressed and before grouting
- Flexure stresses after the bridge is open to traffic before long-term losses
- Flexure stresses in the bridge during operation and after long-term losses

5.2.8.2.2.1   Temporary Tensile Stress Check
The tensile stresses in concrete due to the load combination $DC + PT$ after stressing all tendons and before grouting shall satisfy the following:

$f_{t\_max} \leq 0.0$ ksi (no tension, for no minimum bonded auxiliary reinforcement through the joints)
$f_{t\_max} \leq 0.19 \sqrt{f_c'}$ ksi [in the area (excluding joints) with sufficient reinforcement to resist the tensile force in concrete and $f_s \leq 0.5 \; f_y$]

5.2.8.2.2.2   Permanent Tensile Stress Check
The normal stresses in concrete due to the load combination $DC + 0.8(LL + IM) + FR + TU + 0.5TG$ after losses shall be satisfied:

$f_{t\_max} \leq 0.0$ ksi (no tension, for no minimum bonded auxiliary reinforcement through the joints)
$f_{t\_max} \leq 0.19 \sqrt{f_c'}$ ksi [in the area (excluding joints) with sufficient reinforcement to resist the tensile force in concrete and $f_s \leq 0.5 f_y$]

5.2.8.2.2.3   Principal Tensile Stresses in Webs
The principal tensile stresses at the neutral axis after long-term losses due to service III load combination shall meet

$$f_{t\_max} \leq 0.110 \sqrt{f_c'} \; \text{ksi}$$

5.2.8.2.3   Special Loading Combination

The special load combination defined by Eq. 2-51 for segmental bridges should be verified at the service limit state.

5.2.8.3 Strength

5.2.8.3.1 Flexure Strength Check

The flexure strengths at control sections shall meet Eq. 3-148a and be verified following the procedure shown in Fig. 3-41. Typically, one of the following load combinations may govern the superstructure design:

Strength I:

$$\sum \gamma_i M_i = 1.25 M_{DC} + 1.5 M_{DW} + 1.75 M_{LL+IM} + 0.5 M_{TU} + M_{PS} + 1.25 \left(M_{CR} + M_{SH} \right) \qquad (5\text{-}2)$$

Strength II:

$$\sum \gamma_i M_i = 1.25 M_{DC} + 1.5 M_{DW} + 1.35 M_{LL+IM} + 0.5 M_{TU} + M_{PS} + 1.25 \left(M_{CR} + M_{SH} \right) \qquad (5\text{-}3)$$

Strength IV:

$$\sum \gamma_i M_i = 1.5 M_{DC} + 1.5 M_{DW} + 0.5 M_{TU} + M_{PS} + 1.5 \left(M_{CR} + M_{SH} \right) \qquad (5\text{-}4)$$

5.2.8.3.2 Shear and Torsion Strength Check

The shear and torsion strengths at control sections shall satisfy Eqs. 3-148b and 3-148c and shall be verified following the procedure shown in Fig. 3-54.

The maximum web shear is the sum of the effects induced by the maximum vertical shear and the corresponding torsion of the section or the effects due to the maximum torsion and the corresponding vertical shear. However, it may be more conservative to use the sum of the effects produced by maximum vertical shear and maximum torsion.

5.3   DESIGN OF BOX SECTION COMPONENTS

5.3.1   TOP SLAB

5.3.1.1   Introduction

The top slab should satisfactorily perform its two functions in the longitudinal and transverse directions.

   a. Longitudinal Functions
   - As a tensile or compression chord in resisting the longitudinal axial force caused by the bending moment about its horizontal axis
   - As a shear wall in resisting the longitudinal axial force and in-plane shear caused by sectional torsion moment
   - As a web in resisting the longitudinal axial force and in-plane shear caused the bending moment about its vertical axis.
   b. Transverse functions
   - As a slab in resisting transverse and longitudinal bending moments and vertical shear induced by its dead loads and live loads.
   - As a frame member in resisting transverse bending and vertical shear induced by sectional distortion.
   - As a support slab in resisting transverse bending moments and tensile forces caused by vehicle impact loading applied to the barrier wall.

The longitudinal functions are ensured by the longitudinal capacity checks discussed in Section 5.3.8. The design of the top slab is normally controlled by meeting the transverse functions.

In addition to its dead loads, the deck slab is directly subjected to moving vehicle loads, temperature gradients, creep and shrinkage effects, and in some regions, de-icing chemicals and freeze-thaw action, etc. As it is costly and very difficult to replace the deck slab without closing the entire bridge, it is always a good strategy to be conservative and allow for reserved capacity[5-14]. Normally, post-tensioning tendons are used to prevent deck cracking and deterioration under service loads and to increase its durability even for short overhangs. The minimum deck thickness should not be less than 8 in. Previous experiences show that increasing the top slab thickness by only 0.5 in. will result in a significant improvement in durability with negligible additional cost[5-15]. Thus, it is preferable to use a deck thickness of 9 in.

5.3.1.2 Loading Application and Transverse Analysis

The analytical methods for determining the responses in the top slab due to dead loads and live loads were discussed in Section 4.4. The AASHTO Specification Article 3.6.1.3.3[5-5] indicates that only the design axles of the HL93 and design tandem need to be used in transverse design. The design live loads should be positioned to produce the maximum and minimum effects at control sections, which are illustrated in Fig. 4-48. Some typical loading positions are shown in Figs. 4-49, 4-51 and 4-53. It should be mentioned that it may be necessary to try several loading positions and cases to determine the related maximum moments. Thus, in some cases, using influence lines is an easy way to determine the most unfavorable loading positions. Figure 5-20a shows the sketch of the moment influence line on the interior face of the web in the top slab, developed by treating one HL-93 truck as a unit loading. From

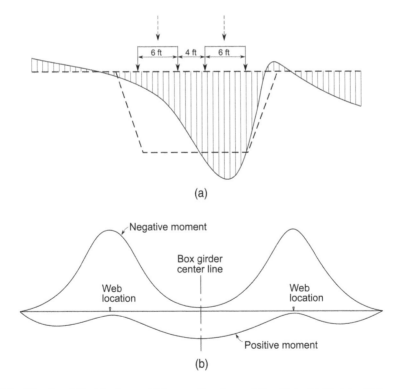

(a)

(b)

FIGURE 5-20 Live Load Positioning and Moment Envelope of Top Slab, (a) Sketch of Moment Influence Line in Interior Face of Right Web, Developed by Treating One HL-93 Truck as a Unit Loading, (b) Typical Moment Envelope of Top Slab.

this figure, we can see that two trucks positioned within the box may induce the maximum absolute negative moment at the section or that the loading case shown in Fig. 4-51 may produce the maximum absolute negative moment if the cantilever is long. A typical moment envelope of the top slab is shown in Fig. 5-20b.

5.3.1.3 Transverse Post-Tensioning Layout

A transverse post-tendon is normally comprised of 3 or 4 of 0.5-in. or 0.6-in.-diameter strands. As the thickness of the top slab is small, the tendons are typically placed in oval-shaped flat ducts as shown in Fig. 1-36b. To effectively use the tendons, the transverse tendons are generally anchored at mid-height of the top slab at the cantilever wing tips and then gradually rise above the neutral axis of the deck over the webs to resist the negative moment. To resist the positive moment at the center region of the box girder, the tendons gradually drop below the neutral axis of the slab in this region. A sketch of a typical tendon layout is illustrated in Fig. 5-21a. The longitudinal spacing of the transverse tendons is determined to meet the service limit and strength limit states requirements and normally range from 2 to 4 ft. To limit the effects of shear lag between anchorages, the maximum spacing is usually not greater than 4 ft (see Fig. 5-21b). Normally, all the transverse tendons will be single end stressed and the stressing end should be alternated to create more uniform prestressing forces. Typical anchorage details for the transverse tendons are shown in Fig. 5-22.

5.3.1.4 Estimation of Transverse Post-Tensioning Tendons

After the maximum and minimum moments of the deck slab at the critical sections due to permanent loads and other design loads have been obtained, the required transverse post-tensioning tendons can be estimated by meeting service limit states and determined by the following equation.

$$A_{ps} = \frac{P}{f_{pe}} \tag{5-5}$$

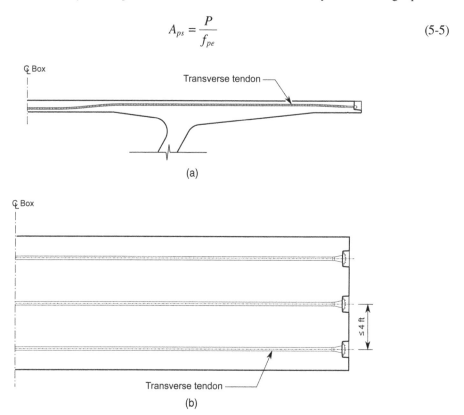

FIGURE 5-21 Transverse Tendon Layout, (a) Elevation View, (b) Plan View.

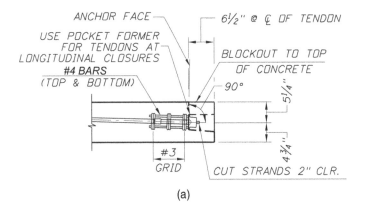

(a)

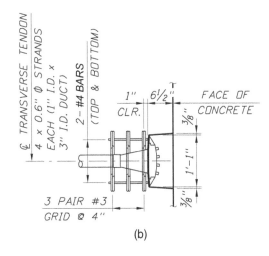

(b)

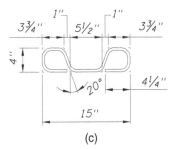

(c)

FIGURE 5-22   Typical Anchor Details for Transverse Tendons, (a) Elevation View, (b) Plan View, (c) #3 Bar Grid.

where

f_{pe} = effective stress of tendon after loss; can be assumed as 16% loss

P = required prestressing force per unit length and can be derived from Eq. 3-27 (in which tensile stress is defined as negative) as

$$P = \frac{6|M| - t^2 f_{ta}}{t + 6e} \tag{5-6}$$

where

$\quad M$ = maximum or minimum moment per unit length
$\quad f_{ta}$ = allowable concrete tensile stress (positive)
$\quad f_{pe}$ = tendon effective stress
$\quad e$ = distance between centroids of post-tensioning tendons and section (positive)
$\quad t$ = thickness of slab

5.3.1.5 Capacity Verifications

After the preliminary layout of the transverse post-tendons has been established, more accurate effects due to the tendons can be evaluated. The losses of prestressing force should include all initial and long-term losses. Secondary post-tensioning effects must be included. The AASHTO specifications require that the capacity at critical sections be checked for service limit states, fatigue and fracture limit state, strength limit states, and extreme event limit states. The fatigue and fracture limit state in the post-tensioned deck is normally deemed to be met.

5.3.1.5.1 Service Limit States

At service limit states, the AASHTO specifications require that the deck and deck system be analyzed as an elastic structure. The current AASHTO specifications require that only service limit state I be checked for tension and compression. The load combination is shown in Table 2-10 and may be reduced as follows for most of the loading situation:

$$M_u = 1.0(DC + DW + EL) + 1.0(PS) + 1.0(LL + IM) + 1.0(CR + SH) + 0.5(TG) + TU \qquad (5\text{-}7a)$$

and

$$M_u = 1.0(DC + DW + EL) + 1.0(PS) + 1.0(CR + SH) + 1.0(TG) + TU \qquad (5\text{-}7b)$$

The maximum concrete compression and tensile stresses calculated from Eqs. 5-7a and 5-7b shall not exceed the values discussed in Section 1.2.5.1.

5.3.1.5.2 Strength Limit States

At strength limit states, the AASHTO specifications require that the decks and deck systems be analyzed as either elastic structures or inelastic structures. For the transverse design, the strength I load combination will normally govern. The strength I load combination is shown in Table 2-10 and can be reduced as follows for typical loading situations:

$$M_u = 1.25(DC) + 1.5(DW) + 1.0(EL) + 1.75(LL + IM) + 1.25(CR + SH) + 0.5TU \qquad (5\text{-}8)$$

The flexural strength at critical sections should be checked following the flow chart shown in Fig. 5-41. The shear strength check is normally ignored for AASHTO design loadings. However, if any large construction loads are anticipated to be placed in the deck, both one-way and two-way action shear should be invested[5-14].

5.3.1.5.3 Extreme Event Limit States

The top slab/bridge deck should be designed for force effects due to vehicle collision and railing loads, i.e., extreme event load combination II. The load combination is shown in Table 2-10 and may be reduced to the following:

$$M_u = 1.25DC + 1.5DW + 0.5(LL + IM) + 1.0CT \qquad (5\text{-}9)$$

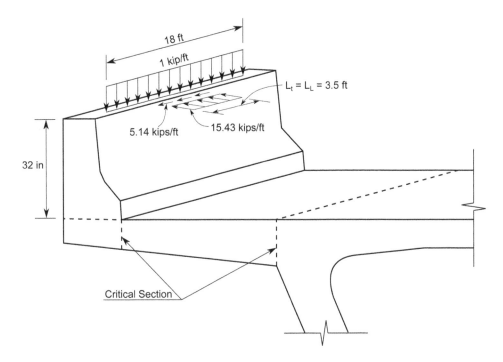

FIGURE 5-23 Vehicle Collision Loadings and Distributions for Test Level TL-4.

The design vehicle collision forces are specified in Article A13.2 of the AASHTO specifications[5-5] and are based on test results that vary with different test levels. The collision loadings and their distributions for Test Level TL-4 which is most often used in the U.S. highway system are illustrated in Fig. 5-23. The vehicle collision loadings for other test levels can be found in the AASHTO specifications[5-5] Table A13.2.1.

Normally, the capacities at two critical sections should be checked: the face of the barrier wall and the root of the cantilever wing (see Fig. 5-23). However, the AASHTO specifications do not clearly address how the collision forces shall be distributed to the aforementioned critical sections. Nevertheless, the AASHTO specifications provide a more direct capacity check method at the extreme event limit state based on the philosophy that the total transverse resistance of the deck is not smaller than the resistance of the barrier wall with the assumption that a yield line failure mechanism will develop (see Fig. 5-24)[5-16], i.e.:

$$M_s \geq M_c \qquad (5\text{-}10)$$

where
M_s = flexural resistance (kips-ft/ft) of the deck determined by acting coincident tensile force T
T = tensile force per unit of deck length (kip/ft)

$$T = \frac{R_w}{L_c + 2H} \qquad (5\text{-}11)$$

R_w = nominal railing resistance to transverse load (kips)
L_c = critical length of yield line failure pattern (ft)
H = height of wall (ft)

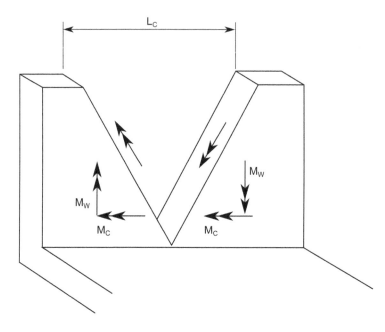

FIGURE 5-24 Yield Line Model for Concrete Barrier Wall of Collision Analysis and Moment Designations[5-5].

For a vehicle hit within a wall segment:

$$R_w = \frac{2}{2L_c - L_t}\left(8M_W + \frac{M_c L_c^2}{H}\right) \tag{5-12a}$$

$$L_c = \frac{L_t}{2} + \sqrt{\left(\frac{L_t}{2}\right)^2 + \frac{8HM_w}{M_c}} \tag{5-13a}$$

For a collision loading applied at the end of a wall or at a joint:

$$R_w = \frac{2}{2L_c - L_t}\left(M_W + \frac{M_c L_c^2}{H}\right) \tag{5-12b}$$

$$L_c = \frac{L_t}{2} + \sqrt{\left(\frac{L_t}{2}\right)^2 + \frac{HM_w}{M_c}} \tag{5-13b}$$

L_t = longitudinal length of distribution of collision force (see Fig. 5-23) =3.5 ft for Test Level TL-4

M_c = flexural resistance of barrier wall about longitudinal direction of wall (kip-ft/ft) (see Fig. 5-24)

M_w = flexural resistance of wall about its vertical axis (kip-ft) (see Fig. 5-24)

M_c and M_w should not vary significantly along the height of the wall to ensure that the results determined by Eqs. 5-12 and 5-13 are correct. For most of the current concrete barrier walls used in the United States, engineers often use the average moments in calculating R_w and L_c to avoid complex analysis.

5.3.2   WEB

The webs of a box girder simultaneously perform two structural functions in the longitudinal and the transverse directions.

   a. Longitudinal functions
      –   As a web in resisting the longitudinal axial force caused by the bending moment about its horizontal axis and vertical shear
      –   As a shear wall in resisting the longitudinal axial force and in-plane shear caused by sectional torsion moment
      –   As a compression and tension chord in resisting the longitudinal axial force caused by the bending moment about its vertical axis
   b. Transverse functions
      –   As a frame member in resisting the bending moments and shear caused by the loading from the top and bottom slabs, deviators, as well as by the section distortion moment
      –   As a slab in resisting bending moment and shear caused by lateral loadings

Theoretically, the web capacity is a function of both the web vertical shear and transverse bending capacities. The capacities of web shear and bending are dependent on each other. However, the web transverse moment normally varies significantly along the web with the maximum moment occurring at the junction between the top slab and the web. The initial cracks of the web caused by the vertical shear and transverse bending will normally develop in different locations. For this reason, engineers often verify the web shear and lateral bending capacities individually without considering their interaction in segmental bridge design. The web bending capacity at the section close to the top slab will be checked after the reinforcement requirement is satisfied in the bridge longitudinal direction. The thickness of the web is mainly governed by detailing consideration and construction requirements, such as to allow proper placement of tendon ducts and concrete. The minimum thickness of the webs is discussed in Section 1.4.2.4.

5.3.3   BOTTOM SLAB

Similar to the top slab and the web, the bottom slab services two types of longitudinal and transverse functions. The longitudinal functions are similar to those for the top slab as discussed in Section 5.3.1. Its transverse functions include (1) as a slab in resisting transverse bending and vertical shear caused by its self-weight, maintenance loadings, and other superposed loads and (2) as a frame member in resisting the transverse moment and shear caused by sectional distortion moment and transverse loadings, such as the forces from deviators and wind loading. In most cases, the effects due to the self-weight and its construction and maintenance loads on the bottom slab are comparatively small. Its design is normally controlled by detailing and construction requirements, such as for proper placement and vibration of concrete and for minimizing differential shrinkage and temperature strains in the girder [5-17], though the capacity of the bottom slab due to a combination of longitudinal and transverse effects should be properly evaluated, including the distortion effect. The AASHTO specifications require that the minimum thickness of the bottom slab is 7 in. and should not less than 1/30 of clear span length between the webs or the haunches. The minimum reinforcement requirement for developing the required crack moment M_r is often governs the design.

5.4   DESIGN OF DIAPHRAGMS AND DEVIATIONS

5.4.1   Design of Diaphragms

5.4.1.1   Introduction

The diaphragms over piers in span-by-span bridges mainly serve two functions: anchoring the longitudinal post-tensioning tendons and transferring the design loadings from the bridge super-structure to the substructure and retaining the super-structure cross-section. The diaphragm is one of the most complex elements in a segmental bridge built by the span-by-span method. Though significant research and analysis has been performed in determining the true behavior of the diaphragm, currently there is no simple, standard, and consistent analytical method available. As the diaphragm is a critical element in segmental bridges, it is normally designed on the rather conservative side with rough analytical methods, such as rules of thumb, and simple equilibrium consideration. With the development of computer software, three-dimensional finite-element models are often used to check its adequacy.

5.4.1.2   Design for Anchoring Post-Tensioning Tendons

The basic analytical methods for the anchorage zone were discussed in Section 3.8. For the diaphragms as shown in Figs. 5-10 and 5-12 with an access opening, there are mainly two issues to be considered in the diaphragm design: (1) ensure that there is enough shear strength to transfer the heavy post-tensioning force to the box slabs and (2) ensure that there is enough reinforcement to resist the bursting forces developed by the post-tensioning forces. The total shear friction force (refer to Section 3.6.4) between the diaphragm and the box slabs should be greater or equal to the total longitudinal post-tensioning forces (see Fig. 5-25), i.e.,

$$\phi \mu f_y A_{vf} \geq \sum_{i}^{n} P_i \cos \alpha_i \tag{5-14}$$

where
P_i =   prestressing force of tendon i ($i = 1$ to n)
α_i =   inclined angle with horizontal direction of tendon i (see Fig. 5-25)
n =   number of post-tensioning tendons in half diaphragm (assuming the box section is symmetrical about its vertical centerline, see Fig. 5-25)

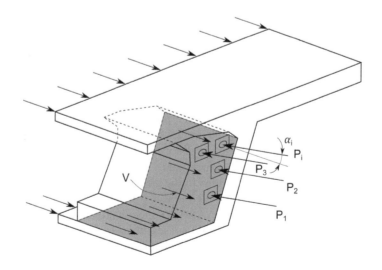

FIGURE 5-25   Post-Tensioning Force Transfer [5-2].

A_{vf} = area of interface shear reinforcement crossing shear plane in half diaphragm as shown in Fig. 5-25

f_y = yield stress of reinforcement

μ = friction factor (see Table 3-2)

The bursting forces in girder vertical and transverse directions are often conservatively estimated by the strut-and-tie models shown in Fig. 5-26.

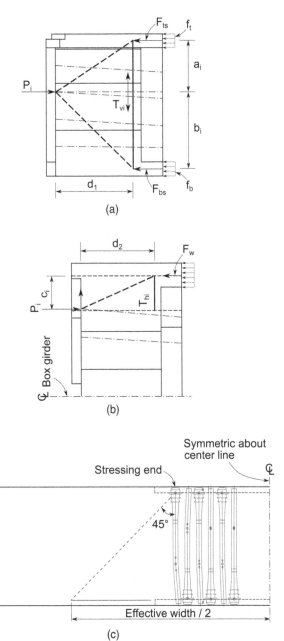

FIGURE 5-26 Analytical Models for Predicating Bursting Forces, (a) Vertical Direction, (b) Transverse Direction, (c) Effective Width.

The total bursting force along the diaphragm's vertical direction may be estimated as

$$T_v = \sum_1^n T_{vi} \qquad (5\text{-}15)$$

where
$\quad T_{vi} \quad$ = vertical bursting force due to tendon i ($i = 1$ to n)

$$T_{vi} = \frac{(F_{ts} + F_{bs})a_i b_i}{d_1(a_i + b_i)} \qquad (5\text{-}16)$$

$\quad F_{ts} \quad$ = total force acting on top slab induced by post-tensioning force (see Fig. 5-26a)

$$F_{ts} = f_t A_{tf} \qquad (5\text{-}16\text{a})$$

$\quad F_{bs} \quad$ = total force acting on bottom flange induced by post-tensioning force (see Fig 5-26a)

$$F_{bs} = f_b A_{bf} \qquad (5\text{-}16\text{b})$$

$\quad f_t, f_b \quad$ = average normal stresses in top and bottom slabs, respectively, can be determined by assuming plane section remaining plane after bending
$\quad A_{tf}, A_{bf}$ = effective areas of the top and bottom slabs, respectively. The effective widths of the top and bottom slabs can be approximately calculated by using a distribution angle of 45° from the tendon stressing end to the opposite face of the diaphragm (see Fig. 5-26c). The designations a_i, b_i, and d_1 are shown in Fig. 5-26a.

The total bursting force along the diaphragm transverse direction may be estimated as

$$T_h = \sum_1^n T_{hi} \qquad (5\text{-}17)$$

where
$\quad T_{hi}$ = transverse tensile force due to tendon i ($i = 1$ to n)

$$T_{hi} = \frac{F_w c_i}{d_2} \qquad (5\text{-}18)$$

where
$\quad F_w$ = total axial force acting on web due to post-tensioning force:

$$F_w = P_i - F_{ts} - F_{bs}$$

The designations c_i and d_2 are shown in Fig. 5-26b.

The required bursting reinforcement can be determined by assuming the allowable stress equal to $0.4 f_y$ at service limit state.

It is apparent that these models are rather rough. Based on limited tested and analytical results, Wollmann[5-17] proposed a refined strut-and-tie model as shown in Fig. 5-27a for predicting the bursting forces. In this figure, the ties are shown by solid lines and the struts are shown by dotted lines as usual. However, it is difficult to develop a general strut-and-tie model that is suitable for all types of diaphragms and different anchorage arrangements as the model may vary with the location

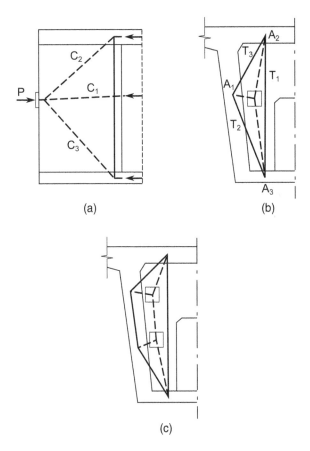

(a)   (b)

(c)

FIGURE 5-27 Refined Analytical Models for Predicating Bursting Forces, (a) Elevation View of Triple Model for Single Anchor, (b)) Side View of Triple Model for Single Anchor, (c) Model for Multiple Anchors.

and number of tendons (see Fig. 5-27b). A more reasonable strut-and-tie model may be developed based on elastic analysis by the finite-element method on a case-by-case basis.

5.4.1.3   Design Considerations for Transferring Super-Structure Loadings to Substructure

The super-structure vertical and horizontal shears as well as torsion moment are transferred to the substructure through the diaphragms. It is important for the designers to clearly know the proper loading paths.

   a. Loading Paths for Vertical Shear

      The loading paths and strut-and-tie models for the diaphragm in vertical shear transferring from the super-structure to the substructure are related to the shapes of the diaphragms and the support conditions. Figure 5-28a and b show two typical loading paths and strut-and-tie models that may serve as a reference in diaphragm design. For a large box girder, to enhance aesthetics and reduce the construction cost, the width of the pier cap and the spacing of bearings are often reduced, which may cause significant tension at the top portion of the diaphragm. The high tensile force can be offset by applying post-tensioning tendons as shown in Fig. 5-28c.

      For a box girder with vertical webs as shown in Fig. 5-29, two critical faces should be investigated[5-16]. One is the shear friction at the interface of the web and diaphragm (see Fig. 5-29a). Another is at the bottom portion of the web. It is important that the web reinforcement at the bottom of the web be designed carefully as it may be subjected to a direct

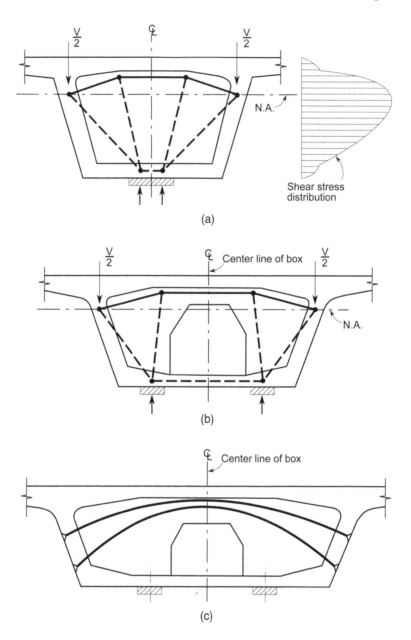

FIGURE 5-28 Loading Paths and Strut-and-Tie Models of Diaphragm Subjected to Vertical Shear, (a) Diaphragm without Opening and with Single Support, (b) Diaphragm with Opening, (c) Transverse Post-Tensioning in Diaphragm.

tension as shown in Fig. 5-29b. The additional reinforcement to resist the tension should be considered in the web design within the limits of the pier and abutment.

b. Loading Path for Torsion and Horizontal Shear

Eccentrically positioned external loadings applied on the bridge super-structure will cause torsion and distortion moments acting on the cross-section of the box. For normal box girder sections in segmental bridges, the effect of the distortion moment may be neglected. The torsion moment is resisted by the shear flow around the box section as discussed in

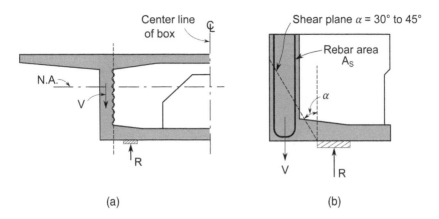

FIGURE 5-29 Vertical Shear Friction Face and Web Direct Tension at Support Area, (a) Shear Friction Face, (b) Web Direct Shear at Support Area.

Section 3.4.3.1 and is transferred to the bearings through the diaphragm. The shear flow in the top slab produces the horizontal force H_t (see Fig. 5-30), which can be written as

$$H_t = \frac{M_T b_t}{2A} \tag{5-19}$$

where

M_T = pure torsion moment
b_t = length of top slab
A = enclosed area of box section (refer to Eq. 3-102 and Fig. 3-26)

The shear stress distribution induced by horizontal force V_h can be determined based on the method discussed in Section 3.4.2.1, and its sketch is shown in Fig. 5-31a. From the shear distribution, the total shear force H_h applied to the top of the slab can be obtained (see Fig. 5-31b). In the diaphragm design, H_h may be roughly estimated using the following equation:

$$H_h = V_h \frac{A_{top}}{A_{top} + A_{bottom}} \tag{5-20}$$

where

V_h = horizontal shear
A_{top} = area of top slab, including cantilever slabs
A_{bottom} = area of bottom slab

The loading paths from the top slab to the bearings in the diaphragm are related to the shapes of the opening in the diagram. Two typical strut-and-tie models for the analysis of diaphragms subjected to torsion moment are shown in Fig. 5-30.

c. Strut-and-Tie Models of Diaphragms Subject to Shear Forces and Torsion
It is apparently difficult to develop a reasonable strut-and-tie model for analyzing the diaphragm subjected to a combined action of shear forces and torsional moment. Though it may be developed based on the loading paths identified by the finite-element method as shown in Fig. 5-32, it may vary with the configurations of the diaphragm and its suitability for practical use can be argued. Engineers can safely use the models shown in Figs. 5-28 to 5-30 to consider the effects of shear forces and torsional moment individually. Either mild reinforcement or post-tensioning tendons can be designed to resist the tensile forces.

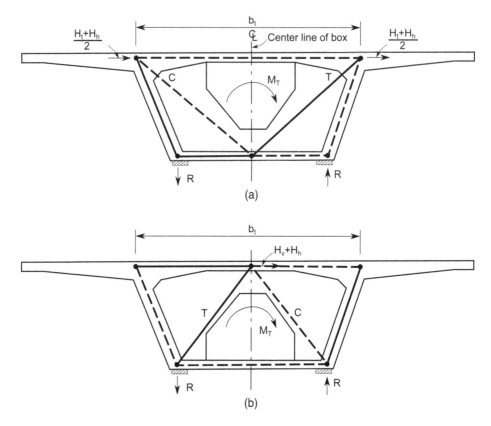

FIGURE 5-30   Strut-and-Tie Models of Diaphragm Subjected to Torsion, (a) V-shape Opening, (b) A-Shape Opening.

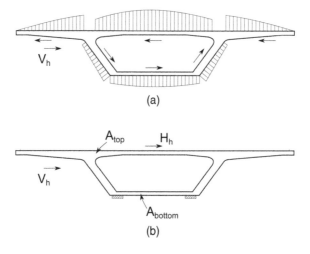

FIGURE 5-31   Shear Stress Distribution Due to Horizontal Shear Force, (a) Shear Stress Distribution, (b) Shear Force Applied to Diaphragm.

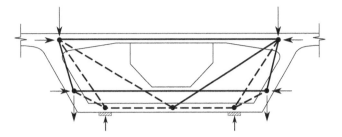

FIGURE 5-32   Strut-and-Tie Model of Diaphragm Subject to Shear Forces and Torsion Moment.

The amount of the top, torsional, web, and bottom reinforcement required can be determined based on the sum of the computed tensile forces. The diaphragm should be sized to adequately resist the compression force.

5.4.2 DESIGN OF DEVIATORS

The deviator is another complex and key element in segmental bridges. Currently, there is no simple and consistent analytical method for its design. It is normally sized on the conservative side as it will not significantly affect the construction cost. In designing the deviator saddles, the following should normally be considered.

1. Provide enough reinforcement to resist the bending moment of deviator diaphragms/ribs as well as to anchor them to the web and bottom slab.
2. The deviators may cause significant bending moment in the web and bottom slab around the deviators. Additional reinforcement may be necessary to resist the bending moment.

The first design consideration listed above is mainly in reference to the deviator itself, and the second design consideration addresses the effect of the deviator on the web and bottom slab around the deviator. An appropriate analytical model to identify the behaviors of the deviator and its effect on the surrounding slabs is largely related to the stiffness of the deviator types selected. Analytical results [5-17] indicate that the tendon deviation forces are largely resisted by bending of the deviator ribs for the full-height diaphragm/rib type, and the surrounding slabs do not have a significant bending moment caused by the deviator. In this case, a two-dimensional frame model can be used to adequately determine the design forces within the deviation rib (see Fig. 5-33). The effective flange width of the rib may be taken as 0.25 times the sum of the web height and the length of the bottom flange or 8 times the rib thickness.

For blister deviators as shown in Figs. 5-34 and 5-36, a significant amount of bending moment may be developed in the webs and bottom flange around the blister[5-4] to resist the deviation forces. In this case, a shell-plate finite-element model can be used to determine the bending moment in the web and bottom flange. Article 5.10.9.3.7 of the AASHTO specifications[5-5] requires that deviation saddles should be designed either using a strut-and-tie model or using methods based on test results. Some typical strut-and-tie models for predicting the direct tension forces and shear are illustrated in Fig. 5-34. The strut-and-tie models for the vertical and horizontal loop bars shown in Fig. 5-34a and b may be developed as illustrated in Fig. 5-34c and d, respectively. A more comprehensive strut-and-tie model may be developed from internal forces of the web and bottom slab based on transverse frame analysis (see Fig. 5-35a). Analytical results indicate that the internal forces in the box web and bottom flange due to the deviation forces can be approximately estimated by assuming the box as a frame structure with fixed supports at the top of the webs (see Fig. 5-35a). Then, a strut-and-tie model can be developed by taking the deviation blister as a free body (see Fig. 5-35b) and the reinforcement for the deviator may be arranged as Fig. 5-35c.

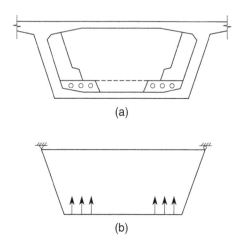

FIGURE 5-33 Analytical Model for Full-Height Diaphragm Deviators, (a) Deviator Segment, (b) Analytical Model.

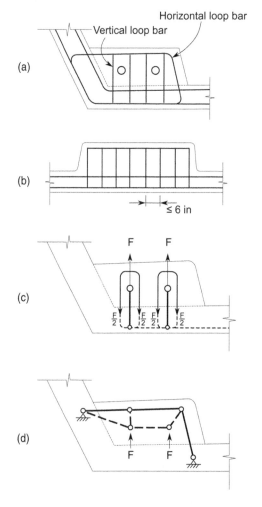

FIGURE 5-34 Strut-and Tie Models for Blister Deviators, (a) Side View of Blister Sketch, (b) Elevation View of Blister Sketch, (c) Strut-and-Tie Model for Vertical Loop Bars, (d) Strut-and-Tie Model for Horizontal Loop Bars.

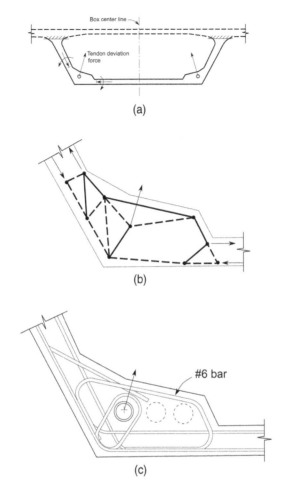

FIGURE 5-35   Refined Strut-and-Tie Model, (a) Transverse Frame Analytical Model, (b) Strut-and-Tie Model, (c) Arrangement of Reinforcement.

Figure 5-36a shows a principal tensile stress contour plot for a typical deviation blister for internal tendons, determined based on finite-element method analysis[5-17]. Based on the analytical and test results, a strut-and-tie model has been proposed by Wollmann,[5-17] and is shown in Fig. 5-36b. The in-plane deviation force due to the change in direction of tendons can be taken as:

$$F = P \sin \alpha \tag{5-21}$$

where
  α = inclined angle of tendon with bottom flange
  P = factored tendon force

From Fig. 5-36a, it can also be observed that high tensile stresses will develop in the bottom flange behind the blister anchorage. To prevent the slab from cracking, AASHTO specifications[5-5] Article 5.10.9.3.4b requires that tie-back reinforcement should be provided behind the intermediate anchor into the concrete section to resist at least 25% of the tendon force at service limit state. The stresses in this reinforcement should not exceed a maximum of 0.6 fy or 36 ksi.

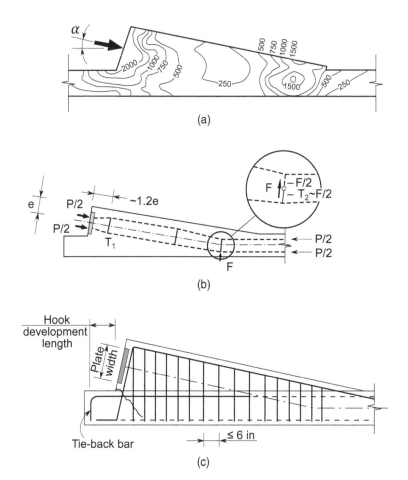

FIGURE 5-36 Strut-and-Tie Model for Internal Tendon Blister Deviator, (a) Stress Contour (psi), (b) Strut-and-Tie Model, (c) Reinforcement Arrangement.

If compressive stresses are developed behind the anchor, the minimum area of the tie-back reinforcement can be calculated as

$$A_{tie} = \frac{0.25P_s - f_{cb}A_{cb}}{0.6\,f_y} \tag{5-22}$$

where

P_s = tendon force (s) at blister anchorage (kip)
f_{cb} = unfactored minimum compressive stress in region behind blister at service limit states (ksi)
A_{cb} = area of web and/or slab within extension of sides of anchor plate or blister as shown in Fig. 5-37 (in.^{2})

Tie-back reinforcement shall be placed within one anchor plate width in the bottom slab from the tendon axis. It shall be fully anchored so that the yield strength can be developed at the base of the blister (see Fig. 5-36c).

Reinforcement should be provided throughout deviators as required for shear friction, corbel action, bursting forces, and deviation forces due to tendon curvature. The required reinforcement can

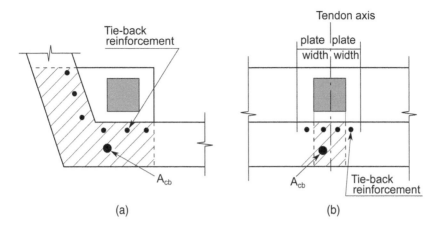

FIGURE 5-37   Tie-Back Reinforcement Arrangement and Effective Compression Concrete Area, (a) Blister Type, (b) Rib Type.

be calculated based on the determined design forces. This reinforcement should extend as far as possible into the flange or web and be bent around transverse bars in the form of ties or U-stirrups (see Figs. 5-34 to 5-36). Spacing should not exceed the smallest dimensions of the deviator or 6.0 in.[5-5]. As a minimum, the following should be checked:

a. Tendon pull-out forces
b. Bending in the deviation rib, diaphragm, and/or surrounding webs and flanges
c. Transverse tension due to deviation forces and shear friction

The maximum deviation force should be factored by 1.7. When a method based on test results is used, the resistance factors for direct tension and shear should be taken as 0.90 and 0.85, respectively.

It should be mentioned that the strut-and-tie models for the type of the blister deviators vary with the different shapes and loading conditions of the blisters. Engineers should use the rule of thumb to identify the basic loading transfer path in the design.

5.5   SUMMARIZATION OF DESIGN PROCEDURES

For clarity, the general design procedures for concrete segmental bridges are summarized as follows:

Step 1: Based on the needs and site conditions, determine the span lengths and bridge width.
Step 2: Based on the AASHTO specifications and the related Department of Transportation (DOT) specific requirements, select types of materials used.
Step 3: Based on the span lengths, AASHTO-ACI-ASBI Design Standards, previous designed bridges and experiences, determine preliminary typical sections.
Step 4: Determine the design loads based on the AASHTO specifications and local DOT special requirements.
Step 5: Perform bridge longitudinal design
a. Assume construction sequences.
b. Calculate the moments, shear at critical sections due to dead loads and live loads, and develop moment and shear envelopes based on the procedures discussed in Sections 5.2.7.2 and 5.2.7.3.

c. Estimate required jacking forces or effective prestressing forces meeting the AASHTO specifications service III by the methods discussed in Section 5.2.7.4, based on the preliminary calculated moments, previous experience, or some bridge design charts.

d. Estimate the required tendon area, the number of tendons, and the layout of the tendons based on the preliminary moment envelopes and previous experiences.

e. Calculate the instantaneous losses of prestressing forces based on the methods discussed in Section 3.2.1.

f. Calculate the secondary moment analysis of post-tensioning forces after instantaneous losses, based on the methods discussed in Section 4.3.2.

g. Determine the secondary effects due to creep, shrinkage, relaxation, temperature, etc., based on the methods presented in Sections 4.3.3. to 4.3.6.

h. Check elastic shortening losses converge on assumed until they are met.

i. Verify service limit states I and III capacities for each loading stage, normally at the end of the construction and 10,000 days. If the limit state cannot be met, revise the tendon area, locations, or the typical cross section as necessary until it is met.

j. Verify flexural strength capacities following the flow chart shown in Fig. 3-41 until they are met by modifying the tendon area or typical cross section as necessary. Normally, Strength Limits I and II will control the design.

k. Size the vertical shear reinforcement based on the factored shear.

l. Verify shear strength capacities following the flow chart shown in Fig. 3-54 until they are met by modifying the shear reinforcement. Normally, strength limits I and II will control the design.

Step 6: Perform bridge transverse design.

a. Calculate the moments, shear at critical sections due to dead loads and live loads using moment and shear influence lines or other methods as discussed in Section 4.4.

b. Based on service state limit I and previous experiences or existing bridge plans, estimate the required deck transverse post-tensioning area per unit length and lay out the transverse post-tensioning tendons.

c. Calculate the losses of prestressing forces.

d. Determine the secondary effects due to post-tensioning forces, creep and shrinkage, temperature, etc.

e. Check elastic shortening losses converge on assumed until they are met.

f. Verify deck capacity for service limit I.

g. Perform flexure reinforcement design based on the maximum factored moments for the deck, web, and bottom flange.

Step 7: Design diaphragm and deviators based on the methods presented in Sections 3.8.2 and 5.4.1.

5.6   DESIGN EXAMPLE I: SPAN-BY-SPAN BRIDGE

To ensure the segmental concrete bridge alternative can compete with other bridge alternatives, the total deck area should normally exceed 110,000 ft^2. The design example is generated based on an existing long bridge owned by FDOT (used with permission of FDOT).

5.6.1   Design Requirements

5.6.1.1   Design Specifications

a. AASHTO LRFD *Bridge Design Specifications,* 7th edition (2014)
b. FDOT *Structures Design Guidelines* (2018)
c. CEB-FIP *Model Code for Concrete Structures,* 1990

5.6.1.2 Traffic Requirements

a. Two 12- ft lanes with traffic flow in the same direction

b. 10- ft- wide right shoulder and 6- ft- wide left shoulder

c. Two 18.5-in.- wide FDOT 32-in. F-shape barrier walls

5.6.1.3 Design Loads

a. Live Load: HL-93 Loading

b. Dead Loads:

- Unit weight of concrete $w_c = 0.145$ kcf
- Unit weight of reinforcement concrete (DC) $w_{dc} = 0.150$ kcf
- 1/2-in. sacrificial deck thickness provided for grooving and planning

c. Thermal Loads

- Uniform temperature: Mean:70°F; Rise: 35°F; Fall: 35°F
- Temperature gradient specified in Section 2.2.10.2 and taking the following values:

Positive Nonlinear Gradient	Negative Nonlinear Gradient
$T_1 = 41°F$	$T_1 = -12.3°F$
$T_2 = 11°F$	$T_2 = -3.3°F$
$T_3 = 0°F$	$T_3 = 0°F$

5.6.1.4 Materials

a. Concrete:

$$f_c' = 6.5 \ ksi, \ f_{ci}' = 5.2 \ ksi$$

$$E_c = 33,000 \ w_c^{1.5} \sqrt{f_c'} = 4645 \ ksi$$

b. Prestressing Strands

ASTM A416 Grade 270 low relaxation

$$f_{pu} = 270 \ ksi, \ f_{py} = 0.9 f_{pu} = 243 \ ksi$$

Maximum jacking stress: $f_{pj} = 216$ ksi

Modulus of elasticity: $E_p = 28,500$ ksi

Strand anchor Set: 3/8 in.

Bar anchor Set: 1/16 in.

Friction coefficient: 0.23 (plastic duct)

Wobble coefficient: 0.00020 (internal), 0 (external)

c. Reinforcing Steel

ASTM 615, Grade 60

Yield stress: $f_y = 60$ ksi

Modulus of elasticity: $E_s = 30,000$ ksi

5.6.2 BRIDGE SPAN ARRANGEMENT AND TYPICAL SECTION

5.6.2.1 Span Arrangement

The first consideration for arranging a bridge spans is to meet the minimum horizontal clearance requirement for traffic below and other site situations. The general considerations for the segmental bridge constructed using the span-by-span method were discussed in Section 5.2.1.2.

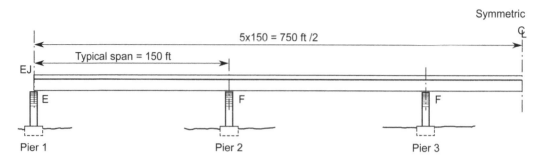

FIGURE 5-38   Elevation View of Example Bridge I.

The span arrangement of the design example is shown in Fig. 5-38. The bridge contains five continuous spans with an equal span length of 150 ft, which is within the normal economic span range as discussed in Section 5.2.1.1. In Fig. 5-38, E represents expansion joint and F represents fixed support. The girder supports are composite elastomeric bearing pads that are actually neither fixed support nor roller support. The supports can be treated as spring supports in the structural analysis.

5.6.2.2   Typical Section and Segment Layout

As the span length of this example bridge is within the span range for the AASHTO-PCI-ASBI segmental box girder standards, it is the first choice to select the standard box section to reduce the construction cost. From the standards (see Appendix A), both 7-ft and 8-ft girder depths for span-by-span construction can meet the minimum girder depth requirement for FDOT and can be considered for the example bridge. However, this bridge contains several units built using both cantilever and span-by-span construction methods. To facilitate precasting and save on construction cost, a girder depth of 9 ft is selected for both span-by-span and cantilever constructions. The total deck width is 43'-1", which accommodates two traffic lanes, two shoulders, and two barrier walls as required. The typical cross-section is shown in Fig. 5-39.

The precast segments are assumed to be transported to the site by trailers on the existing roadway. As discussed in Section 1.4.1, the typical length of segments for this example is designed from 9'-2" to 10'-0" (see Fig. 5-40). The length of the pier segment at the expansion joint is determined as 7'-3". The pier segment over continuous supports is designed to be precasted in two pieces, and the total length of the segment is 10 ft. The maximum segment weight is about 60 tons. The width of the cast-in-place joints ranges from 6" to 7.75". The detail segment arrangement is shown in Fig. 5-40.

The primary dimensions for the deviator segment, the expansion joint pier segment, and the continuous support pier segment are illustrated in Figs. 5-41 to 5-43. To reduce the segment weight of the pier segment over the continuous support, the segment is precast in two halves as shown in Fig. 5-43. The joint between the two half-pier segments is epoxied and connected by 8-1.375" diameter temporary post-tensioning bars which create 40-psf pressure over the cross-section.

5.6.3   Tendon Layout

From the bending moments of the preliminary designed bridge due to the dead and live loads and previous design experiences, the amount of prestressing area can be estimated by assuming the effective prestressing as 0.6 times the tendon ultimate stress. After two or three analysis trials, the required prestressing area can normally be determined. For simplicity, the process is omitted in the example. A total of 10 external tendons, which comprises eight $19 \times 0.6''$ ϕ strand tendons and two $27 \times 0.6''$ ϕ

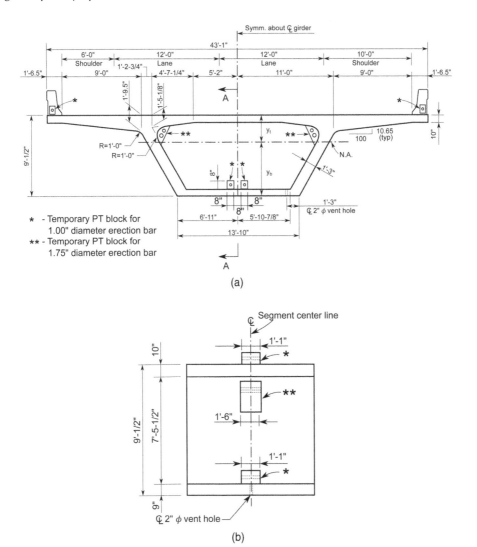

FIGURE 5-39   Typical Section of Example I, (a) Elevation View, (b) Side View.

strand tendons are used for each span. The tendon layout is shown in Fig. 5-44. In the figure, notations T1, T2, T3, and T5 represent $19 \times 0.6''$ φ strand tendons and T4 represents $27 \times 0.6''$ strand tendon for the exterior span. For the interior span, notations T1, T2, T3, and T4 represent $19 \times 0.6''$ φ strand tendons and T5 represents $27 \times 0.6''$ strand tendons. All the tendons will be stressed at one end as shown in Fig. 5-44 in a solid black triangle.

5.6.4   BRIDGE LONGITUDINAL ANALYSIS

5.6.4.1   Assumptions of Construction Sequences

The construction sequences are assumed as follows:

1. Stage I: Gantry erection (see Fig. 5-45a)
   a. Erect expansion joint segment and leading pier segment on temporary support by crane, and provide longitudinal restraint at pier 2 (see Fig. 5-45a).
   b. Install gantry or launch gantry from the previously completed span.
   c. Secure gantry to expansion joint and pier segments.

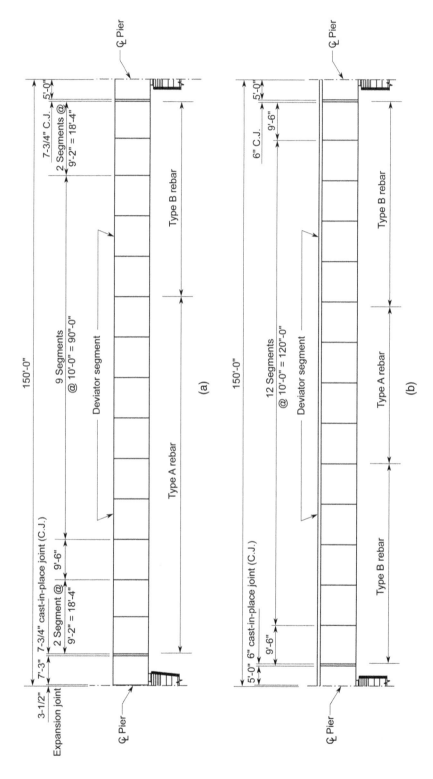

FIGURE 5-40   Segment Layout, (a) End Span, (b) Interior Span.

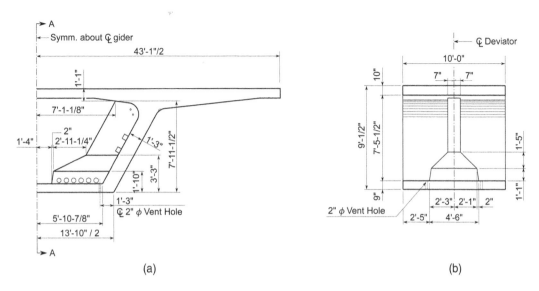

FIGURE 5-41 Primary Dimensions of Deviation Segment, (a) Half of Elevation, (b) Side View.

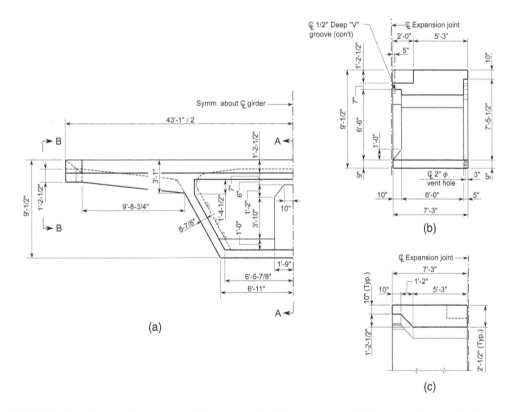

FIGURE 5-42 Primary Dimensions of Expansion Joint Pier Segment, (a) Elevation View, (b) Section *A-A* View, (c) Section *B-B* View.

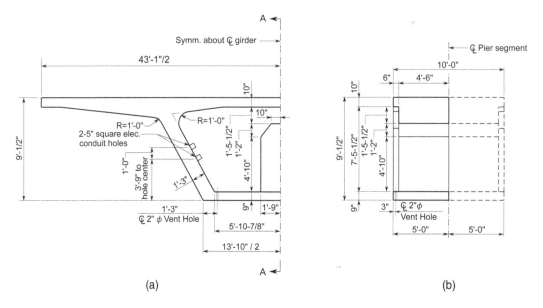

(a)   (b)

FIGURE 5-43 Primary Dimensions of Pier Segment over Continuous Support, (a) Elevation View, (b) Section *A-A* View.

2. Stage II: Segment erection (see Fig. 5-45b)
   a. Deliver precast segments by truck.
   b. Lift each of the segments to its position using a gantry.
   c. After all the segments in the span are lifted, adjust the grade and the alignment of each segment.
3. Stage III: Post-tensioning (see Fig. 5-45c)
   a. Apply epoxy between each segment, and stress together with temporary prestressing bars. After the second segment is stressed to the first segment, a new gap is created between the second and the third segments for applying epoxy.
   b. After all segments between the pier segments are stressed together, install external polyethylene ducts.
   c. Pull longitudinal post-tensioning tendons through the ducts.
   d. Place concrete spacer blocks in the closure joints.
   e. Stress enough tendons in both webs to obtain 10% of total final force for all tendons.
   f. Pour cast-in-place joint concrete.
   g. After the cast-in-place joint concrete has achieved a minimum concrete strength of 3.4 ksi, stress the longitudinal tendons to their final design forces.
   h. Place dry pack mortar plinths at the bearings, and release the pier or expansion joint segment onto the permanent bearings.
4. Stage IV: Move the gantry forward, and install the segments in the next span (see Fig. 5-45d).
   a. Install pier segment at pier 3.
   b. Move the gantry to span 2, and secure the gantry to the pier segments.
   c. Repeat stages II to III until completing the installation of the entire bridge.

5.6.4.2 Analytical Models

The bridge is modeled as a three-dimensional frame structure as shown in Fig. 5-46. The bridge structure is divided into a number of finite elements (see Fig. 5-47) and analyzed by the finite-element method. Each segment is modeled as one element for an easy consideration of time-dependent effects and is assumed to be erected on the 28th day after it was cast. The diaphragms

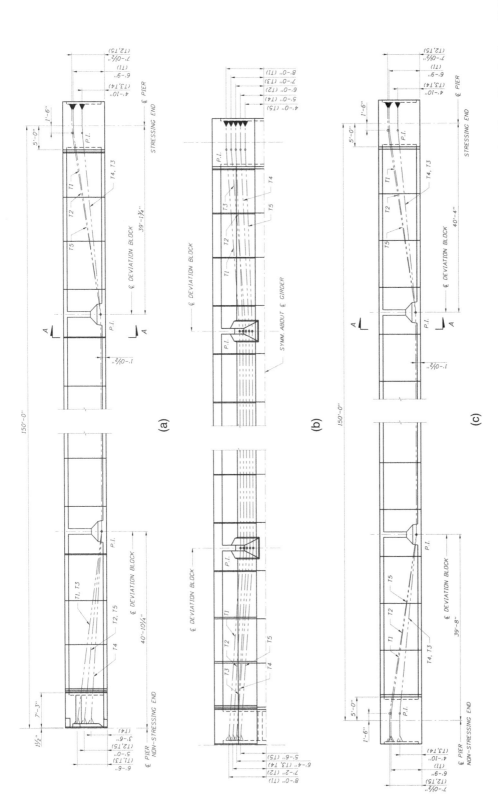

FIGURE 5-44 Tendon Layout and Segment Divisions, (a) Elevation View of Exterior Span, (b) Plan View of Exterior Span, (c) Elevation View of Interior Span, (d) Plan View of Interior Span, (e) Section A-A View. *(Continued)*

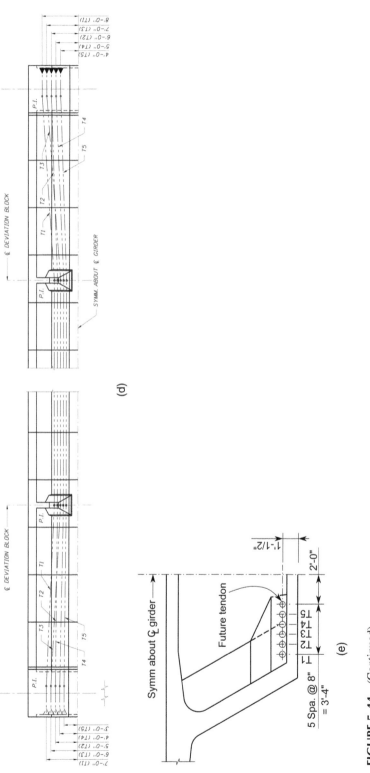

FIGURE 5-44   (Continued)

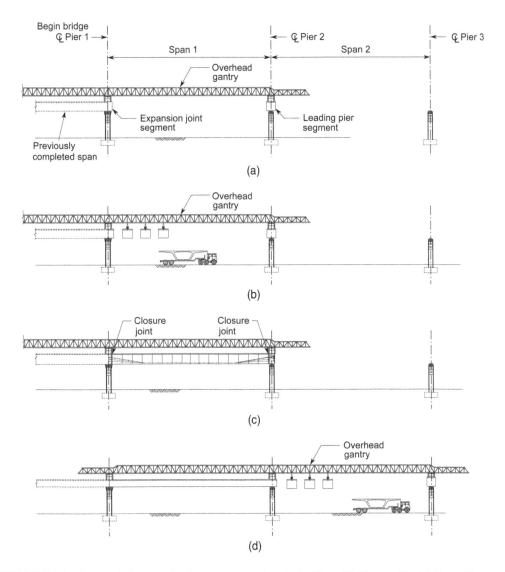

FIGURE 5-45 Assumed Construction Sequences, (a) Stage I, (b) Stage II, (c) Stage III, (d) Stage IV.

and pier caps are modeled with weightless concrete triangle frames with large enough sectional properties for simulating rigid members, and the elastomeric bearings pads are modeled as springs (see Fig 5-46b). The super-structural configurations and support conditions are modeled based on the erection sequences described in Section 5.6.4.1. The structural analysis is performed using the LARSA 4-D computer program.

5.6.4.3 Sectional Properties

The typical sectional properties are calculated as follows:

Section area: $A = 78.28$ ft^2
Area enclosed by shear flow path: $A_0 = 135.42$ ft^2
Moment of inertia about x-axis: $I_z = 759.93$ ft^4
Moment of inertia about y-axis: $I_y = 9061.86$ ft^4
Free torsion moment of inertia: $J = 1565.87$ ft^4

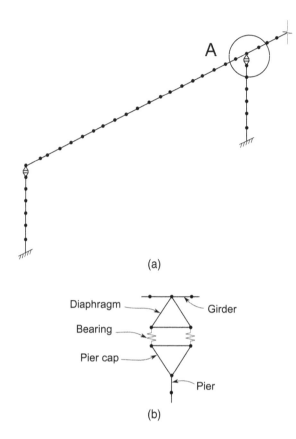

(a)

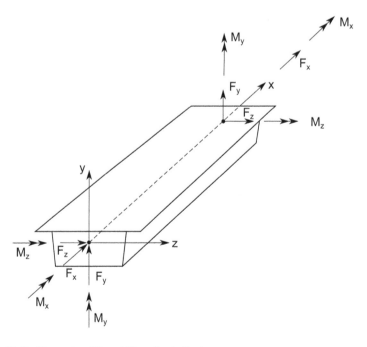

(b)

FIGURE 5-46   Analytical Model of Super-Structure, (a) Three-Dimensional Beam Model, (b) Detail *A* View.

FIGURE 5-47   Finite Element and Local Coordinate System.

Distance from neutral axis to top of the top slab: $y_t = 2.75$ ft
Distance from neutral axis to bottom of the bottom slab: $y_b = 6.25$ ft
Static moment of area above neutral axis: $Q_Z = 106.05$ ft^3
Web orientation angle: $\theta_w = 60.61°$

5.6.4.4 Effects Due to Dead Loads

The moment and shear due to box girder self-weights are analyzed based on the assumed bridge construction sequences as shown in Fig. 5-45. The self-weight of the barrier walls is calculated as 0.84 kips/ft and is applied on the final five-span continuous bridge. The moment and shear envelopes due to the dead loads are illustrated in Figs. 5-48 and 5-49, respectively.

5.6.4.5 Effects Due to Live Loads

The design live loads include a single truck or tandem with a uniform distributed lane loading as discussed in Section 2.2.3. The maximum effects at the related sections are determined by considering the following in the application of the live loads:

    a. In the bridge longitudinal direction, position the live loads based on the shape of the influence line of the section of concern as discussed in Section 5.2.7.3.
    b. In the bridge transverse direction, load one to the maximum allowed loading lanes individually and apply multiple presence factors contained in Table 2-2, respectively, to determine the most unfavorable loading case and effects for design. To produce the maximum torsion of the bridge, the centerline of the exterior wheel is positioned 2 ft from the edge of the travel lane as shown in Fig. 2-2. The number of the maximum loading lanes is determined as

$$N = integer\ part\ of\left(\frac{w}{12} = \frac{40}{12}\right) = 3 \text{ lanes}$$

The braking forces are determined based on the method described in Section 2.2.4.4 and taken as the maximum effects caused by the four loading cases shown in Fig. 5-50. Each of the horizontal forces acting eccentrically from the girder neutral axis is transferred to one horizontal force and one moment acting at the girder neutral axis.

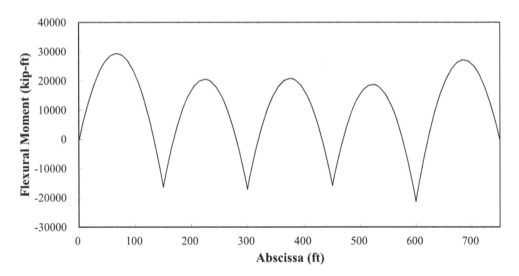

FIGURE 5-48   Moment Envelope Due to Dead Loads.

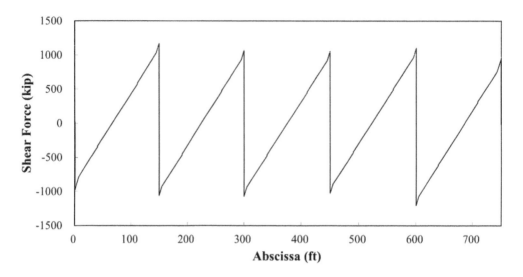

FIGURE 5-49   Shear Envelope Due to Dead Loads.

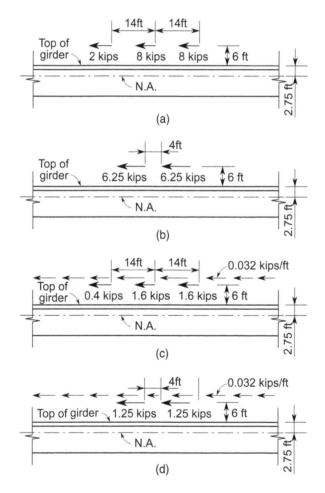

FIGURE 5-50   Braking Loading Cases, (a) 25% Design Truck, (b) 25% Design Tandem, (b) 5% Design Truck Plus 5% Design Lane Loading, (d) 5% Design Tandem Plus 5% Design Lane Loading.

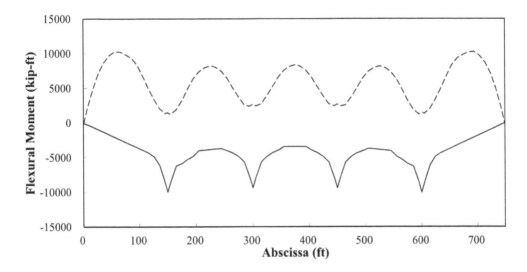

FIGURE 5-51　Moment Envelope Due to Live Loads.

The effects due to live loads are analyzed by the computer program LARSA 4-D and the moment, shear, and torsion envelopes are shown in Figs. 5-51 to 5-53. In these figures, a dynamic load allowance of 1.33 is included. The dotted lines represent the positive responses, and the solid lines indicate the negative responses.

5.6.4.6　Effect Due to Temperatures

As all the supports are elastomeric bearing pads, the effects of uniform temperature rise and fall on the girder axial forces can be neglected. The moment and shear distributions due to the temperature gradient specified in Sections 2.2.10.2 and 5.6.1.3 are illustrated in Figs. 5-54 and 5-55. In Figs. 5-54 and 5-55, the solid and dotted lines represent the responses due to the positive and negative vertical temperature gradients as presented in Section 5.6.1.3, respectively.

5.6.4.7　Effects of Post-Tendons

All the prestressing force losses are determined based on the methods described in Section 3.2 and the assumptions given in Section 5.6.1.4. The detailed analysis is performed using the LASAR

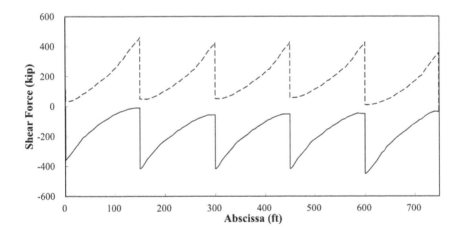

FIGURE 5-52　Shear Envelope Due to Live Loads.

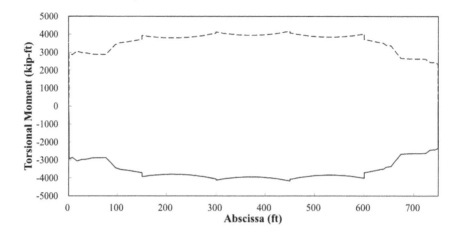

FIGURE 5-53   Torsion Envelope Due to Live Loads.

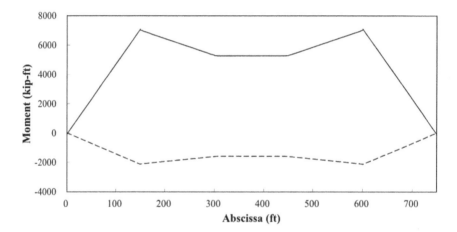

FIGURE 5-54   Moment Distribution Due to Temperature Gradient.

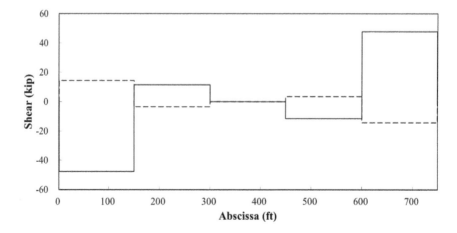

FIGURE 5-55   Shear Distribution Due to Temperature Gradient.

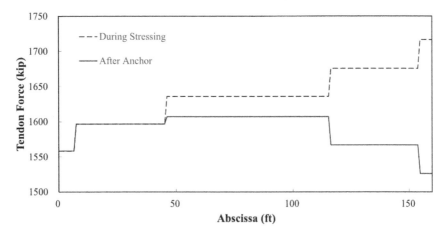

FIGURE 5-56   Variation of Tensile Force in Two Typical Tendons Due to Anchor Set.

4-D computer program. Figure 5-56 illustrates the prestressing force variation of two tendons with 19-0.6″ diameter strands before and after the anchor set. The variation of the effective prestressing stresses of a typical tendon is shown in Fig. 5-57. The moment envelopes due to primary, secondary, and total effects are provided in Fig. 5-58. The shear envelope due to effective post-tensioning forces is given in Fig. 5-59.

5.6.4.8   Secondary Effect of Creep and Shrinkage

The creep and shrinkage effects are evaluated based on the CEB-FIP Model Code for Concrete Structures 1990[5-18], i.e., Eq. 1-9b and Eq. 1-15b. Each of the segments is assumed to be erected on the 28th day after it was cast. The creep coefficient and shrinkage strain are calculated as shown next.

5.6.4.8.1   Creep Coefficient

The perimeter of the box girder in contact with the atmosphere: $u = 28{,}880$ mm

The notational size of the box girder:

$$h = 2\frac{A}{u} = \frac{2 \times 78.28 \times (12 \times 25.4)^2}{28{,}880} = 503.7 \text{ mm}$$

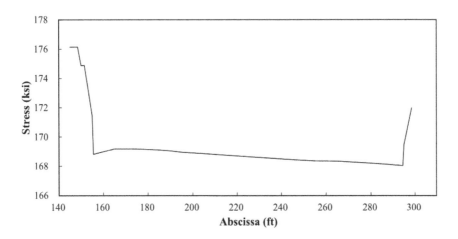

FIGURE 5-57   Variation of Effective Prestressing Stresses of One Typical Tendon at 10,000 days.

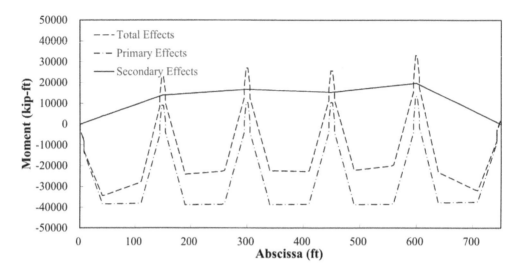

FIGURE 5-58   Moment Envelope Due to Post-Tensioning Tendons at 10,000 Days.

Based on FDOT specifications, the relative humidity:

$$H = 75\%$$

From Eq.1-9b, we have

$$\phi_{RH} = 1 + \frac{1-H}{0.46(0.01 \times h)^{1/3}} = 1 + \frac{1-0.75}{0.46(0.01 \times 503.7)^{1/3}} = 1.317$$

$$\beta(f_c') = \frac{5.3}{(0.1 f_c')^{0.5}} = \frac{5.3}{\left(0.1 \times \frac{6500}{145}\right)^{0.5}} = 2.503$$

$$\beta(t_0) = \frac{1}{1 + t_0^{0.2}} = \frac{1}{1 + 28^{0.2}} = 0.488$$

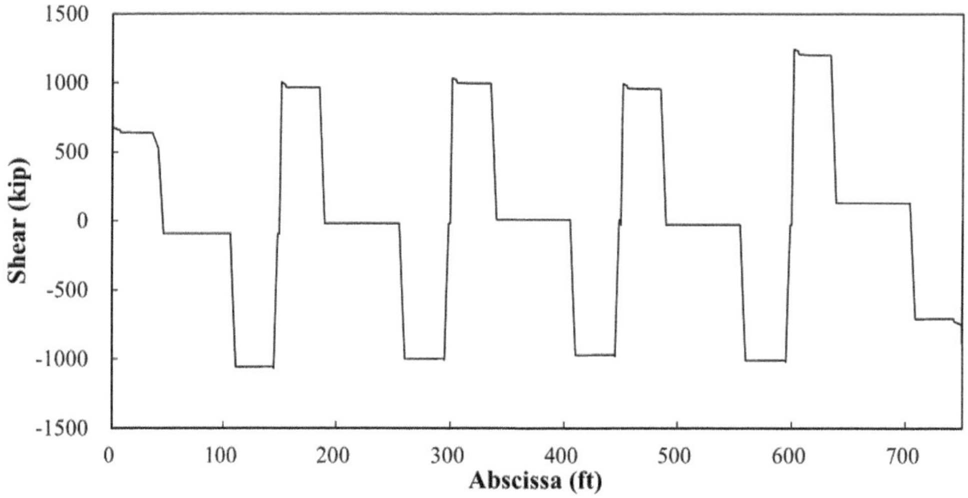

FIGURE 5-59   Shear Envelope Due to Post-Tensioning Tendons at 10,000 Days.

$$\phi_0 = \phi_{RH}\beta(f'_c)\beta(t_0) = 1.317 \times 2.503 \times 0.488 = 1.61$$

$$\beta_H = 1.5\, h\left(1+(1.2\ H)^{18}\right) + 250 = 1.5 \times 503.7 \times \left(1+(1.2 \times 0.75)^{18}\right) = 1119$$

$$\beta_c(t-t_0) = \left[\frac{t-t_0}{\beta_H+(t-t_0)}\right]^{0.3} = \left[\frac{10000-28}{1119+(10000-28)}\right]^{0.3} = 0.969$$

The creep coefficient is

$$\phi(t,\ t_0) = \phi_0\beta_c(t-t_0) = 1.61 \times 0.969 = 1.56$$

5.6.4.8.2 Shrinkage Strain
From Eq. 1-15b, we have:

Shrinkage development coefficient:

$$\beta_s(t-t_0) = \left[\frac{t-t_0}{350(h/100)^2}\right]^{0.5} = \left[\frac{10000-28}{350\left(\frac{503.7}{100}+1000-28\right)}\right]^{0.5} = 0.727$$

$$\beta_{RH} = -1.55\left(1-H^3\right) = -1.55\left(1-0.75^3\right) = -0.896$$

$$\varepsilon_s(f'_c) = \left[16+5(9-0.1f'_c)\right] \times 10^{-5} = \left[16+5\left(9-0.1\frac{6500}{145}\right)\right] \times 10^{-5} = 3.859 \times 10^{-4}$$

$$\varepsilon_{cs0} = \varepsilon_s(f'_c)\beta_{RH} = 3.859 \times 10^{-4} \times (-0.896) = -3.458 \times 10^{-4}$$

The shrinkage strain is

$$\varepsilon_{cs}(t,\ t_0) = \varepsilon_{cs0}\beta_s(t-t_0) = -3.458 \times 10^{-4} \times 0.727 = -2.515 \times 10^{-4}$$

5.6.4.8.3 Results
The moment and shear distributions due to creep and shrinkage at 10,000 days after casting time caused by dead and post-tensioning forces are illustrated in Figs. 5-60 and 5-61, respectively.

5.6.4.9 Effect of Wind Loading
5.6.4.9.1 Lateral Wind Pressure on Super-Structure WS_L
The wind pressure is calculated based on the method described in Section 2.2.5.2.1.

Height of super-structure: $Z = 62$ ft
For service I, design wind velocity at 30 ft: $V_{30} = 55$ mph
Base wind speed: $V_B = 100$ mph
From Table 2-3, base wind pressure: $P_B = 0.05$ ksf
From Table 2-4, for open county
Friction velocity: $V_0 = 8.20$ mph
Friction length: $Z_0 = 0.23$ ft

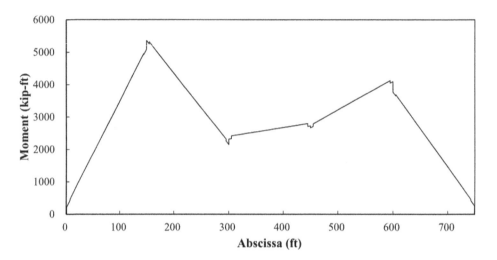

FIGURE 5-60   Moment Distribution Due to Creep and Shrinkage Effects.

From Eq. 2-10, design wind velocity at design elevation is

$$V_{DZ} = 2.5V_0 \frac{V_{30}}{V_B} \ln\left(\frac{Z}{Z_0}\right) = 63.1 \text{ mph}$$

Design wind pressure is

$$P_D = 0.05 \frac{63.1^2}{10000} = 0.02 \text{ ksf}$$

The resulting wind forces per linear feet applied along the box girder neutral axis (see Fig. 5-62) are

$$\text{Horizontal force: } WS_L = P_D D_S = 0.02 \times 12.54 = 0.251 \text{ kips/ft}$$

$$\text{Moment: } M_L = WS_L \left(\frac{D_S}{2} - y_t\right) = 0.251(6.27 - 2.75) = 0.884 \text{ kips-ft/ft}$$

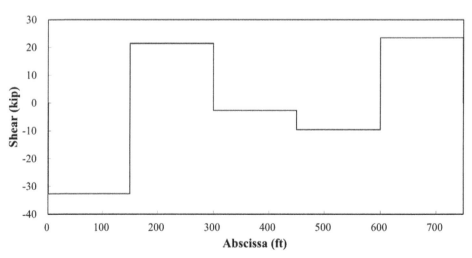

FIGURE 5-61   Shear Distribution Due to Creep and Shrinkage Effects.

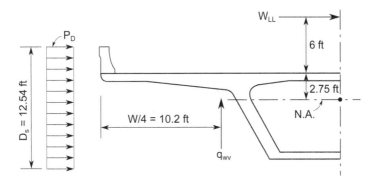

FIGURE 5-62   Design Wind Loadings.

5.6.4.9.2   Vertical Wind Force on Super-Structure WS_V

The vertical wind force applied to the super-structure is calculated based on the method described in Section 2.2.5.2.2. From Eq. 2-11, the upward line wind load can be determined as

$$q_{wv} = 0.02 \times total\ deck\ width = 0.02 \times 43.08 = 0.862\ \text{kips/ft}$$

The resulting moment per linear feet due to the eccentricity of the upward wind load (see Fig. 5-62) is

$$\text{Moment:}\ M_V = q_{wv}\left(\frac{43.08}{4}\right) = 9.28\ \text{kips-ft/ft}$$

5.6.4.9.3   Wind Force on Vehicle W_{LL}

The wind force applied to the vehicles is determined based on the method described in Section 2.2.5.3. The resulting wind forces on vehicles applied along the girder neutral axis (see Fig. 5-62) are

$$\text{Horizontal force:}\ W_{LL} = 0.1\ \text{kips/ft}$$

$$\text{Moment:}\ M_{LL} = W_{LL}\left(6.0 + y_t\right) = 0.1(6.0 + 2.75) = 0.875\ \text{kips-ft/ft}$$

The effects due to the wind load at some control sections are given in Table 5-2.

5.6.4.10   Summary of Effects

The maximum effects of the principal loadings at several critical sections are summarized in Table 5-2.

5.6.5   BRIDGE LONGITUDINAL CAPACITY VERIFICATION

5.6.5.1   Service Limit

For prestressed concrete bridges, two service limit states should be checked, i.e., service I and service III. For segmentally constructed bridges, AASHTO specifications require a special load combination check as indicated in Eq. 2-51. For the bridges constructed by the span-by-span method, the special load combination will not control the design and is therefore omitted in this example. As the effects of the concrete creep and shrinkage varies with time, the bridge capacity verifications may be performed for different loading stages, such as immediately after instantaneous prestressing losses, immediately after construction, 4,000 days, 10,000 days, and final stage. The following verifications are done for 10,000 days after casting time for example.

TABLE 5-2

Summary of Maximum Load Effects in Girder Longitudinal Direction at Control Sections

Section from Left End Support (ft)	Load Effects	Loadings								
		DC	CR + SH	PS	LL(1 + IM)	BR	WS	WL	TU	TG
66.2	Moment (kips-ft)	29,395	2,390	−35,100	10,200	180	1,460	1.53	24.6	3,080
	Axial force (kips)	—	—	−7,440	—	42.6	5.15	0.45		
150.0	Moment (kips-ft)	−16,400	5,360	22,882	−9,960	211	2,070	15	37	7,080
	Axial force (kips)	—	—	−15,300	—	42.6	31	3.63		
150.0	Shear (kips)	−1,170	−33	−127	−461	3.71	78.2	10.6	1.48	−48
	Corresponding torsion (kips-ft)	—	—	—	3,925	—	1,370			
305.0	Torsion (kips-ft)	—	—	—	4,117	—	491	2.79		
	Corresponding shear (kips)	−933	−2.71	1,020	−403	2.14	60.4			

5.6.5.1.1 Service I

Service I check is to investigate concrete compression strengths in prestressed concrete components. Normally, the concrete compression strength in three loading cases should be checked (see Section 1.2.5.1):

Case I: The sum of the concrete compression stresses caused by tendon jacking forces and permanent dead loads should be less than $0.6f_c'$, i.e.,

$$f_j^c + f_{DC+DW}^c \leq 0.6f_c'$$

Case II: Permanent stresses after losses: The sum of the concrete compression stresses caused by tendon effective forces and permanent dead loads should be less than $0.45f_c'$, i.e.,

$$f_e^c + f_{DC+DW}^c \leq 0.45f_c'$$

Case III: The sum of the concrete compression stresses caused by tendon effective forces, permanent dead loads, and transient loadings should be less than $0.60\phi_w f_c'$. The corresponding load combination for the bridge example is:

$$DC + CR + SH + PS + LL\,(1 + IM) + BR + 0.3WS + WL + TU + 0.5TG$$

Take case III and one cross-section located about 66 ft away from the left end support (see Fig. 5-38), for example, to illustrate the calculation procedure for the verification.

From Eq. 3-269, the top slab slenderness ratio is

$$\lambda_w^t = \frac{X_u^t}{t_t} = \frac{124}{10} = 12.4 < 15$$

Thus, the reduction factor $\phi_w^t = 1.0$, and the allowable compressive concrete stress at top fiber $= 0.60\phi_w^t f_c' = 3.9$ ksi.

The bottom slab slenderness ratio is

$$\lambda_w^b = \frac{X_u^b}{t_b} = \frac{141.75}{9} = 15.75 > 15$$

Thus, the reduction factor:

$$\phi_w^b = 1.0 - 0.025(15.75 - 15) = 0.98$$

and the allowable compressive concrete stress at the bottom fiber $= 0.60\phi_w^b f_c' = 3.8$ ksi.

From Table 5-2, the maximum concrete compressive stresses can be calculated as

Top fiber:

$$
\begin{aligned}
f_{top}^c &= \frac{(M_{DC} + M_{CR+SH} + M_{PS} + M_{LL+IM} + M_{BR} + 0.3M_{WS} + M_{WL} + M_{TU} + 0.5M_{TG})y_t}{I} + \frac{P_{PS}}{A} \\
&= \frac{(29395 + 2390 - 35100 + 10200 + 180 + 0.3 \times 1460 + 1.53 + 24.6 + 0.5 \times 3080) \times 2.75}{759.93} \\
&\quad + \frac{7440}{78.28} = 0.888 < 3.9 \text{ ksi (ok!)}
\end{aligned}
$$

Bottom fiber:

$$
\begin{aligned}
f_{bottom}^c &= \frac{-(M_{DC} + M_{CR+SH} + M_{PS} + M_{LL+IM} + M_{BR} + 0.3M_{WS} + M_{WL} + M_{TU} + 0.5M_{TG})y_b}{I} + \frac{P_{PS}}{A} \\
&= \frac{-(29395 + 2390 - 35100 + 10200 + 180 + 0.3 \times 1460 + 1.53 + 24.6 + 0.5 \times 3080) \times 6.25}{759.93} \\
&\quad + \frac{7440}{78.28} = 0.142 < 3.8 \text{ ksi (ok!)}
\end{aligned}
$$

For the remaining sections, the verification of service I can be done in a similar way as above.

5.6.5.1.2   Service III

Service III check is to investigate the maximum tensile stresses in the prestressed concrete components to ensure the maximum tensile stresses are not greater than the limits specified in Sections 1.2.5.1.1.2 and 1.2.5.1.2.2. Three cases should be checked:

Case I: Temporary tensile stresses before losses
Case II: Tensile stresses at Service III load combination
Case III: The principal tensile stress at the neutral axis in the web

The Service III load combination for the bridge example is

$$DC + CR + SH + PS + 0.8 \ LL \ (1 + IM) + 0.8BR + TU + 0.5TG$$

Take the section about 66 ft away from the beginning of the bridge to check its maximum tensile stress at the bottom fiber. Also consider the section at the first interior support to check its principal stress.

5.6.5.1.2.1   Maximum Tensile Stress Check   From Section 1.2.5.1.2.2, no tension is allowed for the section without the minimum bonded auxiliary reinforcement.

From Table 5-2, the maximum tensile stress in the bottom fiber can be calculated as:

$$f_{bottom} = \frac{(M_{DC} + M_{CR+SH} + M_{PS} + 0.8M_{LL+IM} + 0.8M_{BR} + M_{TU} + 0.5M_{TG})y_b}{I} - \frac{P_{PS}}{A}$$

$$= \frac{(29395 + 2390 - 35100 + 0.8 \times 10200 + 0.8 \times 180 + 24.6 + 0.5 \times 3080) \times 6.25}{759.93}$$

$$- \frac{7440}{78.28} = -0.286 \text{ ksi (in compression, ok!)}$$

The maximum tensile stresses for other locations are illustrated in Fig. 5-63. From this figure, we can see, almost all sections are in compression, except those sections within the limits of the pier segments which have tensile stresses that are less than the allowable tensile stress specified in Section 1.2.5.1.2.2, i.e.:

$$f_{ta} = 0.0948\sqrt{f'_c} = 0.242 \text{ ksi}$$

5.6.5.1.2.2 Principal Stress Check   For a given section, the principal stress check should be considered maximum for one of the following four cases: maximum absolute negative shear with corresponding torsion; maximum positive shear with corresponding torsion; maximum absolute negative torsion with corresponding shear; and maximum positive torsion with corresponding shear. Take a section located about 305 ft away from the left end support (see Fig. 5-38), for example, to show the basic procedures for principal stress verification.

Allowable principle stress:

$$f_{ta} = 0.11\sqrt{f'_c} = 280 \text{ psi}$$

Total effective web thickness: $b_w = 2 \times 15 = 30$ in.

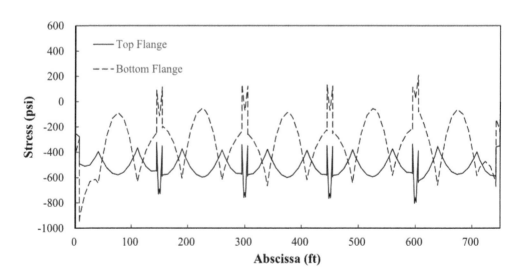

FIGURE 5-63   Variation of Tensile Stresses for Service III.

The maximum factored torsion from Table 5-2:

$$T_{max} = 0.8 \times 4117 = 3293.4 \text{ kips-ft}$$

The corresponding shear force from Table 5-2:

$$V_{max} = -933 - 2.71 + 1020 - 0.8 \times 403 + 0.8 \times 2.14 = -236.4 \text{ kips}$$

Shear stress in the web at the neutral axis location;

$$f_v = \frac{|V_{max}|Q_z}{2I_z b_w} + \frac{|T_{max}|}{2A_0 t_w} = \frac{236.4 \times 106.05}{2 \times 759.93 \times 30 \times 12} + \frac{3293.4}{2 \times 135.42 \times 15 \times 12} = 0.113 \text{ ksi}$$

Normal stress at the neutral axis location:

$$f_x = \frac{-P_{PS}}{A} = \frac{-7440}{78.28 \times 12 \times 12} = -0.660 \text{ ksi}$$
$$f_y = 0$$

From Eq. 3-176, the principal tensile stress can be calculated as

$$f_1 = \frac{f_x + f_y}{2} + \sqrt{\left(\frac{f_x - f_y}{2}\right)^2 + f_v^2} = \frac{-0.66}{2} + \sqrt{\left(\frac{0.66}{2}\right)^2 + 0.113^2} = 18.8 \text{ psi} < f_{ta} \text{ (ok)}$$

The principal stresses for other sections are illustrated in Fig. 5-64. From this figure, it can be seen that all the principal stresses are smaller than the allowable.

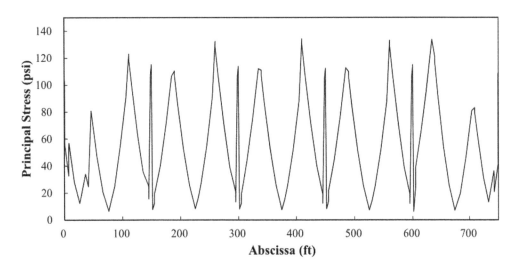

FIGURE 5-64 Variation of Web Principal Stresses for Service III.

5.6.5.2 Strength Limit

Strengths I, II, III, and V and extreme events I and II should be evaluated. Take Strength I for example. The load combination for this example can be written as

$$1.25DC + 1.25\ CR + 1.25SH + 1.75\ LL(1 + IM) + 1.75\ BR + 0.5TU$$

5.6.5.2.1 Flexure Strength

Take the section with maximum positive moment, located about 66 ft away from the left end support (see Fig. 5-38), for example, to illustrate the capacity verification process. The flexure strength check is performed using the method described in Section 3.6.2.2.

From Table 2-14, the flexure residence factor for unbounded tendons

$$\phi_f = 0.9$$

Area of the post-tendons:

$$A_{ps} = 0.217(8 \times 19 + 2 \times 27) = 44.702 \text{ in.}^2$$

The prestressing stress of the tendons can be determined based on Eq. 3-161.

For simplicity, the box section as shown in Fig. 5-39 is simplified as an I-section with uniform top and bottom flange thicknesses. The effective top flange width is determined based on the method described in Section 3.4.2.2 and can be taken as the actual flange width for the side span.

Width of top flange: b $\quad = 517$ in.

Equivalent thickness of top flange: $h_f = 13.16$ in.

Equivalent total web thickness: $b_w \quad = \dfrac{2 \times 15}{\sin(60.61)} = 34.43$ in.

From Eq. 3-149e, the stress block factor: $\beta_1 = 0.85 - 0.05\left(f_c' - 4.0\right) = 0.725$

Using Eq. 3-164 and assuming the box section as a rectangular section and tendon steel has yielded at ultimate strength, the distance from the extreme compression fiber to the neutral axis c can be determined by Eq. 3-165b as

$$c = \frac{A_{ps}f_{py}}{0.85f_c'\beta_1 b} = \frac{44.702 \times 0.9 \times 270}{0.85 \times 6.5 \times 0.725 \times 517} = 5.25 \text{ in.}$$

From Eq. 3-149b, the compression block depth is

$$a = \beta_1 c = 0.725 \times 5.25 = 3.8 \text{ in. (less than the thickness of top flange, assumption is correct)}$$

Effective depth of tendons as defined in Eq. 3-166 (see Figs 1-37 and 5-44) is:

$$d_e = d_p = 9 \times 12 - 12 - .5 - 1.0 = 94.5 \text{ in.}$$

From Fig. 5-57, we can find the effective prestressing force at this section is

$$f_{pe} = 166.443 \text{ ksi}$$

Average tendon length: $l_{ps} = 150.73$ ft

For side span, the number of support hinges (see Fig. 3-39): $N_s = 1$
From Eq. 3-155b, the effective tendon length: $l_e = \frac{2l_{ps}}{2+N_s} = \frac{2\times150.73}{2+1} = 100.49$ ft
From Eq. 3-161, the total tendon stress is

$$f_{ps} = f_{pe} + 900\frac{(d_p - c)}{l_e} = 166.443 + 900\frac{94.75 - 3.8}{100.49\times12} = 234.323 < f_{py} = 243 \text{ ksi}$$

f_{ps} is approximately equal to f_{py}, the assumption is ok!
From Eq. 3-164, the factored flexural resistance is

$$\phi_f M_n = \phi_f A_{ps} f_{ps}\left(d_p - \frac{a}{2}\right) = 0.90\times44.702\times234.323\left(94.075 - \frac{3.8}{2}\right)\frac{1}{12} = 72750.7 \text{ kips-ft}$$

The factored moment:

$$M_u = 1.25DC + 1.25\,(CR + SH) + 1.75\,LL\,(1+IM) + 1.75\,BR + 0.5TU$$
$$= 1.25\times(29395+2390) + 1.75\times10200 + 1.75\times180 + 0.5\times24.6 = 57908.55 \text{ kips-ft (ok!)}$$

Check minimum reinforcement:
The section modulus is

$$S_b = \frac{759.93}{6.25}12^3 = 210105 \text{ in.}^3$$

From Section 1.2.2.3, the modulus of fracture is

$$f_r = 0.20\sqrt{f_c'} = 0.51 \text{ ksi}$$

The compressive stress due to effective prestressing forces at the bottom fiber is

$$f_{cpe} = \frac{M_{ps}\times y_b}{I} + \frac{P_e}{A} = \frac{35100\times6.25}{759.93\times12^2} + \frac{7440}{78.28\times12^2} = 2.665 \text{ ksi}$$

From Eq. 3-168, the section crack moment:

$$M_{cr} = \gamma_3\left(\gamma_1 f_r + \gamma_2 f_{cpe}\right)S_b = 0.67(1.2\times0.51+1.0\times2.665)\frac{210105}{12} = 38442.0 \text{ kips-ft}$$
$$1.33M_{cr} = 51127.9 \text{ kips-ft} < \phi_f M_n \text{ (ok)}$$

The flexure strength checks for other locations can be performed using the approach described above. Figure 5-65 illustrates the variation of factored moments along the bridge longitudinal direction. From the figure, we can see that the flexure strengths of all sections are adequate.

5.6.5.2.2 Shear and Torsion Strength
Take the section located about 305 ft away from the left end support (see Fig. 5-38) and the maximum torsion corresponding to its shear. The shear strength check can be done by following the flow chart shown in Fig. 3-54.

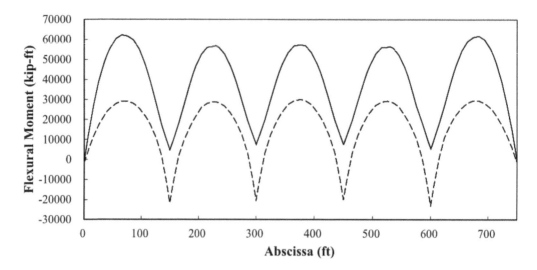

FIGURE 5-65   Variation of Factored Moments for Strength I.

Factored absolute shear per web:

$$V_u = 1.25DC + 1.25\,(CR + SH) + PS + 1.75\,LL\,(1 + IM) + 1.75\,BR + 0.5TU$$
$$= (1.25 \times 933 + 1.25 \times 2.71 - 1020 + 1.75 \times 403 - 1.75 \times 2.14)/2 = 425.57 \text{ kips per web}$$

Factored torsion:

$$T_u = 1.25DC + 1.25\,(CR + SH) + PS + 1.75\,LL\,(1 + IM) + 1.75\,BR + 0.5TU$$
$$= 1.75 \times 4117 \times 12 = 86457 \text{ kips-in}$$

5.6.5.2.2.1 Required Torsion Reinforcement   Effective shear depth as defined by Eq. 3-167:
$d_v = d_p - \frac{a}{2} = 92.6 \ in > 0.9 \ d_e$

Unfactored compressive stress in concrete after loses at the centroid: $f_{pc} = 660$ psi
From Eq. 3-216c, coefficient K is

$$K = \sqrt{1 + \frac{f_{pc}}{0.0632\sqrt{f_c'}}} = 2.26, \ K = 2$$

Effective width of the shear flow path is the minimum thickness of the top and bottom flange and
the web: $b_e = 9.0$ in.
From Eq. 3-216b, the torsion crack moment: $T_{cr} = 0.0632K\sqrt{f_c'}2A_0b_e = 113115.4$ kips-in
From Section 2.3.3.5, resistance factor for shear and torsion: $\phi_v = 0.9$.
The minimum required nominal torsion resistance is

$$T_{n\_req} = \frac{T_u}{\phi_v} = 96063 \text{ kips-in}$$

$T_u > \frac{1}{3}\phi_v T_{cr}$, torsion effect should be considered (3.6.3.4)

$$\text{Required torsion reinforcement per inch: } A_{s\_t} = \frac{T_{n\_req}}{2A_0 f_y} = 0.041 \text{ in.2/in.}$$

The required additional longitudinal reinforcement can be calculated by Eq. 3-218. As the requirement is not critical for segmental bridges, the calculations are omitted in this example.

5.6.5.2.2.2 Required Vertical Shear Reinforcement Effective web width for vertical shear: $b_v = t_w = 15.0$ in.

Depth d_0 is calculated by Eq. 3-220: $d_0 = 0.8\ h = 86.4$ in.

Concrete nominal shear strength (Eq. 3-219b): $V_c = 0.0632K\sqrt{f_c'}b_v d_0 = 417.6$ kips

$V_u > 0.5\phi_v V_c$, transverse shear reinforcement is required (see Section 3.6.3.4.2). The required minimum shear reinforcement per web per inch (Eq. 3-219c):

$$A_{s\_v} = \frac{\frac{V_u}{\phi_v} - V_c}{d_0 f_y} = 0.0107 \text{ in.}^2/\text{in.}$$

5.6.5.2.2.3 The Required Transverse Reinforcement The total required transverse reinforcement per web: $A_s = A_{s\_t} + A_{s\_v} = 0.0517$ in.^{2}/in.

Using #6 rebar at 6-in. spacing (see Fig. 5-66) and conservatively assuming that the one leg of the web shear rebars resists the vertical shear and another resists the transverse bending moment though the bending effect may be neglected as discussed in Section 5.3.2, the web reinforcement for vertical shear provided is

$$A_{s\_provided} = \frac{0.44}{6} = 0.0733 \text{ in.}^2/\text{in. (ok)}$$

5.6.5.2.2.4 Nominal Shear Strength and Cross Section Dimension Check Nominal shear resistance based on the provided shear reinforcement:

$$V_s = A_s f_y d_0 = 380.0 \text{ kips}$$

From Eq. 3-219a, the nominal shear strength is

$$V_{n1} = V_c + V_s = 797.6 \text{ kips}$$

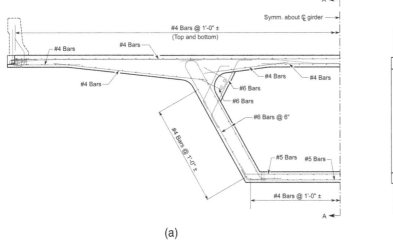
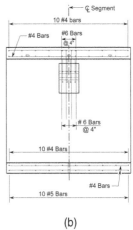

(a)   (b)

FIGURE 5-66   Reinforcement Arrangements of Type B Section, (a) Elevation View, (b) Side View.

From Eq. 3-220, the nominal shear strength is

$$V_{n2} = 0.379\sqrt{f_c'}b_v d_0 = 1473 \text{ kips}$$

Take the lesser of the V_{n1} and V_{n2}, the normal shear strength is

$$V_n = 797.6 \text{ kips}$$

The factored shear is

$$V_u + \frac{T_u}{2A_0} = 425.57 + \frac{86457}{2 \times 135.42 \times 12^2} = 427.79 < \phi_v V_n \text{ (ok!)}$$

Check cross-section dimension:

$$\frac{V_u}{b_v d_v} + \frac{T_u}{2A_0 b_e} = \frac{425.57}{15.0 \times 86.4} + \frac{86457}{2 \times 135.42 \times 9.0 \times 12 \times 12} = 0.575 \; ksi < 0.474\sqrt{f_c'} = 1.208 \text{ ksi (ok)}$$

The required shear reinforcement for other sections can be determined in the same way as described above. The actual shear reinforcement arrangement for typical section Type B is shown in Fig. 5-66.

5.6.6 TRANSVERSE ANALYSIS AND CAPACITY VERIFICATION

5.6.6.1 Effects of Dead Loads

The dead loads include the self-weight of the box section and the barrier walls. A two-dimensional frame model is used for analyzing the effects of dead loads as shown in Fig. 5-67. The structural analysis is performed using the LASAR-4D computer program. As there are no rigid arm elements provided in LASRA 4-D, the region between the top slab and web are modeled as three elements with a relative large stiffness. Some moments at critical sections as shown in Fig. 5-67 are provided in Table 5-3.

5.6.6.2 Effects of Live Loads

The effects of live loads are determined based on the methods described in Sections 4.4 and 5.3.1.2 and are analyzed using the finite-element method with the LASAR 4-D computer program. The box girder is simulated as a thin-wall structure and divided into a number of shell-plate finite elements. The length of the girder is taken as 80 ft. The finite-element meshes in the bridge transverse direction are taken as those shown in Fig. 5-67. The length of each of the finite elements in the girder longitudinal direction is 1 ft. The vertical and transverse displacements at each of the nodes at the

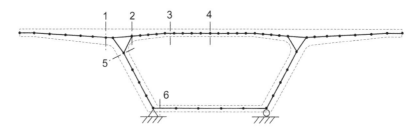

FIGURE 5-67   Transverse Analytical Model for Dead Loads and Post-Tensioning Tendons.

TABLE 5-3

Summary of Maximum Load Effects in Girder Transverse Direction

Loads	Responses	Locations (kips-ft/ft or kips/ft)					
		Section 1	Section 2	Section 3	Section 4	Section 5	Section 6
DL	Moment	−12.842	−6.684	−0.599	1.018	−3.604	−1.255
HL93 truck (1 + I)	Min. moment	−18.137	−13.275	−3.031	−0.703	−11.601	−2.553
	Max. moment	2.468	2.832	4.545	5.683	9.495	3.611
Tandem	Min. moment	−30.138	−23.258	−4.802	−0.833	−16.169	−2.755
(1 + I)	Max. moment	2.279	2.634	5.934	8.581	12.785	4.121
Post-tendons	Primary moment	36.911	24.169	6.819	−7.695		
	Secondary moment	0.780	2.795	−1.574	−1.563		
	Shrinkage creep moment	0	−1.518	−1.533	−1.548	1.180	0.415
	Total moment	37.691	26.964	5.245	−12.353	1.180	0.415
	Axial force	−59.058	−59.261	−59.382	−61.782	0.199	0.199

girder ends are restrained to simulate the end diaphragms. The analytical results at the control sections are given in Table 5-3.

5.6.6.3 Effects of Post-Tensioning Loads

5.6.6.3.1 Estimation of Required Transverse Tendons

As discussed, replacing the bridge deck is very difficult if not impossible. Thus, the deck is designed with zero tensile stresses. Assuming that the profile of the transverse tendons is as shown in Fig. 5-68, the minimum required prestressing force for section 1 can be determined using Eq. 5-6 as

$$P = \frac{6(M_D + M_{LL}) - t^2 f_{ta}}{t + 6e} = \frac{6(12.842 + 30.138) - 0}{\left[21 + 6\left(\frac{21}{2} - 3\right)\right]\frac{1}{12}} = 46.89 \text{ kips/ft}$$

Assuming the effective prestressing force as $0.84 \times f_{jp}$ after losses, i.e.,

$$f_{pe} = 0.84 \times 0.75 \times 270 = 170.1 \text{ kis}$$

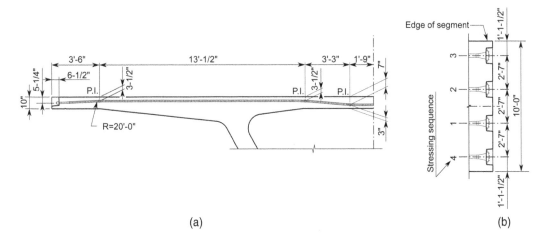

FIGURE 5-68 Arrangement of Transverse Post-Tensioning Tendons, (a) Elevation View, (b) Partial Plan View.

From Eq. 5-5, the required post-tendon area is

$$A_{PS} = \frac{P}{f_{pe}} = 0.2757 \text{ in.}^2/\text{ft}$$

Using 4-0.6″ diameter strands yields the number of tendons required for the 10-ft-length segment:

$$N_{strand} = \frac{A_{ps} \times 10}{4 \times 0.217} = 3.18, \text{ using 4 tendons}$$

The required tendons for the other sections can be determined in the same way as above, and it was found that the required tendon number for section 1 governs the design. The preliminary arrangement of post-tensioning tendons for the 10-ft-long segment is shown in Fig. 5-68.

It should be mentioned that the effect of the deck construction tolerance of 0.25 in. may be considered in the deck design as the deck thickness is small.

5.6.6.3.2   Effects of Post-Tensioning Tendons

Based on the preliminary arrangement of the transverse post-tensioning tendons, their prestressing loss effects due to concrete creep and shrinkage, and secondary effects are performed using the LASAR 4-D computer program. The variation of the effective prestressing stresses for a typical tendon is shown in Fig. 5-69. From this figure, it can be seen that the assumed effective post-tensioning stress approaches the analytical results, and it is not necessary to perform more accurate analysis. The effects of post-tendons at the critical sections are given in Table 5-3. In Table 5-3, the positive moment for the top and bottom slabs indicates that the bottom fibers are in tension. For the web, the positive moment indicates that the outside fibers are in compression.

5.6.6.4   Capacity Verifications

5.6.6.4.1   Top Slab

5.6.6.4.1.1   Service I   The AASHTO specifications require that the post-tensioned deck be checked for service I. The effect of temperature gradient (TG) is comparatively small and neglected in this example: therefore, the load combination is simplified to:

$$DC + (CR + SH) + PS + LL\,(1 + IM)$$

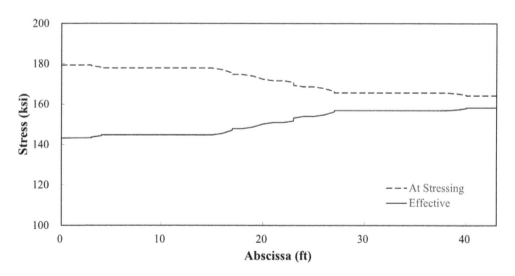

FIGURE 5-69   Variation of Effective Prestressing Stresses of a Typical Transverse Post-Tensioning Tendon.

Taking section 1 (refer to Fig. 5-67), for example, neglecting the deck sacrificial thickness of 0.5 in., the thickness of section 1 is 21 in.

$$\text{The section area per foot: } A = 12 \times 21 = 252 \text{ in.}^2/\text{ft}$$

$$\text{The section moment of inertia: } I = \frac{bt^3}{12} = 9261 \text{ in.}^4/\text{ft}$$

$$\text{The eccentricity of the tendon: } e = \frac{t}{2} - 3 = 7.5 \text{ in.}^2$$

The total moment in the section can be determined from Table 5-3:

$$M_{t01} = M_{DC} + M_{LL} + M_{CR+SH} + M_{PS} = -12.842 - 30.138 + 37.691 = -5.289 \text{ kip-ft/ft}$$

Concrete stress in top fiber: f_{top} $= \dfrac{-P_e}{A} + \dfrac{0.5\ t \times M_{t01}}{I} = -0.162 \text{ kips (compression, <3.9 ksi, ok!)}$

Concrete stress in bottom fiber: f_{bottom} $= \dfrac{-P_e}{A} - \dfrac{0.5\ t \times M_{t01}}{I} = -0.306 \text{ kips (compression, <3.9 ksi, ok!)}$

The capacity for other critical sections can be verified in the same way as above.

5.6.6.4.1.2 Strength Check Take strengths I, for example, to show the general procedure for strength capacity verification. Neglecting the effect of temperature, the load combination can be simplified as:

$$1.25DC + 1.25(CR + SH) + 1.75\ LL\ (1 + IM)$$

From Table 5-3, the factored moment at section 1 can be determined as

$$M_u = -1.25 \times 12.842 - 1.75 \times 30.138 = -68.794 \text{ kips-ft}$$

From Fig. 5-66, we can see:

Area of mild reinforcement: $A_s = 0.2 \text{ in.}^2/\text{ft}$
Distance from centroid of mild reinforcement to bottom fiber: $d_s = 18.75 \text{ in.}$
Area of post-tendons: $A_{ps} = 0.347 \text{ in.}^2/\text{ft}$
Distance from centroid of tendons to bottom fiber: $d_p = 18.0 \text{ in.}$
Neglecting the effect of the mild reinforcement in compression; using Eqs. 3-152b and 3-165a, the depth of the equivalent stress block can be determined as

$$k = 2\left(1.04 - \frac{f_{py}}{f_{pu}}\right) = 0.28$$

$$c = \frac{A_{ps}f_{pu} + A_s f_s}{0.85 f_c'\beta_1 b + kA_{ps}\frac{f_{pu}}{d_p}} = \frac{0.347 \times 270 + 0.2 \times 60}{0.85 \times 6.5 \times 0.725 \times 12 \times +0.28 \times 0.347 \times \frac{270}{18}} = 2.134 \text{ in.}$$

$$a = \beta_1 c = 0.725 \times 2.134 = 1.547 \text{ in.}$$

As the effective stress of the transverse tendons is greater than half of the tendon ultimate strength, the tendon stress at failure can be determined by Eq. 3-152a as

$$f_{ps} = f_{pu}\left(1 - k\frac{c}{d_p}\right) = 270 \times \left(1 - 0.28\frac{1.547}{18}\right) = 263.5 \text{ ksi}$$

From Eq. 3-164, the factored resistance is

$$\phi_f M_n = \phi_f \left(A_{ps} f_{ps} \left(d_p - \frac{a}{2} \right) + A_s f_s \left(d_s - \frac{a}{2} \right) \right)$$

$$= 0.9 \times \left(0.347 \times 263.5 \left(18 - \frac{1.547}{2} \right) + 0.2 \times 60 \left(18.75 - \frac{1.547}{2} \right) \right) = 134.31 \; \frac{kip - ft}{ft} > M_u \; (\text{ok})$$

The minimum reinforcement is checked as follows:

The total section area at section 1: $A_1 = 252$ in.^{2}/ft

The section modulus: $S_1 = 882$ in.^{3}/ft

The net compressive stress at the top fiber:

$$f_{cpe} = \frac{M_{PS}}{S_1} + \frac{P_e}{A} = \frac{37.691 \times 12}{882} + \frac{59.06}{252} = 0.747 \text{ ksi}$$

From Eq. 3-168, the crack moment is

$$M_{cr} = \gamma_3 \left(\gamma_1 f_r + \gamma_2 f_{cpe} \right) S_1 = 0.67 (1.2 \times 0.51 + 1.1 \times 0.747) \times \frac{882}{12} = 70.6 \text{ kip-ft}$$

$$\phi_f M_n > 1.33 \times M_{cr} = 93.9 \text{ kips-ft (ok)}$$

The capacity verifications of other sections can be performed in the same way as shown above. It should be mentioned that the deck capacity to resist its self-weight at critical sections should be checked to determine if the forms can be removed before transverse post-tensioning tendons are installed.

5.6.6.4.2   Web

The webs in transverse direction behavior essentially as a reinforced concrete structure. Both service and strength limit states should be checked. For simplicity, the check of service I limit state is omitted in this example. The webs are subjected to both positive and negative moments. We take section 5, for example, to check its negative moment capacity for strength I limit state.

From Table 5-3, the maximum factored negative moment at section 5 can be calculated as

$$M_{u5} = 1.25 M_D + 1.75 M_{LL} + 1.25 M_{(SH+CR)} = -1.25 \times 3.604 - 1.75 \times 16.169 + 1.25 \times 1.180$$

$$= -31.326 \text{ kip-ft/ft}$$

Assuming #6 rebars at 6 in. used on both sides of the web (see Fig. 5-66) and a clear concrete cover of 2 in., the distance from extreme compression fiber to the rebar centroid can be determined as

$$d_s = 12.625 \text{ in.}$$

By neglecting the effect of the compression reinforcement, from Eqs. 3-149b and 3-165a, the depth of equivalent stress block can be determined as follows:

$$c = \frac{A_s f_s}{0.85 f_c' \beta_1 b} = \frac{0.88 \times 60}{0.85 \times 6.5 \times 0.725 \times 12} = 1.098 \text{ in.}$$

$$a = \beta_1 c = 0.725 \times 1.098 = 0.796 \text{ in.}$$

The factored resistance is

$$\phi_f M_n = \phi_f A_s f_y \left(d_s - \frac{a}{2} \right) = 0.9 \times 0.88 \times 60 \times \left(12.625 - \frac{0.796}{2} \right) \frac{1}{12} = 48.19 \text{ kips-ft (ok)}$$

The minimum reinforcement is checked as follows:

$$\text{Section modulus: } S_5 = 450.0 \text{ in.}^3/\text{ft}$$

From Eq. 5-168, crack moment:

$$M_{cr} = \gamma_3 \gamma_1 f_r S_5 = 0.67 \times 1.2 \times 0.51 \times 450.0 \frac{1}{12} = 15.38 \text{ kip-ft/ft}$$

$$\phi_f M_n > 1.33 M_{cr} = 20.45 \text{ kips-ft/ft (ok)}$$

5.6.6.4.3   Bottom Slab

The bottom slab is subjected to both positive and negative moments. We take section 6, for example, to check its negative moment capacity for strength I limit state.

From Table 5-3, the maximum factored negative moment at section 6 can be calculated as

$$M_{u6} = 1.25 M_D + 1.75 M_{LL} + 1.25 M_{(SH+CR)} = -1.25 \times 1.255 - 1.75 \times 2.755 + 1.25 \times 0.415$$

$$= -5.871 \text{ kip-ft/ft}$$

Assuming #5 rebars at 12 in. used on the top sides of the bottom slab (see Fig. 5-66) and a clear concrete cover of 2 in., the distance from the extreme compression fiber to the rebar centroid can be determined as

$$d_s = 6.688 \text{ in.}$$

By neglecting the effect of the compression reinforcement, from Eqs. 3-149c or 3-165a, the depth of equivalent stress block can be determined as follows:

$$c = \frac{A_s f_s}{0.85 f_c' \beta_1 b} = \frac{0.31 \times 60}{0.85 \times 6.5 \times 0.725 \times 12} = 0.387 \text{ in.}$$

$$a = \beta_1 c = 0.725 \times 0.387 = 0.281 \text{ in.}$$

The factored resistance is

$$\phi_f M_n = \phi_f A_s f_y \left(d_s - \frac{a}{2} \right) = 0.9 \times 0.31 \times 60 \times \left(6.688 - \frac{0.281}{2} \right) \frac{1}{12} = 9.13 \text{ kips-ft (ok)}$$

The minimum reinforcement is checked as follows:

$$\text{Section modulus: } S_6 = 162.0 \text{ in.}^3/\text{ft}$$

From Eq. 5-167, the crack moment is:

$$M_{cr} = \gamma_3 \gamma_1 f_r S_6 = 0.67 \times 1.2 \times 0.51 \times 162.0 \frac{1}{12} = 5.54 \text{ kip-ft/ft}$$

$$\phi_f M_n > 1.33 M_{cr} = 7.37 \text{ kips-ft/ft (ok)}$$

5.6.7 Diaphragm Design

5.6.7.1 Longitudinal Design of Diaphragm

5.6.7.1.1   Properties of Effective Section

Take a typical diaphragm over pier 3 for example. The locations of the tendons at the stressing end are shown in Fig. 5-70. Using the method discussed in Section 5.4.1.2, the average effective width of the top flange can be calculated as (see Fig. 5-70):

$$w_e = 33.0 \text{ ft}$$

The properties of the effective section:

Effective area of top flange:

$$A_{et} = 43.12 \text{ ft}^2$$

Effective area of bottom flange:

$$A_{eb} = 10.67 \text{ ft}^2$$

Effective area of web flange:

$$A_{ew} = 19.02 \text{ ft}^2$$

Total effective area:

$$A_e = 72.83 \text{ ft}^2$$

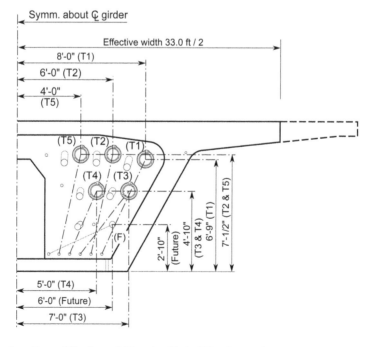

FIGURE 5-70   Locations of Tendons at Stressing End of Pier Segment.

Effective moment of inertia:

$$I_e = 714.64 \text{ ft}^4$$

Distance from neutral axis to top of the top slab: $y_{et} = 2.95$ ft
Distance from neutral axis to bottom of the bottom slab: $y_{eb} = 6.05$ ft

5.6.7.1.2 Normal Forces of Box Components Due to Jacking Forces

In designing the diaphragms, the maximum jacking stress of 216 ksi is only applicable to the last stressing tendon. For simplicity, we conservatively use it for all tendons. Then, the maximum jacking forces for each pair of the tendons can be obtained as

$$\text{Tendons T1 to T4 (pair): } P_1 = P_2 = P_3 = P_4 = 1781 \text{ kips}$$
$$\text{Tendon T5 (pair): } P_5 = 2531 \text{ kips}$$

Take tendon 5, for example, to illustrate how to calculate the normal forces and vertical and horizontal splitting forces.

The eccentricity e_{t5} of tendon T5 from the neutral axis is

$$e_5 = y_{et} - (9 - 7.042) = 0.992 \text{ ft}$$

The moment about the neutral axis due to tendon T5 is

$$M_5 = e_5 \times P_5 = 2510 \text{ kips-ft}$$

The distance between the center of gravity of the top flange and the neutral axis is

$$h_t = 2.338 \text{ ft}$$

The distance between the center of gravity of the bottom flange and the neutral axis is

$$h_b = 5.675 \text{ ft}$$

The average normal stress in the top flange can be calculated as

$$f_5 = \frac{P_5}{A_e} + \frac{M_5 h_t}{I_e} = 42.97 \text{ kips/ft}^2$$

The average normal stress in the bottom flange can be calculated as

$$f_{b5} = \frac{P_5}{A_e} - \frac{M_5 h_b}{I_e} = 14.82 \text{ kips/ft}^2$$

The total axial force in the top flange due to tendon T5 is

$$F_{ts5} = f_5 A_{te} = 1853.0 \text{ kips}$$

The total axial force in the bottom flange due to tendon T5 is

$$F_{bs5} = f_{b5} A_{be} = 158.4 \text{ kips}$$

The axial forces in the top and bottom slabs due to tendons T1 to T4 can be obtained in a similar way as above for tendon T5.

*5.6.7.1.3 Determination of Vertical Splitting Force and Required Prestressing
Bar and Mild Reinforcement*

Assuming the post-tensioning bars that are used to resist part of the split forces are located as shown in Fig. 5-71, we take tendon T5 as an example to illustrate the procedure for determining the vertical split forces. The distance d_1 between the post-tensioning bars to the point of application of tendon T5 is 8 ft (see Fig. 5-71).

From Eq. 5-16, the vertical splitting force due to tendon T5 can be calculated as

$$T_{v5} = \frac{(F_{ts5} + F_{bs5})(a_5 \times b_5)}{d_1(a_5 + b_5)} = 251.47 \text{ kips}$$

Using the same method as above, the vertical splitting forces caused by tendons T1 to T4 can be obtained. Then using Eq. 5-15, we can obtain the total vertical splitting forces as

$$T_v = \sum_{i=1}^{5} T_{vi} = 974.4 \text{ kips}$$

In the vertical direction, there is limited space to arrange the required mild reinforcement. We can use post-tensioning bars to resist part of the splitting forces. Assuming that six Grade 150 post-tensioning bars with a diameter of 1-3/8 in. are used, based on the manufacturer's specifications, the total effective prestressing force provided by the bars can be taken as

$$P_{eff} = 6 \times 0.6 f_{pu} \times A_{ps} = 853.2 \text{ kips}$$

The remaining splitting force is

$$T_{v\_remaining} = T_v - P_{eff} = 121.2 \text{ kips}$$

To control concrete cracks, FDOT requires the maximum tensile stress of Grade 60 cannot exceed 24 ksi at service limit state. Then, the required vertical reinforcement per web is

$$A_{vs\_required} = \frac{T_{v\_remaining}}{2 \times 24.0} = 2.53 \text{ in.}^2 /\text{per web}$$

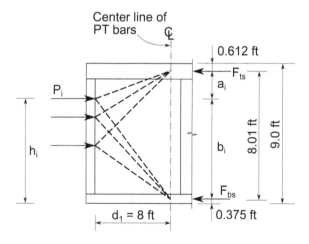

FIGURE 5-71   Analytical Model of Vertical Splitting Forces.

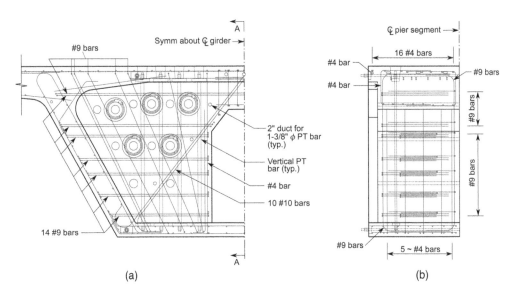

FIGURE 5-72 Primary Reinforcement of Diaphragm, (a) Elevation View, (b) Side View.

From Fig. 5-72, it can be seen that six #9 bars (6.0 in.^{2}) are provided and is more than enough for this project.

5.6.7.1.4 Determination of Horizontal Splitting Force and Required Mild Reinforcement
Taking tendon T5, for example, to illustrate the procedure for determining the horizontal splitting forces, the total axial force in the webs due to tendon T5 is

$$F_{w5} = P_5 - F_{ts5} - F_{bs5} = 519.6 \text{ kips}$$

Assuming that only mild reinforcement bars are used to resist the horizontal splitting force, the distance d_2 from the point of application of P_i to the center of the rebar (see Fig. 5-73) can be estimated as

$$d_2 = 8.792 \text{ ft}$$

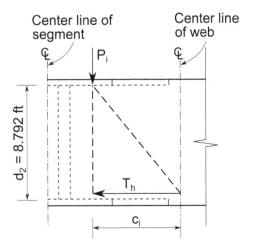

FIGURE 5-73 Analytical Model of Horizontal Splitting Force.

From Eq. 5-18, the horizontal splitting force per web due to tendon T5 can be calculated as

$$T_{h5} = \frac{F_{w5}c_5}{2 \times d_2} = \frac{519.6 \times 6.169}{2 \times 8.792} = 182.29 \text{ kips}$$

In a similar way as described above, we can obtain the horizontal splitting forces caused by tendons T1 to T4. Using Eq. 5-17, the total horizontal splitting force per web can be calculated as

$$T_h = \sum_{i=1}^{5} T_{hi} = 255.62 \text{ kips}$$

Then, the required horizontal reinforcement per web is

$$A_{hs\_required} = \frac{T_h}{24.0} = 10.65 \text{ in.}^2 / \text{per web}$$

From Fig. 5-72, it can be observed that 14-#9 bars (14 in.²) are provided and more than enough.

5.6.7.1.5 Determination of Longitudinal Friction Shear and Required Reinforcement

The longitudinal post-tensioning forces are transferred to the girder through shear friction between the diaphragm and webs, top and bottom flanges. The shear forces can be determined based on the normal stress distributions on the box section as follows:

Shear force on the top flange and the diaphragm interface per web:

$$V_{lt} = \sum_{1}^{5} F_{tsi} = 2942 \text{ kips}$$

Shear force on the bottom flange and the diaphragm interface per web:

$$V_{lb} = \sum_{1}^{5} F_{bsi} = 657 \text{ kips}$$

Shear force on the web and the diaphragm interface per web:

$$V_{lw} = \sum_{1}^{5} F_{wi} = 1229 \text{ kips}$$

The required reinforcement crossing the interfaces can be determined based on Eq. 3-223a. By neglecting the effect of cohesion forces, the required reinforcement can be simplified as

$$A_{vs} = \frac{V_l}{\mu \phi_{sf} f_y} = \frac{V_l}{1.4 \phi_{sf} f_y}$$

Based on the current AASHTO specifications, shear resistance factor ϕ_{sf} may be taken as 0.9. As the jacking force actually has not been factored in this example (load factor $\gamma = 1.2$ for jacking force, see Section 2.3.3.3.3.4) and is similar to a service load, engineers often conservatively take

$$\phi_{sf} = 0.6$$

Thus, the required friction shear reinforcement can be calculated as

In the top flange interface:

$$A_{vs} = \frac{V_{lt}}{\mu \phi_{sf} f_y} = \frac{2942}{1.4 \times 0.6 \times 60} = 58.37 \text{ in.}^2$$

The actual rebar area provided is 66 in.², which is more than enough.

In the bottom flange interface:

$$A_{vs} = \frac{V_{lb}}{\mu \phi_{sf} f_y} = \frac{657}{1.4 \times 0.6 \times 60} = 13.04 \text{ in.2}$$

The actual rebar area provided is 66 in.², which is more than enough.

In the web interface:

$$A_{vs} = \frac{V_{lw}}{\mu \phi_{sf} f_y} = \frac{1229}{1.4 \times 0.6 \times 60} = 24.38 \text{ in.}^2$$

The actual rebar area provided is 42 in.², which is more than enough.

5.6.7.2 Transverse Design of Diaphragm
5.6.7.2.1 Required Diagonal Reinforcement
5.6.7.2.1.1 Tensile Forces Caused by Torsion As an example, consider the section about 150 ft away from the left end support (pier 2) and service I load combination, the maximum shear, and the corresponding torsion to illustrate the design procedures.
From Table 5-2, we have

Total service torque at pier 2: $T_c = 3925$ kips-ft

The shear flow area enclosed by trapezoid *ACDE* as shown in Fig. 5-74a is

$$A_0 = 135.42 \text{ in.}^2$$

From Eq. 3-206, the shear flow can be obtained as

$$q = \frac{T_c}{2A_0} = 14.49 \text{ kips/ft}$$

The shear forces in the components of the box girder can be calculated as

In the top flange:

$$F_{ts} = q \times AB = 14.49 \times 22.12 = 320.52 \text{ kips}$$

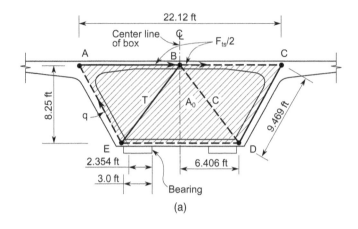

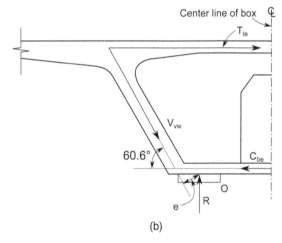

FIGURE 5-74 Model for the Diaphragm Transverse Analysis, (a) Strut-and-Tie Model, (b) Tensile Forces Due to Support Eccentricity.

In the bottom flange:

$$F_{bs} = q \times DE = 14.49 \times 12.81 = 185.62 \text{ kips}$$

In the web:

$$F_{ws} = q \times CD = 14.49 \times 9.45 = 136.93 \text{ kips}$$

To simplify the analysis, we assume that half of F_{ts} is taken by tie AB and that another half is taken by the strut-and-tie model BDE (see Fig. 5-74). The tensile force of tie BE can be calculated as

$$T = \frac{F_{ts}}{2} \frac{BE}{0.5DE} = 261.31 \text{ kips}$$

5.6.7.2.1.2 Determination of Required Diagonal Reinforcement The required diagonal reinforcement can be determined by limiting the reinforcement stress to 24.0 ksi:

$$A_{s-diagnal} = \frac{T}{24} = 10.89 \text{ in.}^2$$

From Fig. 5-72, ten #10 bars (12.7 in.²) are provided, which is more than enough.

5.6.7.2.2 Required Top Flange Reinforcement

From Table 5-2, we have

$$\text{Total absolute service shear at pier 2: } V = 1747.97 \text{ kips}$$

The tensile force of the top flange due to torsion is

$$T_{tt} = \frac{F_{ts}}{2} = 160.26 \text{ kips}$$

The tensile forces caused by the eccentricity of the support (see Fig. 5-74b) can be calculated as follows:

The eccentricity of the web centerline from the center of the support bearing O as shown in Fig. 5-74b can be obtained as

$$e = 1.557 \text{ ft}$$

The total web shear is

$$V_{vw} = \frac{V}{2\sin(\theta_w)} + F_{wt} = 1140.13 \text{ kips}$$

Taking moment about point O, the tensile force in the top flange due to the eccentricity can be computed as

$$T_{te} = \frac{V_{vw}e}{8.25} = 215.17 \text{ kips}$$

The total tensile forces in the top flange are

$$T_{ts} = T_{tt} + T_{te} = 375.43 \text{ kips}$$

The effective prestressing force provided by the transverse post-tendons in the top flange is

$$P_{ts} = 4 \times 4 \times 0.217 \times f_{pu} \times 0.62 = 581.21 \text{ kips}$$

The remaining tensile force in the top flange is

$$T_{tsr} = T_{ts} - P_{ts} = -205.78 \text{ kips}$$

No extra mild reinforcement in the top flange is required.

5.6.7.2.3 Required Web Shear Friction Reinforcement

From Fig. 5-74b, part of the web shear V_{vw} directly transfers to the bearing pad and part of the shear will transfer to the bearing pad through the diaphragm. The shear force in the interface between the web and the diaphragm can be conservatively determined as

$$V_{interface} = V_{vw}\frac{2.354}{3.0} = 894.62 \text{ kips}$$

The required shear friction reinforcement is

$$A_{ws} = \frac{V_{interface}}{\mu \phi_{sf} f_y} = 17.75 \text{ in.}^2$$

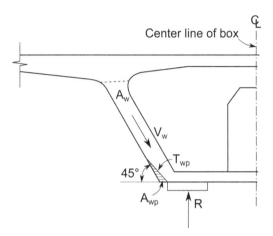

FIGURE 5-75 Analytical Model for Pop-out Force.

The total required shear friction reinforcement in the web is

$$A_{vws} = A_{ws} + A_{vs} = 17.75 + 24.38 = 42.13 \text{ in.}^2$$

The area of the actual reinforcement provided is 44 in.², which is more than enough.

5.6.7.2.4   Required Web Pop-out Reinforcement

Assuming the plane of the pop-out concrete section is oriented at 45° from the horizontal direction as shown in Fig. 5-75, the area of the pop-out section can be calculated as

$$A_{wp} = 0.704 \text{ ft}^2$$

The total effective area of the concrete web is

$$A_w = 10.47 \text{ ft}^2$$

The total force acting on the pop-out section may be estimated as

$$F_{wp} = V_{vw} \frac{A_{wp}}{A_w} = 76.66 \text{ kips}$$

By limiting the maximum tensile stress to 24.0 ksi, the required reinforcement may be approximately calculated as

$$A_{s\_pop} = \frac{F_{wp}}{24} = 3.19 \text{ in.}^2$$

From Fig. 5-72, we can see that the reinforcement provided is more than enough.

5.6.8   DEVIATION DESIGN

5.6.8.1   Determination of Pull-out Reinforcement

Conservatively use maximum jacking stress $0.8 f_{pu}$ to design the deviator. The maximum pull-out force is caused by tendon T5 and is

$$F_{pull} = 206.15 \text{ kips}$$

Load factor for deviator force (AASHTO Article 5.10.9.3.7) [5-5] is

$$\gamma_{pt} = 1.7$$

Resistance factor for tension is

$$\phi_t = 0.9$$

Required pull-out reinforcement for each of the tendons can be calculated as:

$$A_{spo} = \frac{F_{pull}\gamma_{pt}}{\phi_t f_y} = 6.49 \text{ in.}^2$$

The primary dimensions and reinforcement of the deviator are shown in Fig. 5-76. From this figure, it can be seen that 10 pairs of #6 bars (8.8 in.²) are provided and are more than enough.

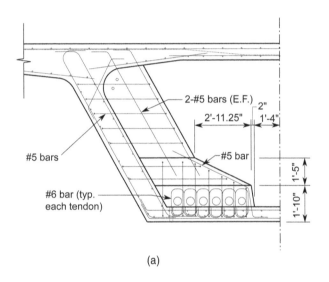

(a)

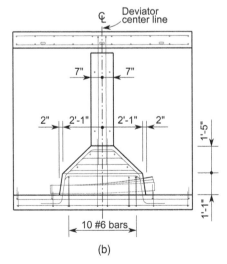

(b)

FIGURE 5-76 Dimensions and Primary Reinforcement of Deviator, (a) Elevation View, (b) Side View.

5.6.8.2   Determination of Minimum Shear Concrete Area and Friction Shear Reinforcement

The interface shear between the deviator and web is calculated as

$$V_{sweb} = 566.78 \text{ kips}$$

From Table 3-2, for normal-weight concrete placed monolithically, we have

$$\text{Cohesion factors: } c = 0.40$$

$$\text{Friction factor: } \mu = 1.4$$

$$K_1 = 0.25, \ K_2 = 1.5$$

Using Eq. 3-223b, it can be found that the minimum shear concrete area is controlled by limiting the interface shear resistance and is determined as

$$A_{cv\_min} = \frac{\gamma_{pt} V_{sweb}}{K_2} = 642.3 \text{ in.}^2$$

The actual provided interface area between the deviator and the web can be conservatively estimated as (see Fig. 5-76)

$$A_{cv} = 1859 \text{ in.}^2 \text{ (ok)}$$

From Section 2.2.3.5, the resistance factor for shear is $\phi_s = 0.9$.

The required shear friction reinforcement can be calculated as (see Eq. 3-223a)

$$A_{sfw} = \frac{\gamma_{pt} V_{sweb} - c \times A_{cv}}{\phi_s \mu f_y} = 2.91 \text{ in.}^2$$

From Fig. 5-76, the reinforcing area that is actually provided is 4.34 in.² (seven pair #5 bars), which is adequate. It should be mentioned that the capacity check is very conservative as the top and bottom slabs will work together with the web in resisting the shear.

5.6.8.3   Required Flexure Reinforcement in Web Due to Deviator Forces

Using a simplified frame model as shown in Fig. 5-33 and Fig. 5-77 and the effective length of the segment of 5.5 ft, which is the width of the deviator at its bottom, the maximum factored bending moment at the web and bottom slab corner can be determined as

$$M_{rib} = 1302.3 \text{ kips-ft}$$

From Eq. 3-150 and Fig. 5-77, the minimum reinforcement should meet the following:

$$M_{rib} = A_s f_y \left(d_s - \frac{\frac{A_s f_y}{0.85 f_c' b_w}}{2} \right)$$

$$1302.3 \times 12 = A_s \times 60 \left(43 - \frac{\frac{60 A_s}{0.85 \times 6.5 \times 14}}{2} \right)$$

Solving the equation above yields the required flexure reinforcement in the web:

$$A_s = 6.43 \text{ in.}^2$$

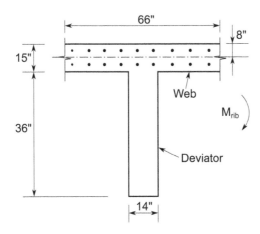

FIGURE 5-77 Simplified Model for Deviator Bending Design.

or rewritten as

$$A_{s\_web} = \frac{A_s}{5.5} = 1.169 \text{ in.}^2 /\text{ft}$$

The reinforcement provided is 1.66 in²/ft and is adequate.

The minimum flexure reinforcement is checked using Eq. 3-168 as follows:

The section modulus is

$$S_c = 18020 \text{ in.}^3$$

From Section 1.2.2.3, the modulus of fracture is

$$f_r = 0.20\sqrt{f_c'} = 0.51 \text{ ksi}$$

From Eq. 3-168, the section cracking moment is

$$M_{cr} = \gamma_3 \gamma_1 f_r S_c = 0.67 \times 1.2 \times 0.51 \times \frac{18020}{12} = 615.74 \text{ kips-ft}$$

$$1.33 M_{cr} = 818.94 \ kips\text{-}ft < M_{rib} \ (\text{ok})$$

5.6.9 REMARKS OF THE DESIGN EXAMPLE

The purpose of the design example is to provide readers with a general procedure for designing a segmental bridge constructed by the span-by-span method. It can be definitely improved. Some of the analytical methods may be very rough and conservative, which is mostly due to the minimal effect on the total construction cost caused by the conservative design. Design and construction experiences show that simplifying the construction method may be more effective than reducing construction materials with regard to the total construction costs. For a large project, the designer should try to use the same details for the entire project as much as possible.

REFERENCES

5-1. ASBI, *Construction Practices Handbook for Concrete Segmental and Cable-Supported Bridges*, American Segmental Bridge Institute, Buda, TX, 2008.

5-2. Tassin, D. M., *Design of Precast Segmental Bridges Built by Spa by Span*, J. Muller International, Tallahassee, FL, 1989.

5-3. Florida Department of Transportation, *Structures Design Guidelines*, Florida Department of Transportation, Tallahassee, FL, 2018.

5-4. Huang, D. Z., and Hu, B., Evaluation of Cracks in a Large Single-Cell Precast Concrete Segmental Box Girder Bridge without Internal Struts, ASCE, *Journal of Bridge Engineering*, 2014.

5-5. AASHTO, *LRFD Bridge Design Specifications*, 7th edition, Washington, D.C., 2014.

5-6. AASHTO-PCI-ASBI, "AASHTO-PCI-ASBI Segmental Box Girder Standards for Span-by-Span and Balanced Cantilever Construction", May 2000.

5-7. Kulka, F., and Thoman, S. J., Feasibility Study of Standard Sections for Segmental Prestressed Concrete Box Girder Bridges, *PCI Journal*, September–October 1983.

5-8. Freyermuth, C. L., AASHTO-PCI-ASBI Segmental Box Girder Standards: A New Product for Grade Separations and Interchange Bridges, *PCI Journal*, September–October 1997.

5-9. Yuan, A., He, Y, Dai, H., and Cheng, L., Experimental Study on Precast Segmental Bridge Box Girder with External Unbounded and Internal Bonded Post-Tensioning under Monotonic Vertical Loading, *Journal of Bridge Engineering*, ASCE, Vol. 20, No. 4, 2015, pp. 04014075-1 to 0401075-9.

5-10. Xu, D., Lei, J., and Zhao, Y., Prestressing Optimization and Local Reinforcement Design for a Mixed Externally and Internally Prestressed Precast Segmental Bridge, *Journal of Bridge Engineering*, Vol. 21, No. 7, 2016, pp. 05016003-1 to 05016003-10.

5-11. Florida Department of Transportation, *New Directions for Florida Post-Tensioned Bridges*, Corven Engineering, Inc., Tallahassee, FL, 2004.

5-12. Florida Department of Transportation, *Structures Detailing Manual*, Tallahassee, FL, 2018.

5-13. Florida Department of Transportation, *Design Standards*, Tallahassee, FL, 2018.

5-14. Theryo, T. S., Precast Balanced Cantilever Construction Cantilever Bridge Design Using AASHTO LRFD Bridge Design Specifications, *Proceedings, Seminar for Design and Construction of Segmental Concrete Bridges*, Orinda, FL, 2004.

5-15. Menn, C., *Prestressed Concrete Bridges*, Springer-Verlag, Wien, 1986.

5-16. Corven, J., *Post-Tensioned Box Girder Design Manual*, Federal Highway Administration, U.S. Department of Transportation, Washington, D.C., 2015.

5-17. Wollmann, G. P., *Anchorage Zones in Post-Tensioned Concrete*, University of Texas, Austin, TX, May 1992.

5-18. Comite Euro-International Du Beton, *CEB-FIP MODEL CODE 1990: Design Code*, Thomas Telford, Great Britain, 1993.

6 Design of Cantilever Segmental Bridges

- Span arrangement and determination of preliminary dimensions
- Typical segments and detailing
- Features of longitudinal analysis and layout of post-tensioning tendons
- Determination of cambers
- Theory and design of curved segmental bridges
- Construction analysis and capacity verification
- Design example: balanced cantilever segmental bridge

6.1 INTRODUCTION

For bridge span lengths greater than 150 ft, span-by-span segmental bridges normally will not be the most economical alternative and balanced cantilever segmental bridges become a more lucrative option. The basic concepts and construction sequences have been discussed in Section 1.5.3. Based on the cast methods, the cantilever segmental bridges can be classified into categories: precast and cast-in-place. Precast cantilever segmental bridges are suitable for span lengths ranging from 150 to 450 ft. They have numerous advantages, such as (1) increased erection speed of the superstructure because the segments are fabricated while constructing the substructure, (2) reduced effects of concrete creep and shrinkage, due to the maturity of the concrete at the time of erection, and (3) high-quality control obtained for factory-produced precast concrete. However, there are also some disadvantages when compared with cast-in-place segmental bridges, such as more difficulties in (1) correcting errors in the bridge geometry during erection and (2) in connecting the longitudinal post-tensioning tendons and making the reinforcing continuous. For span lengths exceeded 400 ft, the weight of the deep haunched segments near the pier increases the difficulty of using precast segments in balanced cantilever construction. In such cases, cast-in-place balanced cantilever bridges can be considered. Cast-in-place balanced cantilever bridges may be suitable for span lengths ranging from 230 to 850 ft. However, the economical span range for this type of bridges is about 230 to 800 ft. Cast-in-place cantilever bridges are especially suitable for bridges over deep valleys and rivers where temporary supports are difficult and expensive to construct[6-1 to 6-3]. The main disadvantage of the cast-in-place segmental bridge construction method is the time required to construct the superstructure, which is typically about 5 days for each segment, much longer than the precast segment construction with an average erection rate of four segments per day.

Though the construction methods for precast and cast-in-place segmental bridges are different, the basic design and analytical methods are essentially the same. Thus, the discussion in this chapter focuses on the precast balanced segmental bridges. The precast balanced cantilever bridges possess many characteristics of span-by-span bridges, which will not be repeated in this chapter. The readers are encouraged to read Chapter 5 before reading this chapter.

6.2 SPAN ARRANGEMENT AND TYPES OF SEGMENTS

The first step in bridge design, as discussed in Chapter 5, is to properly arrange the spans and the locations of piers and abutments based on the site conditions and the opening requirements under the bridge. As discussed in Chapter 1, in balanced cantilever construction, segments are placed in

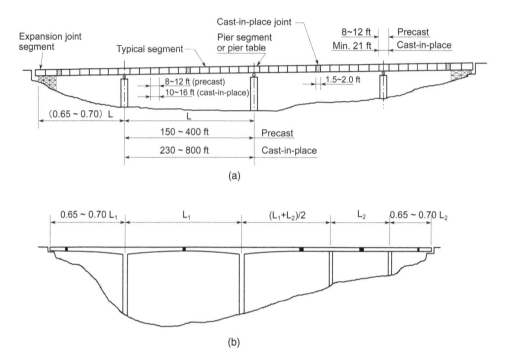

FIGURE 6-1 Optimum Span Arrangements in Balanced Cantilever Construction, (a) Equal Interior Spans, (b) Variable Interior Spans.

a symmetric fashion about a pier. For a bridge with equal interior spans as shown in Fig. 6-1a, the length of side spans is about 0.65 to 0.7 times the length of the interior span. This span arrangement is recommended as it reduces the number of the segments next to the abutment, which normally has to be installed on falsework (see Fig. 6-1a)[6-3, 6-4]. For a bridge with variable span lengths as shown in Fig. 6-1b, an intermediate span with a span length of the average of the two flanking spans is introduced to ensure the bridge can be built in balanced cantilever fashion (see Fig. 6-1b). The cantilever girders can be either simply supported on the piers (see Fig. 6-1a) or rigidly connected with the piers (see Fig. 6-1b). The fixed connection between the superstructure and substructure has the following characteristics:

Advantages

   a. Provides an additional restraint required for ship impact, seismic forces, and significantly tall piers
   b. Can eliminate temporary stabilization of the cantilever required during construction
   c. Eliminates bearing and reduce maintenance work

Disadvantages

   a. Increases the effects of creep, shrinkage, and temperature on the substructure and superstructure
   b. Increases the restraint effect of continuous tendons

The pinned or rolled connection between the super-structure and substructure has opposite characteristics to those of fixed connection. Thus, for the bridges with short piers, the pinned or rolled connection is recommended. For high and relatively flexible piers, priority consideration should be given to fixed connections.

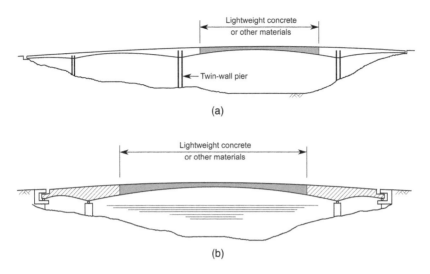

FIGURE 6-2   Unbalanced Span Arrangements, (a) Multi-Unequal Interior Spans, (b) Short End Spans.

In early designed segmental bridges, the connections between two cantilevers were often designed as hinge connections. The performances of such bridges show that the middle hinges will cause many problems, such as more maintenance work, larger long-term deflection, and high dynamic loadings[6-5, 6-6]. This hinge connection is not recommended.

It may not always be possible to provide an optimum span arrangement due to the site limitations, and an unbalanced span arrangement may be used (see Fig. 6-2). To increase the super-structure stability during construction and reduce the temporary supports required, a single solid pier can be separated into two pieces, which is often called a twin wall pier as shown in Fig. 6-2a. Another way of reducing the unbalanced cantilever bending moment and the weight of the super-structure is to use composite materials in the super-structure; such as using normal-weight concrete for the segments close to the pier and lightweight concrete for the segments in the middle of the span (see Fig. 6-2). Also, the most effective way of increasing the span length of segmental girder bridges is to use composite materials in the super-structure. In the situation of a short end span, the abutment can be designed to resist uplift forces to balance the large cantilever moment (see Fig. 6-2b). However, this type of span arrangement is often costly and should be avoided unless necessary.

The determination of segment lengths and weights for precast cantilever segmental bridges is similar to that for span-by-span bridges and is mainly based on two factors: shipping and erection equipment. For on-road hauling, the length and weight of a segment is normally limited to 10 ft and 60 tons, respectively. For crane erections, the weight of a segment is normally limited to 80 tons. For gantry systems and beam-winch systems, the weight of a segment usually ranges from 80 to 100 tons. The length of segments generally ranges from 8 to 12 ft[6-2]. The typical segmental lengths for cast-in-place cantilever bridges range from 10 to 16 ft. Segmental lengths longer than 16 ft may produce large, out-of-balance loadings during construction. The segment over the pier is called a pier table, and its minimum length is about 21 ft for placing form travelers side by side during the beginning of segment construction[6-3].

6.3   SELECTION OF TYPICAL SECTIONS

The general discussion on the selection of typical sections and the determination of preliminary dimensions has been presented in Section 1.4.2 and Section 5.2.2. The segmental box girder standards for balanced cantilever construction developed by AASHTO, PCI, and ASBI[6-7] are valuable

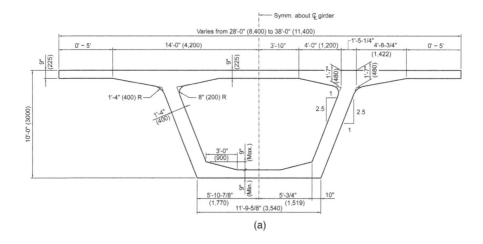

(a)

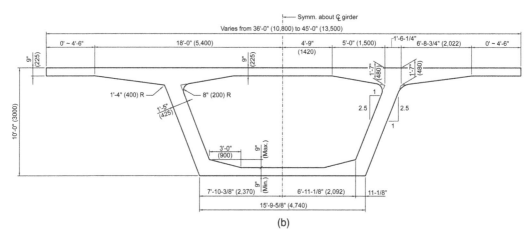

(b)

FIGURE 6-3 Segmental Box Girder Standards for Balanced Cantilever Construction with Girder Depth of 10 ft, (a) Deck Width from 28 to 38 ft, (b) Deck Width from 36 to 45 ft (Used with permission by ASBI).

sources in selecting segmental box girder cross sections. These design standards were developed based on an extensive survey of segmental box girder bridges built in the United States and Canada and other research[6-8, 6-9]. It is recommended that these standards be used as much as possible to allow more reuse of casting equipment. These standards are suitable for bridge span lengths ranging from 100 to 200 ft, bridge widths ranging from 28 to 45 ft, and girder heights ranging from 6 to 10 ft. The typical sections of girders with a height of 10 ft are shown in Fig. 6-3. Appendix A includes other standard sizes and dimensions for different girders.

For span lengths exceeding 200 ft, to achieve a cost-effective segmental bridge alternative, it may be necessary to utilize segments with variable cross sections along the length of the bridge. The variable cross sections can normally be made by changing the thickness of the bottom slab while keeping a constant girder depth for bridge span lengths not greater than 260 ft (see Fig. 6-4a). For longer-span bridges, the variable cross sections may be designed by changing both the thickness of the bottom flange and the girder depth (see Fig. 6-4b). Analytical results indicate that a span-to-depth ratio of about 20 can often achieve the structural optimum and maximum efficiency of prestressing for bridges with constant girder depth. For variable-depth box girders, the span-to-depth ratios generally range from 18 to 20 at pier location and 40 to 50 at midspan. More discussions for preliminary girder depths can be found in Section 1.4.2.2.

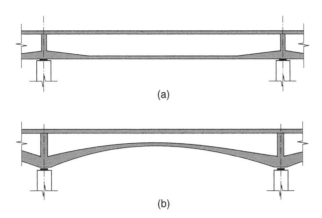

(a)

(b)

FIGURE 6-4 Longitudinal Profiles for Cantilever Segmental Bridges, (a) Constant Girder Depth with Variable Bottom Flange Thickness, (b) Variable Depth.

6.4   ESTIMATION AND LAYOUT OF POST-TENSIONING TENDONS

6.4.1   ESTIMATION AND SELECTION OF TENDONS

After the preliminary span arrangement and cross sections are set, a preliminary structural analysis due to dead and live loads can be performed based on the assumed erection methods and construction sequences. The required post-tensioning tendons can be estimated by assuming the effective prestressing stress of about 60% times the ultimate strength of the strand. Typically, there are four types of post-tensioning tendons in cantilever segmental bridges. They are top slab cantilever tendons, bottom slab continuity tendons, top slab continuity tendons, and draped continuity tendons. The top slab cantilever tendons are used to resist the negative moment during construction and in service. The top and bottom slab continuity tendons are used to connect two cantilevers and often called span tendons. Experiences show that the top cantilever tendons designed for resisting the maximum erection condition are normally sufficient to resist all loads of dead, live, and temperature loadings without additional tendons. Thus, the top cantilever tendons may be estimated based on the dead loads and construction loads. The bottom slab continuous tendons can be estimated based on the preliminary analytical results due to live loads plus the effects due to shrinkage and creep calculated by the approximate method at day 10,000 as discussed in Chapter 3. The top continuous tendons can be estimated as a few percentages of the bottom slab continuous tendons[6-3]. After the required areas of tendons are determined, select proper tendon sizes to determine the required number of tendons. In selecting the tendon sizes, the following factors should be considered:

a. Use commercially available tendon sizes with 0.6 in. diameter.
b. Keep the types of tendon sizes to a minimum to reduce the number of post-tensioning jacks required.
c. Use larger tendon sizes if the minimum tendon numbers required in Sections 6.4.3 to 6.4.5 can be met and are feasible. For typical tendon sizes that are commonly used, refer to Section 5.2.3.6.

In balanced cantilever construction, most tendons are internal tendons, though external tendons are also used. The minimum duct spacing, radius, and tangent length discussed in Section 1.4.3 should be followed.

6.4.2   Top Slab Cantilever Tendons

The top cantilever tendons are used to resist negative moment due to dead loads and erection loads during the cantilever girder construction. The top cantilever tendons are typically anchored in the junction of the top slab and web (see Fig. 6-5a), which is often referred to as face anchored. Normally, one or two interior tendons will be face anchored. The top cantilever tendons can also be anchored in a blister located in the inside of the box girder close to the top slab and the web (see Fig. 6-5b). The typical layouts of the top cantilever tendons with one and two face-anchored tendons per web are illustrated in Figs. 6-6 and 6-7[6-10], respectively.

Segmental duct couplers should be used for all internal tendon ducts at all joints between precast segments and shall be installed normal to the joints to allow stripping of the bulkhead forms. The maximum angle between the internal tendon and the duct coupler should not exceed 6°; this is to ensure the tendons pass through the coupler while theoretically not touching the duct or coupler[6-11] (see Fig. 6-8).

6.4.3   Bottom Span Tendons

The bottom span tendons are installed after the midspan closure segment is cast and has gained enough required strength. They are designed to resist the positive moments due to superimposed dead loads, live loads, and the effects induced by creep and shrinkage and temperature. They are normally provided in approximately the middle two-thirds of the span and are internal tendons located in the bottom slab close to the web (see Fig. 6-9). The bottom span tendons are typically anchored in blisters located in the inside of the box girder at the intersection of the bottom slab and the web as shown in Fig. 6-5a. For a curved variable-depth box girder, the pull-out forces on the bottom slab soffit should be properly designed[6-2, 6-10]. To ensure structural redundancy, a minimum of two bottom span tendons should be provided per web.

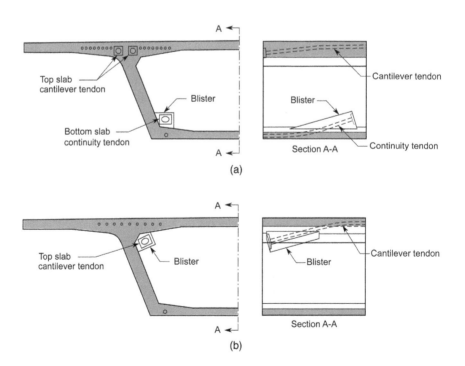

FIGURE 6-5   Typical Internal Longitudinal Tendon Anchorage Locations in Cantilever Segmental Bridges, (a) Face Anchored, (b) Blister Anchored[6-1].

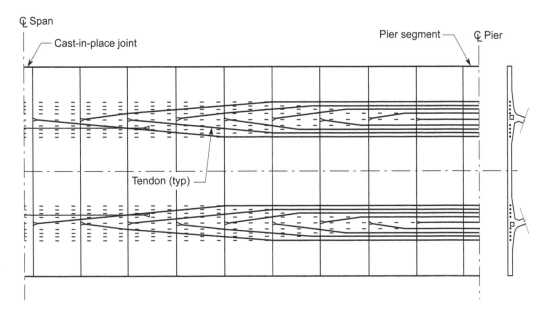

FIGURE 6-6   Typical Cantilever Tendon Layout with One-Face Anchored Tendon.

6.4.4   TOP SPAN TENDONS

The top span tendons are normally small sizes of 4-0.6″ diameter or 7-0.6″ diameter tendons across the midspan closure pours in balanced cantilever bridges to resist mainly the negative moment due to thermal gradient. They can be either external tendons or internal tendons and are generally anchored in the second or third segment from the closure segment (see Fig. 6-10). For easy replacement, the Florida Department of Transportation requires that all top span tendons be externally anchored in blisters. Top span tendons are required even when the calculations show that no top continuity tendons are necessary to meet the allowable stresses. This requirement is necessary because these tendons may be the only reinforcement crossing the closure pour joint. For box girders with cantilever wing lengths not greater than 0.6 times notational box width W (see Fig. 6-10b), the following minimum top span tendons are recommended to be installed at proper locations to produce the minimum required stresses in the cast-in-place top slab[6-11]:

a. For notational box width W ranging from 12 to 20 ft, provide two 4-0.6″ diameter tendons, one adjacent to each web anchored in the third segment back.
b. For notational box width W ranging from 20 to ≤ 25 ft, provide two 7-0.6″ diameter tendons, one adjacent to each web anchored in the third segment back.
c. For notational box width W ranging from 25 to ≤ 30 ft, provide three 7-0.6″ diameter tendons, one adjacent to each web anchored in the second segment back and one at the middle of the cell anchored in third segment back.
d. For notational box width W greater than 30 ft, provide four 7-0.6″ diameter tendons, one adjacent to each web anchored in the third segment back and two evenly spaced across the cell and anchored in the fourth segment back

For boxes with wing lengths greater than 0.6 times the box girder nominal width W, provide enough top span tendons to produce a minimum uniform stress of 75 psi over the top slab in the closure segment. The lateral distribution of tendon force across the top slab can be determined by using 30° distribution model as discussed in AASHTO LRFD Section C4.6.2.6.2. The external

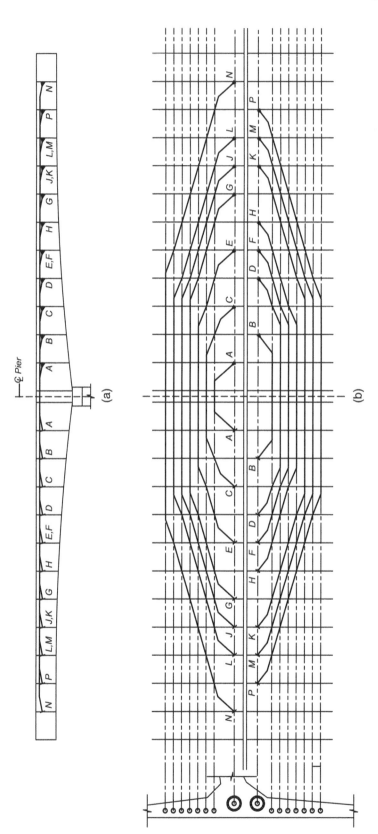

FIGURE 6-7 Typical Cantilever Tendon Layout with Two-Face Anchored Tendons. (a) Elevation View, (b) Plan View with Two-Face Anchored Tendons.

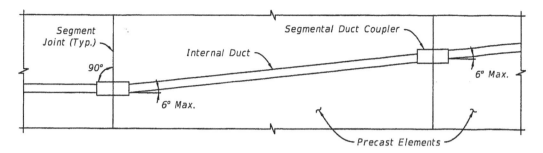

FIGURE 6-8   Layout of Internal Tendons with Segment Duct Couplers (Used with permission of FDOT[6-11]).

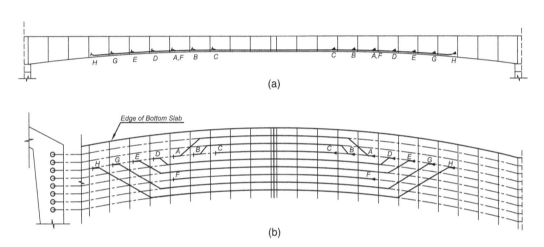

FIGURE 6-9   Typical Layout of Bottom Span Tendons, (a) Elevation View, (b) Part of Plan View.

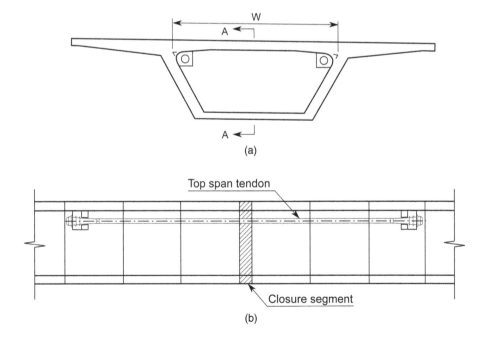

FIGURE 6-10   Typical Arrangement of Top Span Tendons, (a) Cross Section, (b) Elevation View.

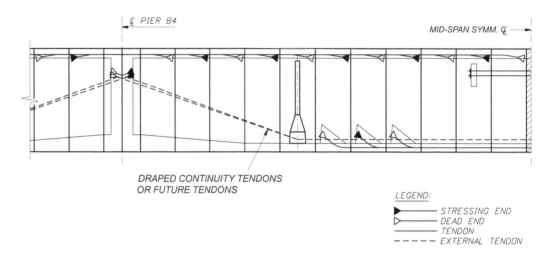

FIGURE 6-11   Draped Continuity Tendons and Future Tendons[6-11].

top slab continuity tendons cannot be anchored in the segments adjacent to the closure pour. To provide a uniform pressure, the tendons may be evenly distributed across each of cell and within the wings.

6.4.5   DRAPED CONTINUITY TENDONS AND FUTURE TENDONS

The draped continuity tendons are installed between piers for special cases and events and also increase the shear capacity due to its inclination (see Fig. 6-11). It is preferred to design the draped continuity tendons as external tendons, though they can be either external or internal tendons. The Florida Department of Transportation[6-11, 6-12] requires that at least two positive moment externally draped continuity tendons be provided per web that extends to adjacent pier diaphragms. The future tendons required by AASHTO are for bridge repair and strengthening purposes and should be designed as external tendons. They should be designed such that any span can be strengthened independently of the adjacent spans. Segments shall have at least one duct/anchorage location for expansion joint diaphragms and two duct/anchorage locations for internal pier segment diaphragms to accommodate future ducts[6-11].

6.5   ERECTION POST-TENSIONING BARS

Erection bars can be temporary or permanent and have been discussed in Section 5.2.3.5. Even though the bottom slab may theoretically have enough uniformly distributed compressive stress due to dead load and those created by the top slab temporary bars, which are required for balanced cantilever construction, it is still necessary to provide bottom erection post-tensioning bars to ensure proper mating along the bottom slab and across the entire match-cast interface. There is always the potential for improper mating along the bottom slab, which may cause an unintentional angle break, leading to serious implications for geometry control of the cantilever[6-2].

   For segmental bridges with a significant curvature, the use of internal temporary bars may not be feasible, as it may not be possible to couple them. In such cases, the designer should compute the required angle break, which should be within the specified limit of the bar anchorage and couple[6-2].

   The importance of avoiding the placement of internal temporary bars within the mid-portion of the box top flange should be mentioned, as it may cause longitudinal cracks to develop in the flanges along the bar ducts due to large splitting forces.

6.6 DESIGN FEATURES OF SEGMENTS

6.6.1 TYPICAL SEGMENT

The typical segment for cantilever-constructed segmental bridges shares many characteristics with those of span-by-span segmental bridges. Most tendons in cantilever segmental bridges are embedded in the box cross section and can be face anchored (see Fig. 6-12) due to their construction sequences. The shape of the tendon ducts may be curved in both the vertical and the horizontal directions in anchorage zones. The minimum duct spacing, radius of curvatures, and the tangent length adjacent to the anchorages should meet the requirements contained in Tables 1-10 and 1-11. The top and bottom blisters should be designed as discussed in Section 5.4.2. In designing the blisters, it is important to provide the minimum clearance for jacking and tendon replacement as shown in Fig. 6-13 and Tables 1-15 and 5-1.

6.6.2 PIER SEGMENT

6.6.2.1 Precast Pier Segment

As discussed in Chapter 5, the pier segment is normally the heaviest segment in a precast segmental bridge. In balanced cantilever segmental bridges, the pier segment is normally subjected to much less longitudinal force than that for a span-by-span segmental bridge during the construction, especially for the pier segment without external tendons installed. To reduce the segmental weight, the length of the pier segment may be reduced to less than 6 ft (see Fig. 6-14), but not less than 4 ft. The pier segment also can be precasted into two pieces as shown in Fig. 5-43 to reduce segment weight.

6.6.2.2 Precast Shell

When the super-structure is designed to be rigidly connected with the substructure, there are typically three ways to accomplish this rigid connection. One way is to use cast-in-place segments. This method eliminates the requirement of lifting heavy pier segments by cranes and stabilization of pier segments during construction. A disadvantage of this method is the requirement of two small cast-in-place closure pours on each side of the pier segment to connect to the first precast segments. The second method is to use a precast pier segment that is grouted to the top of the column and post-tensioned together with the pier by prestressing tendons. This method eliminates the two small cast-in-place pours. However, it is often difficult to lift the heavy pier segment and to properly place sufficient vertical post-tensioning tendons in the congested pier segment[6-2]. The third method is a combination of the first two methods. First, precast the pier segment shell, which is similar to the typical segment without the diaphragm and with the majority of the bottom slab block-out (see Fig. 6-15). The pier segmental shell is then placed above the pier that has reinforcing bars protruding into the shell. The diaphragm reinforcement is then placed, and the segmental shell acts as a form for the cast-in-place diaphragm that provides a rigid connection with the pier after curing. This method greatly reduces the segment weight and eliminates the cast-in-place closure pours between the pier segment and the first precast segments. If it is necessary to further reduce the precast pier segment shell weight, part of the webs also can be block-out.

6.6.3 CLOSURE SEGMENT

Figure 6-16 shows the typical reinforcing for the closure segments located between the tips of the cantilevers in the interior spans or between the tip of the cantilever and the adjacent end-span segments that are erected on falsework. The reinforcement arrangement of the closure segment is essentially the same as that for the typical section without the rebar related to the longitudinal anchorage blisters. It is important to firmly secure the tips of the cantilevers to prevent relative

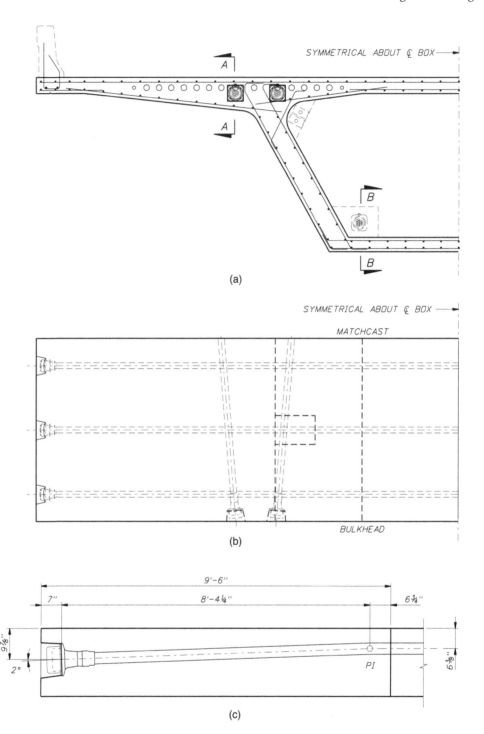

FIGURE 6-12 Typical Segment for Cantilever Segmental Bridges, (a) Elevation View, (b) Plan View, (c) Section *A-A* View, (d) Section *B-B* View.

(Continued)

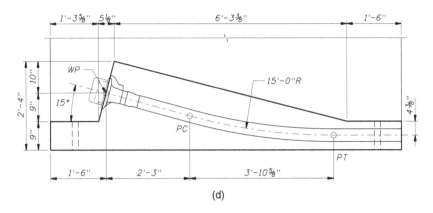

(d)

FIGURE 6-12   (Continued)

displacements between segments during curing. Typically, strongback beams are placed cross the closure joint and secured to the girders using post-tensioning bars (refer to Chapter 12).

6.7   TEMPORARY STABILITY OF CANTILEVERS

6.7.1   INTRODUCTIONS

In balanced cantilever construction, it is impossible to have a pure balanced condition[6-13 to 6-16]. There are always unbalanced moments due to wind loading, erection equipment, accent loadings, etc. (see Fig. 6-17). It is essential to ensure the girder stability during construction and to ensure that

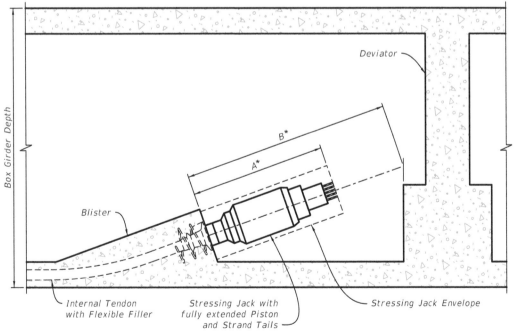

- Variable A* and B* see Table 5-1

FIGURE 6-13   Clearance of Bottom Span Tendon Anchorage Near Deviator[5-12] (Used with permission of FDOT)-Variable A* and B* see Table 5-1.

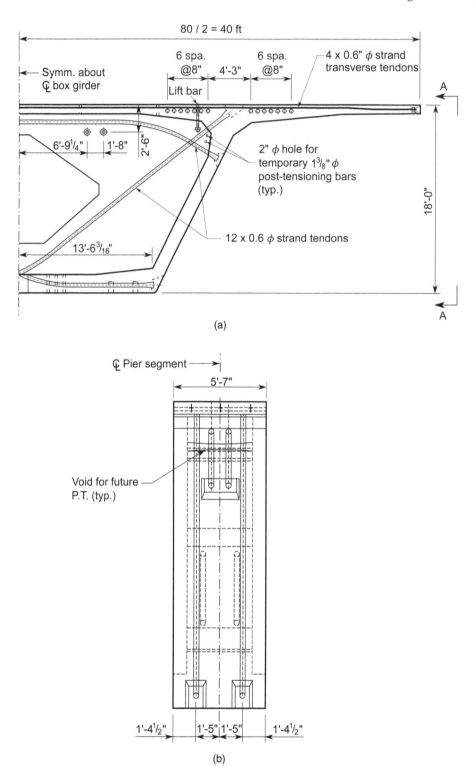

(a)

(b)

FIGURE 6-14 Typical Single-Piece Pier Segment, (a) Elevation View, (b) Side View.

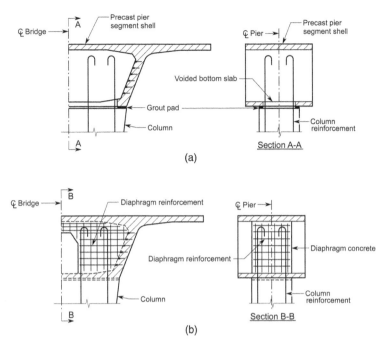

FIGURE 6-15   Typical Precast Pier Segment Shell, (a) Shell over Column Reinforcement, (b) Cast-in-Place Diaphragm and Reinforcement[6-2].

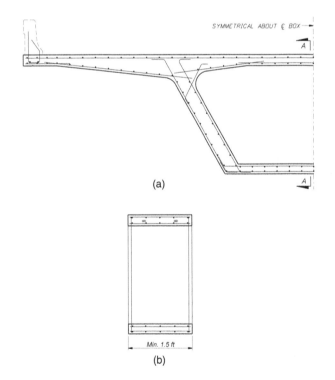

FIGURE 6-16   Typical Closure Segment, (a) Elevation View, (b) Section *A-A* View.

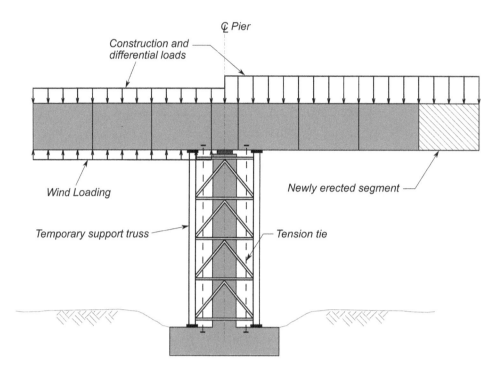

FIGURE 6-17   Out-of-Balance Construction Loadings[6-3, 6-16].

the effects of the construction loadings on the piers do not exceed those of the permanent designed loadings by installing temporary supports or using other methods.

6.7.2   Double Rows of Elastomeric Bearings

If there are two rows of bearings supporting the balanced cantilever girder in the longitudinal direction as shown in Fig. 6-18, it is often adequate to resist the anticipated unbalanced moment during construction with temporary post-tensioning bars to create a fixed condition between the superstructure and pier. First, install temporary steel or reinforced concrete bearings in the positions of permanent bearings. Then, use temporary post-tensioning bars to rigidly connect the pier segment to the pier. After the closure pours at the tips of the cantilevers have been installed and reach the required strength, remove the temporary post-tensioning tendons. Finally, using flat jacks, raise the girder and replace the temporary supports with permanent elastomeric bearings[6-4]. It should be mentioned that the bending capacity of the pier during construction must be checked.

6.7.3   Flexible Piers

Generally, flexible piers cannot resist the anticipated unbalanced moment during construction and temporary supports are necessary to maintain the cantilevers stability. There are typical four types of temporary supports[6-1]: column frame supports (see Fig. 6-17), one-side single-frame supports (see Fig. 6-19a), two-side frame supports (see Fig. 6-19b), and tension ties supports (see Fig. 6-19c). It is often a cost-effective alternative to rest the temporary supports on the pier footing by using column frame supports if possible (see Fig. 6-17). Though the designers may not be required to design the temporary supports, it is critical to specify the locations of the temporary supports and check the capacity of both the super-structures and substructures during the entire construction sequence with assumed locations of the temporary supports.

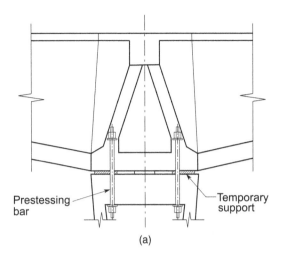

(a)

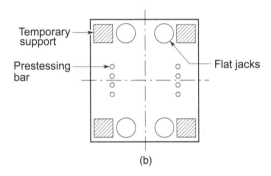

(b)

FIGURE 6-18   Temporary Rigid Connection between Super-Structure and Pier, (a) Elevation, (b) Plan View[6-4].

6.8   FEATURE OF LONGITUDINAL ANALYSIS AND CAPACITY VALIDATIONS

The general method for the longitudinal analysis of balanced cantilever segmental bridges due to dead loads and live loads is similar to that of span-by-span segmental bridges as discussed in Section 5.2.7. As the analysis of the effects due to dead loads and post-tensioning forces is always related to the assumed construction sequences and the analysis of creep and shrinkage effects is time consuming, an accurate structural analysis of a balanced cantilever bridge is typically performed using a computer program based on the theory discussed in Chapter 4. However, based on some simplifications, it is still possible to design balanced cantilever segmental bridges using hand calculations. Considering a three-span balanced cantilever segmental bridge, for example[6-14], an approximate analytical method and procedure for estimating the internal forces due to dead and post-tensioning forces are discussed as follows:

Assuming that a three-span continuous segmental bridge is built based on the construction sequence shown in Fig. 6-20, the longitudinal effects of the segment self-weight $q(x)$ can be estimated as follows:

Step 1: Build segments BD and DF using the balanced cantilever method with necessary temporary supports near the piers and install segments AB and FG on the falseworks as shown in Fig. 6-20a. At this stage, the super-structure is statically determinate and its bending moment distribution $M_{d1}(x)$ is easily calculated and shown in Fig. 6-21a.

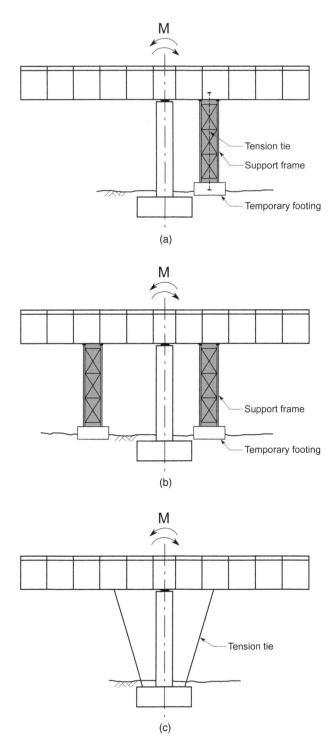

FIGURE 6-19 Temporary Supports for Stability during Construction, (a) One-Side Support, (b) Two-Side Supports, (c) Tension Ties Supports[6-1].

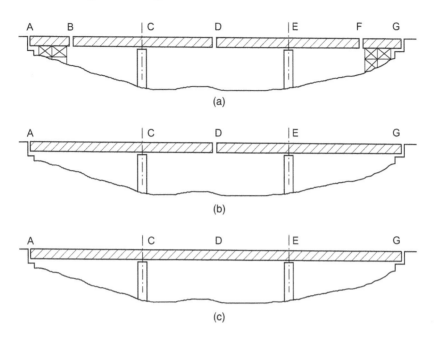

FIGURE 6-20 Construction Sequences of a Typical Three-Span Segmental Bridge, (a) Stage I, (b) Stage II, (c) Stage III.

Step 2: Install closure joints B and F, and remove the falseworks (see Fig. 6-20b). At this stage, the super-structure is still statically determinate. Its moment distribution $M_{d2}(x)$, including $M_{d1}(x)$, is shown in Fig. 6-21b.

Step 3: Install closure joint D, and complete the entire installation of the segments. At this stage, the super-structure is statically indeterminate. Assuming that the same dead load $q(x)$ is applied to the continuous girder, the moment distribution $M_{dc}(x)$ can be determined based on any methods discussed in Chapter 4 and shown in Fig. 6-21c.

Step 4: Assuming that all the segments are cast at the same time, the effects due to creep $M_{dcr}(x)$ can be obtained from Eq. 6-1, which is simplified from Eq. 4-131 based on Dischinger's method and plotted as shown in Fig. 6-21d.

$$M_{dcr}(x) = \left(M_{dc}(x) - M_{d2}(x)\right)\left(1 - e^{-\phi(t,t_0)}\right) \qquad (6\text{-}1)$$

where $\phi(t,t_0)$ is the creep coefficient and can be determined by Eq. 1-9b. Time t can be taken as 10,000 days, and t_0 is the cast time at the change of the static system. For an approximate calculation, $\left(1 - e^{-\phi(t,t_0)}\right)$ may be simply taken as 0.8 (see Eq. 4-135).

Step 5: The total effect due to the dead load is the sum of $M_{d2}(x)$ and $M_{dcr}(x)$ (see Eq. 4-131) and shown in Fig. 6-21e, i.e.

$$M_d(x) = \left(M_{d2}(x) + M_{dcr}(x)\right) \qquad (6\text{-}2)$$

The effects of post-tensioning tendons can also be estimated using a similar method as described above. For simplicity, we assume that only one straight tendon with a constant eccentricity is installed in each of segments BD and DF (see Fig. 6-22a). The total effects of the tendons can be estimated as follows:

Step 1: Determine the moment distribution $M_{p1}(x)$ in the stage I structure (see Fig. 6-22a) due to the post-tensioning tendons as shown in Fig. 6-22b.

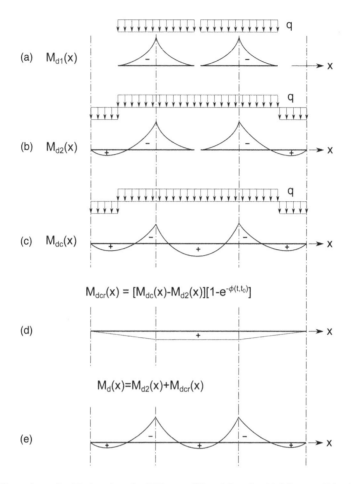

FIGURE 6-21 Procedures for Estimating the Effects of Dead Loads, (a) Moment Distribution in Stage I Static System, (b) Moment Distribution in Stage II Static System, (c) Moment Distribution in Completed Structure Due to Entire Dead Loads, (d) Distribution of Moment Due to Creep Effects, (e) Total Moment Due to Dead Load and Creep.

Step 2: Assume that the post-tensioning forces are applied in the completed three-span continuous structure as shown in Fig. 6-20c, the moment distribution $M_{pc}(x)$ due to the post-
tensioning forces can be determined as shown in Fig. 6-22c by any methods discussed in
Section 4.3, such as Section 4.3.2.1.

Step 3: The creep effects $M_{pcr}(x)$ due to the post-tensioning forces can be calculated as follows and are shown in Fig. 6-22d.

$$M_{pcr}(x) = \left(M_{pc}(x) - M_{p1}(x) \right)\left(1 - e^{-\phi(t,t_0)} \right) \qquad (6\text{-}3)$$

Step 4: Determine the total effects due to the post-tensioning forces as

$$M_p(x) = \left(M_{p1}(x) + M_{pcr}(x) \right) \qquad (6\text{-}4)$$

The capacity verifications for balanced cantilever bridges are the same as those for span-by-span
bridges and were discussed in Section 5.2.8.

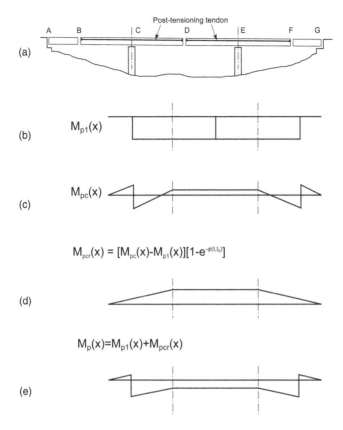

FIGURE 6-22 Procedures for Estimating the Effects of Post-Tensioning Tendons[6-14], (a) Simple Tendon Arrangement in Stage I Static System, (b) Moment Distribution in Stage I Static System Due to Tendons, (c) Moment Distribution in Completed Structure Due to Tendons, (d) Distribution of Moment Due to Creep Effects, (e) Total Moment Due to Tendons and Creep.

6.9   DESIGN OF SHEAR KEYS AT MATCH-CAST JOINTS

6.9.1   SHEAR KEY DETAILING

As discussed in Section 1.4.6.1, multiple small-amplitude shear keys should be provided at the match-cast faces in webs as well as in the top and bottom slabs to transfer section shear forces and facilitate the correct alignment of two match-cast segments during construction (see Fig. 1-39). The shear keys should be distributed over the web as much as possible if they are not in conflict with other details. Typically, the dimensions of the web shear keys and their distribution should meet the following requirements by AASHTO[6-17]:

$$\frac{h_k}{h_0} \approx \frac{b_k}{b_w} = 0.75 \tag{6-5}$$

$$\frac{a}{d} \approx 0.5 \tag{6-6}$$

where the designations of the variables h_0, h_k, a, and d are shown in Fig. 6-23. The depth of the keys should be as

$$1.25'' \le a \le \text{twice the diameter of the top aggragate size.}$$

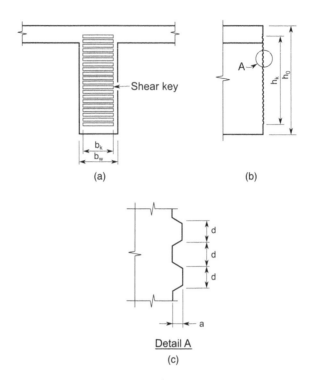

FIGURE 6-23   Typical Shear Key Arrangement in the Web, (a) Elevation View, (b) Side View, (c) Detail *A*.

6.9.2   Determination of Key Shear Strength

Though many tests and analyses in this area have been performed[6-18 to 6-24], the current AASHTO LRFD specifications[6-17] do not provide any recommended methods to determine the shear key strength. Test results show that the shear failure plane for both dry and epoxied match-cast joints is located at the base of the keys (see Fig. 6-24). The shear strength for epoxied joints is generally higher than that for dry joints. Based on test results, the AASHTO *Guide Specifications for Design*

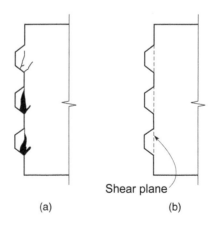

FIGURE 6-24   Shear Key Failure Plane, (a) Tested Failure Plane for Both Dry and Epoxied Joints, (b) Shear Plane for Design.

and Construction of Segmental Concrete Bridges[6-25] recommend the following equation for determining the ultimate shear strength for dry joints:

$$V_{uj} = A_k \sqrt{f_c'} \left(12 + 0.017 f_{pc}\right) + 0.6 A_{sm} f_{pc} \ \text{(psi)} \qquad (6\text{-}7a)$$

where

A_k = area of base of all keys in failure plane (in.²)
f_c' = concrete compression strength at 28 days (psi)
f_{pc} = concrete compression stress at the centroid of the section due to effective prestress (psi)
A_{sm} = area of contact between smooth surfaces on failure plane (in.²)

Recent research results show that Eq. 6-7a may overestimate the shear strength[6-21], and the authors recommend that Eq. 6-7a be revised as

$$V_{uj} = 0.7 A_k \sqrt{f_c'} \left(12 + 0.017 f_{pc}\right) + 0.6 A_{sm} f_{pc} \qquad (6\text{-}7b)$$

Extensive test results and numerical analysis conducted by Shamass et al. and Buyukozturk et al.[6-23, 6-24] indicate that the shear strength for single-keyed epoxied joints can be estimated as

$$V_{uj} = A_k \left(11.1 \sqrt{f_c'} + 1.2 f_{pc}\right) \ \text{(psi)} \qquad (6\text{-}8)$$

The definitions of the notations in Eq. 6-8 are the same as those shown in Eq. 6-7a.

AASHTO *Standard Specifications for Highway Bridges*[6-33] Article 9.20.1.5 requires that possible reverse shearing stresses in the shear keys be investigated, especially in segments near a pier. At the time of erection, the shear stress carried by the shear key shall not exceed $2\sqrt{f_c'}$, i.e.,

$$f_{va} \leq 2\sqrt{f_c'} \ \text{(psi)} \qquad (6\text{-}9)$$

Typically, only web shear keys are considered in resisting vertical shear force.

6.10 CONSTRUCTION ANALYSIS AND CAPACITY VERIFICATIONS

During the construction of a segmental bridge, its structural systems and boundary conditions constantly change from the start of the construction to the completed bridge. It is a critical design issue to ensure the structural stability, overturning and bulking, during each construction stage. The strength and stability of the structure and its foundation must be checked for each of the construction stages, especially for a structural system with a low degree of redundancy and unbalanced loads, such as a free cantilever structure.

Temporary supports are often necessary for keeping the structure stable during construction. It is important for the engineer to clearly state the assumed construction loads and the locations of the temporary supports in the plans.

The construction loads, load factors, and load combinations for both service and strength limits are presented in Sections 2.2.12 and 2.3.3.4. During construction, the concrete stresses should be kept within the allowable stresses at the service limit specified in Section 1.2.5.1.3. The construction load combinations at the strength limit states are discussed in Section 2.3.3.4.2. All of the service and strength limits should be checked and met during construction. The loading applications should be considered to produce the most unfavorable load cases for the related sections. Some typical loading applications for service and strength limits are shown in Figs. 6-25 and 6-26[6-15]. The load designations shown in these figures are the same as those contained in Section 2.3.3.4.1.

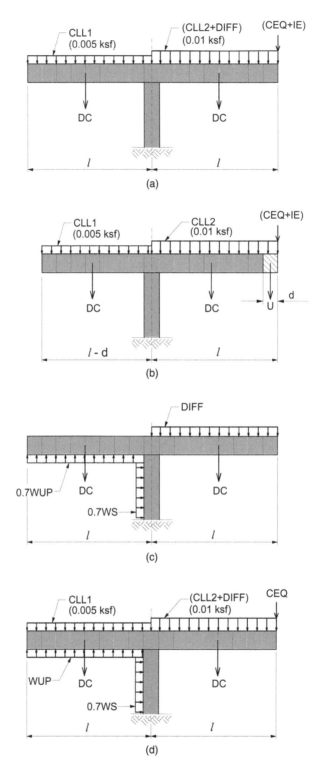

FIGURE 6-25 Load Applications for Service Limit During Construction, (a) Load Combination a, (b) Load Combination b, (c) Load Combination c, (d) Load Combination d, (e) Load Combination e, (f) Load Combination f [6-15].

(Continued)

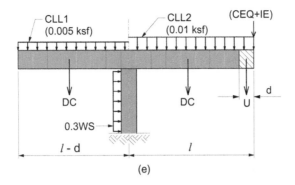

(e)

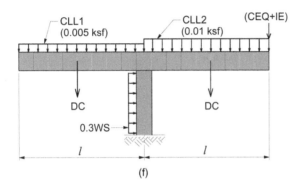

(f)

FIGURE 6-25 (Continued)

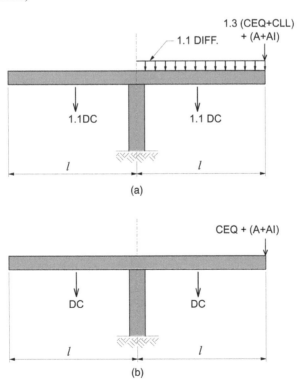

(a)

(b)

FIGURE 6-26 Load Applications for Strength Limit During Construction, (a) Maximum Force Effect, (b) Minimum Force Effect.

6.11 DEFLECTION CALCULATION AND CAMBERS

6.11.1 Determination of Deflection of Cantilevers During Construction

The deflection of the balanced cantilevers during construction is induced by segment self-weights, post-tensioning tendons, concrete creep, and construction loadings, etc. Properly predicting the deflection of the cantilevers during construction is the key to successfully building a balanced cantilever segmental bridge; though accurately determining the deflections may be very difficult as they are affected by many factors, such the material properties, temperature, and humidity. To develop the equation for determining the deflection, let's consider the cantilever as shown in Fig. 6-27[6-4, 6-16].

Assuming that the cantilever girder consists of n segments, the length of segment i is denoted as s_i and the average moment due to the segment self-weight and post-tensioning forces along the segment length is denoted as M_i (see Fig. 6-27a). Using the conjugate beam method as discussed in Section 4.2.1.1 and from Fig. 6-27b, the deflection at an arbitrary section x_j can be written as:

$$y_j = \sum_{i=1}^{j} \frac{M_i(t)}{E_i(t)I_i}(x_j - l_i)s_i \ (i \le j) \tag{6-10}$$

where

$M_i(t)$ = average moment in segment i, at cast time t
I_i = moment of inertia of segment i
$E_i(t)$ = elastic modulus at cast time t
l_i = distance from center of segment i to end of cantilever

From Section 4.2.1.1, we can see that Eq. 6-10 represents the sum of the deflections induced by the average deflection angle of each segment. The deflections calculated by Eq. 6-10 only include the effect of the bending moment $M_i(t)$. From the discussion in Section 4.3.4.2, in statically determinate structures, the deflection induced by concrete creep varies linearly with external loadings and the deflections due to both bending moment $M_i(t)$ and creep effect can be written as

$$y_j = \sum_{i=1}^{j} \frac{M_i(t)}{E_i(t)I_i}(x_j - l_i)s_i \left(1 + \phi(t,t_0)\right) (i \le j) \tag{6-11}$$

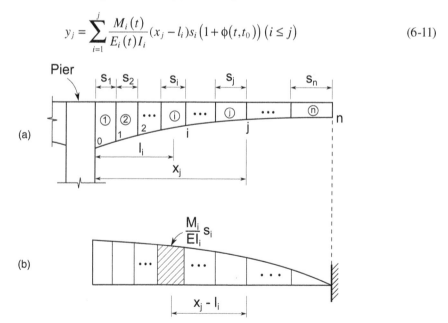

FIGURE 6-27 Analytical Model for Determining Deflections of Cantilever during Construction, (a) Fixed Cantilever Girder, (b) Conjugate Beam.

where $\phi(t,t_0)$ is the creep coefficient.

As the average moment in segment i increases with successive segments erected at different concrete age time, $M_i(t)$ is the function of age time. For an easy calculation, we can write the total average moment in segment i as the sum of the moments induced by related segments, such as the total average moment in segment 1 can be written as the sum of n moments induced by n segments, i.e.,

$$M^1 = M_1^1 + M_2^1 + \cdots + M_i^1 + \cdots M_n^1 = \sum_{i=1}^{n} M_n^1 \tag{6-12}$$

where

M^1 = total average moment in segment 1
M_i^1 = average moment in segment 1 due to segment i

Assuming that each of the segments is cast-in-place and that the time for casting each of the segments is Δt, then the deflection at section j induced by the average moment in segment 1 due to its self-weight and post-tensioning force applied at segment 1 can be written as

$$\Delta y_{j1}^1 = \frac{M_1^1}{E_1(t)I_1}(x_j - l_1)s_1(1+\phi(t,\Delta t)) \tag{6-13a}$$

The deflection at section j induced by the average moment in segment 1 due to the self-weight of segment 2 and the post-tensioning force applied at segment 2 can be written as

$$\Delta y_{j2}^1 = \frac{M_2^1}{E_1(t)I_1}(x_j - l_1)s_1(1+\phi(t,2\times\Delta t)) \tag{6-13b}$$

Thus, the deflection at section j induced by the total moment or deformation in segment 1 at time t can be written as

$$y_j^1 = \frac{(x_j - l_1)s_1}{I_1}\sum_{i=1}^{j}\frac{M_i^1}{E_1(t)}(1+\phi(t,i\Delta t)) \tag{6-13c}$$

Similarly, the deflection at section j induced by the total moment or deformation in segment 2 at time t can be written as

$$y_j^2 = \frac{(x_j - l_2)s_2}{I_2}\sum_{i=1}^{j-1}\frac{M_i^2}{E_2(t)}(1+\phi(t-\Delta t,i\Delta t)) \tag{6-13d}$$

In the same way as above, we can determine the deflections at section j caused by the total average moments or average rotations at sections 3, 4, …, and j. The total deflection at section j can be written as

$$y_j = \frac{(x_j - l_1)s_1}{I_1}\sum_{i=1}^{j}\frac{M_i^1}{E_1(t)}(1+\phi(t,i\Delta t)) + \frac{(x_j - l_2)s_2}{I_2}\sum_{i=1}^{j-1}\frac{M_{i+1}^2}{E_2(t)}(1+\phi(t-\Delta t,i\Delta t)) + \dots$$

$$+ \frac{(x_j - l_k)s_k}{I_k}\sum_{i=1}^{j-(k-1)}\frac{M_{i+(k-1)}^k}{E_k(t)}(1+\phi(t-(k-1)\Delta t,i\Delta t)) + \dots \tag{6-14}$$

$$+ \frac{(x_j - l_j)s_j}{I_j}\frac{M_j^j}{E_j(t)}(1+\phi(t-(j-1)\Delta t,\Delta t))$$

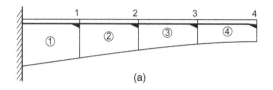

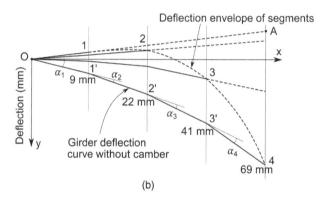

FIGURE 6-28 Variation of Girder Deflections with Segments Installed, (a) Segmental Cantilever Girder, (b) Girder Deflections (in = 25.4 mm).

Equation 6-14 incudes the effects of creep and the variable modulus of elasticity of concrete and can be used for both cast-in-place and precast construction. For precast segments, it is necessary to properly adjust the loading times based on actual casting times.

6.11.2 Camber Determinations

6.11.2.1 Deflections Development During Construction

It is important to clearly understand the development of the various vertical deflections during the installation of segments. For simplicity, take a cast-in-place single cantilever girder consisting of four segments as an example to illustrate the deflection development at the end of segments during construction. Assume that segment 1 is cast in the horizontal direction during construction and post-tensioned; node 1 would deflect upward −5 mm (0.20 in) as shown in Fig. 6-28b and Table 6-1. If segments 2 to 4 were attached to segment 1 when segment 1 was cast and post-tensioned, nodes 2 to 4 would deflect upward −11, −17, and −23 mm (−0.43, −0.67, −0.91 in), respectively (see Fig. 6-28 dotted line 1-*A* and the numbers in round brackets in the third row in Table 6-1). If segment 2 is

TABLE 6-1

Deflections Due to Self-Weights and Post-Tensioning Forces at Different Loading Stages

Segments Cast and Post-tensioned	Locations (mm) (in = 25.4 mm)			
	Node 1	Node 2	Node 3	Node 4
1	−5	(−11)	(−17)	(−23)
2	1	5	(9)	(13)
3	5	10	20	(30)
4	8	18	29	49
Total	9	22	41	69

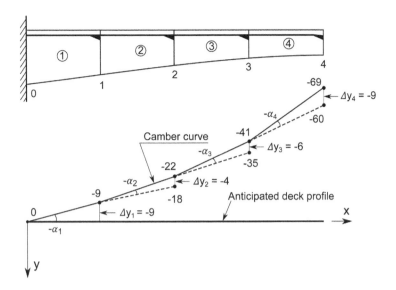

FIGURE 6-29 Camber Curve Setting (unit = mm, in = 25.4 mm).

cast against segment 1 in the orientation of segment 1 and post-tensioned, nodes 1 and 2 would deflect downward by 1 and 5 mm (0.04 and 0.20 in), respectively (see Fig. 6-28b and Table 6-1). Assuming all segments are installed in the same way, at the end of casting and post-tensioning segment 4, the total deflection of each node is the sum of the deflections at each construction stage and is shown in the last row in Table 6-1. If the deflection curve within a segment is assumed as a straight line, the deflection curve of the girder at the end of construction is a broken line 0-1′-2′-3′-4 as shown in Fig. 6-28b. Each of the segments is rotated by an angle relative to the horizontal direction or its previously installed segment by α_1, α_2, α_3, and α_4, respectively (see Fig. 6-28b). The Curve 0-1-2-3-4 shown in Fig. 6-28b represents the deflection of the segments at the end of installation of each of the segments during construction and is called the envelope of deflection curve.

6.11.2.2 Camber Method in Construction

From the discussion in Section 6.11.2.1, to ensure that the deck longitudinal profile is horizontal after construction, it is necessary to set a camber curve to offset the total deflection during construction. In practice, it is common to provide a small opposing angle $-\alpha_i$ or a relative camber Δy_i at the end of each cast segment i (see Fig. 6-29). For example, casting segment 1 in the orientation with an angle of $-\alpha_1$ from the horizontal direction, we yield the deflection of -9 mm (0.35 in) at the end of segment 1. When segment 2 is cast, we rotate an angle $-\alpha_2$ from segment 1 and obtain a relative camber of 4 mm (0.16 in) and total camber of 22 mm (0.87 in) at the end of segment 2, and so on. More discussion on establishing the camber will be discussed in Chapter 12.

When calculating the total deflection at each node during construction, it is important to include all significant effects, such as construction equipment, creep, and pier rotation, except the self-weight of segments and post-tensioning forces. Thus, the camber calculations are typically performed by specialty engineers hired by the contractor based on the contractor's construction sequence and schedule.

6.12 CURVED SEGMENTAL BRIDGE DESIGN FEATURES

The need for a smooth dissemination of congested traffic, the limitation of right-of-way along with economic, aesthetic, and environmental considerations have motivated the increased use of curved box girder bridges. Researchers[6-26] have shown that the effects of curvature should be considered for segmental concrete box bridges that have a center span angle exceeded 11° degrees to ensure the calculated normal and shear stresses are within 10% tolerance. In this section, the concept of concordant tendons in simply supported curved cantilever girder bridges is discussed first. Then, the effects of curvature on continuous segmental box girder bridges are presented.

6.12.1   CONCEPT OF CONCORDANT TENDONS IN CURVED GIRDERS AND DETERMINATION OF TENDON LOCATIONS

6.12.1.1   Concept of Concordant Tendons

As discussed in Sections 3.1 and 4.3.2.3, if the prestressing tendons are properly arranged in a straight girder, it can behave as an axial loaded column. There are no moment, shear, and torsion at any section along the girder, i.e.,

$$M_{yq} + M_{yP} = 0 \qquad (6\text{-}15\text{a})$$

$$Q_{zq} + Q_{zP} = 0 \qquad (6\text{-}15\text{b})$$

$$T_{xq} + T_{xP} = 0 \qquad (6\text{-}15\text{c})$$

where

M, Q, T = moment, shear, and torsion, respectively
q, P = external load and post-tensioning force, respectively

The tendons meeting Eqs. 6-15a to 6-15c are often called concordant tendons. As discussed in Section 3.5, in a curved girder, the vertical bending and torsion are coupled with each other. It is impossible to use prestressing forces to create a pure compression condition for a curved girder. In practice, the tendon locations are arranged based on the bending moment requirement. Then the tendon locations are adjusted to reduce the external torsion and the remaining torsion is resisted by providing torsion reinforcement.

6.12.1.2   Determination of Tendon Locations in Cantilever Girders

6.12.1.2.1   Internal Forces Induced by External Loadings

For simplicity, let's examine a cantilever girder[6-27] as shown in Fig. 6-30. Assuming the curved cantilever girder is subjected to a uniform loading q, has a radius of R, and has a central angle of θ_0, from Fig. 6-30b, the vertical bending moment, torsion, and shear at section θ can be written as

Vertical moment:

$$M_{yq}(\theta) = -\int_0^\theta qR^2 \sin\theta d\theta = -qR^2(1-\cos\theta) \qquad (6\text{-}16)$$

Shear:

$$Q_{zq}(\theta) = -qR\theta \qquad (6\text{-}17)$$

Torsion:

$$T_{xq}(\theta) = \int_0^\theta qR^2(1-\cos\theta)d\theta = qR^2(\theta-\sin\theta) \qquad (6\text{-}18)$$

The sketches for the variations of the moment, torsion, and shear of the curved girder are illustrated in Fig. 6-31.

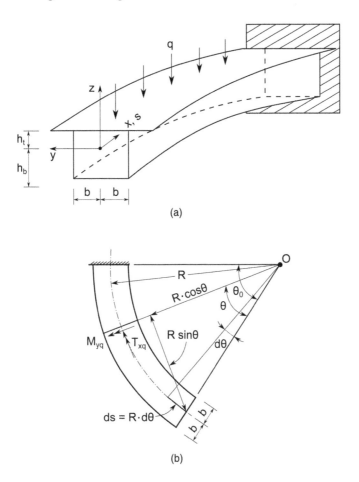

FIGURE 6-30 Curved Cantilever Girder Subjected to a Uniform Loading, (a) Loading Condition and Coordinate System, (b) Analytical Model.

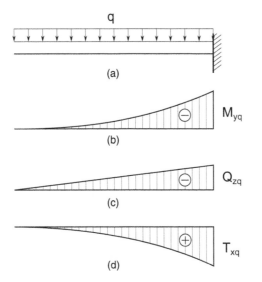

FIGURE 6-31 Variations of Internal Forces in Curved Cantilever Girder, (a) Loading, (b) Moment, (c) Shear, (d) Torsion.

6.12.1.2.2 Tendon Locations for Balancing the Bending Moment

Assume that there are two tendons located in each of the inside and outside webs of the cantilever girder, respectively (see Fig. 6-32c), and that each of the tendons is stressed to P. The moment induced by the post-tensioning forces can be written as

$$M_{yP}(\theta) = 2Pe_z(\theta) \tag{6-19}$$

where

P = one tendon force

$e_z(\theta)$ = eccentricity of tendons to centroid lines, assuming the same for both inside and outside tendons for balancing bending moment

To create concordant tendons, the sum of the bending moments induced by external loading and the post-tensioning forces should be equal to zero. Rewrite Eq. 6-15a as

$$M_{yq}(\theta) + M_{yP}(\theta) = 0 \tag{6-20}$$

Substituting Eqs. 6-16 and 6-19 into Eq. 6-20 yields

$$e_z(\theta) = \frac{qR^2}{2P}(1 - \cos\theta) \tag{6-21}$$

Assuming the tendons are located in the top flange at the fixed end, we have the boundary condition as

$$\theta = \theta_0, \ e_z(\theta_0) = h_t \tag{6-22}$$

Substituting Eq. 6-21 into 6-19 and then using Eq. 6-20, we can obtain

$$P = \frac{qR^2}{2h_t}(1 - \cos\theta_0) \tag{6-23a}$$

$$e_z(\theta) = \frac{(1 - \cos\theta)}{(1 - \cos\theta_0)}h_t \tag{6-23b}$$

From Eqs. 6-23a and 6-23b, we can obtain the required post-tensioning force and the locations of the tendons for balancing the moment caused by external loadings.

If the tendons are arranged according to Eq. 6-23b, the total shear can be written as

$$Q_{zq} + Q_{zP} = -qR\theta + 2P\frac{de_z(\theta)}{Rd\theta} \tag{6-24}$$

Substituting Eq. 6-23b into Eq. 6-24 yields

$$Q_{zq} + Q_{zP} = -qR\theta + \frac{qR^2}{h_t}(1 - \cos\theta_0)\frac{b}{R}\frac{\sin\theta}{(1 - \cos\theta_0)} = -qR(\theta - \sin\theta) \tag{6-25}$$

From Eq. 6-25, it can be seen that the total shear will not be equal to zero, although the total bending moment does equal zero as the shear induced by external loading q is not only related to the bending moment but also related to torque. From Eq. 3-139 and noticing the coordinate system, we have,

$$Q_{zq} = \frac{dM_{yq}}{ds} - \frac{T_{xq}}{R} \tag{6-26}$$

From Eq. 6-20, the shear induced by post-tensioning force is

$$Q_{zP} = \frac{dM_{yP}}{ds} = -\frac{dM_{yq}}{ds} \tag{6-27}$$

From Eqs. 6-26 and 6-27, we have

$$Q_{zq} + Q_{zP} = -\frac{T_{xq}}{R} \neq 0 \tag{6-28}$$

From Eq. 6-28, we can conclude that there are no pure concordant tendons in curved girders because the bending moment and torque are coupled with each other.

6.12.1.2.3 Tendon Locations for Balancing Torsion

From Sections 3.1.2 and 4.3.2.2, we can see that changing the tendon curvature will change the shear in the girder. To offset the torsion induced by the bending moment, we can move the tendon in the inside web down and the tendon in the outside web up by the same amount $\Delta e_z(\theta)$ as shown in Fig. 6-32 to produce an opposite torque, i.e.,

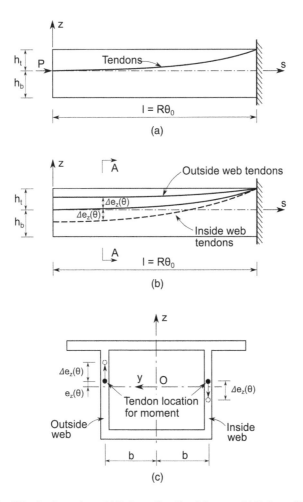

FIGURE 6-32 Idealized Tendon Locations, (a) Balance Bending Moment, (b) Balance Torsion, (c) Tendon Shift.

Torsion produced by the relocated tendons:

$$T_{xP}(\theta) = -2P\frac{d\Delta e_z(\theta)}{Rd\theta}b \tag{6-29}$$

where

b = distance between tendon and z-axis

$\Delta e_z(\theta)$ = amount relocated of tendons (inside tendon down and outside tendon up)

To offset the torsion, the following equation should be met:

$$T_{xq}(\theta) + T_{xP}(\theta) = qR^2(\theta - \sin\theta) - 2P\frac{d\Delta e_z(\theta)}{Rd\theta}b = 0 \tag{6-30}$$

Rearranging Eq. 6-30 yields

$$\frac{d\Delta e_z(\theta)}{d\theta} = \frac{qR^3(\theta - \sin\theta)}{2bP} \tag{6-31}$$

Solving Eq. 6-31 yields

$$\Delta e_z(\theta) = \int_0^{\theta} \frac{qR^3(\theta - \sin\theta)}{2bP}d\theta = \frac{qR^3}{2bP}(\theta^2 - \cos\theta) + c \tag{6-32a}$$

Using boundary condition: $\Delta e_z(\theta_0) = 0$, we have

$$\Delta e_z(\theta) = \pm\frac{qR^3}{2bP}\left(\frac{\theta_0^2 - \theta^2}{2} + \cos\theta_0 - \cos\theta\right) \tag{6-32b}$$

In Eq. 6-32b, $\Delta e_z(\theta)$ is taken as positive for the tendon in the outside web and negative for the tendon in the inside web.

From Eqs. 6-23b and 6-32b, we can see that it is theoretically possible to completely offset the moment and torque induced by external loadings if the tendons are properly located, though there is a remaining shear of $-\frac{T_{xq}}{R}$.

6.12.1.3 Concept of Tendon Layout in Simply Supported Statically Indeterminate Curved Girders

Figure 6-33 shows a simply supported curved girder subjected to uniform loading q. As the vertical loading will cause torsion, the girder is an indeterminate structure. The variations of the moment and torsion along the length of the girder are shown in Fig. 6-33c and d. To balance the bending moment induced by the external loading, the tendons can be arranged in a shape that is similar to that of the moment (see Fig. 6-33e). To balance the torsion, the inside and outside tendons should be curved in the opposite direction (see Fig. 6-33f). To balance both bending moment and torsion, the locations of the inside and outside tendons are the sum of coordinates of the tendons shown in Fig. 6-33e and f (see Fig. 6-33g).

6.12.1.4 Effect of Curvatures in Continuous Girders

It is theoretically possible to completely balance both bending moment and torsion in curved box girder bridges through prestressing forces because the tendon placement is limited and due to the construction considerations. In practice, tendons are typically arranged symmetrically

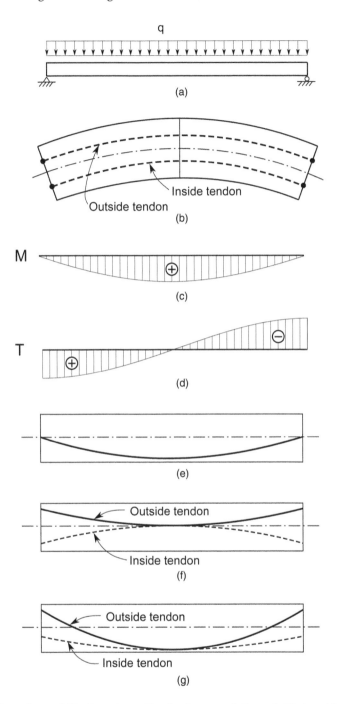

FIGURE 6-33 Locations of Tendons in a Simply Supported Curved Girder, (a) Girder Elevation View, (b) Girder Plan View, (c) Moment Distribution, (d) Torsion Distribution, (e) Locations of Tendons for Balancing Moment, (f) Locations of Tendons for Balancing Torsion, (g) Locations of Tendons for Balancing both Moment and Torsion.

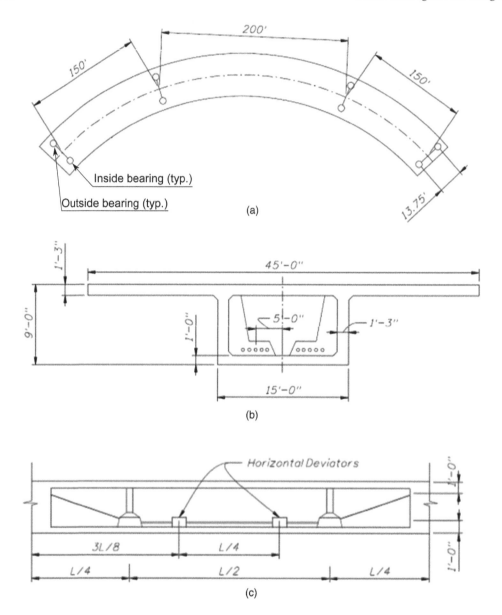

FIGURE 6-34 Prototypical Bridge, (a) Plan View, (b) Typical Section, (c) External Tendon Layout of Center Span.

about the girder's vertical axis. For continuous curved segmental bridges, it is more complicated to balance both bending moment and torsion through prestressing forces. However, the basic concept is the same as that for simply supported curved girders. In this section, the effects of curvature on segmental box girder bridges are briefly discussed[6-26, 6-28].

Figure 6-34 shows a typical three-span continuous box girder bridge that is a simplified version of an existing bridge. At each pier location, two rigid vertical supports are provided underneath the bottom slab of the box at the centerline of the webs. With the exception of one transverse constraint and two longitudinal constraints located at the beginning of the bridge model for stability, all supports are free to rotate. Diaphragms with 6-ft thickness are used at all pier locations to limit local effects and to anchor the post-tensioning tendons. The external prestressing tendons with a yield

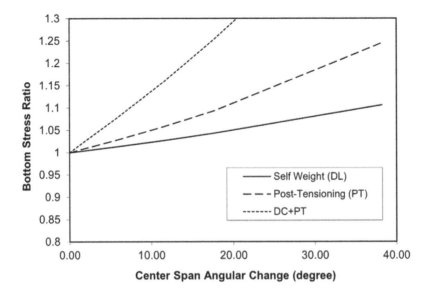

FIGURE 6-35 Variation of Maximum Bottom Stresses in Center Span with Curvatures.

stress of 270 ksi are arranged symmetrically about the box girder's vertical axis in each web. The end span has 6-27ϕ0.6-in.-diameter strands (35.15 in.²), while the interior span has 10-27ϕ0.6-in.-diameter strands (58.59 in.²). The tendon quantity is determined such that no tensile stresses develop at the critical sections, under the combined effects of dead load and live load (around 45% of positive moment due to dead load). Vertical deviators are placed at quarter-span locations for every span. Horizontal deviators are located at midspan for each of the end spans and at 3/8L and 5/8L span locations for the center span. The vertical eccentricities of the external tendons are 1 ft from the bottom of the box at each deviator location and 1 ft from the top of the box at all diaphragm locations. Horizontally, the centroid of the external tendon is located at an offset of 5 ft from the centerline of the box (see Fig. 6-34).

By changing the bridge radii from infinity to 300 ft (the corresponding end span angles varying from 0° to 28.7° and the corresponding center span angles varying from 0° to 38.2°), the variation of the maximum absolute normal stresses, shear stresses, and support reactions with the radii can be obtained. For clarity, assuming the responses for a straight bridge as 1.0, the variations of the ratios of the stresses for the curved bridges to that of the corresponding straight bridge with radii are shown in Figs. 6-35 to 6-37. Figure 6-35 illustrates the variation of the maximum bottom stresses with span angles for the center span. From this figure, the following can be observed: (a) The normal stresses induced by the dead load and post-tensioning do not vary significantly with curvature, while the combined stress is more sensitive to curvature. (b) If a maximum normal stress ratio of 5% is used to determine whether a straight model is adequate to estimate the normal stress of curved ones, the central angle can reach 22° and 11° for self-weight and post-tensioning load cases, respectively, while only 4° for the combined stress results.

The variations of the maximum web shear stress ratios with radii are shown in Fig. 6-36. From these figures, we can observe the following: (a) Under self-weight, the exterior webs are subject to higher shear stresses than the interior webs for curved bridges due to torsion effect. With the center angle changing from 0° to 38°, the maximum absolute shear stresses for the interior web reduces by 13% and that for the exterior web increases by 18%. (b) Under post-tensioning loads, the shear stresses are opposite to that induced by its self-weight. The interior webs are subjected to more shear stresses than the exterior webs. When the center angle changes from 0° to 38°, the maximum absolute shear stresses in the interior web increase by 46% and that of the exterior web decreases

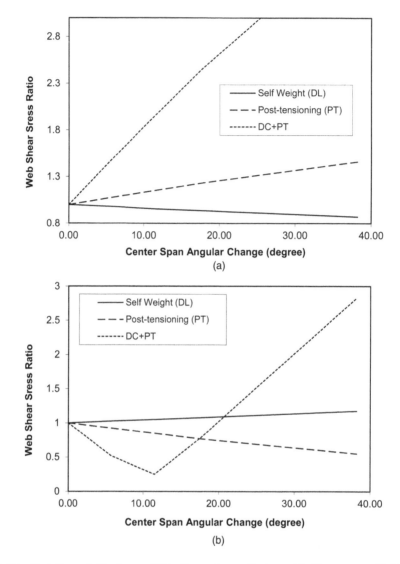

FIGURE 6-36 Variation of Maximum Web Shear Stress Ratios with Curvatures, (a) Interior Web, (b) Exterior Web.

by 45% relative to the straight bridge. (c) For the combined load case, the maximum absolute shear stresses in the interior and the exterior web with a center angle of 38° increases by about 425% and 300% relative to the straight bridge results. For the bridges with a center angle less than 20°, the post-tensioning forces can effectively reduce the exterior web shear stress, without reducing the interior web shear stress.

Figure 6-37 shows the variations in the bearing reactions at the abutment with the radius. From the figures, the following can be observed: (a) Under the dead load, the outside bearings (see Fig. 6-34) will take more reaction than the inside bearings due to torsion. The bearing reaction ratios between the outside and the inside bearings range from 1.00 to 2.0 at the end pier locations (see Fig. 6-37a). (b) With post-tensioned external tendons, the bearing reactions due to post-tensioning are caused by secondary prestressing effects. For the straight bridge model, intermediate piers are subjected to uplifting forces, while end piers take compressive forces. As the curvature increases, the outside bearings tend to take more compressive reaction, while the inside bearings are subjected to more

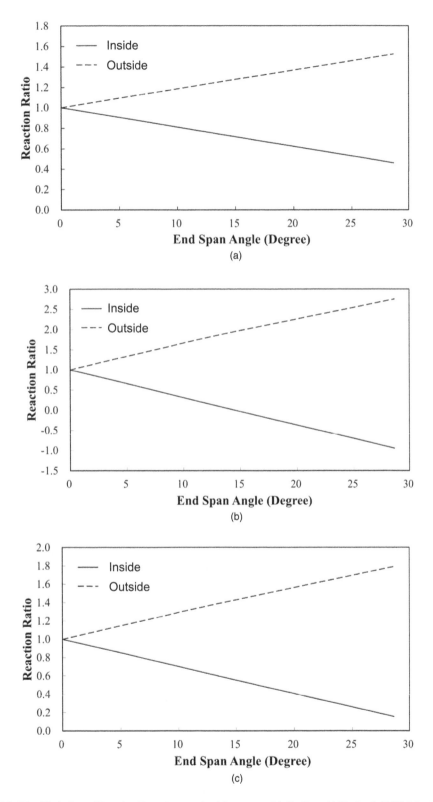

FIGURE 6-37 Variation of Bearing Reactions at the Abutment with Radius, (a) Under Self-Weight, (b) Under Post-Tensioning Forces, (c) Under Combined Loadings of Self-Weight and Post-Tensioning Force.

uplifting reaction. For a bridge with a center angle of 28°, the inside bearings at the end supports start uplifting. The outside bearing reactions increase by about 2.5 times (see Fig. 6-37b). Figure 6-37c illustrates the effects of curvature under the combined loads of self-weight and post-tensioning. This figure indicates that the curvature has significant effects on the bearing reactions. For the bridge with a center angle of about 4.5°, the reaction difference between the outside and inside bearings is about 30% to 40%, while for the bridge with a center angle of about 28°, the outside bearing virtually takes the total pier reaction.

The forgoing discussions about the effects of curvatures have not considered the time-dependent effects of concrete creep and shrinkage. Research[6-26, 6-28] shows the following: (a) The time-dependent effect will reduce the influence of curvature on the bearing reactions, and the reduction in the intermediate piers is more obvious than at the end piers. The final intermediate pier bearing reactions are much less sensitive to the curvature change than that of the end piers. For bridges with tight curvature, the inside bearings in the end piers take a very small portion of the total vertical load and have the highest tendency of uplifting. (b) With 10% tolerance, a straight model can be used to model a curved bridge with a center span angle up to 11° for normal and shear stresses and 3° for bearing reactions.

6.12.2 Reducing Support Uplift Forces and Bearing Arrangement

Per the aforementioned discussion, as the radius decreases, the inside bearings of curved segmental bridges may be subjected to large uplift forces. The uplift force can be determined as

$$V_{up} = \frac{M_T}{d_b} - V_s \tag{6-33}$$

where
  M_T = torsion moment at support
  d_b = transverse distance between bearings
  V_s = reaction due to vertical shear

The uplifting support force often causes more maintenance work or expensive details and should be avoided if possible. Although it is acceptable for the bridge to be designed with uplift-resisting bearing assemblies or structural integration of super-structures and substructures. From Eq. 6-33, it can be seen that there are three possible ways to reduce the support uplifting forces. The first way is to increase the transverse distance between bearings. The second way is to increase the bearing compressive reaction, and the third is to reduce the torsion moments at the supports.

Increasing the bearing distance is an effective way of reducing uplifting forces. However, this means that the length of the pier cap will be increased, which may affect the traffic below and increase construction cost. If the bottom slab soffit is not wide enough, special outriggers should be used, which will increase the complexity of the details and cause additional rebar and/or PT duct congestion in the diaphragm.

Increasing the bearing compressive reaction typically means increasing unnecessary dead load, such as using a heavier material in the end span than in the interior spans[6-27]. However, the complexity of the connection often outweighs the benefits, and therefore this approach is feasible only in certain types of long-span bridges.

There are three ways to reduce the torsion moments at supports. The first one is to adjust tendon locations as discussed in Section 6.12.1. This method is often limited by the available vertical space. The second method is to adjust the torsion-resistant bearing locations, such as using two free end torsion bearings (see Fig. 6-38a) or only two end torsion resistant bearings (see Fig. 6-38b)[6-27]. The layout shown in Fig. 6-38b eliminates the torsion moment, and thus the uplifting issue for the intermediate bearings, and is often used in bridges with tight curvatures. However, this layout often requires

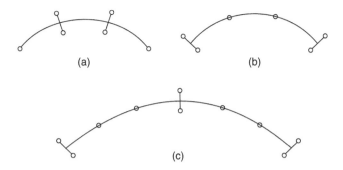

FIGURE 6-38 Bearing Layouts for Reducing Uplifting Forces, (a) Two End Free Torsion Supports, (b) Two End Torsion Resistant Supports, (c) Three Torsion-Resistant Supports.

outriggers to take torsion moments at the end supports. The bearing layout shown in Fig. 6-38a eliminates the torsion moment at the end supports. Analytical results show that this layout will not adversely affect the behavior of the bridge, minimizes the impact of design and construction sequences, and avoids the need for outriggers. However, this layout increases rotation at the end supports and is only suitable for narrow bridges, such as one-lane bridges. The third method is to provide eccentrically located non-torsion-resistant bearings (see Fig. 6-39), which creates opposing torsion forces to resist the torsion due to dead and live loads. Figure 6-40 illustrates the torsion distributions[6-27] before and after adjusting the eccentricities of the single bearings of the interior piers for a three-span continuous bridge with span lengths of 28.1 m + 42.0 m + 23.5 m (92.19 ft + 137.80 ft + 77.10 ft) and a radius of 90 m (295.28 ft). The eccentricities of the single bearings in pier 2 and pier 3 are 0.29 m (0.95 ft) and 0.25 m (0.82 ft), respectively. From this figure, it can be seen that this method is very effective in reducing torsion at the torsion-resistant end supports. Analytical results also show that this method will not cause significant changes to the bending moment and shear[6-27 to 6-29].

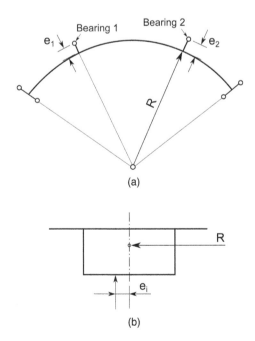

FIGURE 6-39 Eccentric Single Bearing Layout, (a) Plan View, (b) Cross Section.

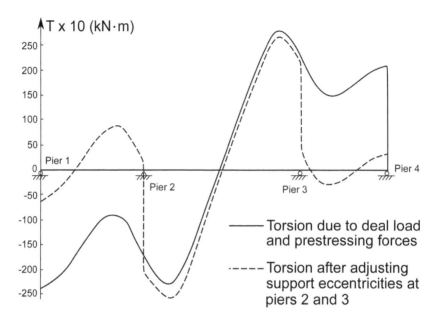

FIGURE 6-40 Torsion Distributions for Symmetrically and Eccentrically Arranged Single Bearings[6-27] (1 kN-m = 0.738 kips-ft).

6.12.3   EQUIVALENT LOADS AND EFFECTS OF CURVED TENDONS

6.12.3.1   Equivalent Loads

Longitudinal tendons in curved segmental bridges are typically curved in both the vertical and the horizontal directions (see Fig. 6-41). There are two common shapes used for the tendon profile in concrete segmental bridges, i.e., parabolic and circular. Parabolic curves are typically used for the vertical tendon deviation in cast-in-place box girder bridges. Circular deviations are typically used

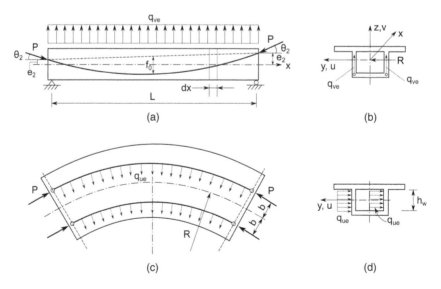

FIGURE 6-41 Equivalent Loadings in Curved Box Girder Bridges, (a) Vertical Equivalent Loadings—Elevation View, (b) Vertical Equivalent Loadings—Side View, (c) Horizontal Equivalent Loadings—Plan View, (d) Horizontal Equivalent Loadings—Side View.

in anchorage blister and at deviation of external tendons[6-30]. From the discussion in Section 3.1.2.2 and assuming the tendon angles at two anchorages as shown in Fig. 6-41a are small, then the vertical equivalent loading of the tendon can be written as

$$\text{Parabolic curve: } q_{ve} = \frac{8f_0}{L^2}P \tag{6-34}$$

$$\text{Circular curve: } q_{ve} = \frac{P}{R} \tag{6-35}$$

The horizontal alignment of a curved segmental bridge is typically along a circular curve, and the tendon equivalent forces in the horizontal direction can be calculated using Eq. 6-35.

6.12.3.2 Effects of Curved Tendons

6.12.3.2.1 Effect of Vertical Curvature

Theoretically, tendons curved in the vertical direction in straight girders do not induce horizontal forces in the girders. However, the vertical tendon force is distributed against the circular wall of the post-tensioning duct (see Fig. 6-42) and the distributed loadings will induce a horizontal force, which is called an out-of-plane force due to the wedging action of the strands. The out-of-plane force may cause web cracks and should be considered in bridge design. AASHTO Article 5.10.4.3.2[6-17] recommends that the out-of-plane force be estimated as

$$F_{v-out} = \frac{P_u}{\pi R_v} \tag{6-36}$$

where
F_{v-out} = out-of-plane force per unit length of tendon due to vertical curvature
P_u = factored tendon force
R_v = radius of curvature of tendon in vertical plane at location being considered

6.12.3.2.2 Effect of Horizontal Curvature

From Fig. 6-41b, it can be observed that the horizontal curvature of a tendon will induce out-of-plane forces. From Eq. 6-35, the-out-of-plane force can be written as

$$F_{u-out} = \frac{P_u}{R_u} \tag{6-37}$$

where
F_{u-out} = out-of-plane force per unit length of tendon due to horizontal curvature (kip/ft)
R_u = radius of curvature of tendon in a horizontal plane at considered location (ft)

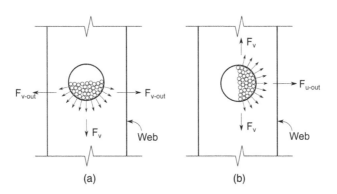

(a)   (b)

FIGURE 6-42   Effects of Curvatures, (a) Vertical Curvature, (b) Horizontal Curvature.

The tendon horizontal curvature will also cause vertical forces (see Fig. 6-42b), which can be neglected in comparison with the vertical forces caused by other loadings.

6.12.3.3 Design Consideration for Out-of-Plane Forces

6.12.3.3.1 Web Regional Bending

Take a free body of a small web segment element[6-30 to 6-32] dx from the curved girder as shown in Fig. 6-43. Assuming that the tendon does not have a vertical profile but remains flat and follows the curvature along the centerline of the web, the web is subjected to a uniform radial force p_t produced by the post-tensioning force and a uniform prestressed circumferential compression force p_c due to the tendon (see Fig. 6-43). The effects of the web due to these forces can be analyzed using shell-plate finite-element methods or other simplified methods in which the web is modeled as a beam supported at the top and bottom slabs as shown in Fig. 6-44. By neglecting the effects of the radial force p_c produced in the compressed concrete web, AASHTO LRFD Article 5.10.4.3.1d[6-17] provides a simplified method to estimate the factored web moment as

$$M_u = \frac{\phi_{cont} F_{out} h_c}{4} \tag{6-38}$$

where

ϕ_{cont} = continuity factor = 0.6 for interior webs and 0.7 for exterior webs
h_c = web length between top and bottom slabs measured along axis of web

$$F_{out} = F_{v-out} + F_{u-out} = \text{out-of-plane force} \tag{6-39}$$

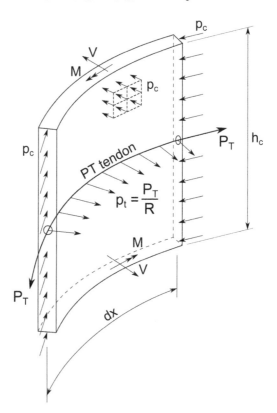

FIGURE 6-43 Free Body of Curve Web Subjected to Post-Tensioning Tendon Force[6-30].

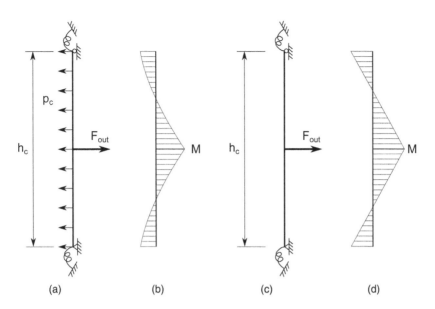

FIGURE 6-44 Simplified Models for Analyzing Web Regional Bending Moment, (a) Simplified Model, (b) Moment Distribution, (c) LRFD Simplified Model, (d) LRFD Simplified Moment Distribution.

The required web vertical reinforcement to resist the moment due to the out-of-plane forces should be added to the reinforcement required for resisting the vertical shear and the transverse moment in the box section.

6.12.3.3.2 Web Local Shear Resistance to Pull-Out

The lateral out-of-plan forces may cause tendons to pull out of the webs toward the center of the girder's curvature if there is not sufficient embedment in the concrete section of the web or enough tie-back reinforcing stirrups in the web. The shear resistance provided by the concrete in the webs on the side of the ducts can be written as[6-17]:

$$V_r = \phi V_n \tag{6-40a}$$

$$V_n = 0.15 d_{eff} \sqrt{f'_{ci}} \tag{6-40b}$$

where

V_n = nominal shear resistance of two shear planes per unit length (kips/in.)

ϕ = resistance factor for shear = 0.75

d_{eff} = half of effective length of failure plane in shear and tension for a curved element (in.) (see Fig. 6-45)

If the clear distance between two adjacent ducts s_{duct} is smaller than the duct diameter d_{duct}, the effective length can be determined based on pull-out path 1, as shown in Fig. 6-45, and can be calculated as

$$d_{eff} = d_c + \frac{d_{duct}}{4}, \text{ for } s_{duct} < d_{duct} \tag{6-41a}$$

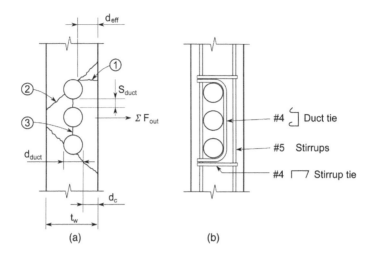

FIGURE 6-45 Effective Length of Shear Failure Planes and Typical Stirrup and Duct Ties, (a) Effective Length of Shear Failure Planes, (b) Typical Arrangement of Stirrup and Duct Ties.

For $s_{duct} \geq d_{duct}$, d_{eff} may be estimated based on the lesser of pull-out paths 2 and 3, as shown in Fig. 6-45, and calculated as

$$d_{eff} = t_w - \frac{d_{duct}}{2} \tag{6-41b}$$

$$\text{or } d_{eff} = d_c + \frac{d_{duct}}{2} + \frac{\Sigma s_{duct}}{2} \tag{6-41c}$$

The definitions of the variables in Eqs. 6-41a to 6-41c can be found in Fig. 6-45. If the factored horizontal out-of-plane force exceeds the factored shear resistance of the concrete determined by Eq. 6-40, fully anchored stirrup and duct ties hooked around the outermost vertical stirrup legs should be provided to resist the out-of-plane forces (see Fig. 6-45b).

6.12.3.3.3 *Cracking of Web Concrete Cover*
When the clear distance between ducts oriented in a vertical column is less than 1.5 in., the cracking resistance should be checked at the ends and at midspan of the idealized unreinforced cover concrete beam[6-17] (see Fig. 6-46). The idealized concrete beam cover can be treated as a beam with two fixed ends as shown in Fig. 6-46b, and the moments at the ends and at midspan can be calculated as:

$$\text{Moment at the ends per unit length: } M_{end} = \frac{\frac{\Sigma F_{u-out}}{h_{ds}} h_{ds}^2}{12} \tag{6-42a}$$

$$\text{Moment at mid-span per unit length: } M_{mid} = \frac{\frac{\Sigma F_{u-out}}{h_{ds}} h_{ds}^2}{24} \tag{6-42b}$$

where
$\quad h_{ds} \quad$ = height of duct stack as shown in Fig. 6-46
$\quad \Sigma F_{u-out}$ = total lateral out-of-plane forces per unit length (see Eq. 6-37)

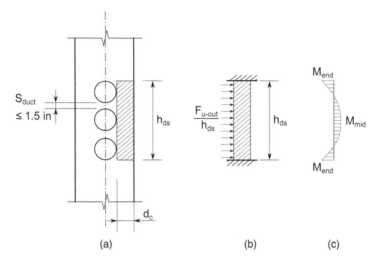

FIGURE 6-46 Analytical Model for Local Bending Moment, (a) Stacked Tendons, (b) Analytical Model, (c) Moment Distribution.

The flexure tensile stresses in the concrete cover is calculated using Eq. 6-42a or Eq. 6-42b, plus the tensile stresses from regional bending of the web as defined in Section 6.12.3.3.1 cannot exceed the cracking stresses per the following equation:

$$f_{cr} = \phi f_r = 0.85 \times 0.24 \sqrt{f_{ci}'} \qquad (6\text{-}43)$$

Otherwise, the ducts shall be restrained by stirrup and duct tie reinforcement.

6.12.4 IN-PLANE DEFORMATIONS OF CURVED BRIDGES AND SUPPORT ARRANGEMENTS

Shortening of the super-structure is caused by temperature, shrinkage, prestressing forces, and creep. The deformations of girders due to temperature and shrinkage are volumetric deformations. The length, width, and depth of the girder are reduced or increased by a given strain. Thus, the radius of curvature is also changed by the same strain, but the span center angles will not be changed. Assuming a statically determinate curved girder as shown in Fig. 6-47a, the radius of the girder due to the effect of temperature and shrinkage can be written as

$$R = R_0 \left(1 - \alpha \Delta t - \varepsilon_{sh} \right) \qquad (6\text{-}44)$$

where
R = changed radius due to temperature and shrinkage
R_0 = original girder radius
α = coefficient of temperature
Δt = drop or raise in temperature, drop temperature as positive
ε_{sh} = shrinkage strain, positive

The displacement at the end support can be written as

$$\Delta_{ts3} = 2 \left(R_0 - R \right) \sin \frac{\theta_0}{2} \qquad (6\text{-}45)$$

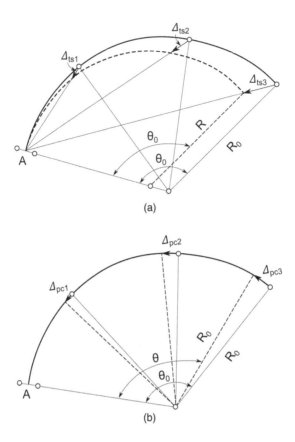

FIGURE 6-47 In-Plane Displacements in Curved Girders, (a) Displacements Due to Temperature and Shrinkage, (b) Displacements Due to Prestressing Forces and Creep.

where

Δ_{ts3} = displacement at support 3 toward fixed support A

θ_0 = initial girder center angle

The axial component of the prestressing force induces a purely axial strain if we neglect the effect of Poisson's ratio. The creep strain is induced by the sustained strain due to the prestressing and will also be purely axial. Thus the girder contracts and shortens along the girder axial direction. The girder radius does not change, but the girder center angle is changed (see Fig. 6-47b). The modified girder center angle due to prestressing force and creep can be written as

$$\theta = \theta_0 \left(1 - \varepsilon_{pc}\right) \tag{6-46}$$

where ε_{pc} = axial strain due to prestress and creep.

The axial displacement at the end support due to prestress and creep can be written as

$$\Delta_{pc3} = R_0 \left(\theta_0 - \theta\right) \tag{6-47}$$

where Δ_{pc3} = axial displacement at support 3.

Temperature change and concrete shrinkage in curved girders will not only induce a displacement along the girder axial direction, but will also induce a displacement along the curved girder radial direction. The volumetric characteristics of temperature and shrinkage deformation must be considered in the layout of the bearings. It is common practice to use guided bearings at expansion

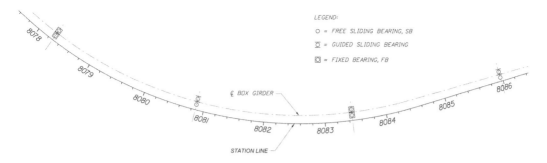

FIGURE 6-48 Typical Bearing Layout.

joints to constrain displacements such that they are parallel with the original alignment of the curved girder. This ensures that the expansion joints will perform properly for the service life of the bridge. The effects of the restraints should be considered in the design. If guided bearings are used at piers between the expansion joints, the guides can be set tangent to the net direction of the total displacements and should be taken into account when orienting the expansion joints[6-31]. To reduce in-plane torque due to temperature change and shrinkage, typically only the inside bearing is guided and the outside bearing is free sliding. Figure 6-48 illustrates a typical bearing layout for a curved segmental bridge.

6.12.5 SEGMENT TREATMENT OF HORIZONTAL CURVATURE

For construction purposes, the bulkhead face is typically perpendicular to the segment chord line (see Fig. 6-49b). The curvature of the girder is attained by rotating the match-cast face (see Fig. 6-49).

6.13 SUMMARIZATION OF DESIGN PROCEDURES FOR BALANCED CANTILEVER BRIDGES

The general design procedures for balanced cantilever bridges are similar to those for span-by-span bridges as discussed in Section 5.5. There are minor differences in the longitudinal design and analysis of balanced cantilever bridges. The design and analysis procedure is summarized below:

Step 5: Bridge longitudinal design
a. Assume construction sequences.
b. Estimate the number and sizes of the cantilever tendons required based on the cantilever segments self-weight and construction loading, by assuming the tendon effective stress of about 0.6 times the ultimate stress.
c. Check the strength of the concrete and the tendons at each of construction stages.
d. Check the stability of the pier and the temporary supports.
e. Preliminarily determine the effects due to additional permanent dead loads and live loads on continuous or partially continuous structures based on the assumed construction sequences, and estimate the required continuous tendons.
f. Determine the dead load moment redistribution due to creep and shrinkage, and check the strength of the concrete and the tendons.
g. Determine the post-tensioning moment redistribution due to concrete creep and shrinkage.
h. Determine prestress losses.
i. Determine the effects due to temperature.
j. Check final strength, at completed construction and at 10,000 days.
k. Revise design as necessary until all design requirements are met.

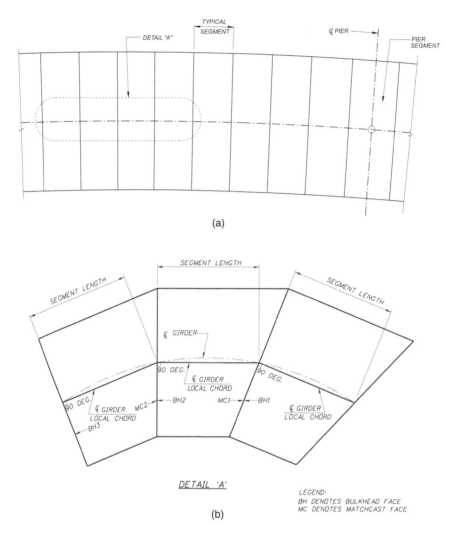

FIGURE 6-49   Treatment of Horizontal Curvature, (a) Partial Plan View, (b) Detail *A* View.

6.14   DESIGN EXAMPLE II—BALANCED CANTILEVER BRIDGE

Take one unit from an existing long bridge owned by FDOT (used with permission by FDOT) as a design example to illustrate the basic design procedures and unique details. The design requirements are the same as those described in Section 5.6.1, except the concrete strength for the precast segments is 8.5 ksi. The bridge transverse design and diaphragm design are similar to those for span-by-span bridges and have been discussed in Section 5.6.6. The readers are encouraged to refer to this section for design procedures.

6.14.1   SPAN ARRANGEMENT AND TYPICAL SECTION

6.14.1.1   Span Arrangement

Based on the minimum horizontal clearance requirement for the traffic below and other site restrictions, the span arrangement is established and is shown in Fig. 6-50. It is apparent that span-by-span bridges are not suitable for the proposed span lengths. The maximum required span length is within the economical span range for balanced cantilever segmental bridges and the ratio of side

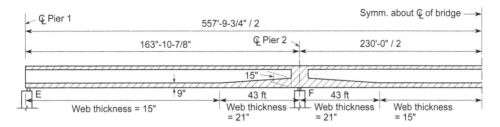

FIGURE 6-50   Span Arrangement of Design Example II.

span length to middle span length is about 0.7. In Fig. 6-50, the capital letters "E" and "F" indicate expansion and fixed joints, respectively. To simplify the segment fabrications, a constant girder depth of 9'-1/2" is used for the entire bridge. To meet the design requirement, variable bottom slab thicknesses are used for this bridge within the first 43 ft on either side of the interior pier supports. The bottom slab thickness varies from 9 to 15 in. as shown in Fig. 6-50.

6.14.1.2   Typical Section and Segment Layout

6.14.1.2.1   Typical Section and Section Properties

The typical web thickness of the segments is 15 in., and the web thickness within the portion with variable bottom slab thickness is designed as 21 in. (see Fig. 6-51). The typical section for the segments with constant bottom slab thicknesses for this bridge is the same as that shown in Fig. 5-39. The typical section for the segments with the variable slab thickness is shown in Fig. 6-51. To allow for core form overlap to the segments with web thickness of 15 in., a blockout of 6 in. by 6 in. is provided in the segments with a web thickness of 21 in. (see Fig. 6-51c).

The precast segments are assumed to be transported to the site by trailers over existing road. Based on the discussion regarding segment length selection in Section 1.4.1, the segment lengths are set in the range 9'-2" to 10'-0". The length of the expansion joint pier segments is 7'-3", and the total length of the pier segments over continuous supports is 10 ft. To reduce the segment weight, the pier segments over the continuous supports are precasted in two 5-ft units. The maximum segment weight is approximately 70 tons. The width of the cast-in-place joints range from 3'-3/8" to 5'-10". There are 18 precast segments in each of the side spans and 25 precast segments in the center span. Each of the cantilevers over pier 2 has 12 precast segments, and each of the cantilevers over pier 3 has 13 precast segments. The detail segment arrangement is shown in Fig. 6-52. The basic dimensions of the deviation and pier segments are similar to those shown in Figs. 5-41 to 5-43.

6.14.2   TENDON LAYOUT

As discussed in Section 6.4 and assuming the effective prestress as 0.6 times the tendon ultimate stress, the required areas of the cantilever tendons for the balanced cantilevers over piers 2 and 3 are estimated to be about 81.59 in.^{2} and 86.80 in.^{2}, respectively. For the cantilevers over pier 2, 8 × 19-ϕ0.6" and 3 × 12-ϕ0.6" tendons per web are used (see Fig. 6-53a and b). For the cantilevers over pier 3, 8 × 19-ϕ0.6" and 4 × 12-ϕ0.6" tendons per web are used (see Fig. 6-54a and b). All the cantilever tendons are internal and designated as C2 to C13 (see Figs. 6-53 to 6-55). Tendons C4, C7, and C11 to C13 are 12-ϕ0.6" tendons, and the remaining cantilever tendons are 19-ϕ0.6" tendons. The cantilever tendons are anchored close to the webs and 11 in. from the top of the deck (see Fig. 6-55b). The detailed locations of the cantilever tendons are shown in Figs. 6-53 to 6-55. The required continuity tendons for the side span are estimated to be about 34.99 in.^{2} and 3 × 27-ϕ0.6" external tendons per web are used (see Figs. 6-53 and 6-55). The required continuity tendons for the center span are estimated about 50.54 in.^{2}, using 3 × 27-ϕ0.6" external and 3 × 12-ϕ0.6" internal tendons per web (see Figs. 6-53 to 6-55). The external continuity tendons are designated as T3, T4,

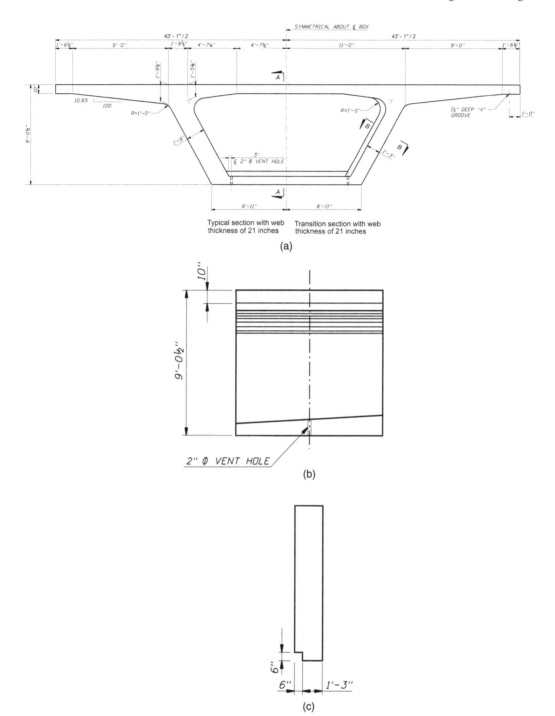

FIGURE 6-51   Dimensions of Segment with Variable Web and Bottom Slab, (a) Cross Section, (b) Section *A-A*
View, (c) Section *B-B*.

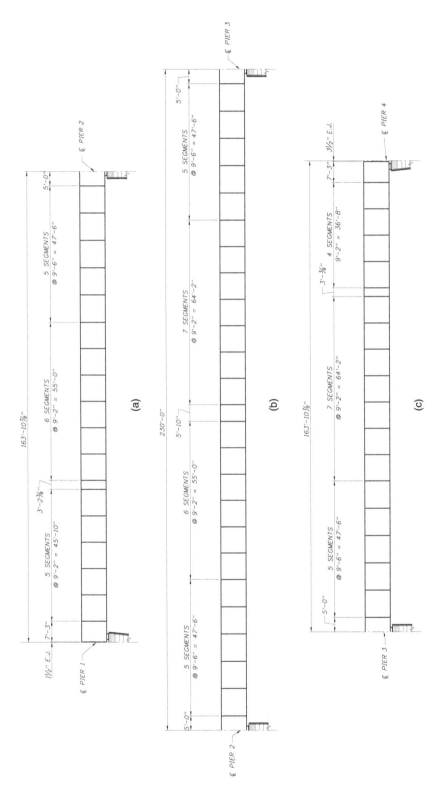

FIGURE 6-52 Segment Layout, (a) First Span, (b) Center Span, (c) Third Span.

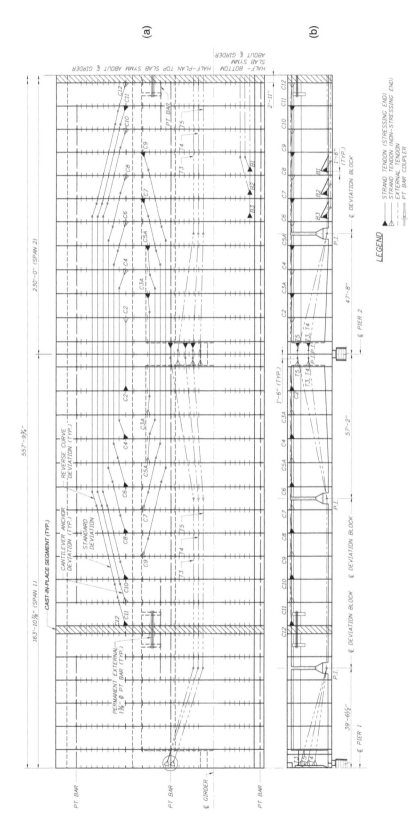

FIGURE 6-53   Tendon Layout for Span 1 and Part of Span 2, (a) Plan View, (b) Elevation View.

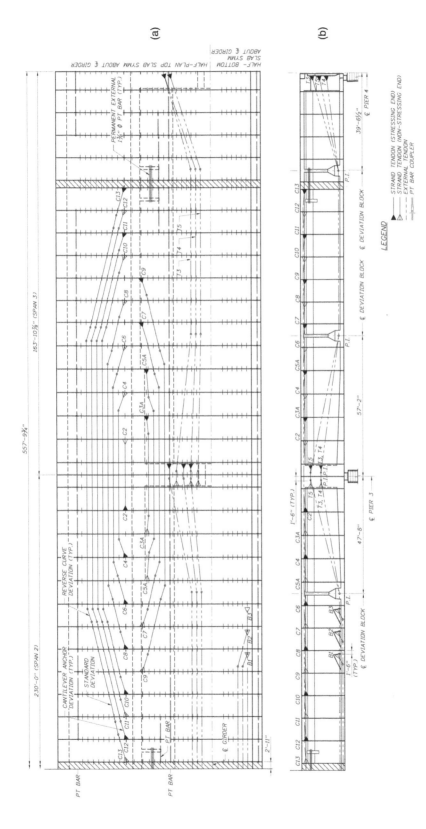

FIGURE 6-54 Tendon Layout for Part of Span 2 and Span 3, (a) Plan View, (b) Elevation View.

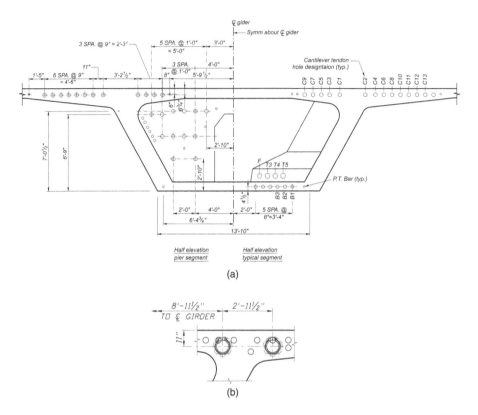

FIGURE 6-55 Tendon Locations in Cross Section, (a) Cross Section, (b) Anchor Locations of Cantilever Tendons.

and T5, respectively. The internal continuity tendons are designated as B1, B2, and B3. respectively. One future external tendon per web is provided and designated as "F" (see Fig. 6-55). The internal continuity tendons are anchored at the bottom corners of the box girder (see Fig. 6-56). Schedule 40 pipes are used for the ducts, and the internal diameters of the ducts are 3.5, 4, and 4.5 in. for 12-ϕ0.6″, 19-ϕ0.6″, and 27-ϕ0.6″ tendons, respectively.

6.14.3 CONSTRUCTION SEQUENCES

The bridge is assumed to be built by the balanced cantilever method, and the construction sequences are described as follows:

Phase I
    a. Construct substructures, abutment, and piers.
    b. Install temporary support at pier 2, and secure it to the pier (see Figs. 6-57a and 6-58).
    c. Place pier segment on top of pier 2.
    d. Erect adjacent segments next to the pier segment. Install and stress the longitudinal erection PT bars.
Phase II
    a. Erect remaining cantilever segments by placing segments on alternating sides of the pier to maintain balance until the cantilever is completed (see Fig. 6-57b). Do not exceed one segment out of balance.
    b. Install and stress cantilever PT tendons after the segments are assembled by means of erection PT bars.

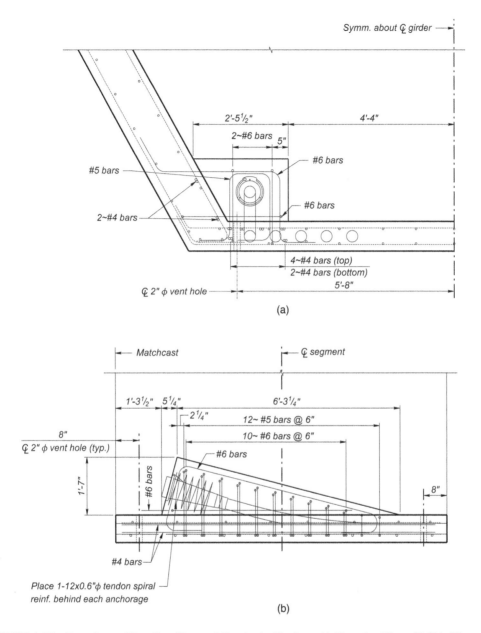

FIGURE 6-56 Location and Details of Internal Continuity Tendons, (a) Elevation View, (b) Side View.

Phase III

    a. Install falsework near abutment 1 and temporary supports at abutment 1 (see Fig. 6-57c).

    b. Place remaining segments in span 1 on the falsework.

    c. Install and stress erection PT bars.

    d. Cast closure segment in span 1.

    e. Install and stress external continuity tendons T3 to T5 and permanent external PT bars in span 1.

    f. Remove falsework near abutment 1.

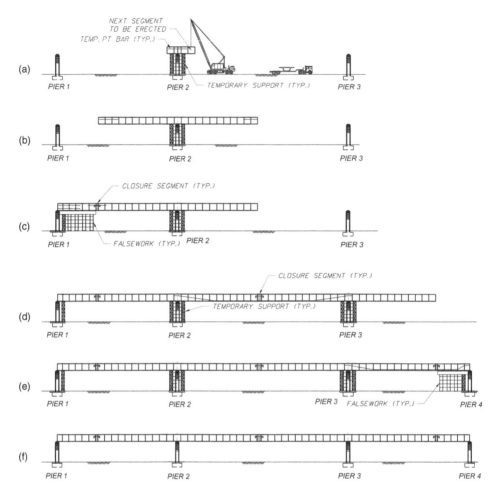

FIGURE 6-57 Construction Sequences, (a) Phase I, (b) Phase II, (c) Phase III, (d) Phase IV, (e) Phase V, (f) Phase VI.

Phase IV
   a.   Erect cantilevers on pier 3 using the same procedures described in phases I and II (see Fig. 6-57d).
   b.   Cast closure segment in Span 2.
   c.   Install and stress continuity tendons T3 to T5, B1 to B3 in span 2, and external PT bars.

Phase V
   a.   Install falsework near abutment 4 and temporary supports at pier 4 (see Fig. 6-57e).
   b.   Place remaining segments in span 3 on the falsework.
   c.   Install and stress PT bars.
   d.   Cast closure segment in span 3.
   e.   Install and stress external continuity tendons T3 to T5 and permanent external PT bars.
   f.   Remove falsework.

Phase VI
   a.   Grout all bearings.
   b.   Install and stress internal continuity tendons in span 3.
   c.   Remove temporary supports (see Fig. 6-57f).

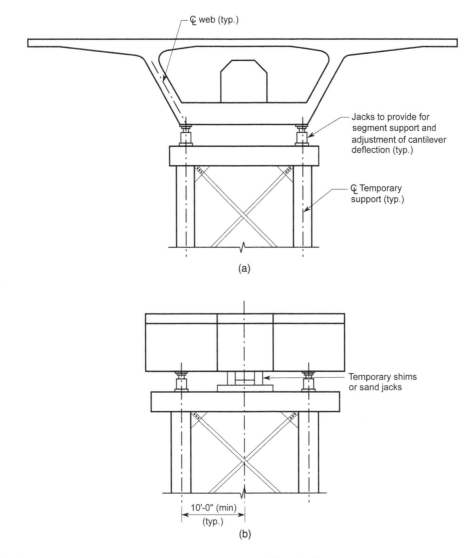

FIGURE 6-58   Sketch of Temporary Tower, (a) Elevation View, (b) Side View.

6.14.4   BRIDGE LONGITUDINAL ANALYSIS AND CAPACITY CHECK

6.14.4.1   Section Properties and Flange Effective Widths

6.14.4.1.1   Section Properties

The section properties for the section with constant bottom slab thickness were presented in Section 5.6.4.3. The section properties at two typical sections are summarized as

A. Section next to the pier segment
   Section area: $A = 93.27$ ft^2
   Moment of inertia about x-axis: $I_z = 959.47$ ft^4
   Area enclosed by the centerlines of box girder plates: $A_0 = 135.42$ ft^2
   Distance from neutral axis to top of the top slab: $y_t = 3.21$ ft
   Distance from neutral axis to bottom of the bottom slab: $y_b = 5.79$ ft
   Static moment of the area above the neutral axis: $Q_z = 135.08$ ft^3

B. Section close to midspan
   Section area: $A = 79.24$ ft^2
   Moment of inertia about x-axis: $I_z = 761.86$ ft^4
   Area enclosed by the centerlines of box girder plates: $A_0 = 135.42$ ft^2
   Distance from neutral axis to top of the top slab: $y_t = 2.74$ ft
   Distance from neutral axis to bottom of the bottom slab: $y_b = 6.26$ ft
   Static moment of the area above the neutral axis: $Q_Z = 106.05$ ft^3

6.14.4.1.2 Flange Effective Widths

The flange effective widths are determined based on the method described in Section 3.4.2.2. The designations for physical flange widths and effective widths are the same as those shown in Fig. 3-24. From Figs. 5-39a and 6-51a, it can be seen that the top flange cantilever length is

$$b_1 = 10.5 \ ft$$

The notional span lengths and ratios of physical flange width to notional span length:
   End Span:

$$l_i = 0.8 \times 163.83 = 131.1 \ ft$$
$$\frac{b_1}{l_i} = 0.080$$

Interior Span:

$$l_i = 0.6 \times 230 = 138.0 \ ft$$
$$\frac{b_1}{l_i} = 0.076$$

Cantilever (maximum cantilever length during construction):

$$l_i = 1.5 \times 107.5 = 161.25 \ ft$$
$$\frac{b_1}{l_i} = 0.065$$

Based on the ratios of physical flange width to notional span length and from Fig. 3-25 or Eq. 3-100, we can obtain the effective flange width coefficients:
   End Span:

$$b_{1s} = 0.80, \ b_{1m} = 1.0$$

Interior Span:

$$b_{1s} = 0.81, \ b_{1m} = 1.0$$

Cantilever (maximum cantilever length during construction):

$$b_{1s} = 0.85$$

Thus, the top flange cantilever effective length in the middle portion of the span is the same as its physical length without the reduction.

The effective widths of the top flange cantilever over supports are:
End Span:

$$b_{1se} = 0.80 \times b_1 = 0.80 \times 10.5 = 8.4 \, ft$$

Interior Span:

$$b_{1se} = 0.81 \times b_1 = 0.81 \times 10.5 = 8.5 \, ft$$

Cantilever (maximum cantilever length during construction):

$$b_{1se} = 0.85 \times b_1 = 0.85 \times 10.5 = 8.9 \, ft$$

Using the same method, we can obtain the effective widths of the top and bottom flanges. The analysis shows that the effective widths for the top and bottom flanges of the box in the middle portion of the span are the same as their physical widths without the reduction. The effective widths of the box top flange over the supports are
End Span:

$$b_{2se} = 0.85 \times b_2 = 0.85 \times 9.42 = 8.0 \, ft$$

Interior Span:

$$b_{2se} = 0.84 \times b_2 = 0.84 \times 9.42 = 7.9 \, ft$$

Cantilever (maximum cantilever length during construction):

$$b_{2se} = 0.88 \times b_2 = 0.88 \times 9.42 = 8.3 \, ft$$

Transition Length of Effective Width:

$$a = 10.5 \, ft \; (< 0.25 \, l_i)$$

The effective widths are typically used to calculate the moment of inertia of the section and the location of the neutral axis. In calculating the uniform normal stress (P/A) caused by post-tensioning forces, it is typical to use the full cross-sectional area[6-14] so as not to overestimate the "P/A" component of the post-tensioning stress. Recent research[6-26] shows that the shear lag effect caused by post-tensioning forces is very small.

6.14.4.2   General Analytical Model and Assumptions

The bridge is modeled as a three-dimensional structure consisting of three-dimensional beam elements and analyzed using the finite-element method. The entire longitudinal analysis of the bridge is performed using the LARSA computer program. The details of the bridge modeling were discussed in Chapter 5. In the analysis, the following assumptions are made:

   a. All superstructure segments are to be erected 28 days after they are cast.
   b. The time for erecting one segment and two balancing segments is one day.
   c. The span closure segment is cast 10 days after the last cantilever segment is erected.
   d. All horizontal bearing springs are removed before stressing tendons to allow for girder elastic deformation. After the tendon has been stressed, the horizontal bearing springs are replaced back to the models.
   e. The connections between the super-structure and the intermediate piers are modeled as fixed to simulate the temporary supports during construction.

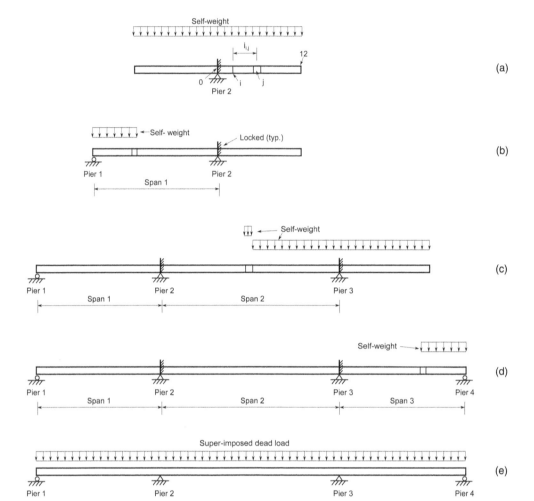

FIGURE 6-59 Sketch of the Analytical Stages, (a) Free Cantilever, (b) One-Span Cantilever, (c) Two-Span Continuous Cantilever, (d) Three-Span Continuous Girder, (e) Three-Span Continuous Girder without Temporary Supports.

The structural analysis is carried out based on the proposed construction sequences. The total effects of the design loads are the sum of the effects determined for each of the following construction stages and their redistributions due to creep and shrinkage:

Stage I: Free balanced cantilever structure with fixed connection at the top of pier 2 (see Fig. 6-59a).

Stage II: One-span cantilever structure with fixed connection at the top of pier 2 (see Fig. 6-59b).

Stage III: Two-span continuous cantilever structure with fixed connections at the top of piers 2 and 3 (see Fig. 6-59c).

Stage IV: Three-span continuous beam structure with fixed connections at the top of piers 2 and 3 (see Fig. 6-59d).

Stage V: Three-span continuous beam structure and release locked joints at piers 2 and 3 (see Fig. 6-59e).

6.14.4.3   Cantilever System Analysis

At this construction stage, the free cantilever structure is subjected to two types of permanent loads: self-weights of the segments and the cantilever tendons. Their effects for the cantilever over pier 2 are calculated as follows:

Step 1: Calculate the moment due to the segment's self-weights as

$$M_{di} = -\sum_{j=i+1}^{12} m_{i,j} = -\sum_{j=i+1}^{12} l_{i,j} w_j \ \left(\text{negative}\right)$$

where
i	=	section location (0-11, see Fig. 6-59a)
j	=	segment number (1-12)
w_j	=	self-weight of segment j
l_{ij}	=	distance between centroid of segment j and section i
$m_{i,j}$	=	moment at section i due to self-weight of segment j

For example, the moment at section 0 due to the cantilever tip segment is

$$m_{0,12} = -l_{0,12} \times w_{12} = -102.92 \times 108.95 = -11213 \ kip\text{-}ft$$

The total moment due to the segment self-weights are

$$M_{d0} = \sum_{j=1}^{12} m_{i,j} = -75484 \ kip\text{-}ft$$

The moment distribution due to self-weight is illustrated in Fig. 6-60a.

Step 2: Determine the moment due to the cantilever tendons.
As aforementioned, there are a total of 22 tendons over pier 2, two for each pair of balanced segments. The moment induced by the post-tensioning tendons can be calculated as

$$M_{pi} = f_{ps} A_{ps} \times e_{pi}$$

where
M_{pi}	=	moment at section i due to post-tensioning tendons
f_{ps}	=	effective stress of tendons
A_{ps}	=	area of tendons
e_{pi}	=	distance between centroid of tendons and neutral axis of section i (see Figs. 6-54 and 6-55)

For example, the moment at section 0 is

$$M_{p0} = f_{ps} A_{ps} \times e_{p0} = 191.91 \times 81.59 \times (3.21 \times 12 - 6) = 42433 \ ft\text{-}kips$$

The moment distribution due to the cantilever tendons is shown in Fig. 6-60b.

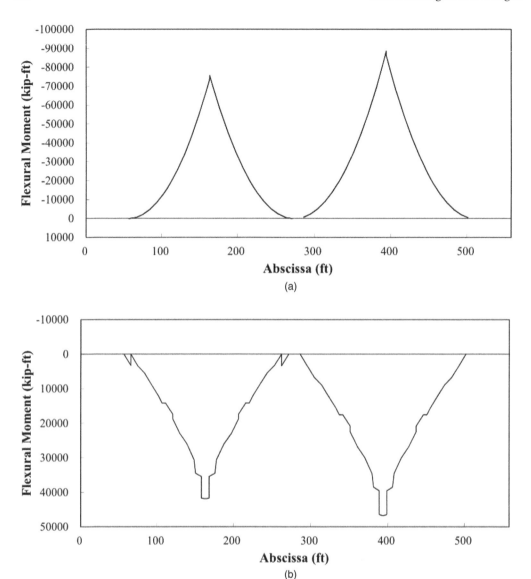

FIGURE 6-60 Moment Distributions in Free Cantilever Systems Due to Segment Self-Weights and Cantilever Tendons, (a) Moment Due to Self-Weight, (b) Moment Due to Cantilever Tendons.

6.14.4.4 One-Span Cantilever System

At this stage, the previous cantilever system has two types of additional loadings, the self-weight of the segments placed on the falsework plus the post-tensioning forces due to external tendons T3 to T5. The additional self-weights and tendon forces are assumed to be applied in the girder with one end simply supported and another rigidly connected with the previously constructed cantilever. The moment distributions due to the additional loads can be determined by any method discussed in Section 4.2 and are shown in Fig. 6-61a. Adding the additional moments shown in Fig. 6-61a to the previously determined cantilever moment shown in Fig. 6-60a and b, we can obtain the total moments in the one-span cantilever girder. The summation of the moments due to the segments self-weight and post-tendons are shown in Fig. 6-61b. In Fig. 6-61, the solid lines represent the moment distribution due to self-weight and the dotted lines represent those due to post-tendons.

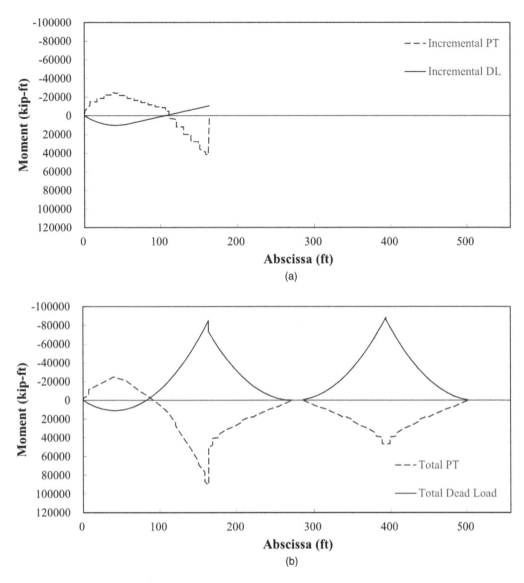

FIGURE 6-61 Moment Distributions in One-Span Cantilever System Due to Segment Self-Weights and Tendons, (a) Moment Distributions Due to Additional Self-Weight and Tendons T3 to T5 in Span 1, (b) Total Moment Distributions Due to Self-Weight and Tendons.

6.14.4.5 Two-Span Continuous Cantilever

The free cantilever over pier 3 is built first and analyzed in the same way as described in Section 6.14.4.3. The self-weight of the midspan closure segment and six internal continuity tendons B1 to B3 are applied to the two-span continuous girder (see Fig. 6-59c). The following procedure can be used for determining the moment distribution:

a. Determine the moment in the cantilever structure over pier 3 due to self-weight and cantilever tendons using the procedures described in Section 6.14.4.3 (see Fig. 6-60 or 6-61b).

b. Determine the moment due to the self-weight of the closure segment and the effects of internal continuity tendons T3 to T5 and B1 to B3 on the two-span continuous cantilever

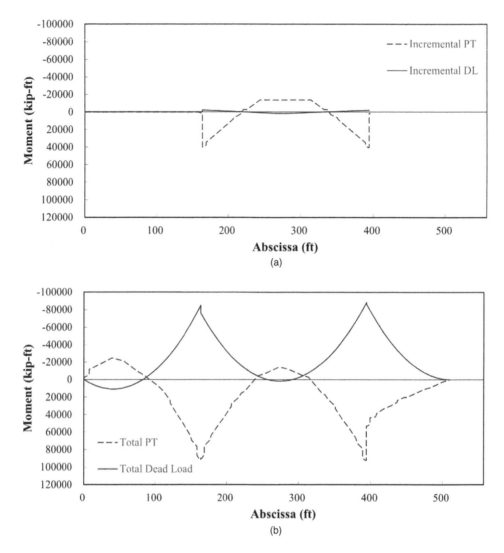

FIGURE 6-62 Moment Distributions in Two-Span Continuous Cantilever Girder Due to Segment Self-Weights and Tendons, (a) Moment Distributions Due to Additional Self-Weights and Continuity Tendons in Span 2, (b) Total Moment Distributions Due to Self-Weights and Tendons.

girder using the methods discussed in Section 4.3. The moment distributions due to the additional self-weight of the segments and the tendons are shown in Fig. 6-62a.

c. The total moment distribution due to the self-weight and the tendons in the two-span continuous cantilever girder can be obtained by adding the moments shown in Fig. 6-62a to the moment distributions shown in Fig. 6-61b and are shown in Fig. 6-62b.

6.14.4.6 Completed Three-Span Continuous Girder

The additional loads applied to the three-span continuous girder with temporary supports at piers 2 and 3 (see Fig. 6-59d) after completing the phase IV girder erection are

a. Self-weights of the remaining segments supported on the falsework near pier 4
b. Six external continuity tendons T3 to T5 applied to span 3

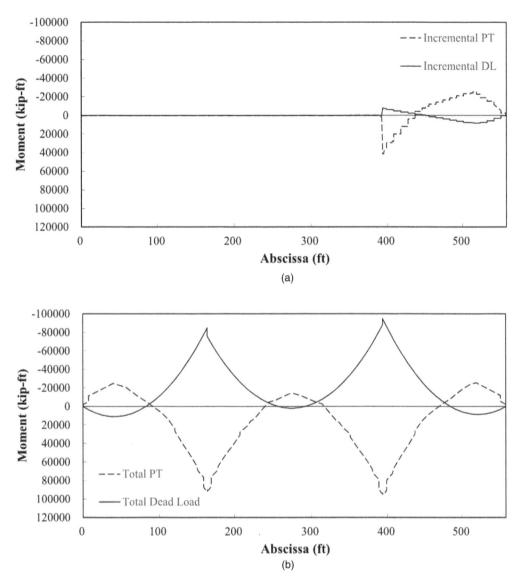

FIGURE 6-63 Moment Distributions in Three-Span Continuous Girder Due to Segment Self-Weights and Tendons, (a) Moment Distributions Due to Additional Self-Weights and External Tendons in Span 3, (b) Total Moment Due to Self-Weights and Tendons.

The moment distributions due to additional self-weights and tendons T3 to T5 in span 3 can be analyzed using any of the methods discussed in Sections 4.2 and 4.3 and are shown in Fig. 6-63a. The distribution of the total moment in phase IV due to self-weight and tendons can be obtained by adding the moments shown in Figs. 6-62b and 6-63a. The total moment distributions in phase V can be determined by releasing the fixed moments at piers 2 and 3 as illustrated in Fig. 6-63b.

6.14.4.7   Final Bridge Analysis

6.14.4.7.1   Effects of Dead Loads

In addition to the box girder self-weights, there is an additional self-weight from the barrier walls that is calculated as 0.84 kips/ft and applied to the final three-span continuous girder (Fig. 6-59e).

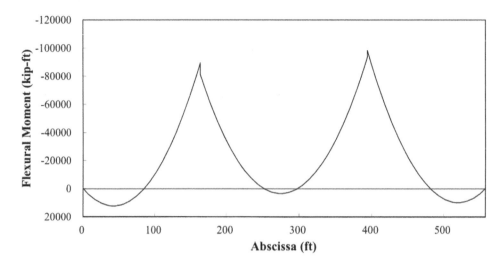

FIGURE 6-64 Moment Envelope of Example Bridge II Due to Dead Loads.

The moment and shear envelopes due to all dead load are determined using the LARSA computer program. The envelopes are illustrated in Figs. 6-64 and 6-65, respectively.

6.14.4.7.2 Effects of Live Loads

The design live loads include a single truck or tandem with a uniform distributed lane loading as discussed in Section 2.2.3. The maximum effects at the related sections are determined based on the procedures discussed in Section 5.6.4.5.

The effects due to live loads are analyzed by finite-element methods and are performed using the LARSA 4-D computer program. The moment, shear, and torsion envelopes are shown in Figs. 6-66 to 6-68, respectively, in which, a dynamic load allowance of 1.33 is included. In these figures, the dotted lines represent the maximum positive responses and the solid lines indicate the maximum negative responses.

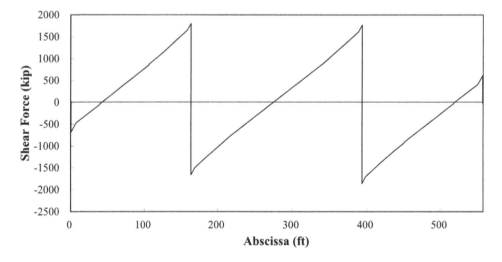

FIGURE 6-65 Shear Envelope of Example Bridge II Due to Dead Loads.

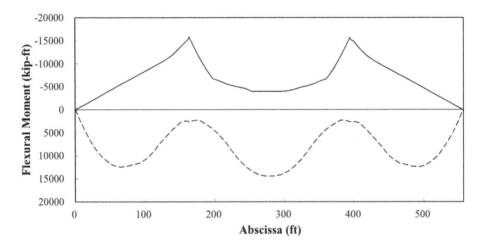

FIGURE 6-66   Moment Envelope of Example Bridge II Due to Live Loads.

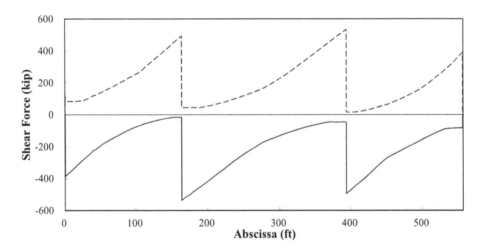

FIGURE 6-67   Shear Envelope of Example Bridge II Due to Live Loads.

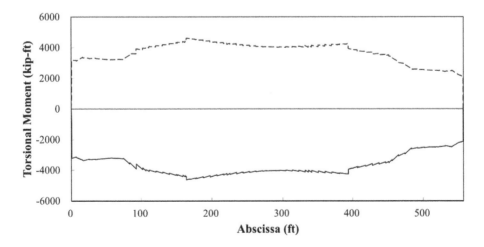

FIGURE 6-68   Torsion Envelope of Example Bridge II Due to Live Loads.

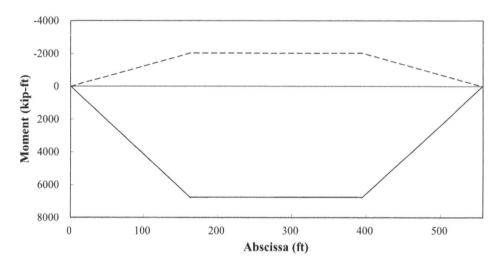

FIGURE 6-69 Moment Distributions of Example Bridge II Due to Gradient Temperatures.

6.14.4.7.3 Effects of Temperature

The mean temperature is assumed to be 70° F, and the rise and fall in temperature is assumed to be 35° F. The temperature gradient specified in Section 2.2.10.2 is used for analyzing the effects of temperature gradient, and the related values are provided in Section 5.6.1.3c. The effects of the uniform temperature change are found to be comparatively small. The moment and shear distributions due to the temperature gradient are illustrated in Figs. 6-69 and 6-70. In these figures, the solid and dotted lines represent the responses due to the positive and negative vertical temperature gradients, respectively.

6.14.4.7.4 Effects of Post-Tendons

All the prestress losses are determined based on the methods described in Section 3.2 and the assumptions given in Section 5.6.1.4. The detailed analysis is done with the LARSA computer program. The moment envelopes due to primary, secondary, and total effects are provided in Fig. 6-71. The shear envelope due to effective post-tensioning forces is given in Fig. 6-72.

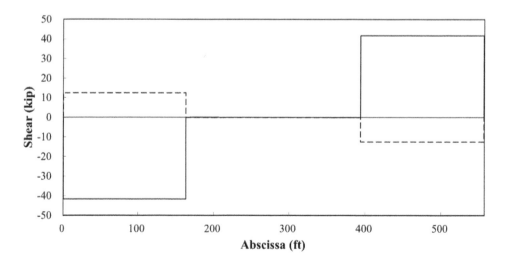

FIGURE 6-70 Shear Distributions of Example Bridge II Due to Gradient Temperatures.

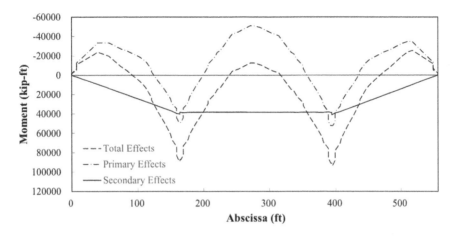

FIGURE 6-71   Moment Envelope of Example Bridge II Due to Post-Tensioning Tendons.

6.14.4.7.5   Effects of Creep and Shrinkage

The creep and shrinkage effects are evaluated based on the CEB-FIP Model Code for Concrete Structures, 1990, i.e., Eqs. 1-9b and 1-15b. Each of the segments is assumed to be erected on the 28th day after it was cast. The creep coefficient and shrinkage strain are obtained in Sections 5.6.4.8.1 and 5.6.4.8.2, respectively. The moment and shear distributions caused by dead and post-tensioning forces due to creep and shrinkage over 10,000 days after the casting date are illustrated in Figs. 6-73 and 6-74, respectively.

6.14.4.7.6   Summary of Effects

The effects of vehicle braking forces, wind loading, etc., are analyzed as described in Section 5.6.4. The maximum effects of the principal loadings at several critical sections are summarized in Table 6-2.

6.14.4.8   Capacity Check

6.14.4.8.1   Service Limit

As mentioned in Section 5.6, for prestressed concrete bridges, two service limit states should be checked: Service I and Service III. For segmentally constructed bridges, the AASHTO specifications require that a special load combination be check as indicated in Eq. 2-51. As the effects of

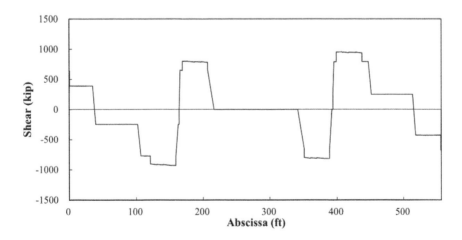

FIGURE 6-72   Shear Envelope of Example Bridge II Due to Post-Tensioning Tendons.

TABLE 6-2
Summary of Maximum Load Effects of Example Bridge II in the Girder Longitudinal Direction at Control Sections

Sections from the Left End Support (ft)	Load Effects	Loadings								
		DC	CR + SH	PS	LL (1 + IM)	BR	WS	WL	TU	TG
53.21	Moment (kips-ft)	11,400	862	−20,500	11,900	202	1,600	44	24	2,190
	Axial force (kips)	—	—	−6,470	—	46	14	2		
277.24	Moment (kips-ft)	3,360	4,640	−12,700	13,820	84	2,925	135	−38	−2,020(6,750)
	Axial force (kips)	—	—	−9540	—	34	36	5		
388.91	Moment (kips-ft)	−85,500	8,500	75,100	−14,100	−234	−4,775	−146	−74	−2,020
	Axial force (kips)	—	—	−20,800	—	46	41	6		
168.48	Torsion (kips-ft)	—	—	—	4,390	—	2,625	304		
	Corresponding Shear (kips)	1,510	35	−648	510	2	99	1		

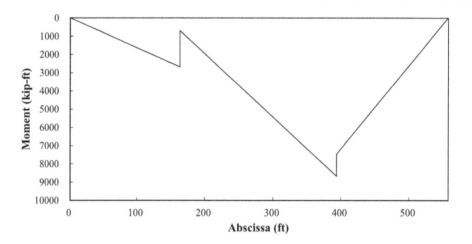

FIGURE 6-73 Moment Distribution of Example Bridge II Due to Creep and Shrinkage Effects (10,000 days).

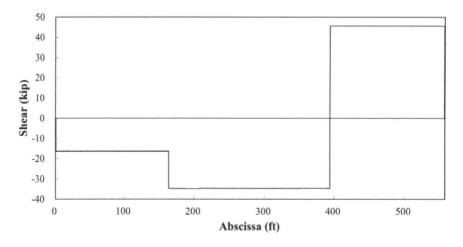

FIGURE 6-74 Shear Distribution of Example Bridge II Due to Creep and Shrinkage Effects (10,000 days).

concrete creep and shrinkage vary with time, the bridge capacity verifications may be performed for different loading stages, such as immediately after the instantaneous prestressing losses, immediately after construction, at 10,000 days, and at the final stage. In this example, the capacity verifications are done at 10,000 days after casting time for the purpose of illustration.

6.14.4.8.1.1 Service I The Service I check is to investigate concrete compression strengths. As discussed in Section 5.6.5.1, the concrete compression strength is typically checked for the following three loading cases:

Case I: The sum of the concrete compression stresses caused by tendon jacking forces and permanent dead loads should be less than $0.6f_c'$, i.e.,

$$f_j^c + f_{DC+DW}^c \le 0.6f_c'$$

Case II: Permanent stresses after losses. The sum of the concrete compression stresses caused by tendon effective forces and permanent dead loads should be less than $0.45f_c'$, i.e.,

$$f_e^c + f_{DC+DW}^c \le 0.45f_c'$$

Case III: The sum of the concrete compression stresses caused by tendon effective forces, permanent dead loads, and transient loadings should be less than $0.60\phi_w f_c'$. The corresponding load combination for the example bridge is

$$DC + CR + SH + PS + LL(1 + IM) + BR + 0.3WS + WL + TU + 0.5TG$$

Consider case III at a cross section located about 389 ft away from the left end support as an example to illustrate the calculation procedures for the verification.

From Eq. 3-269, the top slab slenderness ratio (see Fig. 6-51a) is

$$\lambda_w^t = \frac{X_u^t}{t_t} = \frac{110.25}{10} = 11.03 < 15$$

Thus, the reduction factor $\phi_w^t = 1.0$ and the allowable compressive concrete stress in the top fiber $= 0.60\phi_w^t f_c' = 5.1$ ksi.
The bottom slab slenderness ratio is

$$\lambda_w^b = \frac{X_u^b}{t_b} = \frac{128.0}{15} = 8.53 < 15$$

Thus, the reduction factor: $\phi_w^b = 1.0$ and the allowable compressive concrete stress in the bottom fiber $= 0.60\phi_w^b f_c' = 5.1$ *ksi*.
From Table 6-2, the maximum concrete compressive stresses can be calculated as
Top fiber:

$$f_{top}^c = \frac{(M_{DC} + M_{CR+SH} + M_{PS} + M_{LL+IM} + M_{BR} + 0.3M_{WS} + M_{WL} + M_{TU} + 0.5M_{TG})y_t}{I_{zz}} + \frac{P_{PS}}{A}$$

$$= \frac{(-85500 + 8500 + 75100 - 14100 - 234 - 0.3 \times 4775 - 146 - 0.5 \times 2020 - 74) \times 3.21}{959.47} + \frac{20800}{93.27}$$

$$= 1.113\ (ksi) < 5.1\ \text{ksi}\ (ok!)$$

Bottom fiber:

$$f_{bottom}^{c} = \frac{-(M_{DC}+M_{CR+SH}+M_{PS}+M_{LL+IM}+M_{BR}+0.3M_{WS}+M_{WL}+M_{TU}+0.5M_{TG})y_b}{I_{zz}} + \frac{P_{PS}}{A}$$

$$= \frac{-(-85500+8500+75100-14100-234-0.3\times4775-146-0.5\times2020-74)\times5.79}{959.47}$$

$$+ \frac{20800}{93.27} = 2.341\,(ksi) < 5.1 \text{ ksi (ok!)}$$

For the remaining sections, the verification of Service I can be done in a similar way as shown above. It should be mentioned that the effective section properties discussed in Section 3.4.2.2 should be used based on the AASHTO specifications. As the influence of the effective section properties on the section of concern is small, the shear lag effect is neglected in the analysis above.

6.14.4.8.1.2 Service III   The Service III check is to investigate the maximum tensile stresses in the prestressed concrete components to ensure the maximum tensile stresses are not greater than the limits specified in Sections 1.2.5.1.1.2 and 1.2.5.1.2.2. Three cases should be checked:

Case I: Temporary tensile stresses before losses
Case II: Tensile stresses at the Service III load combination
Case III: Principal tensile stress at the neutral axis of the web

The Service III load combination for the example bridge is

$$DC + CR + SH + PS + 0.8\,LL\,(1+IM) + 0.8\,BR + TU + 0.5TG$$

Consider the section about 389 ft away from the beginning of the bridge to check its maximum tensile stress at the bottom fiber and consider the section at the first interior support to check its principal stress as illustrative design examples.

6.14.4.8.1.2.1 Maximum Tensile Stress Check
From Section 1.2.5.1.2.2, no tension is allowed for the section without the minimum bonded auxiliary reinforcement.

From Table 6-2, the maximum tensile stress at the top fiber can be calculated as

$$f_{top} = \frac{-(M_{DC}+M_{CR+SH}+M_{PS}+0.8M_{LL+IM}+0.8M_{BR}+M_{TU}+0.5M_{TG})y_t}{I_{zz}} - \frac{P_{PS}}{A}$$

$$= \frac{(-85500+8500+75100-0.8\times14100-0.8\times234-0.5\times2020-74)\times3.21}{959.47} + \frac{20800}{93.27}$$

$$= 1.213 \text{ ksi (in compression, ok!)}$$

The maximum tensile stresses for other locations are illustrated in Fig. 6-75. From this figure, we can see that all sections are in compression.

6.14.4.8.1.2.2 Principal Stress Check
For a section, the principal stress check should be considered maximum for one of the following four cases:

• Maximum absolute negative shear with corresponding torsion
• Maximum positive shear with corresponding torsion

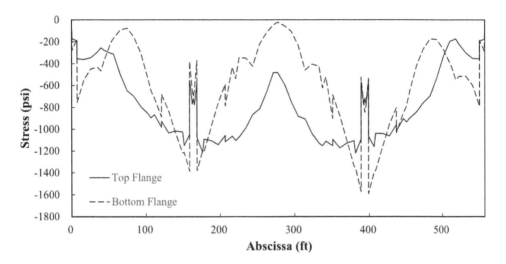

FIGURE 6-75   Variation of Tensile Stresses for Service III of Example Bridge II.

- Maximum absolute negative torsion with corresponding shear
- Maximum positive torsion with corresponding shear

Consider the cross section located about 165 ft away from the left end support as an example to illustrate the calculation procedure for the verification. Analytical results show that the maximum positive torsion with its corresponding shear induces the maximum principal stress for this section. Its principal stress check is shown as follows:

From Section 1.2.5.1.2.2., the allowable principle stress is

$$f_{ta} = 0.11\sqrt{f_c'} = 320 \ psi$$

The equivalent total effective web thickness is

$$b_w = 2 \times 21 = 42.0 \ in$$

The maximum factored torsion from Table 6-2 is

$$T_{max} = 0.8 \times 4390 = 3512 \ kips\text{-}ft$$

The corresponding shear force from Table 6-2 is

$$V_{max} = 1510 + 35 - 648 + 0.8 \times (530 + 2) = 1322.6 \ kips$$

The shear stress in the web at the neutral axis location is

$$f_v = \frac{|V_{max}|Q_z}{2I_{zz}b_w} + \frac{|T_{max}|}{2A_0t_w} = \frac{1322.6 \times 135.08}{2 \times 959.47 \times 42.0 \times 12} + \frac{3512}{2 \times 135.42 \times 21 \times 12} = 0.236 \ ksi$$

The normal stress at the neutral axis location is

$$f_x = \frac{-P_{PS}}{A} = \frac{-20800}{93.27 \times 12 \times 12} = -1.549 \ ksi$$
$$f_y = 0$$

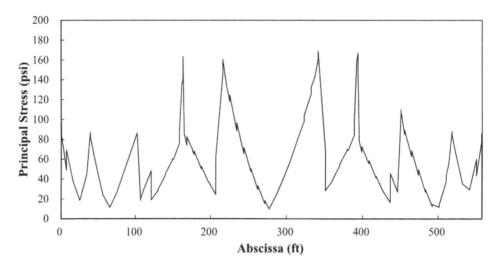

FIGURE 6-76   Variation of Web Principal Stresses for Service III of Example Bridge II.

From Eq. 3-176, the principal tensile stress can be calculated as

$$f_1 = \frac{f_x + f_y}{2} + \sqrt{\left(\frac{f_x - f_y}{2}\right)^2 + f_v^2} = \frac{-1.549}{2} + \sqrt{\left(\frac{1.549}{2}\right)^2 + 0.236^2} = 35 \ psi < f_{ta} \ (\text{ok})$$

The principal stresses for other sections are illustrated in Fig. 6-76. From this figure, it can be seen that all the principal stresses are smaller than the allowable.

6.14.4.8.1.3 Special Load Combination Check   This loading combination may control the design for the locations where the live load effects are small or where they are outside of the pre-compressed tensile zones, such as the tension in the top of the closure pours and compression in the top of box girder over the piers. For this design example, the special load combination as shown in Eq. 2-51 can be written as

$$DC + CR + SH + TG + PS$$

Consider the cross section at the center closure pour about 277 ft away from the left end support as an example to show the analytical procedures.

From Table 6-2, the tensile stress at top fiber is

$$f_{top}^t = \frac{(M_{DC} + M_{CR+SH} + M_{PS} + M_{TG})y_t}{I_z} + \frac{P_{PS}}{A} = \frac{(3360 + 4640 - 12700 - 2020)(-2.74)}{761.86 \times 12 \times 12} + \frac{-9540}{79.24}$$

$$= -668 \ psi \ (\text{in compression, ok!})$$

The tensile stress at bottom fiber is

$$f_{bottom}^t = \frac{(M_{DC} + M_{CR+SH} + M_{PS} + M_{TG})y_b}{I_z} + \frac{P_{PS}}{A} = \frac{(3360 + 4640 - 12700 - 2020)(6.26)}{761.86 \times 12 \times 12} + \frac{-9540}{79.24}$$

$$= -1017 \ psi \ (\text{in compression, ok!})$$

6.14.4.8.2 Strength Limit

Strengths I, II, III, and V and Extreme Events I and II should be evaluated. Take Strength I, for example. The load combination for this example can be written as

$$1.25\ DC + 1.25\ CR + 1.25\ SH + 1.75\ LL\ (1 + IM) + 1.75\ BR + 0.5\ TU$$

6.14.4.8.2.1 Flexure Strength

Consider the section with maximum negative moment that is located about 389 ft away from the left end support pier 1 (see Fig. 6-50) as an example to illustrate the capacity verification process. This section contains both internal and external tendons. Article 5.7.3.1.3 of the AASHTO specifications, as described in Section 3.6.2.3, provides two methods for determining the section flexural capacity. In this example, we neglect the effect of the unbonded tendons on the flexure strength at this section for simplicity.

From Table 2-14, the flexure residence factor for bonded tendons is

$$\phi_f = 0.95$$

Neglecting the unbound continuous tendons, the area of the post-tendons in the top flange is

$$A_{ps} = 0.217(8 \times 19 + 4 \times 12) \times 2 = 86.8\ \text{in.}^2$$

The prestressing stress of the tendons can be determined based on Eq. 3-152a for bounded tendons.

For simplicity, the box section is simplified as an I-section with uniform top and bottom thicknesses. The effective bottom flange width is determined based on the method described in Section 3.4.2.2 and can be taken as the actual flange width:

Width of bottom flange: $b = 174.63$ in.
Thickness of bottom flange: $h_f = 15.0$ in.
Equivalent total web thickness for bending: $b_w = \frac{2 \times 21}{\sin(60.61)} = 48.45$ in.

The distance from the extreme compression fiber to the centroid of the prestressing tendons (see Figs. 1-37 and 6-55):

$$d_p = 9 \times 12 - 6 - 0.75 = 101.25\ \text{in.}$$

From Eq. 3-149e, the stress block factor: $\beta_1 = 0.85 - 0.05(f'_c - 4.0) = 0.625.$
From Eq. 3-152b, $k = 2(1.04 - 0.9) = 0.28.$
Using Eq. 3-165a and assuming the box section as a T-section, the distance from the extreme compression fiber to the neutral axis c can be determined as

$$c = \frac{A_{ps}f_{pu} - 0.85f'_c(b - b_w)h_f}{0.85f'_c\beta_1 b_w + kA_{ps}\dfrac{f_{pu}}{d_p}} = \frac{86.8 \times 270 - 0.85 \times 8.5(174.63 - 48.45) \times 15}{0.85 \times 8.5 \times 0.625 \times 48.45 + 0.28 \times 86.8 \frac{270}{101.25}} = 34.48\ \text{in.}$$

From Eq. 3-149b, the compression block depth is

$$a = \beta_1 c = 0.625 \times 34.48 = 21.55\ \text{in.}$$

From Eq. 3-152a, the average stress in the prestressing steel can be calculated as

$$f_{ps} = f_{pu}\left(1 - k\frac{c}{d_p}\right) = 270\left(1 - 0.28\frac{34.48}{101.25}\right) = 244.22\ \text{ksi}$$

From Eq. 3-164, the factored flexural resistance is

$$\phi_f M_n = \phi_f \left[A_{ps} f_{ps} \left(d_p - \frac{a}{2} \right) + 0.85 f_c' (b - b_w) h_f \left(\frac{a}{2} - \frac{h_f}{2} \right) \right]$$

$$= 0.95 \times \left[86.8 \times 244.44 \left(102 - \frac{21.55}{2} \right) + 0.85 \times 8.5 \times (174.63 - 48.45) \times 15 \times \left(\frac{21.55}{2} - \frac{15}{2} \right) \right] \frac{1}{12}$$

$$= 155315.34 \text{ kips-ft}$$

The magnitude of the factored moment (negative) is

$$M_u = 1.25 DC + 1.25 \, (CR + SH) + 1.75 \, LL \, (1 + IM) + 1.75 \, BR + 0.5 TU$$

$$= 1.25 \times (85500 - 8500) + 1.75 \times 14100 + 1.75 \times 234 + 0.5 \times 74 = 121371.5 \text{ kips-ft} \, \left(< \phi_f M_n, \text{ ok!} \right)$$

Check the minimum reinforcement:
   The section modulus is

$$S_t = \frac{959.47}{3.21} 12^3 = 516499.74 \text{ in.}^3$$

From Section 1.2.2.3, the modulus of fracture is

$$f_r = 0.20 \sqrt{f_c'} = 0.20 \sqrt{8.5} = 0.583 \text{ ksi}$$

The moment M_{pe} and axial force P_e due to effective prestressing force of the bonded tendons are

$$M_{pe} = 45137.74 \text{ kips-ft} \qquad P_e = 14061.6 \text{ kips}$$

The compressive stress due to effective prestressing forces at the bottom fiber is

$$f_{cpe} = \frac{M_{pe} \times y_t}{I_{zz}} + \frac{P_e}{A} = \frac{45137.74 \times 3.21}{959.47 \times 12^2} + \frac{14061.6}{93.27 \times 12^2} = 2.10 \text{ ksi}$$

From Eq. 3-168, the section crack moment is

$$M_{cr} = \gamma_3 \left(\gamma_1 f_r + \gamma_2 f_{cpe} \right) S_b = 0.67 (1.2 \times 0.583 + 1.1 \times 2.10) \frac{516499.94}{12} = 156547.44 \text{ kips-ft}$$

$$1.33 M_{cr} = 208208.1 \text{ kips-ft} > \phi_f M_n$$

Using Eq. 3-163 and considering the effect of unbounded tendons, the factored flexure resistance can be estimated as

$$M_r = \phi_f M_n = 217251.8 \text{ kips-ft}$$

$$M_r > 1.33 M_u \, (\text{ok})$$

   The flexure strength checks for other locations can be performed using the procedure described above. Figure 6-77 illustrates the variation of factored moments along the bridge length. From the figure, we can see that the flexure strengths of all sections are adequate.

6.14.4.8.2.2   Shear and Torsion
6.14.4.8.2.2.1   Factored Shear and Torsion
The shear strength check is performed by following the flow chart shown in Fig. 3-54. Consider a section about 168 ft away from the left end support as an example.

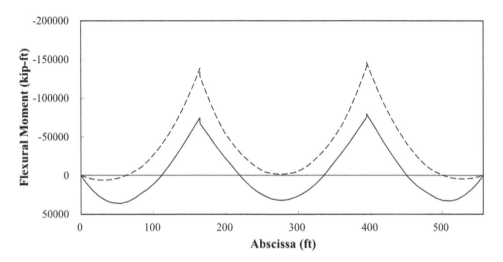

FIGURE 6-77   Variation of Factored Moments for Strength I of Example Bridge II.

Factored shear per web:

$$V_u = 1.25DC + 1.25\ (CR + SH) + PS + 1.75\ LL\ (1 + IM) + 1.75\ BR + 0.5TU$$

$$= \frac{1.25 \times 1510 + 1.25 \times 35 - 648 + 1.75 \times 510 + 1.75 \times 2}{2} = 1089.63 \text{ kips per web}$$

Factored torsion:

$$T_u = 1.25DC + 1.25\ (CR + SH) + PS + 1.75\ LL\ (1 + IM) + 1.75\ BR + 0.5TU$$

$$= 1.75 \times 4390 = 7682.5 \text{ kips-ft}$$

6.14.4.8.2.2.2   Required Torsion Reinforcement

Unfactored compressive stress in concrete after loses at the centroid:

$$f_{pc} = \frac{P_e}{A_g} = 1.54 \text{ ksi}$$

From Eq. 3-216c, coefficient K is

$$K = \sqrt{1 + \frac{f_{pc}}{0.0632\sqrt{f_c'}}} = 3.06,\ K = 2$$

The effective width of the shear flow path is the minimum thickness of the top and bottom flange and the web: $b_e = 10.0$ in.

From Eq. 3-216b, the torsion cracking moment is

$$T_{cr} = 0.0632K\sqrt{f_c'}2A_0 b_e = 119770.7 \text{ kips-ft}$$

From Section 2.3.3.5, the resistance factor for shear and torsion is $\phi_v = 0.9$.

The minimum required nominal torsion resistance is

$$T_{n\_req} = \frac{T_u}{\phi_v} = 8536.1 \text{ kips-ft}$$

$$T_u > \frac{1}{3}\phi_v T_{cr}, \text{ torsion effect should be considered (Section 3.6.3.4.)}$$

The required torsion reinforcement per inch is

$$A_{s\_t} = \frac{T_{n\_req}}{2A_0 f_y} = 0.044 \text{ in.}^2/\text{in.}$$

The required additional longitudinal reinforcement can be calculated by Eq. 3-218. As the requirement is not critical for segmental bridges, the calculations are omitted in this example.

6.14.4.8.2.2.3   Required Vertical Shear Reinforcement
The effective web width for vertical shear is $b_v = t_w = 21$ in.
   Depth d_0 is calculated by Eq. 3-220: $d_0 = 0.8h = 86.4$ in.
   The concrete nominal shear strength (Eq. 3-219b) is

$$V_c = 0.0632 K \sqrt{f_c'} b_v d_0 = 668.64 \text{ kips (per web)}$$

Thus, $V_u > 0.5 f_v V_c$. The transverse shear reinforcement is required (Section 3.6.3.4.2). The required minimum shear reinforcement per web per inch (Eq. 3-219c) is

$$A_{s\_v} = \frac{\frac{V_u}{\phi_v} - V_c}{d_0 f_y} = 0.105 \text{ in.}^2/\text{in.}$$

6.14.4.8.2.2.4   Required Transverse Reinforcement
The total required transverse reinforcement per web is

$$A_s = A_{s\_t} + A_{s\_v} = 0.148 \text{ in.}^2/\text{in.}$$

Use two legs #7 rebars at 6-in. spacing per web to resist the shear and torsion (see Fig. 6-78). Conservatively assume that 0.05-in.² rebar area per inch is for resisting the transverse bending moment, though the effect of web transverse bending on shear may be neglected as discussed in Section 5.3.2. Then the web reinforcement for vertical shear provided is

$$A_{s\_provided} = \frac{2 \times 0.60}{6} - 0.05 = 0.15 \text{ in.}^2/\text{in. (ok)}$$

6.14.4.8.2.2.5   Nominal Shear Strength and Cross-Sectional Dimension Check
From Eq. 3-219c, the nominal shear resistance based on the shear reinforcement provided per inch is

$$V_s = A_s f_y d_0 = 777.60 \text{ kips}$$

From Eq. 3-219a, the normal shear strength is

$$V_{n1} = V_c + V_s = 668.64 + 777.60 = 1446.24 \text{ kips}$$

From Eq. 3-220, the normal shear strength is

$$V_{n2} = 0.379 \sqrt{f_c'} b_v d_0 = 2004.85 \text{ kips}$$

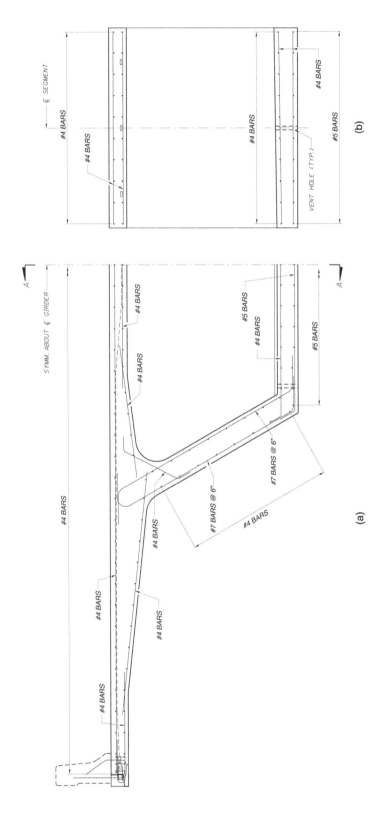

FIGURE 6-78 Reinforcement Arrangements of the Section Close to Interior Pier, (a) Elevation View, (b) Section A-A View.

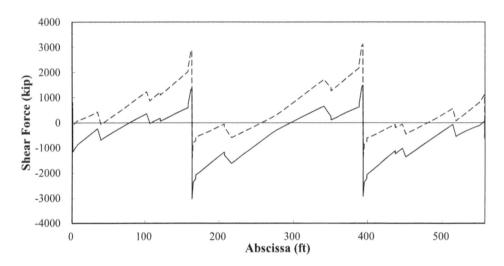

FIGURE 6-79   Variation of Factored Shear for Strength I of Example Bridge II.

Taking the lesser of V_{n1} and V_{n2} yields, the normal shear strength is

$$V_n = 1446.24 \text{ kips}$$

The factored shear is

$$V_u + \frac{T_u}{2A_0} = 1089.63 + \frac{7682.5}{2 \times 135.4 \times 12} = 1091.99 < \phi_v V_n = 0.9 \times 1446.24 = 1301.61 \ kips \ (\text{ok!})$$

Check the cross-sectional dimension:

$$\frac{V_u}{b_v d_v} + \frac{T_u}{2A_0 b_e} = \frac{1089.63}{21 \times 86.4} + \frac{7682.5}{2 \times 135.4 \times 10 \times 12 \times 12} = 0.62 \ ksi < 0.474\sqrt{f_c'} = 1.382 \ ksi \ (\text{ok})$$

The actual used shear reinforcement arrangement for the section close to the interior pier is shown in Fig. 6-78.

The required shear reinforcement for other sections can be determined using the same procedure described above. The variation of factored shear with bridge length for Strength I is shown in Fig. 6-79.

6.14.5   CONSTRUCTION ANALYSIS AND CHECK

6.14.5.1   Capacity Check during Free Cantilever Construction

Though the service and strength limits required by the AASHTO specifications may need to be investigated for each of the construction stages, the longest free cantilever system is the most critical case in this type of bridge construction. The construction check includes the capacity validations for both super-structure and substructure for the loading combinations discussed in Section 6-10. For simplicity, only the strength limit check under maximum force effect is shown below. Figure 6-80 shows a critical loading case for the strength limit check based on the assumed construction sequences. Assuming that segments 1 to 11 have been erected and segment 12 is being erected, check the capacity at section C. The moment capacity at section C should be met the following:

$$\phi M_n \le 1.1 M_{DC} + 1.1 M_{DIFF} + 1.3 M_{CEQ} + 1.0 M_A + 1.0 M_{AI}$$

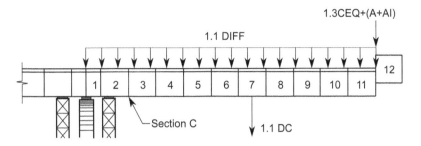

FIGURE 6-80   Analytical Model for Strength Limit Check in Balanced Cantilever Construction.

6.14.5.1.1   Determination of Sectional Nominal Moment Strength M_n

From Section 6.14.4.8.2.1, we have

$$\beta_1 = 0.625 \text{ and } k = 0.28$$

The effective web width for bending capacity is

$$b_w = \frac{2 \times 21}{\sin \frac{60.1^0 \times 2\pi}{360^0}} = 48.45 \text{ in.}$$

The average width of the bottom slab is

$$b = 166 + \frac{13.5}{\tan \sin \frac{60.1^0 \times 2\pi}{360^0}} = 173.76 \text{ in.}$$

The effective depth of tendons is

$$d_p = 108 - 6 - 0.95 = 101.05 \text{ in.}$$

Area of the cantilever tendons at section C for the assumed construction stage (neglecting the effect of permanent PT bars) is

$$A_{ps} = 0.217(16 \times 19 + 4 \times 12) = 76.38 \text{ in.}^2$$

From Eq. 3-165a, the distance from the extreme compression fiber to the neutral axis c can be determined as

$$c = \frac{76.38 \times 270 - 0.85 \times 8.5(173.76 - 48.45)13.5}{0.85 \times 8.5 \times 0.625 \times 48.45 + \frac{0.28 \times 76.38 \times 270}{101.05}} = 30.45 \text{ in.}$$

From Eq. 3-149b, the compression block depth is

$$a = \beta_1 c = 0.625 \times 30.45 = 19.03 \text{ in.}$$

From Fig. 3-152a, we can find the tendon prestressing at failure as

$$f_{ps} = f_{pu}\left(1 - k\frac{c}{d_p}\right) = 270\left(1 - 0.28\frac{30.45}{101.05}\right) = 247.22 \text{ ksi}$$

From Eq. 3-164, the nominal flexural resistance can be approximately calculated as

$$M_n = A_{ps}f_{ps}\left(d_p - \frac{a}{2}\right) = 76.38 \times 247.22 \times \left(101.05 - \frac{19.03}{2}\right) = 144043 \ kip\text{-}ft$$

6.14.5.1.2 *Determination of Factored Moment*

Using the method described in Section 6.13.4.3, the moment at section C due to dead load can be calculated as

$$M_{DC} = 44240 \ kip\text{-}ft$$

The moment at section C due to differential load is

$$M_{DIFF} = 0.02 \ M_{DC} = 0.02 \times 44240 = 884.8 \ kip\text{-}ft$$

The moment at section C due to segment 12 is

$$M_A = 9633.9 \ kip\text{-}ft$$

The moment at section C due to the accidental release of segment 12 is

$$M_{AI} = 1.0 \times M_A = 9633.9 \ kip\text{-}ft$$

There is no special construction equipment assumed, thus

$$M_{CE} = 0$$

The factored moment at section C is

$$M_u = 1.1M_{DC} + 1.1M_{DIFF} + M_A + M_{AI} = 68905.1 \ kip\text{-}ft$$

From Table 2-14, the residence factor for bounded tendons is

$$\phi_f = 0.95$$

The factored resistance is

$$\phi_f M_n = 136841.3 \ kip\text{-}ft > M_u$$

6.14.5.2 Design of Temporary PT Bars

As discussed in Section 5.2.3.5, two loading cases should be evaluated in designing post-tensioning bars. Loading case I is dead load and PT bars, and loading case II is dead load, PT bars, and cantilever tendons. Consider section C as indicated in Fig. 6-81 as an example to show the general design procedures.

6.14.5.2.1 *Check Compression Stress Caused by Dead Load and PT Bars*

Based on the information presented in Sections 6.14.1.2.1 and 6.14.4.1, the section properties of section C can be obtained as:

Physical section area: $A_c = 92.87 \ ft^2$
Effective area: $A_{ce} = 88.52 \ ft^2$

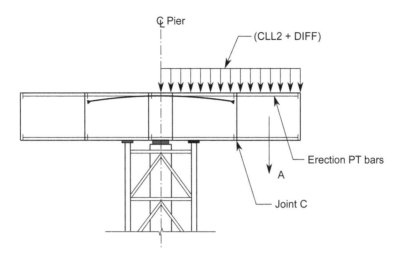

FIGURE 6-81 Analytical Model for Erection PT Bar Design.

Effective moment of inertia: $I_e = 898.76 \ ft^4$
Neutral axis location from top: $y_t = 3.31 \ ft$
Neutral axis location from bottom: $y_b = 5.69 \ ft$
Segment weight: $A = 131.39$ kips
Differential load: $DIFF = 0.02A = 2.63$ kips
Distributed construction load: $CLL = 0.01 \times 43 = 0.43 \ kips/ft$
Moment due to external loads: $M = -(A + DIFF)\frac{9.5}{2} - CLL\frac{9.5^2}{2} = -655.99 \ kip\text{-}ft$
Try using 1-3/8″ diameter Grade 150 Deformed PT bar, 6 in the top flange and 2 in the bottom flange.
Assume the maximum jacking stress as $0.75f_{pu}$, i.e.,
Jacking stress $f_{pj} = 0.75f_{pu} = 0.75 \times 150 = 112.5 \ ksi$
From Eq. 3-49a, the loss due to friction is

$$\Delta f_{pF} = f_{pj}\left(1 - e^{-(0.0002 \times 9.5 + 0)}\right) = 0.214 \ ksi$$

The loss due to an anchor set is

$$\Delta f_{pS} = E_s \times \varepsilon_{set} = 30000 \times \frac{\frac{1}{16}}{9.5 \times 12} = 16.45 \ ksi$$

The effective stress immediately after the setting is

$$f_{pe} = f_{pj} - \Delta f_{pF} - \Delta f_{pS} = 95.84 \ ksi$$

From Table 1-5, the area per bar is

$$A_{ps} = 1.58 \ in^2$$

The total top prestressing force is

$$F_{tps} = 6 \times A_{ps} \times f_{pe} = 908.56 \ kips$$

The total bottom prestressing force is

$$F_{bps} = 2 \times A_{ps} \times f_{pe} = 302.85 \ kips$$

From Figs. 6-54 and 6-55, we have
The distance between the centroid of the top bars and the neutral axis is $e_t = 33.72$ in.
The distance between the centroid of the bottom bars and the neutral axis is $e_b = 63.78$ in.
The normal concrete stress at the top fiber is

$$f_{top} = \frac{\left(F_{tps} + F_{bps}\right)}{A_c} + \frac{\left(F_{tps}e_t - F_{bps}e_b\right)y_t}{I_e} - \frac{My_t}{I_e} = 0.0979 \ ksi > 0.03 \ ksi \ (ok!)$$

The normal concrete stress at the bottom fiber is

$$f_{bottom} = \frac{\left(F_{tps} + F_{bps}\right)}{A_c} + \frac{-\left(F_{tps}e_t - F_{bps}e_b\right)y_b}{I_e} + \frac{My_b}{I_e} = 0.0779 \ ksi > 0.03 \ ksi \ (ok!)$$

The average normal stress is

$$f_{sa} = \frac{f_{top} + f_{bottom}}{2} = 0.0879 \ ksi > 0.04 \ ksi \ (ok!)$$

6.14.5.2.2   Check Compression Stress Caused by Dead Load, PT Bars, and Tendons

From Fig. 6-53, it can be seen that there are two tendons designated as C3 passing through section C and anchored in segment 3.
The total tendon area is

$$A_{ps} = 2 \times 19 \times 0.217 = 8.246 \ in^2$$

The length of tendon C3 is

$$L_{c3} = 4 \times 9.5 + 10 = 48 \ ft$$

From Section 5.6.1.1, the maximum jacking stress is

$$f_{pj} = 0.80 \times 270 = 216.0 \ ksi$$

The loss of friction is

$$\Delta f_{pF} = f_{pj}\left(1 - e^{-(0.0002 L_{c3} + 0)}\right) = 1.93 \ ksi$$

The loss of anchor set is

$$\Delta f_{ps} = E_{ps}\varepsilon_{set} = 30000 \times \frac{3}{8 \times 12 \times L_{c3}} = 19.53 \ ksi$$

The tendon effective stress immediately after set is

$$f_{pe} = f_{pj} - \Delta f_{pF} - \Delta f_{ps} = 194.54 \ ksi$$

The total effective prestressing force is

$$F_{pe} = f_{pe} \times A_{ps} = 1604.2 \ kips$$

From Figs. 6-53 to 6-55, we have
The distance between the centroid of the tendons and the neutral axis is $e_{ps} = 32.97$ in.

The moment due to tendons is

$$M_{ps} = e_{ps} \times F_{pe} = 4407.5 \ kip\text{-}ft$$

The stress at the top fiber due to tendons is

$$f_{top} = \frac{F_{pe}}{A_c} + \frac{M_{ps} y_t}{I_e} = 0.223 \ ksi \ (compression)$$

The stress at the bottom fiber due to tendons is

$$f_{bottom} = \frac{F_{pe}}{A_c} - \frac{M_{ps} y_b}{I_e} = -0.0738 \ ksi \ (tension)$$

The total concrete stress at the top fiber is

$$f_{top} = 0.0979 + 0.233 = 0.3309 \ ksi > 0 \ (ok!)$$

The total concrete stress at the bottom fiber is

$$f_{bottom} = 0.0779 - 0.0738 = 0.0041 \ ksi > 0 \ (ok!)$$

6.14.5.3   Design of Match-Cast Shear Keys

As mentioned in Section 6.9.2, the current AASHTO LRFD specifications do not provide a design method for shear keys. The following design is based on the AASHTO Standard Design Specifications Article 9.20.1.5[6-15, 6-33]. Again, consider section C shown in Fig. 6-82 as an example to illustrate the design procedures.

Based on the AASHTO Standard Specifications Article 9.20.1.5[6-33] and Section 6.14.5.2.1, the factored shear at section C during construction is

$$V_u = 1.1(A + DIFF) = 1.1(131.39 + 2.63) = 147.42 \ kips$$

From the AASHTO Standard Specifications[6-33] Article 9.2, the required shear strength per web is

$$V_c = \frac{V_u}{2\phi_v} = \frac{147.42}{2 \times 0.9} = 81.9 \ kips$$

From Eq. 6-9, the allowable concrete shear stress is

$$f_{va} = 2\sqrt{f_c'(psi)} = 2\sqrt{8500} = 184.4 \ psi$$

The shear area per key is

$$A_k = 3.875 \times 13 = 50.375 \ in^2$$

The shear strength per key is

$$V_{cp} = A_k \times f_{va} = 9.29 \ kips$$

The required number of shear keys is

$$n = \frac{V_c}{V_{cp}} = \frac{81.9}{9.29} = 8.8, \ (use \ 11 \ keys, \ ok!)$$

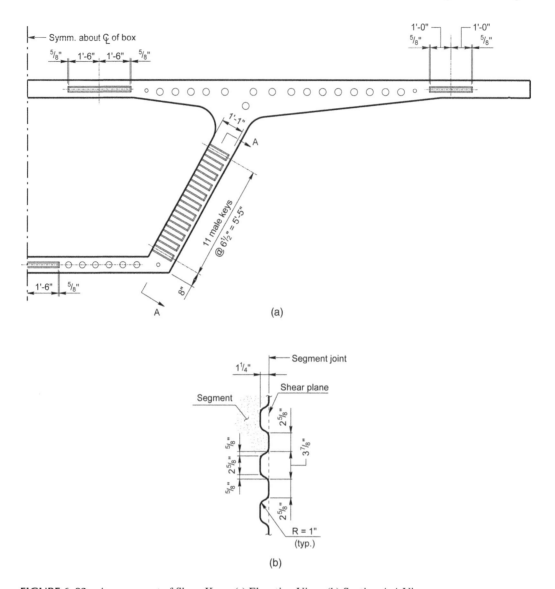

FIGURE 6-82   Arrangement of Shear Keys, (a) Elevation View, (b) Section *A-A* View.

REFERENCES

6-1. ASBI, *Construction Practices Handbook for Concrete Segmental and Cable-Supported Bridges*, American Segmental Bridge Institute, Buda, TX, 2008.
6-2. Jeakle, D., "Precast Segmental Cantilever Bridges Design," *Proceedings, Seminar for Design and Construction of Segmental Concrete Bridges*, American Segmental Bridge Institute, Orlando, FL, 2004.
6-3. Theryo, T. S., "Cast-in-Place Cantilever Bridge Design and Construction," *Proceedings, Seminar for Design and Construction of Segmental Concrete Bridges*, Orlando, FL, 2004.
6-4. Podolny, W., and Muller, J. M., "Construction and Design of Prestressed Concrete Segmental Bridges," A Wiley-Interscience Publication, John Wiley and Sons, New York, 1982.
6-5. Huang, D. Z., and Wang, T. L., "Impact Analysis of Cable-Stayed Bridges," *Journal of Computers and Structures*, Vol. 43, No. 5, 1992, pp. 897–908.
6-6. Wang, T. L., Huang, D. Z., and Shahawy, M., "Dynamic Behavior of Continuous and Cantilevered Thin-Walled Box Bridges," *Journal of Bridge Engineering*, ASCE, Vol. 1, No. 2, 1996, pp. 67–75.

6-7. AASHTO-PCI-ASBI, *AASHTO-PCI-ASBI Segmental Box Girder Standards for Span-by-Span and Balanced Cantilever Construction*, May 2000.

6-8. Kulka, F., and Thoman, S. J., "Feasibility Study of Standard Sections of Segmental Prestressed Concrete Box Girder Bridges," *PCI Journal*, Vol. 28, No. 5, September–October 1983, pp. 54–77.

6-9. Freyermuth, C. L., "AASHTO-PCI-ASBI Segmental Box Girder Standards: A New Product for Grade Separations and Interchange Bridges," *PCI Journal*, Vol. 42, No. 5, September–October 1997, pp. 32–42.

6-10. Huang, D. Z., and Hu, B., "Evaluation of Cracks in a Large Single-Cell Precast Concrete Segmental Box Girder Bridge without Internal Struts," ASCE, *Journal of Bridge Engineering*, 2014.

6-11. Florida Department of Transportation, *Structures Design Guidelines*, Florida Department of Transportation, Tallahassee, FL, 2017.

6-12. Florida Department of Transportation, *Structures Detailing Manuals*, Florida Department of Transportation, Tallahassee, FL, 2017.

6-13. Florida Department of Transportation, *New Directions for Florida Post-Tensioned Bridges*, Corven Engineering, Inc., Tallahassee, FL, 2004.

6-14. PCI and PTI, *Precast Segmental Box Girder Bridge Manual*, Prestressed Concrete Institute and Post-Tensioning Institute, IL, 1978.

6-15. Theryo, T. S., "Balanced Cantilever Construction Cantilever Bridge Design Using AASHTO LRFD Bridge Design Specifications," *Proceedings, Seminar for Design and Construction of Segmental Concrete Bridges*, Orlando, FL, 2004.

6-16. Fan, L. C., "Design of Prestressed Continuous Concrete Bridges," China Communication Press, Beijing, China, 1980.

6-17. AASHTO, *LRFD Bridge Design Specifications*, 7th edition, Washington, D.C., 2014.

6-18. Tassin, D., Dodson, B., Takobayashi, T., Deaprasertwong, K., and Leung, Y. W., *Computer Analysis and Full-Scale Test of the Ultimate Capacity of a Precast Segmental Box Girder Bridge with Dry Joint and External Tendons*, American Segmental Bridge Institute, Phoenix, AZ, 1995.

6-19. Jiang, H., Chen, L., Ma, Z., and Feng, W. (2015). "Shear Behavior of Dry Joints with Castellated Keys in Precast Concrete Segmental Bridges," *Journal of Bridge Engineering*, ASCE, 20(2), 04014062 1/12.

6-20. Turmo, J., Ramos, G., and Aparicio, Á. C., "Shear Behavior of Unbonded Post-Tensioned Segmental Beams with Dry Joints," *ACI Structural Journal*, Vol. 103, No. 3, 2006, pp. 409–417.

6-21. Zhou, X., Mickleborough, N., and Li, Z., "Shear Strength of Joints in Precast Concrete Segmental Bridges," *ACI Structural Journal*, Vol. 102, No. 1, 2005, pp. 3–11.

6-22. Shamass, R., Zhou, X., and Alfano, G. (2015). "Finite-Element Analysis of Shear-Off Failure of Keyed Dry Joints in Precast Concrete Segmental Bridges," *Journal of Bridge Engineering*, ASCE, Vol. 20, No. 6, 2015, 04014084-1/12.

6-23. Shamass, R., Zhou, X., Wu, Z., "Numerical Analysis of Shear-off Failure of Keyed Epoxied Joints in Precast Concrete Segmental Bridges," *Journal of Bridge Engineering*, Vol. 22, Issue 1, ASCE, 2017, 04016108.

6-24. Buyukozturk, O., Bakhoum, M. M., and Beattie, S. M., "Shear Behavior of Joints in Precast Concrete Segmental Bridges," *Journal of Structural Engineering*, Vol. 116, No. 12, 1990, pp. 3380–3401.

6-25. AASHTO, *Guide Specifications for Design and Construction of Segmental Concrete Bridges*, 2nd edition, AASHTO, Washington, D.C., 1992.

6-26. Hu, B., and Huang, D. Z., "Curvature Effects on Post-Tensioned Concrete Box Girder Bridges with External Tendons," *Proceedings, PCI National Bridge Conference*, Orlando, FL, October 5-7, 2008.

6-27. Shao, R. G., and Xia, K., *Curved Concrete Girder Bridges*, People's Communication Publication House, Beijing, China, 1991.

6-28. Hu, B., and Huang, D. Z., "Impact of Construction Methods on Curved Post-Tensioned Concrete Box Girder Bridges," *International Bridge Conference*, Pittsburgh, PA, June 14-17, 2009.

6-29. Hu, B., and Ghali, G. M., *Use of Single Bearings to Reduce Uplifting on Curved Narrow Segmental Bridges*, ASBI, Pittsburgh, PA, June 14-17, 2009.

6-30. Corven, J., *Post-Tensioned Box Girder Design Manual*, Federal Highway Administration, U.S. Department of Transportation, Washington D.C., 2015.

6-31. Post-Tensioning Institute, *Post-Tensioning Manual*, 6th edition, PTI, Phoenix, AZ, 2006.

6-32. Thompson, M. K., et al., "Measured Behavior of a Curved Precast Segmental Concrete Bridge Erected by Balanced Cantilevering," Research Report 1404-2, Center for Transportation Research, University of Texas, Austin, TX, 1998.

6-33. AASHTO, *Standard Specifications for Highway Bridges*, 17th edition, Washington, D.C., 2002.

7 Design of Incrementally Launched Segmental Bridges

- Span arrangement and determination of preliminary dimensions
- Typical launching methods and design
- Analysis and behaviors of bridges during launching
- Layout of post-tensioning tendons
- Design example: incrementally launched segmental bridge

7.1 INTRODUCTION

The basic concepts of incrementally lunched concrete segmental bridges were discussed in Section 1.5.5. Though the principles of launching bridge super-structures have been used in steel bridges for many years, the first attempt at launching concrete segmental bridges was in 1959 in the bridge over the Ager River, designed by Leonhardt and Baur in Austria[7-1, 7-2]. Since the Rio Caroni Bridge in Venezuela was built in 1962 and 1963, many incrementally launched concrete segmental bridges have been built throughout the world[7-1, 7-3, 7-4]. Though this type of bridge is generally economical for span lengths ranging from 100 to 200 ft, many launched bridges with longer span lengths have been successfully built[7-2, 7-8], by using temporary piers and/or launching from the opposite abutments. The total length of this type of segmental bridge is generally not greater than 3000 ft. The advantages and disadvantages of this type of bridge are summarized as follows:

Advantages

- As the entire bridge super-structure is cast behind the abutment on land and moved forward on the piers, no falsework is required for the construction of the super-structure. This feature eliminates the problem in passing over roads, railways, rivers, building, etc.
- The cost of segments transportation is eliminated.
- As there is no epoxy joint required, the construction can be done at low temperatures.
- In comparison to the cast-in-place balanced cantilever construction method, it is much simpler to cast concrete and prestress and provide safer operations.
- The number of construction joints is reduced, which results in faster construction as it is possible to cast much longer segments than those in cast-in-place cantilever construction.
- Heavy crane and launching trusses are eliminated.

Disadvantages

- A high level of dimension control is required at each construction stage. Any dimension control errors could cause difficulty with launching and with correcting geometric misalignment.
- This type of bridge is typically suitable for straight bridges and curved bridges with constant curvatures. The girder depth is constant.

In this chapter, the overview of bridge span arrangement, bridge configuration, limitation, and the basic mechanical behaviors will be discussed first. Then, the typical construction procedures are presented. Finally, some design features for this type of bridge and a design example will be provided.

7.2   GENERAL DESIGN FEATURES OF INCREMENTALLY LAUNCHED SEGMENTAL BRIDGES

7.2.1   BRIDGE ALIGNMENT

To build an incrementally launched segmental bridge, with consecutive sections of a rigid girder being pushed along the same trajectory, the bridge's horizontal and vertical alignment must be: (a) straight (see Fig. 7-1a), (b) curved with constant radius (see Fig. 7-1b), or (c) S-shaped with constant radius (see Fig. 7-1c) when launching on both sides.

7.2.2   BRIDGE SPAN ARRANGEMENT AND SEGMENT DIVISION

7.2.2.1   Span Arrangement

As we will see in Section 7.3, for the best design and detailing, all the spans typically have equal length, except the end spans (see Fig. 7-2a). The ratio of side span to interior span is typically smaller or equal to 0.7 to 0.8. For a bridge launched from both sides or if auxiliary piers are used, the span can be twice that of the standard spans (see Fig. 7-2b and c).

7.2.2.2   Segment Division

Based on previous experience, the following rules should be considered in the division of longitudinal segments[7-2]:

1. Within one week of construction time for one segment, the segment length should be as long as possible. The typical length of a segment ranges from 50 to 100 ft (15 to 30 m).
2. The entire bridge should be divided into as many similar segments as possible to facilitate construction.
3. Avoid having joints between the segments located at midspan and over the piers in the final stage, where there are maximum stresses.

The segments over the piers are often called support segments, which are typically equal-length segments. The first and last cast segments are often called begin and end segments,

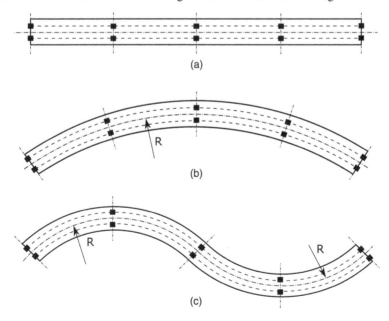

FIGURE 7-1   Bridge Alignment, (a) Straight, (b) Curved, (c) S-shape.

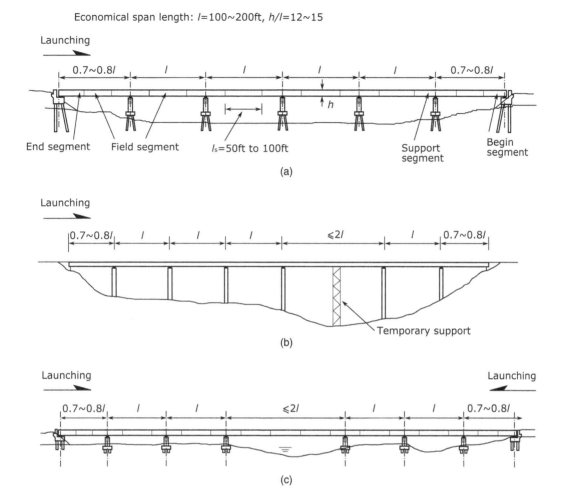

Economical span length: l=100~200ft, h/l=12~15

FIGURE 7-2 Typical Span Arrangement, (a) Equal Interior Span Lengths, (b) Unequal Span Lengths with Auxiliary Pier, (c) Unequal Span Lengths with Launching from Both Abutments.

respectively, and may have different lengths than the typical segments. The remaining segments between the begin and end segments are called field segments (see Fig. 7-2a). If a bridge has equal interior span lengths, the lengths of the field segments are generally equal to the length of the support segments. For a straight bridge, the plan view of the segments is typically rectangular (see Fig. 7-3a). For a curved bridge, the plan view of segments is generally wedge shaped (see Fig. 7-3b). If the curved bridge has a constant profile with a constant vertical slope α (see Fig. 7-3c), the super-structure follows a path of a helix with constant angle α and radius R. The shape of the segments in the elevation view is also typically rectangular (see Fig. 7-3c).

7.2.3　Typical Sections and Estimation of Preliminary Dimensions

7.2.3.1　Typical Sections

The most suitable cross sections for launched bridges are the single-cell box section and the double T-beam (see Fig. 7-4). The box section is more effective in resisting both bending and torsion, and it facilitates easier placement of the prestressing tendons in the section. In current practice, the double T-beam section is used less in launched segmental bridges.

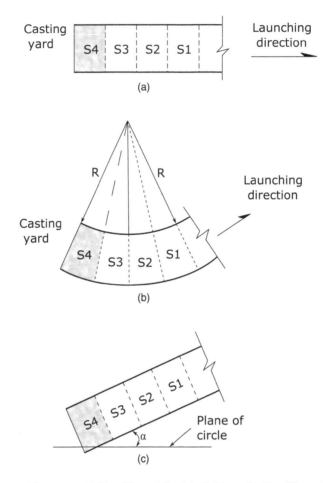

FIGURE 7-3 Shapes of Segments, (a) Plan View of Straight Bridges, (b) Plan View of Curved Bridges with Constant Slope, (c) Elevation View of Curved Bridges with Constant Slope.

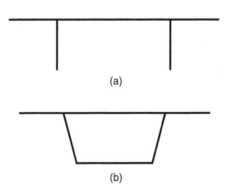

FIGURE 7-4 Typical Sections for Incrementally Launched Segmental Bridges, (a) Box Girder, (b) Double T-beam.

7.2.3.2　Girder Depth

As the uniform moment distribution of the super-structure due to its self-weight during the launching process and the resistance of the super-structure at all sections should be ensured, the girder depth should be constant for incrementally launched segmental bridges. To reduce the amount of construction tendons, the girder's span-to-depth ratio is typically larger than that for other types of concrete segmental bridges was discussed in Chapters 5 and 6. The span-to-depth ratio for incrementally launched segmental bridges normally ranges from 12 to 15. The small value of the ratio is for larger spans, and the larger value is for short span lengths. The span-to-depth ratio typically should not be greater than 17. The AASHTO LRFD specifications[7-5] provide more detailed span-to-depth ratios for incrementally launched segmental bridges, which were presented in Section 1.4.2.2. Based on some statistical analysis[7-1, 7-6, 7-7], the girder depth can be preliminarily estimated as

$$h = 0.94 + \frac{l}{22.7} \ (m) \tag{7-1}$$

where l = span length (m).

7.2.3.3　Estimation of Dimensions for Single-Cell Box Section

In determining the dimensions of the bridge super-structure sections, the requirements of both construction and design limits should be considered. The related minimum dimensions have been discussed in Section 1.4.2. The following approximate methods developed based on some statistical analysis[7-1] can be used for preliminary sizing the box girder sections.

The ratio of the box section cantilever (l_c) to the top slab width (B) (see Fig. 7-5) is

$$\frac{l_c}{B} \approx 0.24 \tag{7-2}$$

The ratio of bottom to top slab lengths is (see Fig. 7-5)

$$\frac{l_b}{B} \approx 0.44 \tag{7-3}$$

The top slab thickness can be estimated as

$$t_t = \frac{l_t}{\lambda} \geq 8.5 \ to \ 10.0 \ in \ (0.22 \ to \ 0.26 \ m) \tag{7-4}$$

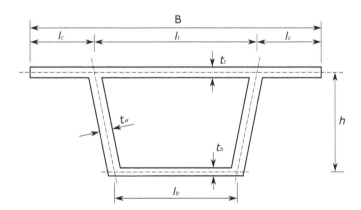

FIGURE 7-5　Box Section Dimensions.

where $\lambda = 25$ to 30, the smaller value is for wide top slabs and the larger value is for narrower top slabs

The bottom slab thickness can be estimated as

$$t_b = \left(l_b - t_w\right)\left(0.07 - \frac{l}{3700}\right) \tag{7-5}$$

where
l_b = bottom slab width (m)
t_w = web thickness (m)
l = span length (m)

For launched bridges, because of the effect of longitudinal prestressing tendons, the minimum bottom slab thickness typically should not be smaller than 8 in. (0.21 m).

The strength of the web is typically governed by the launching shear as it passes over the piers. Thus, web thickness is related to the span length and can be estimated as

$$\text{For bound tendons: } t_w = B\left(\frac{l}{2100} + 0.02\right) \tag{7-6a}$$

$$\text{For unbound tendons: } t_w = \frac{Bl - 500}{2000} + 0.3 \tag{7-6b}$$

where B = total width of the top slab/deck (m). The minimum web thickness should meet the requirements presented in Section 1.4.2.4.

The minimum slab thicknesses above refer to the thicknesses in the slab at the midspan of each of the box section components. In the areas around the intersections of the components, i.e., the corners of the box section, the thickness of the slabs should be increased to effectively resist the anticipated loadings and accommodate the post-tendons. A typical section is illustrated in Fig. 7-6. Generally, the thickness of the top slab is controlled by transverse bending moment and shear, while the thickness of the bottom slab is often determined based on the launching methods. The bottom area of the web has to accommodate the high reaction forces from the launching bearings along the entire bridge length and should be carefully designed. The distance between the edges of the launching pad and the edge of the web should not be less than 3 in. (see Fig. 7-6b and c). If the tendons are embedded in the bottom of the web, the vertical distance from the edge of the ducts to the bottom of the box girder should not be smaller than 6 in. (see Fig. 7-6b). To reduce the eccentricities of the bearing reactions and ensure sufficient distance between the launching pads and the edge of the bottom of the box girder, the bottom 16 to 20 in. (0.4 to 0.5 m) of the outside face of the web is detailed to be more upright than the remaining height of the web. However, this portion should have a minimum slope of 10% from vertical to allow stripping of the shuttering by vertical lowering[7-2].

In the following sections, we will see that each segment of the launched girder will be subjected to both negative and positive bending moment due to self-weights during launching. Normally, the ratio of the minimum negative moment to the maximum positive moment due to self-weights is about 2:1. Thus, it can be more effective to use construction tendons if the selected section has a ratio of top section modulus to bottom section modulus equal to 2:1.

7.3   TYPICAL LAUNCHING METHODS

There are typically three launching systems that use hydraulic devices to move the bridge super-structure forward: self-clamping rear thrust launching, pulling launching, friction launching[7-1, 7-8, 7-9]. Among these systems, the most often used launching systems in segmental bridge construction are pulling and friction launching systems.

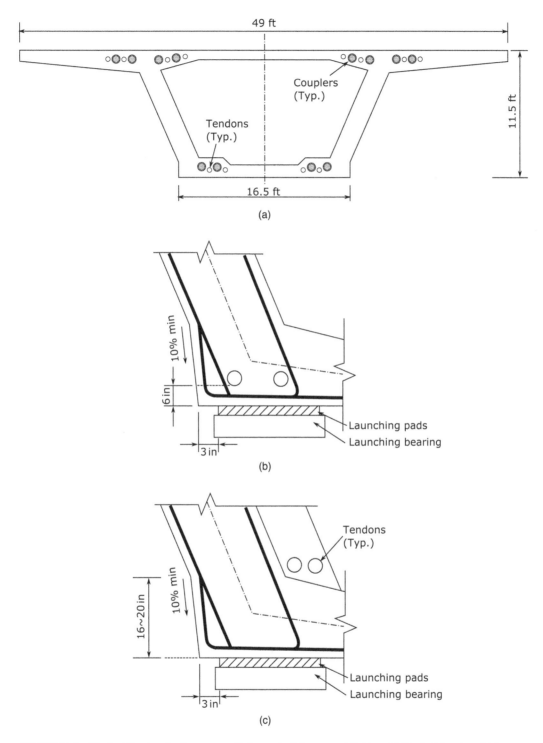

FIGURE 7-6 Typical Section for Incrementally Launched Segmental Bridge, (a) Typical Section, (b) Bottom of Web with Tendons, (c) Bottom Tendons Arranged Outside of Web.

7.3.1 PULLING LAUNCHING SYSTEM

The pulling launching system typically consists of center-hole jacks, prestressing strands or bars, as well as the steel launching pins or other types of launching shoes (see Fig. 7-7). One end of the prestressing strands or bars is connected to the center-hole jack, which is set against a support frame anchored to an abutment, and another end is connected to the steel launching pin that is mounted to the rear end section of the new segment. When the jack pushes against the support frame mounted on the abutment, it pulls the prestressing stands that pull the girder forward. Depending on the types of jacks and pumps, the launching speeds generally range from 10 to 20 ft/h. The maximum launching force may be up to 580 tons (5800 kN) per jack[7-4].

7.3.2 FRICTION LAUNCHING SYSTEM

7.3.2.1 General Launching Principle and Procedures

The friction launching system is often called a friction launcher and is composed of jacks that act vertically and horizontally (see Fig. 7-8). The vertical jacks are also called lifting jacks, and they are seated on a stainless-steel–Teflon base with a low friction coefficient of about 1% to 4%. On the top of the lifting jacks, there are hardened and roughened steel plates that create a friction coefficient against the concrete surface of approximately 0.75. The horizontal jacks are often called pushing jacks and generate launching forces. One end of the pushing jack is mounted to the braking saddle on the bridge abutments or/and piers, and another end is connected to the vertical jack. The braking saddle also supports the super-structure when the vertical jacks retract and is often called the

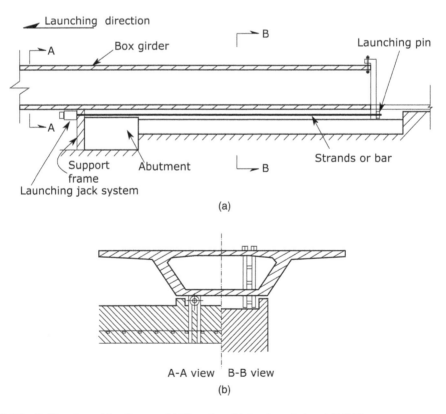

(a)

A-A view   B-B view

(b)

FIGURE 7-7 Pulling Launching System, (a) Elevation, (b) Sections *A-A* and *B-B* Views.

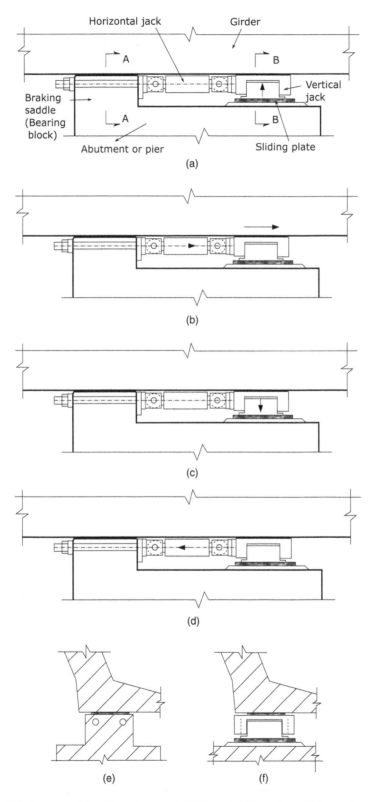

FIGURE 7-8 Friction Launching Procedures, (a) Lifting, (b) Thrusting, (c) Lowering, (d) Retracting, (e) Section *A-A*, (f) Section *B-B*.

bearing block. The friction launching procedure is basically a lift-and-push operation. The launching sequences of the friction launchers are as follows:

Step 1: Install the friction launchers on the abutment and/or piers as shown in Fig. 7-8a. The vertical jack is set on the sliding plate with a low friction surface.

Step 2: Hoist the super-structure about 3/16 in. [not exceeding 0.197 in. (5 mm) to avoid creating an excessive negative moment] around the bearing block by the vertical jack. The friction element at the top of the jack engages the super-structure. Thus, the support reaction of the girder is transferred from the anchor bearing block to the movable vertical jack as shown in Fig. 7-8a.

Step 3: Use the horizontal jack to push the vertical jack. The vertical jack slides on the low friction surface, and the thrust force from the horizontal jack is transferred to the girder by the friction between the bottom surface of the concrete girder and the top surface of the jack and moves the girder forward as shown in Fig. 7-8b.

Step 4: When the piston reaches its limit, the vertical jacks are retracted to lower the girder onto the bearing block as shown in Fig. 7-8c.

Step 5: Retract the longitudinal piston to the initial position, and repeat the cycle until launching is finished.

The friction launching system has the following advantages:

1. It can launch a heavy super-structure by placing synchronized launchers on multiple piers as required.
2. The launching speed can reach 30 ft/h as the complete electric control of the hydraulic system allows synchronization of the action of the launchers.
3. Compared to the pulling launching system, the friction launching system is much safer during downhill launching.

7.3.2.2 Maximum Launching Bridge Length by One-Location Friction Launching

7.3.2.2.1 Minimum Required Pushing Force

The minimum required pushing forces for a launched bridge is related to the friction and the longitudinal slope of the bridge super-structure and can be written as

$$F_{hr} = L_{max} \times q_g \times (\mu + s) \tag{7-7}$$

where
L_{max} = maximum launching length
q_g = girder weight per unit length
μ = friction coefficient between bottom deck and launching pads (%)
s = longitudinal slope of deck at sliding plane in %, negative for launching downhill

7.3.2.2.2 Maximum Pushing Force Provided by One Launcher

The maximum possible pushing force by the friction launchers is related to the friction forces between the bottom slab of the super-structure and the roughed steel plate on the top of the lifting jack as well as between the bottom of the lifting jack and sliding surface. As discussed in Section 7.3.2.1, the friction coefficient between stainless-steel–Teflon base is 1 to 4% and the friction coefficient between the top of the lifting jacks and the concrete surface is about 0.75.

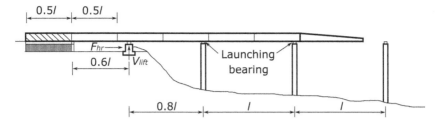

FIGURE 7-9 Typical Launched Bridge Configuration [7-2].

If using these friction coefficients, the maximum pushing force provided by the launcher can be written as

$$F_p = V_{lift} \times \left(\frac{0.75 - 0.04}{s_a} \right)$$ (7-8)

where
V_{lift} = maximum jack lifting force
s_a ≈ 1.4 = safety factor[7-2]

For a typical launched bridge configuration as shown in Fig. 7-9, the maximum jack lifting force can be approximately written as

$$V_{lift} = \frac{0.6 + 0.8}{2} l \times q_g$$ (7-9)

Substituting Eq. 7-9 into Eq. 7-8 yields the maximum pushing force as

$$F_p = 0.35 \times l \times q_g$$ (7-10)

7.3.2.2.3   Maximum Bridge Launching Length per Set of Friction Launching Equipment
The maximum pushing force should not be smaller than the minimum required pushing force, i.e.,

$$0.35 \times l \times q_g \geq L_{max} \times q_g \times (\mu + s)$$

Rearranging the above equation yields

$$L_{max} \leq \frac{0.35 \times l}{\mu + s}$$ (7-11)

For the preliminary design, Fig. 7-10 illustrates the variations of maximum launching lengths with friction and longitudinal slopes developed based on Eq. 7-11. If the maximum length exceeds the required bridge length, the additional sets of friction launching equipment can be mounted at different piers in front of the abutment or using additional pulling bars.

7.3.3   LAUNCHING BEARINGS AND SIDE GUIDES

7.3.3.1   Launching Bearings
During the super-structure launching, the super-structure is supported on launching bearings mounted on the top of the piers (see Figs. 1-44 and 7-9). A typical launching bearing is shown in Fig. 7-11.

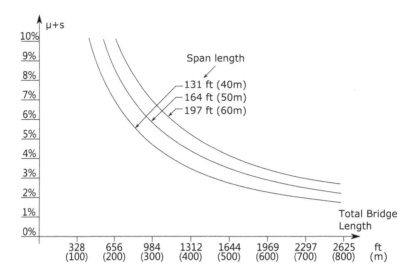

FIGURE 7-10   Variation of Maximum Bridge Launching Length with Friction and Slope[7-2].

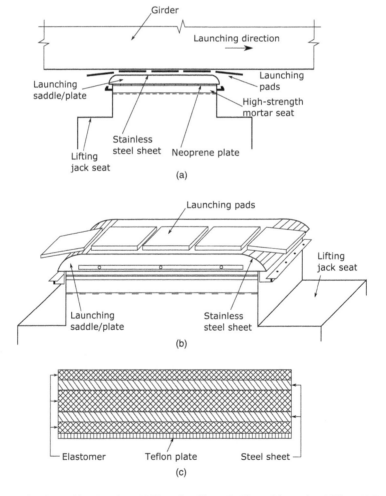

FIGURE 7-11   Typical Launching Bearing, (a) Elevation View, (b) Three-Dimensional View, (c) Launching Pad.

The super-structure sits on launching pads that consist of neoprene sheets with alternating steel sheets and are glued to a bottom Teflon plate (see Fig. 7-11c). The multi-layered elastomeric bearing is used to uniformly distribute the contact stress. The Teflon™ [polytetrafluoroethylene (PTFE)] plate is used to reduce sliding friction between the super-structure and launching supports and ensures that the friction coefficient will not be greater than 4%. The launching pads can be inserted without lifting the super-structure, which makes the launch continuous and increases the launching speed[7-1].

7.3.3.2 Side Guides

To maintain the correct plane alignment of the super-structure, some side guides are placed next to the launching bearings outside of the web (see Fig. 7-12). The side guides create fixed or adjustable lateral forces through rollers or PTFE contacts. The lateral pressure generally is not greater than 1 ksi (7 MPa).

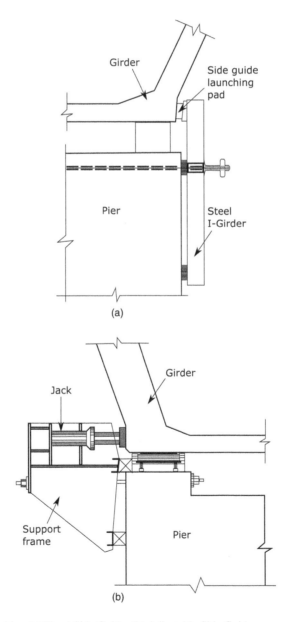

FIGURE 7-12 Side Guides, (a) Fixed Side Guide, (b) Adjustable Side Guide.

7.4 ANALYSIS AND BEHAVIOR OF INCREMENTALLY LAUNCHED BRIDGES

7.4.1 INTRODUCTION

Per the foregoing discussion, an incrementally launched bridge in the final construction stage is typically a continuous girder with equal span length and constant depth. The analytical methods of the effects due to dead loads, live loads, post-tendons, temperature, shrinkage, creep, etc., is the same as those described in Chapter 5 and can be analyzed using any of the methods presented in Chapter 4. However, the super-structure is subjected to continually alternating bending moments during launching. The moment at each of the sections changes from negative to positive or vice versa as shown in the dotted line in Fig. 7-13b. A typical moment envelope due to the girder's self-weight q_g during the bridge launching is shown in Fig. 7-13b (solid lines). From this figure, it can be seen that the negative and positive moments in each of the cross sections are about $q_g l^2/12$ and $q_g l^2/24$, respectively. These moments must be balanced by internal axial prestressing forces, similar to a column. The characteristics of dead load effects in incrementally launched bridges are quite different from those of other types of segmental bridges, and some approximate methods for determining the maximum and minimum moments due to self-weights are discussed in the following.

7.4.2 GENERAL ANALYSIS OF THE EFFECTS DUE TO GIRDER SELF-WEIGHT AND LAUNCHING NOSE

Figure 7-14a shows a typical incrementally launched bridge with equal span length and constant cross section, as well as a cantilevered launching nose. Though its effects due to the dead loads can be analyzed by the methods presented in Chapter 4, the following approximate simple method can be employed in determining its moment distribution[7-10].

Step 1: Determine the moment M_{ic} at support i due to M_c induced by the cantilevered segment of the girder and launching nose as shown in Fig. 7-14a, using the following equation:

$$M_{ic} = \eta_1 M_c \tag{7-12}$$

where

η_1 = coefficient found from Table 7-1 based on the span number n and support number i
M_c = cantilever end moment at support n (see Fig. 7-14b)
n = total span number

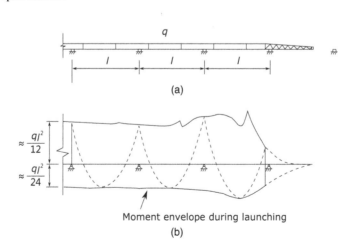

FIGURE 7-13 Typical Moment Envelope for Incrementally Launched Segmental Bridges, (a) Elevation View of Launched Girder, (b) Moment Envelope.

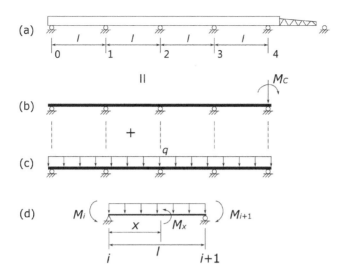

FIGURE 7-14 Analytical Model for Self-Weights of Girder and Launching Nose, (a) Elevation View of Girder and Launching Nose, (b) Continuous Girder with End Moment Due to Cantilever, (b) Continuous Girder with Self-Weight, (c) Free Body of Simply Supported Girder.

Step 2: Determine the moment M_{iq} at support i due to the girder self-weight q as shown in Fig. 7-14c, using the following equation:

$$M_{iq} = \eta_2 q l^2 \tag{7-13}$$

where
η_2 = coefficient found from Table 7-2, determined based on the span number n and support number i
l = span length
q = girder self-weight per unit length

Step 3: Determine the total moment at support i as

$$M_i = M_{ic} + M_{iq} \tag{7-14}$$

TABLE 7-1

Coefficient η_1

Span Number	M0	M1	M2	M3	M4	M5	M6	M7	M8
1	0	−1							
2	0	0.25000	−1						
3	0	−0.06667	0.26667	−1					
4	0	0.01786	−0.07143	0.26786	−1				
5	0	−0.00479	0.01914	−0.07177	0.26794	−1			
6	0	0.00128	−0.00513	0.01923	−0.07180	0.26795	−1		
7	0	−0.00034	0.00137	−0.00515	0.01924	−0.07180	0.26795	−1	
8	0	0.00009	−0.00037	0.00138	−0.00516	0.01924	−0.07180	0.26795	−1

Table 7-2 Coefficient η_2

Span Number	M0	M1	M2	M3	M4	M5	M6	M7	M8
1	0	0							
2	0	−0.12500	0						
3	0	−0.10000	−0.10000	0					
4	0	−0.10714	−0.07143	−0.10714	0				
5	0	−0.10526	−0.07895	−0.07895	−0.10526	0			
6	0	−0.10577	−0.07692	−0.08654	−0.07692	−0.10577	0		
7	0	−0.10563	−0.07747	−0.08451	−0.08451	−0.07747	−0.10563	0	
8	0	−0.10567	−0.07732	−0.08505	−0.08274	−0.08505	−0.07732	−0.10567	0

Step 4: Determine the moment at other sections as (see Fig. 7-14d)

$$M_x = \frac{q(lx - x^2)}{2} + \frac{(M_{i+1} - M_i)x}{l} + M_i \tag{7-15}$$

7.4.3 DETERMINATION OF MAXIMUM AND MINIMUM MOMENTS DUE TO GIRDER SELF-WEIGHT AND LAUNCHING NOSE

As aforementioned, the moment at any of the cross sections will continuously change. Its maximum and minimum moments are related to the girder span length, launching nose length, stiffness, and unit weight of both the girder and the launching nose. The maximum and minimum moments are used to design the construction tendons, which are often called uniform axial posttendons. It may be difficult to use a simple equation to determine the locations of the maximum and minimum moments. These moments can be either determined by computer program or through trial-and-error method. The following approximate methods can be used for preliminary bridge design.

7.4.3.1 Maximum Positive Moment

The maximum positive moment of a launched girder may be analyzed by treating the launched girder as a three-span continuous beam when the front end of the girder just reaches the support (see Fig. 7-15). The location of the maximum moment is approximately equal to 0.4 l from the end (see Fig. 7-15), and the maximum moment can be estimated as

$$M_{max} = \frac{q_g l^2}{12}\left(0.933 - 2.96\gamma\beta^2\right) \tag{7-16}$$

where

q_g = weight of girder per unit length

γ = ratio of weight per unit length of launching nose (assuming a constant) to that of main girder

β = ratio of length of launching nose to that of main girder

7.4.3.2 Minimum Negative Moment

Typically, two loading situations as shown in Fig. 7-16 should be investigated to determine the minimum negative moment. The first loading case is when the launching nose just reaches the end abutment before resting on the temporary support as shown in Fig. 7-16a. In this situation, the cantilever

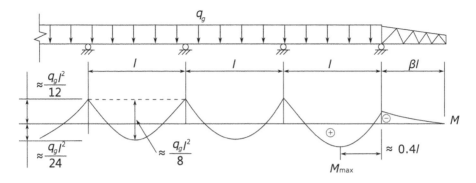

FIGURE 7-15 Analytical Model for Maximum Moment of Launched Girder.

beam including the launching nose is longest. The minimum negative moment of the girder can be estimated using the following equation:

$$M_{min} = -\frac{q_g l^2}{2}\left(\alpha^2 + \gamma\left(1-\alpha^2\right)\right) \tag{7-17}$$

where α = ratio of concrete girder cantilever length to span length.

The second loading case is that of the two-span continuous cantilever as shown in Fig. 7-16b with the end support at about midspan of the launching nose. The negative moment over the interior support can be easily determined by the methods discussed in Sections 4.2.2 and 4.2.3.

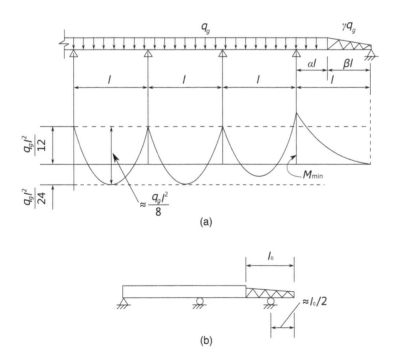

FIGURE 7-16 Models for Analyzing Minimum Moment of Launched Girders, (a) With Longest Cantilever, (b) Two-Span Continuous Cantilever.

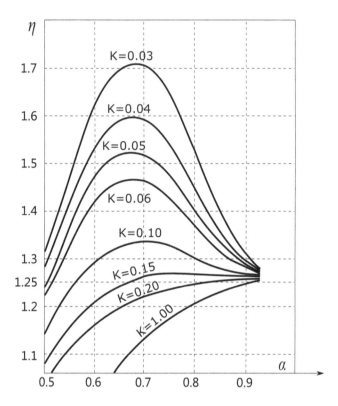

FIGURE 7-17   Variation of Negative Moment Coefficients η.

If the ratio of girder self-weight to that of launching nose is about 0.1, the minimum moment at the support can be estimated as

$$M_{min} = -\eta \frac{q_g l^2}{12} \tag{7-18}$$

where η is a function of coefficients K and α, which can be determined from Fig. 7-17 based on the values of α and K. The variable K is defined as the ratio of launching nose stiffness to main girder stiffness, i.e.,

$$K = \frac{E_s I_s}{E_c I_c} \tag{7-19}$$

where
$\quad E_s, E_c$ = moduli of launching nose and the concrete girder, respectively
$\quad I_s, I_c$ = moments of inertia of launching nose and concrete girder, respectively

7.4.4 ANALYTICAL EXAMPLE

Figure 7-18a shows a four-span continuous incrementally launched segmental bridge. The span length $l = 131$ ft, the girder section and stiffness (EI) are constant, the girder unit weigh $q_g = 1.2$ kips/ft, the length of launching nose $l_0 = 0.65l = 85.15$ ft, and the unit weight of the launching nose $q_0 = 0.12$ kips/ft. Use the approximate methods to estimate the girder maximum and minimum bending moments during girder launching.

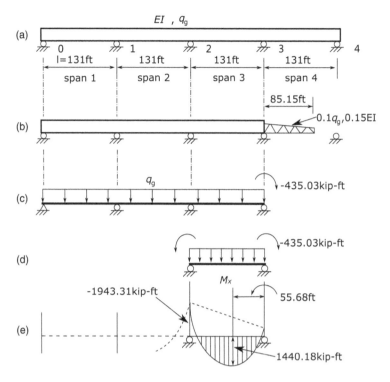

FIGURE 7-18 Loading Cases for Determining Maximum Positive Moment, (a) Span Arrangement, (b) Girder Position, (c) Analytical Model, (d) Free-Body for Span 3, (e) Moment Distribution of Span 3.

7.4.4.1 Determination of Maximum Bending Moment

7.4.4.1.1 Determination by Simplified Analytical Method

As discussed in Section 7.4.3.1, the loading case as shown in Fig. 7-18b will induce the maximum moment for the girder. The cantilever moment at support 3 due to the launching nose is

$$M_c = \frac{0.12 \times 85.15^2}{2} = 435.03 \text{ kip-ft}$$

To determine the moment at support 2 due to the cantilever moment, we can use Eq. 7-12. For span number $n = 3$, from Table 7-1, we have

$$\eta_1 = 0.26667$$

$$M_{2c} = 0.26667 \times (435.03) = -116.01 \text{ kip-ft}$$

From Table 7-2, we have

$$\eta_2 = -0.10000$$

$$M_{2q} = -0.10000 \times 1.2 \times 131^2 = -2059.32 \text{ kip-ft}$$

From Eq. 7-14, we have

$$M_2 = M_{2c} + M_{2q} = -1943.31 \text{ kip-ft}$$

Take span 3 as a free body applied with end moments M_2 and M_c and self-weight q_g, as shown in Fig. 7-18d. Then the maximum positive moment at approximate location $x = 0.425l$ can be determined by Eq. 7-15 as

$$M_{max} = 1440.18 \text{ kip-ft}$$

7.4.4.1.2 Determination by Approximate Equation

Using approximate Eq. 7-16, the maximum positive moment can be estimated as

$$M_{max} = \frac{q_g l^2}{12}\left(0.933 - 2.96\gamma\beta^2\right) = \frac{1.2 \times 131^2}{12}\left(0.933 - 2.96 \times 0.1 \times 0.65^2\right) = 1386.51 \text{ kip-ft}$$

The result is closed to that obtained by the simplified analytical method.

7.4.4.2 Determination of Minimum Negative Moment

As discussed in Section 7.4.3.2, two loading cases as shown in Fig. 7-19a and b should normally be investigated to determine the minimum negative moment. Figure 7-19a illustrates the loading situation of the launching girder with maximum cantilever length. The minimum negative moment occurs at support 2 and can be estimated by Eq. 7-17 as

$$M_{min1} = -\frac{q_g l^2}{12}\left(\alpha^2 + \gamma\left(1-\alpha^2\right)\right) = -\frac{1.2 \times 131^2}{2}\left(0.35^2 + 0.1\left(1 - 0.35^2\right)\right) = -2164.86 \text{ kip-ft}$$

Figure 7-19b shows another construction stage that is a two-span continuous cantilever. It is a one-degree indeterminate structure and can be solved by force methods as discussed in Section 4.2.2. Taking the moment at support 1 as an unknown moment X_1, the moment distributions in the two simply supported beams due to $X_1 = 1$ and the self-weights can be easily calculated and are shown in Fig. 7-19d and e, respectively.

The relative angle at support 1 due to unknown moment can be obtained by Eq. 4-15a, using the diagram multiplication method discussed in Section 4.2.1.2, as

$$\Delta_{11} = X_1\delta_{11} = X_1\left(\int_0^{1.675l} \frac{m_1(x)m_1(x)}{EI}dx + \int_{1.675l}^{2.0l} \frac{m_1(x)m_1(x)}{0.15EI}dx\right) = \frac{95.83}{EI}X_1$$

The relative angle at support 1 due to the self-weight in the primary structure can be obtained by Eq. 4-15e, using the conjugate-beam method discussed in Section 4.2.1.1 (see Fig. 7-19f), as

$$\Delta_{1P} = \int_0^{1.675l} \frac{m_1(x)M_{1p}(x)}{EI}dx + \int_{1.675l}^{2.0l} \frac{m_1(x)M_{1P}(x)}{0.15EI}dx = -\frac{24331.75}{EI}$$

The relative angle at the section over support 1 should be equal to zero, i.e.,

$$\Delta_{11} + \Delta_{1P} = 0$$

Then, we have

$$X_1 = M_{min2} = \frac{24331.75}{95.83} = 2518.33 \text{ kip-ft (the assumed moment direction is correct)}$$

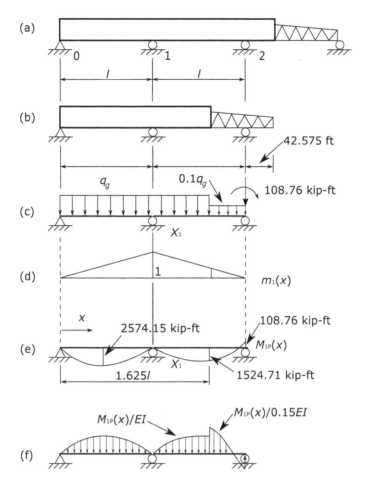

FIGURE 7-19 Construction Stages for Minimum Negative Moment Analysis, (a) Two-Span Continuous Beam with Maximum Cantilever, (b) Two-Span Continuous Beam with Launching Nose Supported at Its Midspan, (c) Analytical Model for Two Span Loading Case, (d) Moment Distribution Due to Unit Unknown Moment $m_1(x)$, (e) Moment Distribution Due to Self-Weights $M_{1p}(x)$, (f) Conjugate Beam.

Thus, the minimum negative moment due to self-weight during launching is

$$M_{min} = -2518.33 \text{ kip-ft}$$

7.5 METHODS FOR REDUCING NEGATIVE MOMENTS OF SUPER-STRUCTURES DURING LAUNCHING

To reduce a large negative moment in the girder during launching over the pier before it reaches the next pier, there are typically two methods used in construction, in addition to using temporary piers. The first method is to use a launching nose that has a much lighter self-weight than the super-structure, as previously discussed, and the second method is to use a tower-and-stays system. Generally, in comparison with the tower-and-stays system, the launching nose method is simpler and provides faster and more economical construction.

7.5.1 Launching Nose

7.5.1.1 Typical Launching Nose

A current commonly used launching nose consists of two steel plate girders connected with lateral bracings at the bottom flanges (see Fig. 7-20). Only bottom flanges are connected by lateral bracings, which are mainly to transfer the side guide force to the super-structure. The top flanges and webs are stiffened by enough vertical stiffeners and longitudinal stiffeners to prevent the top flanges, webs, and girders from bulking (see Fig. 7-20). For ease of transport and the convenience of being able to be reused on different projects, the launching nose is typically divided into several segments and then spliced together as the required nose length varies for different projects.

7.5.1.2 Attachment of Launching Nose to Concrete Girder

The attachment details between the launching nose and the super-structure should ensure that there is enough capacity to transfer moment and shear between them and facilitate easy disassembly after completing the launch. The launching nose is typically assembled in front of the casting bed, and the first segment is cast directly against the base of the nose (see Fig. 7-21). The tensile forces caused by the bending moment from the launching nose are transferred to the concrete super-structure through prestressing bars embedded in the girder. The length of the high-strength bars embedded in the concrete is about 1.5 to 2 times the girder depth, or a 13-ft (4-m) lap with prestressing tendons or bars in the deck (see Fig. 7-21).

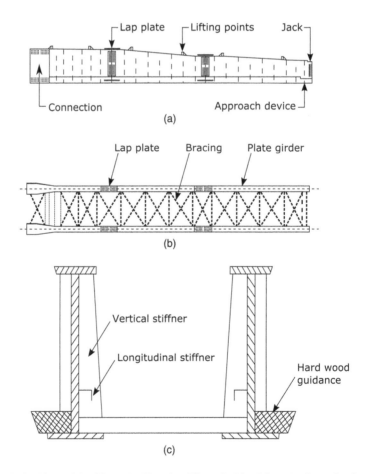

FIGURE 7-20 Typical Launching Nose, (a) Elevation View, (b) Plan View, (c) Cross Section.

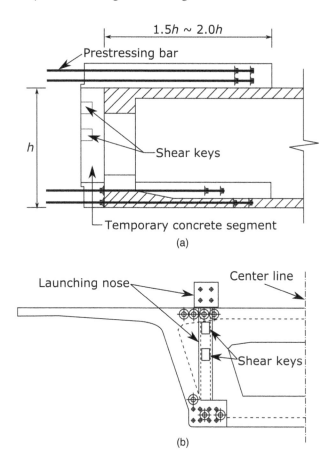

FIGURE 7-21   Typical Connection between Launching Nose and Girder, (a) Elevation View, (b) Side View.

7.5.1.3   Optimum Length and Stiffness of Launching Nose

From Section 7.4, it can be seen that the moments of the super-structure during launching vary with the super-structure self-weight q_g and the following ratios:

γ	=	ratio of weight per unit length of launching nose (assuming a constant) to that of main girder
$\alpha = \frac{x}{l}$	=	ratio of concrete girder cantilever length to span length
$\beta = \frac{l_0}{l}$	=	ratio of length of launching nose to that of main girder
K	=	ratio of launching nose stiffness to main girder stiffness (Eq. 7-19)
x, l, l_0	=	girder cantilever length, span length, length of launching nose (see Fig. 7-22), respectively

To better size the launching nose, let us examine the variations of girder moment with different lengths and stiffnesses of the launching nose. For simplicity, we assume that the super-structure has equal span length and constant section. Then it is not difficult to determine the moment variation of the super-structure during launching by considering two loading stages as shown in Fig. 7-22 [7-1] and using the methods discussed in Sections 4.2 and 7.4.

First, we assume the length of the launching nose is equal to $0.8l$, i.e., $\beta = 0.8$ (long launching nose) and $\gamma = 0.1$. The variations of the girder moment M_B over pier B with α for a different stiffness ratio K[7-1] are shown in Fig. 7-23. In this figure, the coordinate represents the ratio of the moment

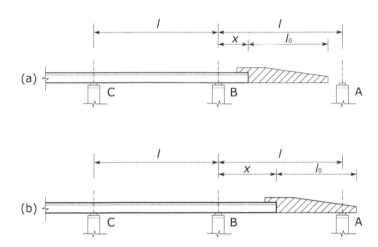

FIGURE 7-22   Loading Stages of Super-Structure during Launching, (a) Before Launching Nose Reaching Next Pier, (b) After Launching Nose Reaching Next Pier.

M_B divided by $q_g l^2$ and the abscissa is the ratio of the concrete girder cantilever length to the span length α. From this figure, we can see that the negative moment of the girder over pier B increases with α until the launching nose reaches pier A ($\alpha = 0.2$). Then, the moment variation is dependent on the stiffness ratio K and the maximum negative moment decreases with the stiffness of the launching nose until the stiffness ratio $K = 0.2$. A larger value of K than 0.2 will not help to reduce the maximum negative moment. Thus, we can conclude that the optimum stiffness ratio K is about 0.2 for this case. From Fig. 7-23, we can also observe that that maximum cantilever moment when $\alpha = 0.2$ (see Fig. 7-23, point E_1) is much smaller than that when the girder reaches pier A, i.e., $\alpha = 1.0$ (see Fig. 7-23, point E_2), which indicates that the length of the launching nose can be reduced without increasing the maximum negative moment over pier B.

Next let us examine a case with a short launching nose where the length of the launching nose is equal to 0.5 times the span length, i.e., $\beta = 0.5$, and $\gamma = 0.1$. The variation of the moment M_B over pier B with girder cantilever length and the stiffness ratio is shown in Fig. 7-24. From this figure, we can observe: (a) the maximum girder cantilever moment M_B over pier B ($\alpha = 0.5$, point E_1) is

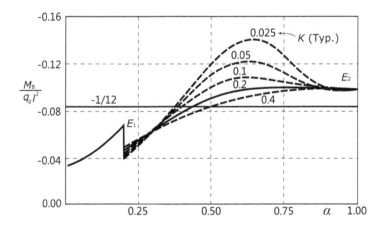

FIGURE 7-23   Variation of M_B with Girder Cantilever Length and Launching Nose Stiffness ($\beta = 0.8$ and $\gamma = 0.1$).

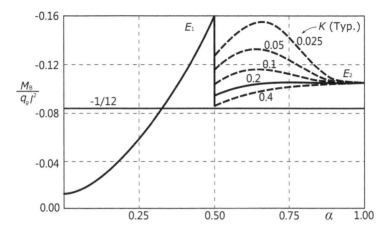

FIGURE 7-24 Variation of M_B with Girder Cantilever Length and Launching Nose Stiffness ($\beta = 0.5$ and $\gamma = 0.1$).

significantly higher than that when the girder reaches pier A ($\alpha = 1.0$, point E_2), which indicates that the length of the launching nose is too short and should be increased to achieve an economical design, and (b) after the launching nose reaches pier A, the moment M_B over pier B decreases as the stiffness of the launching nose increases to $K = 0.2$, which means that it is not beneficial to further increase the launching nose stiffness and that the optimum stiffness ratio K is about 0.2 for this case. In practical bridge design, K values are generally within the range:

$$K \approx 0.085 \sim 0.2$$

From the aforementioned discussion, it can be seen that the optimum length of the launching nose is between $\beta = 0.8$ to 0.5. Through the trial-and-error method[7-1], the optimum length of the launching nose is about 0.65 the span length l, i.e., $\beta = 0.65$ with the stiffness ratio $K = 0.2$, and the variations of moment M_B with α are shown in Fig. 7-25. From this figure, we can see that the maximum cantilever moment M_B ($\alpha = 0.35$, point E_1) is closed to the moment when the girder reaches pier A ($\alpha = 1.0$, point E_2) and that the moment in the intermediate launching stages is smaller.

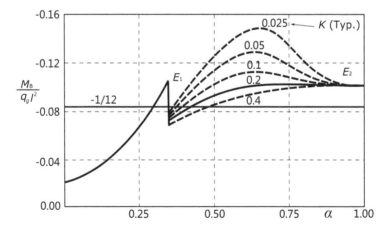

FIGURE 7-25 Variation of M_B with Optimum Length and Stiffness of Launching Nose ($\beta = 0.65$ and $\gamma = 0.1$).

The above discussion regarding the optimum length and stiffness of the launching nose is based on the assumptions of equal super-structure span lengths with constant cross section and under a condition of limiting the minimum moment M_B over pier B during bridge launching only. For other situations and conditions, the optimum length and stiffness of the launching nose may be changed and can be obtained by optimization techniques through computer analysis[7-11].

7.5.2 TOWER-AND-STAY SYSTEM

For longer span lengths, it may not be cost effective to attempt to reduce the cantilever moment by using the optimum launching nose technique. A tower-and-stay system or a combination of the system with launching nose may be used to effectively reduce the cantilever moment in the front part of the super-structure (see Fig. 7-26a and c). The tower-and-stay system is most efficient when the tower is located above a pier. However, when the tower is located near the

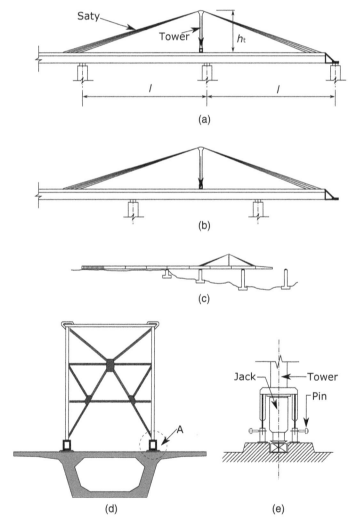

FIGURE 7-26 Tower-and-Stay System and Its Combination with Launching Nose, (a) Tower-and-Stay System, (b) Tower Located Near Midspan, (c) Combination of Tower-Stay System and Launching Nose, (d) Cross Section, (e) Device Adjusting Tower and Stay Forces.

midspan (see Fig. 7-26b), the tower may apply a larger downward force and cause an additional positive moment to the super-structure. For this reason, a long stroke jack is often installed between each of the tower legs and the bridge deck (see Fig. 7-26e) to adjust the forces in the tower legs and stays.

For a wide and heavy super-structure with span lengths ranging from 200 to 230 ft (60 to 70 m), the combined system of the tower-stay system and launching nose can be used. The tower height h_t can be taken as

$$h_t = 0.40(l - l_0) \sim 0.44(l - l_0)$$ (7-20)

and the length of the launching nose can be taken as

$$l_0 = 0.4l \sim 0.6l$$ (7-21)

7.6 ESTIMATION OF LONGITUDINAL TENDONS AND TENDON LAYOUT

7.6.1 INTRODUCTION

As discussed in Section 7.4.1, each of the super-structure sections is subjected to a continual changing of both positive and negative bending moments during launching (see Fig. 7-27b). These moments can be effectively balanced by a uniform axial prestressing, which is often called launching pre-stressing or centric prestressing. The tendons used for the prestressing are called launching tendons or centric tendons. After the completion of bridge launching, the maximum positive moment of

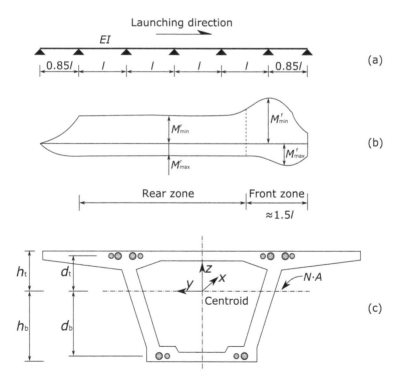

FIGURE 7-27 Typical Moment Envelopes During Launching and Estimation of Launching Tendons, (a) Typical Incrementally Launched Segmental Bridge, (b) Moment Envelope, (c) Locations of Neutral Axis and Launching Tendons.

the super-structure is located in the mid-portion of the span and the maximum absolute negative moment is located over the interior support. These moments should be balanced by variable axial prestressing, which is often called eccentric prestressing. The tendons for this prestressing are often called eccentric tendons or permanent tendons. After launching, some of the launching tendons are no longer necessary and may even be harmful and can therefore be removed. Some of them can remain to resist part of the design loads. The remaining design loads are resisted by providing the eccentrically placed permanent post-tensioning tendons.

7.6.2 LAUNCHING TENDONS

Figure 7-27b illustrates a typical moment envelope for a six-span continuous girder with equal interior spans and constant cross section during launching. From this figure, it can be seen that the moment envelope has two distinguished zones. The rear zone has almost constant envelopes and the moment envelopes in the front zone vary significantly and include both the largest positive moment and largest absolute negative moment. Typically, the launching tendons are determined based on the moment in the rear zone[7-12]. Then, additional tendons are determined for the front zone based on the maximum positive and maximum negative moments in this zone.

7.6.2.1 Estimation of Launching Tendons

7.6.2.1.1 Estimation Launching Tendons in Rear Zone

The required amount of launching tendons is typically determined based on the allowable concrete tensile stress. Assuming the total launching prestressing force as P, we have

$$\text{Tensile stress at top fiber: } f_{t-top} = -\frac{P}{A} - \frac{M^r_{min}h_t}{I} \leq f_a \tag{7-22}$$

where
  A = area of section (see Fig. 7-27c)
  I = moment of inertia of section
  M^r_{min} = minimum negative moment in rear zone (negative for negative moment)
  f_a = allowable concrete tensile stress (positive for tensile stress)
  h_t = distance between top fiber and neutral axis (upward positive)

$$\text{Tensile stress at bottom fiber: } f_{t-bottom} = -\frac{P}{A} + \frac{M^r_{max}h_b}{I} \leq f_a \tag{7-23}$$

where
  M^r_{max} = maximum positive moment in rear zone (positive for positive moment)
  h_b = distance between bottom fiber and neutral axis

From Eqs. 7-22 and 7-23, it can be seen that the minimum required launching prestessing is

$$f_{t-top} = f_{t-bottom} = f_a \tag{7-24}$$

From Eq. 7-24, we can obtain

$$\frac{h_b}{h_t} = \left|\frac{M^r_{min}}{M^r_{max}}\right| \tag{7-25}$$

As previous discussed, for a super-structure with equal span length and constant section,

$$\left| \frac{M^r_{min}}{M^r_{max}} \right| \approx 2 \tag{7-26}$$

That means, the best cross section for the super-structure is the section that has

$$\frac{h_b}{h_t} \approx 2 \tag{7-27}$$

Actually, it is often difficult to satisfy Eq. 7-27 due to the wide top slab required. Thus, the minimum required launching prestressing force P_{min} is determined based on the larger one determined by Eqs. 7-22 and 7-23, i.e.,

$$P_{min} = \max(P_1, P_2) \tag{7-28}$$

where

$$P_1 = -\left(\frac{M^r_{min} h_t}{I} + f_a \right) A$$

$$P_2 = \left(\frac{M^r_{max} h_b}{I} - f_a \right) A$$

If the launching tendons are arranged in the top and bottom slabs as shown in Fig. 7-27c, to ensure that the resulting launching prestressing force passes through the centroid axis, the minimum launching prestressing force can be distributed as

$$\text{Top launching prestressing force: } P_t = \frac{d_b}{d_t + d_b} P_{min} \tag{7-29}$$

where d_t and d_b = distances from the section centroid axis to the centroids of the top and bottom tendons, respectively.

$$\text{Bottom launching prestressing force: } P_b = \frac{d_t}{d_t + d_b} P_{min} \tag{7-30}$$

The required top and bottom launching tendons can be estimated as
Top launching tendon area:

$$A_{t-ps} = \frac{P_t}{0.60 \ f_u} \tag{7-31a}$$

Bottom launching tendon area:

$$A_{b-ps} = \frac{P_b}{0.60 \ f_u} \tag{7-31b}$$

The tendon size used is dependent on the thickness and form of the plate. Typically, 12-0.6″ diameter strands are used.

7.6.2.1.2   Estimation of Launching Tendons in Front Zone

Generally, the maximum absolute negative moment in the front zone is much higher than that in the rear zone. The prestressing force determined based on the moment envelope in the rear zone must be increased, and the positive moment in the front zone may also require additional prestressing force. The required additional post-tensioning force ΔP with an eccentricity Δe (upward positive) can be estimated from the following equations:

$$-\frac{P+\Delta P}{A} - \frac{M_{min}^{f} - \Delta P \times \Delta e}{I} h_t = f_a \tag{7-32a}$$

$$-\frac{P+\Delta P}{A} + \frac{M_{max}^{f} + \Delta P \times \Delta e}{I} h_b = f_a \tag{7-32b}$$

where M_{min}^{f} and M_{max}^{f} are the minimum and maximum moments in the front zone, respectively.

Solving Eqs. 7-32a and 7-32b, we can obtain ΔP and Δe from which the required additional post-tendons and locations can be determined.

7.6.3   PERMANENT TENDONS

After the completion of launching, the bridge super-structure is in its final position. Some of the launching tendons are no longer useful and may be harmful and should be removed if possible. Some of the launching tendons in these areas can remain to carry mainly the super-structure self-weight and become permanent tendons. The super-structure segment over the interior supports will carry more negative moment and shear from mainly live loads and superimposed dead loads. The portion in midspan must carry more positive moments. Additional tendons are required in these areas. The required additional tendons for positive and negative moments can be estimated based on the maximum design negative and positive moments by meeting the service limits or strength limits as discussed in Chapters 5 and 6. It is preferable to locate the additional tendons to obtain the largest eccentricity of the resulting post-tensioning force.

7.7   TENDON LAYOUT

7.7.1   LAUNCHING TENDON LAYOUT

From the aforementioned discussion, the resulting force of the launching tendons should pass through or almost pass through the centroid of the cross section to create a uniform compression over the section. Currently, there are typically three ways to arrange the launching concentric tendons: coupled straight tendons, overlapped straight tendons, and curved and polygonal tendons.

7.7.1.1   Coupled Straight Tendons

With this method, every other launching tendon is joined by couplers at every other construction joint; i.e., only half of the straight launching tendons are coupled at each of the joints (see Fig. 7-28). This method has the following advantages and disadvantages:
Advantages:

- Simple form work and reinforcement
- Low prestress losses due to friction and low secondary stresses

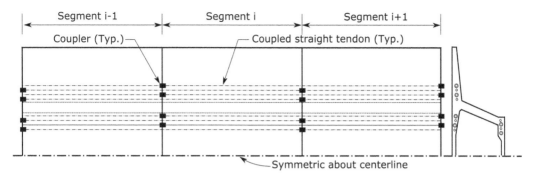

FIGURE 7-28   Coupled Straight Launching Tendon Layout.

Disadvantages:

- It is impossible to remove the unnecessary launching tendons after the completion of launching.
- It is often necessary to increase the deck thickness to accommodate the couplers and increase the super-structure self-weight.
- More post-tensioning tendons are necessary due to the short average tendon length, and couplers are expensive.

7.7.1.2   Overlapped Straight Tendons

With this method, the straight launching tendons run through the top and bottom slabs in the empty ducts and are anchored in blisters inside of the box girder (see Fig. 7-29a). The main advantage of

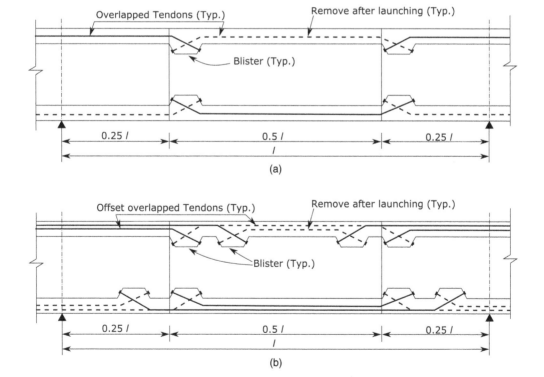

FIGURE 7-29   Overlapped Straight Launching Tendon Layouts, (a) Overlapped Post-Tensioning Tendons, (b) Offset Overlapped Post-Tensioning Tendons.

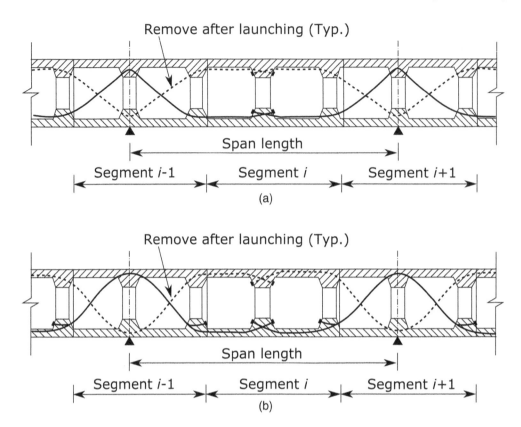

FIGURE 7-30 Polygonal and Curved Launching Tendon Layouts, (a) Polygonal Tendon Layout, (b) Curved Tendon Layout.

this tendon layout is that the tendons that are not required after launching can be removed. The main disadvantage of the method is that many blisters are required, which may increase the conventional reinforcement by about 20%. The coupled straight tendon layout also increases the formwork. To increase the shear resistance from the web, the top and bottom blisters can be offset from each other as shown in Fig. 7-29b.

7.7.1.3 Polygonal and Curved Tendon Layouts

The polygonal launching tendons are typically external tendons (see Fig. 7-30a). This type of tendon layout allows the launching tendons to be removed when they are no longer useful after the completion of launching. This layout also allows for the replacement of damaged tendons in the future. The curved launching tendons are normally designed as internal tendons which run though the webs and are anchored at blisters inside the box girder (see Fig. 7-30b). The shape of the tendons is more effective in resisting the anticipated design loadings in terms of moment and shear.

7.7.2 PERMANENT FINAL TENDON LAYOUT

After the completion of launching, some of the launching tendons are removed and the remaining tendons serve as permanent tendons. The additional tendons needed for resisting mainly the live loading and superimposed dead loadings are typically arranged in polygonal or curved shapes (see Fig. 7-31).

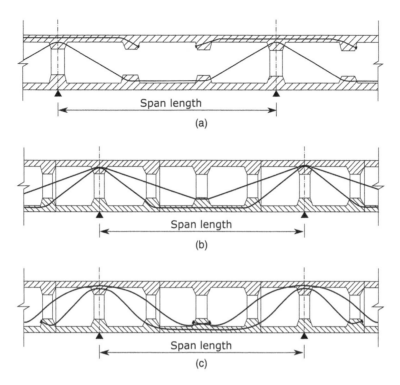

FIGURE 7-31 Final Permanent Tendon Layout, (a) Straight Internal Tendons and Polygonal External Tendons, (b) Polygonal External Tendon Layout, (c) Curved Internal Tendon Layout.

7.8 LONGITUDINAL AND TRANSVERSE ANALYSIS OF SUPER-STRUCTURES

After the preliminary tendon layouts have been finished, refined structural analysis can be performed to consider the effects of creep and shrinkage, prestressing losses, as well as others. Revise the preliminary design until all design limits are met. Currently, the structures longitudinal and transverse analysis of an incrementally launched bridge is typically done though computer software. The general analytical procedures are similar to those discussed in Chapters 5 and 6 and are therefore not repeated in this chapter.

7.9 DESIGN AND DETAILS OF DIAPHRAGMS AND BLISTERS

To reduce the self-weight of segments, diaphragms are typically built after the completion of launching. The basic design method and details of diaphragms are similar to those discussed in Sections 5.2.4 and 5.4.1. The design method and details of blisters can be found in Sections 5.2.4 and 5.4.2.

7.10 LAUNCHING EFFECTS ON SUBSTRUCTURES AND DESIGN CONSIDERATION OF PIER

7.10.1 LONGITUDINAL FORCES ON PIER DURING LAUNCHING

During launching, the sliding bearings are always parallel to the soffit of the girder deck (see Fig. 7-32). The lateral forces on the substructure due to the combined effects of the girders self-weight, its vertical slope/gradient, and friction forces during launching are generally much different from those during service and should be carefully considered. There are two distinct launching

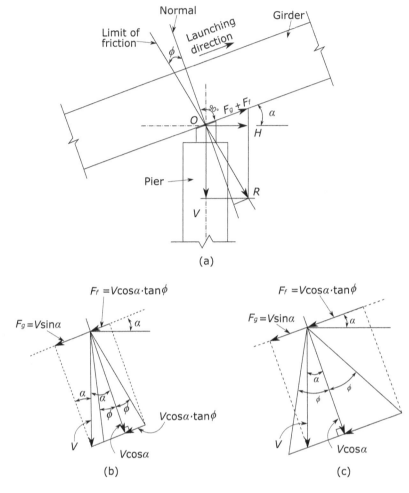

FIGURE 7-32 Launching Force Effect on Pier during Upward Launching, (a) Upward Launching, (b) Forces on Pier Cap for $\alpha > \phi$, (c) Forces on Pier Cap for $\alpha < \phi$.

methods depending on the longitudinal grades: upward launching (see Fig. 7-32) and downward launching (see Fig. 7-33). In these figures, the following notations are used:

α = angle of bridge super-structure about the horizontal (rad)
ϕ = angle of friction of bearing (rad)
R = reaction of super-structure self-weight on pier
V = vertical component of R
H = horizontal component of R

7.10.1.1 Upward Launching

Figure 7-32b illustrates the geometrical relationships between the vertical component V of the reaction R and the forces applied on the sliding face for $\alpha > \phi$. Figure 7-32c shows the geometrical relationships of the forces on the pier cap for $\alpha < \phi$. From these figures, it can be seen that the required launching force due to the super-structure self-weight and friction force for both conditions of $\alpha > \phi$ and $\alpha < \phi$ can be written as

$$F = F_g + F_f \ \left(\text{positive in launching direction}\right) \tag{7-33}$$

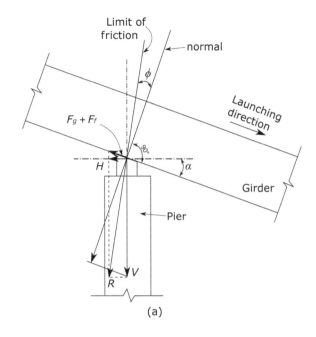

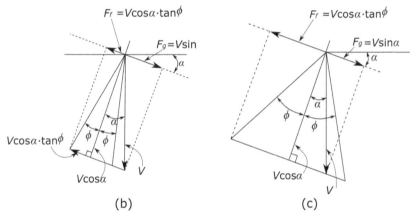

FIGURE 7-33 Launching Force Effect on Pier during Downward Launching, (a) Downward Launching, (b) Forces on Pier Cap for $\alpha > \phi$, (c) Forces on Pier Cap for $\alpha < \phi$.

where
  F_g = upward force due to reaction R:

$$F_g = V \sin \alpha \qquad (7\text{-}34)$$

F_f = upward friction force on pier during launching:

$$F_f = V \cos \alpha \tan \phi \qquad (7\text{-}35)$$

From Fig 7-32a, the lateral force on the pier can be written as

$$H = F \cos \alpha = V \cos \alpha (\sin \alpha + \cos \alpha \tan \phi) \qquad (7\text{-}36)$$

If α and ϕ are small, we have

$$\cos\alpha \approx 1, \quad \sin\alpha \approx \alpha, \quad \text{and} \quad \tan\phi \approx \phi$$

Thus, Eq. 7-36 can be written as

$$F = H = V(\alpha + \phi) \ (\text{positive in launching direction}) \tag{7-37}$$

7.10.1.2 Downward Launching

Figure 7-33b and c show the geometrical relationships between the vertical component V of the reaction R and the forces applied on the sliding face for the conditions of $\alpha > \phi$ and $\alpha < \phi$, respectively. From these figures, we can also see that the force due to component V and the friction force on the sliding surface are in opposite direction. The required launching force and the forces on the pier due to the super-structure component V and the friction force for both conditions of $\alpha > \phi$ and $\alpha < \phi$ can be written as

Launching force:

$$F = F_f - F_g = V(\cos\alpha\tan\phi - \sin\alpha) \ (\text{positive in launching direction}) \tag{7-38}$$

Force on pier:

$$H = F\cos\alpha = V\cos\alpha(\cos\alpha\tan\phi - \sin\alpha) \ (\text{positive in launching direction}) \tag{7-39}$$

If α and ϕ are small, Eqs. 7-38 and 7-39 can be written as

$$F = H = V(\phi - \alpha)(\text{positive in launching direction}) \tag{7-40}$$

From Eq. 7-40, if $\alpha > \phi$, the launching force and the force acting on the pier are opposite to the launching direction. As the possible variation of the angle of friction due to environmental condition, the launching equipment and the pier are typically designed for $F = H = -V\alpha$ by assuming $\phi \approx 0$ for both $\alpha > \phi$ and $\alpha < \phi$. The downward movement of the super-structure is controlled by a restraining jacking force and a braking system.

7.10.2 TRANSVERSE FORCES DURING LAUNCHING

During launching, the transverse friction is theoretically zero due to movement and should not be taken into account in the design. All the anticipated transverse lateral forces should be resisted by the transverse guides (see Fig. 7-34). Transverse forces may be caused by wind, transverse slope, possibly out of tolerance of guides, and curvatures[7-2].

7.10.2.1 Transverse Force Due to Wind and Transverse Slope

The horizontal wind loading as shown in Fig. 7-34 was discussed in Section 2.2.5. The wind load not only generates transverse forces but also additional moment on the pier.

Typically, the soffit of the super-structure is designed to have 1% to 3% transverse slope as shown in Fig 7-34 for a more effective guidance of the super-structure during launching. The lateral force due to the slope can be approximately written as

$$H_t = V_g\beta \tag{7-41}$$

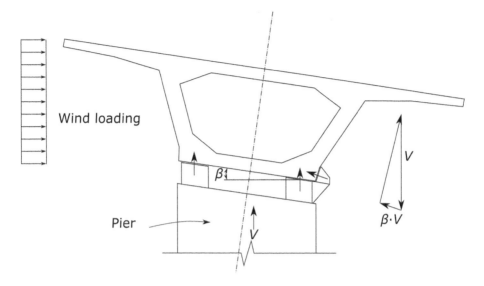

FIGURE 7-34   Forces on the Pier Due to the Wind and Transverse Slope.

where
   V_g = vertical component of reaction on t sliding bearing
   β = transverse slope of soffit

7.10.2.2   Radial Force Due to Curvature

For a curved super-structure, a radial force must be added using guides to make the rigid super-structure turn during launching (see Fig. 7-35). Assuming all supports and piers have the same stiffness, from the discussion in Section 3.5, the radial force on pier n can be approximately written as

$$H_{ri} = H_{li} \frac{l_i}{R} \; (i = 1 \text{ to n})$$

(7-42)

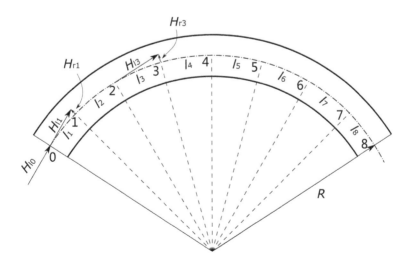

FIGURE 7-35   Radial Forces in Curved Bridge Due to Launching and Analytical Example.

where
H_{ri} = radial force at support i during launching
H_{li} = longitudinal force at support i during launching
l_i = length of span i
R = radius
n = total span number

From Eqs. 7-37 and 7-40, the launching force at the end of the bridge can be written as

$$H_{l0} = \sum_{i=1}^{n} V_i (\mu + \alpha)$$

(7-43)

where
V_i = vertical loading at support i
μ = coefficient of friction
α = longitudinal grade (upward positive)

Assuming the longitudinal force at pier 1 is equal to the total launching force, i.e., $H_{l1} = H_{l0}$, from Eq. 7-42, we have

$$\text{Radial force at pier 1: } H_{r1} = H_{l0}\frac{l_1}{R}$$

The longitudinal and radial forces at pier 2 can be written as

$$H_{l2} = H_{l0} - V_1(\mu + \alpha)$$

$$H_{r2} = \left[H_{l0} - V_1(\mu + \alpha)\right]\frac{l_2}{R}$$

Thus, Eq. 7-42 can be rewritten as

$$H_{ri} = \left[H_{l0} - \sum_{j=1}^{i-1} V_j(\mu + \alpha)\right]\frac{l_i}{R} \quad (j \le i)$$

(7-44)

7.10.2.3 Analytical Example

An eight-span continuous box girder bridge as shown in Fig. 7-35 is to be built using the incremental launching method. Determine the radial forces at piers 1 and 3 during launching, assuming the bridge has radius $R = 800$ ft, span length $l = 123$ ft, total bridge length $L = 984$ ft, coefficient of friction $\mu = 4\%$, longitudinal grade $\alpha = 1\%$, and unit weight per foot $g = 11.7$ kips/ft.

Solve:
From Eq. 7-43, we have the total launching force as

$$H_{l0} = \sum_{i=1}^{n} V_i(\mu + \alpha) = 11.7 \times (8 \times 123)(0.04 + 0.01) = 575.64 \text{ kips}$$

Using Eq. 7-44 for $i = 1$, the radial force at pier 1 can be obtained as

$$H_{ri} = \left[H_{l0} - \sum_{j=1}^{1-1} V_j (\mu + \alpha) \right] \frac{l_1}{R} = (575.64) \frac{123}{800} = 88.5 \text{ kips}$$

The total vertical loading at pier 1 is

$$V_1 = 123 \times 11.7 = 1439.1 \text{ kips}$$

The ratio of radial force at pier 1 to the vertical loading is

$$\frac{H_{ri}}{V_1} = \frac{88.5}{1439.1} = 6.2\%$$

The radial force at pier 3 can be obtained by using Eq. 7-44 and inserting $i = 3$, that is

$$H_{r3} = \left[H_{l0} - \sum_{j=1}^{4-1} V_j (\mu + \alpha) \right] \frac{l_3}{R} = \left[575.64 - 11.7 \times 3 \times 123 \times (0.04 + 0.01) \right] \frac{123}{800} = 53.3 \text{ kips}$$

The ratio of the radial force at pier 3 to the vertical loading is

$$\frac{H_{ri}}{V_1} = \frac{53.3}{1439.1} = 3.8\%$$

From the above example, it can be seen that a side guide is needed to resist the radial force at both piers even with a transverse slope of 3%.

7.10.3 Temporary Pier Staying Cables for Reducing the Effects of Launching Forces

As discussed above, during launching, the bridge slides over the top of the pier, and a large moment in the pier may be induced and may control the pier design. The buckling length of the pier during construction is also larger than that during the bridge service. Additional supports during launching for a longitudinal slope greater than 5% may be necessary to reduce the effect of launching forces. For tall piers, it may be best to use ties anchored at the top of the pier and to the ground to balance the launching forces (see Fig. 7-36). When designing the ties, the following issues should be considered[7-2]:

1. Typically, 1- or 1½-in.-diameter thread bars, grade 150, are used for the staying cables.
2. The reduced modulus of elasticity of the cables E_r due to large sag should be used, i.e.,

$$E_r = \frac{E}{1 + \frac{(\gamma + l)^2}{12 f_s^3}} \tag{7-45}$$

where
E = modulus of elasticity of cables
γ = unit weight per length
l = horizontal length of cable
f_s = tensile stress of cable

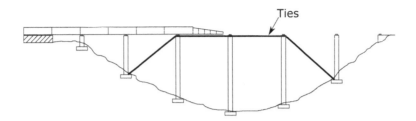

FIGURE 7-36 Typical Ties for Multi-Piers.

3. Because of the effects of temperature and sag, prestressing forces of about 22 ksi or up may be required for the cables.
4. Considering the fatigue effects of the coupler and anchor of the threaded bars and wind-induced oscillation, the design stress of the bars should typically not be greater than 0.5 times the allowable stress.

If a pier is very high, the launching equipment can be directly installed in the pier to eliminate the horizontal force caused by launching.

7.10.4 Pier Cap Dimensioning for Sliding Bearing

The thickness of the pier cap should be able to accommodate the placement of the sliding bearing and the hydraulic jacks as shown in Fig. 7-37. For an average span length and girder dimensions, the length of the sliding bearing is about 47 in. (1200 mm). A total width of 8.2 ft (2.5 m) should be enough for arranging both sliding bearing and jacks.

7.11 DESIGN EXAMPLE III: INCREMENTALLY LAUNCHED SEGMENTAL BRIDGE[1]

This design example is based on an existing bridge built in China[7-13]. The purpose of this design example is to illustrate the basic design procedures for this type of bridge. As the bridge transverse analysis and design are similar to Design Example I presented in Chapter 5, only the bridge's longitudinal design and analysis will be discussed in this design example. The readers should realize that some of the details may not meet the current AASHTO specifications and that the design could be improved. As the original bridge was designed using the metric system, both metric and English units will be shown in this design example for most cases.

7.11.1 Design Requirements

7.11.1.1 Design Specifications

a. AASHTO LRFD *Bridge Design Specifications*, 7th edition, 2014
b. CEB-FIP *Model Code for Concrete Structures*, 1990

7.11.1.2 Traffic Requirements

a. Two 3.66-m (12-ft) lanes with traffic flow in the same direction
b. Right shoulder and left shoulder each 1.83 m (6 ft) wide
c. Two 0.5-m-wide (18.5-in.-wide) FDOT 32 in. F-shape barrier walls

[1] This design example is provided by Prof. Chunsheng Wang, and Dr. Peijie Zhang, School of Highway, Chang'an University, China, revised by Dr. Dongzhou Huang, P.E.

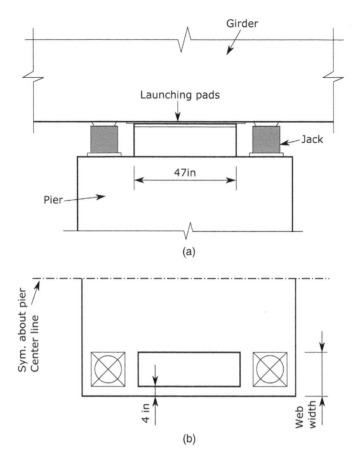

FIGURE 7-37   Cap Dimensioning for Sliding Bearing, (a) Elevation View, (b) Plan View.

7.11.1.3   Design Loads
a. Live load: HL-93 loading
b. Dead loads:
   –Unit weight of reinforced concrete (DC): $w_{dc} = 23.56$ kN/m^3 (0.150 kcf)
   – Asphalt overlay: 8 cm (3.15 in.) thick, 22 kN/m (1.51 kips/ft)
c. Thermal loads:
   – Uniform temperature:
     Mean: 21.11°C (70°F); rise: 19.45°C (35°F); fall: 19.45°C (35°F)
   – Temperature gradient specified as Section 2.2.10.2 with the following values:

Positive Nonlinear Gradient | Native Nonlinear Gradient
$T_1 = 5$°C(41°F) | $T_1 = -24.63$°C (-12.3°F)
$T_2 = -11.67$°C (11°F) | $T_2 = -19.61$°C (-3.3°F)
$T_3 = -17.78$°C (0°F) | $T_3 = -17.78$°C (0°F)

d. Design wind speeds: 25 m/s (55 mph) with traffic; 40.5 m/s (90 mph) without traffic

7.11.1.4   Materials
a. Concrete
   $f'_c = 44.82$ MPa (6.5 ksi), $f'_{ci} = 35.86$ MPa (5.2 ksi)
   $E_c = 32,034.48$ MPa (4645 ksi)
   Relative humidity: 0.75

   b. Prestressing strands
     ASTM A416 Grade 270 low relaxation
     $f_{pu} = 1862$ MPa (270 ksi), $f_{py} = 0.9\ f_{pu} = 1676\ MPa$ (243 ksi)
     Maximum jacking stress: $f_{pj} = 1490$ MPa (216 ksi)
     Modulus of elasticity: $E_p = 196,552$ MPa (28,500 ksi)
     Anchor set: 9.52 mm (0.375 in.)
     Friction coefficient: 0.23 (plastic duct)
     Wobble coefficient: 0.00020 (internal)
   c. Reinforcing steel
     ASTM 615, Grade 60
     Yield stress: $f_y = 413.7$ MPa (60 ksi)
     Modulus of elasticity: $E_s = 206,900$ MPa (30,000 ksi)

7.11.2   BRIDGE SPAN ARRANGEMENT AND TYPICAL SECTION

7.11.2.1   Span Arrangement

The existing bridge consists of an eastbound and westbound pair that cross a river with deep water. Considering the requirements for navigation and site conditions, a five-span continuous bridge was proposed. Based on the comparisons among several feasible bridge alternatives, an incrementally launched segmental alternative was selected for this bridge. The plan and elevation views of the eastbound bridge are shown in Fig. 7-38. The bridge typical section view is illustrated in Fig. 7-39. The interior spans have equal span lengths of 40 m (131.23 ft), and the end spans are 30 m (98.43 ft). In Fig. 7-38, notations "E" and "F" represent expansion and fixed supports, respectively.

7.11.2.2   Typical Sections

As previously discussed, for an incrementally launched segmental bridge, the most suitable cross section for the super-structure is a box section, which is selected for this bridge (see Fig. 7-39). From the discussion in Section 7.2.3.1, the ratio of span length to girder depth generally should not be greater than 17. As the span length is short, a constant girder depth of 2.5 m (8.2 ft) is selected, and the ratio of span to girder depth is 16. The total bridge width is 12 m (39.37 ft) as required. To meet the design requirement, variable bottom slab thicknesses are

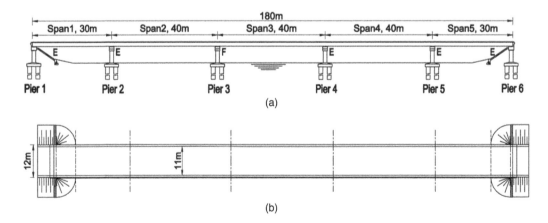

FIGURE 7-38   Elevation and Plan Views of Design Example III, (a) Elevation View, (b) Plan View, (unit = m, 1 m = 3.28 ft).

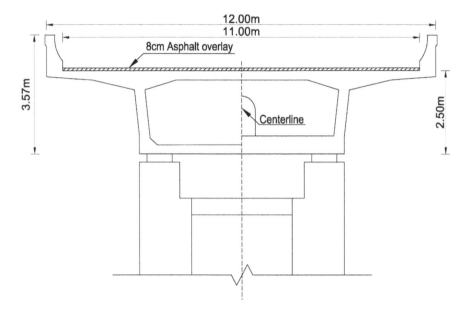

FIGURE 7-39 Side View of Design Example III (1 m = 3.28 ft, 1 cm = 0.3937 in.).

used for this bridge. The bottom slab thickness varies from 55 cm (21.65 in.) at the section over the interior support to 27.5 cm (10.83 in.) at the section 1.05 m (3.44 ft) away from the support. The constant thicknesses of the top slab and web are 25 cm (9.84 in.) and 32 cm (12.6 in.) respectively. There is an asphalt overlay of 8 cm (3.15 in.) over the top slab. The principle dimensions of the box section are given in Fig. 7-40.

7.11.3 SEGMENT LAYOUT AND CONSTRUCTION SEQUENCE

7.11.3.1 Segment Layout

For an easy arrangement of the segments, the length of the typical cast segments is selected as 10 m (32.81 ft) and the length of two short segments in the end spans is set as 5 m (16.40 ft). There are a total of 19 cast segments (see Fig. 7-41). It should be mentioned that a longer segment length may yield a better design as discussed in Section 7.2.2.2.

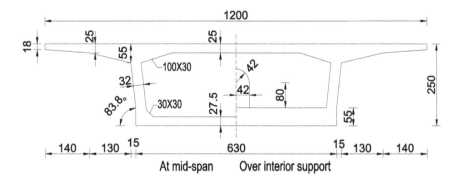

FIGURE 7-40 Typical Section of Design Example III (unit = cm, 1 cm = 0.3937 in.).

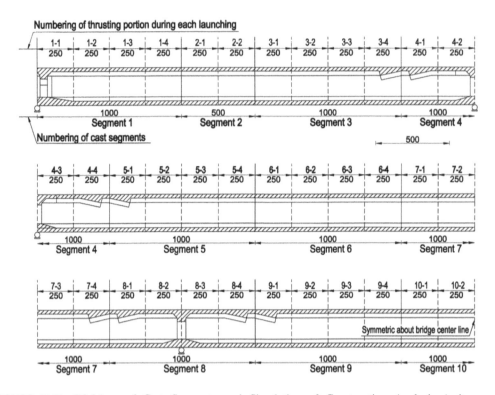

FIGURE 7-41 Divisions of Cast Segments and Simulation of Construction Analysis (unit = cm, 1 cm = 0.3937 in.).

7.11.3.2 Construction Sequences and Assumptions

The example bridge is assumed to be launched from pier 6 to pier 1 (see Fig. 7-38). Theoretically, the bridge undergoes an unlimited number of stress states. However, for actual structural analysis, we need to assume a limited number of construction stages and determine the corresponding structural behavior. For this reason, we assume the girder thrusting length during each launch is 2.5 m (8.2 ft) and then the girder internal forces are analyzed based on the resulting structural configuration at the end of each launching sequence. The notations of the thrusting portions during each launch are illustrated in Fig. 7-41. Each of the notations contains two numbers in which the first one represents the cast segment and the second one indicates the launching number within the segment. From Fig. 7-41, we can see that there are 72 different super-structural configurations from beginning to end of the launching based on the assumptions. To reduce the girder bending moments during construction, a launching nose (see Fig. 7-42) and several temporary supports in span 5 are employed (see Fig. 7-43). The span length of the temporary supports is 5 m (16.4 ft). The length of the launching nose is assumed to be 25 m (82 ft) as shown in Fig. 7-42. The ratio of the length of the launching nose to the interior span length is 0.625, which is close to the optimum length ratio of 0.65 as discussed in Section 7.5.1.3. The launching nose consists of two steel plate girders connected by lateral bracings, and the steel plate girders are bolted together with the first box girder segment through the anchors embedded into it (Fig. 7-42). The typical construction sequences are described as follows:

Step 1: Install temporary supports, cast segment 1, install required post-tendons, and attach the launching nose to the front end of the segment (see Fig. 7-43a).
Step 2: Thrust one-quarter length of segment 1 (portion 1-1, Fig. 7-41) forward as shown in Fig. 7-42b.
Step 3: Thrust portion 1-2 (see Fig. 7-41) forward as shown in Fig. 7-43c.

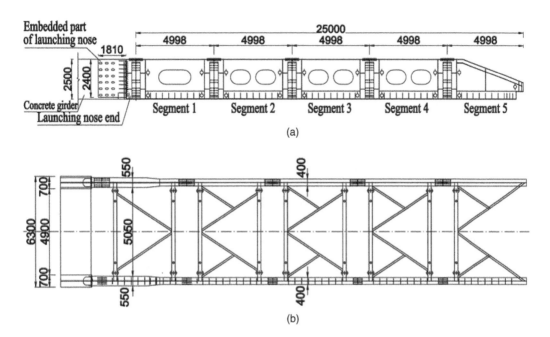

FIGURE 7-42   Launching Nose, (a) Elevation View, (b) Plan View (unit = mm, 1 mm = 0.03937 in.).

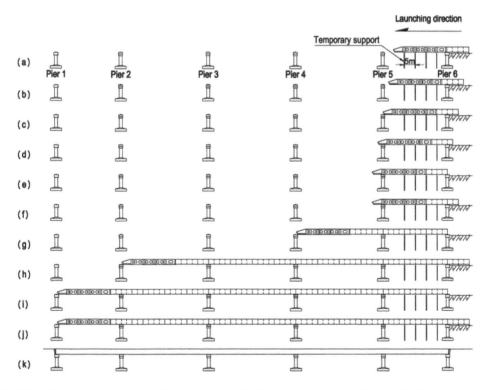

FIGURE 7-43   Assumed Construction Sequences, (a) Step 1, (b) Step 2, (c) Step 3, (d) Step 4, (e) Step 5, (f) Step 6, (g) Step 7, (h) Step 8, (i) Step 9, (j) Step 10, (k) Step 11.

Step 4: Thrust portion 1-3 (see Fig. 7-41) forward as shown in Fig. 7-43d.

Step 5: Thrust portion 1-4 (see Fig. 7-41) forward as shown in Fig. 7-43e.

Step 6: Cast segment 2, and install required post-tendons as designed (see Fig. 7-43f).

Step 7: Thrust portions 2-1 and 2-2 (see Fig. 7-41) forward, and follow steps 2 to 5 to thrust segments 3 to 5 forward as shown in Fig. 7-43g.

Step 8: Following steps 2 to 6, thrust segments 6 to 13 forward and cast segment 14 and install required post-tendons as designed (see Fig. 7-43h).

Step 9: Following steps 2 to 6, thrust segments 14 to 16 forward, as shown in Fig. 7-43i.

Step 10: Use the same procedure as described in steps 8 and 9 (see Fig. 43j) or steps 6 and 7 until the first segment reaches pier 1.

Step 11: Remove temporary supports, disconnect the launching nose, install final tendons, and complete box girder installation (see Fig. 7-43k).

7.11.4 Longitudinal Tendon Layout

After the preliminary dimensions of the bridge super-structure and construction procedures have been determined, the girder moment variations along the bridge length during construction and in the final bridge configuration due to construction loadings, dead loads, and live loads can be calculated (see Section 7.11.5). The required post-tensioning tendons can be estimated to meet the service requirements by assuming the tendon effective stress as 0.6 times the tendon ultimate strength f_{pu}. To simply the stressing operation of the tendons, only 12~ϕ15.4-mm (12~ϕ0.6-in.) tendons will be used for the bridge. As discussed in Section 7.6, there are two types of tendons for incrementally launched segmental bridges. To simplify construction, in this design example, all the launching centric tendons are straight, permanent, and embedded in the top and bottom slabs. The amount of the launching tendons is estimated according to Eqs. 7-31 and 7-32 (see Section 7.11.5.2). The launching tendons layout is to produce an approximate centric prestressing force as discussed in Section 7.6.2 and arranged close to the webs (Fig. 7-44). There are 10 launching tendons in the bottom slab,

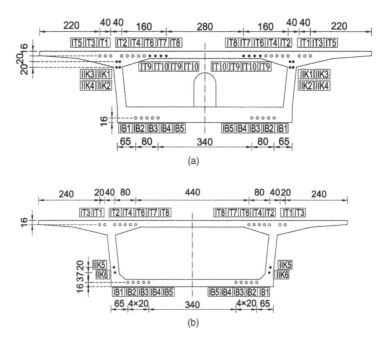

FIGURE 7-44 Tendon Layout at Typical Sections, (a) Section over Pier 2, (b) Section at Bridge Center Section (unit = cm, 1 cm = 0.3937 in.).

TABLE 7-3

Number of Tendons at Typical Sections

Section		Launching Centric Tendons		Permanent Eccentric Tendons		
					Web	
Number of Thrusting Portion	Location	Top Slab	Bottom Slab	Top Slab	Upper Region	Lower Region
1-1	Begin	12	10	—	4	
1-3	End	12	10	—	—	4
2-2	End	16	10	—	—	4
3-3	End	16	10	4	4	4
4-2	End	16	10	8	8	
5-1	End	16	10	—	2	2
6-2	End	16	10	—	—	4
7-2	End	14	10	—	4	4
8-2	End	14	10	8	8	
9-2	End	14	10	—	2	2
10-2	End	14	10	—	—	4

while the number of launching tendons in the top slab varies from 12 to 16 (see Table 7-3). The typical locations of the launching tendons at the section over pier 2 and the bridge center section are illustrated in Fig. 7-44a and b, respectively. Part of the plan view of the top and bottom launching tendons layouts are illustrated in Figs. 7-45 and 7-46, respectively. The top launching tendons are denoted as IT1 to IT8, and the bottom launching tendons are denoted as IB1 to IB5. Both the top and

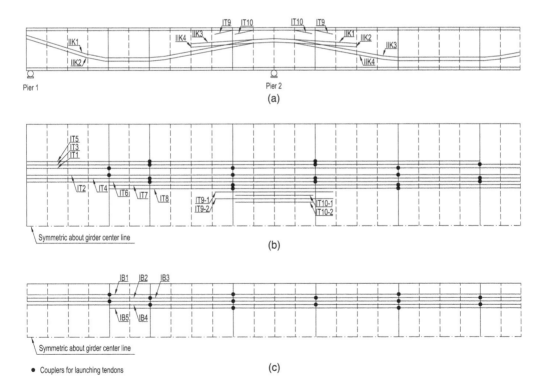

FIGURE 7-45 Tendon Layout in the Bridge Longitudinal Direction from the Beginning of the Bridge to the First Three-Quarters of Span 2 (a) Elevation View, (b) Top Plan View, (c) Bottom Plan View.

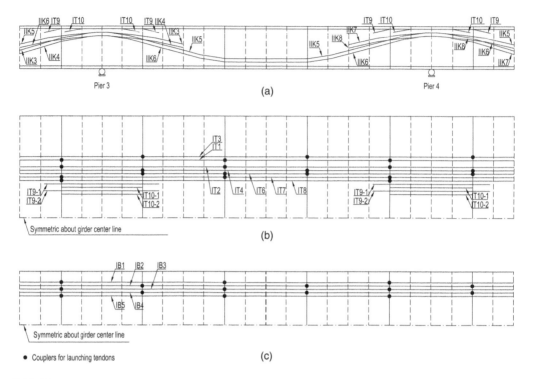

FIGURE 7-46 Tendon Layout in the Bridge Longitudinal Direction from the Last One-Quarter of Span 2 to the First One-Quarter of Span 4 (a) Elevation View, (b) Top Plan View, (c) Bottom Plan View.

bottom tendons are depicted as voided (unshaded) circles in Fig. 7-44. There are 16 12~ϕ15.4-mm (12~ϕ0.6-in.) permanent eccentric tendons in the top slab or in the webs over the interior support to resist any additional negative moment due to live loads and additional dead loads. Four permanent eccentric tendons are used to resist any positive moment due to additional live and dead loads. The numbers of eccentric tendons in the remaining sections are given in Table 7-3. The shape of the eccentric tendons is similar to the shape of the moment envelope in the final bridge structure. Part of the permanent eccentric tendon layout in the bridge longitudinal direction is illustrated in Figs. 7-45 and 7-46. In Figs. 7-44 to 7-46, the notations IT9 to IT10 and IIK1 to IIK8 indicate the permanent eccentric tendons, which are shown as solid circles in Fig. 7-44.

7.11.5 Bridge Longitudinal Analysis

7.11.5.1 Bridge Model and Determination of Section Properties

7.11.5.1.1 Bridge Modeling

As discussed in Section 7.11.3.2, it would be time consuming to perform accurate structural analysis for different construction stages using hand calculations. The computer program MIDAS Civil was used for the structural analysis. The bridge is modeled as a three-dimensional beam structure (see Fig. 7-47) and analyzed using the finite-element method as discussed in Chapter 4. The coordinate system is shown in Fig. 7-47. The length of the typical finite elements is 2.5 m (8.2 ft), which matches the assumed thrusting length. For the segments with variable bottom slab thickness, the length of the elements is reduced to 1.25 m (4.1 ft) to increase analytical accuracy. The launching nose attached to the main girder is divided into 10 elements with an equal element length of 2.5 m (8.2 ft). The super-structure systems with the launching nose and support conditions are modeled

FIGURE 7-47 Analytical Model of the Super-Structure.

based on the erection sequences described in Section 7.11.3.2. To obtain the maximum negative moment of the girder, the girder is assumed to be a cantilever when it just reaches a pier.

7.11.5.1.2 Typical Section Properties
A. Sectional Properties at Midspan

Section area: $A = 6.586 \ m^2 \ \left(70.891 \ ft^2\right)$
Area enclosed by shear flow path: $A_0 = 13.72m^2 \ \left(147.681 \ ft^2\right)$
Moment of inertia about y-axis: $I_y = 6.279m^4 \ \left(727.496 \ ft^4\right)$
Moment of inertia about z-axis: $I_z = 5.726m^4 \ \left(663.425 \ ft^4\right)$
Free torsion moment of inertia: $J = 1.389m^4 \ \left(160.932 \ ft^4\right)$
Distance from neutral axis to top of the top slab: $z_t = 0.963m \ \left(3.159 \ ft\right)$
Distance from neutral axis to bottom of the bottom slab: $z_b = 1.537m \ \left(5.043 \ ft\right)$
Static moment of area above neutral axis: $Q_z = 4.647m^2 \ \left(50.020 \ ft^2\right)$
Web orientation angle: $\theta_w = 83.88°$

B. Properties at Sections over Interior Support

Section area: $A = 8.331m^2 \ \left(89.674 \ ft^2\right)$
Area enclosed by shear flow path: $A_0 = 12.87m^2 \ \left(138.101 \ ft^2\right)$
Moment of inertia about y-axis: $I_y = 7.963m^4 \ \left(922.607 \ ft^4\right)$
Moment of inertia about z-axis: $I_z = 6.188m^4 \ \left(716.953 \ ft^4\right)$
Free torsion moment of inertia: $J = 1.581m^4 \ \left(183.178 \ ft^4\right)$
Distance from neutral axis to top of the top slab: $z_t = 1.193m \ \left(3.914 \ ft\right)$
Distance from neutral axis to bottom of the bottom slab: $z_b = 1.307m \ \left(4.288 \ ft\right)$
Static moment of area above neutral axis: $Q_z = 6.102m^2 \ \left(65.681 \ ft^2\right)$
Web orientation angle: $\theta_w = 83.88°$

C. Launching Nose

Section area: $A = 0.214 \ m^2 \ \left(2.303 \ ft^2\right)$
Moment of inertia about y-axis: $I_y = 0.123 \ m^4 \ \left(14.251 \ ft^4\right)$

7.11.5.2 Construction Analysis and Estimation of Launching Tendons
7.11.5.2.1 Construction Analysis
7.11.5.2.1.1 Estimation by Approximate Equations
A. Front Zone

The maximum and minimum moments in the launched girder due to girder self-weight can be estimated using Eqs. 7-16 to 7-18 as follows. From Section 7.11.5.1.2, we can obtain

Average self-weight of main girder per unit length: $q_g = 164.65$ kN / m (11.283 kips/ft)
Unit weight of launching nose: $q_{ln} = 16.80$ kN / m (1.151 kips/ft)
Ratio of self-weight of launching nose to that of main girder: $\gamma = \frac{q_g}{q_{ln}} = 0.1$
Ratio of length of launching nose to span length: $\beta = \frac{25}{40} = 0.625$
Ratio of concrete girder cantilever length to span length: $\alpha = 1 - \beta = 0.375$

The maximum positive moment can be estimated using Eq. 7-16 as

$$M_{max} = \frac{q_g l^2}{12}\left(0.933 - 2.96\gamma\beta^2\right) = \frac{164.65 \times 40^2}{12}\left(0.933 - 2.96 \times 0.1 \times 0.625^2\right)$$
$$= 17944.1 \ kN - m \ (13135.55 \ \text{kips-ft})$$

By trying both Eqs. 7-17 and 7-18, it was found that the maximum negative moment occurs when the launching nose just reaches the end abutment, i.e.,

$$M_{min} = -\frac{q_g l^2}{2}\left(\alpha^2 + \gamma\left(1 - \alpha^2\right)\right) = -\frac{164.65 \times 40^2}{2}\left(0.375^2 + 0.1\left(1 - 0.375^2\right)\right)$$
$$= -29842.8 \ kN - m \ (-22012.02 \ \text{kips-ft})$$

B. Rear Zone
The maximum and minimum moments of the launched girder due to its self-weight can be estimated as follows (see Fig. 7-13):

Positive moment: $M_{pos} = \frac{q_g l^2}{24} = \frac{164.65 \times 40^2}{24} = 10976.6 \ kN - m \ (8096.33 \ \text{kips-ft})$

Negative moment: $M_{neg} = -\frac{q_g l^2}{12} = \frac{164.65 \times 40^2}{12} = -21953.3 \ kN - m \ (-16192.73 \ \text{kips-ft})$

7.11.5.2.1.2 Analysis by Numerical Method Based on the assumed construction sequences described in Section 7.11.3.2 and the bridge modeling described in Section 7.11.5.1.1, the maximum and minimum moments at different girder locations are determined through analyzing 72 different super-structure configurations developed during the launching process. The moment envelope is illustrated in Fig. 7-48. From Fig. 7-48, it can be observed that Eqs. 7-16 to 7-18 provide relatively accurate results in estimating the maximum and minimum moments in the front zone that the girder develops during the launching process.

7.11.5.2.2 Estimation of Launching Tendons
From Section 1.2.5.1.3.2, the allowable concrete flexure tensile stress excluding "other loads" for construction is

$$f_{ta} = 0.19\sqrt{f_c'(ksi)} = 3.34 \text{ MPa} \ (0.484 \text{ ksi})$$

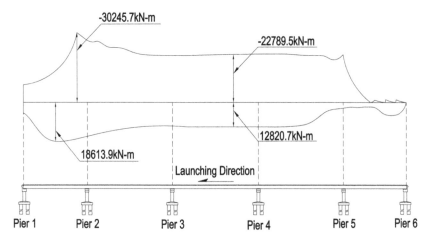

FIGURE 7-48   Envelope of Bending Moments during Launching (1 kN-m = 0.738 kips-ft).

From Eq. 7-28, the minimum launching prestressing force for meeting the minimum negative moment in the rear zone is

$$P_1 = -\left(\frac{M_{min}^r y_t}{I_y} + f_{at} \right) A_g = \left(\frac{22789.5 \times 0.963}{6.279} - 3.34 \right) 6.586 = 22997.3 \; kN$$

The minimum launching prestressing force for meeting the maximum positive moment in the rear zone is

$$P_2 = \left(\frac{M_{max}^r y_b}{I_y} - f_{at} \right) A_g = \left(\frac{12820.7 \times 1.537}{6.279} - 3.34 \right) 6.586 = 20646.9 \; kN$$

The minimum required prestressing force in the rear zone is

$$P_{min} = \max\left(P_1, P_2 \right) = 22997.3 \; kN$$

From Eq. 7-29, the top minimum launching prestressing force in the rear zone is

$$P_{top} = \frac{y_b - 0.16}{y_t + y_b - 0.32} P_{min} = \frac{1.537 - 0.16}{0.963 + 1.537 - 0.32} 22997.3 = 14526.29 \; kN$$

From Eq. 7-30, the bottom minimum launching prestressing force in the rear zone is

$$P_{bottom} = \frac{y_t - 0.16}{y_t + y_b - 0.32} P_{min} = \frac{0.963 - 0.16}{0.963 + 1.537 - 0.32} 22997.3 = 8471.03 \; kN$$

The required top launching tendon area in the rear zone is

$$A_{t-ps} = \frac{P_{top}}{0.6 f_{pu}} = \frac{14526.29 \times 1000}{0.6 \times 1860} = 13016.38 \; mm^2$$

The required bottom launching tendon area in the rear zone is

$$A_{b-ps} = \frac{P_{bottom}}{0.6 f_{pu}} = \frac{8471.03 \times 1000}{0.6 \times 1860} = 7590.53 \ mm^2$$

As aforementioned, using 12φ15.4-mm (12φ0.6-in.) tendons, the minimum number of launching tendons required in the rear zone are

$$\text{Top: } N_{top} = \frac{A_{t-ps}}{140 \times 12} = 7.7$$

$$\text{Bottom: } N_{bottom} = \frac{A_{b-ps}}{140 \times 12} = 4.5$$

Using Eqs. 7-32a and 7-32b, we can estimate the number of launching tendons required in the front zone. Based on the estimated launching tendons in the rear and front zones, as well as considering other factors, the number of bottom launching tendons for both the rear and the front zones is taken as 10 and the numbers of the top launching tendons are selected from 12 to 16 as discussed in Section 7.11.4. It is apparent that the amount of launching tendons selected is more than enough.

7.11.5.3 Bridge Analysis in Service Stage

7.11.5.3.1 Effects Due to Dead Loads

The moment and shear due to the box girder self-weights and other permanent loadings are analyzed based on the assumed bridge construction sequences using the MIDAS Civil computer program. The self-weights of the barrier walls and asphalt overlay are calculated as 37.05 kN/m (2.539 kips/ft) and applied on the final five-span continuous bridge. The moment and shear envelopes due to the dead loads are illustrated in Figs. 7-49 and 7-50, respectively.

7.11.5.3.2 Effects Due to Live Loads

The design live loads include a single truck or tandem with a uniform distributed lane loading as discussed in Section 2.2.3. The maximum effects at the related sections are determined by the considerations discussed in Section 5.6.4.5. The effects due to live loads are analyzed using finite-element methods and performed by the MIDAS Civil computer program. The moment and shear envelopes are shown in Figs. 7-51 and 7-52, respectively. In these figures, a dynamic load allowance

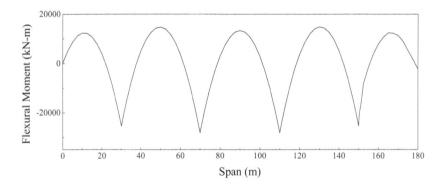

FIGURE 7-49 Moment Envelope Due to Dead Loads (1 m = 3.28 ft, 1 kN-m = 0.738 kips-ft).

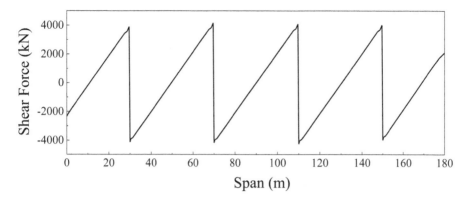

FIGURE 7-50   Shear Envelope Due to Dead Loads (1 m = 3.28 ft, 1 kN = 0.225 kips).

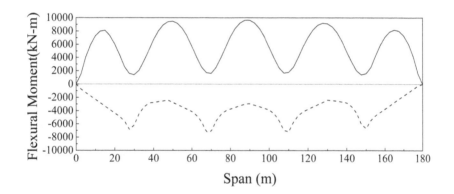

FIGURE 7-51   Moment Envelope Due to Live Loads (1 m = 3.28 ft, 1 kN-m = 0.738 kips-ft).

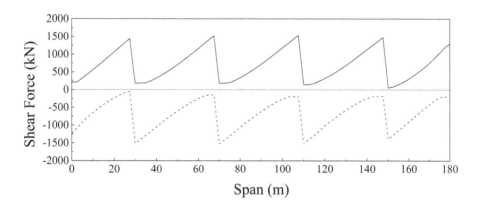

FIGURE 7-52   Shear Envelope Due to Live Loads (1 m = 3.28 ft, 1 kN = 0.225 kips).

TABLE 7-4

Summary of Maximum Load Effects of Design Example III in the Girder's Longitudinal Direction at Control Sections (1 m = 3.28 ft, 1 kN = 0.225 kips, 1 kN-m = 0.738 kips-ft)

Section [Measured from Beginning of Bridge, (m)]	Load Effects	Loadings							
		DC	CR + SH	PS	LL (1 + IM)	BR	WS	WL	TG
90	Positive moment (kN-m)	14,751	−9	−8,260	9,617	497	2,420	—	5,390
	Axial force (kN)	—	—	−39,212	—	178			
70	Negative moment (kN-m)	−28,250	90	29,302	−7,181	538	−4,232	—	4,710
	Axial force (kN)	—	—	−66,516	—	178	−159		
28.5	Torsion (kN-m)	—	—	—	4,168	—	286	956	
	Corresponding shear (kN)	3,902	7	−9,410	1,433	18	657	—	−175(min)/ 50(max)

of 1.33 is included. The dotted lines shown in these figures represent the positive responses, and the solid lines indicate the negative responses.

The braking forces are determined based on the method described in Section 2.2.4.4 and are taken as the maximum effects caused by the four loading cases: 25% of the axle weights of the design truck, 25% of the axle weights of design tandem, 5% of the design truck plus lane load, and 5% of the design tandem plus lane load. Each of the horizontal forces acting eccentrically from the girder neutral axis is transferred to one horizontal force and one moment acting at the girder neutral axis (see Fig. 5-50). The analytical results are summarized in Table 7-4.

7.11.5.3.3   Effects Due to Post-Tensioning Tendons

In determining the prestressing losses of the tendons due to concrete shrinkage and creep, the casting and curing period for each segment is assumed to be 10 days and the launching period for each segment is assumed to be 12 days. The tendon prestress at anchorages immediately after anchor set is assumed to be 1250 MPa (181.25 ksi). All the prestressing force losses are determined based on the methods described in Section 3.2 and the assumptions given in Section 7.11.1.4. The detailed analysis is performed using the MIDAS Civil computer program. The variation of the effective prestressing force along the girder length at 10,000 days is shown in Fig. 7-53. The moment envelopes due to primary, secondary, and total effects at 10,000 days are provided in Fig. 7-54. The shear envelope due to the effective post-tensioning forces at 10,000 days is given in Fig. 7-55.

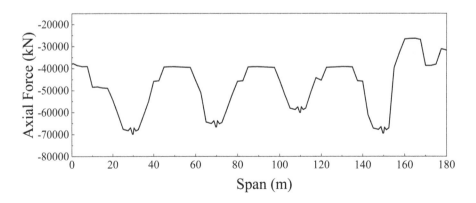

FIGURE 7-53   Variation of Axial Force Due to Post-Tendons (10,000 days) (1 m = 3.28 ft, 1 kN = 0.225 kips).

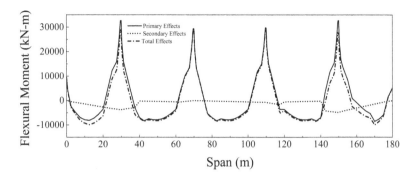

FIGURE 7-54 Variation of Moment Due to Post-Tendons (10,000 days) (1 m = 3.28 ft, 1kN-m = 0.738 kips-ft).

7.11.5.3.4 Effects Due to Temperatures

There is only one fixed pot bearing in the bridge longitudinal direction (see Fig. 7-38). Thus, the uniform temperature rise and fall will not cause any internal forces in the main girder. The effect of the temperature gradient described in Section 7.11.1.3 in the concrete girders are evaluated using the finite-element method discussed in Section 4.3.3 via the MIDAS Civil computer program. The variations of moment and shear due to temperature gradient are illustrated in Figs. 7-56 and 7-57, respectively. In these figures, the solid and dotted lines represent the positive moment and negative moment, respectively.

7.11.5.3.5 Effects Due to Creep and Shrinkage

The perimeter of the box girder in contact with the atmosphere is calculated as follows:

Notational size of box girder (from Eq. 1-9b): $h = 2\frac{A_c}{u} = 578.6\,\text{mm}$

Assumed relative humidity: $H = 75\%$

$$\phi_{RH} = 1 + \frac{1-H}{0.46(0.01 \times h)^{1/3}} = 1.303$$

$$\beta(f_C') = \frac{5.3}{(0.1f_C')^{0.5}} = 2.605$$

$$\beta(t_0) = \frac{1}{1+t_0^{0.2}} = 0.488$$

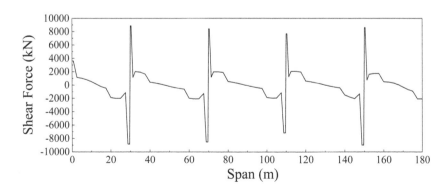

FIGURE 7-55 Variation of Shear Force Due to Post-Tendons (10,000 days) (1 m = 3.28 ft, 1 kN = 0.225 kips).

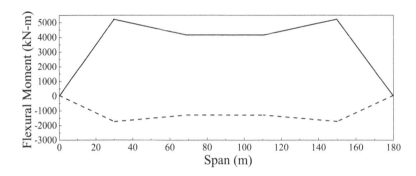

FIGURE 7-56   Moment Distribution Due to Temperatures Gradients (1 m = 3.28 ft, 1 kN-m = 0.738 kips-ft).

$$\phi_0 = \phi_{RH}\beta(f_C')\beta(t_0) = 1.656$$

$$\beta_H = 1.5h\left[1+(1.2H)^{18}\right] + 250 = 1248.21$$

$$\beta_C(t-t_0) = \left[\frac{t-t_0}{\beta_H+(t-t_0)}\right]^{0.3} = 0.965$$

The creep coefficient is

$$\phi(t-t_0) = \phi_0\beta_C(t-t_0) = 1.60$$

From Eq. 1-15b, the shrinkage development coefficient is

$$\beta_s(t-t_0) = \left[\frac{t-t_0}{350(h/100)^2}\right]^{0.5} = 0.922$$

$$\beta_{RH} = -1.55(1-H^3) = -0.896$$

$$\varepsilon_s(f_C') = \left[16+5(9-0.1f_C')\right]\times10^{-5} = 4.031\times10^{-4}$$

$$\varepsilon_{cs0} = \varepsilon_s(f_C')\beta_{RH} = -3.610\times10^{-4}$$

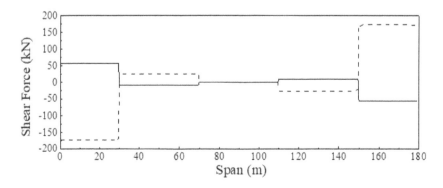

FIGURE 7-57   Shear Distribution Due to Temperatures Gradients (1 m = 3.28 ft, 1 kN = 0.225 kips).

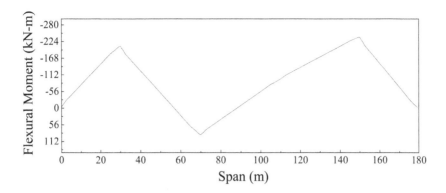

FIGURE 7-58 Moment Distribution Due to Creep and Shrinkage Effects (1 m = 3.28 ft, 1kN-m = 0.738 kips-ft).

The shrinkage strain is

$$\varepsilon_{cs}\left(t-t_0\right)=\varepsilon_{cs0}\beta_s\left(t-t_0\right)=-3.330\times10^{-4}$$

While analyzing the effects of concrete shrinkage and creep, the casting and curing period for each segment is assumed to be 10 days and the launching period for each segment is assumed to be 12 days as previously mentioned. The moment and shear distributions due to creep and shrinkage at 10,000 days after casting as a result of dead load and prestressing force are illustrated in Figs. 7-58 and 7-59, respectively. From these figures it can be observed that the effect of concrete shrinkage and creep is comparatively small in an incrementally launched bridge.

7.11.5.3.6 Effects Due to Wind Loading

7.11.5.3.6.1 Lateral Wind Pressure on Super-Structure WS_L Assuming that the height of the super-structure and design wind speed are the same as those for Design Example I discussed in Chapter 5, the design wind pressure determined in Section 5.6.4.9.1 is

$$P_D =975.61\ Pa\left(0.02ksf\right)$$

The resulting wind forces per linear feet applied along the box girder neutral axis (see Fig. 7-60) are

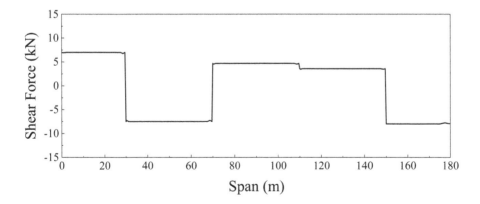

FIGURE 7-59 Shear Distribution Due to Creep and Shrinkage Effects (1 m = 3.28 ft, 1 kN = 0.225 kips).

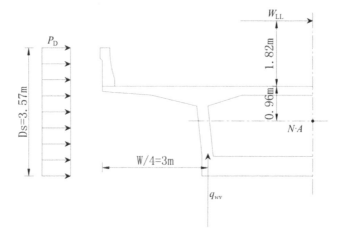

FIGURE 7-60   Design Wind Loadings (1 m = 3.28 ft).

Horizontal force:

$$WS_L = P_D D_S = 3482.93 \frac{N}{m} \left(0.239 \, \frac{kips}{ft} \right)$$

Moment:

$$M_L = WS_L \left(\frac{D_S}{2} - y_t \right) = 2873.42 \, \frac{N-m}{m} \left(0.646 \, \frac{kips\text{-}ft}{ft} \right)$$

7.11.5.3.6.2   Vertical Wind Force on Super-Structure WS_V   The vertical wind force applied to the super-structure is calculated based on the method presented in Section 2.2.5.2.2. From Eq. 2-11, the upward line wind load can be determined as

$$q_{wv} = 11479.89 \text{N/m} \left(0.787 \text{ kips/ft} \right)$$

The resulting moment per linear foot due to the eccentricity of the upward wind load (see Fig. 7-60) is

$$M_V = q_{wv} \left(\frac{12}{4} \right) = 34439.66 \text{N} \cdot \text{m} / \text{m} \left(7.742 \text{ kips-ft/ft} \right)$$

7.11.5.3.6.3   Wind Force on Vehicle W_{LL}   Using a similar method as discussed in Section 5.6.4.9.3, the resulting wind forces on vehicles applied along the girder neutral axis (see Fig. 7-60) can be determined as

Horizontal force:

$$W_{LL} = 1459.32 \frac{N}{m} \left(0.100 \text{ kips/ft} \right)$$

Moment:

$$M_{LL} = W_{LL}\left(1.82 + y_t\right) = 4056.90 \frac{N-m}{m}\left(9.12 \text{ kips-ft/ft}\right)$$

The effects due to the above wind loads at some control sections are given in Table 7-4.

7.11.5.3.6 Summary of Effects
The maximum effects at several critical sections due to the principal loadings are summarized in Table 7-4.

7.11.6 BRIDGE CAPACITY VERIFICATION

The bridge capacity checks include both the bridge longitudinal and transverse directions. Generally, the capacity verifications for incrementally launching box girder bridges are similar to those for Design Examples I and II as discussed in Chapters 5 and 6, respectively. The readers are encouraged to refer to these chapters for the detailed procedures. For simplicity, in this design example, only part of the longitudinal capacity checks will be presented. As previously mentioned, the effects of the concrete creep and shrinkage vary with time, and the bridge capacity verifications may need to be performed for different loading stages, such as immediately after instantaneous prestressing losses, immediately after construction, 4,000 days, 10,000 days, and final stage. The following verifications are done for 10,000 days after casting time in this example.

7.11.6.1 Service Limit
For concrete segmental bridges, three service limit states should generally be checked, i.e., service I, service III, and special load combination. For the bridges constructed by the incrementally launching method, the special load combination will not control the design and is omitted in this example.

7.11.6.1.1 Service I
The service I check is to investigate concrete compression strengths. Normally, the concrete compression strength in three loading cases should be checked.

Case I: The sum of the concrete compression stresses caused by tendon jacking forces and permanent dead loads should not be greater than $0.6f_c'$, i.e.,

$$f_j^c + f_{DC+DW}^c \leq 0.6f_c'$$

Case II: For permanent stresses after losses, the sum of the concrete compression stresses caused by tendon effective forces and permanent dead loads should not be greater than $0.45f_c'$, i.e.,

$$f_{pe}^c + f_{DC+DW}^c \leq 0.45f_c'$$

Case III: The sum of the concrete compression stresses caused by tendon effective forces, permanent dead loads, and transient loadings should not be greater than $0.6\phi_w f_c'$. The corresponding load combination for the example bridge is

$$DC + CR + SH + PS + LL\left(1 + IM\right) + BR + 0.3WS + WL + TU + 0.5TG$$

Consider case III and one cross section located about 90 m away from the left end support as an example to illustrate the calculation procedures for the verification.

From Eq. 3-269, the top slab slenderness ratio is

$$\lambda_w^t = \frac{X_u^t}{t_t} = \frac{396.0}{25} = 15.84 > 15$$

Thus, the reduction factor of the top slab is

$$\phi_w^t = 1.0 - 0.025(15.84 - 15) = 0.979$$

The allowable compressive concrete stress at the top fiber is

$$f_{ca}^t = 0.6\phi_w^t f_c' = 26.33 \ Mpa \ (3.82 \ ksi).$$

The bottom slab slenderness ratio is

$$\lambda_w^b = \frac{X_u^b}{t_b} = \frac{506.0}{27.5} = 18.40 > 15$$

Thus, the reduction factor of the bottom slab is

$$\phi_w^b = 1.0 - 0.025(18.40 - 15) = 0.915$$

The allowable compressive concrete stress at the bottom fiber is

$$f_{ca}^b = 0.6\phi_w^b f_c' = 24.61 \ Mpa(3.57ksi).$$

From Table 7-4, the maximum concrete compressive stresses can be calculated as
Top fiber:

$$
\begin{aligned}
f_{top}^c &= \frac{(M_{DC} + M_{CR+SH} + M_{PS} + M_{LL+IM} + M_{BR} + 0.3M_{WS} + M_{WL} + 0.5M_{TG})y_t}{I} + \frac{P_{PS}}{A} \\
&= \frac{(14751 - 9 - 8260 + 9617 + 497 + 0.3 \times 2420 + 0.5 \times 5390) \times 0.963}{6.279} + \frac{39212}{6.586} \\
&= 9.02 \ Mpa \ (1.31 \ ksi) < 26.33 \ Mpa(3.82 \ ksi)(ok!)
\end{aligned}
$$

Bottom fiber:

$$
\begin{aligned}
f_{bottom}^c &= \frac{-(M_{DC} + M_{CR+SH} + M_{PS} + M_{LL+IM} + M_{BR} + 0.3M_{WS} + M_{WL} + 0.5M_{TG})y_b}{I} + \frac{P_{PS}}{A} \\
&= \frac{-(14751 - 9 - 8260 + 9617 + 497 + 0.3 \times 2420 + 0.5 \times 5390) \times 0.963) \times 1.537}{6.279} + \frac{39212}{6.586} \\
&= 1.05 \ Mpa \ (0.15 \ ksi) < 24.61 \ Mpa \ (3.57 \ ksi)(ok!)
\end{aligned}
$$

For the remaining sections, the verification of service I can be performed in a similar way as above.

7.11.6.1.2 Service III

The service III check is to investigate the maximum tensile stresses in the prestressed concrete components to ensure that the maximum tensile stresses are not greater than the limits specified in Sections 1.2.5.1.1.2 and 1.2.5.1.2.2. Three cases should be checked:

Case I: Temporary tensile stresses before losses
Case II: Tensile stresses at service III load combination
Case III: The principal tensile stress at the neutral axis in the web
The service III load combination for the example bridge is

$$DC + CR + SH + PS + 0.8 \ LL \ (1 + IM) + 0.8BR + 0.5TG$$

Consider the section about 90 m (295.3 ft) away from the left end support of the bridge to check its maximum tensile stress at the bottom fiber, and consider the section at the first interior support to check its principal stress as an example to illustrate the procedures of capacity verifications.

7.11.6.1.2.1 Maximum Tensile Stress Check
From Section 1.2.5.1.2.2, the allowable tensile stress for the section with the minimum bonded auxiliary reinforcement is.

$$f_{ta} = 0.0948\sqrt{f'_c(ksi)} = 1.67 \ Mpa \ (0.242 \ ksi)$$

From Table 5-2, the maximum tensile stress at the bottom fiber can be calculated as

$$
\begin{aligned}
f_{bottom} &= \frac{(M_{DC} + M_{CR+SH} + M_{PS} + 0.8M_{LL+IM} + 0.8 \ M_{BR} + 0.5M_{TG})y_b}{I} - \frac{P_{PS}}{A} \\
&= \frac{(14742 - 9 - 8260 + 0.8 \times 9617 + 0.8 \times 497 + 0.5 \times 5390) \times 1.537}{6.279} - \frac{39212}{6.586} \\
&= -1.73 \ Mpa(-0.25 \ ksi) \ (\text{in compression, ok!})
\end{aligned}
$$

The maximum tensile stresses for other locations can be calculated using the procedure above. It was found that almost all sections are in compression, except for a few sections within the pier segments with tensile stresses that are smaller than the allowable tensile stress.

7.11.6.1.2.2 Principal Stress Check
For a section, the principal stress check should be considered a maximum for one of the following four cases: maximum absolute negative shear with corresponding torsion, maximum positive shear with corresponding torsion, maximum absolute negative torsion with corresponding shear, and maximum positive torsion with corresponding shear. As an example, consider the section 28.5 m away from pier 1 and check the maximum principal stress under the maximum positive torsion with its corresponding shear to illustrate the analytical procedures.

The allowable principal stress from Section 1.2.5.1.2.2 is

$$f_{ta} = 0.11\sqrt{f'_c(ksi)} = 1.93 \ Mpa \ (280 \ psi)$$

The effective total web thickness is

$$b_w = 2 \times 0.32 = 0.64 \ m$$

The maximum torsion from Table 7-4 is

$$T_{max} = 4168 \ kN\text{-}m$$

The corresponding shear force from Table 7-4 is

$$V_{max} = 3902 + 7 - 9410 + 0.8 \times (1433 + 18) + 0.5 \times 50 = -4315 \text{ kN}$$

The shear stress in the web at the neutral axis location is

$$f_v = \frac{|V_{max}|Q_z}{2I_z b_w} + \frac{|T_{max}|}{2A_0 t_w} = \frac{4315 \times 4.647}{2 \times 5.726 \times 0.64} + \frac{4168}{2 \times 13.76 \times 0.32} = 3.209 \; Mpa \; (0.465 \text{ ksi})$$

The normal stress at the neutral axis location is

$$f_x = \frac{-P_{PS}}{A_g} = \frac{-39212}{6.586} = -5.95 \; Mpa \; (0.863 \text{ ksi})$$
$$f_y = 0$$

From Eq. 3-176, the principal tensile stress can be calculated as

$$f_1 = \frac{f_x + f_y}{2} + \sqrt{\left(\frac{f_x - f_y}{2}\right)^2 + f_v^2} = \frac{-5.95}{2} + \sqrt{\left(\frac{-5.95}{2}\right)^2 + 3.209^2}$$
$$= 1.401 \; Mpa \; (0.203 \; ksi) < f_{ta} \; (\text{ok})$$

The principal stress check for other sections and other loading cases can be performed using the same procedure as above. It was found that the principal stresses at all the locations are smaller than the allowable.

7.11.6.2   Strength Limit

Strengths I, II, III, and V and Extreme Events I and II should be evaluated. Take Strength I for example. Load combination for this example can be written as

$$1.25DC + 1.25\,CR + 1.25SH + 1.75\,LL\,(1+IM) + 1.75BR + 0.5TU$$

7.11.6.2.1   Flexure Strength

Consider the section located about 90 m away from the left end support of the bridge as an example to illustrate the capacity verification process. The flexure strength check follows the procedures described in Fig. 3-41.

From Table 2-14, the flexure resistance factor for the bonded tendons is

$$\phi_f = 0.95$$

The area of the post-tendons in the bottom flange is

$$A_{ps} = 140(12 \times (10 + 4) = 23520 \text{ mm}^2$$

The effective top flange width is determined based on the method described in Section 3.4.2.2 and can be taken as the actual flange width for the interior span.

The width of top flange is $b = 12.0$ m.

From Eq. 3-149e, the stress block factor is

$$\beta_1 = 0.85 - 0.05\left(f'_c\,(ksi) - 4.0\right) = 0.725$$

The distance from the centroid of the tendons to the extreme compression fiber can be calculated as

$$d_p = 2.206\ m$$

From Eq. 3-152b, we have

$$k = 2\left(1.04 - \frac{f_{py}}{f_{pu}}\right) = 0.28$$

Using Eq. 3-165a and assuming that the box section is a rectangular section, the distance from the extreme compression fiber to the neutral axis c can be determined as

$$c = \frac{A_{ps}f_{pu}}{0.85f'_c\,\beta_1 b + k\frac{A_{ps}f_{pu}}{d_p}} = \frac{0.02352 \times 1860}{0.85 \times 44.82 \times 0.725 \times 12 + 0.28\frac{0.02352 \times 1860}{2.206}} = 0.13\ m$$

The distance c is less than the thickness of top flange, and the assumption of rectangular section is correct. From Eq. 3-149b, the compression block depth is

$$a = \beta_1 c = 0.725 \times 0.13 = 0.094\ m$$

From Fig. 7-53, we can see

$$f_{pe} = \frac{P_e}{A_{ps}} > 0.5f_{pu}$$

The average stress in the prestressing tendons can be determined by Eq. 3-152a, i.e.,

$$f_{ps} = f_{pu}\left(1 - k\frac{c}{d_p}\right) = 1860\left(1 - 0.28\frac{0.13}{2.206}\right) = 1829.3\ Mpa$$

From Eq. 3-164, the factored flexural resistance is

$$\phi_f M_n = \phi_f A_{ps}f_{ps}\left(d_p - \frac{a}{2}\right) = 0.95 \times 0.02352 \times 1829.3\left(2.206 - \frac{0.094}{2}\right)$$
$$= 88267.1\ \text{kN-m}\ (65090.67\ \text{kips-ft})$$

The factored moment is

$$M_u = 1.25DC + 1.25\,(CR + SH) + 1.75\,LL(1 + IM) + 1.75\,BR + 0.5TU$$
$$= 1.25 \times (14751 - 9) + 1.75 \times 9617 + 1.75 \times 497 = 36127\text{kN-m}\ (26647.24\ \text{kips-ft})\ (< \phi_f M_n,\ \text{ok!})$$

Check the minimum reinforcement. The section modulus is

$$S_b = \frac{I_y}{y_b} = \frac{6.279}{1.537} = 4.085\ m^3$$

From Section 1.2.2.3, the modulus of fracture is

$$f_r = 0.20\sqrt{f'_c} = 0.20\sqrt{6.5(ksi)} = 3.517 \ Mpa \ (0.51 \ \text{ksi})$$

The compressive stress due to effective prestressing forces at the bottom fiber is

$$f_{cpe} = \frac{M_{ps}}{S_b} + \frac{P_e}{A_g} = \frac{8260}{4.085} + \frac{39212}{6.586} = 7975.87 \ Mpa \ (570.11 \ \text{ksi})$$

From Eq. 3-168, the section crack moment is

$$M_{cr} = \gamma_3 \left(\gamma_1 f_r + \gamma_2 f_{cpe}\right) S_b = 0.67(1.2 \times 3.517 + 1.1 \times 7975.87) \times 4.085$$

$$= 24024.06 \ kN\text{-}m \ (17720.12 \ \text{kips-ft})$$

$$1.33 M_{cr} = 31952.01 \ kN-m \ (23567.76 \ \text{kips-ft}) \ (< \phi_f M_n \ (\text{ok}))$$

The flexure strength checks for other locations can be performed using the procedure described above. It is found that the flexure strengths of all sections are adequate.

7.11.6.2.2 Shear and Torsion Strength

The typical shear reinforcement for the box section is shown in Fig. 7-61. The shear and torsion capacity verifications are similar to those presented in Section 6.14.4.8.2.2 (Design Example II). The readers are encouraged to refer to these sections for the detailed procedures. Analytical results show the shear reinforcement provided is quite enough to resist all the anticipated shear forces.

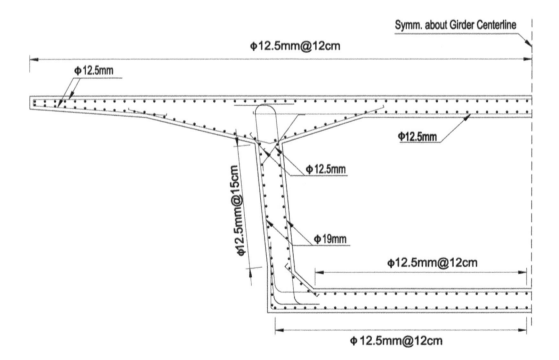

FIGURE 7-61 Reinforcement Arrangements of Midspan Section (1 mm = 0.03937 in., 1 cm = 0.3937 in.).

7.12   CAPACITY CHECK DURING INCREMENTAL LAUNCHING CONSTRUCTION

The service and strength limits required by the AASHTO specifications as discussed in Section 2.3.3.4 should be verified for both the rear and the front zones during construction stages at each of the critical sections. Here, we will consider the section with maximum positive bending moment (see Fig. 7-48) as an example for verifying its flexure capacity. From Eq. 2-52a, the factored moment at this section can be simplified as

$$M_u = 1.1M_{DC} + 1.3\ CLL$$

From Section 2.2.12, the distribution construction live load CLL may be neglected for an incrementally launched bridge; thus, the factored construction load is

$$M_u = 1.1M_{DC} = 1.1 \times 18613 = 20474.3 \text{ kN-m } (15110.03 \text{ kips-ft})$$

The section with maximum positive moment during launching is located close to the midspan of span 1 and has the same geometry as the cross section discussed in Section 7.11.6.2.1. However, there are only 10 launching tendons at the bottom. The area of the post-tendons in the bottom flange is

$$A_{ps} = 140(12 \times 10) = 16800 \text{ mm}^2$$

The distance from the centroid of the tendons to the extreme compression fiber can be calculated as

$$d_p = 2.34 \ m$$

Using Eq. 3-165a and assuming that the box section is a rectangular section, the distance from the extreme compression fiber to the neutral axis c can be determined as

$$c = \frac{A_{ps}f_{pu}}{0.85f_c'\beta_1 b + k A_{ps}f_{pu} / d_p} = \frac{0.0168 \times 1860}{0.85 \times 44.82 \times 0.725 \times 12 + 0.28 \frac{0.0168 \times 1860}{2.34}} = 0.093 \ m$$

The distance c is less than the thickness of the top flange, and the assumption of rectangular section is correct. From Eq. 3-149b, the compression block depth is

$$a = \beta_1 c = 0.725 \times 0.093 = 0.068 \ m \ (2.677 \text{ in})$$

The average stress in the prestressing tendons can be determined using Eq. 3-152a, i.e.,

$$f_{ps} = f_{pu}\left(1 - k\frac{c}{d_p}\right) = 1860\left(1 - 0.28\frac{0.093}{2.34}\right) = 1839.3 \ Mpa \ (266.70 \text{ ksi})$$

From Eq. 3-164, the factored flexural resistance is

$$\phi_f M_n = \phi_f A_{ps} f_{ps}\left(d_p - \frac{a}{2}\right) = 0.95 \times 0.0168 \times 1839.3\left(2.34 - \frac{0.068}{2}\right)$$
$$= 67697.3 \text{ kN-m } (49960.6 \text{ kips-ft})(> M_u, \text{ ok!})$$

The capacity check for other sections and loading conditions during construction can follow the same procedure as described above. It is found that the super-structure has enough capacity to resist all the anticipated loadings during the construction.

7.13   REMARKS ON DESIGN EXAMPLE III

Design Example III is not intended to provide a best design. Readers also can see that this design is conservative and a better design can certainly be made. One of the reasons may be the differences between the Chinese and American bridge design specifications.

REFERENCES

7-1. Rosignoli, M., *Bridge Launching*, Thomas Telford Publishing, London, 2002.
7-2. Göhler, B., and Pearson, B., *Incrementally Launched Bridges: Design and Construction*, Ernst & Sohn, Germany, 2000.
7-3. Podolny, W., and Muller, J., *Construction and Design of Prestressed Concrete Segmental Bridges*, John Wiley & Sons, New York, 1982.
7-4. VSL International LTD, *The Incremental Launching Method in Prestressed Concrete Bridge Construction*, Berne, Switzerland, 1977.
7-5. AASHTO, *AASHTO LRFD Bridge Design Specifications*, 7th edition, Washington, D.C., 2014.
7-6. Rosignoli, M., "Influence of the Incremental Launching Construction Method on Sizing of Prestressed Concrete Bridge Decks," *Proceedings of the Institution of Civil Engineering—Structures and Buildings*, Vol. 122, No. 3. August 1997, pp. 316–325.
7-7. Rosignoli, M., "Presizing of Prestressed Concrete Launched Bridges," *ACI Structural Journal*, Vol. 96, No. 5 September–October 1999, pp. 705–710.
7-8. ASBI, *Construction Practices Handbook for Concrete Segmental and Cable-Supported Bridges*, ASBI, Buda, TX, 2008.
7-9. PCI Committee on Segmental Concrete Construction, "Recommended Practice for Segmental Construction in Prestressed Concrete," *PCI Journal*, Vol. 20, No. 2, March–April 1975, pp. 22–41.
7-10. Sao, X. D., and Gu, A. B., *Bridge Engineering*, 3rd edition, Chinese Communication Publisher of China, Beijing, China, 2014.
7-11. Fontan, A. N., Diaz, J., Baldomir, A., and Hernandez, S., "Improved Optimization Formulation for Launching Nose of Incrementally Launched Prestressed Concrete Bridges," ASCE, *Journal of Bridge Engineering*, Vol. 16, No. 3, 2011, pp. 461–470.
7-12. Rosignoli, M., "Prestressing Schemes for Incrementally Launched Bridges," *Journal of Bridge Engineering*, Vol. 4, No. 2, 1999, Paper No. 18367, pp. 107–115.
7-13. Xu, Y., Zou, C. J., Zhang, L.F., and Zheng, X.Y., *Prestressed Concrete Girder Bridges*, China Peoples, Communication Publishing House, Beijing, China, 2012.

8 Design of Post-Tensioned Spliced Girder Bridges

- Span arrangement and determination of preliminary dimensions
- Segment divisions and detailing
- Typical construction sequences
- Bridge transverse and longitudinal analysis
- Determination and layout of pretensioning and post-tensioning steel
- Design example: spliced girder bridge

8.1 INTRODUCTION

The basic concept for the post-tensioned spliced girder bridges was discussed in Section 1.5.6. A post-tensioned spliced girder bridge is a precast prestressed concrete member fabricated in several relatively long pieces that are assembled into a single girder for the final bridge structure by post-tensioned tendons. The technique is mainly developed for maximizing the span length limit of conventional precast prestressed concrete girder bridges. Because of the limitations of girder transportation and weight, the maximum span length of conventional precast prestressed concrete girder bridges is typically limited to 150 to 175 ft depending on different departments of transportation, though 200 ft may be allowed for a specific situation. The limit of the total weight of a girder generally ranges from 70 to 100 tons. These limitations result in decreased competition between bridge types and materials, which can lead to higher costs for bridge owners. The current maximum length of a post-tensioned spliced girder bridge that has been constructed is 320 ft. In comparison with the conventional precast prestressed concrete girder bridges, post-tensioned spliced concrete girder bridges have the following advantages[8-1]:

1. The number of substructure units, and thus the total project cost, is reduced.
2. The number of girder lines are reduced by increasing the girder spacing, and thus the total project cost is reduced.
3. The structure depth is reduced by using long, continuous members to obtain the required vertical clearance.
4. Piers are placed to avoid water, railroad tracks, roadways, and utilities, and thus reduce the total project cost.
5. Aesthetics are improved by using haunched sections at piers and slender long-span super-structures.

From the viewpoint of the basic construction concept, the post-tensioned spliced girder bridge can be essentially grouped into concrete segmental bridges. However, the post-tensioned spliced girder bridge has distinguishing characteristics that are different from the segmental bridges discussed in Chapters 5 to 7:

1. The segments of a spliced girder are typically long with two to three segments per span and longitudinally prestressed to mainly resist its self-weight and partial dead loads. To reduce the segment self-weight, the cross sections of spliced girders are typically comprised of several individual I-girders or U-beams and the deck is generally cast-in-place on the spliced girders. The segments of a conventional segmental bridge are normally short

with eight or more segments per span that are not longitudinally prestressed. The typical cross section of the conventional segmental bridge is typically box shaped with an integrated bridge deck system.

2. The joints between segments for the spliced girder are typically cast-in-placed, while the joints between segments for conventional segmental bridges are match cast except for the closure pour joints.

3. The fabrication of spliced girder segments can be performed in existing precasting plants with existing or modified forms, while that of the segmental box segments requires the purchase and setup of custom forms. For this reason, the spliced girder bridge is often a cost-effective alternative in comparison with the conventional segmental bridges with a total deck area that is less than 110,000 ft².

4. The erection of a spliced girder bridge typically can be accomplished with standard equipment, and the construction is generally more conventional. This allows bridge contractors to pursue spliced girder projects rather than requiring specialty contractors to perform the work, though a specialty contractor is still typically required to perform the operations related to post-tensioning and grouting. These advantages lead to a more competitive bridge alternative and reduced project cost.

5. Because there are fewer splices and the splices are protected by the composite deck on top of the spliced girders, the corrosion resistance of the spliced girder bridge is better than that of segmental bridges.

Based on the aforementioned distinct features, we can see that the spliced girder bridge possesses the characteristics of both the conventional precast prestressed girder bridges and those of conventional segmental bridges. However, the behaviors of spliced girder bridges more closely match those of conventional precast prestressed girder bridges. For this reason, the AASHTO LRFD specifications[8-2] classify these types of structures as conventional precast girder bridges for design purposes, with the additional requirements that the post-tensioning at splice locations be based on the provisions developed for segmental bridges. The economical maximum span lengths for splice girder bridges generally range from 175 to 300 ft, depending on the span arrangement, type of cross sections, and site conditions.

8.2   TYPICAL BRIDGE CROSS SECTIONS

Two typical bridge sections are often used in post-tensioned spliced girder bridges. One is that the deck system is supported by a number of I-girders (see Fig. 8-1a), and the other is that the deck is supported by a number of U-beams (see Fig. 8-1b). For curved spliced girder bridges, U-beams are typically used. To increase the torsion resistance of the U-beams, a cast-in-place lid slab (see Fig. 8-1b) is required to be poured prior to stressing the post-tendons. To ensure that the bridge has enough redundancy, the minimum number of I-girders is typically not less than four and the minimum number of U-beams should not be less than two. Different departments of transportation in the United States may use different shapes of I-girders or U-beams. The designers should check the design standards for the relevant state when designing a spliced girder bridge to ensure the existing forms in the local precast plants can be used or can be modified.

8.2.1   I-GIRDER

I-girders are most popular in straight spliced girder bridges due to their light weights. Though there are many I-girder shapes, they may be grouped into three types: AASHTO girders[8-3, 8-4], bulb-T beams[8-4], and revised I-girders[8-5] (see Fig. 8-2). The AASHTO girders and bulb-T beams were mainly developed for simple span bridges. Because of the trend toward long span and continuous structures, many states have developed new shapes of I-girders to meet the negative moment

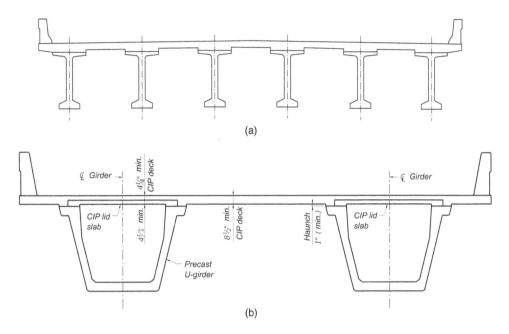

FIGURE 8-1 Typical Bridge Cross Sections in Spliced Girder Bridges, (a) I-girder Bridge Section, (b) U-beam Bridge Section.

and stability requirements, such as the Florida Department of Transportation (FIBs)[8-5], Nebraska Department of Transportation (NU I-girders)[8-6], and Texas Department of Transportation (TxDOT Girders)[8-7 to 8-9]. Some limited comparisons between different types of I-girders with a girder depth of 72 in.[8-1] indicate that the AASHTO Type VI girder has the largest maximum span. However, the AASHTO Type VI girder has a much larger cross section. The bulb-T beam has the lightest weight per foot. The characteristics of the revised I-girders are typically a combination of both. Thus, the most efficient and economical design should be determined based on many factors, such as the site conditions and the preference of different departments of transportation.

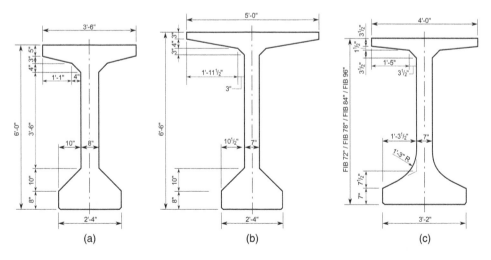

FIGURE 8-2 Typical Sections of I-Girders, (a) AASHTO Type VI Girder, (b) Florida Bulb-T Beam-78, (c) Florida I-Beams.

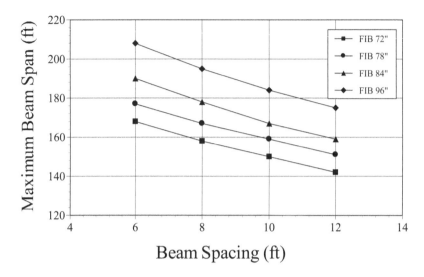

FIGURE 8-3   Maximum Span Lengths for Simply Supported Florida-I Beam[8-5].

The minimum web thickness is generally not smaller than 7 in. for accommodating the post-tensioning ducts. The Florida Department of Transportation (FDOT) design guidelines[8-10] require that the minimum web thickness is not less than 8 in. The girder depths for the spliced I-girder bridges are typically not less than 72 in. For the preliminary design reference, the maximum span lengths for simply supported FDOT I-girders are shown in Fig. 8-3. This chart was developed based on AASHTO LRFD specifications with the following assumptions:

- Girder concrete strength: 8.5 ksi
- Deck concrete strength: 4.5 ksi
- 8-in. composite concrete deck with additional nonstructural ½-in. sacrificial surface
- 20 psf stay-in-place form
- Two 420 plf barrier line loadings

8.2.2   U-Beam

The original development of U-beams was driven by its better aesthetics and stability during construction[8-11, 8-20]. Now, this beam section is preferred for spliced curved girder bridges due to its high level of aesthetics, torsion capacity, and stability. PCI and FDOT have developed two categories of design standards for the spliced U-girder bridges[8-12, 8-22] based on the internal duct sizes of 3 in. and 4 in. (see Fig. 8-4), respectively and are designated as Uxxx-xxx. The first number after the U represents the girder depth, and the second number indicates the maximum duct size in the webs. The designations and dimensions of the typical sections for categories I and II are presented in Fig. 8-4 and Table 8-1. For a preliminary design reference, the maximum possible lengths and the unit weights are shown in Table 8-1. The feasible maximum span lengths were determined by assuming a U-beam concrete strength of 8.5 ksi. The typical sections shown in Fig. 8-4 are mainly for internally bonded tendons imbedded in the webs and bottom slab. To reduce the web thickness and enhance future rehabilitation, FDOT and PCI have recently revised the Category II U-beam design standards and moved the internal tendons in the webs out as external tendons. The typical sections are shown in Fig. 8-5. The maximum diameter of the external ducts is 4 in.

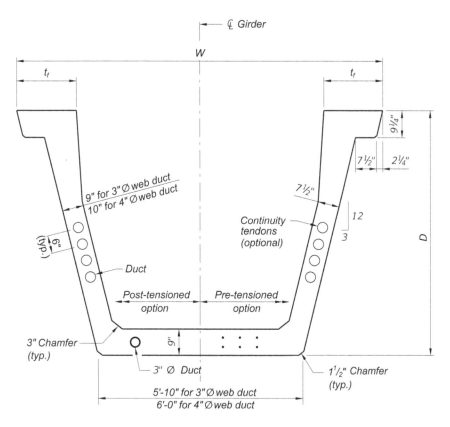

FIGURE 8-4 Typical Sections for PCI Standard U-Beams for Interior Tendons[8-12, 8-22].

8.3 SPAN ARRANGEMENTS AND CONSTRUCTION SEQUENCES

8.3.1 INTRODUCTION

Spliced girders can be used for simple span and continuous-span bridges. One to three spans are most often used in spliced girder bridges. The minimum width of the splice joints is 2 ft. It is generally recommended to provide diaphragms at splice locations to enhance the consolidation of concrete, though permanent diaphragms for straight girder bridges are not required or beneficial for stability or load distribution after the deck has been built.

TABLE 8-1
PCI and FDOT U-Beam Geometry with Internal Tendons

| Category | Girder Type | D | W | t_f | Maximum Span Length | | Weight (plf) |
					Simple	Three-Span	
I	U72-3	6'-0"	10'-1"	1'-8"	175'	205'	2117
(3" φ Web Duct Size)	U84-3	7"-0"	10'-7"	1'-8"	190'	225'	2349
	U96-3	8'-0"	11'-1"	1'-8"	200'	240'	2581
II	U72-4	6'-0"	10'-3"	1'-9"	180'	220'	2271
(4" φ Web Duct Size)	U84-4	7"-0"	10'-9"	1'-9"	200'	265'	2529
	U96-4	8'-0"	11'-3"	1'-9"	220'	280'	2787

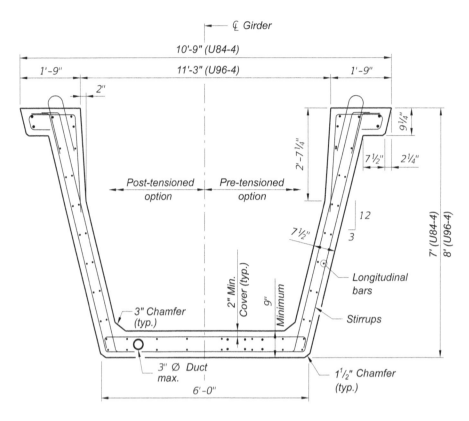

FIGURE 8-5 Typical Sections for FDOT and PCI Standard U-Beams for Bonded and Unbonded Tendons (After Reference 8-22).

For spliced U-beam bridges, access holes should be provided in the diaphragms and in the bottom flanges for maintenance. The spacing of the ingress/egress access openings in the bottom flanges is generally not longer than 600 ft, and a minimum of two access openings per box girder line should be provided. The minimum size of the access opening is 32 in. × 42 in. or 36 in. in diameter.

For curved spliced U-beam bridges, lid slabs must be poured first before stressing the post-tendons as discussed in Section 8.1 and the minimum radius is generally not smaller than 500 ft.

8.3.2 SIMPLE SPAN

A simply supported spliced girder bridge typically consists of three segments per span (see Fig. 8-6b). U-beams are typically used for curved simple spans. The locations of the splice joints are generally determined based on the maximum permitted girder length and the weight, as well as the site limitations. To reduce the moments at the splice joints, the length of the end segments is generally $L/5$ to $L/4$ for spliced I-girders and $L/4$ to $L/3$ for spliced U-girders (see Fig. 8-6). The ratio of girder depth to span is about $0.05L$ and is generally not smaller than $0.045L$ for precast simply supported I-beam bridges based on AASHTO LRFD Article 2.5.2.6.3. However, this limit may be exceeded for spliced girder bridges. A smaller value of the ratio of $0.038L$ may be used for spliced U-girders[8-12]. The typical construction sequences for the simple spliced girder bridges are as follows:

Stage I (see Fig. 8-6a)

1. Construct abutments.
2. Erect temporary supports.

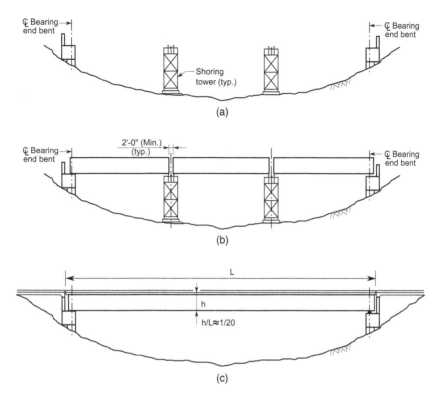

FIGURE 8-6 Simply Supported Spliced Girder Bridge, (a) Construction Stage I, (b) Construction Stage II, (c) Construction Stage III.

Stage II (see Fig. 8-6b)

1. Erect girder segments.
2. Install lateral temporary bracings for girder stability.
3. Cast diaphragms or closure pours at splice locations and lid slab for U-girder.

Stage III (see Fig. 8-6c)

1. Stress partial or all tendons as designed.
2. Cast deck slab.
3. Remove temporary supports.
4. Stress final tendons as designed.
5. Cast barrier walls and approach slabs
6. Install expansion joints, and complete the super-structure construction.

8.3.3   CONTINUOUS SPANS AND SEGMENT ARRANGEMENT

8.3.3.1   General

Spliced girder bridges with two and more continuous spans have been built during the past decades. The three-span continuous sliced girder bridge may be the most commonly used span arrangement in practice. Spliced I-girders are typically used for straight bridges, and the spliced U-girders are generally more suitable for curved bridges, though they can be used for straight bridges. The splice joints are typically located close to inflection points. The locations of inflection points can be determined based on dead loads of girder and deck self-weights, which generally represents 70% of the total load. After the locations of the inflection points are determined, the locations of splice joints can be further determined based on other limitations such as transportation and lifting requirements. Generally, the splice joint in the side span is located at about 0.7 times the side span length away from the end support and the splice joint in the center span is located about 0.22 times the center span length from the nearest interior pier.

8.3.3.2 Continuous Spliced I-Girder Bridges

The girder depth of a continuous spliced I-girder bridge can be either constant or variable. The maximum span length generally ranges from 200 to 300 ft. For a three-span spliced I-girder bridge, each spliced girder is typically divided into five segments: two side segments, two haunch pier segments, and one center segment (see Fig. 8-7). The center segment is often called the drop-in segment. For a long-span continuous spliced I-girder bridge, the girder segments over interior piers are typically haunched (see Fig. 8-7). The ratio of bridge depth over interior piers to the middle span length ranges from 1/16 to 1/20, and the ratio of the bridge depth of the drop-in segment to the middle span length ranges from 1/25 to 1/35. For a spliced I-girder bridge with a variable girder depth, the constant girder depth of the side and drop-in segments is typically first determined based on the discussions contained in Section 8.2.1 and the maximum estimated positive moment. Then, the depth of the section over the pier is determined based on the maximum estimated negative moment. The ratio of the side span length to the interior span is about 0.75 and ranges from 0.7 to 0.85. The individual segments are spliced together through temporary shoring towers (see Fig. 8-6), or strong backs which are commonly used for bridges crossing waterways (Fig. 8-7), or other methods. The strong backs typically consist of several I-section steel members connected to the end of the related segments by post-tensioned threaded bars as shown in Fig. 8-7c and d (see Section 8.4.3 for more details). After the pier segments are erected, the side and drop-in segments are supported on the pier segments using strong backs. The typical construction sequences are as follows:

Stage I (see Fig. 8-7a and b)

1. Construct piers and install temporary supports.
2. Erect pier segments.
3. Tie down the pier segments to temporary supports, and install lateral temporary bracings between pier segments (see Fig. 8-7b).

Stage II (see Fig. 8-7c)

1. Install strong backs to the side girder segments.
2. Erect side girder segments, and connect them to the pier segments using the strong backs.
3. Install lateral temporary bracings for girder stability.

Stage III (see Fig. 8-7d)

1. Install the strong backs for the drop-in segments.
2. Erect drop-in segments, and connect them to the pier segments using the strong backs.
3. Install lateral temporary bracings for girder stability.

Stage IV (see Figs. 8-7e)

1. Cast splices and diaphragms.
2. Stress some or all post-tendons as directed by the design.
3. Remove temporary supports.

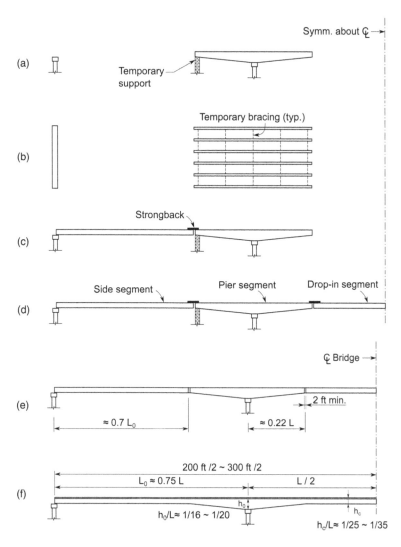

FIGURE 8-7 Typical Arrangement of Three-Span Continuous Spliced I-Girder Bridge and Construction Sequences, (a) Elevation of Stage I, (b) Plan View of Stage I, (c) Stage II, (d) Stage III, (e) Stage IV, (f) Stage V.

Stage V (see Fig. 8-7f)

1. Cast deck slab.
2. Stress final tendons as designed.
3. Cast barrier walls and approach slabs.
4. Install expansion joints, and complete the super-structure construction.

8.3.3.3 Continuous Spliced U-Girder Bridges

The girder depth and bottom slab thickness of a spliced U-girder can be either constant or variable (see Fig. 8-8a and b). The ratio of the bridge depth over the interior piers to the middle span length ranges from 1/18 to 1/20, and the ratio of bridge depth of the drop-in segment to the middle span length ranges from 1/25 to 1/35. The ratio of side span length to interior span ranges from 0.70 to 0.75. For a three-span continuous spliced U-beam bridge, each of the girders is typically divided

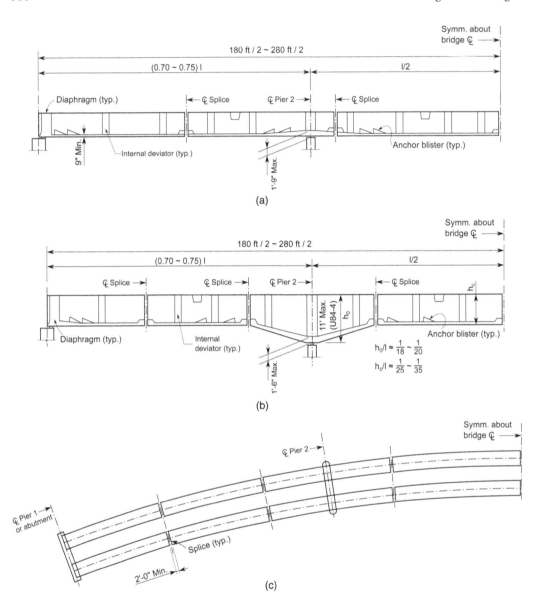

FIGURE 8-8 Typical Span and Segment Arrangements for Spliced U-Beam Bridges, (a) Constant Girder Depth, (b) Variable Girder Depth, (c) Plan View[8-22].

into five or eight segments depending on the span lengths and curvature: two or four side span segments, two pier segments, and one or two center segments (see Fig. 8-8).

The construction method described in Section 8.3.3.2 may not be entirely suitable for the continuous spliced U-beam bridges due to their much larger and heavy segments. Extensive shoring towers may be needed for erecting girder segments. The typical construction sequences for three-span continuous spliced U-beam bridges are as follows (see Fig. 8-9):

Stage I (see Fig. 8-9a)

1. Construct substructures.
2. Erect temporary supports.

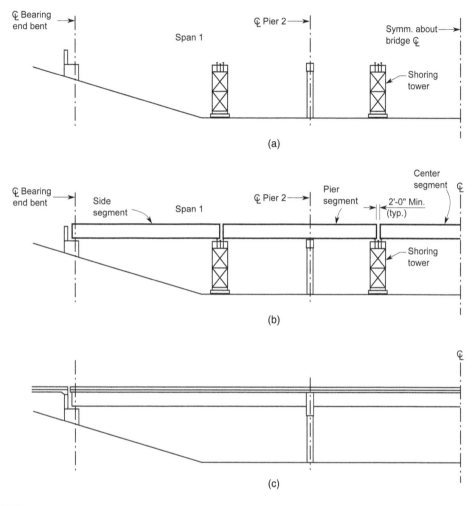

FIGURE 8-9   Typical Construction Sequences for Spliced U-Beam Bridges (a) Stage I, (b) Stage II, (c) Stage III[8-22].

Stage II (see Fig. 8-9b)

1. Erect girder segments.
2. Install transverse bracings between girders.
3. Cast closures.
4. Cast diaphragms over interior piers and splice joints.
5. Cast lid slab over girders.

Stage III (see Fig. 8-9c)

1. Stress continuous tendons.
2. Grout internal tendons.
3. Place flexible filler in external tendons.
4. Remove all shoring towers.
5. Cast deck slab, approach slabs, and bridge rails.
6. Install expansion joints.

FIGURE 8-10 Typical Strongback (Courtesy of Mr. Teddy Theryo).

8.3.4 STRONGBACK AND LATERAL BRACINGS FOR SPLICED I-GIRDERS DURING CONSTRUCTION

8.3.4.1 Strongback

As previously mentioned, the strongback is comparatively simple and typically used to temporally connect the girder segments. It consists of a number of post-tensioning bars and channel or I-shape steel beams (see Fig. 8-10)[8-13]. The post-tensioning forces should be large enough to provide friction forces between the steel beam and the girder segments after the side segments or drop-in segment rest on the pier segments.

8.3.4.2 Temporary Lateral Bracings of Girders

Enough temporary bracings should be provided to ensure the stability of the girder to resist the anticipated wind loading and other construction loads. Before installing the deck forms, a minimum of three lateral bracings shall be provided for girders with depths of less than 72 in.: one at midspan and two at the ends. For girders with depths greater than 78 in., a minimum of five temporary bracings shall be provided: one at midspan and four at the quarter spans and girder ends. The bracings should be designed to resist Strength III horizontal forces.

8.4 BRIDGE ANALYSIS AND CAPACITY CHECK

8.4.1 INTRODUCTION

The analysis of a spliced girder bridge is more complicated than that of a conventional prestressed beam bridge due to the structural static system changes during construction. The effects of permanent and live loads should be performed based on the construction sequences, and the time-dependent effects due to creep and shrinkage should be considered. As the structural analytical procedures of spliced I-girder and U-beam bridges are generally similar, the following discussions are mainly focus on the spliced I-girder bridges.

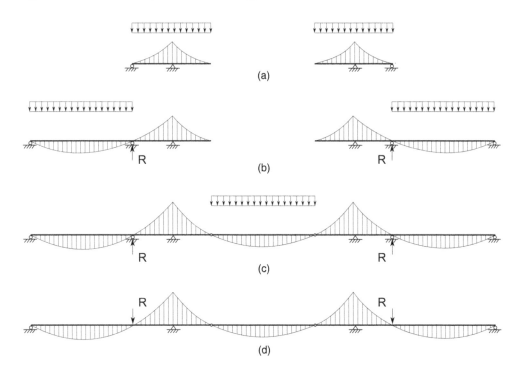

FIGURE 8-11 Analytical Models of Dead Load Effects for Spliced Girders, (a) Cantilever Pier Segments, (b) Simply Supported Side Segments, (c) Simply Supported Center Segment, (d) Three-Span Continuous Girder.

8.4.2 ANALYSIS OF PERMANENT LOAD EFFECTS

The permanent loads applied to the bridge include the girder self-weight, deck slab, barrier walls, and wearing surface. The load due to the deck slab is distributed to each girder based on the center-to-center spacing between girders. The loads due to the barrier walls and wearing surface are applied to the composite section and can be equally distributed to each of the girders. From the construction sequences illustrated in Fig. 8-7, the dead load effects of the spliced I-girders can be analyzed based on the models shown in Fig. 8-11.

8.4.3 ANALYSIS OF LIVE LOAD EFFECTS

8.4.3.1 Introduction

For a straight spliced girder bridge, the effects due to live loads can be either analyzed by load distribution method or the grid model method discussed in Section 4.3.10.1.2. For curved spliced girder bridges with curvature greater than 4°, the AASHTO specifications do not provide any methods for determining the load distribution factors. The methods provided in Reference 8-14 can be used for preliminary design. It is recommended that a finite-element method, such as the grid model method as discussed in Section 4.3.10.1.2, be used for curved spliced girder bridges to ensure accurate analytical results in the final design.

8.4.3.2 Load Distribution Method

8.4.3.2.1 Introduction

The load distribution method has been used in the design of bridge structures for decades[8-15 to 8-18]. As we know, the effects on the girders due to live loads are related to the positions of the vehicles in both the bridge's transverse and longitudinal directions. To simplify the bridge analysis, we assume that the effects of the vehicle transverse and longitudinal locations on each of the girders are not correlated and can be determined separately. First, the effects on the girders due to the vehicle's transverse locations are determined. The transverse effects are defined as transverse load distribution factors, which represent the number of lane loads or the number of wheel loads carried by an individual girder. Using the transverse distribution factor, the bridge can be analyzed based on an individual girder by applying one lane live loading and/or one-line wheel load to the girder along its length. Thus, the total effects on an individual girder can be obtained by multiplying the load effects on a span by the load distribution factors.

8.4.3.2.2 Live Load Distribution Factors

Different counties may have different methods for determining the transverse load factors. The AASHTO specifications[8-2] provide a series of equations for calculating the transverse load distribution factors based on extensive numerical analysis[8-17]. Some of the equations for determining the load factors for I-girder bridges loaded by two or more lanes are given in Eqs. 8-1 to 8-4. The differences between the load distribution factors determined using finite-element methods and Eqs. 8-1 to 8-4 is generally less than 18%[8-18]. However, these equations provided by the AASHTO specifications are only valid for bridges with maximum span lengths of 240 ft. To simplify the time-dependent analysis for long-span bridges, the designer may first determine the load distribution factors using the grid model method or other finite-element methods. After obtaining the distribution factors, the time-dependent analysis of the long-span bridge can be performed on individual girders using any available computer software or hand calculations.

The load distribution factors for moment in interior beams for two or more lanes loaded are

$$g_{Interior}^M = 0.075 + \left(\frac{S}{9.5}\right)^{0.6} \left(\frac{S}{L}\right)^{0.2} \left(\frac{K_g}{12.0Lt_s^3}\right)^{0.1} \qquad (8\text{-}1)$$

where

$\quad S \;\; = \;$ beam spacing (ft) $(3.5 \text{ ft} \leq S \leq 16 \text{ ft})$
$\quad L \;\; = \;$ span length (ft) $(20 \text{ ft} \leq L \leq 240 \text{ ft})$
$\quad t_s \;\; = \;$ thickness of slab (in.) $(4.5 \text{ in.} \leq t_s \leq 12 \text{ in.})$
$\quad K_g \;\; = \; n\left(I + Ae_g^2\right)$
$\quad n \;\; = \;$ modular ratio of girder to slab concrete
$\quad I \;\; = \;$ moment of inertia of girder section (in.^{4})
$\quad A \;\; = \;$ area of girder cross section (in.^{2})
$\quad e_g \;\; = \;$ distance between centroids of girder and slab (in.)

The load distribution factors for moment in exterior beams for two or more lanes loaded are

$$g_{Exterior}^M = e_M g_{Interior}^M \qquad (8\text{-}2)$$

where

$\quad e_M \;\; = \; 0.77 + \frac{d_e}{9.1}$
$\quad d_e \;\; = \;$ distance from exterior web of exterior beam to interior edge of curb or traffic barrier wall (ft) $\left(d_e \leq 3.0 \text{ ft}\right)$

The load distribution factors for shear in interior beams for two or more lanes loaded are

$$g_{Interior}^S = 0.2 + \frac{S}{12} - \left(\frac{S}{35}\right)^{2.0} \qquad (8\text{-}3)$$

The load distribution factors for shear in exterior beams for two or more lanes loaded are

$$g_{Exterior}^{S} = e_S g_{Interior}^{S} \tag{8-4}$$

where $e_S = 0.6 + \frac{d_e}{10}$.

8.4.3.2.3 Dead Load Distribution

It is typically assumed that each of the girders carries the self-weight of the tributary deck area and the weight of barrier walls is uniformly distributed to all the girders.

8.4.3.3 Load Applications

The design live loads discussed in Section 2.2.3 should be positioned to produce maximum force effects at control sections. This can be done through influence lines as discussed in Section 5.2.7.3. For a three-span spliced girder bridge, some sketches of live loading positions for producing extreme force effects are shown in Fig. 8-12.

8.4.4 ESTIMATION OF POST-TENSIONING TENDONS AND PRETENSIONING STRANDS

8.4.4.1 Estimation of Post-Tensioning Tendons

The post-tensioning tendons in a spliced girder bridge are mainly designed to resist live loads and the superimposed dead loads, such as deck slab, traffic barrier walls, and wearing surface. The required post-tensioning force can be estimated using Eq. 3-27. The required area of the post-tensioning tendons can be estimated as

$$A_{ps}^{post} = \frac{P}{f_{pe}^{post}} \tag{8-5a}$$

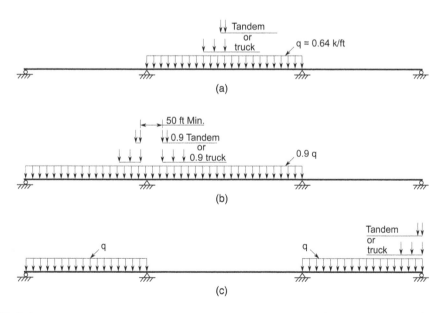

FIGURE 8-12 Most Unfavorable Live Load Positions for Three-Span Continuous Girder Bridges, (a) Maximum Positive Moment in Center Span, (b) Maximum Absolute Negative Moment and Reaction at Interior Support, (c) Maximum Reaction at Exterior Support.

where

P =   minimum required prestressing force determined by Eq. 3-27 in which moment M is mainly induced by live loads and superimposed dead loads at splice locations

f_{pe}^{post} =   effective stress of post-tensioning tendons, which can be estimated by 0.75 times jacking stress

The required amount of post-tensioning tendons may also be estimated based on the required supplementing prestressing force after determining the required pretensioning force discussed in the Section 8.4.4.2 to produce the total net compression stress needed at any point along the span length.

8.4.4.2 Estimation of Pretensioning Strands

The pretensioning strands in a spliced girder bridge are mainly designed to resist anticipated dead loads during construction stage. The Florida Department of Transportation requires that the following conditions be met for spliced I-girder bridges:

a. The pretensioning force should be sufficient to resist all loads applied prior to post-tensioning, including a superimposed dead load equal to 50% the uniform weight of the beam.
b. The initial midspan camber of the precast segment induced by the pretensioning force at release should not smaller 0.5 in.

The minimum required pretensioning strains can be estimated by meeting the service limit states for pretensioned girders based on Eq. 3-27. After determining the maximum positive moments for the side and drop-in segments and the maximum absolute negative moment for pier segments caused by the above dead loads, the required pretensioning forces can be calculated from Eq. 3-27. Then, the required area of the pretensioning strands can be determined as

$$A_{ps}^{pre} = \frac{P}{f_{pe}^{pre}}$$ (8-5b)

where

P =   required pretensioning force determine by Eq. 3-27 in which the moment M is induced by all dead loads applied prior to post-tensioning

f_{pe}^{pre} =   effective stress of prestressing strands, which can be estimated as 0.82 times prestress at release[8-19]

The required amount of pretensioning is often determined based on "supplementing" the post-tensioning determined by Eq. 8-5a to produce the total net compression stress needed at any point along the span length. Also, the use of greater pretensioning can reduce the requirement of post-tensioning tendons.

8.4.5 Capacity Verification

8.4.5.1 Introduction

After the preliminary bridge dimensions and the layout of the post-tensioning tendons have been determined, the capacities of the preliminarily designed bridge should be evaluated for all applicable limit states described in Section 2.3.3.2 for all applicable load combinations presented in Section 2.3.3.3. Generally, the following limit state shall be checked.

8.4.5.2 Service Limit State Stress Check

The service limit states for prestressed concrete bridge super-structures include Service I and Service III as discussed in Section 2.3.3.2.3.

8.4.5.2.1   Service I Limit Check

This limit is to check the concrete compression stresses. For most load cases, the load combination for this limit state can be simplified as

$$\sum \gamma Q_{Service\ I} = DC + (LL + IM) + PS + CR + SH + 0.3WS + WL + FR + TU + 0.5TG \qquad (8\text{-}6)$$

The stresses in concrete should meet the requirements discussed in Section 1.2.5.1. For permanent stresses after losses, the concrete compression stress f_{ta} should satisfy:

$f_{ta} \leq 0.45 f_c'$ ksi (for the sum of effective prestress and permanent load)
$f_{ta} \leq \phi_w 0.6 f_c'$ ksi (for the sum of effective prestress, permanent load, and transient loads)

8.4.5.2.2   Service III Limit Check

Service III Limit State is to check concrete tensile stresses. For most load cases, the load combination can be simplified as

$$\sum \gamma Q_{Service\ III} = DC + 0.8\ (LL + IM) + PS + CR + SH + FR + TU + 0.5TG \qquad (8\text{-}7)$$

For permanent stresses, the concrete tensile stresses should satisfy:

$f_{ta} \leq 0.0$ ksi (no tension, no minimum bonded auxiliary reinforcement through the joints)
$f_{ta} \leq 0.19\sqrt{f_c'}$ ksi (in the area with sufficient reinforcement to resist the tensile force in concrete and $f_s \leq 0.5 f_y$)

8.4.5.2.3   End Zone Splitting Reinforcement Check

From Section 3.8.1, for girder depths equal to or greater than 6.5 ft, the vertical stirrups within a distance of one-quarter of the girder depth, measured from the end of the girder, should resist splitting forces of as much as 8% of the induced prestressing force at transfer[8-20, 8-21], and should meet

$$f_{sv} \leq 20\ \text{ksi}$$

where f_{sv} = stress in vertical stirrups.

8.4.5.3   Strength Check

All applicable strength limits and load combinations shown in Table 2-10 should be investigated.

8.4.5.3.1   Flexure Strength Check

The flexure strengths at control sections should satisfy Eq. 3-148a and should be verified using the procedure shown in Fig. 3-41.

8.4.5.3.2   Shear Strength Check

The shear strengths at control sections should satisfy Eq. 3-148b and should be verified using the procedure shown in Fig. 3-54.

8.5   LAYOUT OF TENDONS AND DETAILS

8.5.1   GENERAL

As previously discussed, in spliced girder bridges, the prestressed girder segments are assembled together using internally bonded and/or external unbounded tendons. The minimum number of tendons per web is three. The minimum duct spacing d as shown in Fig. 8-13b is typically speci-fied as[8-10]:

In the vertical direction, the minimum duct spacing is taken as the greater of the following:

- Outer duct diameter plus 1.5 times maximum aggregate size
- Outer duct diameter plus 2 in.

In the horizontal direction: Outer duct diameter plus 2.5 in.

8.5.2 SPLICED I-GIRDER BRIDGES

In a spliced I-girder bridge, the post-tendons are typically embedded in the webs (see Fig. 8-13). Previous experiences show that a minimum four tendons per girder is recommended for a spliced I-girder bridge with a span length greater than 260 ft. On the spliced face of a precast segment a large shear key and a post-tensioned cast-in-place splice, as shown in Fig 8-13c, is usually provided. Within the cast-in-place joint, the post-tensioning tendon ducts are connected with a short splice between the ends of the ducts projecting from the girder segments. The girder is locally thickened at the anchored zone to transfer highly concentrated compression forces to the girder. The thickened segment is called an anchorage block or an end block. The end block is typically the full height of the girder and the full width of the bottom flange, and its length is approximately 1.5 to 2.5 times the girder depth, plus another transition length of about 4 ft where the web thickness reduces from the width of the flange to the typical web width. The details of a typical anchorage block are illustrated in Fig. 8-14. The analysis of the anchorage zone can be carried out using strut-and-tie models as discussed in Section 3.8.2.3.3.

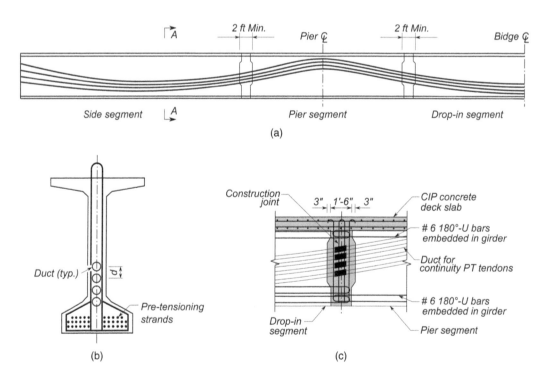

FIGURE 8-13 Typical Layout of Tendons in Spliced I-Girder Bridges, (a) Elevation View, (b) Duct Arrangement at Section *A-A*, (c) Splice Details.

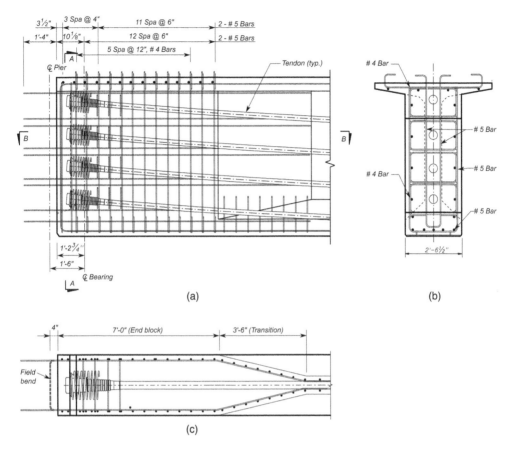

FIGURE 8-14 Typical Anchorage Block, (a) Elevation View, (b) Side View, (c) Plan View.

8.5.3 SPLICED U-GIRDER BRIDGES

The post-tensioning tendons in spliced U-girder bridges can be completely embedded in the webs if the web thickness is wide enough, similar to the spliced I-girder bridges. To reduce the web thicknesses and to facilitate future bridge rehabilitation, FDOT and PCI[8-22] have recently developed some new details for the post-tensioning tendon layout in spliced U-girder bridges. A part of the post-tensioning tendons is internal tendons, which are embedded in the bottom slab and the webs close to the top flanges (see Fig. 8-15). Another part of the tendons is mounted on internal deviators outside of the webs. The external tendons can be made continuous by passing several spans through deviators as shown in Fig. 8-15a. This tendon arrangement is often called the pass through alternative. The external tendons can also be overlapped over the interior piers and anchored in the interior diaphragms (see Fig. 8-15b). This tendon arrangement is often called the laced alternative. The dimension requirements and analysis methods for the end and interior diaphragms, as well as the deviators, are similar to those for typical concrete segmental bridges, discussed in Chapters 5 and 6, and can be found in Sections 3.8 and 5.4. A typical detail of the end diaphragm for spliced U-girder bridges is illustrated in Fig. 8-16a. The diaphragms over the interior piers are typically cast-in-place splices. A typical geometry of the pier segment over the interior piers is shown in Fig. 8-16b.

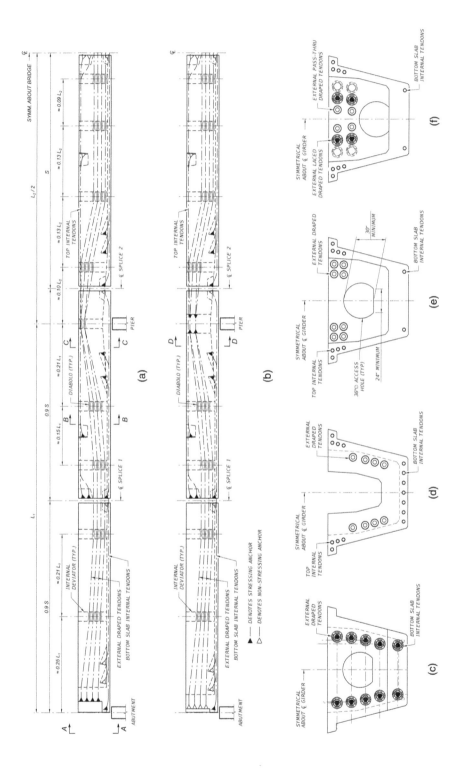

FIGURE 8-15 Typical Tendon Layout of U-Girder, (a) Pass Through Alternative, (b) Laced Alternative, (c) Tendon Layout at End Anchorage (Section *A-A*), (d) Tendon Layout in Deviator (Section *B-B*), (e) Tendon Layout at Interior Pier (Section *C-C*), (f) Tendon Layout at Interior Pier (Section *D-D*)[8-22] (Used with permission by FDOT).

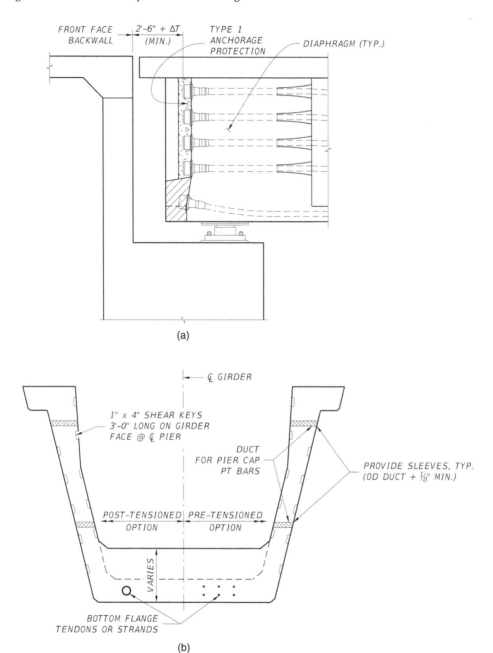

FIGURE 8-16 Typical Details of Diaphragm, (a) End Diaphragm, (b) Cross Section over Pier[8-22] (Used with permision by FDOT).

8.6   DESIGN EXAMPLE IV: SPLICED THREE-SPAN CONTINUOUS I-GIRDER BRIDGE

The design example is generated based on an existing bridge crossing a waterway located in central Florida, owned by FDOT (used with permission of FDOT).

8.6.1   Design Requirements

8.6.1.1   Design Specifications

a. AASHTO LRFD *Bridge Design Specifications*, 7th edition (2014).
b. FDOT *Structures Design Guidelines* (2018).
c. CEB-FIP *Model Code for Concrete Structures* (1990).

8.6.1.2   Traffic Requirements

a. Two 12-ft lanes with traffic flow in the same direction.
b. 10 ft wide for the right shoulder and 6 ft wide for the left shoulder.
c. 5-ft-wide sidewalk on one side.
d. Two 18.5-in.-wide FDOT 32 in F-shape barrier walls and one 12-in.-wide FDOT pedestrian/ bicycle railing.

8.6.1.3   Design Loads

a. Live load: HL-93 loading
Sidewalk pedestrian loading = 0.075 kips/ft²
b. Dead loads:
 – Cast-in-place concrete unit weight (DC) = 0.150 kcf
 – Precast concrete unit weight (DC) = 0.155 kcf
 – Traffic railing barrier wall = 0.421 klf
 – Pedestrian parapet and railing = 0.235 klf
 – Stay-in-place form = 0.02 ksf
 – 1/2-in. sacrificial deck thickness provided for grooving and planning
c. Thermal loads:
 – Uniform temperature: mean: 70°F; rise: 35°F; fall: 35°F
 – Temperature gradient specified as in Section 2.2.10.2 and taking the following values:

Positive Nonlinear Gradient	Negative Nonlinear Gradient
$T_1 = 41°F$	$T_1 = -12.3°F$
$T_2 = 11°F$	$T_2 = -3.3°F$
$T_3 = 0°F$	$T_3 = 0°F$

8.6.1.4   Materials

a. Concrete:
 – Bridge deck, diaphragm, and barrier walls: $f_c' = 5.5$ ksi, $f_{ci}' = 4.4$ ksi
 – Prestressed beams and closure pours: $f_c' = 8.5$ ksi, $f_{ci}' = 6.8$ ksi
 – Modulus of elasticity:

$$\text{Deck: } E_{c-deck} = 33,000 w_c^{1.5}\sqrt{f_c'} = 4273 \text{ ksi}$$

$$\text{Girder: } E_{c-girder} = 33,000 w_c^{1.5}\sqrt{f_c'} = 5312 \text{ ksi}$$

b. Prestressing strands:
ASTM A416, 0.6-in.-diameter, 7-wire, Grade 270 low relaxation
$f_{pu} = 270$ ksi, $f_{py} = 0.9 f_{pu} = 243$ ksi

Maximum jacking stress: $f_{pj} = 216$ ksi
Modulus of elasticity: $E_p = 28,500$ ksi
Anchor set: 0.375 in.
Friction coefficient: 0.23 (plastic duct)
Wobble coefficient: 0.00020
Duct: Maximum outside diameter = 4 in. and minimum inside diameter = 3 1/8 in.
c. Reinforcing steel:
ASTM 615, Grade 60
Yield stress: $f_y = 60$ ksi
Modulus of elasticity: $E_s = 30,000$ ksi

8.6.2   Bridge Span Arrangement and Typical Section

8.6.2.1   Span Arrangement

For a bridge crossing a waterway, the first consideration for arranging the bridge spans is to meet the minimum horizontal and vertical clearance requirements. The horizontal clearance dimension shall include the minimum navigational channel width required by the U.S. Coast Guard and the fender system and the required offset to the centerline of the piers. Based on the bridge site situation, a three-span continuous spliced concrete I-girder bridge is determined. To reduce the construction costs, the piers are set away from the deep water. The side and center span lengths are selected as 225 ft and 285 ft, respectively (see Fig. 8-17). The ratio of the side span to the main span lengths is about 0.79, which is within the typical span ratio of 0.7 to 0.85. In Fig. 8-17, E represents expansion joints and F represents fixed supports. The girder supports are composite elastomeric bearing pads that are actually neither a fixed support nor a roller support. The supports are treated as spring supports in the structural analysis. The bridge is on a tangent alignment with no horizontal coverture, and the angle between the bridge centerline and the centerlines of end bents and piers is 74° (see Fig. 8-17b).

8.6.2.2   Typical Section and Segment Layout

As FDOT's bulb-T girders have been used on many post-tensioned spliced I-girder bridges with great success and the forms for the girder fabrication are readily available in the area where

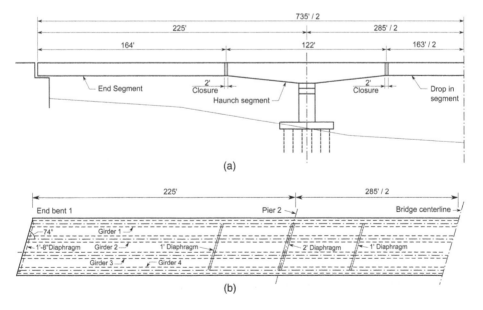

FIGURE 8-17   Example Bridge IV, (a) Elevation View, (b) Plan View.

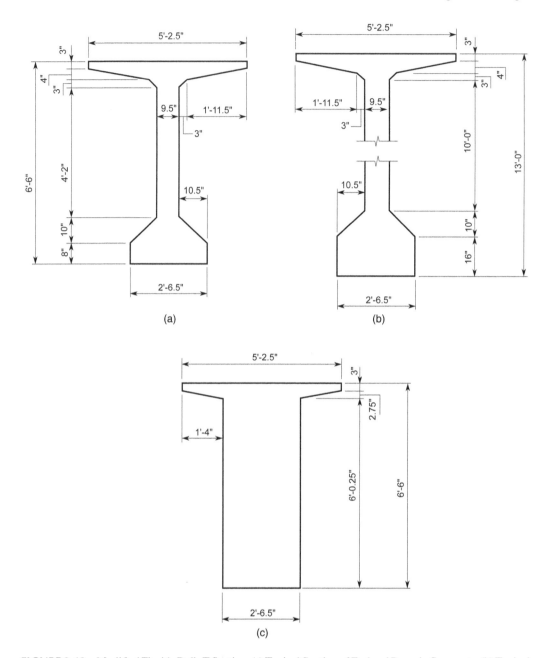

FIGURE 8-18 Modified Florida Bulb-T Section, (a) Typical Section of End and Drop-in Segments, (b) Typical Section of Pier Segments, (c) Section at Anchorage.

the bridge is located, the bulb-T girders are selected for this bridge. From previous experience, the web thickness of the bulb-T girders is increased to 9.5 in. to accommodate 4-in.- diameter post-tensioning ducts by spreading the standard side forms (see Fig. 8-18). The widths of the top and bottom flanges are increased by 2 in. to 5'-2.5″ and 2'-6.5″, respectively. Based on the traffic requirements, the total bridge width is set to 49'-0.5″. The bridge cross section consists of four modified Florida bulb-T girders, and the girder spacing is 13'-6″ with a cast-in-place deck of 8.5 in. thickness, including ½ in. as the sacrificial thickness (see Fig. 8-19). The 1/2-in. sacrificial thickness is not considered to be a structural composite section and is considered as

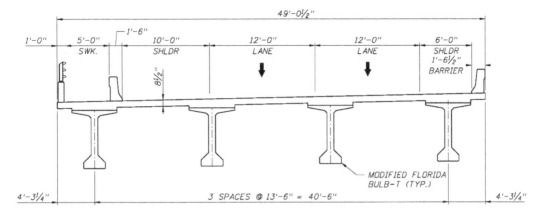

FIGURE 8-19 Typical Bridge Cross Section.

additional superimposed dead load. Each of the girders consists of five segments: two end segments/girders of 162.25 ft, two haunch segments/girders of 120 ft, and one drop-in segments/girders of 161 ft (see Figs. 8-17, 8-22, and 8-23). The segments were connected together by two 2'-0" long cast-in-place closure pours. The splice locations are determined based on the considerations discussed in Section 8.3.3.2. The constant girder depth of the side and drop-in segments is selected as 6.5 ft based on the discussions contained in Section 8.2.1, and the depth of the section over the pier is determined as 13 ft to meet the requirement for the maximum absolute negative moment (see Fig. 8-18).

8.6.3 CONSTRUCTION SEQUENCES

The bridge is analyzed and designed based on the following construction sequences:
   Stage I (see Fig 8-20a)

   a. Construct piers and bents, as well as install bearings.
   b. Erect temporary shoring towers.

Stage II (see Fig 8-20b)

   a. Erect haunch segments, and secure them to the temporary supports before releasing them from the crane.
   b. Install lateral temporary bracings between the haunch segments as shown in Fig. 8-21 when the adjacent haunch segments are erected.
   c. Attach strongbacks to the end segments at the end adjacent to the haunch segments.
   d. Erect end segments, and securely connect them to the haunch segments with strongbacks before releasing them from the cranes.
   e. Install temporary bracings between end segments as shown in Fig. 8-21 when the adjacent end segments are erected.

Stage III (see Fig 8-20c)

   a. Attach steel strongbacks to both ends of the drop-in segments.
   b. Erect the drop-in segments, and securely connect them to the pier segments before releasing them from the cranes.
   c. Install temporary bracings between the drop-in segments as shown in Fig. 8-21 when the adjacent drop-in segments are erected.

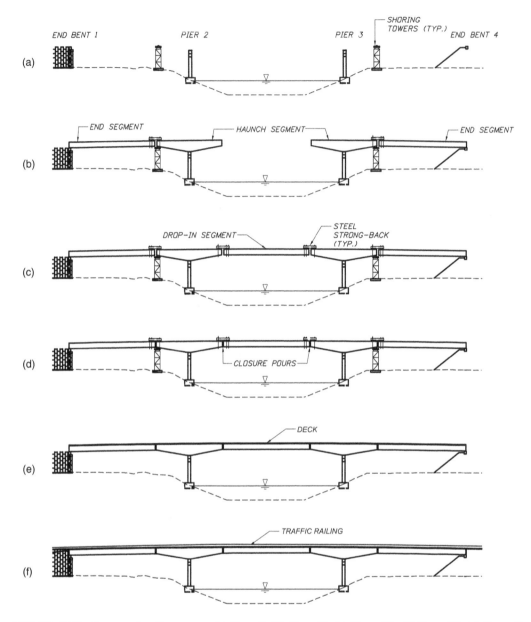

FIGURE 8-20 Construction Sequences, (a) Stage I, (b) Stage II, (c) Stage III, (d) Stage IV, (e) Stage V, (f) Stage VI.

Stage IV (see Fig 8-20d)

a. Cast segment closure pours at the girder splice locations, the intermediate diaphragms, and the pier diaphragms.
b. After the concrete strength in the closure pours reach 6.0 ksi, stress tendon 2 in the following beam order: 2, 3, 4, 1 (refer to Figs. 8-21a, 8-22a, and 8-23a).
c. Stress tendon 1 in the following beam order: 2, 3, 4, 1 (refer to Figs. 8-21a, 8-22a, and 8-23a).
d. Grout tendons 1 and 2.

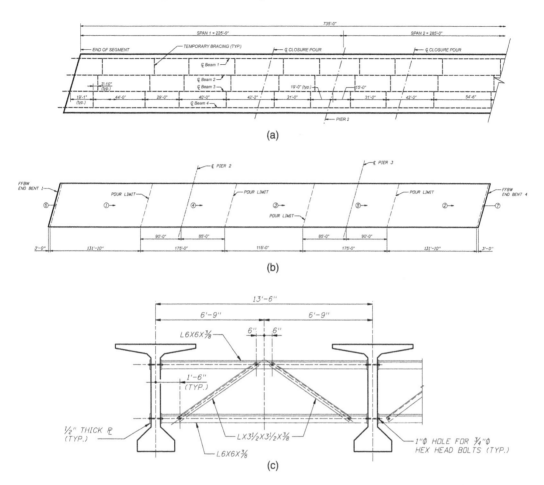

FIGURE 8-21 Arrangement of Temporary Bracings and Deck Pouring Sequences, (a) Temporary Bracing Arrangement, (b) Deck Pouring Sequences, (c) Typical Temporary Lateral Bracing.

Stage V (see Fig 8-20e)

a. Remove strongbacks and temporary supports.
b. Install deck forms and place deck reinforcement.
c. Cast the deck based on the pouring sequences provided in the plans (see Fig. 8-21b).

Stage VI (see Fig 8-20f)

a. After the deck concrete strength reaches 4.5 ksi, stress tendon 3 in the following beam order: 2, 3, 4 and 1 (refer to Figs. 8-21a, 8-22a, and 8-23a).
b. Stress tendon 4 in the following beam order: 2, 3, 4 and 1 (refer to Figs. 8-21a, 8-22a, and 8-23a).
c. Grout tendons 3 and 4.
d. Cast the end diaphragms and the remaining deck pours.
e. Cast the traffic and the pedestrian barriers to complete the bridge.

8.6.4 Layout of Post-Tensioning Tendons and Pretensioning Strands

After the preliminary bridge dimensions and construction sequences have been determined, the moments and shears of each of the girders in different loading stages can be obtained by using any of the methods discussed in Chapter 4 and Section 8.4.3.2 (see Section 8.6.6). Then the required prestressing forces can be estimated.

8.6.4.1 Layout of Pretensioning Strands

As discussed in Section 8.4.4.2, the amount of pretensioning strands is determined mainly to provide enough prestressing force to keep the concrete stresses within certain stress limits at release and during erection. The end segment has 38φ0.6″ strands prestressed to 43.95 kips in the bottom flange and 4φ0.6″ prestressed to 10 kips in the top flange. The drop-in segment has 44φ0.6″ strands prestressed to 43.95 kips in the bottom flange and 4φ0.6″ prestressed to 10 kips in the top flange. The haunch segment has 24φ0.6″ prestressing strands in the top flange and 10 strands in the web. All the strands are straight with some strands deboned at the ends of the segments. The detailed layout of the strands is illustrated in Figs. 8-22 and 8-23.

8.6.4.2 Layout of Post-Tensioning Tendons

From the assumed construction sequences described in Section 8.6.3, we can see that the post-tensioning tendons in this bridge should be provided to resist the live loads and part of the dead loads. Based on the discussions in Section 8.4.4.1, four continuous tendons are used (see Figs. 8-22 and 8-23). Each of the tendons contains 15φ0.6″ low relaxation prestressing strands, with a grade of 270. The duct size is 4 in., and the ratio of the duct diameter to the web width is 0.42, which is a little less than 0.4 required by current AASHTO specifications.

8.6.5 Detailing and Reinforcement

8.6.5.1 Anchorage End

The total length of the end block is 11′-9″ with a transition length of 3′-9″ (see Fig. 8-24). The width of the end block is taken as the full width of the bottom flange. The vertical reinforcement is determined for two loading stages: prestresssing force and post-tensioning force. The vertical reinforcement should be sufficient to resist 8% of the total prestressing forces at release; this reinforcement is distributed within a distance of about one-quarter the girder depth relative to the end of the beam[8-20, 8-21]. The maximum stress of the reinforcement should not exceed one-third of the reinforcement yield stress (20 ksi for this example). The required vertical reinforcement for the post-tensioning forces are determined based on the method discussed in Section 3.8.2.3. The principle reinforcement arrangement is illustrated in Fig. 8-24. The girder end is embedded in the end diaphragm (see Fig. 8-24d).

8.6.5.2 Splice End of Segments

The vertical reinforcement in the splice end of the segments is determined to resist the vertical force caused by the prestressing forces based on the method presented in Section 3.8.1. The reinforcement arrangement for the spliced end of the end segments and the drop-in segments is shown in Fig. 8-25.

8.6.5.3 Splice Joint

The splice joint is 2 ft wide with a shear key (see Fig. 8-26a). Six #5 bars in the top flange and eight #5 bars in the bottom flange are lapped together. Eight #5 bars are provided to transfer shear from the girder to deck. There is a 1-ft-wide diaphragm along each of the splice joints. The principal dimensions and reinforcement arrangement are shown in see Fig. 8-26.

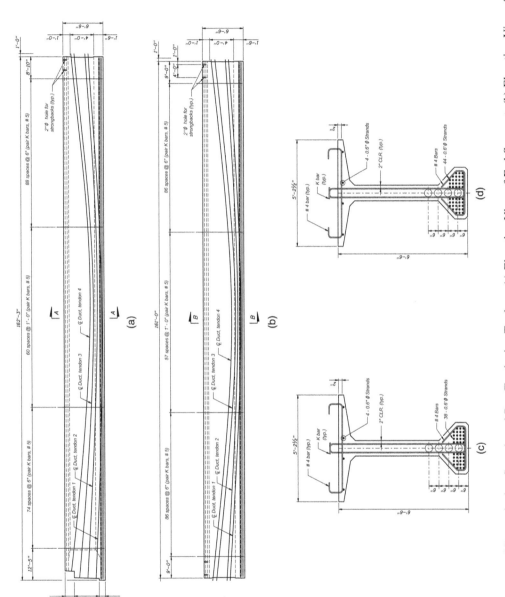

FIGURE 8-22 Arrangements of Pretensioning Strands and Post-Tensioning Tendons. (a) Elevation View of End Segment, (b) Elevation View of Drop-in Segment, (c) Typical Section *A-A* in Side Segment, (d) Typical Section *B-B* in Drop-in Segment.

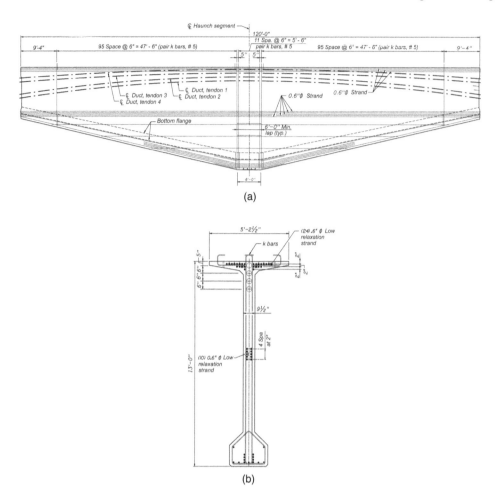

FIGURE 8-23 Arrangements of Pretensioning Strands and Post-Tensioning Tendons in Pier Segment, (a) Elevation View, (b) Section over Pier.

8.6.6 General Analysis of Bridge Super-Structure

8.6.6.1 General

From the assumed construction sequences discussed in Section 8.6.3, it can be observed that: (a) during construction the bridge structural system changes from simply supported girders and cantilevers girders to a three-span continuous girder, (b) the girder cross section is I-shaped when it is first erected and then it becomes a composite T-section after the deck concrete hardens, (c) the prestressing tendons are stressed in three different stages and different cross sections. The accurate time-dependent analysis by hand becomes very difficult and time consuming, though it is still possible. The analysis of the design example has been performed using a computer program based on the analytical theory presented in Chapters 3 and 4. The following discussions are focused on the analytical procedures, assumptions, and results.

8.6.6.2 Section Properties

The cross sections of the girder vary along its length. The section of each of the elements in the analytical model is assumed to be constant, and its section properties are calculated using a computer program based on the information presented in Section 8.6.2.2. The cross section properties

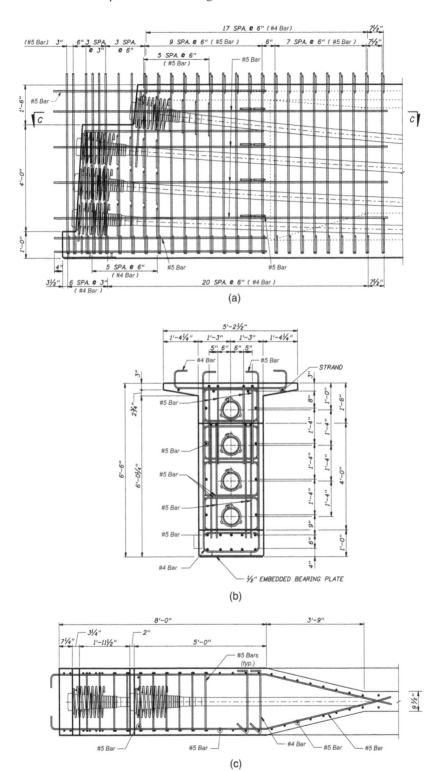

FIGURE 8-24 Post-Tensioning Tendon Anchorage Zone, (a) Elevation View, (b) End View, (c) Plan View, (d) Elevation View of End Diaphragm.

(*Continued*)

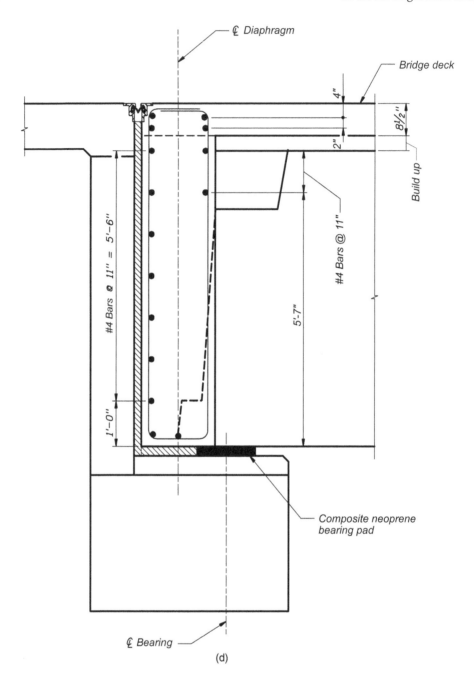

FIGURE 8-24   (Continued)

for each of the girder segments shown in Fig. 8-18a and b and their corresponding composite section properties are summarized in Table 8-2.

8.6.6.3   Assumption of Construction Schedules

To perform a time-dependent analysis, a more detailed construction schedule, in addition to those given in Section 8.6.3, should be assumed to describe the concrete ages for the changing structural systems and stressing tendons. The assumed construction schedule is shown in Table 8-3.

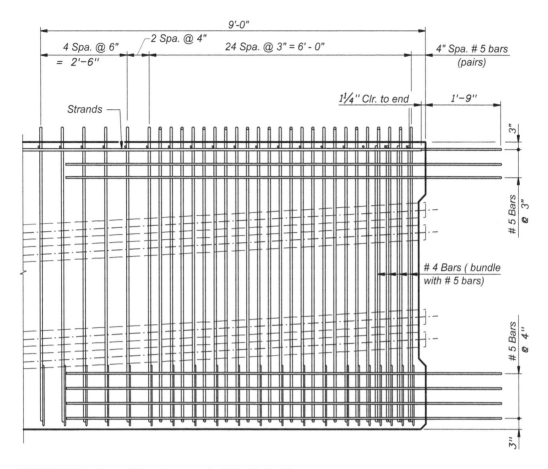

FIGURE 8-25 Typical Reinforcement in Splice End of Segments.

8.6.6.4   Analytical Models

8.6.6.4.1   Introduction

For convenience and from a practical point of view, the time-dependent analysis of post-tensioned girder bridges is typically performed on individual girders using either two- or three-dimensional models. The transverse effects of loads on the girder are accounted for using the load distribution factors discussed in Section 8.4.3.2.

8.6.6.4.2   Transverse Load Distributions

The load distribution factor equations presented in Section 8.4.3.2 are normally suitable for span lengths less than or equal to 240 ft. To ensure more accurate analytical results, the load distribution

TABLE 8-2

Section Properties of Two Typical Sections in Design Example IV

Section	Type	I (ft⁴)	A (ft²)	y_t (ft)	y_b (ft)
Center	Noncomposite	49.648	9.000	3.142	3.358
	Composite	96.701	15.800	2.313	4.853
Pier	Noncomposite	316.749	15.285	6.789	6.211
	Composite	555.790	22.085	5.263	8.403

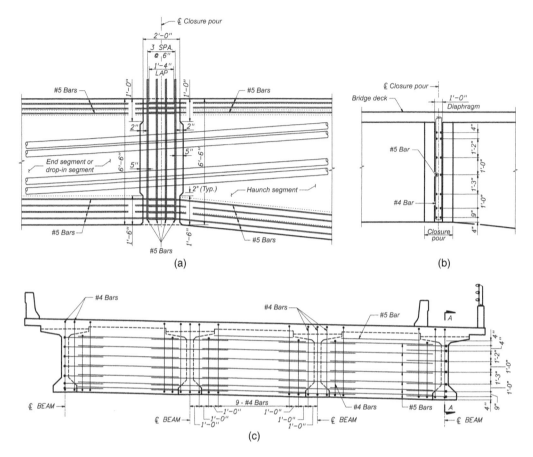

FIGURE 8-26 Typical Reinforcement at Splice Joint, (a) Elevation View, (b) Section *A-A* View, (c) Side View along Diaphragm.

TABLE 8-3
Assumed Construction Schedule for Design Example IV

Sequence	Description	Girder Age (days)
1	Release of pretensioning strands	1
2	Erection of girder segments	80
3	Casting of splices and curing	100
4	Stressing of post-tendons 1 and 2; removal of temporary support and strongbacks	105
5	Installation of stay-in-place deck form; casting of side span segment deck concrete, assuming 2-in. uniform buildup	155
	Casting of center span segment deck concrete	165
	Casting of pier segment deck concrete	175
6	Stressing of post-tendons 3 and 4	200
7	Casting of barrier walls and install utilities	250
8	Open to traffic	500
9	Middle condition	5,000
10	Final condition	10,000

TABLE 8-4

Load Distribution Factors of Design Example IV*

Loading	Internal Force	Range (L = Bridge Length)	Girder 1	Interior Girders	Girder 4
HL-93 (lane)	Positive moment	Entire bridge	0.665	0.820	0.905
	Negative moment	$0' \leq L < 170'$ and $565' \leq L < 735'$	0.670	0.750	0.840
		$280' \leq L < 455'$	0.800	0.800	1.035
		$170' \leq L < 280'$ and $455' \leq L < 565'$	0.670	0.770	0.85
	Shear	Entire bridge	0.780	1.017	1.025
Pedestrian (lb/ft)		Entire bridge	291	176	0
Railings (lb/ft)		Entire bridge	370	320	235

* Note: The load distribution factors should be multiplied by multiple presence factor 0.85 when both HL-93 and pedestrian live loads are considered in the design.

factors can be determined using three-dimensional beam models or plane grid models as discussed in Chapter 4. The maximum live load distribution factors for moment and shear for this bridge were determined based on three-dimensional beam models by applying two- and three-lane loadings including the multiple presence factors as specified in Table 2-2. The maximum load factors are defined as

$$g_i = \frac{F_i}{\sum_{j=1}^{n} F_j}$$ (8-8)

where

g_i = load distribution factor of girder i

F_i = maximum moment or shear of girder i at control section

n = total number of girders supporting deck

$\sum_{j=1}^{n} F_j$ = sum of total moment or shear of girders, determined based on a three-dimensional model, at the same cross section as section for F_i

Theoretically, the load distribution factors vary along the length of the girder. For simplicity, it is common practice to use the load distribution factors determined by Eq. 8-8 at some control sections. The load distribution factors used for the bridge design are determined using a three-dimensional finite-element model and are summarized in Table 8-4.

8.6.6.4.3   Analytical Model for Longitudinal Analysis

Each girder is modeled as a two-dimensional beam and analyzed using the finite-element method. Each neoprene bearing pad supporting the girder is modeled as two springs at each support (see Fig. 8-27). Each end segment is divided into 18 elements with three elements for the thickened web segment. Each pier segment is divided into 22 elements with two elements for the constant girder depth segment. The drop-in segment is divided into 18 elements. Each of the closure pour segments is treated as one element. There are 102 elements in this model (see Fig. 8-27).

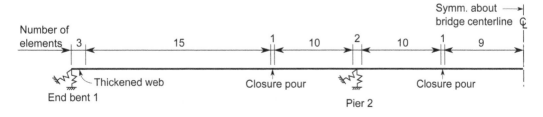

FIGURE 8-27   Analytical Model for Longitudinal Analysis of Design Example IV.

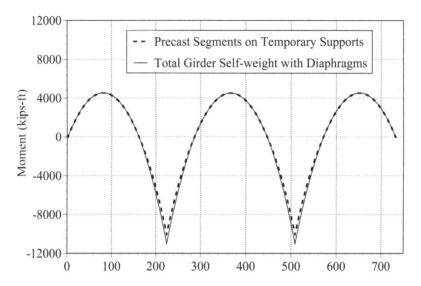

FIGURE 8-28   Distribution of Moment Due to Girder Self-Weight.

8.6.6.5   Analytical Results

Consider the interior girder, for example, to illustrate the analytical results. The results for other girders are similar.

8.6.6.5.1   Effect of Dead Loads

8.6.6.5.1.1   Effects Due to Girder Self-Weight   Based on the construction sequence presented in Section 8.6.3, the moment and shear distributions due to the girder self-weight before installing the closure pours can be determined by treating each of the segments as simply supported beams. The reactions of the temporary supports and self-weights of the closure pours, including the diaphragms, are applied to the three-span continuous structure. The moment and shear distributions due to girder self-weight, including closure pours and diaphragms, are shown in Figs 8-28 and 8-29, respectively. The moment and shear at this stage are resisted by the girder's noncomposite section.

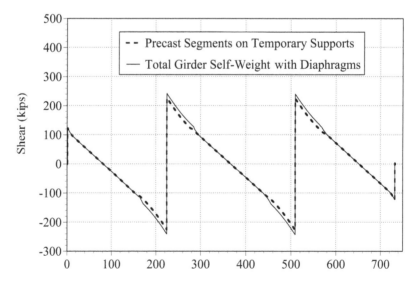

FIGURE 8-29   Distribution of Shear Due to Girder Self-Weight.

8.6.6.5.1.2 Effects Due to Deck Self-Weight The deck concrete pouring sequences are shown in Fig. 8-21b. It is assumed that the deck concrete within the limits of pours 1 and 2 are composited with the girder when starting the third deck pour and that the girder sections within the limits of pours 1 to 3 have become composite sections when pouring the deck concrete over the pier segments. The tributary width of the deck for the interior girder is equal to the girder spacing of 13.5 ft, and the deck weight per unit length is

$$q_{deck} = \frac{0.15 \times 13.5 \times 8.5}{12} = 1.434 \ kips/ft$$

The effects of the deck concrete self-weights for each of the concrete pouring stages are determined by computer analysis and shown in Figs. 8-30 and 8-31.

8.6.6.5.1.3 Effects Due to Traffic and Pedestrian Railings The traffic barrier walls and pedestrian parapets and railings are installed after all the deck concrete has hardened. The distributed uniform loading on each of the interior girders is 320 lb/ft as shown in Table 8-4 and resisted by girder composite sections. The distributions of moment and shear due to the railings are shown in Fig. 8-32.

8.6.6.5.1.4 Total Effects of Dead Loads The total effects due to the dead loads can be obtained by summing the related effects shown in Figs. 8-28 to 8-32 and are illustrated in Fig. 8-33.

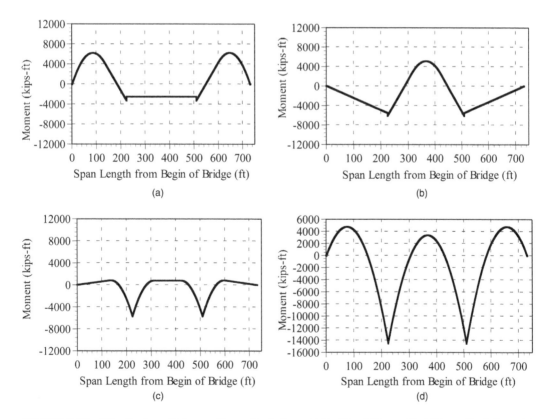

FIGURE 8-30 Distribution of Moment Due to Deck Pour, (a) End Deck Pours, (b) Midspan Deck Pour, (c) Pier Deck Pours, (d) Total Deck Self-Weight.

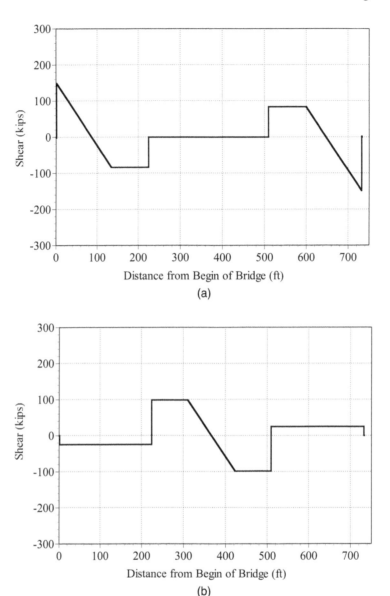

FIGURE 8-31 Distribution of Shear Due to Deck Pour, (a) End Deck Pours, (b) Midspan Deck Pour, (c) Pier Deck Pours, (d) Total Deck Self-Weight.

(Continued)

8.6.6.6 Effects of Live Loads

8.6.6.6.1 *Effects Due to Vehicle Live Loads*

The effects of the following three types of vehicle live loads are analyzed for this design example:

    a. HL-93 design truck and design lane loading as discussed in Section 2.2.3.1
    b. Tandem and design lane loading presented in Section 2.2.3.1
    c. FDOT permit trucks

The live load applications follow the specifications discussed in Section 2.2.3.1.3 and load positions presented in Section 8.4.3.3. It was found that the HL-93 design truck and design lane control

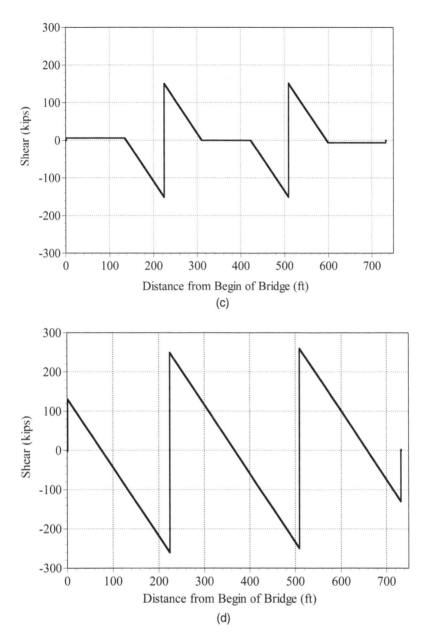

FIGURE 8-31   (Continued)

the design. The moment and shear envelopes due to one lane HL-93 live loadings are shown in Figs. 8-34 and 8-35.

8.6.6.6.2   Effects Due to Pedestrian Live Loads

The distributed pedestrian loading on the interior girder is 176 lb/ft as given in Table 8-4. The effects of the uniform pedestrian loading are shown in Figs. 8-36 and 8-37.

8.6.6.7   Effect of Prestressing Forces

The prestressing forces are applied to the girders in three stages: pretensioned strands, first-stage post-tensioning forces and second-stage post-tensioning forces. The losses of prestressing forces

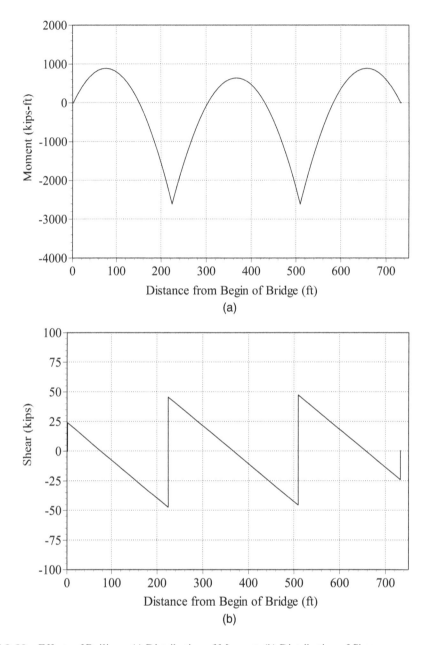

FIGURE 8-32   Effects of Railings, (a) Distribution of Moment, (b) Distribution of Shear.

due to elastic shortening, creep and shrinkage, etc., are determined based on different assumed construction stages and concrete ages and are determined according to the methods discussed in Section 3.2. The detailed analysis is performed using a computer program. The effects of the prestressing forces in several typical construction stages and concrete ages are shown in Figs. 8-38 to 8-44. Figure 8-38 illustrates the distributions of moment and shear due to pretensioning strands before removing temporary supports. The effects of first-stage and second-stage post-tensioning forces are shown in Figs. 8-39 and 8-40, respectively. The variations of axial forces due to the first- and second-stage post-tensioning tendons are presented in Fig. 8-41. The effects of total prestressing forces at 10,000 days are shown in Figs. 8-42 to 8-44.

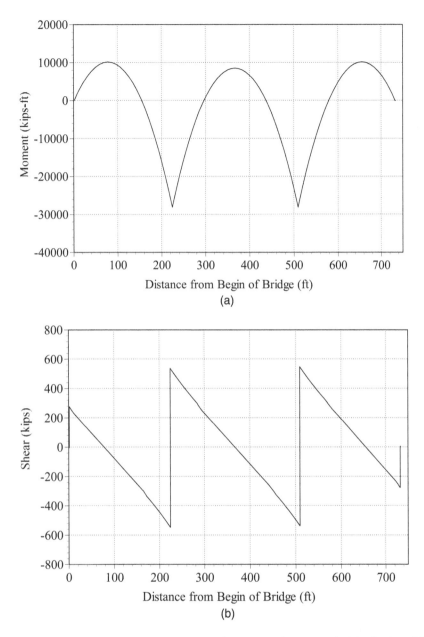

FIGURE 8-33 Total Effects Due to Dead Loads, (a) Distribution of Moment, (b) Distribution of Shear.

8.6.6.8 Effects of Concrete Creep and Shrinkage

The creep and shrinkage effects are evaluated step by step based on the assumed construction sequences and schedules shown in Table 8-3 according to CEB-FIP *Model Code for Concrete Structures*, 1990[8-23], i.e. Eq. 1-9b and Eq. 1-15b, and by the method presented in Section 4.3.4.4.3. The moment distributions due to creep and shrinkage at 10,000 days after casting time caused by dead and prestressing forces are illustrated in Fig. 8-45.

8.6.6.9 Effect of Temperatures

The mean temperature is assumed as 70°F, and the rise and fall in temperature is assumed to be 35°F. The temperature gradient specified in Section 2.2.10.2 is used for analyzing the effects of the

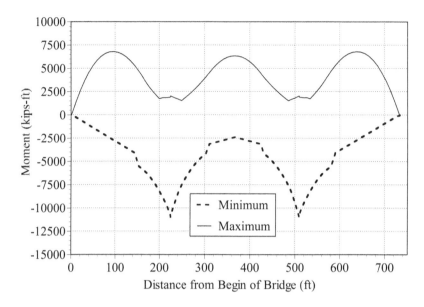

FIGURE 8-34 Moment Envelope of Interior Girder Due to HL-93 Live Load Including Impact Factor.

temperature gradient, and the related values are provided in Section 8.6.1.3.c. The effects of the uniform temperature change are found to be comparatively small. The moment and shear distributions due to the temperature gradient are illustrated in Figs. 8-46 and 8-47. In these figures, the solid and dotted lines represent the responses due to the positive and negative vertical temperature gradients, respectively.

8.6.6.10   Summary of Effects at Some Control Sections

For clarity and capacity verifications, the effects due to different loadings for some control sections are summarized in Table 8-5. The effects due to the HL-93 design loading shown in the table are

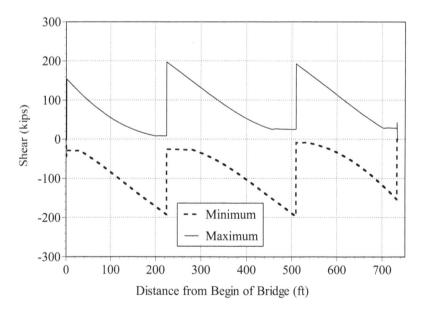

FIGURE 8-35   Shear Envelope of Interior Girder Due to HL-93 Live Load Including Impact Factor.

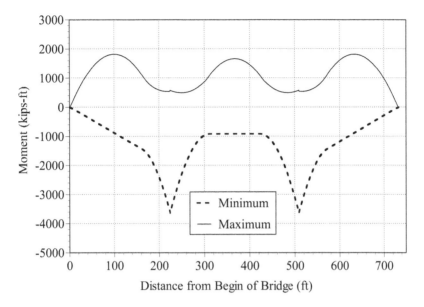

FIGURE 8-36   Moment Envelope of Interior Girder Due to Pedestrian Loading.

calculated by multiplying the results shown in Figs. 8-34 and 8-35 by the load distribution factors shown in Table 8-4 and by the multiple presence factor of 0.85. The effects due to both the HL-93 and the pedestrian live loads are slightly larger than the effects due to HL-93 live load alone.

8.6.7   CAPACITY VALIDATIONS

8.6.7.1   Service Limit

For spliced girder bridges, two service limit states should be checked, i.e., Service I and Service III. As the bridge structural system changes during construction, and the effects of concrete creep

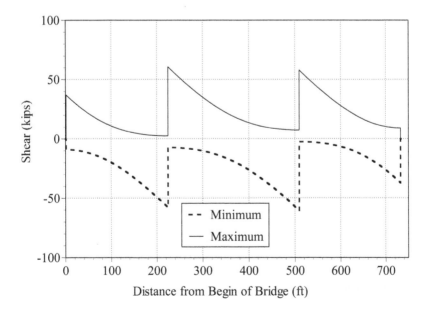

FIGURE 8-37   Shear Envelope of Interior Girder Due to Pedestrian Loading.

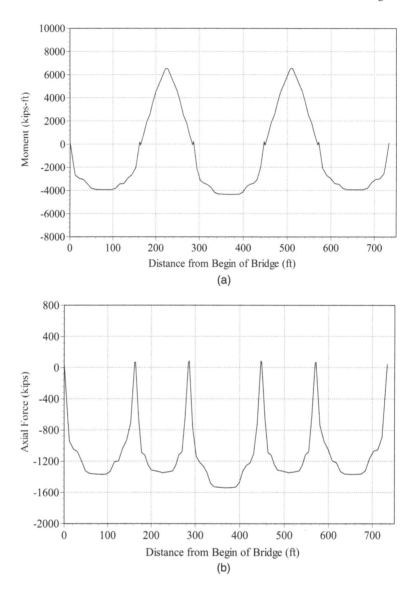

FIGURE 8-38 Effects of Prestressing Forces Due to Pretensioning Strands after Loses, (a) Moment, (b) Axial Force.

and shrinkage vary with time, the bridge capacity verification should be performed for different loading stages to ensure the adequacy of the structure during construction, such as when it is open to traffic and during its final stage. The following verifications are performed at 10,000 days after casting for example.

8.6.7.1.1 Service I

Service I check is to investigate concrete compression strengths. Normally, the concrete compression strength in three loading cases should be checked (see Section 1.2.5.1):

Case I: The sum of the concrete compression stresses caused by temporary forces before losses and permanent dead loads should be less than $0.6 f_c'$, i.e.,

$$f_j^c + f_{DC+DW}^c \leq 0.6 f_c'$$

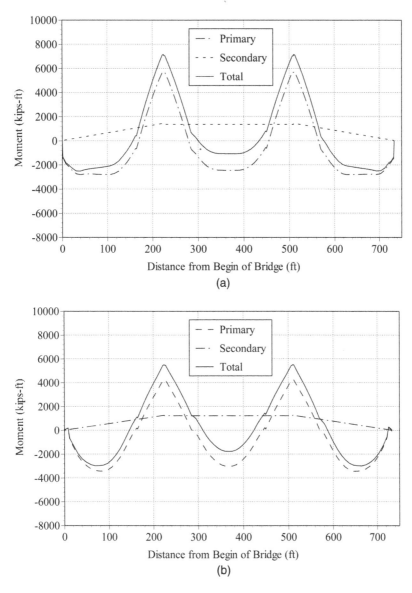

FIGURE 8-39 Moment Distribution Due to Post-Tensioning Tendons after Initial Loses, (a) First-Stage Tendons (Tendons 1 and 2), (b) Second-Stage Tendons (Tendons 3 and 4).

Case II: Permanent stresses after losses: The sum of the concrete compression stresses caused by effective prestressing forces and permanent dead loads should be less than $0.45f'_c$, i.e.,

$$f^c_{pe} + f^c_{DC+DW} \leq 0.45f'_c$$

Case III: The sum of the concrete compression stresses caused by effective prestressing forces, permanent dead loads, and transient loadings should be less than $0.60\phi_w f'_c$. The corresponding load combination for this example is:

$$DC + CR + SH + PS + LL(1+IM) + BR + 0.3WS + WL + TU + 0.5TG$$

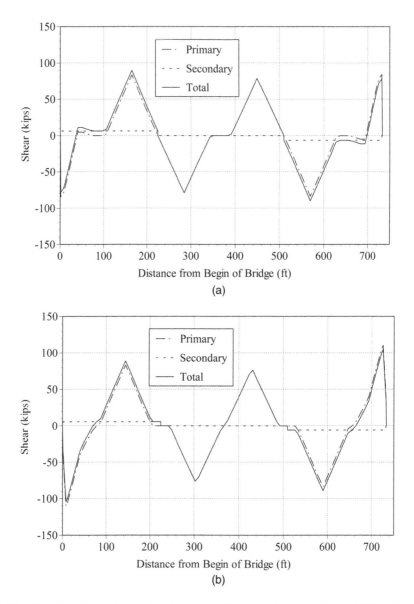

FIGURE 8-40 Shear Distribution Due to Post-Tensioning Tendons after Initial Loses, (a) First-Stage Tendons (Tendons 1 and 2), (b) Second-Stage Tendons (Tendons 3 and 4).

Consider Case II at a cross section located at the midspan of the center span, approximately 367 ft away from the left end support as an example to illustrate the procedures for the capacity verification.
    Allowable concrete compression stress:

Deck concrete: $f_{allow}^c = 0.45 f_c = 0.45 \times 5.5 = 2.475 \ ksi$
Girder concrete: $f_{allow}^c = 0.45 f_c = 0.45 \times 8.5 = 3.825 \ ksi$

From Table 8-5, we have
    The moment due to dead loads on the girder noncomposite section is

$$Md_{non} = 7856 \ kips\text{-}ft$$

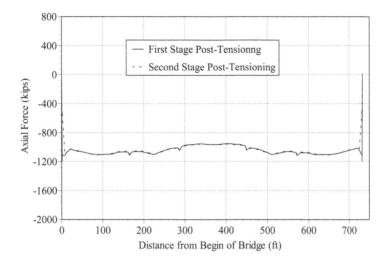

FIGURE 8-41 Axial Force Distribution Due to Post-Tensioning Tendons after Loses (10,000 Days).

The moment due to dead loads on the girder composite section is

$$Md_{com} = 642 \ kips\text{-}ft$$

The moment due to prestressing forces on the girder noncomposite section is

$$Mp_{non} = -6771 + 1376 = -5395 \ kips\text{-}ft$$

The moment due to prestressing forces on the girder composite section is

$$Mp_{com} = -3004 + 1239 = -1765 \ kips\text{-}ft$$

The effective prestressing force on the noncomposite section is

$$N_{non} = -2507 + 3 = -2504 \ kips$$

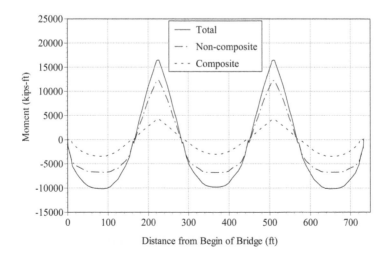

FIGURE 8-42 Moment Distributions Due to Total Prestressing Forces after Loses (10,000 Days).

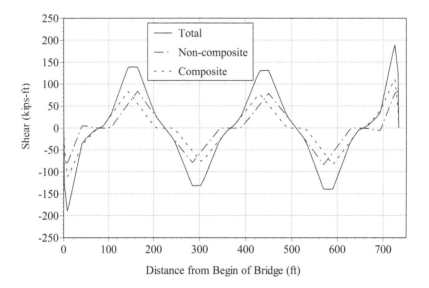

FIGURE 8-43 Shear Distributions Due to Total Prestressing Forces after Loses (10,000 Days).

The effective prestressing force on the composite section is

$$N_{com} = -966 + 2 = -964 \ kips$$

The moment due to creep and shrinkage on the girder noncomposite section is

$$Mcreep_{non} = 24 \ kips\text{-}ft$$

The moment due to creep and shrinkage on the girder composite section is

$$Mcreep_{com} = 194 \ kips\text{-}ft$$

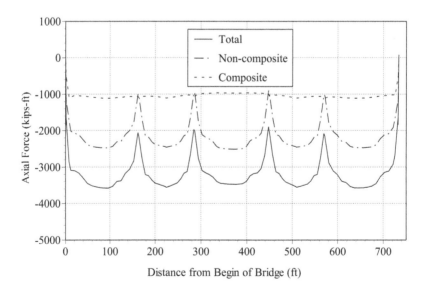

FIGURE 8-44 Axial Force Distributions Due to Total Prestressing Forces after Loses (10,000 Days).

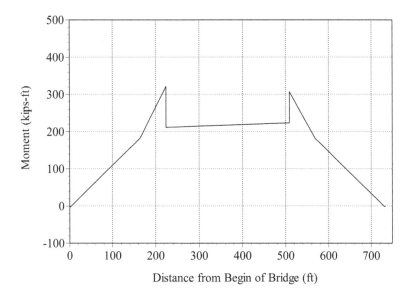

FIGURE 8-45   Distribution of Moment Due to Creep and Shrinkage (10,000 days).

The compressive stress at the top of the girder caused by loads on the noncomposite section is

$$f_{t-non}^{c} = \frac{\left(Md_{non} + Mp_{non} + Mcreep_{non}\right)yt_{non}}{I_{non} \times 12^2} + \frac{N_{non}}{A_{non} \times 12^2} = -3.024 \text{ ksi}$$

The compressive stress at the top of the girder caused by the loads on the composite section is

$$f_{t-com}^{c} = \frac{\left(Md_{com} + Mp_{com} + Mcreep_{com}\right)\left(yt_{com} - 8\right)}{Icom \times 12^2} + \frac{N_{com}}{A_{com} \times 12^2} = -0.314 \text{ ksi}$$

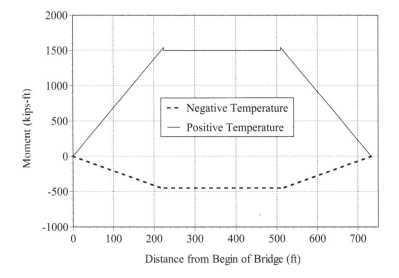

FIGURE 8-46   Distribution of Moment Due to Gradient Temperature.

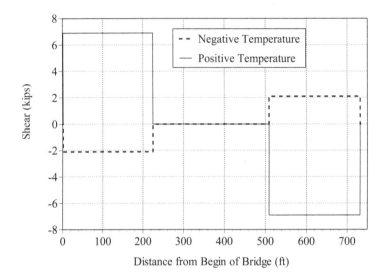

FIGURE 8-47   Distribution of Shear Due to Gradient Temperature.

The total compressive stress at the top of girder is

$$f_t^c = \left| f_{t-non}^c + f_{t-com}^c \right| = 3.338 \ ksi < 3.825 \ ksi \ (ok!)$$

The compressive stress at the bottom of the girder induced by loads on the noncomposite section is

$$f_{b-non}^c = \frac{\left(Md_{non} + Mp_{non} + Mcreep_{non} \right) yb_{non}}{I_{non} \times 12^2} + \frac{N_{non}}{A_{non} \times 12^2} = -0.765 \ ksi$$

The compressive stress at the bottom of the girder caused by loads on the composite section is

$$f_{b-com}^c = \frac{\left(Md_{com} + Mp_{com} + Mcreep_{com} \right) \left(yb_{com} \right)}{Icom \times 12^2} + \frac{N_{com}}{A_{com} \times 12^2} = -0.747 \ ksi$$

The total compressive stress at the bottom of the girder is

$$f_b^c = \left| f_{b-non}^c + f_{b-com}^c \right| = 1.513 \ ksi < 3.825 \ ksi \ (ok!)$$

The compressive concrete stress at the top of the deck is

$$f_{t-deck}^c = \left| \frac{\left(Md_{com} + Mp_{com} + Mcreep_{com} \right) \left(yt_{com} \right)}{I_{com} \times 12^2} + \frac{N_{com}}{A_{com} \times 12^2} \right| = 0.269 \ ksi < 2.475 \ ksi \ (ok!)$$

For the remaining loading cases and sections, the Service I verification can be performed using the same procedure as shown above.

8.6.7.1.2   Service III

Service III check is to investigate the maximum tensile stresses in the prestressed concrete components to ensure the maximum tensile stresses are not greater than the limits specified in Sections 1.2.5.1.1.2 and 1.2.5.1.2.2. Two cases should be checked:

Case I: Temporary tensile stresses before losses
Case II: Tensile stresses at the Service III load combination

TABLE 8-5
Maximum Effects at Several Control Sections

Loads Type	Section	78 ft (0.35L)			224 ft (Pier)			285 ft (Closure Pour)			367.5 ft (Center of Span 3)		
		M (kip-ft)	V (kip)	N (kip)	M (kip-ft)	V (kip)	N (kip)	M (kip-ft)	V (kip)	N (kip)	M (kip-ft)	V (kip)	N (kip)
Dead	Noncompo.	9,336	2.4	0	-25,551	-503.1	0	-2,875	-272.5	0	7,856	0	0
	Total	12,207	1.9	0	-28,165	-550.5	0	-3,321	-298.9	0	8,498	0	0
HL-93	Max.	4,601	63	0	1,423	170	0	2,136	126	0	4,427	66	0
	Min.	-1,389	-56	0	-7,466	-166	0	-3,250	-24	0	-1,555	-66	0
Pedestrian	Max.	689	6	0	216	1	0	267	16	0	666	8	0
	Min.	-272	-6	0	-1,452	-3	0	-475	-4	0	-365	-8	0
Creep	Noncompo.	-2.3	0	0	87	1.2	0	18.9	0.1	0	23.7	0.1	0
	Compo.	87	0.8	0	234	0.8	0	195	0	0	194	0	0
Gradient Temper.	Pos.	533.6	6.9	0	1542	6.9	0	1,499	0	-1	1,499	0	-3.1
	Neg.	-160.1	-2.1	0	-462.7	0	0.3	-449.6	0	0.3	-449	0	0.3
Primary prestress	Noncompo.	-7,123	0	-2,603	12,338	0	-2,452	-704	-78.9	-1,015	-6,771	-78.9	-2,507
	Compo	-3,213	0	-1,077	4,245	0	-1,102	75	-52.8	-1,020	-3,004	-52.8	-966
	Total	-10,335	0	-3,681	16,583	0	-3,552	-629	-131.7	-2,035	-9,775	-131.7	-3,473
Secondary prestress	Noncompo.	532	6.35	0	1,461	6.35	0	1,375	0	3.35	1,376	0	3.35
	Compo.	452.6	5.65	0	1,139	5.65	0	1,239	0	1.85	1,239	0	1.85
	Total	984.6	12	0	2,601	12	0	2,614	0	5.2	2,615	0	5.2

Longitudinal Locations (Distance from Beginning of Bridge, ft)

The Service III load combination for the example bridge is

$$DC + CR + SH + PS + 0.8 \, LL \, (1 + IM) + TU + 0.5TG$$

Consider case II and a cross section located at the midspan of the center span, about 367 ft away from the left end support as an example to illustrate the verification procedure.

From Table 8-5, the total maximum moment due to both the HL-93 and the pedestrian loadings at this section is

$$MLL = 4427 + 666 = 5093 \text{ kips-ft}$$

$$M_{tg} = 1499 \text{ kips-ft}$$

The maximum tensile stress at the bottom of the girder is equal to the stress f_b^c calculated above plus that due to live loads and temperature, i.e.,

$$f_t^c = f_b^c - \frac{(0.8MLL + 0.5M_{tg})yb_{com}}{I_{com} \times 12^2} = 1.513 - \frac{(0.8 \times 5093 + 0.5 \times 1499) \times 4.853}{96.701 \times 12^2} = 0.169 \ ksi$$

From Section 8.6.1.4, the ratio of the modulus of elasticity of steel to that of girder concrete is about 5.4. Thus, the maximum stress of the steel is

$$f_s \approx 5.4 \times 0.169 = 0.913 \ll 30 \ ksi$$

From Section 1.2.5.1.2.2, the concrete allowable tensile stress is

$$f_{ta} = 0.19\sqrt{f_c'} = 0.554 \ ksi > f_t^c = 0.169 \text{ ksi } (ok!)$$

The maximum tensile stress validation for other locations can be performed using the same procedure as above. It was found that all the sections are adequately designed for this limit state.

8.6.7.2   Strength Limit

Strengths I, II, III, V and Extreme Events I and II may need to be evaluated. Take Strength I for example. The load combination for this design example can be simplified as

$$1.25DC + 1.25 \, CR + 1.25SH + 1.75LL(1 + IM) + 0.5TU$$

8.6.7.2.1   Flexure Strength

Consider the same section as above with its maximum positive moment located about 367 ft away from the left end of the bridge as an example to illustrate the capacity verification process. For the tension-controlled prestressed concrete specified in Section 2.3.3.5, the flexure residence factor is

$$\phi_f = 1.0$$

The area of the prestressing steel in the top flange is

$$A_{ps} = 0.217(4 \times 15 + 44) = 22.568 \text{ in}^2$$

From Section 3.4.2.2, the effective width of the top flange can be taken as the center-to-center spacing of the beam, which is also the actual tributary width for the girder, i.e.,

$$b = 13.5 \times 12 = 162 \text{ in}$$

Web thickness: $b_w = 9.5$ in

From Section 1.4.3.3, the distance between the centers of gravity of the duct and tendons $Z = 0.75$ in. for 4-in.-diameter ducts and the distance from the extreme compression fiber to the centroid of the prestressing tendons can be calculated as

$$d_p = \frac{(4 \times 15 \times 62.25 + 44 \times 71.45)}{104} = 66.142 \text{ in}$$

Assuming the section as a rectangular and from Eq. 3-149e, the stress block factor is

$$\beta_1 = 0.85 - 0.05(5.5 - 4.0) = 0.775$$

From Eq. 3-152b,

$$k = 2(1.04 - 0.9) = 0.28$$

Using Eq. 3-165a, the distance from the extreme compression fiber to the neutral axis c can be determined as

$$c = \frac{A_{ps} f_{pu}}{0.85 f_c' \beta_1 b + k A_{ps} \frac{f_{pu}}{d_p}} = \frac{22.57 \times 270}{0.85 \times 5.5 \times 0.775 \times 162 + 0.28 \times 22.57 \frac{270}{66.14}} = 9.94 \text{ in}$$

From Eq. 3-149b, the depth of the compression block is

$$a = \beta_1 c = 0.775 \times 9.94 = 7.71 \text{ in}$$

From Eq. 4-152a, the average stress in the prestressing steel can be calculated as

$$f_{ps} = f_{pu}\left(1 - k\frac{c}{d_p}\right) = 270\left(1 - 0.28\frac{9.94}{66.14}\right) = 258.64 \text{ ksi}$$

From Eq. 3-164, the factored flexural resistance can be approximately determined by treating the section as rectangular as

$$\phi_f M_n = \phi_f A_{ps} f_{ps}\left(d_p - \frac{a}{2}\right) = 1.0 \times 22.57 \times 258.64\left(66.14 - \frac{7.71}{2}\right)\frac{1}{12} = 30299 \text{ kip-ft}$$

The factored moment is

$$M_u = 1.25 M_d + 1.25\ M_{creep} + 1.75\ M_{LL}$$
$$= 1.25 \times (7856 + 642) + 1.25 \times (194 + 23.7) + 1.75 \times (4427 + 666) = 19807 \text{ kips-ft}$$
$$(< \phi_f M_n, \text{ ok!})$$

Check the minimum reinforcement. The section modulus is

$$S_b = \frac{96.701}{4.853} 12^3 = 34432 \text{ in}^3$$

From Section 1.2.2.3, the modulus of fracture is

$$f_r = 0.20\sqrt{f_c'} = 0.20\sqrt{8.5} = 0.583 \text{ ksi}$$

The compressive stress due to the effective prestressing forces at the bottom fiber is

$$f_{cpe} = \left| \frac{Mp_{non} \times yb_{non}}{I_{non}} + \frac{N_{non}}{A_{non}} + \frac{Mp_{com} \times yb_{com}}{I_{com}} + \frac{N_{com}}{A_{com}} = \frac{-5395 \times 3.358}{49.65 \times 12^2} + \frac{-2507}{9.0 \times 12^2} + \frac{-1765 \times 4.853}{96.701 \times 12^2} \right.$$

$$\left. + \frac{-966}{15.8 \times 12^2} \right| = 5.51 \text{ ksi}$$

From Eq. 3-168, the section crack moment is

$$M_{cr} = \gamma_3 \left(\gamma_1 f_r + \gamma_2 f_{cpe} \right) S_b = 0.67(1.6 \times 0.583 + 1.1 \times 5.51) \frac{34432}{12} = 13445 \ kips\text{-}ft$$

$$1.33 M_{cr} = 17882 \ kips - ft < \phi_f M_n \ (ok)$$

The flexure strength checks for other locations can be performed using the same procedure described above. Analytical results show that all the sections are adequately sized.

8.6.7.2.2 Shear Strength Verification

The shear strength check is performed by following the flow chart shown in Fig. 3-54 and using the semi-empirical methods discussed in Section 3.6.3.1.3. Consider a section near the interior support approximately 224 ft away from the left end support for example.

The location of the critical section for shear near the support is determined following Section 3.6.3.1.7. Conservatively, take the shear over the support to check the shear strength. The factored shear is

$$V_u = \left| 1.25 V_{DC} + 1.25 V_{CRSH} + PS + 1.75 V_{LL+I} \right|$$
$$= 1.25 \times 550.5 - 1.25 \times 2 - 12 + 1.75 \times 5 \times 166 = 964.1 \text{ kips}$$

For simplicity, the noncomposite girder section is only considered for the shear capacity. The normal stress at the neutral axis is

$$f_{pc} = \left| \frac{N_{non}}{A_{non}} + \frac{N_{com}}{A_{com}} - \frac{Mp_{com} \times (yt_{non} + 0.667 - yt_{com})}{I_{com}} \right|$$

$$= \frac{2452}{15.285} + \frac{1102}{22.085} + \frac{5384 \times (5.263 - 6.789 - 0.667)}{555.79} = 1.313 \ ksi$$

The effective shear depth is first calculated using Eq. 3-167. However, this result is smaller than 0.9 times the girder effective depth d_e. Thus, the effective shear depth is

$$d_v = 0.9 \times d_e = 0.9 \times 145.53 = 130.98 \ in$$

The shear strength based on Eq. 3-182b is

$$V_{cw} = \left(0.06\sqrt{f_c'} + 0.3 f_{pc} \right) b_w d_v = \left(0.06\sqrt{8.5} + 0.3 \times 1.313 \right) 130.98 \times 9.5 = 707.80 \text{ kips}$$

The section modulus of the girder top fiber for the noncomposite section is

$$S_{t-non} = \frac{I_{non}}{yt_{non}} = 46.66 \ ft^3$$

The section modulus of the girder top fiber for the composite section is

$$S_{t-com} = \frac{I_{com}}{yt_{com}} = 105.60 \ ft^3$$

The compressive stress in concrete due to the effective prestress forces at the top of the girder is

$$f_{cpe} = \frac{2452}{15.285} + \frac{1102}{22.085} + \frac{13799}{46.66} + \frac{5384}{105.60} = 3.868 \ ksi$$

From Eq. 3-184, the moment causing flexural cracking at the top of the girder due to externally applied loadings can be obtained as follows:

$$M_{cr} = S_{t-com}\left(f_r + f_{cpe} - \frac{M_d}{S_{t-non}}\right) = 105.6\left(0.695 + 3.868 - \frac{25551}{46.66 \times 12 \times 12}\right) \times 12^2 = 11560 \ kip\text{-}ft$$

From Eq. 3-185b, the concrete shear capacity considering the shear and moment combination can be obtained as follows:

$$V_{ci} = 0.02\sqrt{f_c'}b_w d_v + V_d + \frac{V_i M_{cr}}{M_{max}} = 0.02\sqrt{8.5} \times 9.5 \times 130.98 + 550.5 + \frac{1.75 \times 169 \times 11560}{1.75 \times 8918} = 842.12 \ kips$$

As $V_{cw} < V_{ci}$, the shear strength contributed by concrete is

$$V_c = V_{cw} = 707.80 \text{ kips}$$

$$\cot\theta = 1.0$$

Form Fig. 8-23a, one pair of #5 vertical bars spaced at 6 in. is used. The shear strength provided by the steel can be determined by Eq. 3-186b as

$$V_s = \frac{A_v f_y d_v \cot\theta}{s} = \frac{0.62 \times 60 \times 130.98 \times 1}{6} = 812.1 \ kips$$

From Section 2.3.3.5., the shear resistance factor is $\phi_v = 0.9$.
   The total factored shear strength is

$$\phi_v V_n = \phi_v\left(V_c + V_s\right) = 0.9(707.80 + 812.1) = 1367.9 \ kips > V_u \ (ok)$$

8.6.7.2.3 Longitudinal Reinforcement Check
The AASHTO specifications require that the longitudinal reinforcement be checked using Eq. 3-201c. However, this requirement is not critical for splice girders, as the entire capacity of the post-tensioning tendon is available up to the end of the girder. Thus, in this design example, the computation is not shown.

8.6.7.3 End Zone Splitting Vertical Reinforcement Check
Consider the side segment for example, the total pretensioning force at transfer is

$$PS = (44 - 12) \times 43.95 = 1406.4 \ kips$$

From Section 3.8.1, the required vertical reinforcement within one-quarter of the girder height from its ends is

$$A_{sv-req} = \frac{PS \times 0.08}{20} = 5.626 \ in^2$$

TABLE 8-6
Live Load Deflections

Span	Maximum Live Load Deflection (in.)	Allowable Live Deflection (in.)
Side span	2.56	2.70
Center span	3.29	3.42

From Fig. 8-25, twenty #5 bars are provided within about one-quarter of the girder height from its end. The total area of the vertical reinforcement is

$$A_{sv-pro} = 20 \times 0.31 = 6.1 \ in^2 > A_{sv-req} \ (\text{ok})$$

8.6.7.4   Live Load Deflection Check

From Section 1.4.2.1 (AASHTO Section 2.5.2.6.2), the maximum deflection caused by vehicular and pedestrian loads is limited to span length/1000, i.e.,

$$\text{For side span: } y_{a-side} = 225 \times 12/1000 = 2.7 \ in$$

$$\text{For center span: } y_{a-center} = 285 \times 12/1000 = 3.42 \ in$$

The calculated maximum deflections caused by live loads are shown in Table 8-6. From this table, it can be seen that the requirement of live load deflection is met.

REFERENCES

8-1. Castrodale, R. W., and White, C. D., "Extending Span Ranges of Precast Prestressed Concrete Girders," NCHRP Report 517, National Cooperative Highway Research Program, Transportation Research Board, Washington, D.C., 20001, 2004.

8-2. AASHTO, *AASHTO LRFD Bridge Design Specifications*, 7th edition, Washington, D.C., 2014.

8-3. Heins, C. P., and Lawrie, R. A., *Design of Modern Concrete Highway Bridges*, John Wiley & Sons, New York, 1984.

8-4. FDOT, *Design Standards*, Florida Department of Transportation, Tallahassee, FL, 2004.

8-5. FDOT, *Design Standards*, Florida Department of Transportation, Tallahassee, FL, 2017.

8-6. Tadros, M. K., Girgis, A., and Pearce, R., *Spliced I-Girder Concrete Bridge System*, Department of Civil Engineering, University of Nebraska–Lincoln, Omaha, NE, 2003.

8-7. Abdel-Karim, A. M., and Tadros, M. K., *State-of-the-Art of Precast/Prestressed Concrete Spliced I-Girder Bridges*, PCI, Chicago, IL, 1995.

8-8. Hueste, M. B. D., Mander, J. B., and Parker, A. S., *Continuous Prestressed Concrete Girder Bridges*. Vol. I: *Literature Review and Preliminary Design*, Texas Department of Transportation, Austin, TX, 2011.

8-9. Hueste, M. B. D., Mander, J. B., and Parker, A. S., *Continuous Prestressed Concrete Girder Bridges*. Vol. II: *Analysis, Testing, and Recommendations*, Texas Department of Transportation, Austin, TX, 2016.

8-10. Florida Department of Transportation, *Structures Design Guidelines*, Florida Department of Transportation, Tallahassee, FL, 2017.

8-11. Florida Department of Transportation, *Curved Precast Spliced U-Girder Bridges*, Florida Department of Transportation, Tallahassee, FL, 2017.

8-12. Reese, G., "Precast Spliced U beams PCI Zone 6 Standards," *PCI National Bridge Conference*, Orlando, FL, 2010.

8-13. Theryo, T. S., and Binney, B., "Spliced Bulb Tee Bridge Design and Construction," *Virginia Concrete Conference*, Richmond, VA, March 2010.

8-14. Zhang, H. L., and Huang, D. Z., "Lateral Load Distribution in Curved Steel I-Girder Bridges," *Journal of Bridge Engineering*, Vol. 10, No. 3, 2005, pp. 281–290.

8-15. Li, G. H., and Si, D., *Transverse Load Distributions of Highway Bridges*, People's Transportations' Publish House, Beijing, China, 1984, pp. 1–21.

8-16. Li, G. H., Si, D., and Huang, D. Z., "Transverse Load Distributions of Arch Bridges," *Journal of Tongji University*, Vol. 98, No. 3, Shanghai China, 1984.

8-17. Zokaie, T., "AASHTO-LRFD Live Load Distribution Specifications," *Journal of Bridge Engineering*, Vol. 5, No. 2, 2000, pp. 131–138.

8-18. Shahawy, M., and Huang, D. Z., (2001). "Analytical and Field Investigation of Lateral Load Distribution in Concrete Slab-on-Girder Bridges," *ACI Structural Journal*, Vol. 98, No. 4, 2000, pp. 590–599.

8-19. AASHTO, *Standard Specifications for Highway Bridges*, 17th edition, Washington, D.C., 2002.

8-20. Huang, D. Z., and Shahawy, M., "Analysis of Tensile Stresses in Transfer Zone of Prestressed Concrete U-Beams," *Transportation Research Record: Journal of the Transportation Research Board*, No. 1928. Transportation Research Board of the National Academies, Washington, D.C., 2005, pp. 134–141.

8-21. Huang, D. Z., Scott, A., and Hu, B., "Evaluation of Cracks in a Spliced Prestressing Concrete I-Girder Bridge," TRB, *Journal of the Transportation Research Board*, No. 2313, Transportation Research Board of the National Academies, Washington, D.C., 2012, pp. 83–91.

8-22. FDOT and PCI, "Example Drawings—Curved Precast Spliced U-Girder Bridges," Florida Department of Transportation and PCI Zone 6, Tallahassee, FL, 2017.

8-23. Comite Euro-International Du Beton, *CEB-FIP Model Code 1990—Design Code*, Thomas Telford, Great Britain, 1993.

9 Design of Concrete Segmental Arch Bridges

- Types of arch bridges and components
- Span arrangement and determination of preliminary dimensions
- Static analysis and behaviors of arch bridges
- Construction methods
- Buckling analysis
- Dynamic analysis
- Design example: segmental arch bridge

9.1 INTRODUCTION

Arch bridges may be considered one of the most appealing of all types of bridges in terms of aesthetics; the curved shape in a scenic setting is elegant and graceful. This type of bridge not only has a unique shape, but also has quite a different mechanical behavior from girder bridges. Under external loadings and self-weight, the arch rib (see Fig. 9-1a) is subjected to a larger horizontal force which greatly reduces its bending moment. Thus, the arch rib is mainly an eccentrically loaded compression member, while a beam is a bending member (see Fig. 9-1).

Though there are several definitions for the arch bridge, it may be more generally defined as a structure shaped and supported in such a manner that its vertical loads are transmitted to the supports primarily by axial compressive thrust in its arch ribs[9-1]. This type of bridge has the following advantages and disadvantages:

Advantages

- Because an arch rib is subjected to a larger compressive axial force and a comparatively small bending moment caused by external loads, the concrete material can be more effectively used in arch bridges.
- In most cases, it is possible to eliminate the prestressing tendons and simplify the construction.
- It is esthetically pleasing with elegant shapes.
- Suitable for long-span bridges and is often the preferred choice for crossing long and deep valleys.

Disadvantages

- The substructure must have the ability to resist high horizontal forces, and its construction cost is generally high.
- For multispan arch bridges, the interior piers must be designed to resist the horizontal thrust caused by one arch span to avoid the arch collapsing due to the collapsing of the adjacent span.

Arch bridges can be suitable for a large range of span lengths from short to long. The economic span range for concrete segmental arch bridges generally ranges from 200 to 1300 ft.

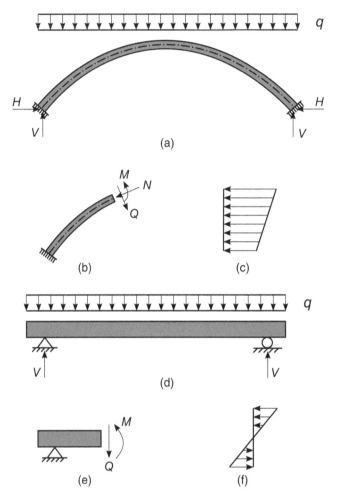

FIGURE 9-1 Difference of Mechanical Behaviors between Arch and Beam, (a) Arch, (b) Internal Forces of Arch, (c) Normal Stress Distribution in Arch Section, (d) Beam, (e) Internal Forces of Beam, (f) Normal Stress Distribution in Beam Section.

9.2 PRELIMINARY DESIGN OF CONCRETE SEGMENTAL ARCH BRIDGES

9.2.1 PRIMARY COMPONENTS AND TERMINOLOGIES OF ARCH BRIDGES

The super-structure of an arch bridge generally consists of the arch ring or rib and spandrel. Though the arch ring and arch rib may be used interchangeably[9-1, 9-2], the arch rib is often designated as an individual member in an arch ring. The arch ring is the primary member which resists the external loadings, and the spandrel is mainly to transfer loading from the deck to the arch ring. The spandrel can be completely or partially filled with proper materials. For concrete segmental arch bridges, the spandrel structure generally comprises columns and the deck system (see Fig. 9-2). The forces in the arch ring are transferred to the foundation through the abutments. The interface between the arch rib and the abutment is often called the "skew back." The top face of the arch rib is generally called the "extrados" or "back," and the bottom face of the rib is called the "intrados" or "soffit." The line connecting the centroids of each of the arch rib sections is often referred to as the "arch axial line,"

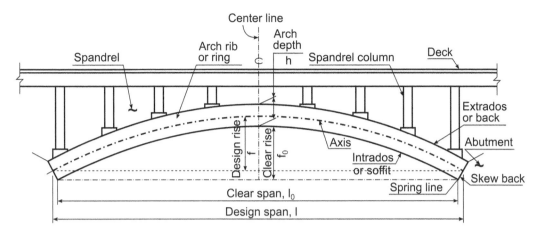

FIGURE 9-2 Primary Components and Terminologies of Arch Bridges.

while the intersection between the intrados and skew back is called a "spring line." The following terminologies are often used in arch bridge design:

Design span (l): Distance between two points where the arch axial line intersects with the skew backs (see Fig. 9-2)

Design rise (f): Distance from the crown of the arch axial line to the line connecting the centroids of the arch rib at the ends of the rib (see Fig. 9-2)

Design rise-to-span ratio: $\lambda = \frac{f}{l}$

Clear span (l_0): Distance between two spring lines (see Fig. 9-2)

Clear rise (f_0): Distance from the crown of the intrados to the line connecting the two spring lines (see Fig. 9-2)

Clear rise-to-span ratio: $\lambda_0 = \frac{f_0}{l_0}$

9.2.2 TYPES OF ARCH BRIDGES

The types of arch bridges can be classified in many ways, such as based on the deck positions, mechanical behaviors, structural systems, curve shape of the rib, and construction material. According to the deck positions, the arch bridges can be classified as a deck arch bridge in which the bridge deck is located above the arch ring (see Fig. 9-3a), a half-through arch bridge in which the bridge deck is approximately located in the middle of the arch design rise (see Fig. 9-3b), and a through arch bridge in which the bridge deck is located below the arch ring (see Fig. 9-3c). The most commonly used type in concrete segmental bridges is the deck arch bridge. Based on the static systems, the arch bridges can be classified as hingeless or fixed, two-hinged, and three-hinged arch bridges (see Fig. 9-4). Both ends of the rib in a hingeless arch bridge are fixed (see Fig. 9-4a). In a two-hinged arch bridge, the arch rib is hinged at both of its ends (see Fig. 9-4b). The three-hinged arch bridge has three hinges, which are generally located at the ends and crown of the arch rib (see Fig. 9-4c), and is rarely used currently in concrete segmental bridges due to its large deformations caused by live loadings. The arch bridges may be classified based on the structural systems as a simple arch system or a generalized arch system. The simple arch system means that the arch rib is a primary member in resisting external loads and its thrusts are directly transfer to the piers or abutments, while the spandrel structures mainly transfer the loadings from the deck to the arch rib (see Fig. 9-4). All the arch types shown in Fig. 9-3 can be designed as a simple arch system. The generalized arch system means that the arch rib and the spandrel structure work together in resisting external loads. The spandrel structure not only transfers the

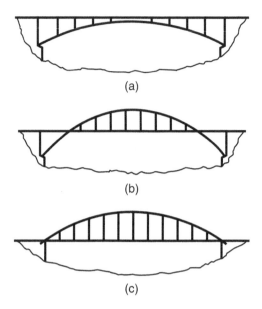

FIGURE 9-3 Types of Arch Bridges Based on Deck Positions, (a) Deck Arch Bridge, (b) Half-Through Arch Bridge, (c) Through Arch Bridge.

loadings from the deck to the arch rib, but it also shares part of the external loads. Bridges with this type of structural system may include tied-arch bridges (Fig. 9-5a), inclined leg frame bridges (Fig. 9-5c and d), arch-beam bridges (Fig. 9-5b), and truss arch bridges (Fig. 9-5e and f).

9.2.3 TYPICAL SECTIONS OF ARCH RIB AND ARCH RING

For concrete segmental arch bridges with short or medium span lengths, it is often more cost effective to use solid cross sections for the arch ribs. The arch ribs are connected by lateral bracings to form an entire arch ring (see Fig. 9-6a). For long-span concrete segmental bridges, it is often more cost effective to use box sections. The cross section of the arch ring can be formed by two or more

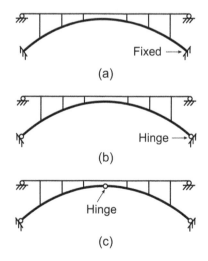

FIGURE 9-4 Types of Arch Bridges Based on Static Systems, (a) Hingeless Arch Bridge, (b) Two-Hinged Arch Bridge, (c) Three-Hinged Arch Bridge.

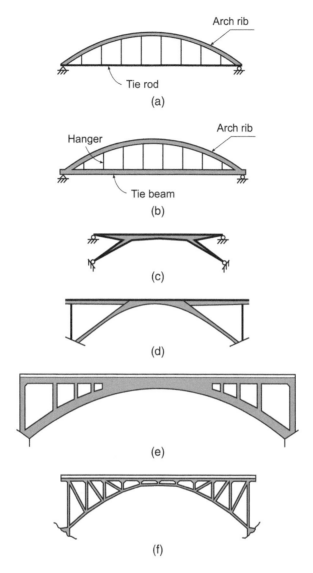

FIGURE 9-5   Generalized Arch Bridge Systems, (a) Tie-Arch System, (b) Arch-Beam System, (c) Inclined Short Leg Frame System, (d) Inclined Long Leg Frame System, (e) Deck-Stiffened Truss Arch System, (f) Braced Truss Arch System.

separated box arch ribs connected by lateral bracings (see Fig. 9-6b). The cross section of the arch ring also can be formed by side-by-side multiple precast arch ribs with the I-section or box section connected by cast-in-place joints (see Fig. 9-6c and d). For cast-in-place arch ribs, a monolithic multi-cell box section can be used (see Fig. 9-6e). The cross section of the arch ring shown in Fig. 9-6d has been successfully used in many long arch bridges. Some typical dimensions are shown in Fig. 9-7 for the readers' reference[9-3], though it may not be suitable for all western countries.

9.2.4   TYPICAL DETAILS AT ARCH RIB ENDS

As discussed, arch ribs can be hinged or fixed supported at the abutments. For a long- or medium-span arch bridge, an arc shape support as shown in Fig. 9-8a may be used. The ratio of the radius of the concave (R_2) to that of the convex (R_1) ranges from 1.2 to 1.5 (see Fig. 9-8a). Enough

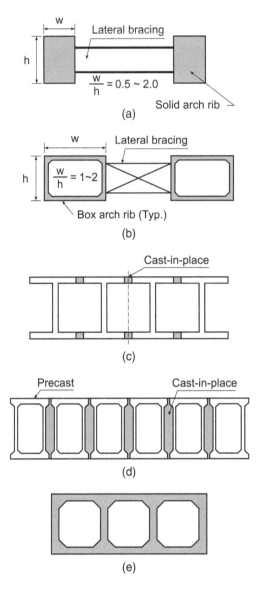

FIGURE 9-6 Typical Sections of Arch Ring, (a) Separated Solid Section Rib, (b) Separated Box Section Ribs, (c) Side-by-Side I-Shape Ribs, (d) Side-by-Side Box Section Ribs, (e) Monolithic Box Section.

reinforcement should be provided to resist large splitting force in the end zone, which is typically about 1.15 to 2.0 times the arch rib depth measured from the end of the arch rib (see Fig. 9-8b). A typical detail of a fixed support condition is illustrated in Fig. 9-8c[9-3, 9-4]. The fixed end of the arch rib is often completed after the entire rib is assembled to eliminate some secondary effects and facilitate construction.

9.2.5 COMPONENTS AND PRELIMINARY DIMENSIONS OF DECK ARCH BRIDGES

9.2.5.1 Introduction

A deck arch bridge typically consists of its deck system, including deck, longitudinal beams, transverse beams or cap beams, columns, and arch ring (see Fig. 9-9). After the span length has been

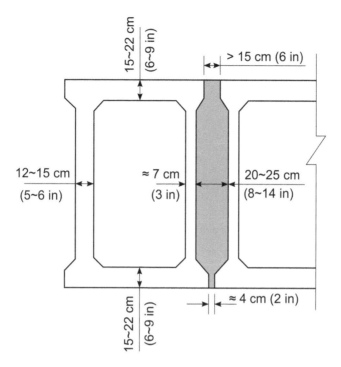

FIGURE 9-7 Typical Dimensions for Precast Side-by-Side Box Section Ribs.

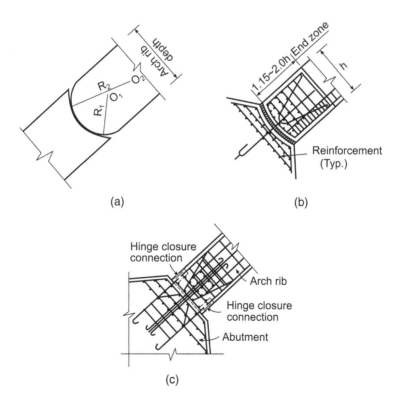

FIGURE 9-8 Typical Details at Arch Rib Ends, (a) Hinge Support Geometry, (b) Typical Reinforcing in Hinge Support, (c) Fixed Support.

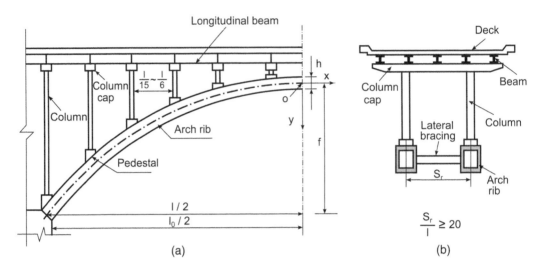

FIGURE 9-9 Preliminary Dimensions of Deck Arch Bridge, (a) Elevation View, (b) Cross Section.

determined based on the requirements of traffic below and the site conditions, the design considerations may include the determination of the rise-to-span ratio, shape of the arch axis, depth of the
arch rib, and arrangement of the spandrel structure.

9.2.5.2 Rise-to-Span Ratio

For an arch bridge with a small ratio of rise to span, a larger thrust force will be developed at the
ends of the arch ribs, which greatly reduces the bending moment in the ribs. However, the abutments
should be designed to resist this large thrust force and may be costly. An arch bridge with a large
rise-to-span ratio may have a smaller thrust force, and the abutments can be cheaper and easier to
build. However, there may be a larger bending moment in the arch ribs. Thus, an optimal rise-to-
span ratio should be selected based on site situations. Generally, for concrete segmental bridges, the
economic rise-to-span ratios may range from 1:10 to 1:3, i.e.,

$$\frac{f_0}{l_0} \approx \frac{1}{10} \sim \frac{1}{3} \tag{9-1}$$

Based on existing bridges, the most frequently used rise-to-span ratios range from 1:6 to 1:3.

9.2.5.3 Profile of Arch Rib

9.2.5.3.1 Shapes of Arch Axis

Different shapes for the arch axis have been used in existing bridges, such as circular, elliptical,
parabolic, and inverse catenary curves. Two important factors to consider when selecting the type
of arch profile are the minimization of the moment in the arch ring and aesthetics. In concrete
segmental arch bridges, the most commonly used curve shapes are parabolic and catenary. If the
coordinate system of the arch bridge is selected as shown in Fig. 9-9, the shapes of the arch axis can
be written as

$$\text{Parabolic: } y = \frac{4f}{l^2} x^2 \tag{9-2}$$

$$\text{Catenary: } y = \frac{f}{m-1}(\operatorname{csch}k\xi - 1) \tag{9-3}$$

where

    f = arch rise

    m = ratio of unit dead loading at arch spring to that at crown (often called arch axis parameter and will be discussed in later sections)

$$k = \ln\left(m + \sqrt{m^2 - 1}\right)$$

$$\xi = \frac{2x}{l}$$

An arch rib will behave as a concentrated loaded compression member under uniformly distributed vertical loading if the arch axis can be described by Eq. 9-2. Under a distributed vertical loading proportional to $y(x)$, there will be no bending moment in the arch rib whose axis is a catenary curve described by Eq. 9-3. These types of arch axis functions are often called perfect or optimum arch axis corresponding to their related loading patterns and will be discussed in later sections.

9.2.5.3.2 Depth-to-Span Ratio and Width-to-Span Ratio of Arch Ring

The depth of an arch rib can be constant or variable along the bridge's length. For an arch rib with constant depth, the rib depth can be preliminarily estimated as[9-3]

$$h = \frac{l_0}{100} + \Delta h \tag{9-4}$$

where

    l_0 = clear span length

    Δh = 2.0 to 2.3 ft (0.6 to 0.7 m) for multi-cell box section

        = 2.6 to 3.3 ft (0.8 to 1.0 m) for two or more separated box ribs

The arch rib depth can also be simply estimated as 1/55 to 1/75 of the span length.

For the hingeless deck arch bridges with variable rib depth, the depth of arch rib can be estimated as[9-2]

$$h = \frac{l_0}{29} \sim \frac{l_0}{75}, \text{ at arch spring}$$

$$= \frac{l_0}{44} \sim \frac{l_0}{75}, \text{ at arch crown} \tag{9-5}$$

To ensure the lateral stability of arch bridges, the width–to–span-length ratio of an arch ring is generally not smaller than 1/20.

9.2.5.4 Spandrel Structure

For long-span arch bridges, the spacing of the columns over the arch ribs generally range from 1/15 to 1/6 of the arch span length. For small- and middle-span arch bridges, the spacing of the columns ranges from 1/11 to 1/6 of the arch span length. The locations of columns should be arranged to properly transfer loads from the deck to the arch rib, reduce the self-weight of the spandrel structure, and facilitate construction. Generally, placing the columns directly over the arch crown should be avoided (see Fig. 9-9). The bridge deck can be either completely supported by the columns, or it can be partially fused together with the arch ring in the middle portion over the arch crown; for a distance ranging from about 1/5 to 1/3 the span length (see Fig. 9-10). The longitudinal beams can be simply supported on the column caps (Fig. 9-10a) or be made continuous over the caps or rigidly connected to the cap and the column (see Fig. 9-10b). For a long column, the secondary effect due to large axial

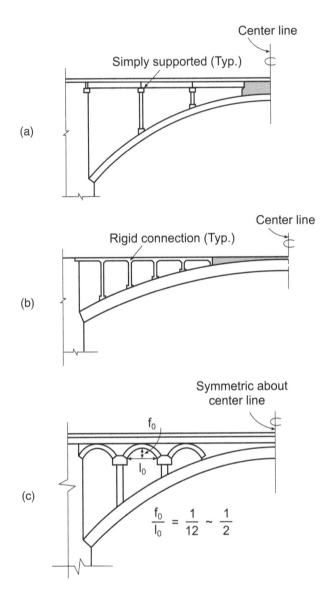

FIGURE 9-10 Typical Arrangements of Spandrel Structures, (a) Simply Supported Longitudinal Beams, (b) Continuous Longitudinal Beams, (c) Arch Spandrel Structure.

forces on column capacity should be considered. To increase the aesthetics for short- to medium-span arch bridges, the longitudinal beams may be replaced with small arch rings (see Fig. 9-10c). The rise-to-span ratio of the small aches typically ranges from 1/12 to 1/2 column spacing. Depending on the spacing of the columns, the deck can be a cast-in-place flat slab, precast prestressed slabs with cast-in-place concrete topping, or cast-in-place deck on concrete girders (see Fig. 9-10a).

9.2.6 COMPONENTS AND PRELIMINARY DIMENSIONS OF THROUGH AND HALF-THROUGH ARCH BRIDGES

9.2.6.1 Introduction

In a comparatively flat area, a through or half-through arch bridge may reduce the height of the bridge deck and shorten the span length of the bridge approaches (see Fig. 9-11). Both through and

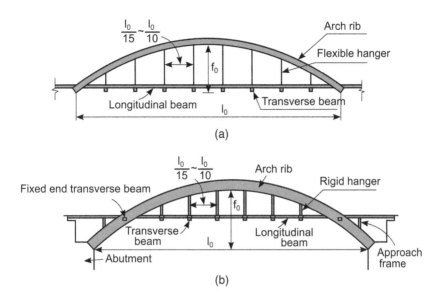

FIGURE 9-11 Typical Components of Through and Half-Through Arch Bridges, (a) Through Arch Bridge, (b) Half-Through Arch Bridge.

half-through arch bridges consist of three portions: arch ring, hangers, and deck system. The arch ring is the primary member that resists the external loads. The hangers typically transfer the loads from the deck to the arch ring. Then, the arch ring transfers the loads to the foundation. The deck system is supported by the hangers, end transverse beams connecting the arch ribs and approach frames for a half-through arch bridge (see Fig. 9-11b).

9.2.6.2 Arch Ring

The arch ring in a through or half-through arch bridge typically consists of two separate arch ribs connected by lateral bracings (see Fig. 9-12). The two ribs may be arranged in parallel and perpendicular to the deck transverse direction (see Fig. 9-12a) or in a triangle and inclined toward each other (see Fig. 9-12b). The inclined arch ribs often offer a more elegant configuration and can reduce lateral bracings, though this arrangement of arch ribs may increase construction costs.

The dead loads of a through or half through arch bridge is comparatively uniformly distributed along the length of the bridge. Thus, the second-order parabolic curve described by Eq. 9-2 is often used for the shape of the arch rib axis, though the catenary curve (Eq. 9-3) can also be used. The rise-to-span ratio of the ribs typically range from 1/7 to 1/4. For short- and medium-span arch bridges, the solid rectangular section shown in Fig. 9-6a is typically used for the arch ribs. The width-to-depth ratio of the rib section ranges from 0.5 to 1.0. The depth-to-span ratio ranges from 1/70 to 1/40. For long-span arch bridges, the sections of the arch ribs are typically a rectangular box shape or other thin-wall shapes. The depth at the arch crown may be estimated based on Eq. 9-4[9-3] in which

$$\Delta h = 2.0 \ ft \ to \ 3.28 \ ft \ (0.6 \ m \ to \ 1.0 \ m), \ \text{if span length } l_0 \leq 100 \ m \ (328 \ ft)$$
$$= 3.94 \ ft \ to \ 8.2 \ ft \ (1.2 \ m \ to \ 2.5 \ m), \ \text{if span length } 328 \ ft \ (100 \ m) < l_0 \leq 984 \ ft \ (300 \ m)$$

9.2.6.3 Hangers

The hangers may be classified into two categories; rigid hangers and flexible hangers. The rigid hangers can be prestressed or post-tensioned concrete columns that are rigidly connected with the arch ribs and the transverse floor beams (see Fig. 9-13). The cross section of the rigid hangers is typically rectangular.

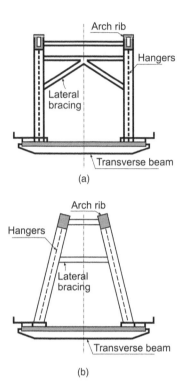

FIGURE 9-12   Typical Arrangements of Arch Ribs, (a) Side View of Vertical Arch Ribs, (b) Inclined Arch Ribs.

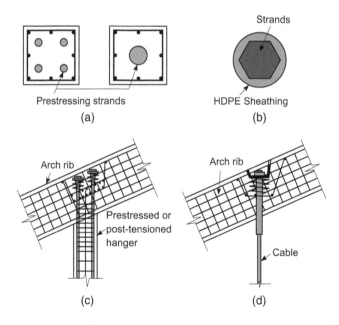

FIGURE 9-13   Typical Sections and Details of Hangers, (a) Typical Sections of Rigid Hangers, (b) Typical Section of Flexible Hanger, (c) Typical Connection between Arch Rib and Rigid Hanger, (d) Typical Connection between Arch Rib and Flexible Hanger[9-4].

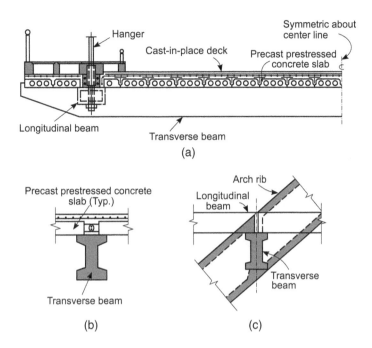

FIGURE 9-14 Typical Details of Deck System, (a) Cross Section of Deck System, (b) Cross Section of Transverse Beam, (c) Connection Details at the Intersection between Arch Ribs and Longitudinal Beams.

To increase the lateral rigidity and reduce the longitudinal moment of the hanger, the longitudinal dimension of the hanger cross-section may be smaller than its transverse one. The flexible hangers are typically made of high-strength strands and are more suitable for short- or medium-span arch bridges due to vibration considerations. The hanger spacing generally ranges from 1/15 to 1/10 span length.

9.2.6.4   Deck System

The deck system of a through or half-through arch bridge typically consists of transverse beams, longitudinal beams, and deck slabs. The longitudinal beams may be simply supported on the floor beams. Depending on the spacing of the transverse beams, the longitudinal beams can be precast prestressed I-girders with cast-in-place concrete deck, cast-in-place reinforced flat slab, or precast prestressed concrete void plates with concrete topping. The transverse beams at the intersections of deck and arch ribs are rigidly connected with the arch ribs. To increase deck smoothness and reduce vehicle dynamic loadings, it is a good practice to use two larger exterior continuous longitudinal beams and rigidly connect them with the interior transverse beams and/or use continuous deck slab[9-13, 9-15] (see Fig. 9-14). The longitudinal beams should be simply supported on the fixed-end transverse beam connected to the arch ribs (see Fig. 9-14c) to allow the arch ribs to transfer their thrusts to the foundation.

9.2.7   Typical Span Arrangement of Arch Bridges

For a long-span arch bridge, it is typical to arrange one long arch span together with multiple short or middle continuous or simple girder spans (see Fig. 9-15a). An arch bridge can be arranged as multiple continuous spans with equal or unequal span lengths (see Fig. 9-15b and c). To avoid the collapsing of one span causing the collapsing of all spans in a continuous multiple-span arch bridge, some or all of the interior piers should be designed to resist the thrust induced by a single span. For a continuous multiple span arch bridge with unequal span lengths, to reduce the bending moment of the pier under permanent loadings, the rise-to-span ratio of the shorter span can be set to be smaller than that of the

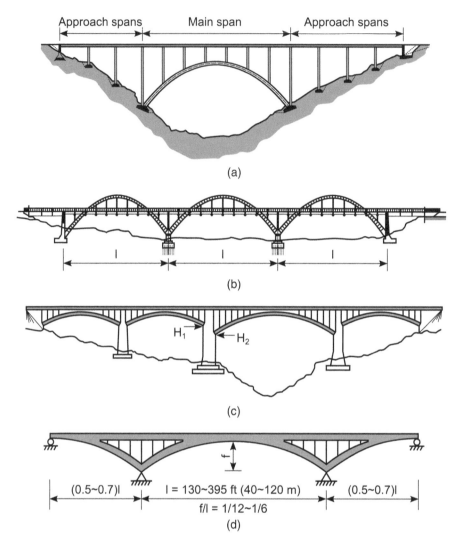

FIGURE 9-15 Typical Span Arrangements of Arch Bridges, (a) Single Span, (b) Continuous Multiple Equal Spans, (c) Continuous Multiple Unequal Spans, (d) Cantilever Spans.

longer span or the arch spring of the shorter span can be set higher than that for the longer span (see Fig. 9-15c). An arch bridge can also be arranged in a double cantilever configuration (see Fig. 9-15d).

9.2.8 INCLINED LEGGED FRAME BRIDGES

An inclined legged frame bridge has its super- and substructures as one unit with its interior piers rigidly connected with the super-structure and moved away from the lower roadway or deep canyon. This type of bridge may provide a safer, economic, and elegant structure under certain site conditions. Similar to general arch bridges, this type of bridge can be suitable for short-, medium-, and long-span concrete segmental bridges. Figure 9-16 shows a typical configuration of the inclined leg frame bridge. For a short- or medium-span inclined legged frame bridge, the girder depth can be constant and be taken as about 1/24 to 1/18 the middle-span length. For a long-span inclined legged frame bridge, the girder depth is typically variable. The girder depth may be estimated based on the guides described in Section 1.4.2.2 for continuous girder bridges. Based on existing inclined legged

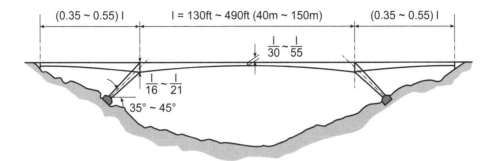

FIGURE 9-16 Typical Configuration of Inclined Legged Frame Bridges.

frame bridges and considering the effect of axial forces in the girder, the girder depth at the location of the legs may range from 1/16 to 1/21 of the middle-span length, while the girder depth at midspan may range from 1/30 to 1/55 of the middle-span length. Depending on the construction methods and the site condition, the ratio of the side span to the middle span may be taken as 0.35 to 0.55. The incline angles of the legs typically range from 35° to 45°. For concrete segmental bridges, the box sections shown in Fig. 1-33 are often used.

The balanced cantilever construction method discussed in Chapter 6 is typically used in the construction of inclined legged frame bridges (see Fig. 9-17a). A typical layout of the post-tensioning tendons is illustrated in Fig. 9-17b.

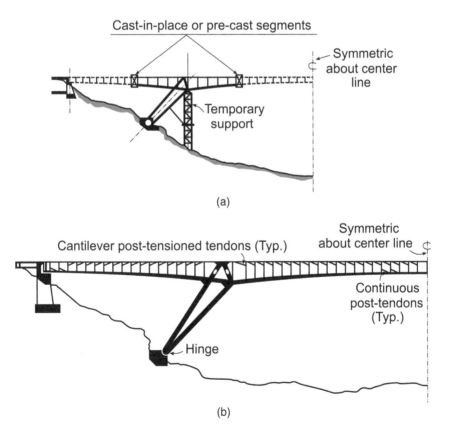

FIGURE 9-17 Typical Construction Method and Tendon Layout of Inclined Legged Frame Bridges, (a) Sketch of Balanced Cantilever Construction, (b) Sketch of Tendon Layout.

9.3 BEHAVIORS AND ANALYSIS OF ARCH BRIDGES

9.3.1 INTRODUCTION

Though there are many powerful commercial computer programs available for analyzing any arch bridge, it is still valuable for bridge engineers to fully understand the basic behaviors of arch bridges under dead and live loadings. As the basic behaviors of all types of arch bridges are similar, in this section, we will mainly focus on the analysis of the deck arch bridges. First, we discuss the important behaviors of arch bridges and the differences from the well-known simply supported beam, by taking a three-hinge arch as an example. Then, the distinguishing behaviors of two-hinged and fixed arch bridges are presented. Finally, we will discuss the buckling and vibration analysis of arch bridges.

9.3.2 ANALYSIS OF THREE-HINGED ARCHES AND BASIC BEHAVIORS OF ARCH BRIDGES

9.3.2.1 Distinguished Behaviors of Arches

Let us examine a three-hinged arch and a simply supported beam as shown in Fig. 9-18. Both the arch and beam have the same span length and loading conditions. From Fig. 9-18a, by taking the moment about point B, we obtain the vertical reaction of the arch at support A, i.e.,

$$\sum M_B = 0: V_A = \frac{\sum_1^3 P_i b_i}{l} = V_A^0 \tag{9-6}$$

$$\sum Y = 0: V_B = \sum_1^3 P_i = V_B^0 \tag{9-7}$$

where
$\quad V_A, V_B$ = vertical reactions at supports A and B of arch, respectively
$\quad V_A^0, V_B^0$ = vertical reactions at supports A_0 and B_0 of beam, respectively

From Eqs. 9-6 and 9-7, we can see that the vertical reactions at the supports of the arch are equal to those of the corresponding simply supported beam.

From $\sum X = 0$, the horizontal thrust at the arch end supports can be written as

$$H_A = H_B = H \tag{9-8}$$

From the free-body shown in Fig. 9-18c and taking moment about middle hinge C, i.e.,

$$\sum M_C = 0: H_A = \frac{V_A \frac{l}{2} - P_1\left(\frac{l}{2} - a_1\right) - P_2\left(\frac{l}{2} - a_2\right)}{f} = \frac{M_C^0}{f} \tag{9-9}$$

where
$\quad M_C^0$ = moment at midspan of simply supported beam
$\quad f$ = arch rise

From Eq. 9-9, we can observe:

    a. The horizontal thrust of a three-hinged arch is equal to the moment at the middle span of its corresponding simply supported beam divided by the arch rise.
    b. The smaller the arch rise, the larger the horizontal thrust will be. This behavior is applicable for all types of arches.

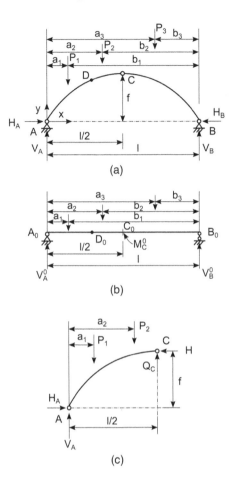

FIGURE 9-18 Determination of Support Reactions of Three-Hinged Arches, (a) Three-Hinged Arch, (b) Simply Supported Beam, (c) Free Body for Determining Arch Support Reactions.

After the determination of the support reactions, we can obtain the internal forces at any location of the arch.

From Fig. 9-19a, we have

$$\sum X = 0: H_D = H \qquad (9\text{-}10)$$

$$\sum Y = 0: Q'_D = Q_D^0 = V_A - P_1 \qquad (9\text{-}11)$$

$$\sum M_A = 0: M_D = V_A x_D - P_1(x_D - a_1) - Hy_D = M_D^0 - Hy_D \qquad (9\text{-}12)$$

where

H_D, Q_D, M_D = horizontal force, vertical force, and moment at point D of arch, respectively
Q_D^0, M_D^0 = shear and moment at point D of the corresponding simply supported beam, respectively

Assume the tangential angle at point D as φ_D, i.e.,

$$\frac{dy}{dx} = \tan \varphi_D$$

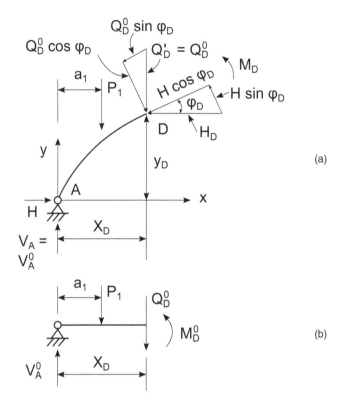

FIGURE 9-19 Determination of Internal Forces in Arches, (a) Free Body for Determining Internal Force in Arch, (b) Free Body for Determining Internal Force in Beam.

Then, the shear Q_D and axial force N_D in the arch at section D can be written as (see Fig. 9-19a)

$$Q_D = Q_D^0 \cos\varphi_D - H \sin\varphi_D \qquad (9\text{-}13)$$

$$N_D = Q_D^0 \sin\varphi_D + H \cos\varphi_D \qquad (9\text{-}14)$$

From Eqs. 9-12 to 9-14, we can observe:

a. In comparison to a beam, the moment and shear in an arch bridge are greatly reduced due to the horizontal thrust.

b. The internal forces M_D, Q_D, and N_D in an arch are not only related to the span length and the rise but also to the shape of arch rib $y(x)$.

9.3.2.2 Selection of the Shapes of Arch Rib Axis

If the shape of an arch rib axis coincides with the pressure line of the arch under a certain dead loading, the arch axis is often referred to as an optimum or perfect arch axis as mentioned in Section 9.2.5.3. There is no bending moment in a perfect arch axis under this dead loading. Based on this condition, we can develop the ideal shapes of arch rib axes for certain dead load distributions. Equation 9-12 shows the moment at any location D of the arch and can be rewritten as

$$M(x) = M^0(x) - Hy(x) \qquad (9\text{-}15)$$

In a perfect arch axis, the moment at any location should be equal to zero, i.e., $M(x) = 0$, and we have

$$y(x) = \frac{M^0(x)}{H} \tag{9-16a}$$

Taking the second derivatives with respect to x on both sides of Eq. 9-16a yields

$$y(x)'' = \frac{M^0(x)''}{H} \tag{9-16b}$$

Assuming that the arch is subjected to a uniform loading q, from Eq. 4-6 and noting the sign definition in Fig. 4-1, we have

$$y(x)'' = -\frac{q}{H}$$

Integrating the above equation twice yields

$$y(x) = -\frac{qx^2}{2H} + c_1 x + c_2 \tag{9-17}$$

where c_1 and c_2 are two integrating constants that can be determined from boundary conditions, i.e.,

$$x = 0, \ y = 0: \ c_2 = 0; \tag{9-18a}$$

$$x = l, \ y = 0: \ c_1 = \frac{ql}{2H}; \tag{9-18b}$$

From Eq. 9-9, with a uniform loading q, we have

$$H = \frac{M_C^0}{f} = \frac{ql^2}{8f} \tag{9-19}$$

Substituting Eqs. 9-18 and 9-19 into Eq. 9-17 yields

$$y = \frac{4fx}{l^2}(l - x) \tag{9-20}$$

Note that Eq. 9-20 is developed based on the coordinate system as shown in Fig. 9-19. From Eq. 9-20, we can conclude that the perfect arch axis of a three-hinged arch subjected to a uniform loading is a second-order parabola. This conclusion can be applicable to two-hinged and fixed arch bridges if neglecting the effect of its elastic deformation.

Next, let us examine the perfect axis of an arch with a solid spandrel. Take the coordinate system as shown in Fig. 9-20 and assume that the dead loading, caused by the self-weight of the arch ring, filling materials, and deck, can be described as

$$q(x) = q_c + \gamma y(x) \tag{9-21}$$

where
   q_c = weight per unit length at arch crown
   γ = unit weight

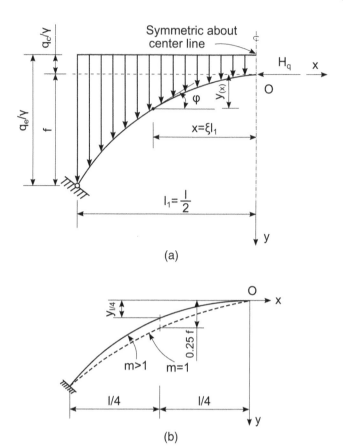

FIGURE 9-20 Determination of Catenarian Arch Axis, (a) Analytical Model, (b) Effect of *m* on Shape of Catenary.

Then the weight per unit length q_e at the arch end can be written as

$$q_e = q_c + \gamma f \tag{9-22}$$

From Eq. 9-22, we have

$$\gamma = \frac{q_e - q_c}{f} = q_c \frac{m-1}{f} \tag{9-23}$$

where *m* is the ratio of the weight per unit length of the bridge at the arch support end to that at the arch crown and is often called the parameter of arch axis, i.e.,

$$m = \frac{q_e}{q_c} \tag{9-24}$$

Assuming the arch axis is a perfect axis and the bridge is symmetric about its center line (see Fig. 9-20a), then there is only a horizontal axial force H_q without vertical shear at the crown. From Eqs. 9-16b and 4-6, we have

$$\frac{d^2y}{dx^2} = \frac{q(x)}{H_q} \tag{9-25}$$

Let

$$x = \xi l_1, \ (0 \le \xi \le 1) \tag{9-26}$$

where $l_1 = \frac{l}{2} =$ half of span length.

Substituting Eqs. 9-21 and 9-26 into Eq. 9-25, we can obtain

$$\frac{d^2 y}{dx^2} - k^2 y = \frac{l_1^2 q_c}{H_q} \tag{9-27}$$

where $k = \frac{l_1^2 q_c}{H_q f}(m-1)$.

Solving Eq. 9-27 yields the function of the perfect axis as

$$y(x) = \frac{f}{m-1}(\cosh k\xi - 1) \tag{9-28}$$

where $\cosh k\xi = \frac{e^{k\xi} + e^{-k\xi}}{2}$, a hyperbolic cosine function.

Equation 9-28 is often called the equation of catenary, which represents the perfect axis for the arches with external loading that can be expressed as Eq. 9-21, including two hinged and fixed arches if neglecting the effect of elastic deformation.

If $\xi = 1$, $y = f$. From Eq. 9-29, we have

$$\cosh k = m \tag{9-29}$$

$$k = \cosh^{-1} m = \ln\left(m + \sqrt{m^2 - 1}\right) \tag{9-30}$$

When $m = \frac{q_e}{q_c} = 1$, the arch is subjected a uniform loading and we can obtain from Eq. 9-27:

$$y(x) = f\xi^2 = \frac{4f}{l^2} x^2 \tag{9-31}$$

Thus, we can see that Eqs. 9-20 and 9-31 represent the same second-order parabola in different coordinate systems.

After the determination of the ratio of rise to span, the shape of the catenary is only related to the parameter of arch axis m. Some important characteristics of the catenary can be discussed by the ratio of the coordinate at the fourth point of the arch to the rise, i.e.,

$$\frac{y_{l/4}}{f} = \frac{1}{m-1}\left(\cosh\frac{k}{2} - 1\right) = \frac{1}{m-1}\left(\sqrt{\frac{\cosh k + 1}{2}} - 1\right) = \frac{1}{\sqrt{2(m+1)} + 2} \tag{9-32}$$

where $y_{l/4} =$ coordinate at fourth point of arch.

From Eq. 9-32, we can see that $y_{l/4}$ decreases when m increases. When m increases, the arch axis moves up (see Fig. 9-20b). Under a uniform loading, $m = 1$ and

$$y_{l/4} = 0.25 \ f \tag{9-33}$$

In current concrete segmental bridges, open spandrel structures are always used (see Figs. 9-9 and 9-10). The actual compressive line of the arch consists of discontinuous segments and is a funicular polygon that is rarely used for aesthetic consideration. If the distribution of the dead loads is close to a uniform loading, the arch axis can be chosen as a second-order parabola. Otherwise, it is typical

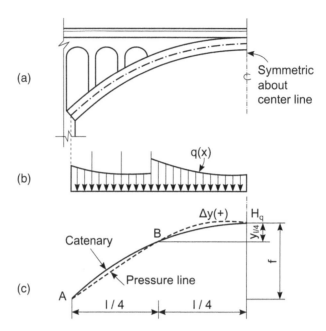

FIGURE 9-21 Determination of Arch Axis with Open Spandrel, (a) Typical Arch Bridge with Open Spandrel, (b) Distribution of Dead Loads, (c) Optimum Arch Axis.

to use a catenary that passes through the actual pressure line of the three-hinged arch under its self-weight at five points of skewbacks, crown, and fourth points of the arch. This method for determining the shape of the arch is often referred to as the five points matching method. The parameter of the arch axis can be determined using the following procedures.

First, we determine the ratio of the rise to span and assume the parameter of the arch axis m $(m \geq 1.0)$. Then, we can preliminarily arrange the arch bridge as shown in Fig. 9-21a. Based on the preliminary bridge arrangement, we can calculate the distribution of the dead loads as shown in Fig. 9-21b. Assuming that the dead load is symmetrical about the centerline, there is only horizontal force H_q at the arch crown (see Fig. 9-21c).

From $\sum M_A = 0$, we have

$$H_q = \frac{\sum M_{Aq}}{f} \tag{9-34}$$

where $\sum M_{Aq}$ = sum of the moment about point A due to dead loadings on half of the arch (see Fig. 9-21b).

From $\sum M_B = 0$, we have

$$H_q = \frac{\sum M_{Bq}}{y_{l/4}} \tag{9-35}$$

where $\sum M_{Bq}$ = sum of the moment about point B, i.e., fourth point of the arch, due to dead loading on half of the arch (see Fig. 9-21b).

From Eqs. 9-34 and 9-35, we can obtain

$$\frac{y_{l/4}}{f} = \frac{\sum M_{Bq}}{\sum M_{Aq}} \tag{9-36}$$

From Eq. 9-32, we have

$$m = \frac{1}{2}\left(\frac{f}{y_{l/4}} - 2\right)^2 - 1 \tag{9-37}$$

If the parameter of the arch axis m calculated by Eq. 9-37 is close enough to the previously assumed m, then the catenary is a suitable arch axis with which there is no bending moment at the aforementioned five points in the three-hinged arch bridge. This conclusion is applicable for two-hinged and fixed arches under the dead loads if the effect of elastic deformation is neglected. If the calculated m is different from the assumed one, we can modify the arch profile and the dead load distribution by using the newly calculated parameter of the arch axis m as assumed until the m determined by Eq. 9-37 closely matches the previously assumed value.

It should be mentioned that there is no perfect arch axis existing under live loadings as the moving vehicles continuously change their loading positions. However, it is possible to select an optimum arch axis to minimize the moment in the arch due to live loadings.

9.3.2.3 Influence Lines of Three-Hinged Arches

It is important to know the influence lines of the internal forces of arch bridges for live load applications and construction sequences. An influence line of an internal force at a section indicates the variation of the internal force as a unit load moves from the beginning of the bridge to the end. As the three-hinged arch is a statically determined structure, its influence lines can be easily determined by the method used for a simply supported beam and Eqs. 9-12 to 9-14 discussed in Section 9.3.2.1. Some typical influence lines at end A and the fourth point of the arch are shown in Fig. 9-22.

9.3.3 ANALYSIS OF TWO-HINGED ARCH BRIDGES

In Section 9.3.2, the basic behaviors of arch bridges were discussed based on the three-hinged arch bridges. In this section, we will discuss a simplified analytical method for two-hinged arch bridges, which are often used in concrete segmental bridges.

If the deck system of an arch bridge is not rigidly connected with its arch ring as shown in Figs. 9-9 to 9-11, the effect of the arch spandrel structures or hangers can be conservatively neglected and the analysis model of the arch as shown in Fig. 9-23 can be used. The two-hinged arch bridges are statically indeterminate structures with one unknown and can be analyzed by any methods discussed in Chapter 4.

Using the force method and taking the horizontal thrust H at support A as an unknown (see Fig. 9-23b), from Section 4.2.2, we have

$$X_1 = H = -\frac{\Delta_{1P}}{\delta_{11}} \tag{9-38}$$

where Δ_{1P} and δ_{11} are the horizontal displacements at support A due to external loads and unit unknown force, respectively, and can be written as follows:

$$\Delta_{1P} = \int \frac{\bar{M}_1(x)M_P(x)}{EI}ds + \int \frac{\bar{N}_1(x)N_P(x)}{EA}ds + \int \frac{k\bar{Q}_1(x)Q_P(x)}{GA}ds \tag{9-39a}$$

where
$\bar{M}_1(x), \bar{N}_1(x), \bar{Q}_1(x)$ = moment, axial force, and shear of arch due to $X_1 = 1$ on primary structure (see Figs. 9-24c, 9-24e, and 9-24g), respectively

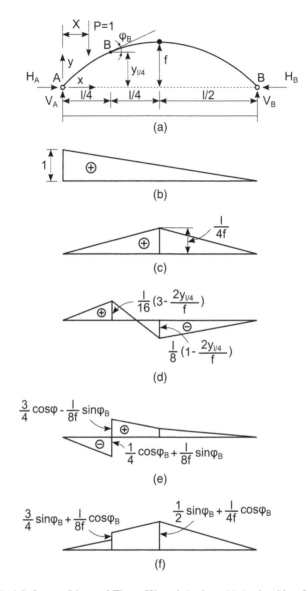

FIGURE 9-22 Typical Influence Lines of Three-Hinged Arches, (a) Arch with a Unit Load, (b) Vertical Reaction at End A, (c) Thrust at End A, (d) Moment at Fourth Point, (e) Shear at Fourth Point, (f) Axial Force at Fourth Point.

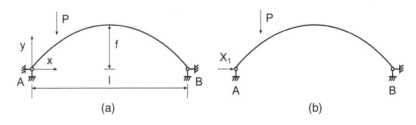

FIGURE 9-23 Analytical Model of a Two-Hinged Arch Bridge, (a) Simplified Analytical Model, (b) Primary Structure.

$M_P(x)$, $N_P(x)$, $Q_P(x)$ = moment, axial force, and shear of arch due to external loads on primary structure (see Figs. 9-24d, 9-24f, and 9-24h), respectively, assuming only one external load for simplicity

I, A = moment of inertia and area of arch rib, respectively
E, G = modulus of elasticity and shear modulus of elasticity, respectively

$$ds = \text{infinitive element along arch axis}$$

$$= \frac{dx}{\cos\varphi} \tag{9-40}$$

In Fig. 9-24,

Q_0 = shear in corresponding simply supported beam caused by external force P and can be determined by Eq. 9-11
φ = tangential angle of arch axis at location x
k = shape factor for cross-sectional area: $k = 1.2$ for rectangular sections; $k \approx 1$ for box sections

$$\delta_{11} = \int \frac{\bar{M}_1(x)^2}{EI}\,ds + \int \frac{\bar{N}_1(x)^2}{EA}\,ds + \int \frac{k\bar{Q}_1(x)^2}{GA}\,ds \tag{9-41a}$$

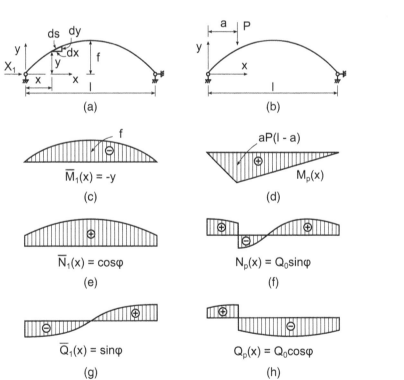

FIGURE 9-24 Variations of Moment, Axial Force, and Shear on the Primary Arch Structure, (a) Arch with Unit Unknown Force, (b) Arch with External Force, (c) Moment Due to Unit Unknown Force, (d) Moment Due to External Load, (e) Axial Force Due to Unit Unknown Force, (f) Axial Force Due to External Load, (g) Shear Due to Unit Unknown Force, (h) Shear due to External Load.

Analytical results show that the effect of shear on Δ_{1P} and δ_{11} is small and can be neglected. The effect of axial force N on Δ_{1P} is also small and can be neglected when $\frac{f}{l} < \frac{1}{5}$. Thus, Eqs. 9-39a and 9-41a can be rewritten as

$$\Delta_{1P} = \int \frac{\bar{M}_1(x) M_P(x)}{EI} ds = -\int \frac{y M_P(x)}{EI} ds \tag{9-39b}$$

$$\delta_{11} = \int \frac{\bar{M}_1(x)^2}{EI} ds + \int \frac{\bar{N}_1(x)^2}{EA} ds = \int \frac{y^2}{EI} ds + \int \frac{\cos^2 \varphi}{EA} ds \tag{9-41b}$$

where φ = tangential angle of the arch axis at location x.

Substituting Eqs. 9-39b and 9-41b into Eq. 9-38 yields the horizontal thrust:

$$X_1 = H = \frac{\int \frac{y M_P(x)}{EI} ds}{(1+\mu) \int \frac{y^2}{EI} ds} \tag{9-42a}$$

where

$$\mu = \text{coefficient of axial force effect} = \frac{\int \frac{\cos^2 \varphi}{EA} ds}{\int \frac{y^2}{EI} ds} \tag{9-42b}$$

After the horizontal thrust force H is obtained by Eq. 9-42a, the internal force at any sections of the arch rib can be calculated using Eqs. 9-12 to 9-14 as discussed in Section 9.3.2.1. The sketches of some typical influence lines for two-hinged arch bridges are shown in Fig. 9-25. In statically indeterminate structures, the elastic shortening due to axial forces, concrete shrinkage, and temperature variation will cause additional internal forces in the arch. Their effects will be discussed in Section 9.3.4.

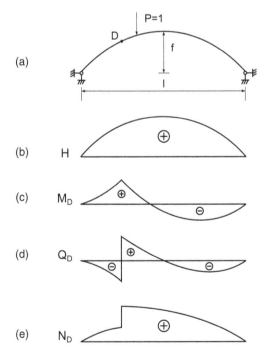

FIGURE 9-25 Typical Influence Lines of Two-Hinged Arch Bridges, (a) Two-Hinged Arch, (b) Horizontal Thrust, (c) Moment at Section D, (d) Shear at Section D, (e) Axial Force at Section D.

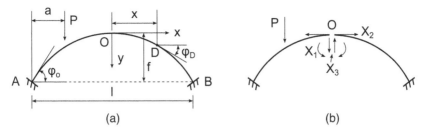

FIGURE 9-26   Analytic Model of Hingeless Arch Bridges, (a) Coordinate System, (b) Primary Structure.

9.3.4   ANALYSIS OF HINGELESS ARCH BRIDGES

9.3.4.1   General Equilibrium Equations by Force Method

A one-span hingeless arch bridge has three unknowns. For an easy integration, the origin of the arch coordinate system is set at the crown as shown in Fig. 9-26. Cutting the arch at the crown and inserting three redundant forces of moment X_1, axial force X_2, and shear X_3, the hingeless arch becomes a statically determined structure. Using the conditions of displacement compatibility (refer to Section 4.2.2), we have

$$\begin{cases} X_1\delta_{11} + X_2\delta_{12} + X_3\delta_{13} + \Delta_{1p} = 0 \\ X_1\delta_{21} + X_2\delta_{22} + X_3\delta_{23} + \Delta_{2p} = 0 \\ X_1\delta_{31} + X_2\delta_{32} + X_3\delta_{33} + \Delta_{3p} = 0 \end{cases} \tag{9-43a}$$

where

Δ_{ip} = displacements caused by external loads in primary structures shown in Fig. 9-26b ($i = 1$ to 3)

δ_{ii} = displacements along direction of X_i due to $X_i = 1$ ($i = 1$ to 3)

δ_{ik} = displacements along direction of X_i due to $X_k = 1$ ($i = 1$ to 3, $k = 1$ to $3, i \neq k$)

$$\Delta_{ip} = \int \frac{\bar{M}_i M_P}{EI} ds + \int \frac{\bar{N}_i N_P}{EA} ds + \int \frac{k\bar{Q}_i Q_P}{GA} ds \; (i = 1 \; to \; 3) \tag{9-44}$$

$$\delta_{ii} = \int \frac{\bar{M}_i^2}{EI} ds + \int \frac{\bar{N}_i^2}{EA} ds + \int \frac{k\bar{Q}_i^2}{GA} ds \; (i = 1 \; to \; 3) \tag{9-45}$$

$$\delta_{ik} = \int \frac{\bar{M}_i \bar{M}_k}{EI} ds + \int \frac{\bar{N}_i \bar{N}_k}{EA} ds + \int \frac{k\bar{Q}_i \bar{Q}_k}{GA} ds \; (i = 1 \; to \; 3, \; k = 1 \; to \; 3, \; i \neq k) \tag{9-46}$$

In Eqs. 9-44 to 9-46,

$M_P(x)$, $N_P(x)$, $Q_P(x)$ = moment, axial force, and shear caused by external forces, respectively (see Fig. 9-27)

$\bar{M}_i(x)$, $\bar{N}_i(x)$, $\bar{Q}_i(x)$ = moment, axial force, and shear caused by unit redundant forces, respectively (see Fig. 9-27)

From Fig. 9-27, it can be seen that

$$\delta_{13} = \delta_{31} = \int \frac{\bar{M}_1 \bar{M}_3}{EI} ds + \int \frac{\bar{N}_1 \bar{N}_3}{EA} ds + \int \frac{k\bar{Q}_1 \bar{Q}_3}{GA} ds = 0$$

$$\delta_{23} = \delta_{32} = \int \frac{\bar{M}_2 \bar{M}_3}{EI} ds + \int \frac{\bar{N}_2 \bar{N}_3}{EA} ds + \int \frac{k\bar{Q}_2 \bar{Q}_3}{GA} ds = 0$$

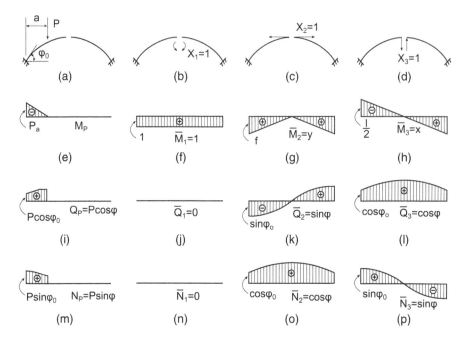

FIGURE 9-27 Moments, Axial Forces, and Shears Due to External Loads and Unit Redundant Forces, (a) Primary Structure with External Loads, (b) Primary Structure with Unit Redundant $X_1 = 1$, (c) Primary Structure with Unit Redundant $X_2 = 1$, (d) Primary Structure with Unit Redundant $X_3 = 1$, (e) Moment Due to External Loading, (f) Moment Due to $X_1 = 1$, (g) Moment Due to $X_2 = 1$, (h) Moment Due to $X_3 = 1$, (i) Shear Due to External Loads, (j) Shear Due to $X_1 = 1$, (k) Shear Due to $X_2 = 1$, (l) Shear Due to $X_3 = 1$, (m) Axial Force Due to External Loads, (n) Axial Force Due to $X_1 = 1$, (o) Axial Force Due to $X_2 = 1$, (p) Axial Force Due to $X_3 = 1$.

Thus, the set of Equations 9-43a can be written as

$$\begin{cases} X_1\delta_{11} + X_2\delta_{12} + \Delta_{1p} = 0 \\ X_1\delta_{21} + X_2\delta_{22} + \Delta_{2p} = 0 \\ X_3\delta_{33} + \Delta_{3p} = 0 \end{cases} \tag{9-43b}$$

9.3.4.2 Elastic Center Method

From Eq. 9-43b, it can be seen that these equations can be easily solved individually if we have $\delta_{12} = \delta_{21} = 0$. To let $\delta_{12} = \delta_{21} = 0$, we can move the redundant forces down a distance y_s from O to C by inserting two rigid arms, which will not affect the structure behavior (see Fig. 9-28a). The new application point C of the redundant forces only changes the distribution of the moment, which is caused by redundant X_2 as shown in Fig. 9-28c and has both positive and negative moment regions. The distance y_s can be determined by the condition of $\delta_{12} = \delta_{21} = 0$, i.e.,

$$\delta_{12} = \int \frac{\bar{M}_1 \bar{M}_2}{EI} ds = \int \frac{1(y_s - y)}{EI} ds = 0 \tag{9-47}$$

Rearranging Eq. 9-47, we have

$$y_s = \frac{\int \frac{y}{EI} ds}{\int \frac{1}{EI} ds} \tag{9-48}$$

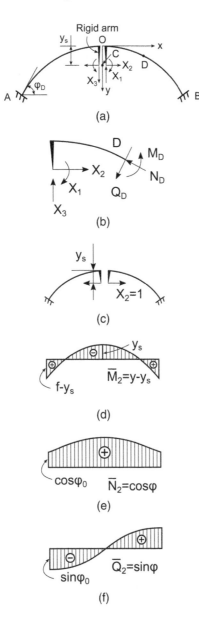

FIGURE 9-28 Elastic Center and Internal Forces Due to Unit Redundant Forces in Primary Structure, (a) Elastic Center, (b) Internal Forces at Section D, (c) Unit Redundant Force X_2 Applied at the Elastic Center, (d) Moment due to $X_2 = 1$, (e) Axial Force due to $X_2 = 1$, (f) Shear Due to $X_2 = 1$.

From Eq. 9-48, it can be seen that y_s indicates the location of the centroid of the arch rib with unit width. Point C is often called elastic center. After the redundant forces are moved to the elastic center C, Eq. 9-43b can be simplified as

$$\begin{cases} X_1\delta_{11} + \Delta_{1p} = 0 \\ X_2\delta_{22} + \Delta_{2p} = 0 \\ X_3\delta_{33} + \Delta_{3p} = 0 \end{cases} \tag{9-43c}$$

The redundant forces can be easily obtained as

$$X_1 = -\frac{\Delta_{1p}}{\delta_{11}} \qquad (9\text{-}49a)$$

$$X_2 = -\frac{\Delta_{2p}}{\delta_{22}} \qquad (9\text{-}50a)$$

$$X_3 = -\frac{\Delta_{3p}}{\delta_{33}} \qquad (9\text{-}51a)$$

Per the discussions in Section 9.3.3, the effect of shear force can often be neglected. Analytical results show that the effect of axial forces on the redundant X_2 is only required to be considered when $\frac{f}{l} < \frac{1}{5}$. Then, Eqs. 9-49a, 9-50a, and 9-51a can be written as

$$X_1 = -\frac{\int \frac{\bar{M}_1 M_p}{EI} ds}{\int \frac{\bar{M}_1^2}{EI} ds} \qquad (9\text{-}49b)$$

$$X_2 = -\frac{\int \frac{\bar{M}_2 M_p}{EI} ds}{(1+\mu) \int \frac{\bar{M}_2^2}{EI} ds} \qquad (9\text{-}50b)$$

where μ = coefficient of axial force effect (see Eq. 9-42b)

$$X_3 = -\frac{\int \frac{\bar{M}_3 M_p}{EI} ds}{\int \frac{\bar{M}_3^2}{EI} ds} \qquad (9\text{-}51b)$$

In Eqs. 9-49b, 9-50b, and 9-51b, $\bar{M}_1$ and $\bar{M}_3$ can be found in Fig. 9-27f and h, respectively. Functions $\bar{M}_2$ and $\bar{N}_2$ can be found in Fig. 9-28d and e, respectively. The method of determining the redundant forces using the elastic center concept is often called the elastic center method. After the redundant forces X_1, X_2, and X_3 due to the external loadings are obtained, the internal forces at any location D can be obtained as

$$M_D = M_D^0 + X_2 y + X_3 x \qquad (9\text{-}52)$$

$$N_D = N_D^D + X_2 \cos\varphi_D + X_3 \sin\varphi_D \qquad (9\text{-}53)$$

$$Q_D = Q_D^D + X_2 \sin\varphi_D + X_3 \cos\varphi_D \qquad (9\text{-}54)$$

In Eqs. 9-52 to 9-54,

M_D^0 = moment at section D due to self-weight and external loads in corresponding simply supported beam (see Fig. 9-19b)

N_D^D, Q_D^D = axial force and shear at section D due to self-weight and external loads in primary structure (can be calculated by Eqs. 9-13 and 9-14)

φ_D = tangential angle at section D

The sign convention is shown in Fig. 9-28b, i.e., positive compression axial force, positive moment inducing tensile stress at bottom fiber, positive shear causing a free body turning clockwise.

9.3.4.3 Influence Lines of Hingeless Arches

It is common to use influence lines in determining the most unfavorable loading positions of live loads. After the redundant forces X_1, X_2, and X_3 due to a unit external loading in a hingeless arch

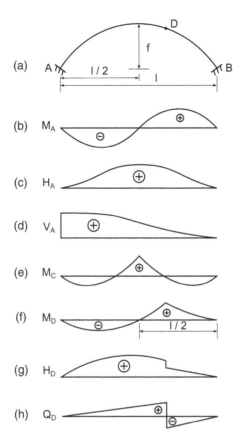

FIGURE 9-29 Typical Influence Lines of Hingeless Arch Bridges, (a) Hingeless Arch, (b) Moment at Arch End A, (c) Horizontal Thrust at Arch End A, (d) Shear at Arch End A, (e) Moment at Arch Crown, (f) Moment at Section D, (g) Axial Force at Section D, (h) Shear at Section D.

bridge are determined, the influence lines at section D can be obtained using Eqs. 9-52 to 9-54. Some typical influence lines are shown in Fig. 9-29.

9.3.4.4 Effect of Elastic Shortening of Arch Ribs

Let us examine an arch subjected to a uniform loading q as shown in Fig. 9-30. Assuming the axis of the arch is a parabola, then there is no moment and shear in the arch rib and there is only horizontal force H_q at the arch crown (see Fig. 9-30a) if the effect of arch elastic deformation is neglected. The axial force in the arch rib will cause the arch rib to shorten, and additional forces will be induced in the arch rib due to its deformation being restricted. The additional forces can be determined by the force method using the elastic center concept as discussed in Section 9.3.4.2. Considering the primary structure as shown in Fig. 9-30a, there is only one unknown at the elastic center X_{Hq}. Using the condition of deformation compatibility, we have

$$X_{Hq}\delta_{22} - \Delta l = 0 \qquad (9\text{-}55)$$

where

X_{Hq} = horizontal force at elastic center due to elastic shortening caused by H_q

Δl = horizontal displacement at arch crown due to axial force N in primary structure (see Fig. 9-30a)

δ_{22} = horizontal displacement due to $X_{Hq} = 1$

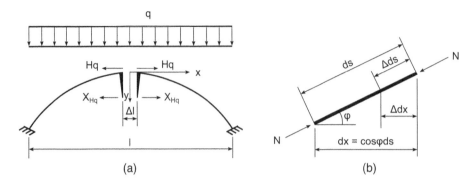

FIGURE 9-30 Effect of Elastic Shortening Due to Axial Force, (a) Primary Structure, (b) Infinitesimal Element.

Taking an infinite length ds along the arch axis (see Fig. 9-30b), we have

$$\Delta l = \int_{-\frac{l}{2}}^{\frac{l}{2}} \Delta dx = \int \cos\varphi \Delta\, ds = \int \frac{\cos\varphi N ds}{EA} \tag{9-56}$$

As there are no moment and shear, the axial force can be written as

$$N = \frac{H_q}{\cos\varphi} \tag{9-57}$$

where H_q = horizontal thrust at arch ends.

Substituting Eq. 9-57 into Eq. 9-56 yields

$$\Delta l = H_q \int_0^l \frac{dx}{EA\cos\varphi} \tag{9-58}$$

From Eq. 9-41b, we have

$$\delta_{22} = (1+\mu)\int \frac{y^2}{EI}\, ds \tag{9-59}$$

Substituting Eqs. 9-58 and 9-59 into Eq. 9-55 yields

$$X_{Hq} = \frac{H_q}{(1+\mu)} \frac{\int_0^l \frac{dx}{EA\cos\varphi}}{\int \frac{y^2}{EI}\, ds} \tag{9-60}$$

The additional forces due to external loadings at any location D of the arch can be written as

$$\Delta N_D = \cos\varphi_D X_{Hq} \tag{9-61}$$

$$\Delta M_D = (y - y_s) X_{Hq} \tag{9-62}$$

$$\Delta Q_D = \sin\varphi_D X_{Hq} \tag{9-63}$$

The total internal forces due to external loadings can be obtained by summing the results calculated by Eqs. 9-52 to 9-54 and Eqs. 9-61 to 9-63, respectively.

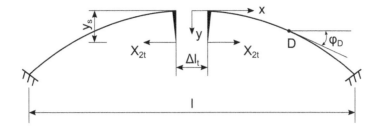

FIGURE 9-31 Effect of Uniform Temperature.

9.3.4.5 Effect of Uniform Temperature

Temperature rise and fall will cause secondary forces in indeterminate arch structures. The general analytical method was discussed in Section 4.3.3. Taking a primary structure as shown in Fig. 9-31, there is only one redundant at the elastic center X_{2t}. Assuming the horizontal displacement at the elastic center due to temperature change as Δl_t, and using the condition of deformation compatibility at the elastic center, the horizontal force due to the temperature change can be written as

$$X_{2t} = \frac{\Delta l_t}{\delta_{22}} \tag{9-64}$$

where

$$\Delta l_t = \alpha l \Delta T \tag{9-65}$$

α = coefficient of temperature
ΔT = temperature change from time when arch rib closes, positive for temperature rise
δ_{22} = horizontal displacement at elastic center due to $X_{2t} = 1$, positive for compressive axial force (see Eq. 9-59)

After the horizontal force X_{2t} due to the temperature change is obtained, the internal forces at any section D can be calculated as

$$M_{Dt} = X_{2t}\left(y - y_s\right) \tag{9-66a}$$

$$N_{Dt} = X_{2t}\cos\varphi_D \tag{9-66b}$$

$$Q_{Dt} = X_{2t}\sin\varphi_D \tag{9-66c}$$

9.3.4.6 Effect of Support Displacements

Non-uniform support displacements of an indeterminate arch bridge will have a significant effect on its internal forces. The displacements include horizontal, vertical, and rotation (see Fig. 9-32).

9.3.4.6.1 Effect of Horizontal Displacement

Assuming support B of a fixed arch (see Fig. 9-32) is displaced horizontally by an amount Δh and considering the primary structure as shown in Fig. 9-32a, there is only one redundant force X_{2h} applied at the elastic center. Using the condition of deformation compatibility at the elastic center, the horizontal force due to the horizontal displacement can be written as

$$X_{2h} = -\frac{\Delta h}{\delta_{22}} \tag{9-67}$$

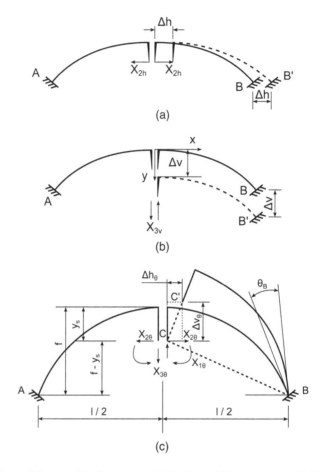

FIGURE 9-32 Effects of Support Displacements, (a) Horizontal Displacement, (b) Vertical Displacement, (c) Rotation.

where
$\quad \Delta h$ = horizontal displacement at support B or relative horizontal displacement between supports A and B, positive as shown in Fig. 9-32a
$\quad \delta_{22}$ = horizontal displacement at the elastic center due to a unit X_{2h}, can be calculated by Eq. 9-59

After X_{2h} is obtained, the internal forces due to the horizontal displacement at the support can be determined using the procedure discussed in Section 9.3.4.5.

9.3.4.6.2 Effect of Vertical Displacement

Assuming support B of a fixed arch (see Fig. 9-32) is displaced vertically by an amount Δv and considering the primary structure as shown in Fig. 10-32b, there is only one redundant force X_{3v} applied at the elastic center. Using the condition of deformation compatibility at the elastic center, the vertical shear force due to the horizontal displacement can be written as

$$X_{3v} = \frac{\Delta v}{\delta_{33}} \qquad (9\text{-}68)$$

where

Δv = vertical displacement at support B or relative vertical displacement between supports A and B, positive, as shown in Fig. 9-32b

δ_{33} = vertical displacement at elastic center due to unit X_{3v} can be calculated using Eq. 9-45 or simplified as

$$\delta_{33} = \int \frac{x^2}{EI} ds \qquad (9\text{-}69)$$

After X_{3v} is obtained, the internal forces due to the vertical displacement at the support can be determined using a similar procedure to the one discussed in Section 9.3.4.5.

9.3.4.6.3   Effect of Support Rotation

Assuming support B of a fixed arch (see Fig. 9-32) is rotated by an angle θ_B as shown in Fig. 9-32c, its elastic center C is moved to C' in the primary structure (see Fig. 9-32c). The horizontal and vertical displacements at the elastic center can be written as

$$\text{Horizontal: } \Delta h_\theta = \theta_B \left(f - y_s \right) \qquad (9\text{-}70)$$

$$\text{Vertical: } \Delta v_\theta = \frac{\theta_B l}{2} \qquad (9\text{-}71)$$

Thus, there are three redundant forces at the elastic center: $X_{1\theta}$, $X_{2\theta}$, and $X_{3\theta}$. From the conditions of deformation compatibility at the elastic center, we have

$$\begin{cases} X_{1\theta}\delta_{11} + \theta_B = 0 \\ X_{2\theta}\delta_{22} + \Delta h_\theta = 0 \\ X_{3\theta}\delta_{33} + \Delta v_\theta = 0 \end{cases} \qquad (9\text{-}72)$$

From Eq. 9-72, we can obtain the redundant forces at the elastic center due to the rotation θ_B as

$$X_{1\theta} = -\frac{\theta_B}{\int \frac{ds}{EI}} \qquad (9\text{-}73a)$$

$$X_{2\theta} = -\frac{\theta_B \left(f - y_s \right)}{\int \frac{y^2 ds}{EI}} \qquad (9\text{-}73b)$$

$$X_{3\theta} = -\frac{\theta_B l}{2 \int \frac{x^2 ds}{EI}} \qquad (9\text{-}73c)$$

The internal forces at any section D in the arch can be written as

$$\text{Moment: } M_{D\theta} = X_{1\theta} - X_{2\theta} \left(y - y_s \right) + X_{3\theta} x \qquad (9\text{-}74a)$$

$$\text{Axial force: } N_{D\theta} = X_{2\theta} \cos \varphi_D + X_{3\theta} \sin \varphi_D \qquad (9\text{-}74b)$$

$$\text{Shear: } Q_{D\theta} = X_{2\theta} \sin \varphi_D + X_{3\theta} \cos \varphi_D \qquad (9\text{-}74c)$$

9.3.5   BUCKLING AND STABILITY OF ARCH BRIDGES

Per the aforementioned discussion, the arch rib of an arch bridge is mainly subjected to axial loading and is basically a compression member. When its external loadings reach a certain magnitude, the arch ribs may buckle in-plane or out-of-plane and the arch bridge loses stability. In this section, we will focus on bifurcation buckling analysis, which is often called the Category I stability of arch bridges which are assumed to have the perfect arch axis.

9.3.5.1 In-Plane Buckling of Arch Bridges

9.3.5.1.1 Basic Theory of Arch In-Plane Buckling

To simplify the theoretical analysis, let us examine a two-hinged circular arch with a radius of R_0 subjected to a uniform radial loading q as shown in Fig. 9-33. Neglecting the effect of the elastic deformation of the arch, the arch axis is coincident with its pressure line and there is only axial

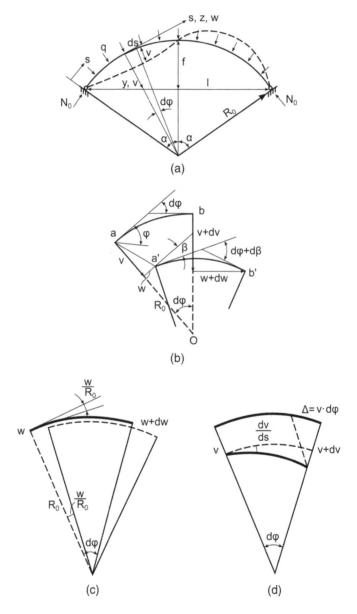

FIGURE 9-33 Circular Arch Buckling with Radial Loading, (a) Circular Arch Subjected to Radial Loading, (b) Displacements of an Infinitesimal Element, (c) Displacements of Element Due to Deformation along the Arch Axis, (d) Displacements of Element Due to Deformation along the Arch Radius.

force N_0 in the rib. When the loading q reaches the arch buckling load, the arch may undergo a small deformation (see dotted line in Fig. 9-33a). The buckling differential equation of the arch is established based on this small deformation status[9-5 to 9-11].

Taking an infinitesimal element $ds = \widehat{ab}$ from the arch, the infinitesimal element $\widehat{ab}$ is displaced to a new position $\widehat{a'b'}$ when buckling as shown in Fig. 9-33b. From this figure, the rotation angle of the element can be written as

$$\beta = \frac{(v+dv)\cos d\varphi + (w+dw)\sin d\varphi - v}{ds}$$

Considering that $d\varphi$ is very small, $\cos d\varphi \approx 1$, $\sin d\varphi = d\varphi$, $dwd\varphi \approx 0$, the above equation above be simplified as (see Figs. 9-33c and 9-33d)

$$\beta = \frac{dv}{ds} + \frac{w}{R_0} \tag{9-75}$$

The axial strain of the element can be written as

$$\varepsilon_s = \frac{(w+dw)\cos d\varphi - (v-dv)\sin d\varphi - w}{ds} \approx \frac{dw}{ds} - \frac{v}{R_0} \tag{9-76a}$$

When the arch buckles, the initial axial force N_0 can be treated as having no change, i.e.,

$$\varepsilon_s = \frac{dw}{ds} - \frac{v}{R_0} = 0 \tag{9-76b}$$

Thus, we have

$$\frac{dw}{ds} = \frac{v}{R_0} \tag{9-76c}$$

From Eqs. 9-75 and 9-76c, we can obtain the curvature of the arch as

$$k_x = \frac{d\beta}{ds} = \frac{d^2v}{ds^2} + \frac{v}{R_0^2}$$

Assuming the compressive axial force as positive and from Eq. 4-4, the equation above can be rewritten as

$$\frac{d^2v}{ds^2} + \frac{v}{R_0^2} = -\frac{M}{EI_x} \tag{9-77a}$$

where M is the moment due to the initial axial force $N_0 = qR_0$ and can be written as

$$M = N_0v = qR_0v$$

Substituting the above equation into Eq. 9-77a, we have

$$\frac{d^2v}{d\varphi^2} + \omega^2 v = 0 \tag{9-77b}$$

where

$$\omega^2 = 1 + \frac{qR_0^3}{EI_x} \tag{9-77c}$$

The general solution of Eq. 9-77b is

$$v = C \sin k\varphi + D \cos k\varphi$$

From the boundary conditions of the arch:

$$\varphi = 0, \ v = 0: \ D = 0$$
$$\varphi = 2\alpha, \ v = 0: \ C \sin 2\omega\alpha = 0$$

As the constant C cannot be equal to zero if the arch buckles,

$$\sin 2\omega\alpha = 0$$

From the equation above, we have

$$2\omega\alpha = n\pi, \ (n = 1, 2, 3, \dots)$$

As there is no movement at the arch ends and from Eq. 9-76b, we have

$$w_A - w_B = \int_0^{2\alpha} v(\varphi) \, d\varphi = 0 \tag{9-78}$$

To satisfy Eq. 9-78, the minimum value of n should be equal to 2, i.e.,

$$\omega_{min} = \frac{\pi}{\alpha}$$

Substituting the equation above into Eq. 9-77c, we can obtain the buckling loading as

$$q_{cr} = \frac{EI_X}{R_0^3} \left(\frac{\pi^2}{\alpha^2} - 1 \right) \tag{9-79a}$$

Equation 9-79a can be rewritten in terms of the axial buckling load as

$$N_{cr} = q_{cr} R_0 = \frac{\pi^2 EI_x}{R_0^2 \alpha^2} \left(1 - \frac{\alpha^2}{\pi^2} \right) = \frac{\pi^2 EI_x}{s_0^2} \tag{9-79b}$$

where s_0 = effective buckling length:

$$s_0 = \frac{R_0 \alpha}{\sqrt{1 - \left(\frac{\alpha}{\pi}\right)^2}} = \frac{s/2}{\sqrt{1 - \left(\frac{s}{2\pi R_0}\right)^2}} = k \frac{s}{2} \tag{9-80a}$$

where
$\quad s$ = length of arch rib axis

$$k = \frac{1}{\sqrt{1 - \left(\frac{s}{2\pi R_0}\right)^2}} \tag{9-80b}$$

k is often called the effective length factor.

9.3.5.1.2   Practical Method for Determining Arch In-Plane Buckling Axial Loading and Design Check

9.3.5.1.2.1   In-Plane Buckling Axial Loading
From Eq. 9-79b, it can be observed that the buckling analysis of an arch bridge is similar to that for a concentrically loaded column by using the concept of effective buckling length. However, the effective buckling length of an arch is not only related to the boundary condition of the arch, but it is also related to its shape. Table 9-1 provides some effective length factors for different types of axis, rise-to-span ratios, and boundary conditions, obtained based on numerical analysis. From this table, we can observe that the types of arch axis—circular, parabolic, and catenary—do not significantly affect the buckling loading of arches. For this reason, AASHTO recommends the effective length factors shown in Table 4-4 be used for bridge design in approximating the second-order effects of arch bridges. Then, the in-plane buckling loading can be determined by Eq. 9-79b.

9.3.5.1.2.2   Design Check of In-Plane Buckling
In general, after the entire bridge has been constructed, the in-plane buckling check of the completed bridge is generally not required due to the interaction of the arch ribs and the spandrel structure. The in-plane buckling of the arch rib or ring, before the deck system has been constructed, should be checked. Currently, the AASHTO LRFD specifications do not provide any detailed method for checking the in-plane buckling capacity of the arch bridges. However, the following equation[9-3, 9-4] may be used for checking arch in-plane buckling:

$$\eta_I N_u \le \phi_a \lambda_s \left(0.85 f_c' A_c + f_y A_s\right) \tag{9-81}$$

$$N_u = \frac{H_u}{\cos \alpha}$$

$$\cos \alpha = \frac{1}{\sqrt{1 + 4\left(f/l\right)^2}}$$

where
N_u = factored axial force at arch end
H_u = factored horizontal force at arch end
η_I = operation important factor (see Section 2.3.3.1)
ϕ_a = resistance factor for compression member = 0.75

TABLE 9-1
Effective Length Factors *k*

Boundary Condition	Shape of Axis	Rise-to-Span Ratio (*f/l*)				
		0.1	0.2	0.3	0.4	0.5
Two-hinged	Circular	1.01	1.07	1.06	1.11	1.15
	Parabolic	1.02	1.04	1.10	1.12	1.15
	Catenary	1.01	1.04	1.10	1.17	1.24
Fixed	Circular	0.70	0.70	0.70	0.71	0.71
	Parabola	0.70	0.69	0.70	0.71	0.72
	Catenary	0.70	0.69	0.68	0.72	0.73
Three-hinged	Circular	1.14	1.15	1.15	1.15	1.15
	Parabolic	1.14	1.11	1.10	1.12	1.15

TABLE 9-2

Stability Coefficient λ_s

s_0/r	≤ 28	35	42	48	55	62	69	76	83	90	97
λ_s	1.0	0.98	0.95	0.92	0.87	0.81	0.75	0.70	0.65	0.60	0.56
s_0/r	104	111	118	125	132	139	146	153	160	167	174
λ_s	0.52	0.48	0.44	0.40	0.36	0.32	0.29	0.26	0.23	0.21	0.19

Note: s_0 = effective buckling length determined by Eq. 9-80a. $r = \sqrt{\frac{I}{A}}$ = radius of gyration (see Eq. 3-9).

λ_s = stability coefficient (see Table 9-2)
α = angle from horizontal to line connecting arch spring to arch crown (see Fig. 9-34)
f_c, f_y = concrete strength and steel yield strength, respectively
A_c, A_s = areas of concrete section and steel, respectively

9.3.5.2 Out-of-Plane Buckling of Arch Bridges

9.3.5.2.1 Lateral Buckling Differential Equations

Assuming an arch is subjected to radial loading q as shown in Fig. 9-35, when q reaches a certain magnitude, the arch may buckle laterally as shown in Fig. 9-35b and c. Consider a moving rectangular coordinate system as shown in Fig. 9-35a. Section S in the arch will be displaced by u, v, and w and rotated by β, γ, and θ along x-, y-, and z-directions, respectively, when the arch buckles laterally.

Taking an infinitesimal element ds from the arch as shown in Fig. 9-35d and considering $\cos d\varphi \approx 1$, $\sin d\varphi = d\varphi$, $d\varphi d\theta \approx 0$, the lateral bending curvature about the y-axis can be written as

$$k_y = \frac{(\gamma + d\gamma)\cos d\varphi - \gamma + (\vartheta + d\vartheta)\sin d\varphi}{ds} = \frac{d\gamma}{ds} + \frac{\theta}{R} \tag{9-82a}$$

where γ is the lateral deflection angle when the arch is buckling (see Fig. 9-35c) and can be written as

$$\gamma = -\frac{du}{ds}$$

Substituting the above equation into Eq. 9-82a yields

$$k_y = \frac{\theta}{R} - \frac{d^2 u}{ds^2} \tag{9-82b}$$

FIGURE 9-34 Simplified In-Plane Buckling Check.

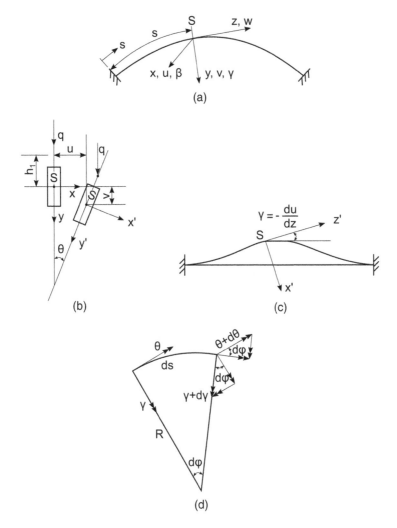

FIGURE 9-35 Lateral Buckling of Arches, (a) Moving Coordinate System, (b) Side View of Arch after Buckling, (c) Plan View of Arch after Buckling, (d) Geometric Relationships after Buckling.

The torsion curvature about the z-axis can be written as

$$k_z = \frac{(\theta + d\theta)\cos d\varphi - \theta - (\gamma + d\gamma)\sin d\varphi}{ds} = \frac{d\theta}{ds} - \frac{\gamma}{R} = \frac{d\theta}{ds} + \frac{du}{R\,ds} \qquad (9\text{-}83)$$

When an arch has buckled, the lateral bending moment M_y, the torsion about the z-axis M_z, and the lateral shear Q_x will be occurring in the arch (see Fig. 9-36). In addition, the external vertical loading q will induce lateral force q_x and torsion M_z along the principal axes of the section (see Fig. 9-35b), i.e.,

$$q_x = \theta q$$
$$M_z = h_1 q_x$$

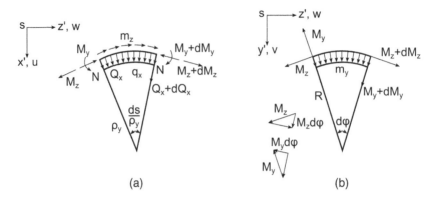

FIGURE 9-36 Force Equilibriums of an Infinitesimal Element, (a) Out-of-Plane, (b) In-Plane.

The initial in-plane axial force N will have little change when buckling and can be assumed to remain constant. The distributed bending moment m_y induced by the external loading due to buckling is also very small and can be neglected. Thus, from the equilibrium conditions and Fig. 9-36, we have

$$\frac{dQ_x}{ds} - k_y N + q_x = 0 \tag{9-84}$$

$$\frac{dM_y}{ds} + \frac{M_z}{R} + Q_x = 0 \tag{9-85}$$

$$\frac{dM_z}{ds} - \frac{M_y}{R} + m_z = 0 \tag{9-86}$$

Eliminating Q_x from Eqs. 9-84 and 9-85 yields

$$M_y'' + \left(\frac{M_z}{R}\right)' + k_y N - q_x = 0 \tag{9-87}$$

From Eqs. 4-4 and 3-85a, we have

$$\begin{cases} M_y = EI_y k_y \\ M_z = GJk_z - \left(EI_\omega k_z'\right)' \end{cases} \tag{9-88}$$

Substituting Eqs. 9-82b and 9-83 as well as Eq. 9-88 into Eqs. 9-86 and 9-87, we can obtain the arch lateral buckling differential equations:

$$\begin{cases} -\left[EI_y\left(u'' - \frac{\theta}{R}\right)\right]'' + \left[\frac{GJ}{R}\left(\theta' + \frac{u'}{R}\right)\right]' - \left[\frac{EI_\omega}{R}\left(\theta' + \frac{u'}{R}\right)'\right]'' + N\left(-u'' + \frac{\theta}{R}\right) - q_x = 0 \\[4mm] \left[\frac{GJ}{R}\left(\theta' + \frac{u'}{R}\right)\right]' - \left[EI_\omega\left(\theta' + \frac{u'}{R}\right)'\right]'' + \left[\frac{EI_y}{R}\left(u'' - \frac{\theta}{R}\right)\right] + m_z = 0 \end{cases} \tag{9-89a}$$

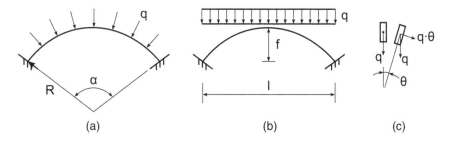

FIGURE 9-37 Lateral Buckling of a Single Rib, (a) Circular Arch Rib with Radial Uniform Loading, (b) Parabolic Rib with Uniform Vertical Loading, (c) Lateral Buckling Deformation of Arch Rib.

For a circular arch with constant cross section, Eq. 9-89a can be simplified as

$$\begin{cases} -E\left(I_y+\dfrac{I_\omega}{R}\right)u''''+\left(\dfrac{GJ}{R^2}-N\right)u''-\dfrac{EI_\omega}{R}\theta''''+\dfrac{EI_y+GJ}{R}\theta''+\dfrac{\theta}{R}N-q_x=0 \\ -\dfrac{EI_\omega}{R}u''''+\dfrac{EI_y+GJ}{R}u''-EI_\omega\theta''''+GJ\theta''-\dfrac{EI_y}{R^2}\theta+m_z=0 \end{cases}$$

(9-89b)

9.3.5.2.2 Lateral Buckling of Single Rib

9.3.5.2.2.1 Circular Arch Ribs with Radial Loading   First, let us examine a simple circular arch rib with a center angle of α, subjected to a uniform radial loading as shown in Fig. 9-37, for developing a general equation for determining the arch lateral buckling loading. Assuming the cross section of the rib is rectangular and constant, we have

$$I_\omega \approx 0$$
$$N = qR$$
$$q_x = q\theta$$
$$m_z = 0$$

Substituting the equations above into Eq. 9-89b and eliminating variable u, we can obtain

$$\frac{d^4\theta}{d\varphi^4}+(2+\omega)\frac{d^2\theta}{d\varphi^2}+(1-\omega\xi)\theta=0$$

(9-90a)

The general solution to Eq. 9-90a can be written as

$$\theta = A\sin k_1\varphi + B\cos k_1\varphi + C\sin k_2\varphi + D\cos k_2\varphi$$

where

$$k_1 = \sqrt{\frac{2+\omega}{2}+\sqrt{\left(\frac{2+\omega}{2}\right)^2+\omega\xi-1}}$$

(9-90b)

$$k_2 = \sqrt{-\left(\frac{2+\omega}{2}\right)+\sqrt{\left(\frac{2+\omega}{2}\right)^2+\omega\xi-1}}$$

(9-90c)

$$\xi = \frac{EI_y}{GJ}$$

$$\omega = \frac{qR^3}{EI_y}$$

(9-90d)

Assume that the arch is hinged at both ends and can rotate about the x-axis and the y-axis but is restrained from rotation about the z-axis (see Fig. 9-35 for coordinate system), i.e.,

$$\varphi = 0: \theta = 0, \ \theta'' = 0$$
$$\varphi = \alpha: \theta = 0, \ \theta'' = 0$$

From the above boundary conditions, we can obtain

$$B = C = D = 0$$
$$A \sin k\alpha = 0 \tag{9-91}$$

As A cannot be equal to zero, the minimum value of k_1 should be

$$k_{1min} = \frac{\pi}{\alpha}$$

From the equation above, we can obtain

$$\omega = \frac{\left(\pi^2 - \alpha^2\right)^2}{\alpha^2 \left(\pi^2 + \xi\alpha^2\right)}$$

Substituting Eq. 9-90d into the equation above, we can obtain the arch lateral buckling axial force as

$$N_{cr} = q_{cr}R = \beta \frac{\pi^2 EI_y}{\left(\kappa S\right)^2} \tag{9-92}$$

where $S = R\alpha$ = length of arch axis

$$\kappa = \text{effective length coefficient} = \begin{cases} 1.0: \text{ for two-hinged arches} \\ 0.5: \text{ for fixed arches} \end{cases}$$

$\beta = \dfrac{\left[1-\left(\frac{\alpha}{\pi}\right)^2\right]^2}{1+\xi\left(\frac{\alpha}{\pi}\right)^2}$, often called the coefficient of arch shape.

9.3.5.2.2.2 Parabolic Arches with Vertical Loading In Section 9.3.5.2.2.1, we developed the equation for determining the buckling loading of circular arches with a special loading case. However, for other shapes of arches, it is difficult to develop a simple formula for determining the arch lateral buckling loading, and is generally necessary to use numerical methods to determine the buckling loadings. For fixed parabolic arch bridges (see Fig. 9-37b), the lateral buckling loading can be calculated using the following equation[9-5]:

$$q_{cr} = K \frac{EI_y}{l^3} \tag{9-93}$$

where
   l = arch span length
   K = lateral buckling coefficient

TABLE 9-3
Lateral Buckling Coefficients for
Parabolic Fixed Arches, K

f/l	ξ		
	0.7	1.0	1.3
0.1	28.5	28.5	28.0
0.2	41.5	41.0	40.0
0.3	40.0	38.5	36.5

K is related to the arch rise-to-span ratio, boundary conditions, and ratio of arch bending stiffness to torsion stiffness ξ (see Eq. 9-90c) and can be determined from numerical analysis. The K values for frequently used arch bridges with fixed ends are given in Table 9-3.

The vertical loading q in practical arch bridges is actually not uniform. The following equivalent vertical loading can be used to evaluate the arch bridge lateral buckling:

$$q_e = \frac{8f}{l^2} H_{max} \tag{9-94}$$

where H_{max} is the maximum horizontal thrust at the arch end or the axial force at the arch crown, including both dead and live loadings.

From Eqs. 9-94 and 9-93, the lateral buckling horizontal force can be written as

$$H_{cr} = K \frac{EI_y}{8fl} \tag{9-95}$$

The lateral buckling axial force can be written as

$$N_{cr} = \frac{H_{cr}}{\cos\alpha} \tag{9-96}$$

where $\cos\alpha = \frac{1}{\sqrt{1+4(f/l)^2}}$

9.3.5.2.3 Lateral Buckling of Twin Ribs

To facilitate construction and increase lateral buckling loads, ribs are often connected with parallel transverse members (see Fig. 9-38). Their axial buckling loads can be determined based on the principle of energy[9-3 to 9-11] and can be written as

$$N_{cr} = \frac{\pi^2 EI_y}{l_e^2} \tag{9-97}$$

where
  I_y = lateral bending moment of inertia of combined two arch ribs about y-axis
  l_e = effective length of arch axis

$$l_e = k_s k S$$

  k = effective coefficient of arch: $k = 1.0$ for two-hinged-arches; $k = 0.5$ for fixed arches
  S = length of arch axis

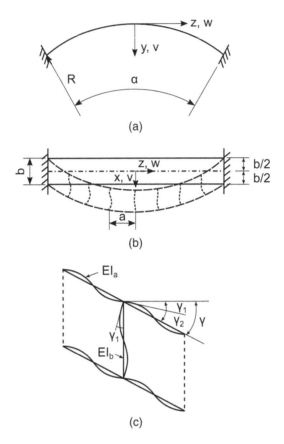

FIGURE 9-38   Lateral Buckling of Twin Arch Ribs, (a) Elevation, (b) Lateral Buckling Mode, (c) Buckling Deformation of Arch Ribs and Bracing.

k_s = reduction factor for shear effect

$$k_s = \sqrt{1 + \frac{\pi^2 EI_y}{(kS)^2} \left(\frac{ab}{12EI_b} + \frac{a^2}{24EI_a (1-\beta)} + \frac{na}{bA_bG} \right)}$$

a = spacing of transverse beams
b = distance between centroids of arch ribs
I_a = lateral bending moment of inertia of single arch rib (about y-axis)
I_b = moment of inertia of transverse beam about y-direction
n = coefficient of section shape of transverse beam: $n = 1.20$ for rectangular; $n = 1.11$ for circular
β = panel buckling coefficient, can be assumed to be zero for initial calculation

$$\beta = \frac{N_{cr}a^2}{2\pi^2 EI_a}$$

9.3.5.2.4   Lateral Buckling of Tied Arches

Figure 9-39a shows a typical tied arch bridge. When the vertical loading q reaches a certain critical value, the arch may buckle laterally as shown in Fig. 9-39c. The hangers are inclined and create a pulled back force F, which is often called a stabilizing force, and it increases the arch lateral

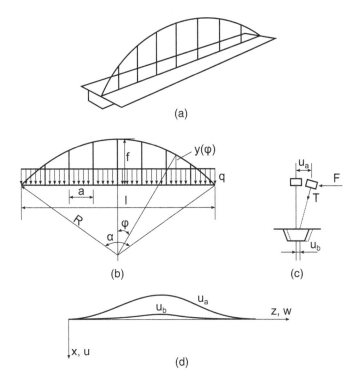

FIGURE 9-39 Buckling of Tied Arch Bridges, (a) Typical Tied Arch Bridge, (b) Elevation View, (c) Side View, (d) Plan View.

buckling loading. Assuming the arch axis is circular and using the principle of energy, we can obtain the arch lateral buckling loading as[9-5]

$$q_{crt} = \eta q_{cr} \qquad (9\text{-}98a)$$

where
q_{cr} = buckling loading without hangers, being calculated using Eq. 9-92 or 9-93
η = hanger effect coefficient

$$\eta = \frac{1}{1-C}$$

$$C = \frac{3}{4}\left(\frac{\alpha}{\pi}\right)^2 \frac{R}{f}$$

Numerical analysis shows that the value of η varies from 2.8 to 3.5 for different shapes of the arch axis. To be conservative, engineers may simply calculate the lateral buckling loading of a tied arch bridge using Eq. 9-93 and multiply it by 2.5, i.e.,

$$q_{crt} = 2.5 \; K \frac{EI_y}{l^3} \qquad (9\text{-}98b)$$

9.3.6 Arch Bridge Analysis by Finite-Element Method

The analytical methods discussed in Section 9.3.5 are generally suitable for simple arch bridges or for a quick estimation of internal forces and buckling loadings of arch bridges. Actual arch bridges

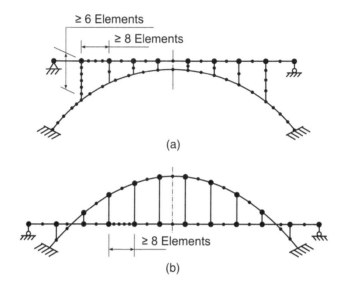

FIGURE 9-40 Modeling of Typical Arch Bridges, (a) Deck Arch Bridge, (b) Half-Through Arch Bridge.

may have variable rib sections, or their spandrel structures and arch rib may function as combined structural elements. The analysis of arch bridges by hand calculation becomes time consuming and very difficult, especially for construction, nonlinear, and time-dependent analyses. Currently, the analyses of arch bridges are typically performed using computer programs developed based on finite-element methods. Most of the currently available commercial computer programs have very powerful structural analysis and design functions, including linear, nonlinear, and buckling analyses. The arch ribs may be divided into a number of straight frame elements. The number of finite elements in each of the ribs should not be less than 12. Generally, 20 elements in each rib are enough to ensure accurate analytical results. If the internal forces of the deck longitudinal beams are needed for their design from the entire bridge structural analysis, the number of finite elements in each of the spans is recommended to be at least 8. If the bending moment of the column is required, its number of finite elements is recommended to be at least 6. The flexible hangers can be treated as a single element (see Fig. 9-40).

9.3.7 Effect of Large Deflections

As arch bridges are comparatively flexible with large axial forces, the AASHTO specifications[9-12] require that the use of large deflection analysis of arches with longer spans be considered. The large deflection theory was discussed in Section 4.3.7, and its analysis can be performed with a computer program. For most practical arch bridges, the approximate method discussed in Section 4.3.7.3 can generally satisfy the design requirement.

9.3.8 Dynamic Behaviors and Dynamic Loading of Arch Bridges

9.3.8.1 General

Comparatively speaking, less is known about the dynamic behavior of arch bridges due to moving vehicles than that for girder bridges. The AASHTO LRFD specifications[9-12] specify a dynamic allowance of 0.33 for all types of bridges, based on limited research and tests on girder bridges. Recent research[9-13, 9-15] indicates that: (1) The vertical stiffness of an arch bridge significantly affects the dynamic loading, especially for vehicle speeds greater than 55 mph (88.5 km/h). For half-through and tied-arch bridges, the most effective way of reducing dynamic loading is to use a deck system with stronger vertical stiffness, such as two stronger continuous longitudinal girders connected with floor beams. (2) Fixed arch bridges have better dynamic behaviors than two-hinged arch bridges, and the dynamic loading of fixed arch bridges due to moving vehicles is generally less than that of two-hinged arch bridges, especially for vehicle speeds greater than 55 mph. (3) Dynamic loadings of axial forces are generally much smaller than those of the bending moment.

9.3.8.2 Dynamic Behaviors and Dynamic Loading Estimation of Deck Arch Bridge

Figure 9-41 shows some typical vibration modes of deck arch bridges. It can be observed that the first torsion vibration mode is symmetric and that the first vertical vibration mode is asymmetric. Considering the vehicle and bridge interaction as well as the road surface roughness effect, Huang[9-14] performed large amounts of numerical analysis and concludes that the dynamic equation contained in

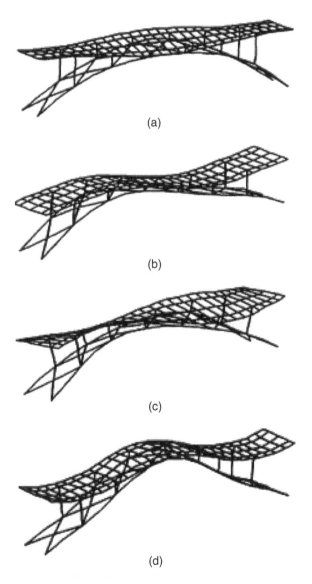

(a)

(b)

(c)

(d)

FIGURE 9-41 Typical Vibration Modes of Deck Arch Bridges, (a) Vibration Mode 1, (b) Vibration Mode 2, (c) Vibration Mode 3, (d) Vibration Mode 4.

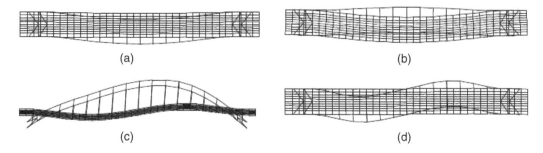

FIGURE 9-42 Typical Vibration Modes of Half-Through Arch Bridge, (a) Vibration Mode 1, (b) Vibration Mode 2, (c) Vibration Mode 3, (d) Vibration Mode 4.

the AASHTO standard specifications[9-16] can be safely used in estimating the dynamic loadings of the arch members, though it may be conservative; i.e., the impact factor can be written as

$$IM = \frac{50}{L_0 + 125} \tag{9-99}$$

where L_0 is the loaded length in feet. For dynamic axial forces, L_0 is the total bridge span length. For the dynamic moment at arch ends, L_0 is equal to half the span length.

9.3.8.3 Dynamic Behaviors and Dynamic Loading Estimation of Half-Through Arch Bridge

Figure 9-42 shows some typical vibration modes of half-through arch bridges. It can be observed that the first lateral bending and torsion vibration mode is symmetric and that the first vertical vibration mode is asymmetric. Extensive numerical dynamic analyses due to moving vehicles have been done by Huang[9-13], including the vehicle and bridge interaction, effect of road surface roughness, effect of span length, and effect of vehicle speed. Based on the numerical analysis, the following impact factor equations are proposed for estimating tie and half-through arch bridges.

Impact factor of axial force at all sections and moment at midspan:

$$\text{If } L > 260 \ ft \ (80 \ m): IM = 0.15 \tag{9-100a}$$

$$\text{If } L \le 260 \ ft \ (80 \ m): IM = 0.25 - 000166(l - 20) \ (\le 0.25) \tag{9-100b}$$

Impact factor of moment at arch rib ends:

$$\text{If } L \ge 460 \ ft \ (140 \ m): IM = 0.08 + 0.21\frac{f}{L} \tag{9-100c}$$

$$\text{If } L \le 260 \ ft \ (80 \ m): IM = 0.33 + 0.21\frac{f}{L} \ (\le 0.40) \tag{9-100d}$$

$$\text{If } 460 \ ft < L < 260 \ ft: IM = 0.6636 - 0.00417L + 0.21\frac{f}{L} \tag{9-100e}$$

In Eqs. 9-100a to 9-100e, f = rise and L = span length. The impact factors of the bending moment at the span quarter point can be taken as the average of those at the arch end and at midspan.

9.4 GENERAL CONSTRUCTION METHODS OF ARCH BRIDGES

There are many erection methods developed in arch bridge constructions, such as falsework method, cable supported cantilever method, and swing method. For concrete segmental arch bridges, the falsework method and cable supported cantilever method are most often used.

9.4.1 FALSEWORK METHOD

The falsework method is the simplest and earliest method used in arch bridge construction. For arch bridges with a span length greater than 260 ft (80 m), steel arch trusses supported at arch ends and/or two or more interior towers may be used as the falsework (see Fig. 9-43a). For arch bridges with a span length less than 260 ft (80 m), a three-hinged steel arch truss may be used (see Fig. 9-43b).

The arch ribs can be either precast or cast-in-place. The sequences of segment installation or casting will significantly affect the internal force and deformation of the falsework. The designer should specify the segment installation sequences in the plan to reduce the bending moment and deformation of the falsework, A typical segment cast sequence is shown in Fig. 9-43a.

The falsework should be designed to facilitate removal. Cambers should be provided in setting the top elevations of the form. The arch cambers may be determined by one of the following methods:

a. Calculate the camber Δ at the arch crown, and determine the cambers Δ_x at other locations by assuming the camber varying with x is a parabolic function or horizontal thrust influence line (see Fig. 9-44a).
b. Calculate the camber Δ at the arch crown, and determine the cambers Δ_x at other locations using Eq. 9-20 or Eq. 9-28, the revised rise-to-span ratio $\frac{f+\Delta}{l}$ and a coefficient of arch axis of $(m - 0.2)$ (see Fig. 9-44b).
c. Determine the camber amounts at all related locations based on actual conditions.

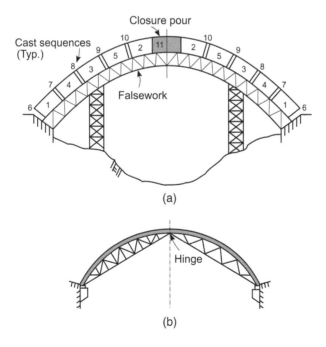

(a)

(b)

FIGURE 9-43 Falsework Method and Segment Cast Sequences, (a) Steel Arch Truss, (b) Three-Hinged Steel Arch Truss.

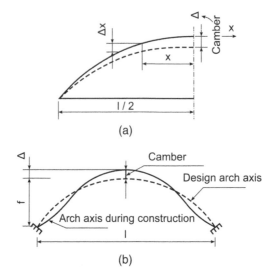

FIGURE 9-44   Camber Setting, (a) Parabolic Distribution, (b) Catenary Distribution.

9.4.2   CABLE-STAYED CANTILEVER METHOD

The cable-stayed cantilever method is the most popular construction method in long-span arch bridges as mentioned in Section 1.5.3.2. Two halves of an arch rib are erected separately from each of the arch springs and then closed at the arch crown. The arch ribs can be built by either the cast-in-place method (see Fig 9-45a) or precast method (see Fig. 9-45b). Temporary cables are used to maintain the arch rib stability during construction. The cables are anchored in piers or temporary

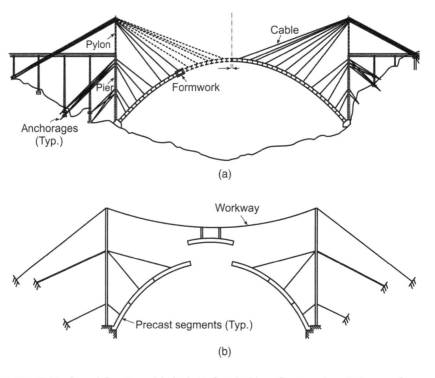

FIGURE 9-45   Cable-Stayed Cantilever Method, (a) Cast-in-Place Construction, (b) Precast Construction.

pylons that are built first and stabilized with some back-tie cables anchored to the ground. During construction, lateral cables are necessary to prevent the arch ribs from buckling. Generally, the arch rib lateral elastic buckling safety factor should not be less than 4. It is common practice to use cableway attached to a trolley to transport precast segments and other materials from the end banks to the anticipated locations (see Fig. 9-45b).

9.4.3 Adjustment and Control of Arch Rib Forces

As discussed, additional forces will be induced due to elastic shortening of the arch rib, concrete creep and shrinkage, temperature, etc. To reduce these effects, it may be necessary to use temporary hinges or jacking methods.

9.4.3.1 Temporary Hinge Method

To avoid additional forces due to elastic shortening, creep and shrinkages, temperature, etc., we may place three temporary hinges, one at each of the skew backs and one at the arch crown, to convert a fixed arch into a statically determined three-hinged arch. After the completion or partial completion of the spandrel structures, the three hinges are keyed and fixed. In this way, the aforementioned effects can be partially or completely eliminated. In general, the three-hinge method may be considered if one of the following conditions is met:

a. Arch ribs or rings with a low rise-to-span ratio for which high deformation forces are expected.
b. A high movement of the arch support is expected.
c. Significantly enlarged rib depth at the arch skew back, which may result in the accentuation of deformation stresses.

Some previously used temporary hinges are shown in Fig. 9-46. The temporary hinge should have enough compression strength to resist the anticipated axial force. Steel spiral ties may be used to increase its compression strength (Fig. 9-46b).

9.4.3.2 Jacking Method

Let's examine a fixed arch that is built on falsework. If the structure is closed while resting on its centering, theoretically, there is no stress in the arch. If we remove the falsework, the loading due to its self-weight will be transferred to the rib and produce axial force that cause elastic shortening by an amount Δ. The arch axis must move down as shown by the dotted line in Fig. 9-47 to maintain its arch shape. This movement induces a negative moment at the skew backs and a positive moment at the crown. If the arch is completely cut at the crown and some hydraulic jacks are installed, the jacks can be engaged to lengthen the arch by an amount equal to the shortening Δ. Then, the arch will be raised back to its original position and the moment due to the elastic shortening will be eliminated. If a lengthening equal to the anticipated shortening is introduced in the arch before engaging the falsework, the load will be transferred to the rib and the span will be decentered without settling or bending. This method of eliminating part of or all the effects due to elastic shortening, shrinkage and creep, temperature, etc., is also referred to as the method of compensation and adjustment[9-1]. The amount of arch compensation stretch provided by the hydraulic jacks may be calculated as

$$\Delta = \Delta_e + \Delta_t + \Delta_{sh} + \Delta_{cr} \qquad (9\text{-}101)$$

where
Δ_e = horizontal shortening due to axial thrust, can be calculated by Eq. 9-58
Δ_t = horizontal shortening due to uniform temperature, can be calculated by Eq. 9-65
Δ_{sh}, Δ_{cr} = horizontal shortening due to concrete shrinkage and creep, respectively

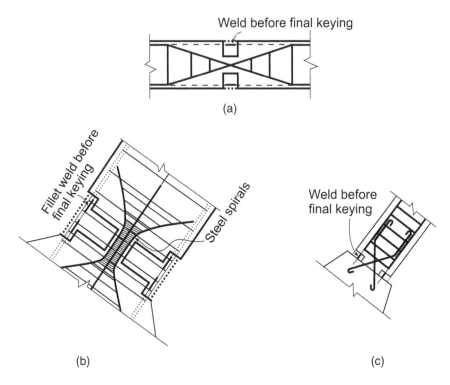

FIGURE 9-46 Typical Temporary Hinges, (a) at Arch Crown, (b) at Skew Back with Steel Spirals, (c) at Skew Back with Cross Steel Bars.

9.5   SUMMARY OF GENERAL ANALYSIS AND DESIGN PROCEDURES FOR SEGMENTAL ARCH BRIDGES

For clarity, the basic procedures for designing and analyzing arch bridges are summarized as follows:

Step 1: Based on the site conditions, opening requirement under the bridge, etc., determine the minimum span length and type of arch bridge as discussed in Section 9.2.2.

Step 2: Select the rise-to-span ratio based on the discussion in Section 9.2.5.2.

Step 3: Determine the shape of the arch rib axis as discussed in Section 9.2.5.3.

Step 4: Select typical section of the arch ring, and estimate the depth of the arch ring based on the discussion in Section 9.2.3 and previously built similar arch bridges.

Step 5: Arrange spandrel structures and deck systems as discussed in Section 9.2.5.

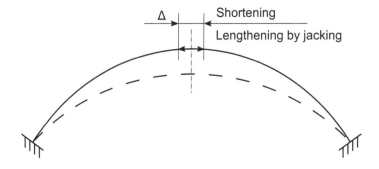

FIGURE 9-47   Deformation Due to Elastic Shortening and Lengthening by Jacking.

Step 6: Select and assume a construction method and construction sequence.

Step 7: Determine the effects due to dead loads using the methods discussed in Section 9.3 by either hand calculations or computer programs. It is typical to calculate the responses at the arch spring, ¼ point of the arch, and at the arch crown. For long-span arch bridges with variable arch rib sections, it may be necessary to calculate the internal forces at each of 1/8 points along the span, including the arch springs and the crown.

Step 8: Determine the maximum and minimum live load effects at the critical locations mentioned in step 7 by applying the design live loads on the influence lines of the internal forces with the same sign (positive or negative) as discussed in Section 9.3.

Step 9: Determine other effects due to elastic shortening, temperature, shrinkage and creep, support settlements, nonlinear deformations, etc.

Step 10: Check in-plane buckling, especially for the arch ribs/ring during construction, based on Eq. 9-81 or using computer programs.

Step 11: Check lateral buckling of the arch based on Eq. 9-95 or Eq. 9-97 or using computer analysis.

Step 12: Check other service capacities of the arch ring.

Step 13: Check the arch rib compression, bending, and shear strengths.

Step 14: Check the capacities for the arch spandrel and/or deck systems.

Step 15: If any of the steps between steps 10 to 14 is not met, revise the relevant members until all requirements are met.

9.6 DESIGN EXAMPLE V[1]

9.6.1 INTRODUCTION

This design example is modified based on the Na Li You River Bridge located in Guang Xi Providence, China, designed by Guangxi Transportation Planning Surveying and Designing Institute (see Fig. 9-48). The bridge consists of a Northbound and a Southbound pair with a total length of 1654.86 m (5429.33 ft) and 1644.03 (5393.80 ft), respectively. The main bridge is an arch bridge with a clear span length of 160 m (524.93 ft). The approach spans are continuous prestressed concrete girders with span lengths ranging from 30 to 45 m (98.43 to 147.64 ft). In this example, we will take the Northbound main bridge for the discussion of the basic design issues of arch bridges. The original bridge was designed using the Chinese highway bridge design specifications. For simplicity and consistency, the *AASHTO LRFD Highway Bridge Design Specifications* will be used in this example. However, some of the details presented may not be frequently used in the United States. As the design of spandrel structures is similar to other types of bridges, the analysis and design in this example will be focused on the arch ring.

9.6.2 DESIGN REQUIREMENTS

9.6.2.1 Design Specifications

a. *AASHTO LRFD Bridge Design Specifications*, 7th edition, 2014[9-12]

b. CEB-FIP *Model Code for Concrete Structures*, 1990

9.6.2.2 Traffic Requirements

a. Two 3.6-m (11.8-ft) lanes with traffic flow in same direction

b. 3.0 m (9.84 ft) wide for right shoulder, and 1.55 m (5.09 ft) wide for left shoulder

c. 1.57-m (5.2-ft) pedestrian sidewalk

d. Two 0.46-m (18.11-in.) wide barrier walls

e. 0.26-m (10.24-in.) pedestrian railing

[1] This design example is provided by Prof. Baochun Chen, Dr. Xiaoye Luo and Fuzhou University; Mr. Zenghai Lin and Mr. Guisong Tu, Institute of Guangxi Communication Planning Surveying and Designing Co., Ltd.; and Dr. Jianjun Wang, Guangxi Road and Bridge Engineering Group Co., Ltd, China; and revised by Dr. Dongzhou Huang, P.E.

FIGURE 9-48 Na Li You River Bridge, Guang Xi, China.

9.6.2.3 Design Loads

a. Live load:
   Vehicle: HL-93 loading
   Pedestrian loading: 2.5 kN/m² (52.2 lb/ft²)
b. Dead loads:
   – Unit weight of reinforcement concrete (DC): $w_{DC} = 2500$ kg/m³ (0.156 kcf)
c. Thermal loads:
   – Uniform temperature: mean: 20°C (68°F); rise: 25°C (45°F); fall: 20°C (36°F)

9.6.2.4 Materials

a. Concrete:

$$f_c' = 34.5 \ Mpa \ (5.0 \ ksi) \ f_{ci}' = 27.6 \ Mpa \ (4.0 \ ksi)$$

$$E_c = 30000 \ Mpa \ (43500 \ ksi)$$

b. Reinforcing steel:
   ASTM 615, Grade 60
   Yield stress: $f_y = 413.8$ MPa (60 ksi)
   Modulus of elasticity: $E_s = 206,897$ MPa (30,000 ksi)

9.6.3 Bridge Span Arrangement and Determination of Principal Dimensions

9.6.3.1 Bridge Span Arrangement and Typical Section

Based on the navigation requirement and site conditions, the arch clear span length was determined to be 160 m (524.93 ft) and the rise was determined to be 24.61 m (80.74 ft). The arch rise-to-span ratio is 1/6.5. Using the five-points matching method discussed in Section 9.3.2.2, the arch axial coefficient m is determined to be 1.78. The spandrel structure consists of 17 spans of precast reinforced void concrete slabs with equal span lengths of 10 m (32.81 ft). The bridge elevation and typical section are shown in Figs. 9-49 and 9-50, respectively.

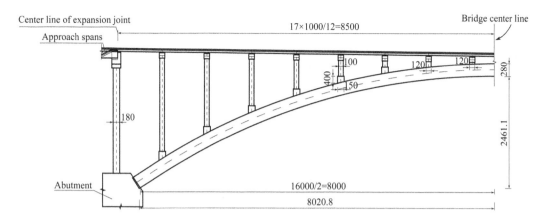

FIGURE 9-49 Bridge Elevation View of Design Example V (unit = cm, 1 in. = 2.54 cm).

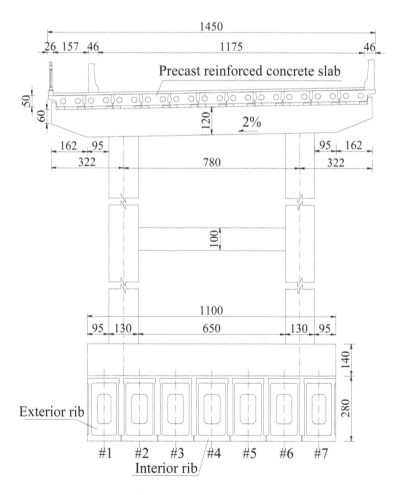

FIGURE 9-50 Bridge Typical Section of Design Example V (unit = cm, 1 in. = 2.54 cm).

9.6.3.2　Arch Ring and Ribs

The arch ring consists of six side-by-side precast concrete box ribs with cast-in-place reinforced concrete topping of 10 cm (3.94 in) thick (see Figs. 9-50 and 9-51). The total depth and width of the arch ring are 2.8 m (9.19 ft) and 11 m (36.09 ft), respectively. The depth of the arch ribs is 2.7 m (8.86 ft). The widths of the interior and exterior box ribs are 1.56 m (5.12 ft) and 1.48 m (4.86 ft), respectively. The interior box ribs have equal web thickness of 8 cm (3.15 in.), and the exterior box ribs have a thicker outside web of 14 cm (5.51 in.). The thicknesses of the top and bottom slab of the arch ribs are 20 cm (7.87 in.) and 25 cm (9.84 in.), respectively. There are six longitudinal cast-in-place joints of 24 cm (9.45 in) thick between the ribs. As the thickness of the box rib webs is comparatively small, some interior diaphragms with a spacing of about 2.4 m (7.87 ft) along the longitudinal direction are provided for each of the arch ribs. The thickness of the interior diaphragms is 10 cm (3.94 in.), except that those attached hooks for pickup is 14 cm (5.51 in.). The remaining dimensions of the arch ring are shown in Fig. 9-51.

There are twenty-six 25-mm-diameter (#8) bars in each of the top and bottom slabs of the box arch rib. The typical reinforcing details of the interior box arch rib are shown in Fig. 9-52. In this figure, the reinforcing bar identifications 1, 1′, 2, and 2′ represent diameter 25-mm-diamter (#8) bars and the remaining rebars are 12-mm-diameter (#4) bars.

From Fig. 9-51 and the above discussions, the thicknesses of the web and diaphragms are comparatively small and their self-weights are significantly reduced for easy erection during the construction. However, the fabrication of the arch ribs is more difficult due to the use of diaphragms and thin webs. To facilitate the fabrication, two stages was used to precast the arch rib. First, the interior diaphragms and the web segments are precast in the yard or the field. The length of precast web segment is about 2 m (6.56 ft). The rebars in each of the precast diaphragms and web slabs are extruded out from its four edges (see Fig. 9-53a and b). The precast diaphragms and web segments are assembled together by casting the top and bottom slabs as well as the joints between the diaphragms and web segments (see Fig. 9-53a).

For erection, each of the arch ribs is divided into seven segments in the bridge longitudinal direction. The total weight of the precast arch rib segment is 78.5 metric tons (86.5 tons) (see Fig. 9-54). After all segments of a rib are in place as designed, the segments are connected by bolting the steel angles embedded in the top and bottom slabs of the precast arch rib segments (see Fig. 9-55). The steel angles are welded to the rebars. The details for connecting the interior

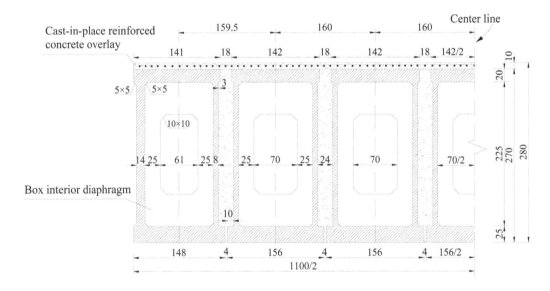

FIGURE 9-51　Typical Section of Arch Ring (unit = cm, 1 in. = 2.54 cm).

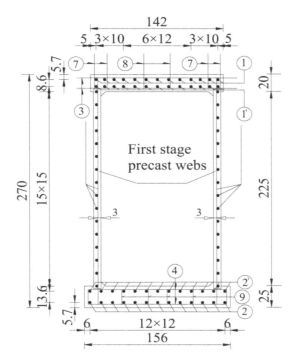

Note: Reinforcing identifications 1, 1', 2, and 2' indicate diameter 25-mm (#8) bars, and the remaining rebars are diameter 12-mm (#4) bars

FIGURE 9-52   Reinforcing Details of the Interior Box Arch Rib (unit = cm, 1 in. = 2.54 cm).

arch rib segments are shown in Fig. 9-55. The connection details for the exterior arch rib segments are similar. The typical reinforcing details at the end of the segments of the interior box arch rib are illustrated in Fig. 9-56.

After all the seven box arch ribs are erected in place as anticipated, six interior transverse diaphragms and two end transverse diaphragms are cast in place along the segment joints (see Figs. 9-54, 9-57, and 9-58). The width of the interior transverse diaphragms (i.e., the cast-in-place joints between the segments) is 71.6 cm (2.35 ft). The details of the interior diaphragm reinforcing are illustrated in Fig. 9-57. The width of the end diaphragms is 80.9 cm (2.65 ft). The connection details between the end of the box arch rib and the abutment are similar to those between the box arch rib segments (see Fig. 9-58). The entire ends of the arch ring are embedded into the abutment by 25 cm (10.0 in.) and connected together with the abutment by bolts to form fixed ends.

9.6.3.3 Spandrel Structures

The spandrel is an open structure with 17 equal spans of 10 m (32.81 ft) supported by reinforced concrete columns or short walls (see Fig. 9-59). The columns are rectangular in cross section with 1.00 × 1.30 m (3.28 × 4.27 ft) dimensions and all the columns are built on their corresponding footing beams with a minimum height of 1.4 m (4.59 ft) (see Fig. 9-60). Each cap is rigidly connected with its corresponding columns to form a frame structure with the footing beams. Because of the limited heights at locations of columns 7 to 10, short wall caps are used instead of vertical columns. Columns 1 to 6 and 11 to 16 as well as their caps are precast. They are connected by cast-in-place joints. The deck system consists of simply supported precast reinforced voided beams with four-span or five-span continuous reinforced concrete topping of 15 cm thick (5.91 in.) (see Fig. 9-61). The remaining principal dimensions of the spandrel can be seen in Figs. 9-59 to 9-61.

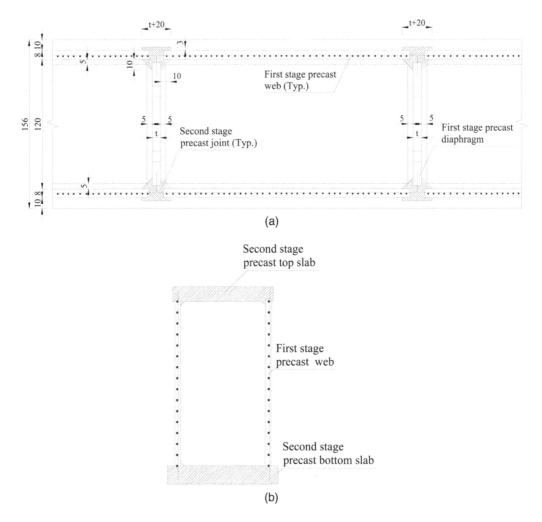

FIGURE 9-53   Assembling of Box Arch Rib, (a) Plan View, (b) Cross Section (unit = cm, 1 in. = 2.54 cm).

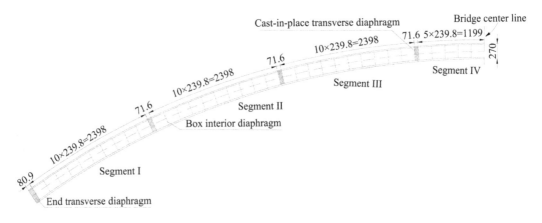

FIGURE 9-54   Segment Division of Box Arch Rib (unit = cm, 1 in. = 2.54 cm).

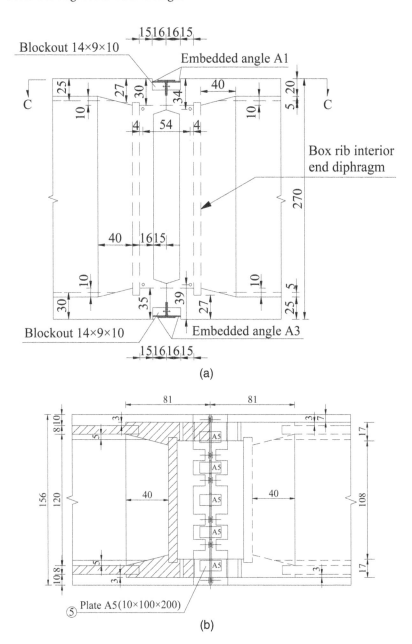

FIGURE 9-55 Longitudinal Connection Details of Interior Box Arch Rib Segments, (a) Elevation View of Arch Rib Connection, (b) Plan View of Arch Rib Connection (*C-C* View), (c) Details of Steel Angle A1 [Plan View (top), Elevation (bottom)], (d) Details of Steel Angle A3 [Plan View (top), Elevation (bottom)] (unit = cm, 1 in. = 2.54 cm, except the unit in callout *A1*, A3, and A5 are mm, in. = 25.4 mm).

(*Continued*)

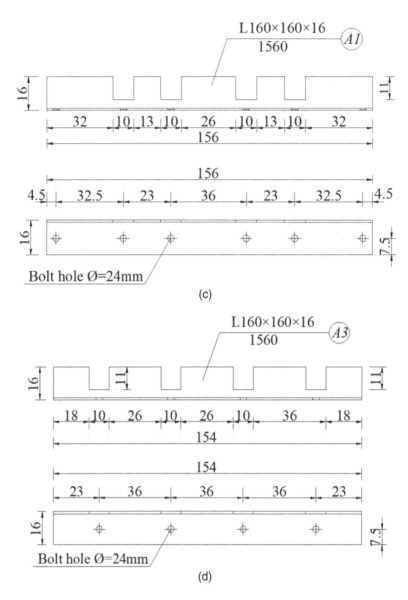

FIGURE 9-55   (Continued)

9.6.3.4   Skew Back and Abutment

Both ends of the arch rings are fixed at the abutments. The south side abutment is built on rock foundation (see Fig. 9-62), and the north side abutment is supported on a pile-supported foundation with a strut beam behind to resist the horizontal thrust (see Fig. 9-63).

9.6.4   BRIDGE ANALYSIS

9.6.4.1   Analytical Models

The bridge is modeled as a three-dimensional structure as shown in Fig. 9-64. The bridge structure is divided into a number of three-dimensional frame elements and analyzed using the finite-element method. All the arch ribs are assumed to be working together, and the columns are assumed to be

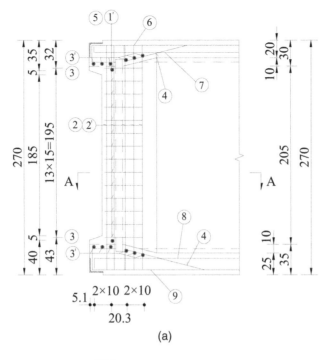

(a)

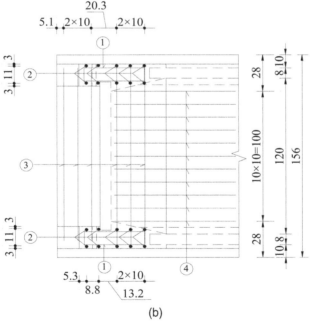

(b)

Note: See Fig. 9-52 for rebar numbering.

FIGURE 9-56 Reinforcement Details of Interior Box Arch Rib Segment Connection at Arch Ends, (a) Elevation View, (b) Plan View (*A-A* View) (unit = cm, 1 in. = 2.54 cm).

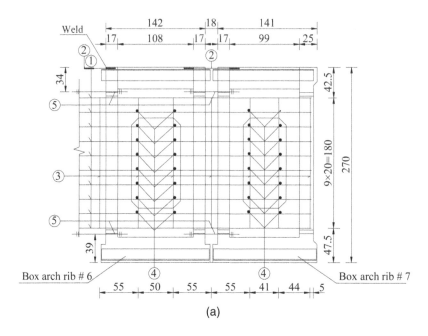

(a)

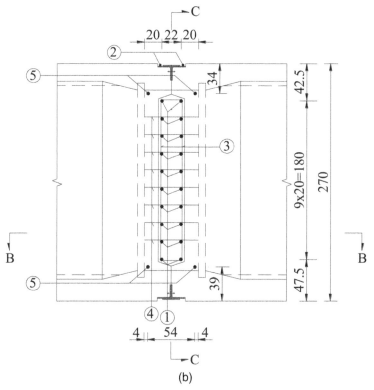

(b)

Note: See Fig. 9-52 for rebar numbering.

FIGURE 9-57 Details of the Interior Diaphragm of the Arch Ring, (a) Partial Elevation View (*C-C* View), (b) Side View (*A-A* View), (c) Plan View (unit = cm, 1 in. = 2.54 cm).

(*Continued*)

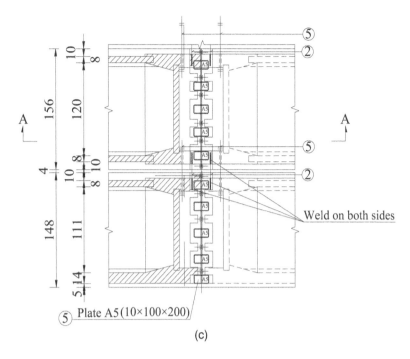

Weld on both sides

(5) Plate A5(10×100×200)

(c)

FIGURE 9-57   (Continued)

rigidly connected with the arch ring that is fixed to the abutments. The precast reinforced concrete slabs with cast-in-place reinforced concrete topping are treated as simply supported on the related caps, which are rigidly connected to the columns. Analytical results show the nonlinear effects are small and can be neglected for the completed arch bridge. The structural analysis is performed using the MIDAS Civil computer program.

9.6.4.2   Sectional Properties
The properties of some typical arch ring sections are calculated as follows:

Gross section area: $A_g = 12.08$ m^2 $\left(130.03 \text{ ft}^2\right)$
Reinforcement area $A_s = 1660.19$ cm^2 $\left(253.73 \text{ in.}^2\right)$
Moment of inertia about x-axis: $I_z = 12.195$ m^4 $\left(1412.93 \text{ ft}^4\right)$
Moment of inertia about y-axis: $I_y = 64.613$ m^4 $\left(7486.18 \text{ ft}^4\right)$
Distance from the neutral axis to top of the top slab: $y_t = 1.361$ m (4.465 ft)
Distance from neutral axis to bottom of the bottom slab: $y_b = 1.439$ m (4.720 ft)

9.6.4.3   Effects Due to Dead Loads
The effects due to dead loads were analyzed based on the assumed construction sequences as shown in Section 9.6.6.3. The permanent dead loads include the self-weight of the arch ring, column footing beams, columns, column caps, precast concrete void beams, concrete topping, traffic barrier walls, and pedestrian railings. The variations in the axial force, shear, and moment along the bridge span as a result of these dead loads are illustrated in Fig. 9-65.

9.6.4.4   Effects Due to Live Loads
The design live loads include a single truck or tandem with a uniform distributed lane loading as discussed in Section 2.2.3. The maximum absolute effects at the related sections can be

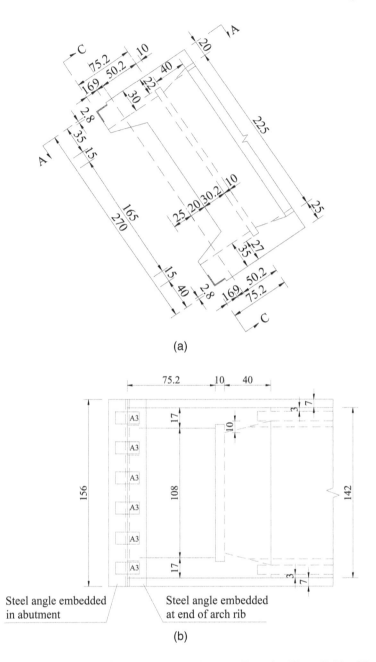

(a)

Steel angle embedded
in abutment

Steel angle embedded
at end of arch rib

(b)

FIGURE 9-58 Details of the End Diaphragm of the Arch Ring, (a) Elevation View, (b) Plan View (*A-A* View)
(unit = cm, 1 in. = 2.54 cm).

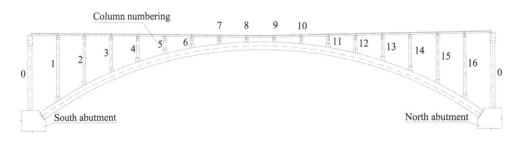

FIGURE 9-59 Arrangement of Spandrel and Column Numbering.

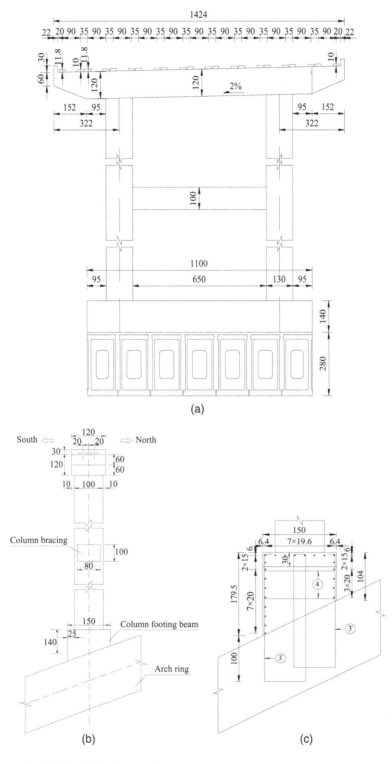

Note: See Fig. 9-52 for rebar numbering.

FIGURE 9-60 Typical Column Frame, (a) Elevation View, (b) Side View, (c) Connection of Footing Beam and Arch Rib (unit = cm, 1 in. = 2.54 cm).

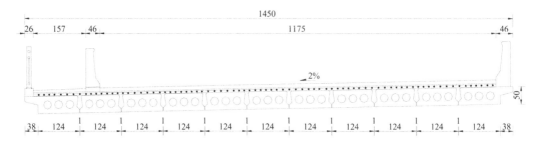

FIGURE 9-61 Typical Section of Bridge Deck (unit = cm, 1 in. = 2.54 cm).

determined by applying the live loads based on the shape of the influence line at the section of concern as shown in Fig. 9-29. From Fig. 9-29, it can be seen that we should only apply the live load on the first half of the span for obtaining maximum negative moment at the left end support of the arch and that we should apply the live load on the entire span for obtaining the maximum axial force at the skew back. The effects due to live loads are analyzed using the MIDAS Civil computer program, and the moment envelopes and the corresponding axial force and shear are shown in Fig. 9-66. The axial force envelopes and the variations of the corresponding moment and shear are illustrated in Fig. 9-67. In these figures, a dynamic load allowance of 1.33 is included. The dotted lines in these figures represent the maximum responses, and the solid lines indicate the minimum responses.

9.6.4.5 Effect Due to Temperatures

Generally, the effects of temperature gradient are small in deck arch bridges, and only the effects of uniform temperature rise and fall on arch bridges should be considered. The axial force, shear, and moment distributions due to temperature rise and fall given in Section 9.6.2.3 are illustrated in Fig. 9-68. In this figure, the solid and dotted lines represent the responses due to the rise and fall, respectively.

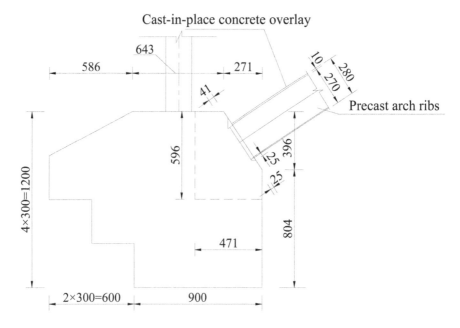

FIGURE 9-62 South Side Abutment (unit = cm, 1 in. = 2.54 cm).

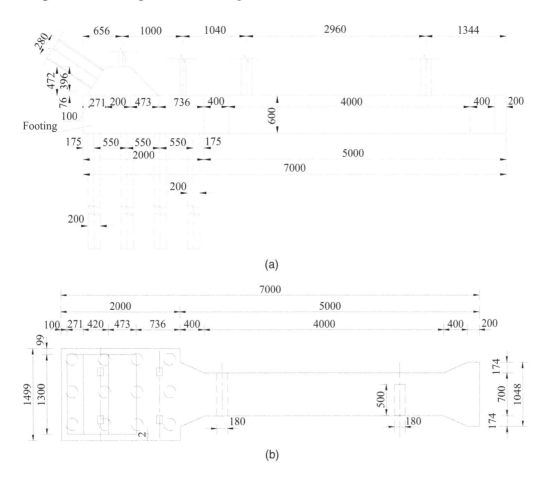

(a)

(b)

FIGURE 9-63 North Side Abutment, (a) Elevation, (b) Plan (unit = cm, 1 in. = 2.54 cm).

9.6.4.6 Secondary Effect of Creep and Shrinkage

The creep and shrinkage effects are evaluated based on the CEB-FIP *Model Code for Concrete Structures*, 1990, i.e., Eqs. 1-9b and 1-15b. To simplify the calculation, the entire arch ring is assumed to be erected on the 60th day after it was cast. The relative humidity is assumed as $H = 75\%$. The axial force, shear, and moment distributions due to creep and shrinkage at 10,000 days after casting time caused by dead loads are illustrated in Fig. 9-69.

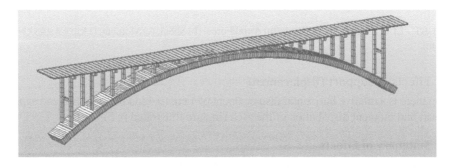

FIGURE 9-64 Bridge Analytical Model.

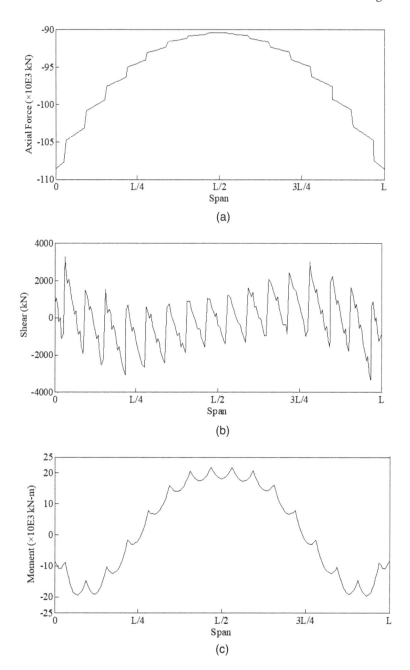

FIGURE 9-65   Effects Due to Dead Loads, (a) Axial Force, (b) Shear, (c) Moment, (1 kip = 4.448 kN, 1 kip-ft = 1.356 kN-m).

9.6.4.7   Effects of Support Displacement

Assuming there is a relative horizontal displacement of 1 cm (0.4 in.) between two end supports, the axial, shear, and moment distribution of the arch ring are illustrated in Fig. 9-70.

9.6.4.8   Summary of Effects

The maximum effects of the principal loadings at several critical sections are summarized in Table 9-4.

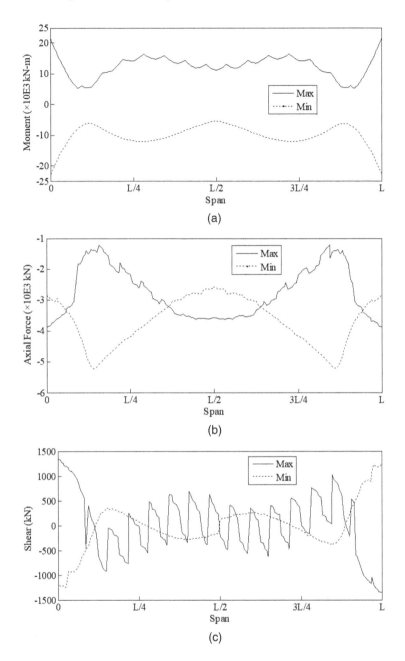

FIGURE 9-66 Moment Envelopes and the Corresponding Axial and Shear Due to HL-93 Live Loads, (a) Moment, (b) Axial Load, (c) Shear (1 kip = 4.448 kN, 1 kip-ft = 1.356 kN-m).

9.6.5 VERIFICATION OF REINFORCEMENT LIMITS AND CAPACITY OF ARCH RING

9.6.5.1 General

The arch ring is generally checked as an eccentrically loaded compression member as there are no prestressing and post-tensioning strands used for the arch ring. The capacity check of the arch ring is essentially the same as that for reinforced concrete structures. The minimum and maximum reinforcement check is to reduce the possibility of brittle failure. Its capacity should be checked at different construction stages, especially when the arch ring has not been closed and the deck system has not been built.

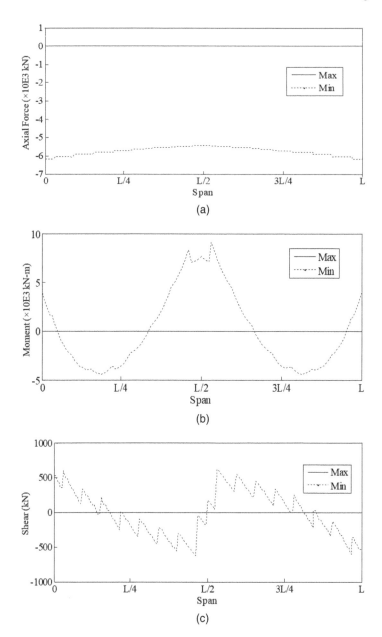

FIGURE 9-67 Axial Force Envelopes and the Corresponding Moment and Shear Due to HL-93 Live Loads, (a) Axial Force, (b) Moment, (c) Shear (1 kip = 4.448 kN, 1 kip-ft = 1.356 kN-m).

For simplicity, the capacity verification of the arch ring is focused on the completed structure. Different limit state requirements as described in Section 2.3.3.2 should be satisfied. Here, we consider Service I and Strength I limit states as an example to illustrate the capacity verification procedures.

9.6.5.2 Verification of the Reinforcement Limits

The ratio of the longitudinal reinforcement to the arch rib gross section is

$$\rho_s = \frac{A_s}{A_g} = \frac{254 \left(in^2 \right)}{18724 \left(in^2 \right)} = 0.014$$

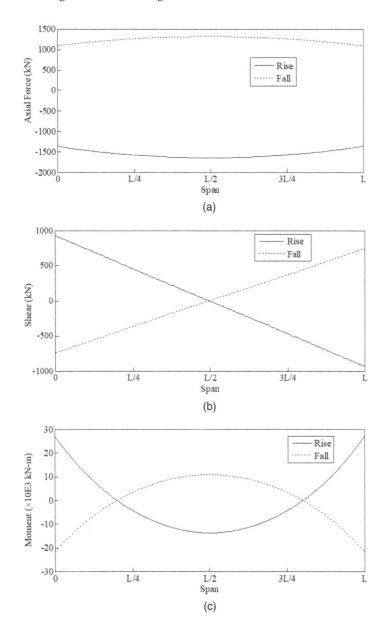

FIGURE 9-68 Effects of Uniform Temperature, (a) Axial Load, (b) Shear, (c) Moment (1 kip = 4.448 kN, 1 kip-ft = 1.356 kN-m).

From Eq. 3-231, the maximum reinforcement ratio is

$$\rho_s < \rho_{smax1} = 0.08 \ (\text{ok!})$$

From Eq. 3-233a, the minimum reinforcement ratio is

$$\rho_{smin} = 0.135 \frac{f_c'}{f_y} = 0.011$$

$$\rho_s > \rho_{smin} \ (\text{ok!})$$

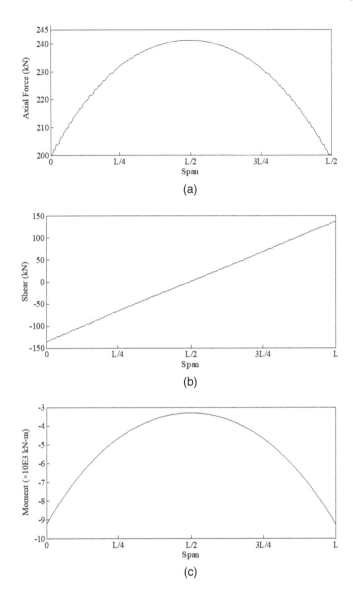

FIGURE 9-69 Effects of Concrete Shrinkage and Creep, (a) Axial Load, (b) Shear, (c) Moment (1 kip = 4.448 kN, 1 kip-ft = 1.356 kN-m).

As the permanent axial force of the arch rib is $N_D = 24409.4 \, (kips) < 0.4A_g f_c' (37448.1 \, kips)$, from Eq. 3-233b, the maximum reinforcement ratio is

$$\rho_{smax2} = 0.015$$

$$\rho_s < \rho_{sman2} \, (ok!)$$

9.6.5.3 Capacity Verification

9.6.5.3.1 Service Limit I

The Service I check is generally to control the crack width for the reinforced concrete segmental arch bridge. As discussed in Section 3.6.2.5.2, the AASHTO LRFD specifications limit the concrete

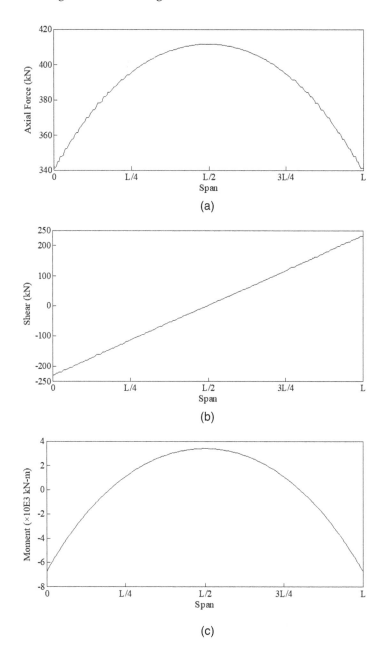

FIGURE 9-70 Effects of Support Displacement, (a) Axial Load, (b) Shear, (c) Moment (1 kip = 4.448 kN, 1 kip-ft = 1.356 kN-m).

crack width by determining the spacing of reinforcement based on the reinforcement tensile stress induced by the Service I load combination.

Considering the arch end section as an example, the maximum absolute bending moment and the corresponding maximum axial load at Service I are:

Maximums absolute Service I Moment

$$M_{s-I} = M_{DC} + M_{CR+SH} + M_{LL(1+IM)} + M_{SE} + 1.0M_{TU} + M_{PL} = 51522.9 \ kips\text{-}ft$$

TABLE 9-4

Summary of Maximum Load Effects in Girder Longitudinal Direction at Control Sections

Section	Internal Force	Dead Loads	HL-93			Shrinkage and Creep	Temperature		Support Displacement	Pedestrian	
			M_{max}	M_{min}	N_{min}		Rise	Fall		Max	Min
End	Moment	-8,497	21,421	-22,670	-3,888	-9,244	26,911	-21,529	-6,728	1,291	-1,197
	Axial	-108,573	-3,884	-2,829	-6,166	200	-1,365	1,092	341	-255	-187
	Shear	840	1,349	-1,265	532	-135	920	-736	-230	84	-67
L/8	Moment	-18,855	5,820	-6,291	-3,667	-6,490	8,105	-6,484	-2,026	223	-357
	Axial	-100,072	-1,511	-4,988	-5,921	219	-1,494	1,195	374	-85	-330
	Shear	-812	-615	134	135	-101	691	-553	-173	-15	15
L/4	Moment	-1,842	14,217	-11,842	-3,573	-4,675	-4,285	3,506	1,071	649	-707
	Axial	-94,515	-2,178	-4,181	-5,707	232	-1,582	1,295	396	-129	-265
	Shear	-1,284	-15	69	-179	-67	454	-372	-113	-10	2
3L/8	Moment	14,255	13,519	-9,801	2,045	-3,645	-11,319	9,055	2,830	618	-542
	Axial	-91,392	-3,305	-3,056	-5,535	239	-1,632	1,305	408	-194	-187
	Shear	-975	288	-258	-412	-32	218	-174	-54	9	-17
L/2	Moment	18,030	11,130	-5,331	7,627	-3,311	-13,599	10,879	3,400	476	-338
	Axial	-90,489	-3,580	-2,561	-5,436	241	-1,646	1,317	412	-198	-179
	Shear	-433	186	110	176	1	-8	6	2	-1	-1

Uni: Axial force and shear: kN; moment: kN-m; 1 kip = 4.448 kN, 1 kip-ft = 1.356 kN-m

Maximums Service I Axial Force

$$N_{s-I} = N_{DC} + N_{CR+SH} + N_{LL(1+IM)} + N_{SE} + 1.0N_{TU} + N_{PL} = 24720.3 \ kips \ (\text{compression})$$

The maximum tensile stress in the top layer of the reinforcement is

$$f_{ts} = \frac{M_{s-I}y_t}{I_z} + \frac{N_{s-I}}{A_g} = \frac{-51522.9 \times 53.58}{29298604.69} + \frac{24720.3}{18724.04} = 0.190 \ ksi \ (\text{compression})$$

The reinforcing bars are in compression, and no crack checking is necessary.

For the remaining sections and other load cases, the service checking can follow the same way as above. It was found that there is no tension in all the sections at the service loading.

9.6.5.3.2 Strength Limit

Strengths I to V and Extreme Events I and II need to be evaluated. Take Strength I, for example:

$$\text{Load combination: } 1.25DC + 1.25 \ CR + 1.25SH + 1.75 \ LL \ (1 + IM) + 1.75 \ BR + 0.5TU$$

9.6.5.3.2.1 Axial and Bending Strength

The axial and bending strengths of the arch ring can be obtained by following the procedures shown in Fig. 3-61. For the Strength I limit state, there is no lateral moment; the strength check can be done directly by developing the $P_n - M_n$ interaction diagram of the arch rib through available computer programs as follows (see Section 3.7):

1. Determine the maximum axial load capacity based on Eq. 3-229.

$$P_n = 0.8 \times \left[0.85f_c'(A_g - A_s) + f_y A_s \right] = 0.8 \times 0.85 \times 5.0 \times (18724.04 - 253.73)$$
$$+60 \times 253.73) = 74978.2 \ kips$$

The factored axial strength is

$$P_r = \phi P_n = 0.75 \times 74978.2 = 56233.65 \ kips$$

2. Calculate P_n and M_n for general loading cases by assuming a number of strain distributions.

Assume the yield concrete strain at extreme fiber as 0.003 and the strains of exterior rebar in tensile side as 0 (zero tension), 0.25, 0.5, 0.75, 1.0 (balanced failure) times of steel yield strain, respectively. Then we can determine the $P_n - M_n$ interaction diagram by the equilibrium conditions discussed in Section 3.7.2 (see Fig. 9-71).

3. Calculate ϕP_n and ϕM_n.

After determining resistance factor ϕ based on Eq. 2-53b, we can obtain the $\phi P_n - \phi M_n$ interaction diagram (see Fig. 9-71).

Taking the arch end section, for example, from Table 9-4, the maximum factored axial loads and moments for strength I are

Maximum Factored Axial Force

- Corresponding to the minimum absolute moment:

$$N_{u1} = 1.25N_{DC} + 1.25N_{CR+SH} + 1.75 \ N_{LL(1+IM)} + N_{SE} + 0.5N_{TU} + 1.75N_{PL} = 31442.7 \ kips$$

- Corresponding to the maximum absolute moment:

$$N_{u2} = 1.25N_{DC} + 1.25N_{CR+SH} + 1.75 \ N_{LL(1+IM)} + N_{SE} + 0.5N_{TU} + 1.75N_{PL} = 32160.7 \ kips$$

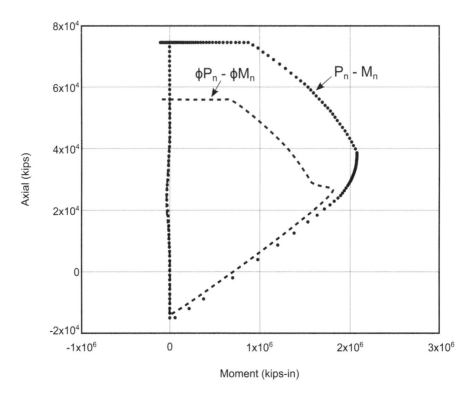

FIGURE 9-71   Interaction Diagram of the Arch Rib.

Maximum Absolute Moment

$$M_{u1} = 1.25M_{DC} + 1.25M_{CR+SH} + 1.75\ M_{LL(1+IM)} + M_{SE} + 0.5M_{TU} + 1.75M_{PL} = 720672\ kips - in$$

Minimum Absolute Moment

$$M_{u2} = 1.25M_{DC} + 1.25M_{CR+SH} + 1.75\ M_{LL(1+IM)} + M_{SE} + 0.5M_{TU} + 1.75M_{PL}$$
$$= 215020\ kips\text{-}in$$

From Fig. 9-71, we can see that the factored axial loads and moments, i.e., points $(7.2 \times 10^5, 3.2 \times 10^4)$ and $(2.2 \times 10^5, 3.1 \times 10^4)$, are within the ϕP_n - ϕM_n curve. The strength capacity of the arch rib is enough. The capacities for other sections can be done in a similar way. It was found the arch rib is adequately designed.

9.6.5.3.2.2 Shear Capacity   Take the arch rib end section, for example. The factored shear force at the arch end is

$$V_u = 1.25V_{DC} + 1.25V_{CR+SH} + 1.75\ V_{LL(1+IM)} + V_{SE} + 0.5V_{TU} + 1.75V_{PL} = 813.6\ kips$$

From the above calculation, we can see that the shear is comparatively small. From Eqs. 3-182b and 3-185b, the minimum concrete shear strength can be calculated as

$$V_{cw} = 0.06\sqrt{f_c'}d_v b_w = 0.06\sqrt{5.0} \times 102.37 \times 105.51 = 1449.1\ kips$$

From Section 2.3.3.5, the shear resistance factor is

$$\phi_v = 0.9$$

The factored shear resistance is

$$V_r = \phi V_{cw} = 1304.2 \ kips > V_u \ (\text{ok!})$$

It was found that the shear capacities of the arch rib at the remaining sections are enough.

9.6.5.3.3   Stability Check

9.6.5.3.3.1   In-Plane Stability

As mentioned in the aforementioned discussion, the AASHTO LRFD specifications do not provide clear requirements for the stability check. Based on the Chinese specification, the in-plane stability for the completed bridge can be checked using Eq. 9-81 as follows:

Conservatively neglecting the effects of the spandrel structure, the effective length of the arch rib can be calculated by Eq. 9-80 and Table 9-1 as

$$s_0 = k\frac{s}{2} = 0.69\frac{6733.95}{2} = 2323.21 \ (\text{in})$$

Radius of gyration:

$$r = \sqrt{\frac{I_x}{A_g}} = 39.56 \ \text{in}$$

The slenderness ratio is

$$\frac{s_0}{r} = \frac{2323.21}{39.56} = 58.73$$

From Table 9-2, we can find the stability coefficient

$$\lambda_s = 0.81$$

Thus, the allowable factored axial force at the end of the arch is

$$N_{cr} = \phi_a \lambda_s \left(0.85 f_c' A_c + f_y A_s\right) = 0.75 \times 0.81 \left(0.85 \times 5.0 \times (18724.04 - 253.73) + 60 \times 253.73\right)$$
$$= 56936.5 \ \text{kips}$$

The Strength I factored axial force at the end of the arch is

$$N_u = 1.25 N_{DC} + 1.25 N_{CR+SH} + 1.75 \ N_{LL(1+IM)} + N_{SE} + 0.5 N_{TU} + 1.75 N_{PS} = 32160.7 \ kips$$

Assuming the operation factor $\eta_I = 1.05$, we have

$$\eta_I N_u < N_{cr} \ (\text{ok!})$$

9.6.5.3.3.2   Out-of-Plane Stability

During construction, the lateral stability of the arch rib is critical and is typically provided by temporary supports. Based on previous experience, the out-of-plane stability of the completed arch bridges is usually adequate and it is not necessary to be checked if the ratio of the arch ring width to the span length is greater than $1/20$[9-3, 9-4]. The ratio for this bridge is

$$r_{bl} = \frac{B_{rib}}{l} = \frac{11.0}{160.16} = \frac{1.38}{20} > \frac{1}{20} \text{ (ok!)}$$

9.6.6 CONSTRUCTION

9.6.6.1 General Construction Sequences

The general construction sequences of an arch bridge is as follows:

Step 1: Construct abutments on both sides of the arch.
Step 2: Construct a temporary cable erection system.
Step 3: Erect the arch box ribs in the sequences as shown in Fig. 9-72.
Step 4: Cast the transverse diaphragms.
Step 5: Cast the longitudinal joints between the precast arch box ribs. Cast the full length of the longitudinal joints simultaneously from both skew backs to the arch crown, and cast the joints symmetrically about the crown. To avoid damaging the precast web during construction, the longitudinal joint should be poured in two layers, pouring the second pour after the first concrete pour has been preliminary consolidated.
Step 6: Pour the reinforced concrete topping of the arch ring symmetrically from both skew backs to the arch crown.
Step 7: Construct the column footing beams symmetrically from both abutments to the crown.
Step 8: Build the columns in the following sequences: (i) columns 2 and 15, (ii) columns 5 and 12, (iii) columns 7 and 10, (iv) columns 3 and 14, (v) columns 6 and 11, (vi) columns 4 and 13, (vii) columns 8 and 9, and (viii) columns 1 and 16. The column numbering can be found in Fig. 9-59. The construction sequences are to reduce the arch bending moment during construction.
Step 9: Construct the column caps using the sequences as discussed in step 8.
Step 10: Erect the precast reinforced concrete void slabs symmetrically from both bridge ends to the arch crown. First, the center three precast reinforced concrete slabs in the bridge cross section are installed for all the spans. Then, place one slab on each side of the previously installed center three slabs until the erection of all deck slabs are complete.
Step 11: Cast reinforced concrete deck topping.
Step 12: Install the traffic barrier walls and the pedestrian railing, and open the bridge to traffic.

9.6.6.2 Arrangement of Erection Cables System

The arch ribs are erected by the cable-supported cantilever method. The span length of the main cable is 316.3 m (1037.7 ft), and the height of towers is about 78 m (255.9 ft). The horizontal angle

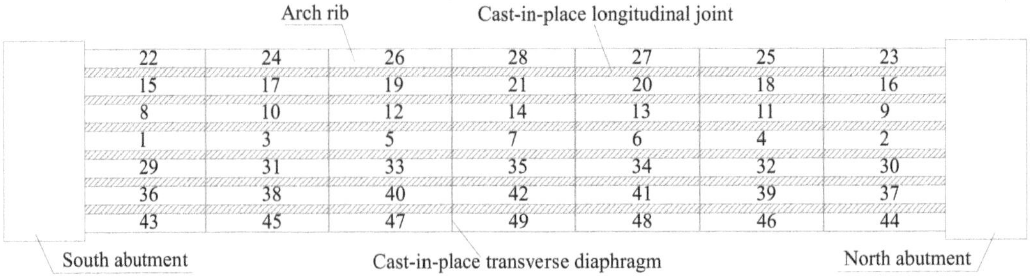

FIGURE 9-72 Erection Sequences of Arch Box Rib Segments.

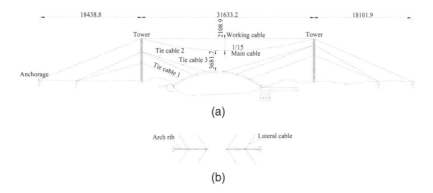

FIGURE 9-73 Erection Cable System, (a) Elevation View, (b) Part of Plan View (unit = cm, 1 in. = 2.54 cm).

of the tie cables generally ranges from 20° to 60°. The maximum erection weight of the main cable is 85 metric tons (93.7 tons), and the maximum design tension of the tie cables is 10 metric tons (11 tons). The design safety factors for all the cables are greater than 3.0. The principal dimensions of the erection cable system are illustrated in Fig. 9-73.

9.6.6.3 Erection Sequences of Arch Ribs

Consider a single arch rib as an example to describe the erection sequences.

Step 1: Erect box rib segments 1 and 7 (see Fig. 9-54 for segment numbering), temporarily connect the segments to the abutments, and use tie cable 1 to adjust the segments to the anticipated positions (see Fig. 9-74a).

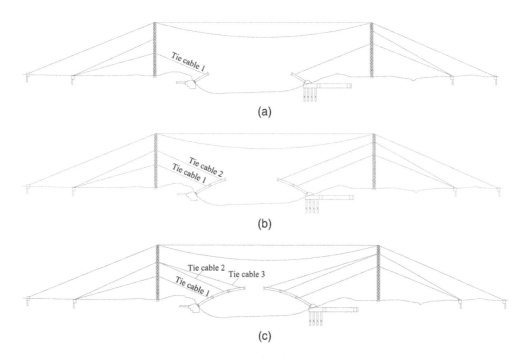

FIGURE 9-74 Typical Sequences for Erecting Arch Ribs, (a) Step 1, (b) Step 2, (c) Step 3, (d) Step 4, (e) Step 5.

(Continued)

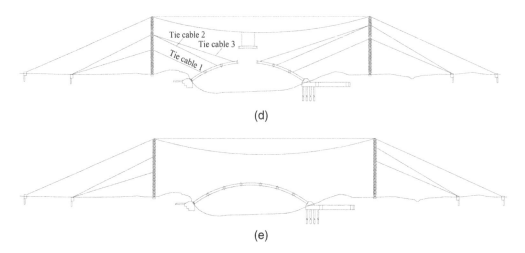

(d)

(e)

FIGURE 9-74   (Continued)

Step 2: Erect box rib segments 2 and 6. After the segments are placed in their anticipated positions, connect them to segments 1 and 7, respectively, by bolts and stabilize them with tie cable 2 and lateral cables (see Fig. 9-74b).

Step 3: Erect box rib segments 3 and 5. After they are placed in their anticipated positions, connect them to segments 2 and 6, respectively, by bolts and stabilize them with tie cable 3 and lateral cables (see Fig. 9-74c).

Step 4: Erect the closure box rib segment 4. After it is placed in the anticipated position, connect its ends to segments 3 and 5, respectively, by bolts (see Fig. 9-74d and Fig. 9-75).

Step 5: Adjust the final arch rib axis using the tie and lateral cables. Gradually release the tie cables and close the arch rib (see Figs. 9-74e, 9-76, and 9-77).

Step 6: Repeat steps 1 to 5 until completing the erection of all ribs (see Fig. 9-78).

FIGURE 9-75   Erection of the Closure Segment.

FIGURE 9-76   Adjusting Rib Axis by Tie and Lateral Cables.

FIGURE 9-77   Releasing Tie Cables.

FIGURE 9-78   Construction of Caps.

9.6.7 EXAMPLE REMARKS

The purpose of the design example is to illustrate the basic design procedures. It is not intended to provide the best possible design. The readers should also keep in mind that a good bridge alternative varies with different site conditions and locations.

REFERENCES

9-1. Xanthakos, P. P., *Theory and Design of Bridges*, John Wiley & Sons, New York, 1994.

9-2. Chen, W. F., and Duan, L., *Bridge Engineering Handbook*, CRC Press, Boca Raton, FL, 2000.

9-3. Shao, X. D., and Gu, A. B., *Bridge Engineering*, 3rd edition, China Communication Press, Beijing, China, 2014.

9-4. Liu, L. J., and Xu. Y., *Bridge Engineering*, 3rd edition, China Communication Press, Beijing, China, 2016.

9-5. Li, G. H., Xian, H. F., Sheng, Z. Y., Dong, S., and Huang, D. Z., *Stability and Vibration of Bridges*, 2nd edition, China Rollway Publishing House, Beijing, China, 1981.

9-6. Huang, D. Z., "Lateral Bending and Torsion Buckling Analysis of Jiujiang Yangtze River Arch-Stiffened Truss Bridge," *IABSE, 19th Congress*, Stockholm, 2016.

9-7. Huang, D. Z., "Lateral Stability of Arch and Truss Combined System," Master Degree Thesis, Tongji University, Shanghai, China, 1985.

9-8. Huang, D. Z., "Elastic and Inelastic Stability of Truss and Arch-Stiffened Truss Bridges" Ph.D. Degree Thesis, Tongji University, Shanghai, China, 1989.

9-9. Huang, D. Z., and Li, G. H., "Inelastic Lateral Stability of Truss Bridges," *Journal of Tongji University*, Vol. 10, No. 4, 1988, pp. 405–419.

9-10. Huang, D. Z., Li, G. H., and Xiang H. F., "Inelastic Lateral Stability of Truss Bridges with Inclined Portals," *Journal of Civil Engineering of China*, Vol. 24, No. 3, 1991, pp. 27–37.

9-11. Huang, D. Z., and Li, G. H., (1991) "Inelastic Lateral Stability of Arch-Stiffened Truss Bridges," *Journal of Tongji University*, Vol. 10, No. 4, 1991, pp. 1–11.

9-12. AASHTO, *LRFD Bridge Design Specifications*, 7th edition, Washington D.C., 2014.

9-13. Huang, D. Z., "Dynamic and Impact Behavior of Half-Through Arch Bridges," *Journal of Bridge Engineering*, ASCE, Vol. 10, No. 2, March 1, 2005, pp. 133–141.

9-14. Huang, D. Z., "Vehicle-Induced Vibration of Steel Deck Arch Bridges and Analytical Methodology," *Journal of Bridge Engineering*, ASCE, Vol. 17, No. 2, March 1, 2012, pp. 241–248.

9-15. Huang, D. Z., Wang, T-L, and Shahawy, M., "Dynamic Loading of a Rigid Frame Bridge," *Proceedings of papers presented at the Structures Congress '94*, Atlanta, GA, April 24–28, 1994, pp. 85–90.

9-16. AASHTO, *Standard Specifications for Highway Bridges*, 17th edition, Washington, D.C., 2002.

10 Design of Concrete Segmental Cable-Stayed Bridges

- Types and characteristics of cable-stayed bridges
- Determination of preliminary dimensions for main girders and pylons
- Static behaviors
- Stability and dynamic analysis
- Capacity verification
- Design example: segmental cable-stayed bridge

10.1 INTRODUCTION

A cable-stayed bridge consists of three basic load-bearing elements: cables, main girder, and pylons, which are also called towers (see Fig. 10-1). Though the basic concept of cable-stayed bridges in which a beam or bridge is supported by inclined cable stays may date back to the 18th century[10-1], the first concrete cable-stayed bridge was the Tempul Aqueduct crossing the Guadalete River in Spain[10-2, 10-3], built in 1925. In the last several decades, concrete cable-stayed bridges have been rapidly developed and have become a primary choice for long-span bridges. These types of bridges have been successfully used for span lengths ranging from 300 to 3340 ft (91 to 1018 m)[10-4], but they may be best suitable for span lengths ranging from 500 to 3000 ft (150 to 900 m).

Though there is no one particular type of bridge best suitable for all site conditions, environments, and problems, designers may consider the following advantages of segmental concrete cable-stayed bridges in selecting bridge alternatives:

a. Because the main girder is supported by a number of cables, except at the piers and abutments, girder moments are comparatively small and vary slightly along the longitudinal direction (see Fig. 10-2). The girder depth is typically designed as constant. The depth of the main girder can be very shallow, and the ratio of girder span length to depth can reach 250 or more. This system is especially suitable for long-span bridges due to the light self-weight.

b. The main girder is subjected to compression forces caused by the horizontal components of the inclined cable forces, which increase girder bending capacity (see Fig. 10-2e).

c. In comparison with steel deck systems, concrete segmental bridges have favorable damping behaviors and are less susceptible to aerodynamic vibrations.

d. It is comparatively simple to adjust the main girder internal forces to reduce the effects of concrete shrinkage, creep, construction errors, etc., through adjustment of the cable forces.

e. The erection of the super-structure is comparatively easy with current bridge construction techniques.

10.2 CATEGORIES AND STRUCTURAL SYSTEMS OF CABLE-STAYED BRIDGES

Cable-stayed bridges are typically classified into two categories: conventional and special. Conventional cable-stayed bridges are defined as those in which the ratios of the height of the pylon to the center span length of the main girder typically range from one-fifth to one-fourth and are greater than one-eighth. In addition, their cables are connected to the main girder and pylons.

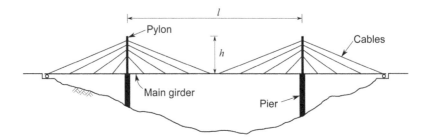

FIGURE 10-1 Components of Cable-Stayed Bridges.

The remaining cable-stayed bridges are considered special. Cable-stayed bridges can be further classified based on structural systems or arrangements of stay cables. Based on their distinguishing mechanical behavior, cable-stayed bridges may be categorized into the following five systems: floating, semi-floating, pylon-beam, frame, partial anchored, and extradosed systems[10-1, 10-5, and 10-6]. Each system has its advantages and disadvantages. Designers should select them based on the site conditions, environment, needs, etc. Types of cable-stayed bridges defined based on cable arrangement will be discussed in Section 10.3.2.

10.2.1 CONVENTIONAL CABLE-STAYED BRIDGES

10.2.1.1 Floating System

In a floating system, the main girder is completely supported by the stay cables and the end abutments (see Fig. 10-3a). The pylons are rigidly connected to the piers. There are no connections between the main girder and the pylons, except for lateral elastomeric bearing supports provided to limit girder lateral displacements. In the bridge longitudinal direction, spring supports with a high damping ratio are typically provided to limit girder longitudinal displacements. This system is often

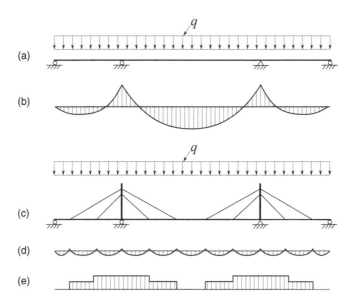

FIGURE 10-2 Comparison of Moment Distribution in Continuous Beams versus Cable-Stayed Bridges, (a) Three-Span Continuous Beam, (b) Moment Distribution in Continuous Beam, (c) Two-Pylon Cable-Stayed Bridge, (d) Moment Distribution in Cable-Stayed Bridge, (e) Distribution of Axial Force in Cable-Stayed Bridge.

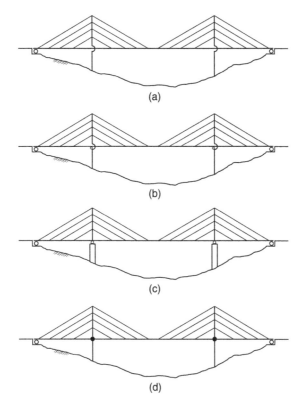

(a)

(b)

(c)

(d)

FIGURE 10-3   Systems of Conventional Cable-Stayed Bridges, (a) Floating System, (b) Semi-Floating System, (c) Pylon-Beam System, (d) T-Frame System.

used for bridges with span lengths greater than 1300 ft (400 m) and has the following advantages and disadvantages:

Advantages

a. Removes large negative moment in the main girder at the locations of the pylons.
b. Reduces the effects of temperature changes, shrinkage, and creep of concrete.
c. The variation of the moment in the main girder in the longitudinal direction is comparatively small and uniform.
d. Can absorb large earthquake loads due to the ability of the entire super-structure to move longitudinally with a long vibration period, thus having a decisive advantage in seismic resistance.

Disadvantages

a. Requires a temporary fixed connection between the main girder and the pylons if the cantilever construction method is used.
b. Because of the inevitable unbalanced forces experienced during construction, removing the temporary fixed connections may cause longitudinal movement of the main girder.

10.2.1.2   Semi-Floating System

In comparison with the floating system, an additional vertical support at each of the pylon locations (see Fig. 10-3b) is provided in the semi-floating system. The main girder as shown in Fig. 10-3b is a three-span continuous girder supported on some spring supports. Compared to the floating system, it possesses the following advantages and disadvantages:

Advantages

a. Stronger stiffness to resist the deformations induced by live loads.
b. Can have similar advantages to the floating system if the vertical supports on the pylons are designed as vertically adjustable or as spring supports.

Disadvantages

a. Higher negative moments in the main girder at the locations of pylons.
b. The effects of temperature, creep, and shrinkage will be higher if the supports at the locations of pylons are not vertically adjustable.

10.2.1.3 Pylon-Beam System

In the pylon-beam system, the pylons are rigidly connected to the main girder. The entire structure consisting of girder, pylons, and cables is similar to a beam simply supported on the piers and abutments (see Fig. 10-3c). The cables are like the external post-tensioning tendons in the aforementioned continuous segmental bridges. This system has the following advantages and disadvantages:

Advantages

a. Significantly reduces the tensile force in the middle region of the main girder.
b. Significantly reduces the forces caused by temperature changes.
c. Comparatively simple for structural analysis.

Disadvantages

a. Large displacement at the top of pylons may result due to the inclined deformation of the pylons when fully loaded at the midspan. This increases the deflection at midspan and the negative moment in side spans.
b. There are large reactions at the piers for long-span cable-stayed bridges, which may require high capacity supports and cause difficulty for future maintenance.

10.2.1.4 Frame System

In the frame system, the piers, pylons, and main girder are rigidly connected (see Fig. 10-3d). This system is often used in single-pylon cable-stayed bridges and has the following advantages and disadvantages:

Advantages

a. Eliminates high-capacity supports at the piers and facilitates maintenance.
b. Comparatively easy for construction.
c. Possesses higher stiffness resulting in a small deflection of the main girder.

Disadvantages

a. High negative moment at the rigidly connected joints between the pylon and the main girder.
b. The piers may need to be designed comparatively flexible to reduce the temperature forces.

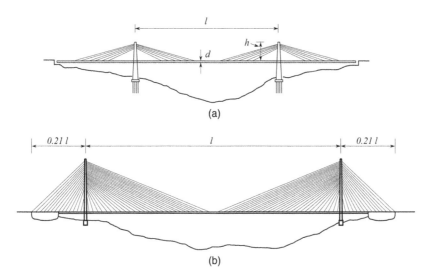

FIGURE 10-4   Special Cable-Stayed Bridges, (a) Extradosed Prestressed Bridge, (b) Partial Anchored Bridge.

10.2.2   SPECIAL CABLE-STAYED BRIDGES

10.2.2.1   Extradosed Prestressed System

Extradosed prestressed bridges are sometimes called partial cable-stayed bridges. Their economic span lengths range from 200 to 900 ft (60 to 270 m). Compared to a conventional cable-stayed bridge, this type of bridge has a smaller ratio of the height of pylon to the center span length (see Fig. 10-4a), which typically ranges from 1/15 to 1/8, i.e.,

$$\frac{h}{l} = \frac{1}{15} \ to \ \frac{1}{8} \tag{10-1}$$

where
  h = height of pylon
  l = center span length

Typically, h/l is about 1/10.

The behavior of this type of bridge lies between that of the conventional cable-stayed bridge and the external post-tensioned continuous girder bridge. The extradosed system has the following characteristics:

a. The girder depth of an extradosed bridge is greater than that for a conventional cable-stayed bridge, but smaller than a conventional girder bridge. The girder depth to midspan length ratio typically has a range of

$$\frac{d}{l} = \frac{1}{50} \ to \ \frac{1}{20} \tag{10-2}$$

  where d = girder depth and l = midspan length.
  At the towers, d/l is about 1/30.
b. Its cables are flatter than those in conventional cable-stayed bridges. The cables share the vertical live loadings, which are typically less than 30% of the total load.

c. Because of the larger stiffness of the main girder, the live load stress range of the cables is generally less than 7.5 ksi (52 MPa). The fatigue effect is comparatively small, and the cables can be designed as the external tendons in post-tensioned girder bridges and sized to prestress the deck.

d. The ratio of side span length to middle span length may range from 0.45 to 0.69[10-2] and is generally larger than 0.5 and typically 0.6.

10.2.2.2 Partial Anchor System

When the site conditions require that the ratio of side span length to midspan length be very small, some of the stay cables in the side spans may be anchored directly to the ground (see Fig. 10-4b). This type of cable-stayed bridge is often called a partial anchor system cable-stayed bridge.

10.3 GENERAL LAYOUT AND COMPONENTS DETAILING OF CONVENTIONAL CABLE-STAYED BRIDGES

10.3.1 SPAN ARRANGEMENTS

10.3.1.1 Arrangement of Two-Pylon Three-Span Bridges

10.3.1.1.1 Ratio of Side Span to Main Span

Most current cable-stayed bridges consist of three spans with two pylons as shown in Fig. 10-5a. This arrangement can provide a large main span and is often used for long-span cable-stayed bridges. To reduce the deflections and bending moment of the pylons due to live loads applied in the middle span, backstays (the cables connecting the tips of pylons to rigid points at the ends of the stayed spans, such as first piers or abutments) are often used (see Fig. 10-5). The backstays are also called end anchor cables (see Figs. 10-5 and 10-6). Live loads in the side spans decrease the tensile stress in the backstay cables, and live loads applied in the main span increase their tensile stress. Thus, the ratio of the side span to main span will significantly affect the stress range due to live load in the backstay cables and the cable fatigue life. The ratio of side span length to center span length for concrete segmental bridges typically ranges from 0.4 to 0.45 and does not exceed 0.5.

10.3.1.1.2 Ratio of Pylon Height to Main Span Length

The stay cable force decreases with the height of the pylon. A good ratio of pylon height to the center span length generally ranges from $\frac{1}{5}$ to $\frac{1}{4}$.

10.3.1.1.3 Ratio of Girder Depth to Main Span Length

The ratio of girder depth to main span length is related to the spacing of the cables and may range from $\frac{1}{200}$ to $\frac{1}{40}$. The smaller ratio corresponds to a small cable spacing, and the larger ratio corresponds to a large cable spacing.

10.3.1.2 Arrangement of Single-Pylon Two-Span Bridges

Single-pylon two-span cable-stayed bridges are also often used and are typically arranged in two unequal spans (see Fig. 10-5b). The longer span is called the main span, and the short one is called the side span. The ratio of the side span to main span may range from 0.5 to 1.0 and often ranges from 0.6 to 0.7.

10.3.1.3 Arrangement of Cable-Stayed Bridges with Four or More Spans

Cable-stayed bridges with four or more spans (see Fig. 10-5c) are rarely used, especially for long spans, due to the bridge stiffness consideration for the interior pylons without end anchor cables. To increase the stiffness, it often must incorporate a stiffened center pylon (see Fig. 10-5c). The ratio

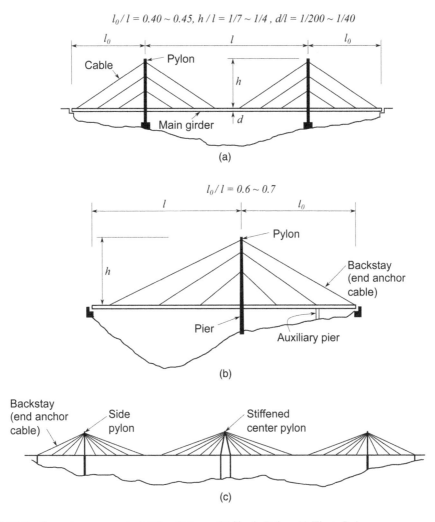

FIGURE 10-5   Span Arrangements, (a) Two Pylons, (b) Single Pylon, (c) Three Pylons.

of side span to interior spans is similar to that for the three-span cable-stayed bridges described in Section 10.3.1.1.

10.3.1.4   Auxiliary Piers and Approach Spans

To increase the vertical stiffness of the cable-stayed bridges as well as to reduce the bending moment of the side spans and rotation deformations at the end of the main girder, the use of auxiliary piers (see Fig. 10-6a) or extension of the main girder to an approach span on each of the side spans (see Fig. 10-6b) are often implemented. These arrangements are especially useful for long-span cable-stayed bridges in reducing the stress ranges and increasing the fatigue life of the cables. This transition span can also be employed, by its weight, to counteract the uplift forces of the backstays.

10.3.2   Layout of Cable Stays

10.3.2.1   Layout in the Longitudinal Direction

There are several types of longitudinal layouts of the cables. The harp, fan, and semi-harp patterns are comparatively more popular (see Fig. 10-7).

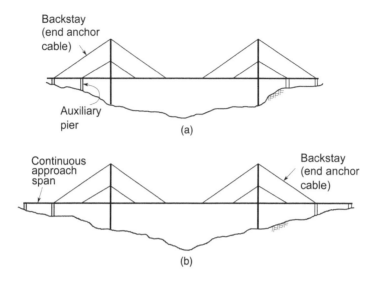

FIGURE 10-6 Auxiliary Piers and Continuous Approach Spans, (a) Arrangement of Auxiliary Piers, (b) Arrangement of Continuous Approach Spans.

10.3.2.1.1   Harp Pattern

The harp pattern is often used in cable layouts (see Fig. 10-7a). The cables are parallel to each other and cross the pylons at a constant angle. This pattern has the following advantages and disadvantages:

   Advantages

   a. Attractive aesthetic appearance
   b. Better distribution of the axial force in the pylons
   c. Simpler details of the connections between pylon and cables

   Disadvantage

   The angle between the cable and the main girder is comparatively small. Thus, the cables may not be most effective to resist the vertical loadings.

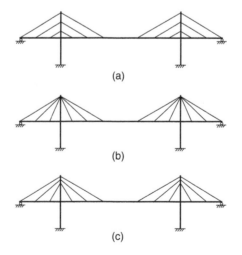

FIGURE 10-7   Typical Longitudinal Layout of Cable Stays, (a) Harp Pattern, (b) Fan Pattern, (c) Semi-Harp Pattern.

10.3.2.1.2   Fan Pattern

In the fan pattern, all cables are anchored at the top of the pylon (see Fig. 10-7b). This pattern offers the following advantages and disadvantages:

Advantages

a. Provides a large mean slope of the stays, which is more effective in resisting bridge vertical loadings. As a result, the total weight of the cables required is substantially lower than that for a harp pattern.
b. Reduces the horizontal forces induced by the cable
c. Increases the stability of the bridge against seismic activity, due to the greater flexibility of the structure

Disadvantages

a. Less attractive from the aesthetics point of view than the harp pattern.
b. High local concentrated stresses at the top of the pylon and complex design of the anchorages
c. Less stability of the pylon and the entire bridge structure

10.3.2.1.3   Semi-Harp Pattern

The semi-harp pattern falls somewhere between the harp and fan patterns (see Fig. 10-7c). It possesses the advantages of both the harp and fan patterns and is currently an often-used pattern for long-span bridges. By spreading the stays in the upper part of the pylon, it not only increases the cable efficiency, but significantly simplifies the design of the cable anchorages in the pylon.

10.3.2.1.4   Cable Spacing and Angle

The cable spacing for concrete segmental cable-stayed bridges typically ranges from 16 to 33 ft (5 to 10 m). The minimum cable anchorage spacing in the main girder is typically not less than 20 ft (6 m). The angle between the cables and the main girder ranges from 25° to 65°, but not smaller than 22°.

10.3.2.2   Layout in the Transverse Direction

In the transverse direction, the stay cables can be arranged in one plane (see Fig. 10-8a), two planes (see Fig. 10-8b), or three and more planes as needed. Theoretically, the stay cables in one-plane arrangements do not have the capacity to resist torsion and the main girder must have greater torsion capacity. The majority of the torsion can be resisted by the stay cables in a two-plane arrangement, and the cross sections of the main girder may have a smaller torsion capacity.

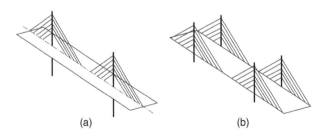

(a)   (b)

FIGURE 10-8   Typical Transverse Layout of Cables, (a) One Plane, (b) Two Planes.

10.3.3 General Design Considerations and Detailing of Main Girders

10.3.3.1 Functions and General Dimensioning of Main Girders

The principal functions of the main girder are to:

a. Transfer its dead and live loadings to the stay cables. The larger the stiffness of the girder, the larger the bending moment in the girder.
b. Resist the large longitudinal forces induced by the stay cables. The main girder must have enough stiffness to resist its buckling.
c. Transfer the forces due to the lateral wind and earthquake loadings to the substructure.

The main girder should have enough stiffness and strength to account for cable replacement and some unexpectedly cables.

Based on previous experience, the girder depth can be selected within the following ranges:
The ratio of girder depth d to bridge width b for the cable layout of two planes is

$$\frac{d}{b} \geq \frac{1}{10} \tag{10-3}$$

The ratio of girder depth to center span length generally ranges as

$$\text{For the cable layout of two planes: } \frac{d}{l} = \frac{1}{200} \ to \ \frac{1}{100} \tag{10-4}$$

$$\text{For the cable layout of a single plane: } \frac{d}{l} = \frac{1}{100} \ to \ \frac{1}{50} \tag{10-5}$$

10.3.3.2 Typical Sections

The typical section of a concrete segmental cable-stayed bridge may be classified as one of the following five types: solid slab (see Fig. 10-9), slab on girders (see Fig. 10-10), two-box (see Fig. 10-11), single-cell box (see Fig. 10-12), and multi-cell box (see Fig. 10-13)[10-5 to 10-8].

10.3.3.2.1 Solid Slab

Solid slab sections are suitable for cable-stayed bridges with comparatively short main span lengths up to about 492 ft (150 m), narrow deck widths up to about 66 ft (20 m), and with stay cables

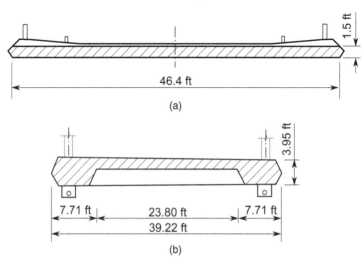

FIGURE 10-9   Solid Slab Sections, (a) Constant Slab Depth, (b) Stiffened.

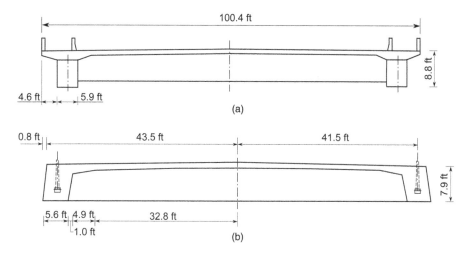

FIGURE 10-10   Slab on Girder Sections, (a) T-Girder, (b) Inclined Trapezoidal Girder.

arranged in two planes (see Fig. 10-9). This type of cross section is simple and has a favorable behavior in wind resistance. However, its torsional rigidity is weak and the sectional efficiency is low. To reduce its self-weight and increase its sectional efficiency, the slab may be designed as a stiffened slab (see Fig. 10-9b).

10.3.3.2.2   Slab on Girders

This type of cross section is suitable for cable-stayed bridges with two-plane stay cables. It consists of the girders, transverse beams/floor beams and deck (see Fig. 10-10). In comparison with the solid slab, it can be suitable for a wide deck and has better sectional efficiency. Most of all, it offers a greater simplicity in cantilever construction with much lighter weight of the traveling scaffold. The girders are built first, and then the cables are installed to support the final cast or precast transverse beam and deck.

10.3.3.2.3   Two-Box Section

This type of section consists of two separated boxes connected by the transverse beams and deck (see Fig. 10-11). The wind nose and slight inclined bottom face provide the lowest wind coefficients

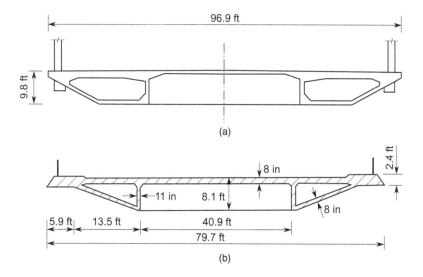

FIGURE 10-11   Two-Box Sections, (a) Four-Sided Box, (b) Three-Sided Box.

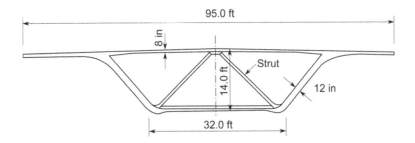

FIGURE 10-12   Single-Cell Box Section.

and the best provision for wind stability. It has comparatively large torsional rigidity and a light self-weight. This type of section is best suitable for long-span two-plane cable-stayed bridges.

10.3.3.2.4   Single-Cell Box

The single-cell box section as shown in Fig. 10-12 is often used for cable-stayed bridges with single-plane stay cables. The inside struts are employed to resist the distortional deformation of the large box section and to transfer the cable force to the main girder. The box section has large torsional rigidity and is most suitable for cable-stayed bridges with single-plane stay cables. The inclined webs not only offer good wind resistance, but also greatly reduce the pier cap width.

10.3.3.2.5   Multi-Cell Box

To increase the distortional rigidity, the cross section of the main girder can be designed as a two-cell box as shown in Fig. 10-13a. This section is suitable for the cable-stayed bridges with both single-plane or two-plane stay cables. For a bridge with a wide deck width, the girder section may be designed as a three-cell box section as shown in Fig. 10-13b. This type of section is more suitable for the cable-stayed bridges with single-plane stay cables. In the construction, the center cell may be cast first, and then the cables installed. Finally, the side cells are cast. In this way, the loading of the traveler scaffolds can be greatly reduced.

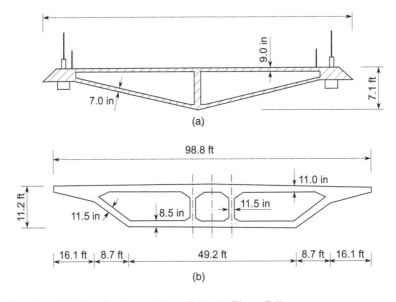

FIGURE 10-13   Multi-Cell Box Sections, (a) Two Cells, (b) Three Cells.

10.3.3.3 Preliminary Determination of Cross Sections

After the girder depth and type of typical section are determined, the preliminary dimensions of the girder section can be determined based on previous experience, such as previously designed bridges and the minimum required dimensions discussed in Section 1.4.2. The preliminary dimensions of the main girder and floor beams may also be estimated by meeting the requirements of bending moments and axial forces, which can be determined by some simplified analytical methods, such as using isolated local analytical models by applying local loadings. Further refinement of the cross-sectional dimensions can be done by the trial-and-error method until the section meets all requirements of service, strength, and buckling. The analysis of cable-stayed bridges will be discussed in Section 10.5.

10.3.3.4 Arrangement of Segments and Details

The determination of the length of segments is dependent on the construction methods, lifting equipment, etc. As discussed in Sections 1.4.1 and 6.2, for precast segments, the length typically ranges from 8 to 12 ft. For cast-in-place box segments, the segment length generally ranges from 10 to 16 ft using the balanced cantilever construction method.

For cable-stayed bridges, the weight of super-structure segments often ranges from 130 to 300 tons. For such heavy segments, easy access to the site must exist, such as by barge. Moreover, a significant initial investment for casting yard, forms, and lifting equipment to manufacture large precast segments is generally required. Thus, a sufficient number of segments are essential to ensure cost effectiveness of the precast construction method. For this reason, most current precast concrete cable-stayed bridges are comparatively long, or the cross section of the approach spans is the same as that of the main span.

10.3.3.4.1 Segment Arrangement and Detailing of Precast Cable-Stayed Bridges

Figure 10-14 shows a typical arrangement of the precast segments of a cable-stayed bridge with a main span length of 1200 ft. The typical arrangement is revised from an existing bridge located in Florida. There are three approach spans with equal span lengths of 240 ft at both ends of the main bridge. The typical section of the approach spans is the same as that of the main bridge. There are primarily four types of segments: typical segment, cable anchor segment, pier segment, and closure segment. The length of the typical segments is 12 ft, and the maximum weight of the segments is about 200 tons. The main dimensions and reinforcing of the typical segment and cable anchor segment are shown in Figs. 10-15 and 10-16, respectively.

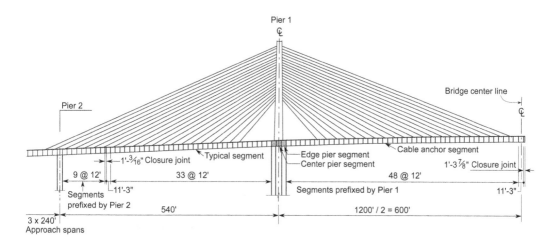

FIGURE 10-14   Segment Arrangement of a Precast Cable-Stayed Bridge.

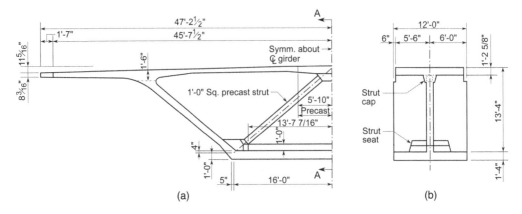

FIGURE 10-15 Cross Section of the Typical Segment of a Precast Cable-Stayed Bridge, (a) Elevation View, (b) Section *A-A* View.

There are three pier segments over pier 1 (see Fig. 10-14). These three pier segments transfer the large forces from the pylon to the pier and were constructed in two phases. The precast pier shell segments were first erected (see Fig. 10-17), and then the concrete in the core area was cast in place (see Fig. 10-18). The edge pier segments rest directly on the piers and are rigidly connected to the piers by post-tensioning tendons. The center segment rests on the piers through two steel beams as shown in Fig. 10-19. The three pier segments are connected by post-tensioning tendons after the strength of the core concrete reaches the required value.

The main girder is supported by stay cables arranged in one plane, and the typical connection details with the stay cable are shown in Fig. 10-20. Some post-tensioning tendons are used to resist the large tensile force from the cables (see Fig. 10-20). To reduce the weight of the segments, the top slab, bottom slab, and the webs are also post-tensioned (see Fig. 10-21). In the areas of piers 1 and 2, as well as at midspan, some longitudinal post-tensioning tendons are arranged to resist the bending moments. The locations of the tendons in cross section are shown in Fig. 10-16a. Their locations in the longitudinal direction can be arranged similar to the locations shown in Figs. 5-44, 6-7, and 6-9.

10.3.3.4.2   Segment Arrangement and Detailing of Cast-in-Place Cable-Stayed Bridges

Figure 10-22 shows a typical arrangement of segments for a cast-in-place concrete cable-stayed segmental bridge. The typical arrangement is revised from an existing bridge located in Florida. The center span length of the cable-stayed bridge is 1300 ft with a typical section of slab-on-beams (see Fig. 10-23). The entire bridge is divided into 76 segments with a typical segment length of 17.5 ft. There are mainly three types of segments: typical segment, stay cable anchor segment, and pier segment. Each of the typical segments has two floor beams, and each of the stay cable anchor segment has one floor beam. The edge beams are cast in place, and the floor beams are precast. The floor beams are connected to the edge beams by post-tensioning tendons (see Fig. 10-23). After the related stay cables are installed, then cast the deck. The details of the stay cable anchor segment are shown in Fig. 10-24.

10.3.3.5   Anchor Types of Stay Cables in Main Girders

Stay cables can be anchored inside of the box girder through anchor blocks or diaphragms (see Fig. 10-25a and b). For cable-stayed bridges with a two-plane cable arrangement, the stay cables may be directly anchored at the bottom of the main girder through anchor blocks or diaphragms (see Fig. 10-25c to f).

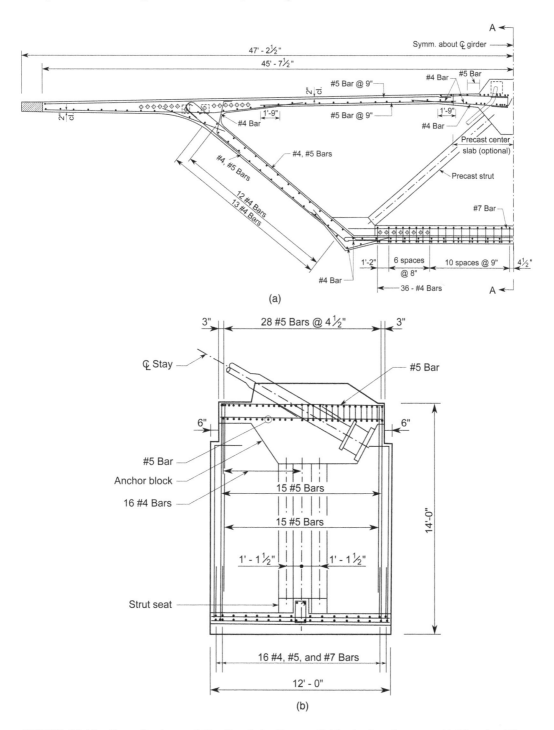

FIGURE 10-16 Cross Section and Details of the Precast Cable Anchor Segment, (a) Elevation View, (b) Section *A-A* View.

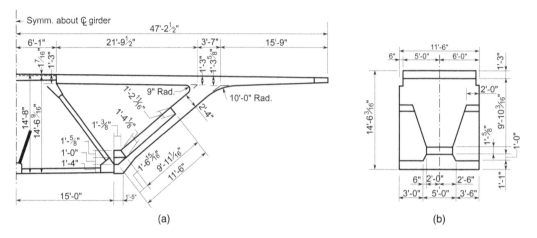

FIGURE 10-17 Typical Section of the Exterior Pier Segment under Pylon of Pier 1, (a) Elevation View, (b) Section *A-A* View.

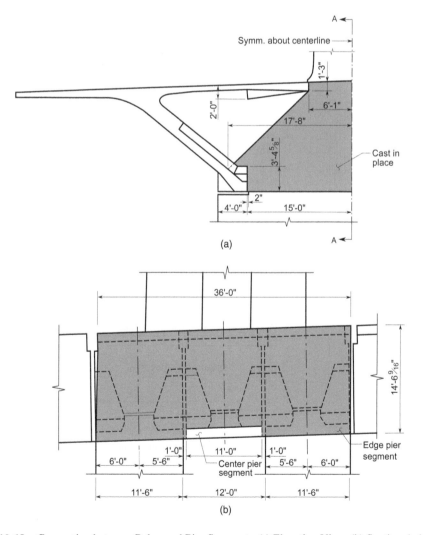

FIGURE 10-18 Connection between Pylon and Pier Segments, (a) Elevation View, (b) Section *A-A* View.

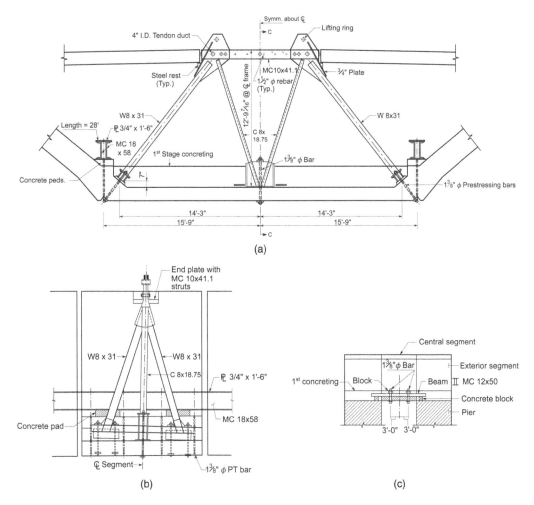

FIGURE 10-19 Typical Section of the Center Pier Segment under Pylon of Pier 1, (a) Elevation View, (b) Section *C-C* View, (c) Erection of the Center Pier Segment.

10.3.4 Layout and Detailing of Pylons

10.3.4.1 Introduction

The pylons or towers of a cable-stayed bridge are subjected not only to large axial forces but also to large bending moments due to their self-weights, forces from inclined cables, transverse and longitudinal forces from the deck system, as well as from wind and earthquakes. The tower may be the most important part in a cable-stayed bridge to display its aesthetic feature, and various types and shapes of towers have been developed.

10.3.4.2 Layout in Longitudinal Direction

To avoid a large bending moment being transferred to the foundation and reduce the size of the substructure, the pylon should be slender and typically designed as a column shape in the bridge longitudinal direction (see Fig. 10-26a). This type of pylon is simple, but is relatively weaker in resisting longitudinal moments. If it is necessary to increase pylon longitudinal stiffness, A-shapes or inverted Y-shapes may be used (see Fig. 10-26b and 10-26c) and a twin-pier system (employing dual columns) may be designed.

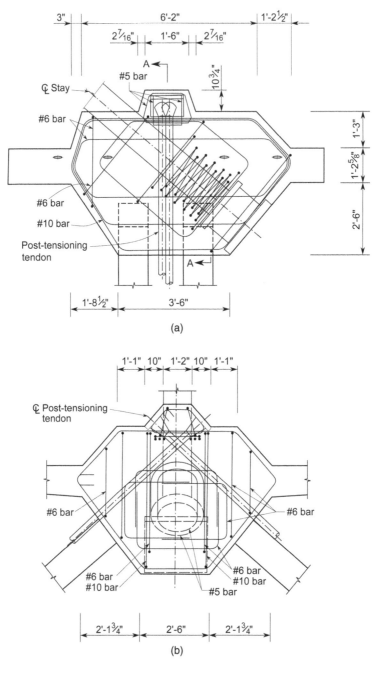

FIGURE 10-20 Principal Dimensions and Details of Single-Plane Stay Cable Anchorage, (a) Elevation View, (b) Section *A-A* View.

10.3.4.3   Layout in Transverse Direction

Though there are many shapes of pylons used in the bridge transverse direction, they may be classified into three types: column shape, inverted Y-shape, and diamond shape for bridges with single-plane stay cables (see Fig. 10-27). For cable-stayed bridges with two-plane cables, they may be categorized as double-column, frame, "H," "A," diamond, and revised diamond shapes (see Fig. 10-28). For long

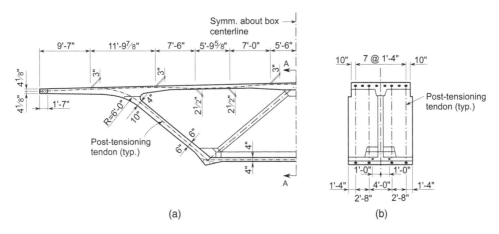

FIGURE 10-21   Arrangement of Post-Tensioning Tendons in Box Section Components, (a) Elevation View, (b) Section *A-A* View.

span cable-stayed bridges in regions with strong winds, the "A," diamond, and revised diamond shapes are often used to resist the large wind forces.

10.3.4.4   Typical Section of Pylons

The typical sections of pylons may be divided into two types: solid and hollow sections (see Fig. 10-29). The shapes of the cross sections can be square, octagonal, elliptical, etc., as needed for aesthetic considerations. For long-span bridges, hollow sections are typically used.

10.3.4.5   Stay Cable Anchor Types in Pylons

There are typically four types of cable anchors in the pylon: intercross external anchor (see Fig. 10-30a and b), internal web anchor (see Figs. 10-30c and 10-30d), steel beam anchor (see Fig. 10-30e), and stay saddle anchor (see Fig. 10-31).

10.3.4.5.1   External Anchor Block

The intercross external anchor method is comparatively simple and mainly suitable for cable-stayed bridges with short- or medium-span lengths and solid towers. As both the horizontal and vertical forces produced by the pair of front and back stay cables are compression (see Fig.10-30a), it is not necessary to provide horizontal post-tensioning tendons, and the design of the tower is comparatively simple.

10.3.4.5.2   Internal Anchor Block

The internal anchor method means the anchor blocks are built inside of the pylon box section and suitable for long-span cable-stayed bridges. The large horizontal tensile force in the pylon web is resisted by post-tensioning tendons (see Fig. 10-30d).

10.3.4.5.3   Steel Anchor Beam

Figure 10-30e illustrates a typical steel anchor beam device. Two steel cable anchorages are mounted at each end of the steel beam, which resists the horizontal forces induced by the inclined cables. The steel beam is supported on the corbels in the tower interior web. This type of anchor device greatly reduces the tensile force in tower web and is suitable for long-span cable-stayed bridges.

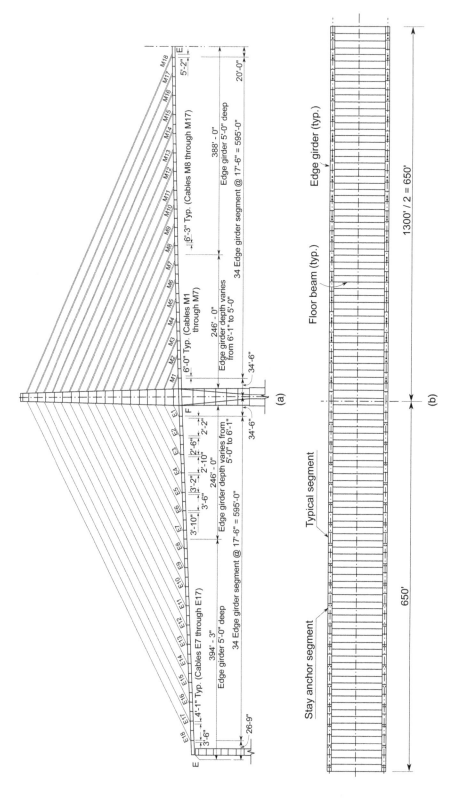

FIGURE 10-22   Segment Arrangement of a Cast-in-Place Cable-Stayed Bridge, (a) Elevation View, (b) Plan View.

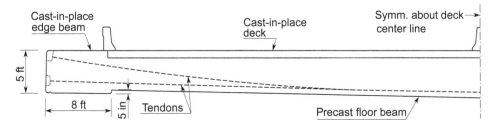

FIGURE 10-23   Typical Section of Main Girder.

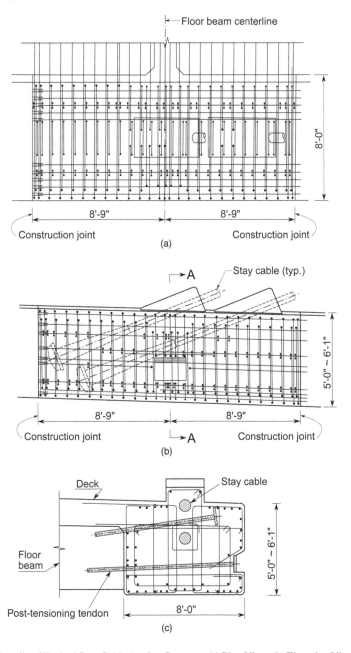

FIGURE 10-24   Details of Typical Stay Cable Anchor Segment, (a) Plan View, (b) Elevation View, (c) Section *A-A*.

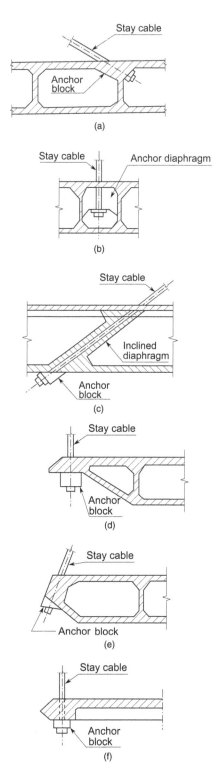

FIGURE 10-25   Types of Stay Cable Anchors in Main Girders, (a) Anchor Block under the Bottom of Top Slab, (b) Anchor Block with Diaphragm, (c) Anchor Block at Bottom of Box Girder through Diaphragm, (d) Anchor Block on Exterior of Box, (e) Anchor Block in Thickened Exterior Web, (f) Anchor Block under Edge Beam [10-5].

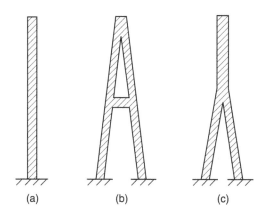

FIGURE 10-26 Pylon Shapes in Bridge Longitudinal Direction, (a) Column Shape, (b) A-Shape, (c) Inverted Y-Shape.

10.3.4.5.4 Stay Saddle

Figure 10-31 shows a typical stay saddle configuration. The stay cables run through the steel pipes that are embedded in the pylon. It generally requires equal numbers of strands on both sides of the pylon. Though this type of anchorage is simple, it requires the pace of cantilever construction on both sides of the tower to always be symmetrical before a cable may be installed and increases construction costs.

10.3.4.5.5 Steel Link Box

Many manufacturers have developed semi-standard steel link boxes as shown in Fig. 10-32 (DYNA Grip Link Box)[10-9]. The stay cables are anchored at each ends of the box outside of the pylon (see Fig. 10-32c and d). The forces are transferred by the shear studs welded to the outside surface of the anchor box flanges (see Fig. 10-32b).

10.3.4.6 Design Examples of Pylons and Aesthetics

As previous mentioned, the pylons have a governing influence on the overall aesthetics effect of a cable-stayed bridge. Though the aesthetics of a pylon may increase the construction cost, in many cases only minor modifications are needed to achieve satisfactory aesthetics.

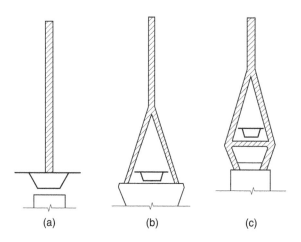

FIGURE 10-27 Shapes of Pylons for Cable-Stayed Bridges with Single-Plane Stays (a) Column Shape, (b) Inverted Y-Shape, (c) Diamond Shape.

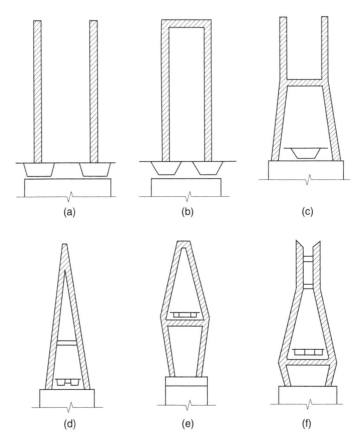

FIGURE 10-28 Shapes of Pylons for Cable-Stayed Bridges with Two-Plane Stays, (a) Double-Column Shape, (b) Frame Shape, (c) H-Shape, (d) A-Shape, (e) Diamond Shape, (f) Revised Diamond Shape.

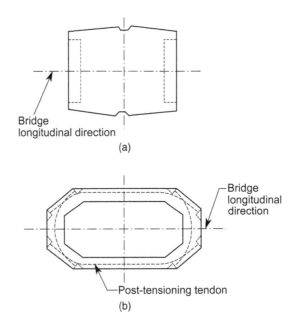

FIGURE 10-29 Types of Pylon Cross Sections, (a) Solid, (b) Hollow.

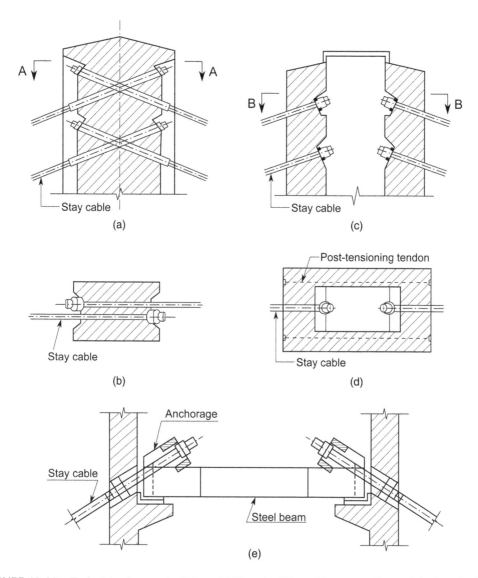

FIGURE 10-30 Typical Anchorages in Pylons, (a) Elevation View of Intercross External Anchor, (b) Plan View of Intercross External Anchor (Section *A-A* View), (c) Elevation View of Internal Web Anchor, (d) Plan View of Internal Web Anchor (Section B-B) View, (e) Steel Beam.

Figure 10-33 shows the arrangement of a pylon for the cable-stayed bridge shown in Fig. 10-14. The typical section of the pylon is a hollow octagonal with a constant width of 11 ft and the length varying from 14 ft at the top to 23 ft at the bottom (Fig. 10-33c). The pylon is rigidly connected to the pier segments and twin piers by post-tensioning tendons. The pier shaft has a hollow elliptical cross section (Fig. 10-33d). The stay cables are anchored on the saddles mounted in the pylon (see Fig. 10-31).

Figure 10-34 illustrates the arrangement of a pylon for the cable-stayed bridge shown in Fig. 10-22. The pylon is a frame-type structure consisting of two columns and three struts. The cross section of the column is rectangular with a constant width and a length varying from 15 ft at the top to 32 ft at the bottom.

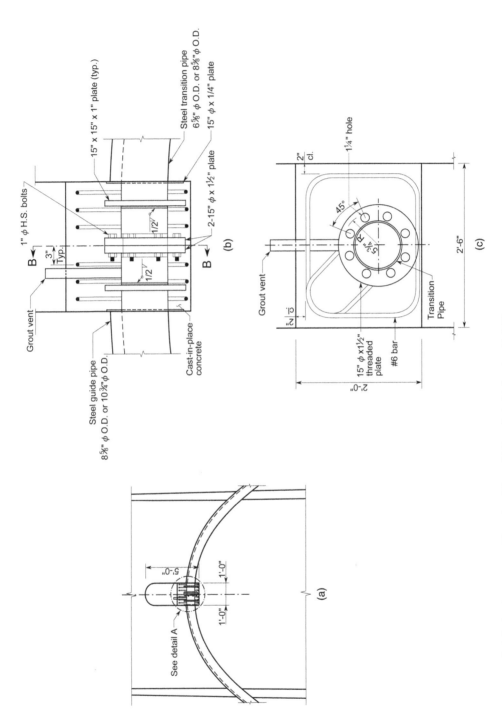

FIGURE 10-31   Stay Saddle Anchorages, (a) Elevation View, (b) Detail A, (c) Section *B-B* View.

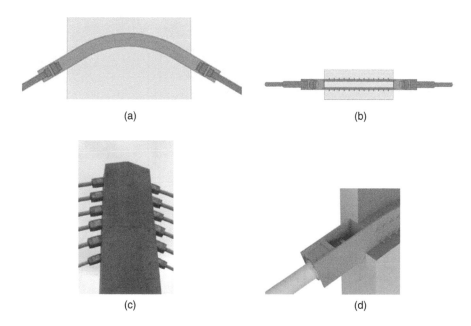

(a)   (b)

(c)   (d)

FIGURE 10-32 Steel Link Box, (a) Elevation View, (b) Plan View, (c) External Installation, (d) Opening for Installation (Courtesy of DYWIDAG International).

Using inclined arms and curved outlines of the towers may greatly enhance the aesthetic features (see Fig. 10-35). However, these may increase the construction difficulty and cost.

10.3.5 Types and Detailing of Stay Cables

10.3.5.1 Types of Stay Cables and General Material Requirements

The stay cables are the most important members in cable-stayed bridges. They must be designed to perform safely against fatigue, to be durable, and to be protected against corrosion. The stay cables most often used currently in cable-stayed bridges may be grouped into three types: parallel-bar cables, parallel-wire cables, and strand cables (see Fig. 10-36). The selection of the cable type may depend on the mechanical properties required and on the structural and economic criteria. The stay cable material should conform to ASTM related specifications. The minimum ultimate tensile stresses recommended by PTI[10-10] for bars, wires, and strands cables are 150 ksi (1035 MPa), 240 ksi (1655 MPa), and 270 ksi (1860 MPa), respectively. The specimens of bars, wires, and strands shall be tested at an upper stress of $0.45f_u$ and $0.55f_u$ for fatigue loads. At least 30% of the specimens shall be tested at $0.55f_u$. The stress ranges of bars, wires, and strands shall be 23.6 ksi (105 MPa), 43.6 ksi (194 MPa), and 35.7 ksi (159 MPa), respectively, for a fatigue life of 2 million cycles[10-10]. After the fatigue life requirement is reached, a static test to failure shall be conducted to ensure that at least a minimum value of 0.95 ultimate tensile strength (MUTS) is provided.

10.3.5.1.1 Parallel-Bar Cables

Parallel-bar cables consist of high-strength steel bars parallel to each other and enclosed in metal ducts (see Fig. 10-36a). The high-strength steel bars should conform to ASTM Standard Specifications A722 for prestressing concrete. Some polyethylene spacers are used to keep them in position in the ducts. Cement grout is injected into the ducts after erection to keep the ducts and

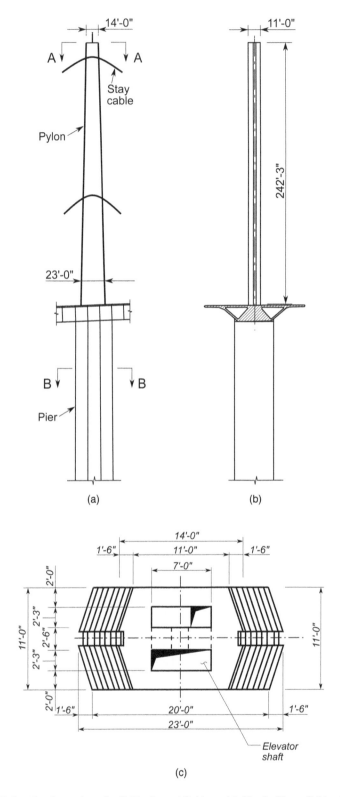

FIGURE 10-33 Pylon Configuration of a Cable-Stayed Bridge with Single-Plane Cable, (a) Elevation View, (b) Side View, (c) Section *A-A* View, (d) Section *B-B* View.

(Continued)

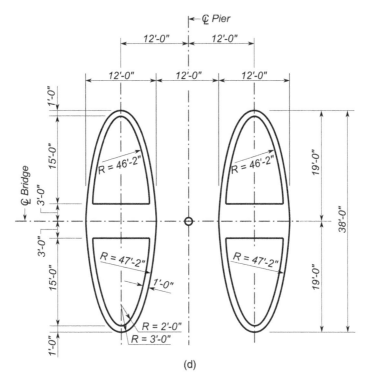

(d)

FIGURE 10-33　(Continued)

the bars working together. The available sizes and properties of Grade 150 high-strength bars are shown in Table 1-5. For large-diameter bars, due to transportation difficulty, the length of the bar is generally not greater than 60 ft (18 m). Thus, couplers must be used to extend the bar lengths in the field, which greatly reduces the fatigue capacity. For this reason, parallel-bars are currently rarely used in long-span cable-stayed bridges.

10.3.5.1.2　Parallel-Wire Cables

Parallel-wire cables are formed of high-strength, drawn steel wires, placed parallel to each in metal or polyethylene ducts. The ducts are typically injected with cement grout after erection. Currently, in the United States, the wire cables are replaced by strand cables for both post-tensioning and stay cable applications, and U.S. manufacturers have stopped making these types of cables. A few manufacturers in other countries produce ASTM A421 wires. The diameters of the wire are generally 5 mm and 7 mm, and the number of wires in the cable typically ranges from 50 to 421. The typical dimensions and mechanical properties of parallel-wire cables with a wire diameter of 7 mm are provided in Table 10-1.

TABLE 10-1

Capacities of Typical Parallel-Wire Cables (diameter = 7 mm)[10-1]

Number of wire	1	61	91	121	163	211	253	313
Nominal area (mm²)	38.5	2,348.5	3,503.5	4,658.5	6,275.5	8,123.5	9,740.5	12,050.5
Ultimate strength (N/mm²)	1,670	1,670	1,670	1,670	1,670	1,670	1,670	1,670
Ultimate load (kN)	64.3	3,922	5,850.8	7,779.7	10,480.1	13,566.3	16,266.6	20,124.3

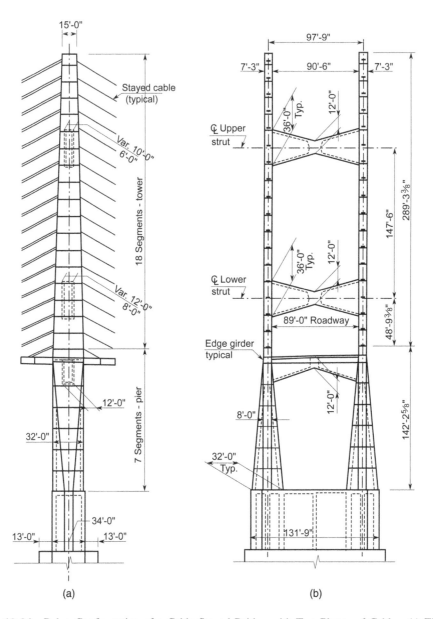

FIGURE 10-34 Pylon Configuration of a Cable-Stayed Bridge with Two Planes of Cables, (a) Elevation View, (b) Side View.

10.3.5.1.3 Strand Cables

Currently, strand cables are most often used in cable-stayed bridges. The strand cables should conform to ASTM A416 standard specifications for steel strand, uncoated seven-wire for prestressing concrete, and low-relaxation grade. Each strand consists of seven parallel or twisted wires with a diameter of 0.5 to 0.7 in. (12.7 to 17.78 mm) as discussed in Section 1.2.3.2.1. Each manufacturer may develop its own types of stay cables and anchors. Most stay cable systems use Grade 270 (1860 MPa) with a strand diameter of 0.6 in. (15.2 mm). Cables with up to 169 strands are available as standard products. Table 10-2 shows the capacities of typical multi-strand stay cables with a strand diameter of 0.6 in. (15.2 mm).

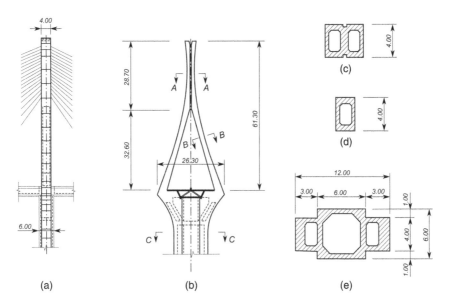

FIGURE 10-35 Curved Arms of a Pylon, (a) Elevation View, (b) Side View, (c) Section *A-A* View, (d) Section *B-B* View, (e) Section *C-C* View (unit = m, 1 m = 3.281 ft)[10-1].

10.3.5.2 Anchorages

The basic working principle of anchorages for the stay cables is the same as that for the prestressing tendons discussed in Section 1.7. However, the stay cables act as vertical supports and take up directly and completely the dead, live, and wind loads acting on the structures. The anchorage is vitally important to ensure structural durability, serviceability, and safety. Anchorages for stay cables are typically proprietary products of cable suppliers and are generally not fully detailed on the contract plans. The bridge designers must provide enough space and access for installing the stay cables and jacking equipment. Thus, designers should obtain the design information provided in the suppliers' product literature before sizing the related details. During construction, the designers will receive shop drawings and test information provided by the selected cable supplier for review and approval. Currently, the stay cable anchor systems provided by the cable suppliers generally have high levels of performance, with good protection against corrosion, high resistance to axial and bending fatigue loads, simple tensioning and re-tensioning, easy to maintain and replace and stay cable vibration control, etc. There are two types of stay cable anchorages, adjustable and fixed anchorages, corresponding to stressing and dead ends, respectively. Adjustable anchorages allow regulation of loads whenever needed, even during the service life of the bridge. The stay cables

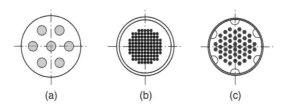

FIGURE 10-36 Types of Stay Cables, (a) Parallel-Bar Cables, (b) Parallel-Wire Cables, (c) Strand Cables.

TABLE 10-2

Capacity of Typical Parallel Strand Cables [15.2-mm (0.6-in.) Diameter Strand]

Number of Strands	Nominal Cross Section (mm²)	Ultimate Load at 100 GUTS (kN)	Service Load at 50% GUTS for Stay Cable	Service Load at 60% GUTS for Extradosed Cable
4	600	1,116	558	670
7	1,050	1,953	977	1,172
12	1,800	3,348	1,674	2,009
19	2,850	5,301	2,651	3,181
27	4,050	7,533	3,767	4,520
31	4,650	8,649	4,325	5,189
37	5,550	10,323	5,162	6,194
43	6,450	11,997	5,999	7,198
55	8,250	15,345	7,673	9,207
61	9,150	17,019	8,510	10,211
73	10,950	20,367	10,184	12,220
91	13,650	25,389	12,695	15,233
109	16,350	30,411	15,206	18,247
127	19,050	35,433	17,717	21,260
169	25,350	47,151	23,576	28,291

Note: 1 mm = 0.0394 in., 1 kN = 0.225 kips. GUTS = guaranteed ultimate tensile strength.

can be either stressed in the pylon or under the deck as required. As previously mentioned, cable suppliers may develop different stay cable systems. However, most systems have eight basic components: anchor heads, ring nuts, bearing plates, transition tubes, deviators, recess tubes, neoprene collars, and caps, although the actual names used may be different. Figure 10-37 shows the principal components of the strand cables anchorages developed by DYWIDAG[10-9]. The main dimensions of the anchorage systems are provided in Table 10-3. The information of the anchorages developed by other manufacturers, such as Freyssinet and TENSA, can be easily found in their manufacturers' websites (see Appendix B). The anchor head transfers forces from individual tension elements of wires, strands, and bars to the ring nut. For strand stay cables, the anchor heads are the wedge plates as discussed in Section 1.7. A ring nut is typically provided at the stressing end anchorage to permit fine adjustments without unseating and reseating the wedges. The bearing plate transfers the cable forces into the structure. The transition tube encloses the flared stay cable bundle and corrosion protective material. The transition tube allows the strands to flare from a tight cable bundle to a wider spacing (see Fig. 10-37). The deviator is typically used for resisting lateral forces due to angle changes of the stay cables in the front end of the transition tube. The recess pipe provides an opening in the structure for cable installation. The neoprene collars or dampers are typically at the front end of the recess tubes. The neoprene provides a flexible lateral support of the cable, acts as a vibration damper, and seals the recess tube. The cap, as previously discussed, is provided to mainly seal the exposed anchor head and prevent water ingress.

10.3.5.3 Dampers

Long stay cables are susceptible to vibration due to wind, rain, etc. Large vibration amplitudes may result in damage to the cables due to bending and fatigue loads. To increase the aerodynamic stability, the outside surface of the cable sheaths is roughened with helical or strip ribs. For long-stay cables, additional dampers or cross ties may be used. Although some cables more

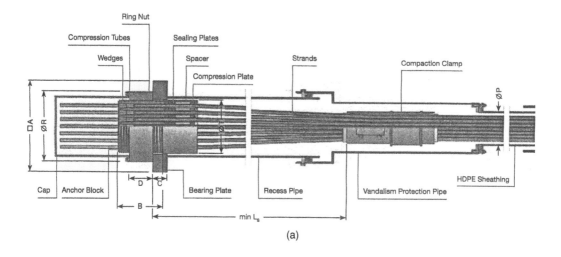

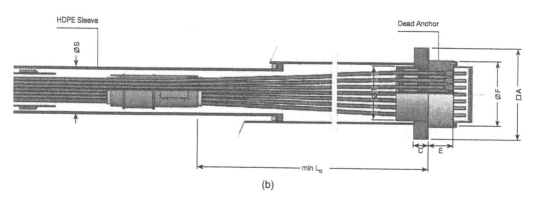

FIGURE 10-37   Components of DYNA Grip Anchorages for Strand Stay Cables, (a) Adjustable Anchorage, (b) Fixed Anchorage (Courtesy of DYWIDAG-Systems International).

than 656 ft (200 m) long have been installed without additional dampers and do not have any vibration problems, it is recommended to use additional damping devices for cable lengths of more than 262 ft (80 m) due to the respective cable parameters. Currently, many manufacturers provide both internal and external dampers (see Fig. 10-38). Both internal and external dampers increase the damping ratios of the cables and are effective in resisting cable vibrations caused by wind, rain, and support excitations. In some exceptional cases, the dampers may be insufficient. Cross ties may be installed to increase the cable stiffness (see Fig. 10-39). The cross ties are often called stabilizing cables. The stabilizing cables are installed in each of the stay cable planes and connected to each of the cables. The cross ties can be arranged in a straight line (Fig. 10-39a) or curved as shown in Fig. 10-39b. The stabilizing cables may be terminated at a main cable or at the deck or pylon. It is recommended that flexible cables be used for the cross ties rather than the prestressing strands or steel elements. The design of the stabilizing cables and their connections should consider fatigue effects. The effect of the cross ties on the sags and inclinations of the main cables should be considered by including the ties in bridge analytical models. Though the cross ties may not be desirable aesthetically, they are generally very effective in resisting cable vibrations.

TABLE 10-3
Technical Data of DYNA Grip Strand Stay Cable Anchorage (unit = mm, 1 in = 25.4 mm)

Number of Strands	Bearing Plate, □A	Bearing Plate, C	Bearing Plate Opening, ØT	Thread, B	Ring Nut, D	Ring Nut, ØR	Dead Anchor, E	Dead Anchor, ØF	Clamp Distance, Stressing End, min L_S	Clamp Distance, Dead End, min L_D	HDPE, ØP	HDPE, ØS
12	300	30	183	200	90	244	120	215	880	770	110	200
19	370	35	219	220	110	287	120	261	1080	970	125	225
31	460	40	267	230	120	350	135	324	1350	1240	160	250
37	500	45	293	240	130	378	135	354	1500	1390	180	250
43	600	55	329	250	140	420	150	398	1690	1580	180	315
55	600	60	341	270	160	440	170	420	1750	1640	200	315
61	640	65	371	275	165	480	170	450	1920	1810	225	315
73	715	70	403	290	180	536	185	490	2070	1960	250	355
85	780	75	429	300	190	554	190	522	2170	2050	280	355
91	780	80	455	310	200	600	195	550	2340	2230	280	355
109	855	85	479	340	230	636	210	586	3020	2910	315	450
127	910	90	531	350	240	700	220	645	3390	3280	315	450

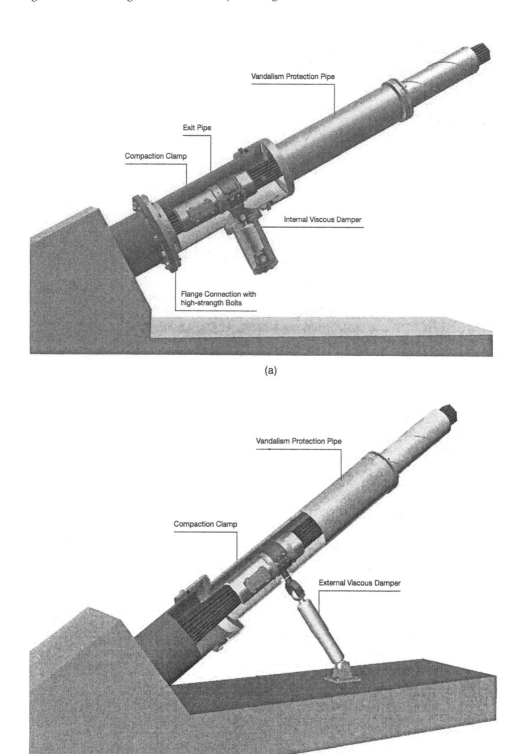

(a)

(b)

FIGURE 10-38 Stay Cable Dampers, (a) Internal Damper, (b) External Damper (Courtesy of DYWIDAG-Systems International).

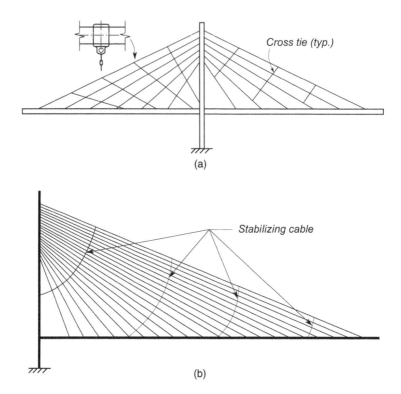

FIGURE 10-39 Cross Tie Cables Arrangement, (a) Straight, (b) Curved Arrangement.

10.4 TYPICAL CONSTRUCTION PROCEDURE

There are several erection methods previously used in cable-stayed bridges. The most often used method may be the cantilever construction method similar to that discussed in Chapter 6. The basic construction procedure is described as follows:

Step 1: Construct abutments or/and piers as shown in Fig. 10-40a.

Step 2: Construct pier segments on falsework (see Fig. 10-40b). If the main girder is designed as simply supported on the pier, temporarily fix the segments to the pier with PT bars (see Fig. 10-41). The pylon can be built first or built together with the main girder (see Fig. 10-40b).

Step 3: Construct the girder segments on both sides of the pier segments on falsework or the traveling scaffoldings (see Fig. 10-40c) and install post-tensioning tendons and temporary post-tensioning bars as required. Install the first pair of stay cables, and stress them to predetermined tensile forces. The cable can be stressed either in the main girder or in the pylon. However, it is typical to stress in the pylon.

Step 4: Move the traveling scaffoldings forward, and construct the new main girder segments and pylon segments next to the previously cast or erected segments. Install a second pair of the stay cables (see Fig. 10-40d).

Step 5: Install the third pair of the stay cables after their anchor segments have been constructed (see Fig. 10-40e). Stress the third pair of cables to the required forces, and adjust the forces of the cables previously installed as necessary.

Step 6: Construct the next segments on the sides of the previously built segments, and install the fourth pair of stay cables until all cantilever-built segments and the stay cables have been installed (Fig. 10-40f).

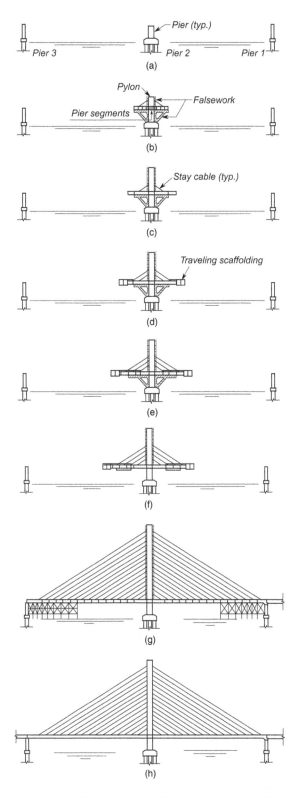

FIGURE 10-40 Typical Construction Procedures, (a) Step 1, (b) Step 2, (c) Step 3, (d) Step 4, (e) Step 5, (f) Step 6, (g) Step 7, (h) Step 8.

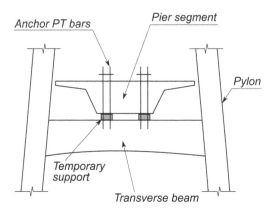

FIGURE 10-41   Temporary Fixed Pier Segment.

Step 7: Construct the segments near piers 3 and 1 on falsework (see Fig. 10-40g). Install the stay cables, stress the cables to the required tensile forces, and install the longitudinal post-tensioning tendons as designed.

Step 8: Remove all falsework and construct barrier walls. Perform the final adjustment of stay cable forces and open them to traffic (Fig. 10-40h).

10.5   ANALYSIS, BEHAVIORS, AND CAPACITY VERIFICATIONS OF CABLE-STAYED BRIDGES

10.5.1   General

The analysis of cable-stayed bridges generally includes: (1) the static analysis, including linear and nonlinear; (2) stability analysis, including entire bridge structural stability, and the stability of pylons and main girders; and (3) dynamic analysis due to wind, earthquake, and moving vehicles. In this section, some typical analytical methods and important behaviors of cable-stayed bridges are first presented. Then, the design verifications are discussed.

10.5.2   Static Analysis

10.5.2.1   Modeling and Analysis by Finite-Element Method

10.5.2.1.1   Modeling

Cable-stayed bridges are generally highly redundant complex structures, and their analysis in the final stage is typically done by the finite-element method through computer programs. The basic theory of the finite-element method was discussed in Section 4.2.5. In general, for routine cable-stayed bridges, two-dimensional frame models (see Fig. 10-42a) with some approximate assumptions, such as transverse load distribution concept and partial models for the transverse analysis, should be adequate, especially for preliminary designs. In certain cases, such as for important bridges and in final design, three-dimensional frame models (see Fig. 11-42b) are often used because there are many powerful commercial computer programs available. In the finite-element model, the main girder, piers, and pylons are divided into a number of frame elements and the cables are treated as cable elements with negligible bending stiffness in the global structural analysis. The nonlinear effect of the cable can be approximately considered by the effective modulus of elasticity (see section 10.5.2.1.2).

10.5.2.1.2   Effective Modulus of Elasticity of Cables

The inclined stay cables are relatively flexible and will sag due to their self-weight. The deflection of the cable varies nonlinearly with the cable axial force. The axial deformation of the cable due to the

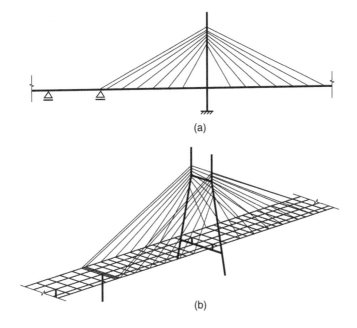

FIGURE 10-42 Typical Analytical Models for Cable-Stayed Bridges, (a) Plane Model, (b) Three-Dimensional Model.

axial force includes the elastic deformation and that due to its sag. Strictly speaking, it is necessary to perform nonlinear analysis for the stay cables. For convenience, the concept of effective modulus of elasticity is often used in the analysis of cable-stayed bridges.

Let us examine an inclined stay cable AB subjected a tensile force F due to self-weight q, as shown Fig. 10-43a. Assuming that the upper support of the cable is on a roller and the lower support is fixed, the cable elongation along the chord AB consists of two portions of Δl_e due to elastic deformation and Δl_s due to the cable sag. As discussed in Chapter 9, the deflected shape of the cable is a catenary curve. Considering its deflection is small, the length of the sagged cable can be determined by assuming it is a parabolic curve and is calculated as

$$l_s = l + \frac{8f^2}{3l}$$

The cable elongation due to its sag is

$$\Delta l_s = l_s - l = \frac{8f^2}{3l} \tag{10-6}$$

where
  f = radial deflection at its midspan (see Fig. 10-43a)
  l = cable length along chord (see Fig. 10-43a)

Using the condition of $\Sigma M = 0$ about the midspan, we have

$$F \times f = \frac{ql^2 \cos\theta}{8}$$

$$f = \frac{ql^2 \cos\theta}{8F} \tag{10-7}$$

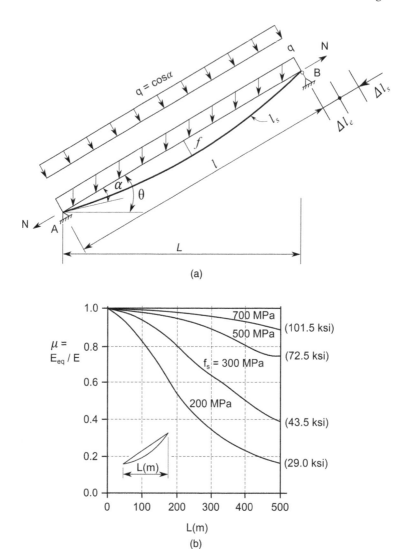

FIGURE 10-43 Effect of Cable Sag and Effective Modulus of Elasticity, (a) Cable Deformation, (b) Effect of Cable Sag.

where

    q = weight per unit horizontal length

    θ = cable angle between horizontal and chord (see Fig. 10-43a)

Substituting Eq. 10-7 into Eq. 10-6 yields

$$\Delta l_s = \frac{q^2 l^3 \cos\theta^2}{24F^2} \tag{10-8}$$

From the definition of modulus of elasticity, we can define the cable effective modulus of elasticity as

$$E_{eq} = \frac{f_s}{\varepsilon_e + \varepsilon_s} \tag{10-9a}$$

where

f_s = $\frac{F}{A_0}$ = cable axial stress

ε_e = $\frac{f_s}{E}$ = elastic strain

ε_s = $\frac{\Delta l_s}{l}$ = generalized strain due to cable sag

A_0 and E = area and modulus of elasticity of the cable, respectively
   Thus, Eq. 10-9a can be rewritten as

$$E_{eq} = \mu E \qquad (10\text{-}9b)$$

where

μ = $\dfrac{1}{1+\frac{(\gamma L)^2}{12 f_s^3}E}$ = reduction factor of modulus of elasticity

γ = density of cable

L = cable horizontal projection (see Fig. 10-43a)

Assuming that the stay cables have a modulus of elasticity of 205,000 MPa (29,725 ksi) and a density of 80 kN/m³ (0.51 kips/ft³), the variations of the reduction factor with cable length for different cable tensile stresses are illustrated in Fig. 10-43b.

10.5.2.2   Determination of Cable Tension and Rotation Angles at Cable Ends
The cable sag effects should be considered in determining the anchor forces and the orientations of the transition tubes. After the forces at the cable ends are determined from the static analysis, the actual cable axial force at the anchorages can be determined as (see Fig. 10-44)

$$F_i = \sqrt{H^2 + V_i^2} \qquad (10\text{-}10)$$

where

i = cable ends = A or B

H = horizontal reaction force of cable at anchorages A and B

V_i = vertical reaction forces of cable at anchorages A or B, respectively

Assuming the shape of sagged cable as a parabolic curve and using the conditions of equilibrium (see Fig.10-44), we have:
   The horizontal reaction force of the cable is

$$H = \frac{qL^2 (\cos\theta)^2}{8f} \qquad (10\text{-}11)$$

The vertical reaction forces at A and B are

$$V_A = H\tan\theta + \frac{qL}{2} \left(\text{upward positive}\right) \qquad (10\text{-}12a)$$

$$V_B = H\tan\theta - \frac{qL}{2} \left(\text{downward positive}\right) \qquad (10\text{-}12b)$$

The angle between the chord and the cable tangent at anchorages is

$$\alpha_i = \beta_i - \theta \qquad (10\text{-}13)$$

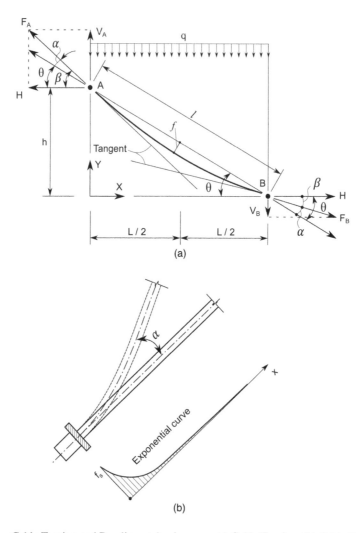

FIGURE 10-44   Cable Tension and Bending at Anchorages, (a) Cable Tension, (b) Cable Bending.

where

$$i = \text{cable ends} = A \text{ or } B$$

$$\tan\theta = \frac{h}{L}$$

$$\tan\beta_i = \frac{V_i}{H}$$

For cables with lengths not greater than 985 ft (300 m), Eqs. 10-10 to 10-13 have good accuracy.

10.5.2.3   Cable Bending at Anchorage

The effect of stay cable bending is neglected in the global structural analysis. However, the effect of the cable bending at the anchorage locations may need to be considered. The bending effect in anchorages is evaluated by testing, and the anchorage design is considered by the cable system suppliers. The bending moments in the cable are related to the magnitude of the angle (α) between the fixed anchorage and the cable tangent at the anchorage (see Fig. 10-44b). As discussed in sections 10.5.2.1.2 and 10.5.2.2, the angle varies with the cable length, unit weight of the cable, cable tension forces, etc., and typically ranges from 0.001 to 0.01 rad[10-10]. The bending moment can be determined as

$$M(x) = \alpha e^{-kx} \sqrt{EIF} \tag{10-14}$$

where
$k = \sqrt{\frac{F}{EI}}$
x = distance from anchorage along chord
I = moment of inertia of cable section.

The remaining notations are designated in Sections 10.5.2.1 and 10.5.2.2.

For typical circular cross section of the stay cables, the flexible stress at the extreme fiber caused by the moment can be written as

$$f_{bs} = 2\alpha \sqrt{Ef_s}\, e^{-kx} \tag{10-15a}$$

The maximum bending stress at the anchorage is

$$f_{bs\_max} = 2\alpha \sqrt{Ef_s} \tag{10-15b}$$

10.5.2.4   Bending in Cable Free Length

Bending stresses in the cable should be avoided or reduced by proper detailing and special devices. The bending stress can be calculated as

$$f_{bs} = \frac{rE}{\chi} \tag{10-16}$$

where
r = radius of cable
χ = local radius of curvature in cable

The bending stress should be considered in the axial stress verification.

10.5.2.5   Geometrical Nonlinear Effects

Cable-stayed bridges are comparatively flexible, and the geometrical nonlinear effects may need to be considered, especially for long-span bridges. The nonlinear effects mainly include the following:

a. Effect of cable sag, which was discussed in Section 10.5.2.1.2; this effect is considered by using the cable equivalent modulus of elasticity.
b. P-Δ effect of the main girder; the main girder is subjected to large axial forces from the inclined cables and bending moment. This P-Δ effect may be approximately determined by Eq. 4-167a.
c. P-Δ effect of the pylon; the pylons of a cable-stayed bridge are generally flexible and subjected to large axial forces and bending deformations. The P-Δ effect may be approximately determined by Eq. 4-167a.

Accurate geometrical nonlinear analysis can be done through computer programs by the finite-element method discussed in Section 4.3.7.2.

10.5.2.6 Basic Static Behaviors of Cable-Stayed Bridges

10.5.2.6.1 Introduction

The static behaviors of cable-stayed bridges are generally more complex than that of girder bridges with the interaction between the main girder, stay cables, and pylons. It is useful to know the basic static behaviors in bridge design although there are many powerful computer programs to perform the analysis of complex bridges. In this section, we take a semi-harp bridge as shown in Fig. 10-45[10-1] as an example to discuss the basic static behaviors of typical cable-stayed bridges. The bridge has a main span of 204.6 m (671.3 ft) and two equal side spans with a span length of 99.2 m (325.5 ft). The height of the pylon is 51 m (167.3 ft). The main girder is fixed at the pylons. The other principal dimensions are shown in Fig. 10-45. Assuming the average moment of inertia of the main girder is 0.07 m^4 (8.11 ft^4) and the moment of inertia of the pylon and pier is 45.56 m^4 (5279 ft^4), the basic static behaviors of the cable-stayed bridge are discussed in the following sections. For ease of discussion, the live load effects are calculated by assuming a uniform loading of 44 kN/m (3.02 kips/ft) loaded to produce maximum effects.

10.5.2.6.2 Typical Influence Lines

The maximum live load effects can be calculated by using influence lines to determine the most unfavorable loading locations. The sketches of some typical influence lines are shown in Fig. 10-46. The axial force influence lines of cables 1 and 2 are illustrated in Fig. 10-46b and c in which the positive and negative signs indicate the tensile and compressive force, respectively. For these figures, it can be observed that applying live loadings in the center span can induce large tensile force in cable 1 which should be used in the design. The moment influence lines at sections 2 and 3 of the main girder are shown in Fig. 10-46d and e. From these figures, we can see the effect of the live loadings on the moment of the main girder is relatively localized. The vertical force and moment influence lines at section 1 of the pier are given in Fig. 10-46f and g. It can be seen from these figures that the maximum bending moment at this section is caused by loading the entire center span.

10.5.2.6.3 Internal Force Distributions in Pylon and Main Girder

The internal force envelopes of the pylon and main girders are illustrated in Figs. 10-47 and 10-48, respectively. Figures 10-47a and 10-48a shows the compression force distributions for the pylon and main girder, respectively. From these figures, we can observe that both the pylon and main girder are subjected to large compression forces, which are located at the bottom of the pylon. The P-Δ effects of the axial forces and buckling should be considered in the bridge analysis and design. The moment envelope of the pylon is shown in Fig. 10-47b. From this figure, it can be seen that the pylon

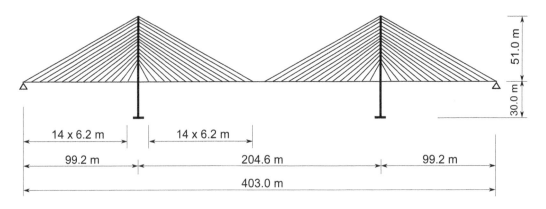

FIGURE 10-45 Principal Dimensions of a Semi-Harp Cable-Stayed Bridge (1 ft = 0.3048 m)[10-1].

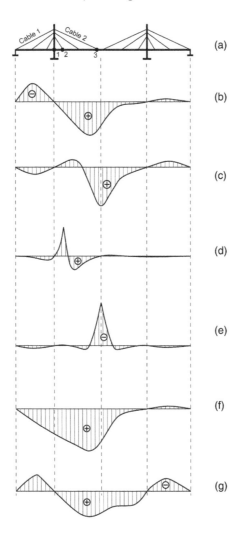

FIGURE 10-46 Typical Influence Lines of Cable-Stayed Bridges, (a) Locations, (b) Axial Force at Cable 1, (c) Axial Force at Cable 2, (d) Moment at Section 2, (e) Moment at Section 3, (f) Reaction at Pier Section 1, (g) Moment at Pier Section 1[10-1].

moment increases greatly from its top to bottom and the maximum bending moment occurs at the bottom of the pylon due to frame action. The moment envelopes of the main girder are illustrated in Fig. 10-48b. From this figure, we can see that the moments due to dead loads for this bridge are comparatively small. We also can see from this figure that the maximum moments due to dead and live loads are located in the side spans, in the range close to the pylon, and in the area near the mid-span. To ensure the cross sections of the main girder are not significantly changing, post-tensioning tendons may be necessary in these locations.

10.5.2.6.4 Main Girder Deflections under Different Loading Cases

Figure 10-49 shows deflections of the main girder under three different loading cases. From this figure, we can observe: (a) maximum deflection occurs in the center span under full loading in the span; (b) the maximum deflection in the side span is much smaller than that in the center span and is induced by loading both side spans (see Fig. 10-49b); (c) the center span behaves similar to an arch with opposite deflections in the span under asymmetric loading as shown in Fig. 10-49c.

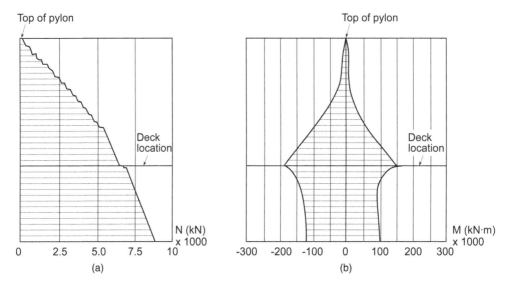

FIGURE 10-47   Internal Force Distributions in Pylon under Dead and Live Loadings, (a) Axial Forces, (b) Moments (1 kips = 4.448 kN, 1 ft = 0.3048 m, 1 kips-ft = 1.356 kN-m)[10-1].

10.5.2.6.5   Effect of Main Girder Stiffness

Figure 10-50a shows the variation of main girder moments with the girder moments of inertia I_{deck} changing from 0.07 to 3.5 m⁴ (8.11 to 405.52 ft⁴). From this figure, it can be seen that increasing the main girder moment of inertia will significantly increase the bending moment in the girder in highly stressed areas. However, the reduction of the bending moment in the pylon is comparatively small. Thus, from this point of view, a girder stiffness that is too large may not yield an economical design.

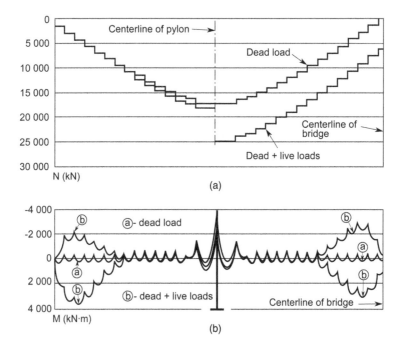

FIGURE 10-48   Internal Force Distribution in Main Girder, (a) Axial Forces, (b) Moments (1 kips = 4.448 kN, 1 ft = 0.3048 m, 1 kips-ft = 1.356 kN-m)[10-1].

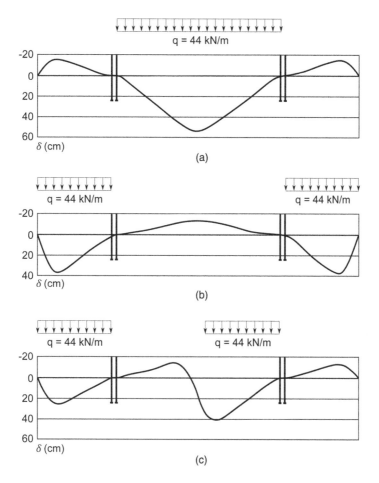

FIGURE 10-49 Main Girder Deflections Due to Different Loading Cases, (a) Loading in Midspan, (b) Loading in Side Spans, (c) Asymmetrical Loading (1 in. = 2.45 cm)[10-1].

10.5.2.6.6 Effect of Pylon Stiffness

The effects of the pylon stiffness by changing its moment of inertia from 12 to 300 m⁴ (1390 to 34,758 ft⁴) are illustrated in Fig. 10-51. It can be observed from Fig. 10-51a that the maximum moments in the main girder decrease with the increase of pylon moment of inertia. However, the extent of the heavily stressed zones in the main girder is hardly affected. Figure 10-51b indicates that the moments in the pylon increase significantly with the increase of the pylon moment of inertia. Thus, for three-span cable-stayed bridges, the selection of a very stiff pylon is not advisable.

10.5.2.6.7 Effect of Types of Connections between Pylon and Main Girder

It is apparent that the types of connections between the main girder and the pylons will play an important role in the bridge behaviors. Let us examine the two extreme cases of fixed connection and free connection with floating deck. Figure 10-52 shows the deformations of cable-stayed bridges with the two different types of connections for two different loading cases. From Fig. 10-52a, it can be seen that the maximum deflection at the midspan will increase over 50% if the connection is changed from fixed to no connection between the main girder and the pylons when the center span is fully loaded. Under the asymmetrical loading as shown in Fig. 10-52b, the maximum deflection without connection between the pylons and the main girder is over 3 times more than that with a fixed connection. Also, the asymmetrical loading case causes significant longitudinal displacement. Thus, the

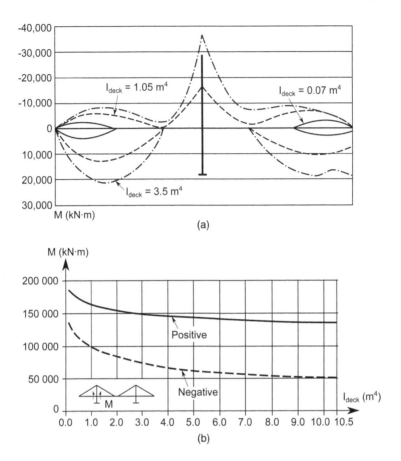

FIGURE 10-50 Effect of Main Girder Stiffness on Moments, (a) Girder Moment Distribution, (b) Variation of Pylon Moments with Girder Moment of Inertia (1 kips-ft = 1.356 kN-m, 1 ft = 0.3048 m)[10-1].

floating system is generally used for long spans in earthquake zones with special devices restraining the bridge deformations in normal service conditions. The increasing deformation is accompanied by an increasing bending moment as shown in Fig. 10-53, which presents the moment envelopes for the main girder and pylon. It can be observed from Fig. 10-53a that the maximum moment in the highly stressed range of the girder will increase over 100% if the connection is changed from fixed to no connection between the main girder and the pylons. From Fig. 10-53b, we can see that a very high bending moment will be induced in the foundation if the floating system is used.

10.5.2.6.8 Geometrical Nonlinear Effects

The geometric nonlinear effects are related to the stiffness of the bridge. The longer the bridge or the smaller the bridge stiffness, the larger the nonlinear effect will be. Figure 10-54 shows the variations of the nonlinear effects on moments and deflections with deck stiffness for this bridge. The superscripts I and II in the figure indicate the responses determined based on first-order and second-order methods, respectively. From this figure, we can observe that the nonlinear effects on both moment and deflection for this bridge discussed is about 10%.

10.5.2.7 Determination of Cable Forces

10.5.2.7.1 Effect of Cable Force on the Behaviors of the Main Girder

One of the most distinguishing characteristics of cable-stayed bridges is that the girder behaviors, including the forces and geometry, are related to the supporting cable forces and can be adjusted

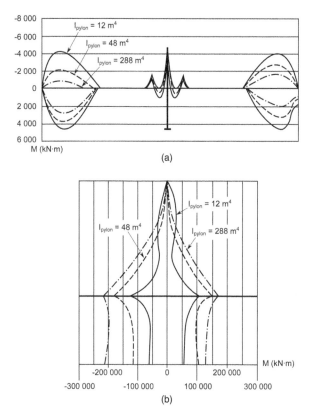

FIGURE 10-51   Effect of Pylon Stiffness on Moments, (a) Girder Moment Distributions, (b) Pylon Moment Distributions (1 kips-ft = 1.356 kN-m, 1 ft = 0.3048 m)[10-1].

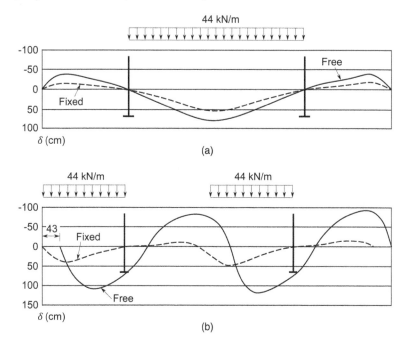

FIGURE 10-52   Effect of Types of Connections on Deflections, (a) Loading in Center Span, (b) Asymmetrical Loadings (1 kips = 4.448 kN, 1 ft = 0.3048 m, 1 in. = 2.54 cm)[10-1].

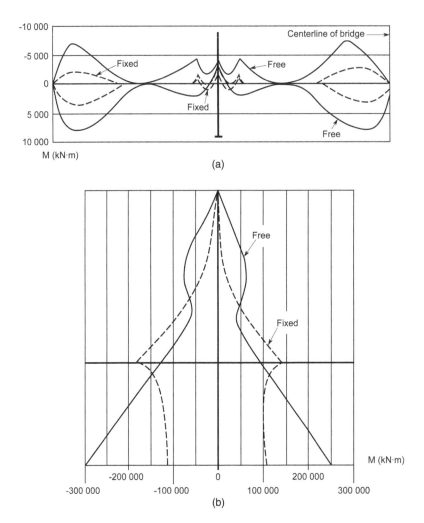

FIGURE 10-53 Effect of Types of Connections on Moments, (a) Moment Envelopes of Main Girder, (b) Moment Envelopes of Pylon (1 kips-ft = 1.356 kN-m)[10-1].

through stressing the stay cables. For ease of understanding, let us examine a simple cable structure as shown in Fig. 10-55. Assuming a simply supported weightless beam AB tied by a cable DC at its midspan (see Fig. 10-55a), then a uniform loading q is applied on the beam. This is a first-order redundant structure. Taking the coordinate system as shown in Fig. 10-55a and denoting cable axial force as N, the bending moments of the beam can be written as

$$M(x) = \frac{q}{2}\left(lx - x^2\right) - \frac{N}{2}x \tag{10-17}$$

Using the displacement compatibility at location C and the force method discussed in Section 4.2.2, the cable axial force N can be obtained as

$$N = \frac{5ql^4/384EI}{l^3/48EI + h/EA} \tag{10-18a}$$

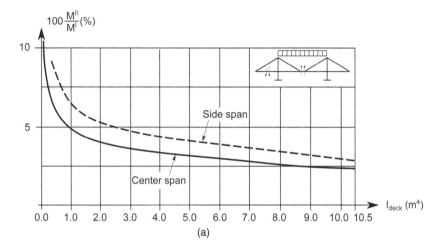

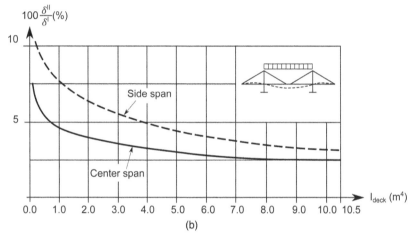

FIGURE 10-54 Nonlinear Effects, (a) Moments, (b) Deformations (1 ft = 0.3048 m)[10-1].

where

E = modulus of elasticity, assuming same for both cable and beam

A and I = area of cable cross section and moment of inertia of beam, respectively

For ease of discussion, assuming $EI/l^3 = 1$ and $EA/h = 192$, the cable axial force can be written as

$$N = \frac{ql}{2} \qquad (10\text{-}18b)$$

The girder behaviors are similar to those for two simply supported beams, and the girder moment distribution is shown in Fig. 10-55c. Assuming the displacement at point C due to the external loading q is equal to zero, similar to adding a rigid support at point C (see Fig. 10-55b), then the reaction at point C can be calculated as

$$R_C = \frac{5ql}{8} \qquad (10\text{-}19)$$

From Eq. 10-19, it can be seen that the beam behaviors will be the same as those of a two-span continuous beam if the cable can be stressed to $N = \frac{5ql}{8}$ without changing its total length of h. Thus,

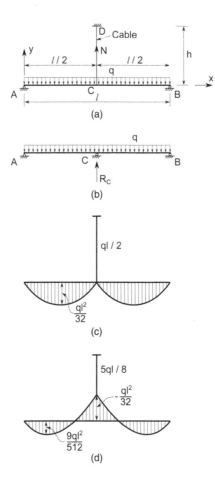

FIGURE 10-55 Effect of Cable Force, (a) Cable Supported Beam, (b) Continuous Beam, (c) Moment Distribution Determined Based on Deformation Compatibility, (d) Moment Distribution Determined Based on Zero Displacement at Cable Location.

the moment distribution of the main girder will be as shown in Fig. 10-55d. From Fig. 10-55c and d, we can observe that the girder behaviors vary with the cable forces. The question is what the optimum cable force is for the beam. To answer this question, we must first set an optimum objective function, such as the sum of the girder moments squared, i.e.,

$$f(N) = \int_0^l M^2(x)\,dx \qquad (10\text{-}20)$$

Substituting Eq. 10-17 into Eq. 10-20 and using the condition of $\frac{df(N)}{dN} = 0$, we can obtain the cable force that meets the optimum objective function Eq. 10-20 as

$$N = \frac{5ql}{8}$$

The above result is the same as the support reaction indicated in Eq. 10-19. Thus, we can conclude that the cable force stressed to the magnitude of the rigid support reaction can make the flexural strain energy minimum for this special case.

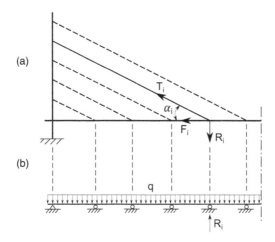

FIGURE 10-56 Determination of Target Cable Forces by Rigid Support Method, (a) Cable-Supported Main Girder, (b) Continuous Girder on Rigid Supports.

10.5.2.7.2 Determination of Optimum Cable Forces

Typically, stay cables are stressed in stages to predetermined tensile forces while simultaneously maintaining anticipated elevations and camber of the supported structure. After a bridge is completely erected, it is often necessary to adjust cable forces to optimum values to meet the requirements for cable forces and geometry. The optimum cable forces or target cable forces are defined as those meeting the state of stress distribution in the bridge as expected by the designer. There are several methods to determine the optimum cable forces. Two often used methods are discussed in following sections.

10.5.2.7.2.1 Rigid Support Method   Currently, the most often used method to determine target cable forces is to set the cable force to balance the dead load shear in the main girder at the cable locations. Take a cable-stayed bridge as shown in Fig. 10-56 for example. First, we assume the cables as rigid support without displacements (see Fig. 10-56b) to determine the support reactions due to dead load q. Then, the tensile force in cable i can be written

$$T_i = \frac{R_i}{\sin \alpha_i} \qquad (10\text{-}21)$$

where
   R_i = reaction determined based on model shown in Fig. 10-56b
   α_i = inclined angle of cable i

10.5.2.7.2.2 Optimum Objective Function Method   The designers may set one or more optimum objective functions and use the optimum method as discussed in Section 10.5.2.7.1 to obtain the target cable forces. For ease of discussion, we take the minimum flexure strain energy of the structure as the optimum objective function. The total structural flexure strain energy can be written as

$$U = \int_0^s \frac{M^2(s)}{2EI} ds \qquad (10\text{-}22a)$$

where s indicates the entire structure.

In the finite-element method, Eq. 10-22a can be written as

$$U = \sum_{i=1}^{m} \frac{l_i}{4E_iI_i}\left(M_{Li}^2 + M_{Ri}^2\right)$$ (10-22b)

where

m = total number of beam elements
E_i, I_i, l_i = modulus of elasticity, moment of inertia, and length of element i, respectively
M_{Li}, M_{Ri} = moments at left and right joints, respectively, of element i

Equation 10-22b can be written in matrix form as

$$U = \{M_L\}^T[B]\{M_L\} + \{M_R\}^T[B]\{M_R\}$$ (10-22c)

where

$\{M_L\}, \{M_R\}$ = vectors of moments at left and right nodes of elements, respectively

$$[B] = \begin{bmatrix} b_{11} & \cdots & 0 \\ \vdots & \ddots & \vdots \\ 0 & \cdots & b_{mm} \end{bmatrix}$$

$b_{ii} = \frac{l_i}{4E_iI_i} (i = 1, 2, \ldots m)$

$\{M_L\}$ and $\{M_R\}$ are the functions of cable tensile forces and can be written as

$$\begin{cases} \{M_L\} = \{M_L^0\} + [C_L]\{T\} \\ \{M_R\} = \{M_R^0\} + [C_R]\{T\} \end{cases}$$ (10-23)

where

$\{M_L^0\}, \{M_R^0\}$ = initial moments at left and right nodes, respectively, of elements before applying cable tensile forces $\{T\}$
$\{T\}$ = vector of cable tensile force
$[C_L], [C_R]$ = matrices of node moments due to unit cable forces = influence matrices of cable forces for left and right nodes, respectively

Substituting Eq. 10-23 into Eq. 10-22c yields

$$U = C_0 + \{M_L^0\}^T[B][C_L]\{T\} + \{T\}^T[C_L]^T[B][M_L^0] + \{T\}^T[C_L]^T[B][C_L]\{T\}$$
$$+ \{M_R^0\}^T[B][C_R]\{T\} + \{T\}^T[C_R]^T[B][M_R^0] + [T]^T[C_R]^T[B][C_R]\{T\}$$ (10-22d)

where C_0 is constant.

Making the total flexure strain energy minimum yields

$$\frac{\partial U}{\partial T_j} = 0 \ (j = 1, 2, \ldots n)$$ (10-24)

where j = number of cables considered.

Substituting Eq. 10-22d into Eq. 10-24 yields

$$\left([C_L]^T[B][C_L] + [C_R]^T[B][C_R]\right)\{T\} = -[C_L]^T[B][M_L^0] - [C_R]^T[B][M_R^0])$$ (10-25)

Solving Eq. 10-25, we can obtain the optimum cable forces $\{T\}$. If the initial moments include the effects of initial cable forces $\{T^0\}$, the optimum cable forces are the sum of $\{T^0\} + \{T\}$.

The optimum objective functions may be taken as strain energy induced by both bending and axial deformations, considering both dead load and live load effects based on the requirements of the structure [10-11].

10.5.2.8 Determination of Cable Initial Stressing Forces during Cantilever Construction

In cantilever construction, the cable forces will change with the subsequently erected segments and cables. The initial cable tensile forces and the segment elevations during each of the construction stages must be determined.

10.5.2.8.1 Determination of Initial Cable Tensile Forces

Assuming that each pair of stay cables are stressed one time to reach their designed tensile forces after completing the entire erection, the predetermined cable forces can be written as

First stressed cables: $T_1 = b_{11}T_{1S} + b_{12}T_{2S} + \cdots + b_{1n}T_{nS} + T_{1q}$

Second stressed cables: $T_2 = b_{22}T_S + b_{23}T_{2S} + \cdots + b_{2n}T_{nS} + T_{2q}$

$\vdots$

nth stressed cables: $T_n = b_{nn}T_{nS} + T_{nq}$

The designed cable tensile forces can be written in matrix form as

$$\{T\} = [B]\{T_S\} + \{T_q\} \tag{10-26}$$

where

$\{T\}$ = target cable tensile forces = $\begin{bmatrix} T_1 & T_2 & \cdots & T_n \end{bmatrix}^T$

$\{T_S\}$ = cable initial stressed forces = $\begin{bmatrix} T_{1S} & T_{2S} & \cdots & T_{nS} \end{bmatrix}^T$

$\{T_q\}$ = cable tensile forces due to system changes, additional dead loads, creep and shrinkage, etc. = $\{T_q\} = \begin{bmatrix} T_{1q} & T_{2q} & \cdots & T_{nq} \end{bmatrix}^T$

$[B]$ = coefficient matrix

$$= \begin{bmatrix} b_{11} & b_{12} & \dots & b_{1n} \\ 0 & b_{22} & \dots & b_{2n} \\ & & & \\ 0 & 0 & \dots & \dots \\ 0 & 0 & 0 & b_{nn} \end{bmatrix}$$

b_{ij} = increment of cable force in cable i due to unit stressing force in cable j $(i, j = 1,2, \dots, n)$

n = total number of cables

From Eq. 10-26, the cable initial stressed forces can be written as

$$\{T_S\} = [B]^{-1}\left(\{T\} - \{T_q\}\right) \tag{10-27}$$

10.5.2.8.2 Determination of Segment Elevations

The initial elevations of segments during construction can be written as

$$\{Y_0\} = \{Y\} - \{\Delta Y\} - \{\Delta Y_q\} \tag{10-28}$$

where

$\{Y_0\}$ = initial elevations (cambers) of segments = $\begin{bmatrix} y_{01} & y_{02} & \cdots & y_{0m} \end{bmatrix}^T$

$\{Y\}$ = anticipated final road profile = $\begin{bmatrix} y_1 & y_2 & \cdots & y_m \end{bmatrix}^T$

$\{\Delta Y\}$ = vector of displacement increments due to segment self-weight, post-tensioning forces, and cable stressed forces = $\begin{bmatrix} \Delta y_1 & \Delta y_2 & ... & \Delta y_m \end{bmatrix}^T$

Δy_1 = $\Delta y_{11} + \Delta y_{12} + ... + \Delta y_m$

Δy_2 = $\Delta y_{22} + \Delta y_{23} + ... + \Delta y_{2m}$

$\vdots$

Δy_m = Δy_{mm}

Δy_{ij} = displacement increment of segment i due to self-weight, post-tensioning forces, and cable stressed forces of segment j ($i, j = 1, 2, ..., m$), upward positive

m = number of segments

$\{\Delta Y_q\}$ = vector of displacement increments due to system changes, additional dead loads, creep and shrinkage, etc. = $\begin{bmatrix} \Delta y_{q1} & \Delta y_{q2} & ... & \Delta y_{qn} \end{bmatrix}^T$

Δy_{qi} = displacement increment of segment i due to system changes, additional dead loads, creep and shrinkage, upward positive, etc.

10.5.2.9 Adjustment and Measurement of Cable Forces

10.5.2.9.1 Adjustment of Cable Forces

Because of many factors, the cable forces and the elevations of the bridge deck after completing erection of all segments and cables are typically not the same as the predetermined values. The cable forces and the elevations after finishing the erection must be adjusted. For concrete segmental bridges, the primary adjustment may be given to cable forces first if the girder stiffness is comparatively large. Then, further adjustments of the elevations to meet both cable forces and deck geometric requirements are needed. The general relationship between the elongations and the increments of the cable tensile force can be written as

$$[K]_{n\times n}\{\Delta\delta\}_{1\times n} = \{\Delta T\}_{1\times n} \tag{10-29}$$

where

n = total numbers of stay cables

$[K]_{n\times n}$ = global stiffness matrix representing forces induced by unit elongations of the cables

$\{\Delta\delta\}_{1\times n}$ = elongations of cables

$\{\Delta T\}_{1\times n}$ = differences between predetermined and measured cable forces

As it is very difficult to meet the design cable forces and design deck profile at the same time, the cable design should consider at least an additional 10% of the design capacity required.

10.5.2.9.2 Measurement of Cable Forces

After completing the main girder erection, it is important to measure the actual cable forces for adjustment to satisfy the design requirements. The stay cable forces can be directly measured using pressure meters or strain gages. The forces also can be measured by the relationship between cable lateral vibration frequency and tensile force, i.e.,

$$F = \frac{4ml^2 f^2}{n_c^2} \tag{10-30a}$$

where

F = cable tensile force

f = cable lateral vibration frequency corresponding to n_c

n_c = cable vibration mode number (see Fig. 10-57)

l = cable length

m = $\frac{w}{g}$ = mass per unit cable length, w = weight per unit cable length, g = gravity acceleration

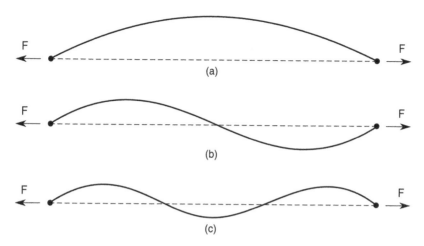

FIGURE 10-57   Stay Cable Vibration Modes, (a) First Vibration Mode, (b) Second Vibration Mode, (c) Third Vibration Mode.

The frequency for $n_c = 1$ is easily measured. If the time t_{50} (seconds) taken for 50 times of the vibrations is measured, then, $f = \frac{50}{t_{50}}$ and Eq. 10-30a can be written as

$$F = \frac{10^4 \, ml^2}{t_{50}} \tag{10-30b}$$

where t_{50} = time (s) taken for 50 times of vibration in the first vibration mode as shown in Fig. 10-57a.

10.5.3   Stability Analysis

10.5.3.1   Introduction

As aforementioned, the main girders and pylons in cable-stayed bridges are subjected to large compression forces and the entire bridge may be buckled in plane or out of plane with increased external loadings. Under static wind loads, the main girder may be buckled out of plane and may lose its lateral bending and torsional stability if the torsion induced by the upraise static wind loading exceeds its torsional capacity. Currently, the buckling analysis of cable-stayed bridges is typically done using finite-element methods through computer programs, as discussed in Section 4.3.9. In this section, some approximate methods to estimate the buckling loads of main girders and pylons will be presented. The torsional stability analysis of main girders due to static wind loading is a Category II stability issue and can be analyzed by the methods discussed in Section 4.3.7.2.

10.5.3.2   Approximate Analytical Methods

Although many computer programs are available for determining the buckling loadings of cable-stayed bridges, approximate methods are always valuable in preliminary design and for checking computer-generated results.

10.5.3.2.1   Estimation of In-Plane Buckling Loading of Main Girders

10.5.3.2.1.1   Buckling Loads of Spring-Supported Beams   Before developing an approximate method for estimating the buckling loading of cable-stayed bridges, let us examine a spring-supported

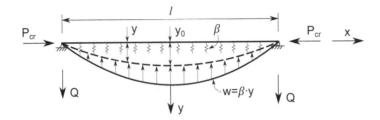

FIGURE 10-58 Buckling of Spring-Supported Beam.

beam. Assuming that the buckling model of a spring-supported beam, as shown in Fig. 10-58, is a half-circle of cosine function which yields the lowest buckling load, i.e.,

$$y = y_0 \cos \frac{\pi x}{l} \tag{10-31}$$

Then the reaction of the spring is

$$w = \beta y \tag{10-32}$$

where β = spring stiffness.

The shear at the end supports can be written as

$$Q = \int_0^{\frac{l}{2}} w dx = \frac{\beta y_0 l}{\pi} \tag{10-33}$$

The moment at the midspan is

$$M_0 = P y_0 - \frac{Ql}{2} + \int_0^{\frac{l}{2}} xw dx = P y_0 - \frac{\beta y_0 l^2}{2\pi} + \beta y_0 l^2 \left(\frac{1}{2\pi} - \frac{1}{\pi^2} \right) \tag{10-34}$$

Using the boundary condition:

$$x = 0, \ EI \frac{d^2 y}{dx^2} = -M_0 = EI y_0 \frac{\pi^2}{l^2} \tag{10-35}$$

where EI is the bending stiffness of the beam.

Solving Eq. 10-34 yields

$$P = \frac{\pi^2 EI}{l^2} + \frac{\beta l^2}{\pi^2} \tag{10-36}$$

From $\frac{dP}{dl} = 0$, we can obtain l, which makes the P minimum as

$$l = \pi^4 \sqrt{\frac{EI}{\beta}} \tag{10-37}$$

Substituting Eq. 10-37 into Eq. 10-36, we can obtain the buckling load of a spring-supported beam as

$$P_{cr} = 2\sqrt{\frac{EI}{\beta}} \qquad (10\text{-}38)$$

10.5.3.2.1.2 Buckling Loadings of Main Girders The main girder in a cable-stayed bridge (Fig. 10-59a) can be simulated as a girder supported by individual elastic supports provided by the stay cables. The vertical stiffness of the individual cable is assumed to be uniformly distributed along the cable spacing (see Fig. 10-59b), i.e., the equivalent stiffness of the spring supports per unit length is

$$\beta(x) = \frac{k(x)}{a(x)} \qquad (10\text{-}39)$$

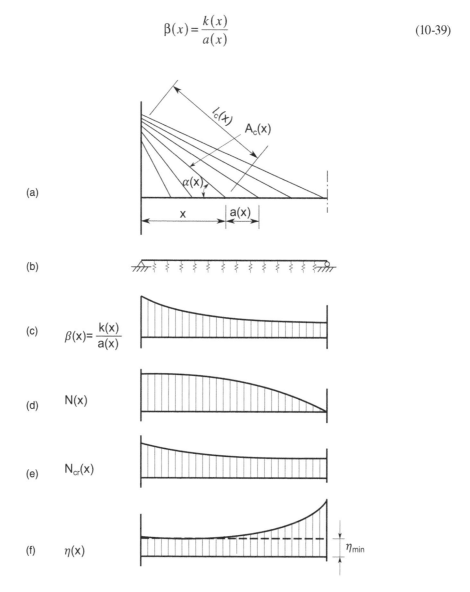

FIGURE 10-59 In-Plane Buckling Model of Main Girder, (a) Cable-Supported Girder, (b) Girder on Elastic Supports, (c) Stiffness of Spring Supports, (d) Variation of Axial Load, (e) Generalized Buckling Loading, (f) Variation of Buckling Safety Factors.

where
$\quad k(x)$ = equivalent stiffness of cable located at x
$\quad a(x)$ = stay cable spacing along girder longitudinal direction

Then, the model shown in Fig. 10-59b is essentially equal to that shown in Fig. 10-58, except that the equivalent spring stiffness and the girder axial loading vary along the girder longitudinal direction (see Fig. 10-59c and d). Thus, from Eq. 10-38, the generalized buckling loading $N_{cr}(x)$ of the main girder can be written as

$$N_{cr}(x) = 2\sqrt{\frac{E(x)I(x)}{\beta(x)}} \tag{10-40}$$

where
$\quad \beta(x)$ can be calculated from Eq. 10-39

$$k(x) = \frac{1}{\delta_1 + \delta_2} = \mu \frac{E_c A_c \sin^2\alpha(x)}{l_c}$$

δ_1, δ_2 = vertical displacements at point A induced by the unit force F due to cable elongation and pylon bending, respectively (see Fig. 10-60a and b).
E_c, A_c, l_c = modulus of elasticity, area, and length of the stay cable at location x, respectively

$$\mu = \frac{1}{1 + \frac{\cos\alpha(x)\eta h}{3l_c}}$$

$$\eta = \frac{E_c A_c h^2}{E_t I_t} = \text{stiffness ratio of cable to pylon}$$

E_t, A_t = modulus of elasticity and area of the pylon, respectively
h = height of pylon from its fixed end to cable in consideration (see Fig. 10-60b)

We define the buckling safety factor of the main girder as

$$\eta(x) = \frac{N_{cr}(x)}{N(x)} \tag{10-41}$$

where $N(x)$ = service axial loading of the main girder.

As the buckling safety factors vary along the girder longitudinal direction (see Fig. 10-59f), we can take the minimum value as the buckling factor of the main girder in bridge design[10-11].

10.5.3.2.2 Buckling Loadings of Pylons

Buckling of a pylon in a cable-stayed bridge may occur during construction and in service. During construction, the buckling loading of a pylon for cable-stayed bridges with harp or semi-harp cable patterns can be estimated by treating it as a uniformly centrically loaded column fixed at its bottom with an equivalent constant cross section and using the following equation:

$$q_{cr} = \frac{7.837\ EI}{h^2} \tag{10-42}$$

where
$\quad EI$ = equivalent in-plane or out-of-plane stiffness of pylon
$\quad h$ = height of pylon from its fixed end to its top
$\quad q_{cr}$ = buckling loading per unit length along the pylon height

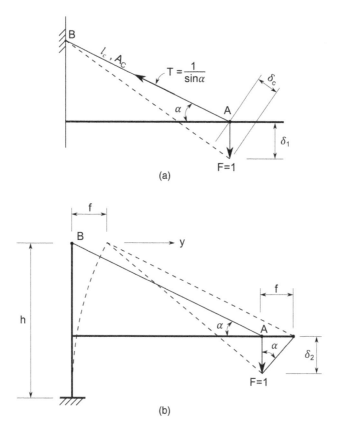

FIGURE 10-60 Deformation Relationships of Stay Cable and Girder, (a) Deformations Induced by Unit Force F at Point A due to Cable Elongation, (a) Deformations Induced by Unit Force F at Point A due to Pylon Bending.

In service, the in-plane buckling loading of a pylon is typically larger than that of the main girder. When its out-of-plane buckling occurs, a stabilizing force provided by the out-of-plane inclined cables will increase the pylon buckling loading, which can be determined by the elastic energy method[10-11 to 10-19].

10.5.4 WIND ACTION ON CABLE-STAYED BRIDGES AND WIND-INDUCED VIBRATION

The effects of wind on cable-stayed bridges are generally more important than those on other types of bridges discussed in previous chapters as the cable-stayed bridges are typically long-span light-weight bridges with relatively small stiffness. The wind action on bridges is very complex and depends on the velocity and direction of wind; the size, shape, and motion of the bridge; as well as the environmental situation. Research indicates that natural wind can be divided into two portions: steady and turbulent. If the stiffness of a bridge is large, the bridge remains static under steady wind and only the static effect of the wind on the bridge should be considered. If the stiffness of a bridge is comparatively small, both steady and turbulent winds will induce oscillations of the bridge. In current bridge design practice, the wind effects are typically considered in two parts: static action and dynamic action, which are discussed in following sections.

10.5.4.1 Static Wind Action

Assuming that a bridge structure is acted upon by a steady wind with constant wind speed V and angle of attack α and that the bridge structure is a static fixed structure or that the bridge

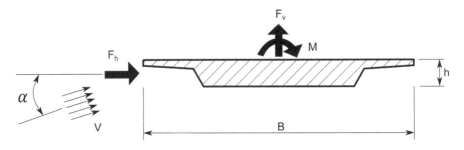

FIGURE 10-61 Wind Static Loadings.

deformations have no effect on the wind action, the forces exerted by the wind on the bridge are called static wind loads. These loads can be expressed as consisting of three components: drag (horizontal) force F_h, lift (vertical) force F_v, and pitching (torsional) moment M (see Fig. 10-61), which can be denoted as

$$F_h = \frac{\rho V^2}{2} C_D A \tag{10-43}$$

$$F_v = \frac{\rho V^2}{2} C_L A \tag{10-44}$$

$$M = \frac{\rho V^2}{2} C_M AB \tag{10-45}$$

In Eqs. 10-43 to 10-45,
ρ = density of air
V = design wind speed
A = exposed area of bridge in longitudinal direction
B = bridge width
C_D, C_L, C_M = dimensionless drag, lift, and torsion coefficients, respectively

Coefficients C_D, C_L, and C_M vary with the wind angles of attack and the shape of the main girder cross sections and should be determined through wind tunnel tests. Some test results are shown in Fig. 10-62.

10.5.4.2 Wind Dynamic Action

10.5.4.2.1 Auto-Excited Oscillation—Flutter

A cable-stayed bridge absorbs energy from steady wind and may move with time as shown in Fig. 10-63. When the steady wind speed reaches a certain speed and the energy received cannot be dissipated by the damping, the bending and torsional movements of the bridge will increase rapidly and the bridge structure becomes unstable and will be destroyed. This phenomenon is called flutter, and the wind speed is called the flutter critical speed. The critical speed of a cable-stayed bridge must be higher than any possible wind speed at the bridge site with a proper safety factor determined by the pertinent authority. The critical wind speed for flutter is dependent on the bridge dynamic properties and is generally determined based on wind tunnel test results. For preliminary design, based on many test results, the critical wind speed for flutter may be estimated as

$$V_{cr} = 2\xi \eta \pi f_B b \tag{10-46}$$

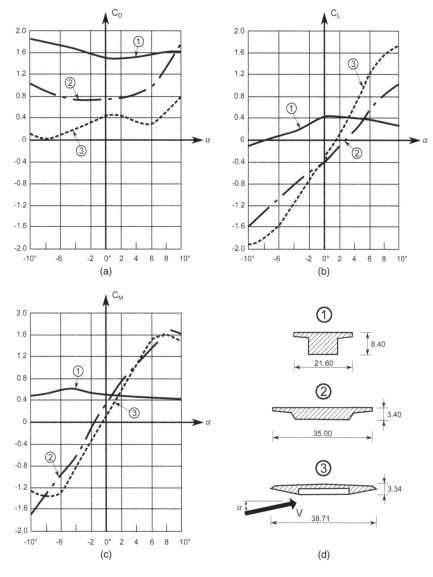

FIGURE 10-62 Variations of Drag, Lift, and Torsion Coefficients with Wind Angle of Attack for Typical Sections, (a) Drag Coefficient C_D, (b) Lift Coefficient C_L, (c) Torsion Coefficient C_M, (d) Legends[10-1].

where

f_B = vertical bending frequency of bridge (Hz)

b = half of bridge width

ξ = ratio of theoretical critical wind speed for a flat plate to $2\pi f_B b$, which can be found in Fig. 10-64

η = correction coefficient for shape of main girder cross section, which can be found in Fig. 10-65

From Fig. 10-64, it can be seen that ratio ξ is a function of the logarithmic decrement of damping ψ, ratio of torsion to bending frequencies ε, ratio μ of deck mass to the density of air, ratio of deck radius of gyration to half deck width $\frac{r}{b}$. These dimensionless coefficients are defined as

$$\varepsilon = \frac{\omega_T}{\omega_B} = \frac{f_T}{f_B}, \ \mu = \frac{m}{\pi \rho b^2}, \ \frac{r}{b} = \frac{1}{b}\sqrt{\frac{I}{A}}$$

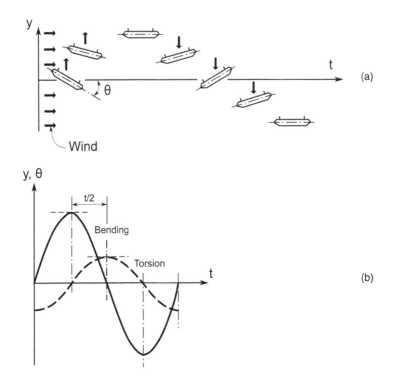

FIGURE 10-63 Flutter of Bridge Deck, (a) Movement Sketch of Deck under Wind, (b) Displacements of Deck with Time.

where

ω_T, ω_B	=	torsional and bending circular frequencies of bridge, respectively
f_T	=	torsional frequency of bridge (Hz)
m	=	mass of bridge per unit length
r	=	radius of gyration
I, A	=	moment of inertia and area of main girder, respectively
ρ	=	density of air

It also can be observed from Fig. 10-64 that (1) the larger the coefficient ε, the larger the critical wind speed will be; i.e., increasing torsional stiffness can increase the capacity of the bridge in resisting wind loading; (2) the heavier the deck, the larger value of μ and critical wind speed will be; and (3) the larger the radius of inertia, the larger the critical wind speed will be.

From Fig. 10-65, we can see that the correction coefficient η is a function of the shape of the main girder cross section and the ratio of torsional to bending frequencies. It can also be observed that a girder cross section with a wind nose will significantly increase the bridge capacity for wind resistance.

10.5.4.2.2 Galloping

Under a steady wind, members with square-shaped, or close to square-shaped, cross sections, such as pylons with a square cross section and stay cables with approximate elliptical cross sections due to rain and ice, may undergo bending-dominated lateral vibrations. When the wind speed reaches a certain value, called the galloping critical wind speed, the vibration becomes unstable. This type of vibration is called galloping. The critical wind speed for stay cables may be estimated by

$$V_{cr}^g = Cfd\sqrt{\frac{m\psi}{\rho d^2}} \tag{10-47}$$

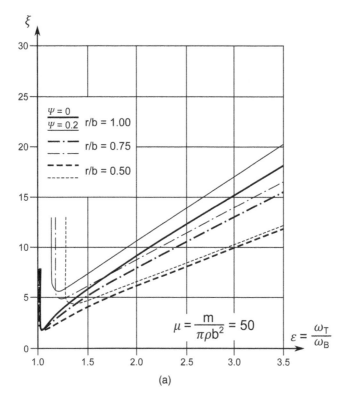

(a)

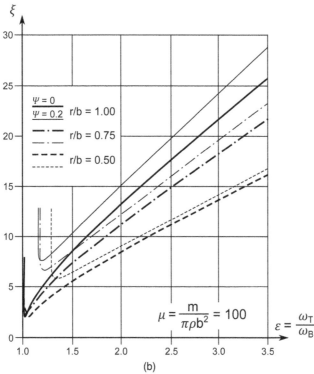

(b)

FIGURE 10-64 Ratio of Theoretical Critical Wind Speed for Flat Plate Deck to ω_B, ξ, (a) $\mu = 50$, (a) $\mu = 100$.

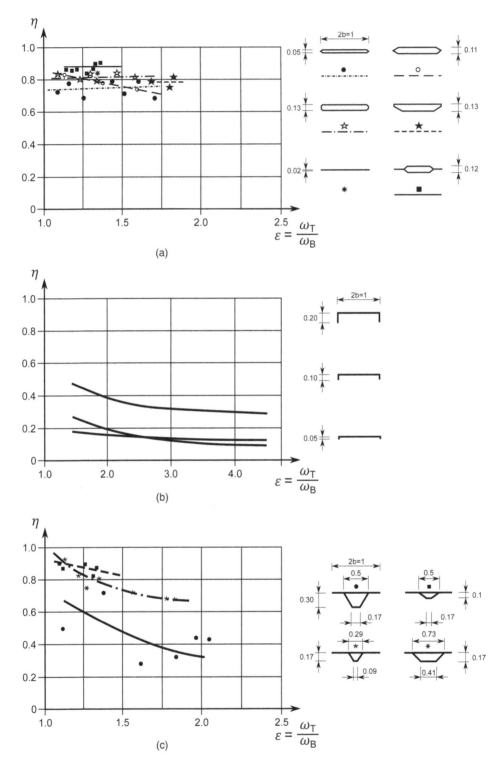

FIGURE 10-65   Correction Coefficients for Non–Flat Plate Decks, η, (a) Decks with Wind Noses, (b) Decks with Two Rectangular Edge Beams, (c) Single Box Sections with Inclined Webs.

FIGURE 10-66 Vortices Shedding.

where

$C \approx 40$, for circular cables

d = diameter of cable

m = mass of cable per unit length

ψ = damping ratio of stay cables

$f = \frac{1}{2l_c}\sqrt{\frac{N_C}{A_C m}}$ = first frequency of cable, ranging from 0.5 to 2 Hz for a cable length of 165 to 655 ft (50 to 200 m)

l_c, A_C, N_C = length, area, and axial force of the cable, respectively

10.5.4.2.3 Forced Oscillations

10.5.4.2.3.1 Vortices or Vortex-Shedding When the steady air flow passes around a bridge deck, eddies may be shed periodically from the deck (see Fig. 10-66). This phenomenon is often called vortices or vortex shedding. The periodic shedding of vortices alternately from the upper and lower surfaces of the bridge deck induces a periodic fluctuation of the aerodynamic forces on the bridge that impose on the bridge a forced vibration. The frequency of the forced vibration is equal to the eddy frequency and can be estimated by

$$f = \frac{SV}{h} \tag{10-48}$$

where

h = depth of deck

V = velocity of wind

S = dimensionless parameter called Strouhal number (see Table 10-4), which may be approximately taken as 0.20 for all types of cross sections

If the frequency is close to one of the natural frequencies of the bridge, such as bending, there is a risk of resonance, because the bridge displacements are only limited by the damping of the bridge.

The periodic force developed by vortices can be estimated by

$$F(t) = F_0 \sin \omega t \tag{10-49}$$

where

$$F_0 = \frac{\rho V^2}{2} C_V h$$

TABLE 10-4
Strouhal Numbers for Different Shapes of Cross Sections

Cross Section	Circular	Inverted U	Single Box with Inclined Webs	Square
S	0.20	0.22	0.23	0.25–0.30

C_V = coefficient of aerodynamic force determined by test

$\omega = \omega(R_e, S)$ = circular frequency of vortices = function of Reynolds number R_e and Strouhal number S

$$R_e = \frac{BV}{\nu}$$

where
- $\nu = 0.15$ cm^2/s = kinematic viscosity of the air
- B = bridge width
- V = velocity of wind

10.5.4.2.3.2 Buffeting As discussed, natural wind is not steady but turbulent. The nonsteady wind can cause dynamic excitation through fluctuation of the local wind speed. In long-span bridge decks, the turbulence of natural wind can induce bridge buffeting, which can cause a local fatigue issue. In design, the concept of impact factors is used to consider the effect of buffeting. The accurate analysis of bridge buffeting may be carried out by modeling the wind action as random processes through computers. The analytical methods can be found in References 10-19 and 10-22.

10.5.4.3 Remark and Summary

The wind effects on cable-stayed bridges are complex and are difficult to determine through pure theoretical analysis. Wind tunnel tests may be necessary for properly evaluating the wind effects, especially for long-span cable-stayed bridges. However, based on previous experience and research, there is generally no danger of wind-induced failures of concrete segmental bridges with two planes of cables if the following conditions are met[10-20]:

1. B (bridge width) $\geq 10h$ (bridge girder depth)
2. Girder cross section with wind noses, if $B < 10h$
3. $B \geq \frac{1}{30}l$ (main span length)

10.5.5 CAPACITY VERIFICATIONS

10.5.5.1 Introduction

In general, similar to the design of other types of structural elements, the designers, together with the cable suppliers and contractors, should ensure the structural durability, serviceability, constructability, and maintainability of a cable-stayed bridge. All bridge components should be designed to resist anticipated loads described in Chapter 2 and meet all the related limit states discussed in Section 2.3.3.2 and Eq. 2-50. The capacity verifications for the main girders and pylons are similar as those presented in Chapters 5 to 9. The behaviors of the stay cables and their capacity verifications are a little different than those previously discussed for concrete members and are discussed in this section.

10.5.5.2 Design Considerations of Stay Cables and Capacity Validations

10.5.5.2.1 Introduction

The stay cables should be designed to resist static loads, dynamic loads, and fatigue as well as to employ appropriate anchorage details, corrosion protection, and cable erection procedures. The bridge designers are typically expected to perform and responsible for all design-related aspects, including:

a. Arranging stay cables and anchorages, including constructability considerations
b. Sizing stay cables and determining their forces
c. Designing connection details and ensuring forces transfer from the anchorages to the main structural components.

10.5.5.2.2 Capacity Validations

10.5.5.2.2.1 General The stay cables should be designed to satisfy Eq. 2-50 for each of the related strength, service, extreme event, and fatigue limit states described in Section 2.3.3.2 and those discussed in the following sections. As the redundant design of the stay cables will be considered based on the loss of one cable, the load modifier contained in Eq. 2-50 is taken as 1.0, i.e.,

$$\eta_i = 1.0$$

Some special design requirements for the stay cables are discussed in the following sections.

10.5.5.2.2.2 Resistance Factors The resistance factors for stay cables are different from those for prestressed and post-tensioned concrete elements. The factors are calibrated to correspond to the earlier allowable stress design approach. The stay cable resistance factors should be taken as:

a. Strength limit state with axial load only: The resistance factor varies with the ratio β_s of unfactored live load plus wind load $(LL + W)_{axial}$ to minimum ultimate tensile stress (MUTS), as shown in Fig. 10-67, i.e.,

$$\beta_s = \frac{(LL + W)_{axial}}{MUTS} \qquad (10\text{-}50)$$

where

$(LL + W)_{axial}$ = unfactored axial force due to live load and wind load
$MUTS$ = minimum ultimate tensile stress of the cable

$$\phi = 0.65, \ if\,\beta_s \geq 0.075 \quad \phi = 0.75, \ if\,\beta_s \leq 0.025$$

The resistance factor varies linearly with $\beta_s = 0.025$ to 0.075.

b. Strength limit state with both axial and bending: $\phi = 0.78$
c. Extreme event limit state: $\phi = 0.95$
d. Fatigue limit state: $\phi = 1.0$

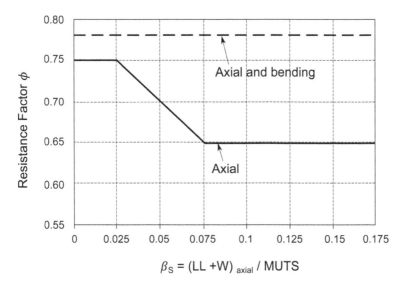

FIGURE 10-67 Strength Resistance Factors for Stay Cables.

10.5.5.2.2.3 Wind Loads on Stay Cables   For preliminary design or for short-span cable-stayed bridge design, the wind load on the stay cables may be estimated as

$$\text{Static lateral wind load per unit length: } q_s = p_D C_D d \qquad (10\text{-}51)$$

$$\text{Total lateral wind load including dynamic effect per unit length: } q_L = q_s g_B g_g \qquad (10\text{-}52)$$

In Eqs. 10-51 and 10-52,

p_D = design wind pressure at middle cable elevation determined by Eq. 2-10.

$C_D \cong 1.2$ = drag coefficient

d = cable diameter

g_B = 1.5 to 2.0 = turbulent gust impact factor

g_g = 1.1 to account for galloping effect

For a single stay cable, $g_g = 1.5$. The dynamic responses in any direction for a single cable can be estimated by

$$q_D = q_s(g_B - 1)g_g \qquad (10\text{-}53)$$

10.5.5.2.2.4 Lateral Loads on Stay Cable Anchorages   At the entrance of the stay cable into the anchorage or at other locations in the anchorage, some angular deviations of the stay cable may occur during construction before the guide deviators are installed or during service when the guide deviator is replaced. This angular deviation will cause lateral forces, which should be considered in designing the anchorage and the guide deviators. Based on actual survey results, the angular deviation generally does not exceed 0.025 rad. Thus, PTI[10-10] recommends that a minimum lateral force of 2.5% of the maximum static cable force should be applied to the anchorage for evaluating its connection or service conditions.

10.5.5.2.2.5 Fatigue Capacity Verification   The stay cables should be designed to meet following fatigue limit equation:

$$\gamma(\Delta F) \le (\Delta F)_n \qquad (10\text{-}54)$$

where

γ = 1.05 = load factor

(ΔF) = stress range due to passage of fatigue loading specified in Section 2.2.3.1.3.2

$(\Delta F)_n$ = nominal fatigue resistance

The nominal fatigue resistance can be determined as follows.

$$(\Delta F)_n = G, \text{ if } \gamma(\Delta F) \le G \qquad (10\text{-}55a)$$

$$\text{Else } (\Delta F)_n = B, \text{ if } \gamma(\Delta F) \le B \qquad (10\text{-}55b)$$

where

$$G = \frac{(\Delta F)_{TH}}{2} \qquad (10\text{-}56a)$$

$$B = \left(\frac{A}{N}\right)^{\frac{1}{3}} \qquad (10\text{-}56b)$$

Table 10-5 Fatigue Constants

Constants	Parallel Strand	Parallel Wire	Uncoupled Bar	Epoxy-Coated Bar
$A \times 10^{11}$, MPa	39.3	74.3	39.3	7.2
$(\Delta F)_{TH}$, MPa	110	145	110	48

$(\Delta F)_{TH}$, A = fatigue constants, which can be found in Table 10-5

$N = 365 \times N_y \times ADTT$ = number of cycles due to passage of the fatigue design truck during the design life of the stay cables

N_y = design life of cables in years

$ADTT$ = average daily truck in one direction and in one lane

If Eqs. 10-55a and 10-55b are not applicable, $\gamma(\Delta F)$ is too high and measures must be taken to reduce it.

10.5.5.2.2.6 Design for Cable Replacement   It is important for the designers to consider how to replace any possible damaged individual cable during the design process and provide a general recommendation in the plans. PTI recommends the following load combination and load factors be used for cable replacement design:

$$1.2DC + 1.4DW + 1.5(LL + IM) + Cable\ Exchange\ Force$$

10.5.5.2.2.7 Design for Loss of Cable   Based on the PTI's recommendation that all cable-stayed bridges shall be designed to be capable of withstanding the loss of any one cable without causing any structural instability, the following load combination and load factors should be checked:

$$1.1DC + 1.35DW + 0.75\ (LL + IM) + 1.1CLDF$$

where CLDF = cable loss dynamic forces. The impact dynamic force due to the sudden rupture of a cable can be conservatively estimated as 2 times the static force in the cable. If a nonlinear dynamic analysis is performed, the dynamic force should not be less than 1.5 times the cable static load.

10.5.5.2.2.8 Design for Construction   PTI[10-10] recommends that the following limit states be used for the construction load combination for cable design, instead of Eqs. 2-52a and 2-52b:

Strength limit state: $1.2(DC + DW + CE + CLE) + 1.2(WS + WE + WUP) + 1.4CLL$

Extreme limit state: $1.1(DC + DW + CE + CLE) + 1.0AI$

10.5.5.2.2.9 Minimum Radius of Stay Cables   For a single strand, the minimum radius of cable bend is 10 ft (3 m). For a multi-strand stay cable, the minimum radius is 13 ft (4 m).

10.5.5.2.2.10 Design for Wind- and Rain-Induced Resonance   To avoid any wind- and rain-induced resonance, the mass-damping parameter or Scruton number S_C for smooth circular cables should meet the following equation:

$$S_C = \frac{m\psi}{\rho d^2} \geq 10 \tag{10-57}$$

where ψ = damping ratio.

Equation 10-57 implies that for typical cable mass densities and diameters, damping ratios of 0.5% to 1% should be sufficient to suppress wind- and rain-induced vibration.

10.5.5.2.2.11 Preliminary Design of Stay Cables In preliminary design, the sizes of stay cables may be estimated by the allowable stress design method. The allowable stress may be taken as $0.45f_{pu}$ for permanent dead load and live load plus impact and $0.56f_{pu}$ for construction or cable replacement.

10.6 DESIGN EXAMPLE VI[1]

10.6.1 INTRODUCTION

This design example is modified from the Xiaogan Bridge located in Shanghai crossing over the Xiaogan Channel (see Fig. 10-68). Based on the minimum opening requirement for navigation, a 300-m (984.25-ft) main span was proposed for its main bridge. Among the types of bridges suitable for this range of span length, three alternatives are investigated: an arch bridge, a cable-stayed bridge with steel and concrete composite main girder, and a cable-stayed bridge with post-tensioned concrete girder. The cable-stayed bridge with a post-tensioned concrete main girder alternative is found to be the most suitable bridge type for this site, after a comparison of the following aspects:

- Construction cost
- Suitability for accommodating large barges and potential ship impact
- Constructability and construction time
- Durability and ease of maintenance
- Resistance to wind and earthquake loadings

FIGURE 10-68 Xiaogan Bridge Overview.

---

[1] This design example is provided by Mr. Xudong Shen and Mr. Fangdong Chen, Planning and Design Institute of Zhejiang Province, Inc., revised by Dr. Dongzhou Huang, P.E.

As the metric system is used in the original design, both metric and English units will be provided in many cases.

10.6.2 DESIGN CRITERIA

10.6.2.1 Design Specifications
    a. *AASHTO LRFD Bridge Design Specifications*, 7th edition (2014)
    b. *CEB-FIP Model Code for Concrete Structures*, 1990

10.6.2.2 Minimum Opening
Minimum vertical clearance for navigation: 30.5 m (100 ft)
    Minimum clearance for navigation: 200 m (656 ft) plus 50 m (164 ft) for small ship dock

10.6.2.3 Traffic Requirements
    a. Two 3.66-m (12-ft) traffic lanes
    b. 3.66-m-wide (12-ft-wide) right shoulder and left shoulder
    c. Two 0.47-m-wide (two 18.5 in. wide) barrier walls

10.6.2.4 Design Loads
    a. Live load: HL-93 Loading
    b. Dead loads:
        – Unit weight of reinforced concrete (DC) $w_{DC} = 23.56$ kN/m^3(0.15 kcf)
        – Other dead load, including barrier wall, 12.7-mm (½-in.) sacrificial deck thickness and utilities: 63.5 kN/m (4.35 kips/ft)
    c. Thermal loads:
        – Uniform temperature: Mean: 18°C (64.4°F); rise: 21°C (37.8°F); fall: 23°C (41.4°F)
        – Temperature difference: Between cables and other members: ±15°C(±27°F)
        Between left and right sides of pylons: ±5°C (±9°F)
        – Main girder temperature gradient as specified in Section 2.10.2 and taking the following values:

Positive Nonlinear Gradient	Negative Nonlinear Gradient
$T_1 = 5°C(41°F)$	$T_1 = -24.61°C(-12.3°F)$
$T_2 = -11.67°C(11°F)$	$T_2 = -19.61°C(-3.3°F)$
$T_3 = -17.78°C(0°F)$	$T_3 = -17.78°C(0°F)$

    d. Design wind speeds: 25 m/s (55 mph) with traffic; 40.5 m/s (90 mph) without traffic; flutter check wind speed: 84.3 m/s (188.6 mph)
    e. Earthquake: horizontal peak ground acceleration (PGA) = 0.1 g
    f. Nonuniform pier settlement: 20 mm (0.79 in.)

10.6.2.5 Materials
    a. Concrete:
        $f_c' = 44.8$ MPa $(6.5$ ksi$)$, $f_{ci}' = 35.9$ MPa $(5.2$ ksi$)$
        $E_c = 32,034$ MPa (4645 ksi)
        Relative humidity: 0.8
    b. Stay cables:
        ASTM A416 Grade 240 low relaxation parallel-wire cables diameter = 7 mm
        $f_{pu} = 1670$ MPa (242 ksi), $f_{py} = 0.9\ f_{pu} = 1503$ MPa (218 ksi)
        Modulus of elasticity: $E_p = 200,000$ MPa (29,000 ksi)

c. Prestressing strands:
   ASTM A416 Grade 270 low relaxation
   $f_{pu} = 1860$ MPa (270 ksi), $f_{py} = 0.9 f_{pu} = 1676$ MPa (243 ksi)
   Maximum jacking stress: $f_{pj} = 1490$ MPa (216 ksi)
   Modulus of elasticity: $E_p = 196{,}600$ MPa (28,500 ksi)
   Anchor set: 9.52 mm (0.375 in.)
   Friction coefficient: 0.23 (plastic duct)
   Wobble coefficient: 0.00020 (internal), 0 (external)
d. Reinforcing steel:
   ASTM 615, Grade 60
   Yield stress: $f_y = 413.8$ Mpa (60 ksi)
   Modulus of elasticity: $E_s = 206{,}900$ MPa (30,000 ksi)

10.6.3 General Arrangement of the Bridge

The main bridge is a three-span cable-stayed bridge with a two-plane semi-harp pattern of cable arrangement (see Fig. 10-69). The lengths of the side and center spans are 130 m (426.51 ft) and 300 m (984.25 ft), respectively. The ratio of side to center span lengths is 0.433. To reduce the uplift force at end piers and eliminate the auxiliary piers, a relative larger ratio of side to center span lengths is used. The typical section of the main girder is a three-cell box girder with inclined outside webs and wind noses (see Fig. 10-70). The total bridge width is 19 m (62.34 ft), including two barrier wall widths and two cable zones and maintenance areas. The girder depth is 2.75 m (9.02 ft), and the ratio of girder depth to the center span length is about 1/109.09. The ratio of girder depth to girder width is about 1/6.91, and the ratio of bridge width to span length is about 1/15.8. The total pylon height from the footing to the top of the pylon is 113.5 m (372.38 ft). The height of the pylon from the girder support to the top of the pylon is about 85 m (278.87 ft) (see Fig. 10-71). The ratio of the pylon height to center span length is about 0.283, and the minimum angle of the stay cables with the main girder is about 27.5°.

To reduce the bridge longitudinal forces due to earthquakes, the semi-floating support system is used for this bridge. The main girder is allowed to move freely in a longitudinal direction to release large energy due to earthquakes. To reduce the bridge deformation during the bridge's normal service life and reduce the physiological and psychological effects on users of the bridge, two longitudinal dampers are installed at each of the pylons under the main girder (see Fig. 10-72). The maximum

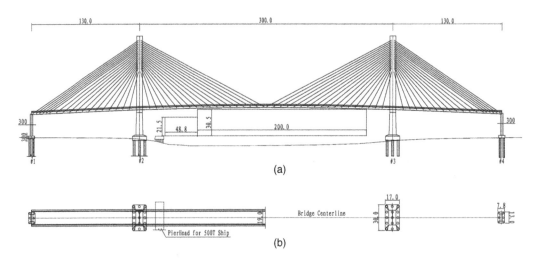

FIGURE 10-69   Plan and Elevation of Example VI, (a) Elevation View, (b) Plan View (unit = m, 1 ft = 0.3048 m).

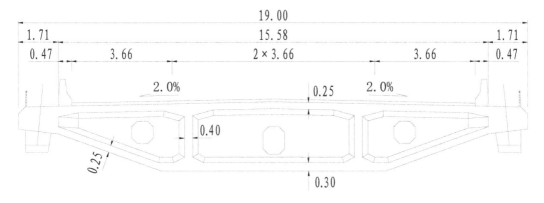

FIGURE 10-70 Typical Section (unit = m, 1 ft = 0.3048 m).

damping force of the dampers is 750 kN (168.62 kips), and their maximum displacement is 350 mm (13.8 in.). There are two lateral supports in each of the pylons to restrain the girder lateral displacements due to wind loadings. The general arrangements of the bridge supports are shown in Fig. 10-72.

The stay cables are arranged in two planes. There are 18 pairs of stay cables in elevation view at each pylon, and the total number of stay cables is 144. The cable spacing in the pylon is 2.0 or 2.1 m based on the requirements for stressing the stay cables. The cable spacing on the main

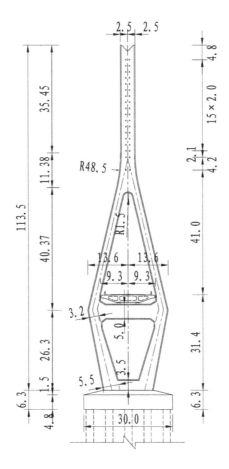

FIGURE 10-71 Pylon and Pier Side View (unit = m, 1 ft = 0.3048 m).

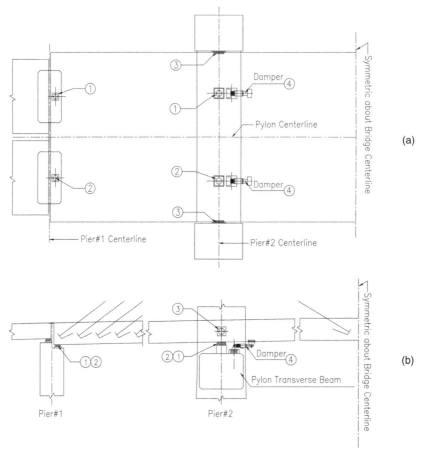

Legend: ① Vertical support, longitudinal guided sliding

② Vertical support, free sliding

③ Horizontal support, free sliding in longitudinal and vertical directions

④ Damper, maximum damping force is 750 kN (169 kips)

FIGURE 10-72 Layout of Supports, (a) Plan View, (b) Elevation View.

girder is typically 8 m. In the range where the concrete counterweight is necessary for reducing uplift force at the end supports in the side spans, the cable spacing is 4 m (see Fig. 10-73). Grade 240 ϕ7-mm parallel-wire stay cables are used for this bridge, and each of the cables has dampers to reduce the cable vibrations. There are seven different cable sizes used in the bridge.

1. Cable numbers SC1, SC2, MC1, and MC2 have 91 ϕ7-mm (0.276-in.) wires.
2. Cable numbers SC3, SC4, MC3, and MC4 have 109 ϕ7-mm wires.
3. Cable numbers SC5 to SC8 and MC5 to MC7 have 127 ϕ7-mm wires.
4. Cable numbers SC9 to SC12 and MC8 to MC10 have 139 ϕ7-mm wires.
5. Cable numbers SC13, SC14, and MC11 to MC14 have 163 ϕ7-mm wires.
6. Cable numbers SC15, SC16, and MC15 to MC18 have 187 ϕ7-mm wires.
7. Cable numbers SC17 and SC18 have 199 ϕ7-mm wires.

The cable identification numbers can be seen in Fig. 10-73.

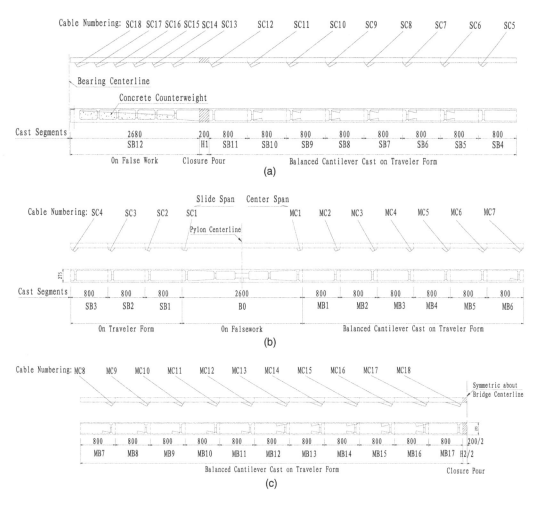

FIGURE 10-73 Notations of Stay Cables and Cast Segment Layout of Main Girder, (a) Part of Side Span, (b) Parts of Side and Center Spans, (c) Part of Center Span (unit = cm, 1 in. = 2.54 cm).

10.6.4 PRINCIPAL DIMENSIONS AND CAST SEGMENT LAYOUT OF MAIN GIRDER

10.6.4.1 Principal Dimensions of Main Girder

The typical thicknesses of the girder top and bottom slabs are 25 cm (10 in.) and 30 cm (12 in.), respectively. The thicknesses of the exterior inclined and interior webs are 25 cm (10 in.) and 40 cm (16 in.) (see Fig. 10-70), respectively. There is a transverse diaphragm 40 cm (16 in.) thick at each of the stay cable anchorages (see Figs. 10-73 and 10-74). Manholes are provided in each of the diaphragms and interior webs. The principal dimensions of the typical girder segments are illustrated in Figs. 10-74b and c. Stay cables are anchored on both sides of the box girder under the thickened top flange through the anchor blisters. All stay cables are stressed at the ends in the pylon, and the cable anchorages in the main girder are fixed.

10.6.4.2 Layout of Cast Segments

The main girder is cast in place, and the segment divisions and numbering are shown in Fig. 10-73. Segments B0 and SB12 are cast in place on falsework, because there are no stay cables in the range

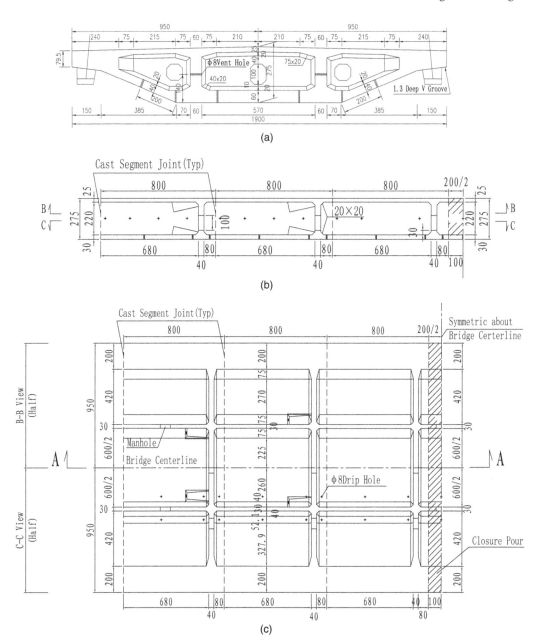

FIGURE 10-74 Typical Section of Segment B0 and Typical Segment, (a) Typical Section of Segment B0, (b) Elevation of Typical Segment, (c) Plan of Typical Segment (unit = cm, 1 in. = 2.54 cm).

of segment B0 and the stay cables in the range of segment SB12 are not symmetric with the cables in the center span. The counterweight concrete used to reduce the support uplift force in pier 1 in segment SB12 is gradually added based on construction requirements. There are three closure segments located in each of the side and center spans designated as H1, H2, and H3 (not shown in the plan as it is symmetric with H1). The remaining segments are called typical segments and are cast in place by the balanced cantilever construction method through the use of traveler forms. The length of the typical segment is 8 m (26.25 ft), and the segment weight is about 320 metric tons (352.64 tons).

10.6.4.3 Layout of Post-Tensioning Tendons and Bars

Post-tensioning forces are provided to resist the loadings during construction and service. The amount of required post-tensioning forces and steel areas can be estimated by Eqs. 3-21 and 3-23 multiplying by a coefficient of 1.1 to 1.3, based on the moments preliminarily determined. Around the pylon area of the main girder and in the segments constructed using the balanced cantilever method, there are twenty ϕ32-mm (1.25-in.) and ten ϕ32-mm post-tensioning bars in the top and bottom flanges, respectively. There are an additional twelve 19-ϕ15.2 mm (0.6 in.) and four 19-ϕ15.2 mm (0.6 in.) tendons in the top and bottom flanges in the area close to the pylons of the main girder. The layouts of the post-tensioning bars and tendons at the pier section of the main girder are shown in Fig. 10-75a. There are twelve 12-ϕ15.2 mm (0.6 in.) continuous tendons in the top flange and fourteen 19-ϕ15.2 mm continuous tendons in the bottom flange in the side spans. In the center span, there are twenty-two 19-ϕ15.2 mm tendons in the bottom flange and ten 19-ϕ15.2 mm tendons in the top flange of the main girder. The locations

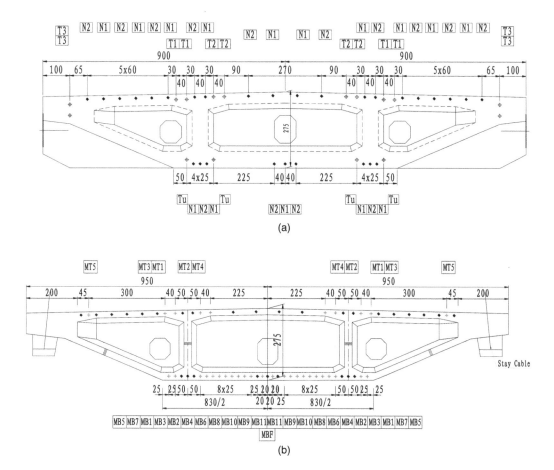

FIGURE 10-75 Layout of Post-Tensioning Tendons and Bars, (a) At Pier Section, (b) At Midsection (unit = cm, 1 in. = 2.54 cm).

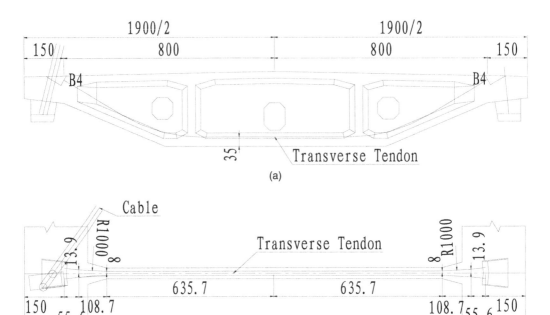

FIGURE 10-76 Layout of Transverse Post-Tensioning Tendons, (a) Vertical View, (b) Plan View (unit = cm, 1 in. = 2.54 cm).

of the tendons at the bridge midsection are shown in Fig. 10-75b. In Fig. 10-75a, the designations N and T represent the post-tensioning bars and tendons, respectively. In Fig. 10-75b, the designations MT and MB represent the top and bottom post-tensioning tendons, respectively. All post-tensioning tendons are anchored at the blister anchor blocks, as shown in Fig. 10-74b and c. The design of the blisters can be found in Chapters 5 and 6. At each of the transverse diaphragms, a post-tensioning tendon of 19-ϕ15.2 mm or 15-ϕ15.2 mm is used to resist its positive moment. The typical transverse tendon layouts are shown in Fig. 10-76.

10.6.4.4   Stay Cable Anchorages in Main Girder

The stay cables are designed to be stressed in the pylon and fixed in the main girder. A sketch of the anchor block in the main girder is shown in Fig. 10-77, and its reinforcement layout is illustrated in Fig. 10-78.

10.6.5   Design and Principal Dimensions of the Pylon

10.6.5.1   Principal Dimensions of the Pylon and Pier

The pylon has an inverted Y-shape. The pylon and pier are integrated through middle and bottom transverse beams. The principal dimensions of the pylon and the pier are illustrated in Fig. 10-79. The cross section of the top segment of the pylon is a 5- × 7-m (16.40- × 22.97-ft) rectangular hollow section (see Fig. 10-79c). The cross section of the legs of the bottom segment of the pylon is a 3- × 7-m (9.84- × 22.97-ft) rectangular hollow section (see Fig. 10-79d). The middle transverse beam is a single-cell rectangular box beam with two intermediate diaphragms (see Figs. 10-79 and 10-82). A manhole is provided in each of the diaphragms for maintenance. The height and width of the box is

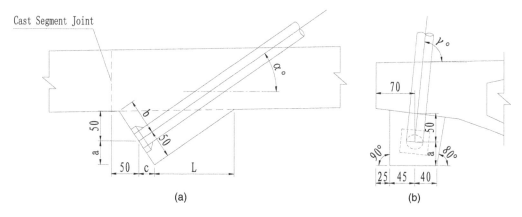

FIGURE 10-77 Layout of Stay Cable Anchorage in Main Girder, (a) Elevation View, (b) Side View (unit = cm, 1 in. = 2.54 cm).

5 m (16.40 ft) and 6 m (19.69 ft), respectively. The wall thickness of the box beam is 0.6 m (1.97 ft). The bottom transverse beam has a solid rectangular section, and its height is 5 m (16.40 ft). Each of the corners of the pylon cross sections is chamfered 0.3 × 0.3 m (0.98 × 0.98 ft) to reduce the wind loading.

10.6.5.2 Stay Cable Anchorages in Pylon

Two types of stay cable anchorages are employed in the pylon. In the top portion of the pylon with a constant cross section, anchor beams are used (see Fig. 10-80a). Within the transition segment from the constant section to the section with two legs, the stay cables are directly anchored to the inside of the pylon box walls. The tensile forces induced by the stay cables are balanced by the circumferential post-tensioning tendons (see Fig. 10-80b). The anchor beams are connected to the brackets with bolts. The brackets are mounted on the pylon concrete walls using studs as shown in Fig. 10-81. For clarity, a three-dimensional view of the anchor beam is illustrated in Fig. 10-81b.

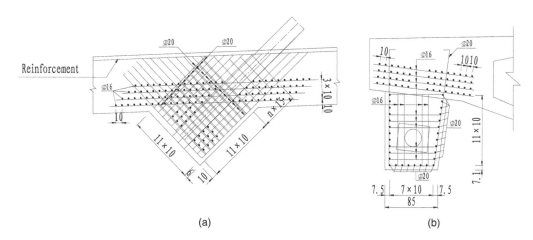

FIGURE 10-78 Reinforcement Layout of Stay Cable Anchorage in Main Girder, (a) Elevation View, (b) Side View (unit = cm, 1 in. = 2.54 cm, rebar diameter = mm, 1 in. = 25.4 mm).

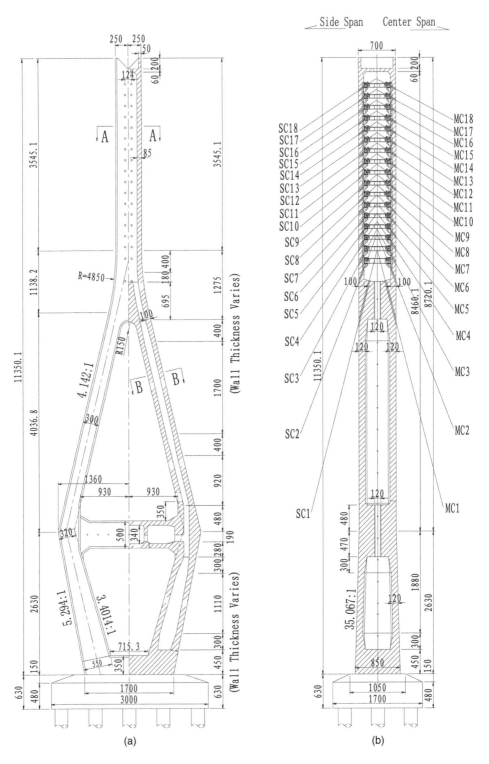

FIGURE 10-79   Principal Dimensions of Pylon and Pier, (a) Elevation View, (b) Side View, (c) Section *A-A*, (d) Section *B-B* (unit = cm, 1 in. = 2.54 cm).

(*Continued*)

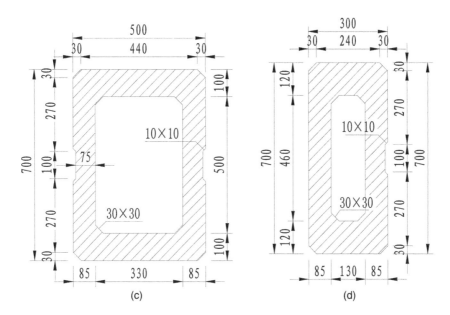

FIGURE 10-79 (Continued)

10.6.5.3 Layouts of Typical Post-Tensioning Tendons in Pylon

Other than the circumferential post-tensioning tendons to resist the tensile stresses induced by the stay cables, some post-tensioning tendons are used in the pylon section transition area and in the middle transverse beam (see Fig. 10-82). The stress distribution in the transition range from one hollow column to the two-leg split is complex. A nearly solid segment of 6.95-m (22.8-ft) height is provided in the transition area. In the top of the segment, there are six 12φ15.2-mm (0.6-in.) tendons in the bridge longitudinal direction. At the bottom of the segment, there are twelve 15φ15.2-mm tendons in the bridge transverse direction (see Fig. 10-82a). The middle transverse beams support the main girder and together with other members act to resist lateral wind loading. There are seventy 15φ15.2-mm post-tensioning strands provided in the beam's bottom flange, top flange, and webs to balance its large bending moments (see Fig. 10-82b).

10.6.6 Construction Sequence

The proposed construction sequence is as follows:

Step I
a. Construct the bridge substructure piers 1 to 4 as shown in Fig. 10-83a.
b. Construct the pylons.
Step II
a. Erect the temporary falsework on the footings under the pylons, and preload it to eliminate nonelastic deformations (see Fig. 10-83b).
b. Build the formwork on the falsework, and cast segment B0 as designated in Fig.10-73b.
c. After the concrete strength in segment B0 reaches 90% of its design strength, install temporary supports over the middle transverse beam and fix segment B0 to the piers.
d. Install longitudinal and transverse post-tensioning tendons and bars as designed.
e. Install and stress the first pair of stay cables SC1 and MC1 as designated in Fig. 10-73b.

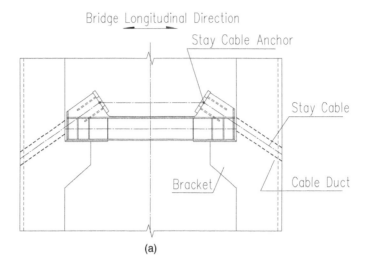

(a)

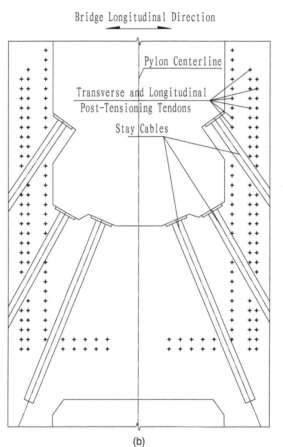

(b)

FIGURE 10-80 Cable Anchorages in Pylon, (a) Anchor Beam, (b) Anchor on Inside of Pylon Hollow Column (unit = cm, 1 in. = 2.54 cm).

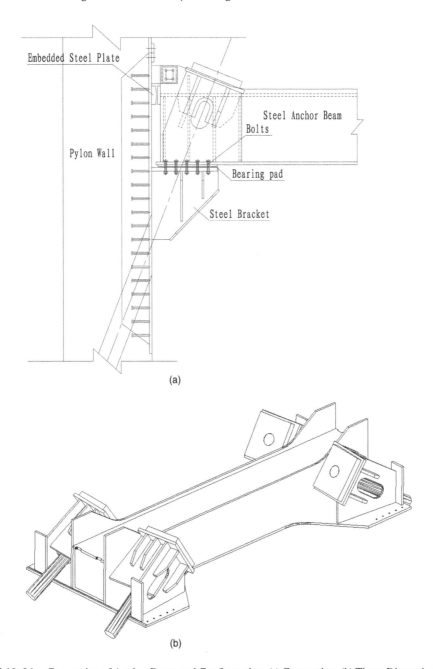

FIGURE 10-81 Connection of Anchor Beam and Configuration, (a) Connection, (b) Three-Dimensional View.

Step III
a. Symmetrically install the traveler forms on both ends of the main girder segment B0 (see Fig. 10-83c).
b. Install the second pair of cables SC2 and MC2 (see Fig.10-73 for designations), and tempo-rarily anchor them in the traveler forms.
c. Initially stress cables SC2 and MC2 to one-third of the final designed cable tension forces.
d. Cast half of the concrete in segments SB1 and MB1.

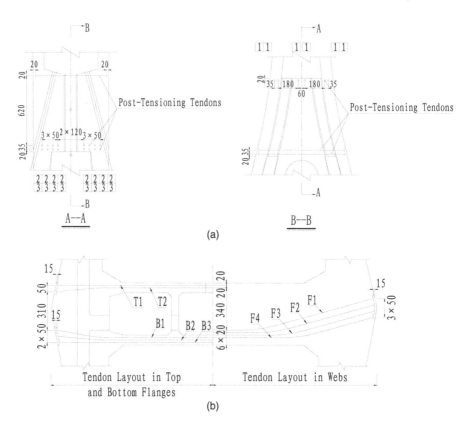

FIGURE 10-82 Layouts of Typical Post-Tensioning Tendons in Pylons, (a) At Corner of Section Transition, (b) Middle Transverse Beam (unit = cm, 1 in. = 2.54 cm).

Step IV

a. Stress cables SC2 and MC2 a second time to another one-third of the final design cable tension forces (see Fig. 10-83d).

b. Complete the concrete casting for segments SB1 and MB1.

c. After the concrete strength reaches 90% of its design strength, install and stress the post-tensioning tendons and bars as designed.

d. Fully stress cables SC2 and MC2 to their designed tensile forces.

Step V

a. Move the traveler forms forward, install the third pair of stay cables SC3 and MC3, and cast segments SB2 and MB2. Follow steps I through IV, until finishing the installations of segments SB11 and MB11 (see Fig. 10-83e).

b. Install temporary falsework next to piers 1 and 4 (see Fig. 10-83e).

c. Cast segment SB12 (see Fig. 10-73 for designation).

d. After the concrete strength reaches 90% of its design strength, install and stress the post-tensioning tendons and bars as designed.

Step VI

a. Cast the side span closure pour segment H1 (see Fig. 10-83f and Fig. 10-73 for designation).

b. After the concrete strength reaches 90% of its design strength, install and stress post-tensioning tendons as designed and the remaining side span stay cables SC12 through SC18, adding concrete counterweights as designed.

c. Remove the traveler form and the form under the closure pour in side spans.

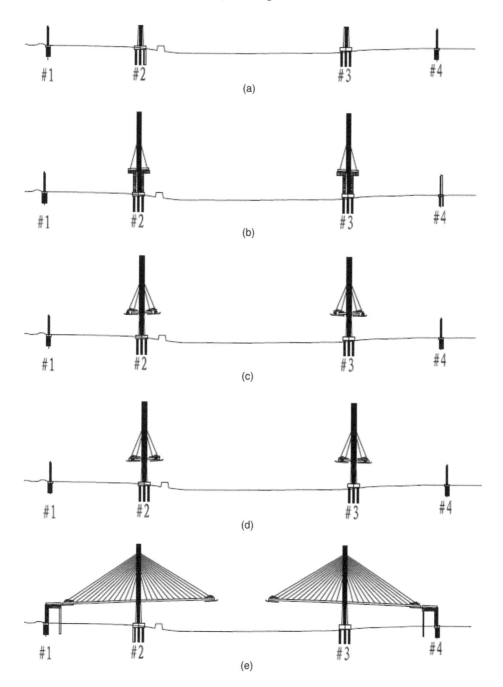

FIGURE 10-83 Construction Sequences, (a) Step I, (b) Step II, (c) Step III, (d) Step IV, (e) Step V, (f) Step VI, (g) Step VII, (h) Step VIII.

(Continued)

Step VII

a. Move the traveler form in the center span forward, and finish the installation of segments MB12 through MB18 as well as stay cables MC12 to MC18.

b. Remove the falseworks around piers 1 and 4.

c. Install brackets in the center span closure pour.

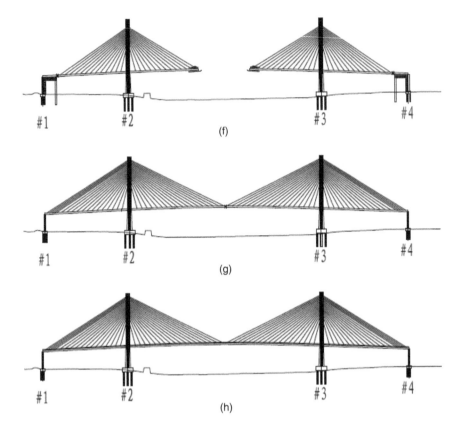

FIGURE 10-83   (Continued)

  d. Remove the falseworks around piers 2 and 3.
  e. Install formwork and reinforcement for the center closure pour.
  f. Cast concrete in the closure pour.
  g. After the concrete strength reaches 90% of its design strength, stress the continuous post-
    tensioning tendons as designed.
  Step VIII
  a. Install barrier walls and perform other related miscellaneous work.
  b. Check and adjust cable tensile forces until the cable tensile forces and deck geometry meet
    design requirements.

10.6.7   Bridge Analysis

10.6.7.1   Properties of Cross Sections

The properties of the cross sections of each of the members are determined by the computer program. The properties of some of the controlling sections are given in Table 10-6.

10.6.7.2   Analytical Model

The bridge is modeled as a three-dimensional beam system (see Fig. 10-84) and analyzed by the
finite-element method. The main girder is divided into 150 elements, pylons are divided into
164 elements, and each of the stay cables is treated as a cable element. The structural analysis is
done by the commercial computer program MIDAS.

TABLE 10-6
Properties of Some of the Controlling Sections

Member	Location	Area A (m²)	Moment of Inertia I_x (m⁴)	Distance from Neutral Axis to Top Fiber y_t (m)	Static Moment about Neutral Axis S_x (m³)	Box Enclosed Area A_0 (m²)
Main girder	Pylon	18.7	15.8	1.3077	5.81	32.94
	Midspan	13.6	12.3	1.1351	7.89	30.78
Pylon	Bottom	25.58	96.5	2.75		
Cable	Side (SC18)	7.658E-03	4.7E-06	4.94E-02		

10.6.7.3 Effects Due to Dead Loads

The axial force, moment, and shear due to bridge self-weight are analyzed based on the assumed bridge construction sequence as shown in Fig. 10-83. The self-weight of the barrier walls and utilities are calculated as 63.5 kN/m (4.35 kips/ft) and applied on the final three-span continuous bridge. The axial force, moment, and shear distributions due to the dead loads for the main girder and pylon are illustrated in Figs 10-85 and 10-86, respectively. The distribution of the stay cable tensile force is shown in Fig. 10-103a.

10.6.7.4 Effects Due to Live Loads

The design live loads include a single truck or tandem with a uniform distributed lane loading as discussed in Section 2.2.3. The maximum effects at the related sections are determined by considering the following in the application of the live loads:

a. In the bridge longitudinal direction, position the live loads based on the shape of the influence line of the relevant section, as discussed in Section 10.5.2.6.2.
b. In the bridge transverse direction, load one lane up to the maximum number of allowed loading lanes individually and apply the appropriate multiple presence factors from Table 2-2 to determine the most unfavorable loading case and effects for design. To produce the maximum torsion on the bridge, the centerline of the exterior wheel is positioned 2 ft from the edge of the travel lane as shown in Fig. 2-2. The number of the maximum loading lanes is determined as

$$N = integer\ part\ of\left(\frac{w}{12} = \frac{48.03\ (ft)}{12} \right) = 4\ \text{lanes}$$

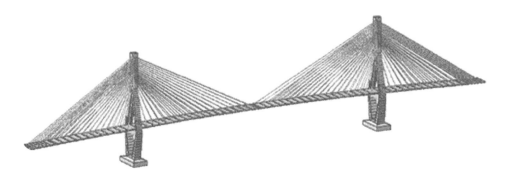

FIGURE 10-84 Analytical Model.

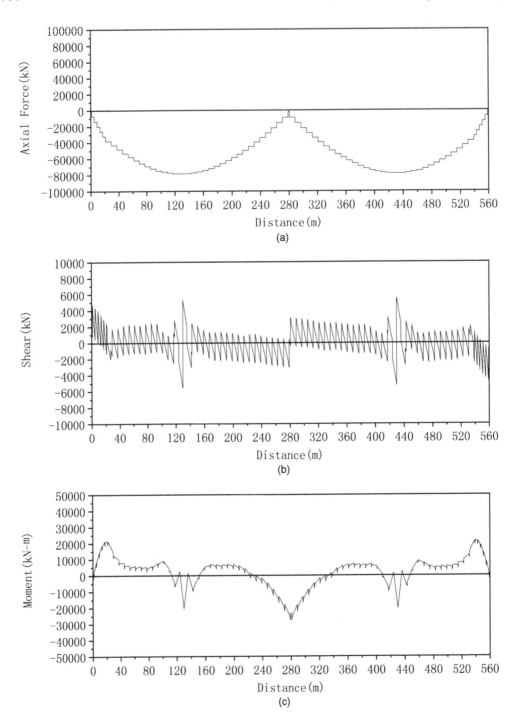

FIGURE 10-85   Effects of Dead Loads on Main Girder, (a) Axial Force, (b) Shear, (c) Moment (1 kip-ft = 1.356 kN-m, 1 ft = 0.3048 m, 1 kip = 4.448 kN).

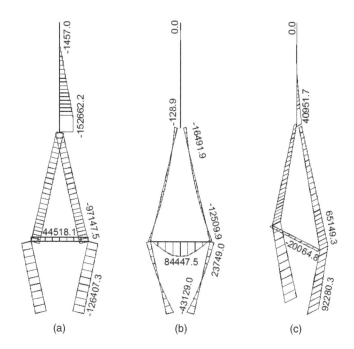

FIGURE 10-86 Effects of Dead Loads on Pylon, (a) Axial Force, (b) Transverse Moment, (c) Longitudinal Moment (unit: kN-m for moment, kN for axial force, 1 kip-ft = 1.356 kN-m, 1 kip = 4.448 kN).

The effects due to live loads are analyzed by computer program MIDAS and the axial force, moment, and shear envelopes for the main girder and pylon are shown in Figs. 10-87 and 10-88, respectively. The distribution of stay cable tensile force is shown in Fig. 10-103b.

10.6.7.5 Effects of Prestressing Force

All prestressing force losses are determined based on the methods described in Section 3.2 and the assumptions given in Section 10.6.2.5. The detailed analysis is done by the aforementioned computer program. The total effects of the effective post-tensioning forces on the main girder and pylon, including primary and secondary effects, are illustrated in Figs. 10-89 and 10-90, respectively. The effects on stay cable tensile forces are comparatively small and are not shown for simplicity.

10.6.7.6 Effects of Creep and Shrinkage

The creep and shrinkage effects are evaluated based on the *CEB-FIP Model Code for Concrete Structures*, 1990, i.e., Eqs. 1-9b and 9-15b. Each of the segments is assumed to be erected on the seventh day after it was cast. The creep coefficient and shrinkage strain can be determined similarly to those shown in Section 5.6.4.8.1. The effects of creep and shrinkage on the main girder and pylon at 10,000 days after casting time caused by dead and post-tensioning forces are illustrated in Figs. 10-91 and 10-92, respectively. The distribution of stay cable tensile force is shown in Fig. 10-103g.

10.6.7.7 Effect of Wind Loading

The static wind pressures on the main girder, pylons, and cables are calculated based on the method described in Sections 2.2.5.2 and 10.5.5.2.2.3. The design wind speeds for the main girder, pylons, and cables are determined by Eqs. 2-10, 10-51, and 10-52 based on their related design elevations. The detailed calculation procedures are similar to those presented in Section 5.6.4.9. The effects of the wind loading on the main girder and pylon are illustrated in Figs. 10-93 and 10-94, respectively. The distribution of stay cable tensile force due to longitudinal wind is shown in Fig. 10-103e.

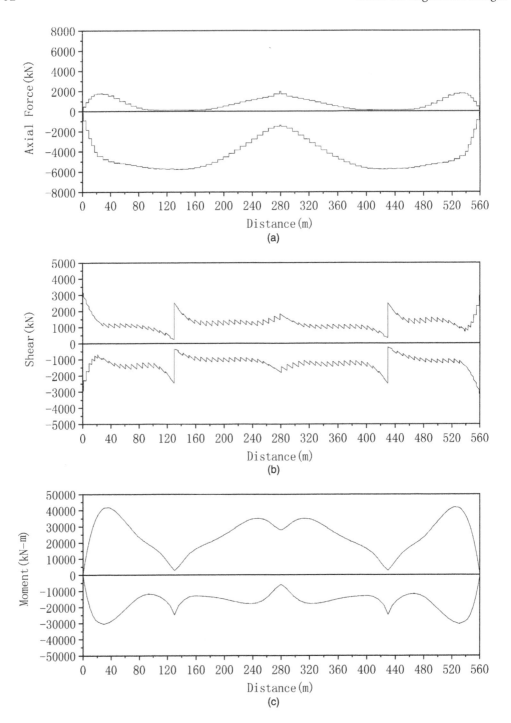

FIGURE 10-87   Effects of Live Loads on Main Girder, (a) Axial Force, (b) Shear, (c) Moment (1 kip-ft = 1.356 kN-m, 1 ft = 0.3048 m, 1 kip = 4.448 kN).

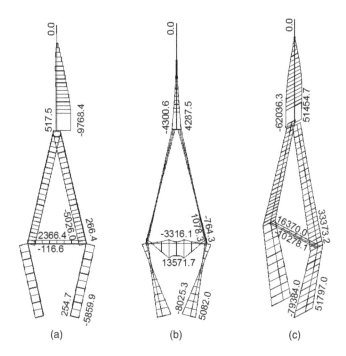

(a)   (b)   (c)

FIGURE 10-88   Effects of Live Loads on Pylon, (a) Axial Force, (b) Transverse Moment, (c) Longitudinal Moment (unit: kN-m for moment, kN for axial force, 1 kip-ft = 1.356 kN-m, 1 kip = 4.448 kN).

10.6.7.8   Effect of Earthquakes

The effects of earthquakes on the bridge are determined by the multimode spectral method (MM) as described in Section 2.2.6.2. The horizontal peak ground acceleration is taken as 0.1 g. The short spectral acceleration is taken as 0.2 g. The long period spectral acceleration is taken as 0.06 g. The maximum axial forces, shear, and moment of the main girders and pylon are illustrated in Figs. 10-95 and 10-96, respectively. The distribution of stay cable tensile force is shown in Fig. 10-103f.

10.6.7.9   Effects of Nonuniform Support Settlements

Through settling every other supports by 20 mm (0.79 in.) assumed sequentially, we can obtain the envelopes of axial forces, shears, and moments of the main girder and pylon as shown in Figs. 10-97 and 10-98. The distribution of stay cable tensile force is shown in Fig. 10-103h.

10.6.7.10   Effects of Temperature

The effects of temperature changes described in Section 10.6.2.4 on the main girder and pylons are determined by computer program MIDAS. The axial forces, shear, and moment distributions due to the uniform temperature changes for the main girder and pylons are shown in Fig. 10-99a to c and Fig. 10-100a to c, respectively. In Fig. 10-99, the solid lines represent the effects due to temperature rise and the dotted lines indicate temperature fall. Figure 10-99d to f and Fig. 10-100d to f illustrate the distributions of the axial forces, shears, and moments of the main girder and pylon induced by the main girder gradient temperatures, respectively. In Fig. 10-99, the solid lines represent the effects due to positive temperature gradient and the dotted lines indicate negative temperature gradient. The effects of the temperature differences between the cables and the remaining members on the main girder and pylon are displayed in Figs. 10-101 and 10-102, respectively. The effects of cable temperature differences on stay cable tensile force are shown in Fig. 10-103c and d.

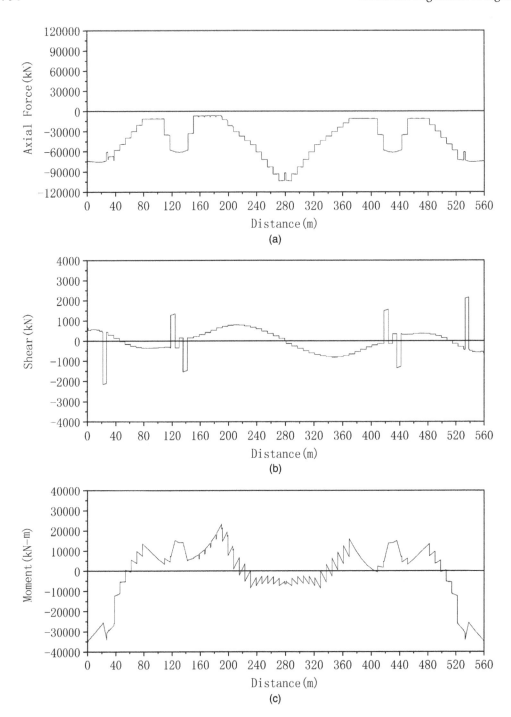

FIGURE 10-89 Effects of Post-Tensioning Tendons and Bars on Main Girder, (a) Axial Force, (b) Shear, (c) Moment (1 kip-ft = 1.356 kN-m, 1 ft = 0.3048 m, 1 kip = 4.448 kN).

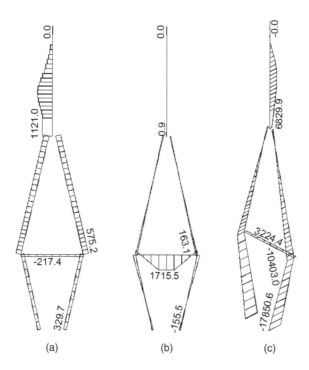

FIGURE 10-90   Effects of Post-Tensioning Tendons and Bars on Pylon, (a) Axial Force, (b) Transverse Moment, (c) Longitudinal Moment (unit: kN-m for moment, kN for axial force, 1 kip-ft = 1.356 kN-m, 1 kip = 4.448 kN).

10.6.7.11   Summary of Effects

The maximum effects of the principal loadings at several critical sections are summarized in Table 10-7.

10.6.7.12   Buckling Analysis

The bridge buckling analysis is done by simulating the bridge as a three-dimensional beam model and using the finite-element method discussed in Section 4.3.9. Bridge buckling analysis should be performed for all construction stages and the completed bridge. For the final completed structure, the design live loadings are placed to produce maximum axial forces for the main girder and pylon, based on their influence lines. The first two buckling modes are shown in Fig. 10-104. From this figure, we can see that the first buckling mode is in-plane buckling and the second buckling mode corresponds to lateral buckling. The corresponding elastic buckling safety factors obtained by solving Eq. 4-185 for the in-plane buckling and out-of-plane buckling are

$$\lambda_{k1} = 11.35 \ \left(\text{first in-plane buckling}\right)$$

$$\lambda_{k2} = 22.19 \ \left(\text{first out-of-plane buckling}\right)$$

10.6.7.13   Vibration Analysis

10.6.7.13.1   Free Vibration Behaviors

The bridge dynamic behaviors are analyzed by the finite-element method. The first several frequencies of the bridge are given in Table 10-8.

10.6.7.13.2   Wind-Induced Vibration

From the Table 11-6, it can be seen that the first torsion frequencies are much larger than the first bending frequencies, which indicates the bridge has good dynamic behaviors in resisting

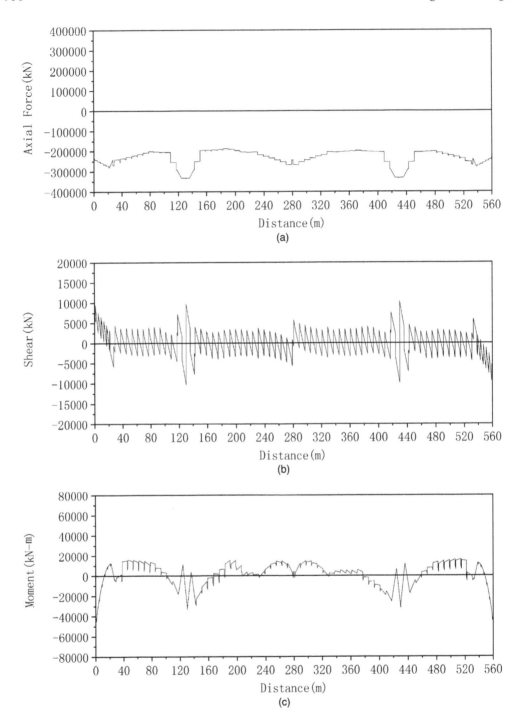

FIGURE 10-91 Effects of Shrinkage and Creep on Main Girder, (a) Axial Force, (b) Shear, (c) Moment (1 kip-ft = 1.356 kN-m, 1 ft = 0.3048 m, 1 kip = 4.448 kN).

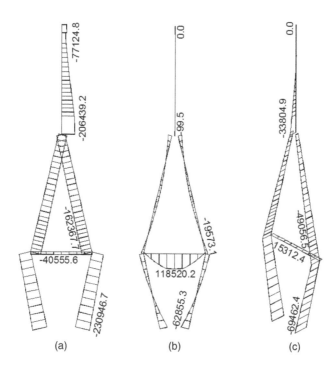

FIGURE 10-92 Effects of Shrinkage and Creep on Pylon, (a) Axial Force, (b) Transverse Moment, (c) Longitudinal Moment (unit: kN-m for moment, kN for axial force, 1 kip-ft = 1.356 kN-m, 1 kip = 4.448 kN).

wind-induced vibration (see Section 10.5.4.2). The critical wind speed for bridge flutter is determined from the model tests:

The critical wind speed for flutter during construction is

$$V_{cr} = 112\frac{m}{s}(250.59mph) \gg required : 70.8\,\frac{m}{s}\,(158.41mph)\,(ok)$$

The critical wind speed for flutter for the completed bridge is

$$V_{cr} = 104\frac{m}{s}(232.69mph) \gg required : 84.3\,\frac{m}{s}(188.61mph)\,(ok)$$

The test results on vortices also show the vibration amplitudes due to bending and torsion vortex shedding are within allowable limits.

10.6.7.14   Determination of Cable Forces in the Completed Bridge

The cable forces in the completed bridge are determined based on the assumption of a continuous beam supported on rigid supports at the cable anchor locations with zero vertical displacement under dead loads. Then, the cables forces are adjusted to an optimum state that meets all requirements of stress and geometry. The adjusted cable forces in the completed bridge are given in Table 10-9.

10.6.7.15   Determination of the Elevations of the Top Face of the Bottom Formwork for Main Girder Cast Segments

The elevations of the top face of the bottom formwork for the main girder segments are calculated as

$$El_{i(formwork)} = El_{i(design)} + f_{i(camber0)} - h_{i\,(girder\ depth)} - \Delta h_{i\,(overlay)} + f_{i\,(elastic)}$$

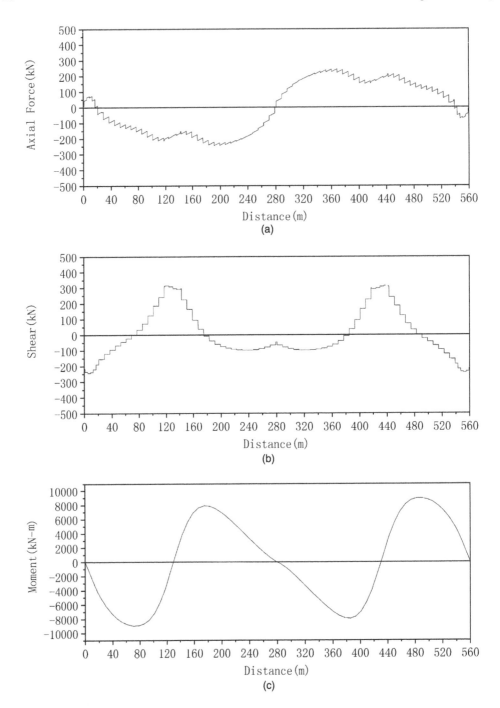

FIGURE 10-93 Effects of Wind Loadings on Main Girder, (a) Axial Force, (b) Shear, (c) Moment (1 kip-ft = 1.356 kN-m, 1 ft = 0.3048 m, 1 kip = 4.448 kN).

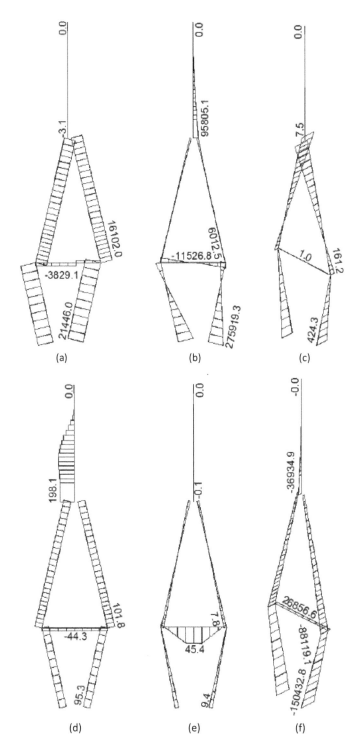

FIGURE 10-94 Effects of Wind Loadings on Pylon, (a) Axial Force Due to Transverse Wind, (b) Transverse Moment Due to Transverse Wind, (c) Longitudinal Moment Due to Transverse Wind, (d) Axial Force Due to Longitudinal Wind, (e) Transverse Moment Due to Longitudinal Wind, (f) Longitudinal Moment Due to Longitudinal Wind (unit = kN-m for moment, kN for axial force, 1 kip-ft = 1.356 kN-m, 1 kip = 4.448 kN).

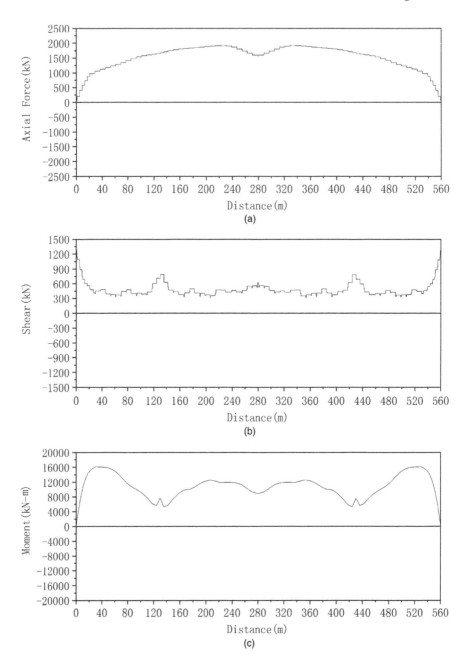

FIGURE 10-95 Effects of Earthquake Forces on Main Girder, (a) Axial Force, (b) Shear, (c) Moment (1 kip-ft = 1.356 kN-m, 1 ft = 0.3048 m, 1 kip = 4.448 kN).

where

$i = 1,...,n$ = cast segment number; n is total number of segments

$El_{i(formwork)}$ = top face elevation of bottom formwork

$El_{i(design)}$ = design elevation of deck

$f_{i(camber0)}$ = required camber after completing segment erection = deflection due to shrink-age and creep + deflection due to half of design live load, which is calculated by assuming that deflection varies in a parabolic curve and that maximum deflection is located at midspan

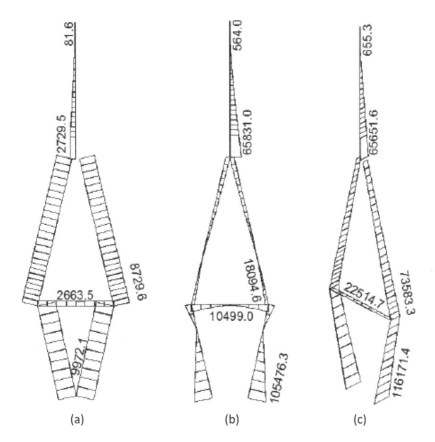

FIGURE 10-96 Effects of Earthquake Forces on Pylon, (a) Axial Force, (b) Transverse Moment, (c) Longitudinal Moment (unit = kN-m for moment, kN for axial force, 1 kip-ft = 1.356 kN-m, 1 kip = 4.448 kN).

$h_{i(girder\ depth)}$ = girder depth
$h_{i(overlay)}$ = deck overlay thickness
$h_{i(elastic)}$ = deflection induced by formwork, newly cast concrete, and post-tensioning forces

10.6.8 CAPACITY VERIFICATIONS

10.6.8.1 Capacity Check of Main Girder

The capacity check for the main girder is similar to that for other types of concrete segmental bridges discussed in Chapters 5 to 7 and 9. For simplicity and clarity, only a brief description of the required capacity check items is presented in this section. Refer to Design Examples I, II, V, and VII for detailed procedures of the verification of girder capacity, including the anchorages.

10.6.8.1.1 Service Limit States

For prestressed concrete bridges, two service limit states should be checked, i.e., Service I and Service III. For the segmentally constructed main girder, the AASHTO specifications require a special load combination check as indicated in Eq. 2-51. As the effects of the concrete creep and shrinkage vary with time, the bridge capacity verifications may be performed for different loading stages. These include immediately after instantaneous prestressing losses, immediately after construction, 4000 days, 10,000 days, and final stage.

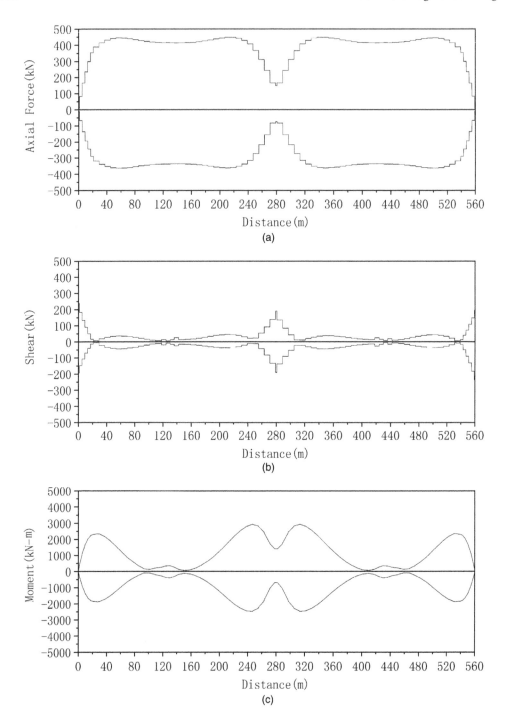

FIGURE 10-97 Effects of Nonuniform Support Settlements on Main Girder, (a) Axial Force, (b) Shear, (c) Moment (1 kip-ft = 1.356 kN-m, 1 ft = 0.3048 m, 1 kip = 4.448 kN).

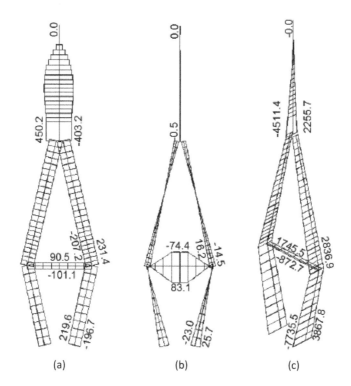

FIGURE 10-98 Effects of Nonuniform Support Settlements on Pylon, (a) Axial Force, (b) Transverse Moment, (c) Longitudinal Moment (unit = kN-m for moment, kN for axial force, 1 kip-ft = 1.356 kN-m, 1 kip = 4.448 kN).

10.6.8.1.1.1 Service I Limit State The Service I check investigates concrete compression stresses. Normally, the concrete compression stresses in three loading cases should be checked (see Section 1.2.5.1):

Case I: The sum of the concrete compression stresses caused by tendon jacking forces and permanent dead loads should be less than $0.6f_c'$, i.e.,

$$f_j^c + f_{DC+DW}^c \leq 0.6f_c'$$

Case II: Permanent stresses after losses. The sum of the concrete compression stresses caused by the tendon effective forces and permanent dead loads should be less than $0.45f_c'$, i.e.,

$$f_e^c + f_{DC+DW}^c \leq 0.45f_c'$$

Case III: The sum of the concrete compression stresses caused by tendon effective forces, permanent dead loads, and transient loadings should be less than $0.60\phi_w f_c'$. The corresponding load combination for the example bridge is

$$DC + CR + SH + PS + LL\,(1+IM) + BR + 0.3WS + WL + 1.0TU + 0.5TG$$

The detailed calculation procedures are similar to those discussed in Sections 5.6 and 6.14.

Analytical results indicate the actual concrete compression stresses at all locations are less than the allowable concrete stresses for all the above three loading cases.

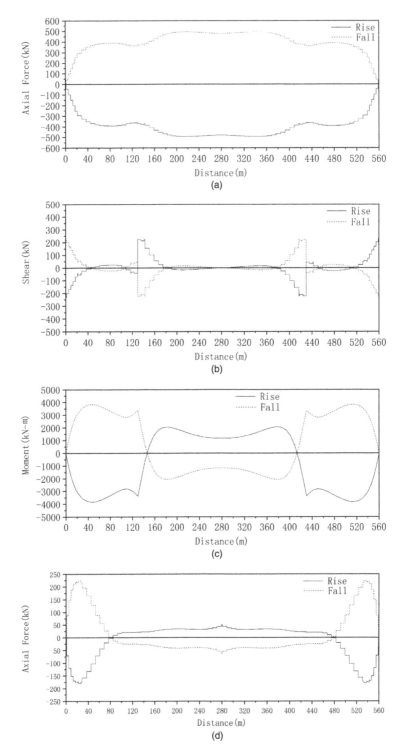

FIGURE 10-99 Effects of Temperature Changes on Main Girder, (a) Axial Force Due to Uniform Temperature, (b) Shear Due to Uniform Temperature, (c) Moment Due to Uniform Temperature, (d) Axial Force Due to Temperature Gradient, (e) Shear Due to Temperature Gradient, (f) Moment Due to Temperature Gradient (1 kip-ft = 1.356 kN-m, 1 ft = 0.3048 m, 1 kip = 4.448 kN).

(*Continued*)

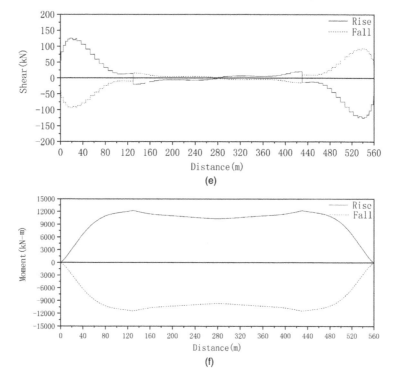

FIGURE 10-99 (Continued)

10.6.8.1.1.2 Service III Limit State The Service III check investigates the maximum tensile stresses in the prestressed concrete components to ensure those stresses are not greater than the limits specified in Sections 1.2.5.1.1.2 and 1.2.5.1.2.2. Three cases should be checked:

Case I: Temporary tensile stresses before losses
Case II: Tensile stresses at the Service III load combination
The allowable tensile stress for this design example is

$$f_{ta} = 0.0948\sqrt{f_c'(ksi)} = 1.67 \ Mpa \ (0.242 \ ksi)$$

Case III: The principal tensile stress at the neutral axis in the web

For a section, the principal stress check should be considered the maximum of the following four cases: maximum absolute negative shear with corresponding torsion, maximum positive shear with corresponding torsion, maximum absolute negative torsion with corresponding shear, and maximum positive torsion with corresponding shear.
The allowable principal stress for this design example is

$$f_{ta} = 0.11\sqrt{f_c'(ksi)} = 1.93 \ Mpa \ (280 \ psi)$$

The Service III load combination for the example bridge is

$$DC + CR + SH + PS + 0.8 \ LL \ (1 + IM) + BR + 1.0TU + 0.5TG$$

The detailed calculation procedures are similar to those discussed in Sections 5.6 and 6.14.

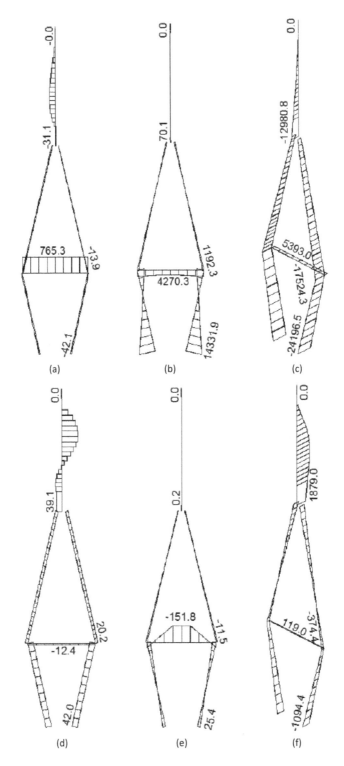

FIGURE 10-100 Effects of Temperature Changes on Pylon, (a) Axial Force Due to Uniform Temperature Rise, (b) Shear Due to Uniform Temperature Rise, (c) Moment Due to Uniform Temperature Rise, (d) Axial Force Due to Positive Temperature Gradient, (e) Shear Due to Positive Temperature Gradient, (f) Moment Due to Positive Temperature Gradient (unit = kN-m for moment, kN for axial force, 1 kip-ft = 1.356 kN-m, 1 kip = 4.448 kN).

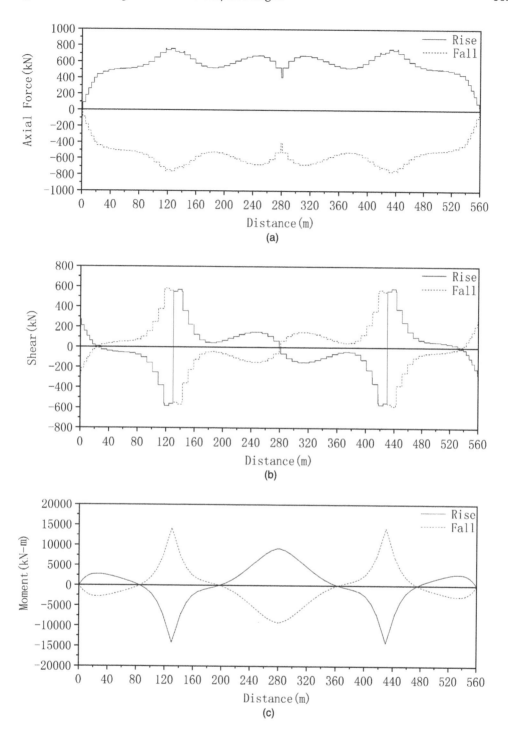

FIGURE 10-101 Effects of Cable Temperature Difference on Main Girder, (a) Axial Force, (b) Shear, (c) Moment (1 kip-ft = 1.356 kN-m, 1 ft = 0.3048 m, 1 kip = 4.448 kN).

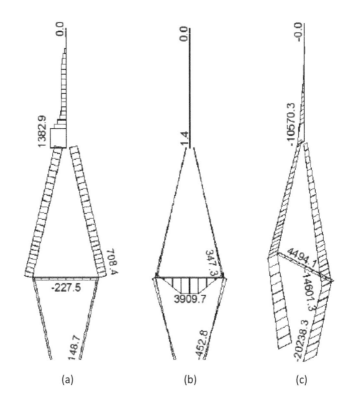

FIGURE 10-102   Effects of Cable Temperature Difference on Pylon, (a) Axial Force, (b) Transverse Moment, (c) Longitudinal Moment (unit: kN-m for moment, kN for axial force, 1 kip-ft = 1.356 kN-m, 1 kip = 4.448 kN).

10.6.8.1.1.3 Special Load Combination Check   This loading combination may control the design for locations where the live load effects are small or where the locations are outside of the precompressed tensile zones, such as the tension in the top of closure pours and compression in the top of the box girder over the piers. For this design example, the special load combination as shown in Eq. 2-51 can be written as

$$DC + CR + SH + TG + DW + WA + PS$$

The detailed calculation procedures are similar to those discussed in Section 6.14. For this bridge, the special load combination does not control the design.

10.6.8.1.2 Deflection Check

The bridge maximum deflections at the side and center spans due to live load are calculated by computer program. They are
   At the center span:

$$\Delta_{center\ span} = 136.77\ mm\ (5.38\ in)$$

   The ratio of deflection to span length: $\frac{\Delta_{center\ span}}{l_{center\ span}} = \frac{136.77}{300 \times 1000} = 1/2195 < 1/800\ (ok)$
At the side span:

$$\Delta_{side\ span} = 69.09\ mm\ (2.72\ in)$$

   The ratio of deflection to span length: $\frac{\Delta_{side\ span}}{l_{side\ span}} = \frac{69.09}{130 \times 1000} = 1/1882 < 1/800\ (ok)$

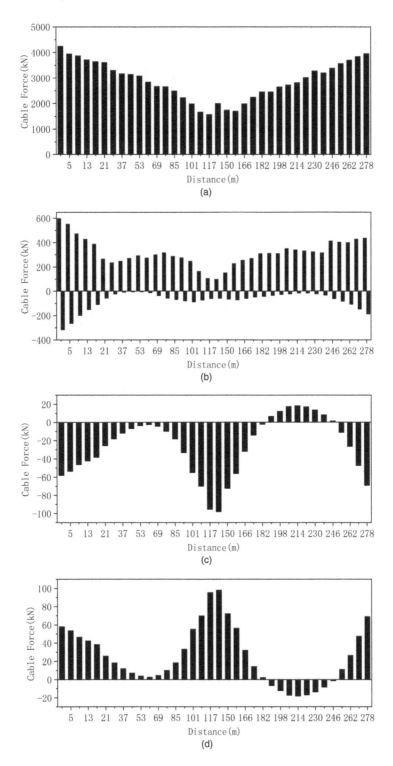

FIGURE 10-103 Variations of Cable Tensile Forces, (a) Effects of Dead Load, (b) Effects of Live Loading, (c) Effects of Cable Temperature Rise, (d) Effects of Cable Temperature Fall, (e) Effects of Longitudinal Wind Loading, (f) Effects of Earthquake, (g) Effects of Shrinkage and Creep, (h) Envelope of Cable Tensile Forces Due to Nonuniform Support Settlements (1 kip = 4.448 kN).

(Continued)

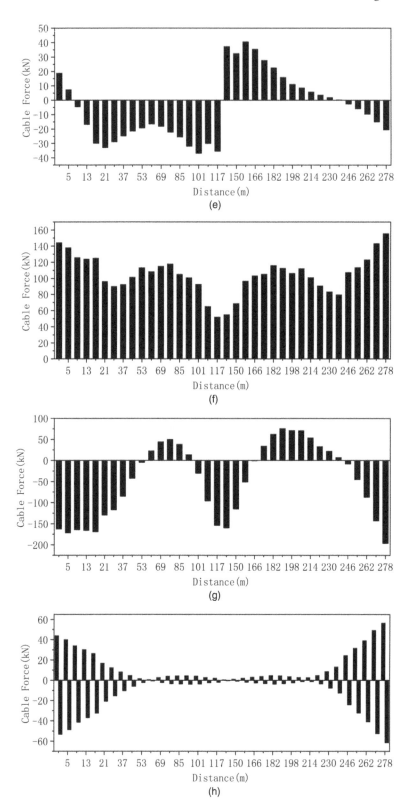

FIGURE 10-103   (Continued)

Table 10-7
Summary of Maximum Effects at Typical Control Sections

Type of Loadings	Type of Forces	Member and Location			
		Main Girder		Pylon	Cable
		Interior Support	Midspan	Bottom	SC18
Dead loads	M (kN-m)	−20,479	−23,312	−43,139	
	T (kN-m)	−32.4	0	−7,028.7	
	V (kN)	5,552	0	−4,187	
	N (kN)	−78,261	−1,945	−126,406	4,243
Live loads	M (kN-m)	−24,783	−6,113	−8,032	
	T (kN-m)	12,253.7	7,205.2	5,784.2	
	V (kN)	2,477	−1,796	−436	
	N (kN)	−5,712	−1,396	−5,860	598
Post-tensioning	M (kN-m)	13,954	−6,368	−155.5	
	T (kN-m)	0.9	0	1,219.7	
	V (kN)	347	−0	−15.8	
	N (kN)	−60,935	−91,406	329.7	−97.5
Shrinkage and creep	M (kN-m)	−28,837	3,869	−62,843	
	T (kN-m)	16.7	0	5,178.0	
	V (kN)	10,305	0	−6,390	
	N (kN)	−342,536	−255,491	−230,947	−162.6
Uniform temperature	M (kN-m)	−3,377 (3,377)	1,161 (−1,161)	−14,395 (14,395)	
	T (kN-m)	−105.5 (105.4)	0	−1,872.2 (1,872.2)	
	V (kN)	−42(42)	0	−722.9 (722.9)	
	N (kN)	−363 (363)	−479 (479)	46 (−46)	31.3 (−31.3)
Gradient temperature	M (kN-m)	12,144 (−6,072)	10,331 (−5,166)	25.3 (−12.6)	
	T (kN-m)	−0.3(0.3)	0	56.6 (−28.3)	
	V (kN)	16.9 (8.5)	0	2 (−1)	
	N (kN)	21.6 (−10.8)	50.6 (−25.3)	41.8 (−21)	39.2 (−19.6)
Cable temperature	M (kN-m)	−14,223,(14,223)	9,159 (−9,159)	−453 (453)	
	T (kN-m)	−0.3 (0.3)	0	1,562.1 (−1,562.1)	
	V (kN)	560 (−560)	0	−41.1 (41.1)	
	N (kN)	759 (−759)	405 (−405)	148.7 (−148.7)	−58.1 (58.1)
Wind loading	M (kN-m)	−140	0	10.7	
	T (kN-m)	9.4	−9.8	10,152.7	
	V (kN)	308	−46.6	0.36	
	N (kN)	182	0	95.5	18.8
Earthquake	M (kN-m)	7,589	8,995	105,393	
	T (kN-m)	2,767.1	52.2	7,967.7	
	V (kN)	694	620	4,951	
	N (kN)	1,653	1,567	9,972	144
Support settlement	M (kN-m)	−412	−702	−23	
	T (kN-m)	0.2	0.1	603.2	
	V (kN)	8.2	−190	−0.9	
	N (kN)	415	−74	−197	43.8

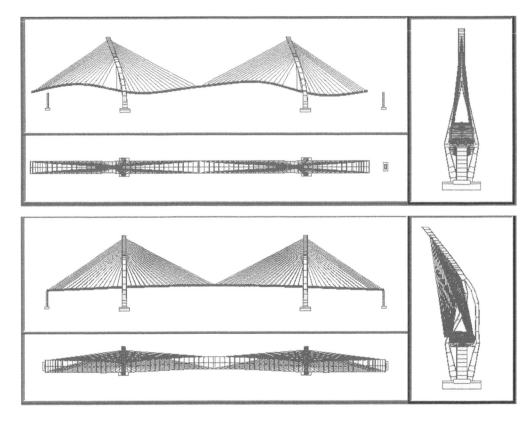

FIGURE 10-104 Buckling Modes, (a) First Buckling Mode, (b) Second Buckling Mode.

10.6.8.1.3   Strength Limit States

Strengths I, II, III, and V and Extreme Events I and II need to be evaluated. The strength checks include biaxial flexural strength, shear, and torsional strength.

10.6.8.1.3.1   Biaxial Flexural Strength   The main girder in a cable-stayed bridge is subjected to not only large bending movements but also to large axial forces as previously discussed. Its capacity check can be done by following the procedures shown in Fig. 3-61 and verified by Eqs. 3-230a and 3-230b discussed in Section 3.7.3.2, i.e.:

TABLE 10-8
Natural Frequencies (Hz)

Structural System	Vertical Bending		Lateral Bending		Torsion	
	First	Second	First	Second	First	Second
Completed bridge	0.3396	0.4911	0.5029	1.3804	1.4435	1.8412
During construction, with single maximum cantilever	0.3675	0.7576	0.3987	1.8630	1.5350	2.5888
During construction, with double maximum cantilevers	0.8360	1.1357	0.8733	1.2243	2.1104	2.1620

TABLE 10-9
Cable Forces in the Completed Bridge

Side Span, Cable Number	Cable Force (kN)	Center Span, Cable Number	Cable Force (kN)
1	1879	1	1910
2	1613	2	1705
3	1560	3	1733
4	2238	4	2053
5	2499	5	2332
6	2659	6	2554
7	2619	7	2544
8	2763	8	2705
9	2954	9	2748
10	2976	10	2799
11	3103	11	2955
12	2962	12	3205
13	3525	13	3100
14	3479	14	3251
15	3618	15	3415
16	3803	16	3535
17	3910	17	3651
18	4096	18	3746

If factored axial force $P_u \geq 0.10\phi_a f_c' A_g$:

$$P_{rxy} \geq P_u$$

Otherwise:

$$\frac{M_{ux}}{M_{rx}} + \frac{M_{uy}}{M_{ry}} \leq 1.0$$

See Eqs. 3-230a and 3-230b for more details.

The procedures for checking the biaxial flexural capacities can be found in Section 11.7 (Design Example VII). Analytical results show the biaxial flexural capacities at all cross sections of the main girder are adequate.

10.6.8.1.3.2   Shear and Torsional Strength   The main girder shear and torsional strength should meet (see Eq. 3-217b)

$$V_u + \frac{T_u}{2A_0} \leq \phi_v V_n$$

From Section 2.3.3.5, the resistance factor for shear and torsion is

$$\phi_v = 0.9$$

The enclosed area A_0 of the box section is shown in Table 10-6. The detailed procedure for checking the shear and torsional capacities can be found in Fig. 3-54 and Design Examples I and II (sections 5.6 and 6.14). Analytical results show the shear and torsional strength of the main girder is adequate.

The main girder cross-sectional dimension also should meet the following requirement (see Eq. 3-221):

$$\frac{V_u}{b_v d_v} + \frac{T_u}{2 A_0 b_e} \leq 0.474 \sqrt{f'_c (ksi)}$$

The detailed procedure for checking the cross-sectional dimension can be found in Design Examples I and II (sections 5.6 and 6.14). The analytical results indicate that the requirement is satisfactory.

10.6.8.2 Capacity Check of Pylon

10.6.8.2.1 General

The transverse beams of the pylons and the portion around the pylon section transition area are post-tensioned, and their capacity should be verified following the same procedures discussed in Design Examples I and II (sections 5.6 and 6.14). As there are not any prestressing and post-tensioning strands used in the vertical direction of the pylon, the capacity check of the horizontal sections of the pylon is essentially the same as that for reinforced concrete structures. The pylon is subjected to large axial forces as well as large bending moments. Its capacity is generally checked as an eccentrically loaded compression member. The capacity checks of the pylon include the minimum and maximum reinforcement checks, service capacity check, and strength capacity check. As the basic calculation procedures are the same as those presented in Sections 9.6 and 11.7 (Design Examples V and VII), only a brief description of the items to be checked is presented as follows.

10.6.8.2.2 Minimum and Maximum Reinforcement Check

The minimum and maximum reinforcement check is performed to reduce the possibility of brittle failure. The ratio of the pylon reinforcement should not be greater than the values determined based on Eqs. 3-231 and 3-232.

The reinforcement ratio of the pylon should not be smaller than that determined by Eq. 3-233a or greater than that determined by Eq. 3-233b.

Analytical results show the pylon reinforcement ratios meet both maximum and minimum reinforcement requirements.

10.6.8.2.3 Capacity Verification of Pylon

10.6.8.2.3.1 Service I Limit State
The Service I check is generally to control the crack width for the reinforced concrete segmental bridge. As discussed in Section 3.6.2.5.2, the AASHTO LRFD specifications limit the concrete crack width by determining the spacing of reinforcement based on the reinforcement tensile stress induced by the Service I load combination.

Analytical results indicate that all the cross sections meet the Service I requirement.

10.6.8.2.3.2 Strength Limit State
Strengths I through V and Extreme Events I and II may need to be evaluated. The strength checks include axial and bending strength as well as shear strength. The detailed calculation procedures can be found in Sections 9.6 and 11.7 (Design Examples V and VII). Analytical results show all required strength capacities are satisfactory.

10.6.8.3 Stability Check

As previously mentioned, the AASHTO LRFD specifications do not have clear requirements for stability checks. Based on Chinese specifications, the elastic buckling safety factor of a cable-stayed bridge should not be less than 4. From the buckling analysis of the bridge shown in Section 10.6.7.12, we have the minimum buckling safety factor:

$$\lambda_k = 11.35 > 4 \ (ok!)$$

10.6.8.4 Capacity Check of Cables

10.6.8.4.1 General Required Check

Based on the discussions in Section 10.5.5.2.2, the following strength and fatigue limits should be checked for this design example (refer to Table 2-10 and Section 10.5.5.2.2). The factored cable tensile forces can be calculated as:

1. Strength I

$$T_{u1} = 1.25DC + 1.75(LL + IM) + PS + 1.25(CR + SH) + 0.5TU + SE$$

2. Strength II

$$T_{u2} = 1.25DC + 1.35(LL + IM) + PS + 1.25(CR + SH) + 0.5TU + SE$$

3. Strength III:

$$T_{u3} = 1.25DC + 1.4WS + PS + 1.25(CR + SH) + 0.5TU + SE$$

4. Strength IV

$$T_{u4} = 1.5DC + PS + 1.5(CR + SH) + 0.5TU$$

5. Strength V

$$T_{u5} = 1.25DC + 1.35(LL + IM) + 0.4WS + WL + PS + 1.25(CR + SH) + FR + 0.5\ TU + SE$$

6. Extreme Event I

$$T_{u6} = 1.25DC + (LL + IM) + PS + 1.25(CR + SH) + FR + EQ$$

7. Fatigue

$$T_{u7} = 1.05\Delta F(LL)$$

8. Cable Replacement Event

$$T_{u8} = 1.2\text{DC} + 1.4\text{DW} + 1.5(\text{LL} + \text{IM}) + \text{Cable Exchange Force}$$

9. Cable Loss Event

$$T_{u9} = 1.1DC + 1.35DW + 0.75\ (LL + IM) + 1.1CLDF$$

10. Construction Strength Limit State

$$T_{u10} = 1.2(DC + DW + CE + CLE) + 1.2(WS + WE + WUP) + 1.4CLL$$

11. Construction Extreme Limit State

$$T_{u11} = 1.1(DC + DW + CE + CLE) + 1.0AI$$

10.1.8.4.2 Strength I Check

Take the side stay cable SC18, for example, to check its Strength I and Fatigue capacities:
Strength I:
Factored Tensile Force:

$$
\begin{aligned}
T_{u1} &= 1.25DC + 1.75(LL + IM) + PS + 1.25(CR + SH) + 0.5TU + SE \\
&= 1.25 \times 4243 + 1.75 \times 598 \times 1.33 - 97.5 - 1.25 \times 162.6 + 0.5 \times 31.3 + 0.5 \times 58.1 + 43.8 \\
&= 6483.0 \ kN
\end{aligned}
$$

Stay Cable SC18 has 199 ϕ7-mm wires, and its nominal tensile strength is (see Table10-1).

$$
T_n = 64.3 \times 199 = 12796.0 \ kN
$$

The ratio of live load effect plus wind to minimum ultimate strength is

$$
\frac{(LL + W)}{T_n} = \frac{598 \times 1.33 + 18.8}{12796} = 0.0636
$$

From Fig. 10-67, we have the resistance factor

$$
\phi_a = 0.66
$$

Thus,

$$
T_{u1} < \phi_a T_n = 8445.4 \ kN \ (\text{ok})
$$

10.1.8.4.3 Fatigue Check

The stress range due to one lane of the fatigue load, including the dynamic allowance 0.15 (see Section 2.2.4.2), was calculated as

$$
\Delta F_{LL} = 51.67 \ Mpa
$$

From Eq. 10-54, the factored stress range is

$$
\Delta F_u = 1.05 \times \Delta F_{LL} = 54.25 \ Mpa
$$

From Table 10-5, for parallel wire cables, we have the fatigue constant as

$$
(\Delta F)_{TH} = 145 \ Mpa
$$

From Eq. 11-56a, we have

$$
G = \frac{(\Delta F)_{TH}}{2} = 72.5 \ Mpa
$$

$$
\Delta F_u < G
$$

$$
(\Delta F)_n = G = 72.5 \ Mpa \ (\text{ok})
$$

The capacity checks for other state limits can be done in a similar way as above, and it is found that all limit state requirements meet the AASHTO specifications.

REFERENCES

10-1. Walther, R., Houriet, B., Isler W., and Moia, P., *Cable Stayed Bridges*, Thomas Telford Ltd, London, 1988.

10-2. Torroja, E., *Philosophy of Structures*, English version by Polivka, J. J., and Polivka, M., University of California Press, Berkeley and Los Angeles, 1958.

10-3. Podolny, W., and Muller, J. M., *Construction and Design of Prestressed Concrete Segmental Bridges*, A Wiley-Interscience Publication, John Wiley & Sons, New York, 1982.

10-4. ASBI, *Construction Practices Handbook for Concrete Segmental and Cable-Supported Bridges*, American Segmental Bridge Institute, Buda, TX, 2008.

10-5. Shao, X. D., and Gu, A. B., *Bridge Engineering*, 3rd edition, China Communication Press, Beijing, China, 2014.

10-6. Liu, L. J., and Xu. Y., *Bridge Engineering*, 1st edition, China Communication Press, Beijing, China, 2017.

10-7. Ling, Y. P., *Cable-Stayed Bridges*, 3rd edition, China Communication Press, 2014, Beijing, China, 2003.

10-8. Fan, L. C., *Bridge Engineering*, China Communication Press, Beijing, China, 1989.

10-9. DSI, "DYWIDAG Multi-Strand Stay Cables Systems," DYWIDAG-Systems International, www.dywidag-systems.com.

10-10. PTI, *Recommendations for Stay Cable Design, Testing, and Installation*, 6th edition, Post-Tensioning Institute, Farmington Hills, MI, 2012.

10-11. Xiang, H. F., *Advanced Theory of Bridge Structures*, China Communication Press, Beijing, China, 2001.

10-12. Menn, C., *Prestressed Concrete Bridges*, Birkhauser Verlag AG Basel, Berlin, Germany, 1990.

10-13. Huang, D. Z., "Lateral Stability of Arch and Truss Combined System," Master Degree Thesis, Tongji University, Shanghai, China, 1985.

10-14. Huang, D. Z., "Elastic and Inelastic Stability of Truss and Arch-Stiffened Truss Bridges," Ph.D. Degree Thesis, Tongji University, Shanghai, China, 1989.

10-15. Huang, D. Z., and Li, G. H., "Inelastic Lateral Stability of Truss Bridges," *Journal of Tongji University*, Vol. 10, No. 4, 1988, pp. 405–420.

10-16. Huang, D. Z., Li, G. H., and Xiang, H. F., "Inelastic Lateral Stability of Truss Bridges with Inclined Portals," *Journal of Civil Engineering of China*, Vol. 24, No. 3, 1991, pp. 27–36.

10-17. Huang, D. Z., and Li, G. H., "Inelastic Lateral Stability of Arch-Stiffened Truss Bridges," *Journal of Tongj University*, Vol. 10, No. 4, 1991, pp. 1–11.

10-18. Huang, D. Z., "Lateral Bending and Torsion Buckling Analysis of Jiujian Yangtze River Arch-Stiffened Truss Bridge," *Proceedings, IABSE Congress*, Stockholm, 2016, pp. 150–157.

10-19. Li, G. H., Xiang, H. F., Chen, Z. R., Shi, D., and Huang D. Z., *"Stability and Vibration in Bridge Structures,"* China Railway Publisher, Beijing, China, 1992.

10-20. Leonhardt, F., "Cable Stayed Bridges with Prestressed Concrete," *PCI Journal*, Special Report, Vol. 32, No. 5, 1987, pp. 52–80.

10-21. Stroh, S. L., "Extradosed Prestressed Concrete Bridges," *ASPIRE*, Summer 2015, pp. 28–30.

10-22. Xiang, H. F., "Modern Theory and Practice of Bridge Wind Resistance," China Communications Press, Beijing, China, 2005.

11 Design of Substructure

- Types and arrangement of substructures
- Analysis of flexure piers
- Analysis and selection of bridge bearings
- Expansion joints
- Design example: pier

11.1 INTRODUCTION

In concrete segmental bridges, greater savings may often be achieved from optimization of sub-structure design, including the foundation, than from the super-structure itself. However, in this book we do not intend to cover the foundation design as it is related to geotechnical information. In this chapter, the basic types of piers, abutments, and bearings are introduced first. Then, the analytical methods and design of bridge substructures are discussed. After that, the analysis and design of typical expansion joints are presented. Finally, a design example generated based on the existing pier is given.

11.2 TYPES OF PIERS

11.2.1 TYPICAL PIER SHAPES

Piers may be classified as solid and box piers based on their cross sections or as moment-resisting piers and flexible piers based on their mechanical behaviors.

11.2.1.1 Solid Piers

One of the primary advantages of concrete segmental bridges is that the width of supporting pier caps is significantly smaller than that for multi-girder bridges. For piers with short and medium column heights, single columns with solid sections to resist large compression forces are typically used (see Fig. 11-1). The shape of the pier section can be rectangular (see Fig. 11-2a), circular (see Fig. 11-2b), rectangular with two circular ends (see Fig. 11-2c), diamond (see Fig. 11-2d), etc. Generally, circular shapes are more suitable for variable directions of stream or tidal flow or acute angles less than 75° between the water flow direction and the bridge longitudinal direction (see Fig. 11-2). The diamond-shaped section is more suitable for acute angles greater than 85° between the water flow direction and the bridge longitudinal direction and reduces the effect on the bridge foundation scour. The shape of the rectangle with two semi-circular ends is generally more suitable for acute angles greater than 75° between the water flow direction and bridge longitudinal direction.

11.2.1.2 Box Piers

For tall piers, it may be cost effective to use thin-walled box piers (see Fig. 11-3). This type of pier can have a significantly reduced self-weight and more effectively uses the concrete members. Moreover, this type of pier may be precast due to its lighter self-weight (see Figs. 11-3 and 11-4). The precast tubular segments (see Fig. 11-5) and the foundation (see Fig. 11-9) are post-tensioned together. Thus, the construction time may be significantly reduced. The box pier cross sections can have different shapes, as shown in Fig. 11-6. To increase the wall buckling capacity, stiffening ribs can be used (see Fig. 11-4a).

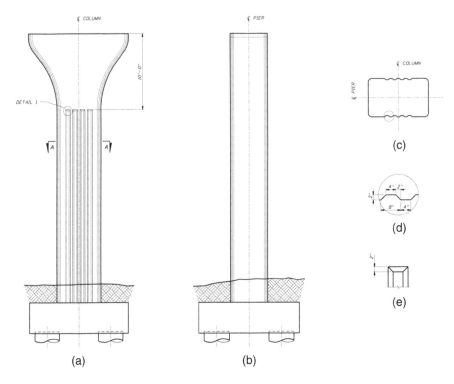

FIGURE 11-1 Typical Solid Pier, (a) Elevation View, (b) Side View, (c) Section *A-A*, (d) Rustication Detail, (e) Detail I.

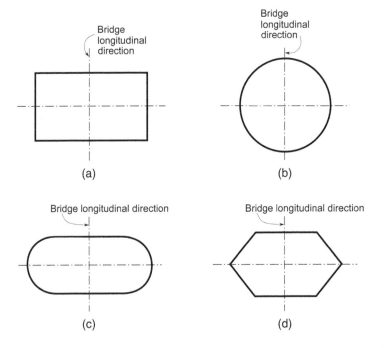

FIGURE 11-2 Typical Shapes of Solid Piers, (a) Rectangular, (b) Circular, (c) Rectangle with Semi-Circular Ends, (d) Diamond[11-1, 11-2].

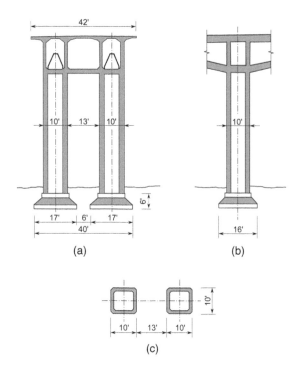

FIGURE 11-3 Precast Square Box Pier, (a) Elevation View, (b) Side View, (c) Plan View[11-3].

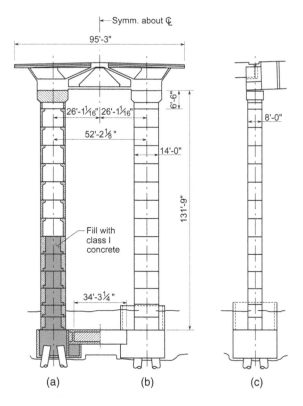

FIGURE 11-4 Precast Ellipse Box Pier, (a) Transverse Half-Section, (b) Transverse Half-Elevation, (c) Side View[11-3].

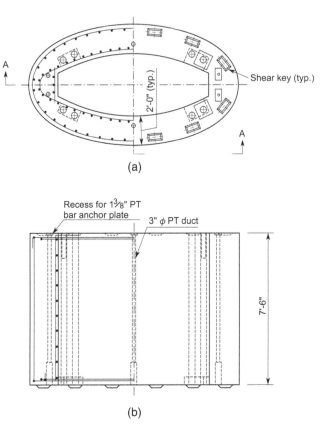

FIGURE 11-5 Typical Pier Segment, (a) Plan View (Half Cross Section on Left; Half Top View on Right), (b) Section *A-A* View.

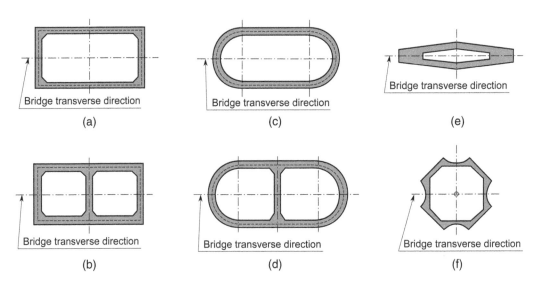

FIGURE 11-6 Typical Sections of Box Piers, (a) Single-Cell Rectangular Box, (b) Two-Cell Rectangular Box, (c) Single-Cell Box with Semi-Circular Ends Box, (d) Two-Cell Box with Semi-Circular Ends Box, (e) Single-Cell Long Diamond, (f) Single-Cell Diamond[11-1, 11-2].

11.2.1.3 AASHTO-PCI-ASBI Standard Segmental Box Piers and Details

Based on previous design experience, AASHTO-PCI-ASBI[11-4 to 11-6] developed a set of standard rectangular box pier segments for span-by-span and cantilever segmental bridges with span lengths ranging from 100 to 200 ft. Figure 11-7 shows a typical pier segment with a cross section of 15 ft by 8 ft. The corresponding cap segment is illustrated in Fig. 11-8. The dimensions for other typical segments can be found in Appendix A. The precast pier segments are match-cast and utilize Type A epoxy joints. Post-tensioning tendons are used to connect the segments together. The post-tensioning tendons can be anchored in the cast-in-place footing (see Fig. 11-9a) or in the cast-in-place column stub (see Fig. 11-9b). More details and requirements for the standard precast box piers can be found in Appendix A.

11.2.1.4 Transverse Layouts

11.2.1.4.1 Single Boxes and Single Piers

If a bridge super-structure consists of one single box girder and if the pier location does not interfere with the roadway or structures below, the pier cap can be symmetrically arranged as shown in Fig. 11-10a. If a larger or a multi-cell box is used for the super-structure, a V-shape pier may be used (see Fig. 11-10b). In some cases, the proposed pier cannot be placed along the proposed overhead roadway alignment due to the limitations of the roadway or structures below. In those cases, the pier may be asymmetrically arranged as a cantilever cap pier as shown in Fig. 11-10c. For this type of pier, it is often necessary to provide post-tensioning tendons to resist large bending moments, especially for long cantilever caps (see Fig. 11-10d).

11.2.1.4.2 Multi-Box Multi Piers

For deck widths larger than 85 ft, a large single-box girder is seldom used in concrete segmental bridges due to construction considerations. A common practice is to use two smaller similar box

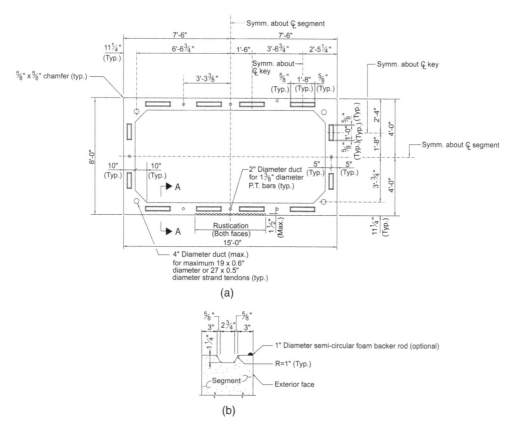

FIGURE 11-7 Typical Standard Box Pier Segment, (a) Plan View, (b) Section *A-A* View[11-4, 11-5].

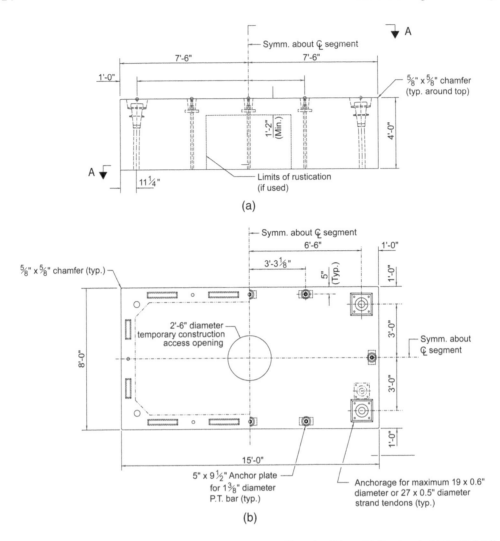

FIGURE 11-8 Typical Standard Box Pier Cap Segment, (a) Elevation View, (b) Section *A-A* View[11-4, 11-5].

girders that are connected by a cast-in-place longitudinal joint (see Fig. 11-11). For this case, similar separate piers are typically used to support each of the box girders as shown in Fig. 11-11.

11.2.1.4.3 Straddle Bents

In highly urbanized areas, it is often difficult to place the piers to exactly accommodate the requirements of the proposed roadway alignment. If a cantilever pier cap is not practical to resolve the limitation under the proposed roadway, a straddle bent is typically used (see Fig. 11-12). The box girder pier segments are integrated with the cap beam of the straddle bent. The cap beam may be simply supported on the bent columns (Fig. 11-12a) or fixed with the bent columns (Fig. 11-12b). Post-tensioning tendons are used to resist large bending moments induced by the longitudinal box girders. A typical detail of the integral cap is shown in Fig. 11-13.

11.2.1.5 Connections between Super-Structures and Piers

11.2.1.5.1 Moment-Free Connections

For short- and medium-height piers, hinge connections between the piers and super-structures are typically used to reduce the effects on the substructures due to super-structure deformations induced by temperature, shrinkage, and creep as well as to avoid large moments being transferred

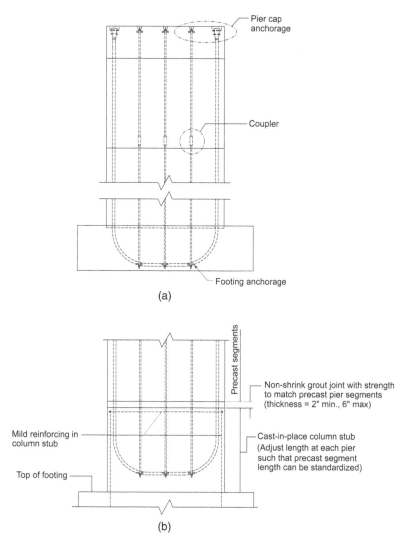

FIGURE 11-9   Precast Pier Post-Tensioning Anchorages, (a) Anchor in Footing, Anchor in Column Stub[11-4, 11-5].

to the foundations. During balanced cantilever construction, temporary falsework can be used to resist unbalanced moments (see Fig. 6-58).

11.2.1.5.2   Moment-Resistant Connections

It is desirable to choose piers and connections that have adequate stability without temporary aids during construction, especially for tall piers. There are typically three methods to achieve this: fixed connections with double flexible piers (Fig. 10-33), fixed connection with two parallel or inclined flexible legs (Figs. 11-3 and 11-14a), and partially fixed connections with two rows of neoprene bearing pads (Fig. 11-14b). The fixed connections between the super-structure and the double flexible piers or two flexible legs can provide excellent stability during construction with little temporary equipment, except for bracing between the slender walls to prevent elastic instability. In the final structure, the tall flexible piers make it possible to develop frame action between the super-structure and piers without impairing the free expansion and contraction of the structure, while sufficiently accommodating the longitudinal braking forces. The behavior of the moment resistant piers with a double row of neoprene bearings are similar to that of double flexible piers. During

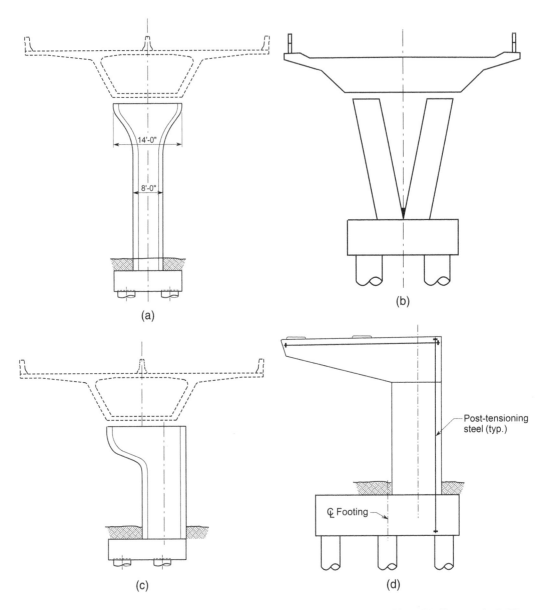

FIGURE 11-10 Transverse Pier Layout for Single-Box Single Column Pier, (a) Symmetrical Pier, (b) V-Shape Pier (c) Asymmetrical Pier, (d) Typical Layout of Post-Tensioning Tendons.

construction, temporary supports and post-tensioning bars are used to fix the super-structure to the pier (see Fig. 6-18). In the finished structure, the neoprene bearings allow free expansion and contraction of the continuous super-structure. The bending moment transferred to the substructure can be controlled by properly selecting the dimensions of the neoprene bearings.

11.3 ABUTMENTS

Bridge abutments are provided at both ends of the bridge and have two main functions. The first function is to support the bridge super-structure and allow a smooth transition from the bridge deck to the roadway as well as free expansion and contraction of the super-structure. The second function is to retain the fill of the approach embankment where required due to geometric conditions. The

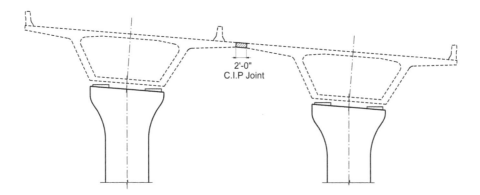

FIGURE 11-11   Transverse Pier Layout for Multi-Box Girders.

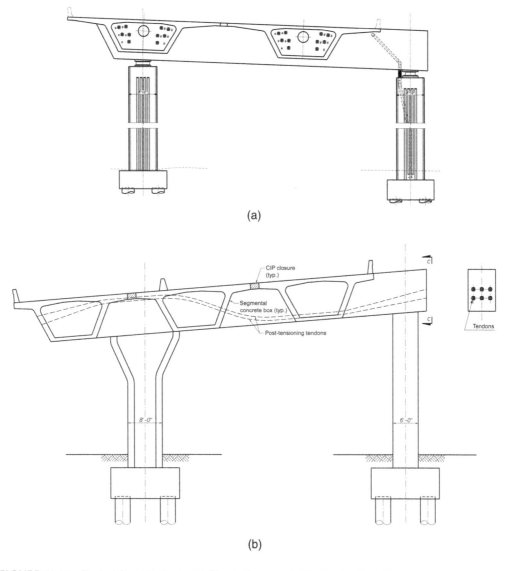

(a)

(b)

FIGURE 11-12   Typical Straddle Bents, (a) Simply Supported, (b) Fixed with Columns.

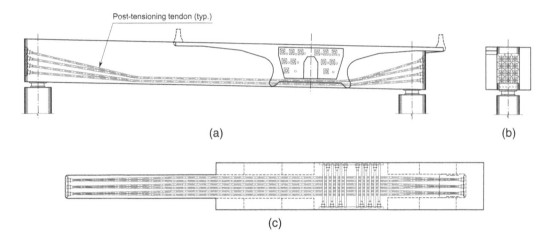

FIGURE 11-13 Typical Post-Tensioning Tendon Layout in Straddle Bent, (a) Elevation View, (b) Side View, (c) Plan View.

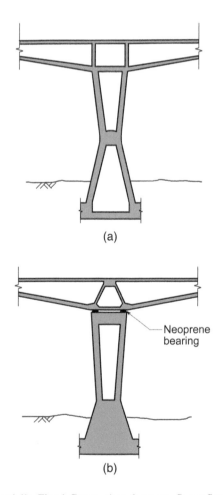

FIGURE 11-14 Fixed and Partially Fixed Connections between Super-Structure and Piers, (a) Pier with Double Inclined Flexible Legs, (b) Pier with Double Row of Neoprene Bearings[11-3].

two functions may be accomplished by using an integrated structure or two separate structures (end bent and retaining wall). In some cases, the retaining wall may be greatly minimized by allowing the approach fill to be sloped under the structure.

11.3.1 U-Shape Abutments

The U-Shape abutment is one of the earliest abutment types and consists of side walls, front wall, back wall, cap, and footing (see Fig. 11-15a). The cap and the front wall together support the loadings from the super-structure. The side walls and the front wall are anchored to the footing and together

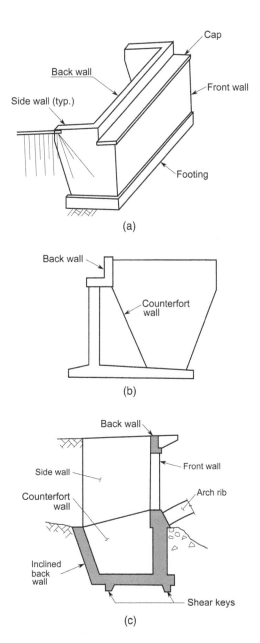

FIGURE 11-15 Typical U-Shape Abutment, (a) Components of U-Shape Abutment, (b) Cross Section of Reinforced Concrete U-Shape Abutment, (c) U-Shape Abutment for Arch Bridges.

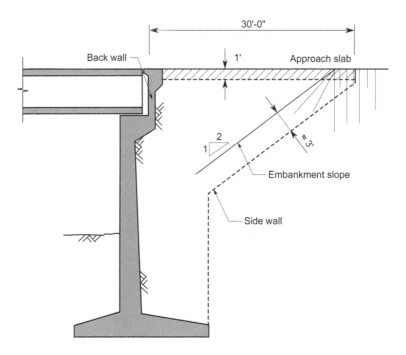

FIGURE 11-16   Ear Wall Abutment.

with the back wall retain the approach embankment fill. The walls can be gravity walls or rein-forced concrete retaining walls, depending on the wall height. To reduce abutment self-weight, rein-forced concrete walls are typically used for U-shape abutments (see Fig. 11-15b). For arch bridges, a large horizontal force should be resisted by its abutments. Shear keys and an inclined bottom back wall are typically provided (see Fig. 11-15c). The top back wall receives an approach slab typically 30 ft long to avoid roadway profile discontinuity between the rigid deck and flexible pavement over the approach embankment. In many cases, the side walls do not necessarily have the same length from the top to the bottom and can be designed as triangular shapes as shown in Fig. 11-16. These types of abutments are often called ear abutments.[11-3].

11.3.2   MINI-ABUTMENTS

If the girder depth is not large and the prevailing conditions allow the sloping fill to be placed below and around the super-structure, the abutment can be designed as a simple less-expensive structure as shown in Fig. 11-17.

11.3.3   END BENT AND MSE WALL COMBINED ABUTMENTS

With the development of mechanically stabilized earth (MSE) walls, the abutments of concrete segmental bridges often consist of two separate structures, an end bent with back wall and a wrap-around MSE wall (see Fig. 11-18). This type of abutment has been proven to be cost effective for many site conditions and is prevalent in current bridge construction in the United States. The end bent is built first, and then the wraparound MSE wall is constructed. The MSE wall consists of precast reinforced concrete panels (see Fig. 11-19) and metallic or nonmetallic (geosynthetic) soil reinforcing strips (see Fig. 11-20). The typical reinforced concrete square panel size is 5 ft by 5 ft and generally does not exceed 30 ft^2 in area. The concrete panels are anchored to the strips embed-ded in the soil (see Fig. 11-20). If the end bent back wall is comparatively high, strips are used to reduce its bending moment due to the earth pressure (see Fig. 11-20).

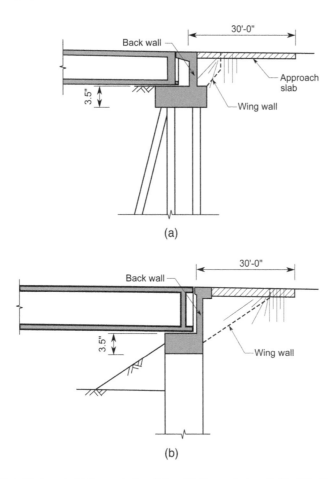

FIGURE 11-17 Mini-Abutment, (a) Supported on Piles, (b) Supported on Drilled Shafts.

11.4 BRIDGE BEARINGS

11.4.1 Introduction

The basic function of bridge bearings is to transfer loadings from the super-structure to the substructure. A bearing may be required to resist up to three types of axial loadings and to accommodate anticipated super-structure deformations. Though the cost of bridge bearings is small in comparison with the total cost of bridge construction, it is essential to choose a proper type of bearing as its failure may damage the bridge or cause long-term deterioration. Bearings are typically classified as fixed or movable/expansion. Fixed bearings will transmit longitudinal and transverse lateral loads, while expansion bearings generally only resist small friction forces from deformation of the bearing during longitudinal expansion and contraction. Movable bearings may include guides to control the direction of translation. Fixed and guided bearings should be designed to resist and restrain all unwanted translations. There are numerous types of bearings, such as elastomeric bearings, plain sliding bearings, spherical bearings, disc bearings, pot bearings, and rocker bearings. Three types commonly used in segmental bridge construction are neoprene bearings, disk bearings, and pot bearings. These will be discussed in this section. The resistance factor for bearings is generally taken as

$$\phi_b = 1.0$$

For multi-rotational bearing, its vertical bearing capacity should not be larger than 5 times the design vertical loads[11-8].

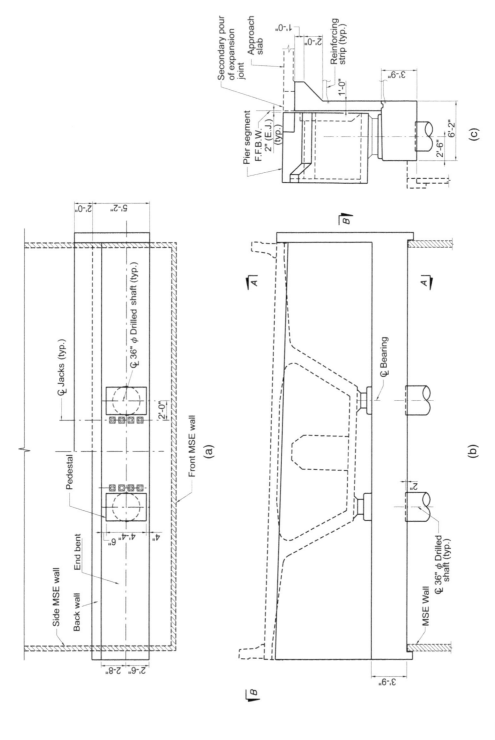

FIGURE 11-18 End Bent and MSE Wall Combined Abutment, (a) Plan View, (b) Elevation View, (c) Side View.

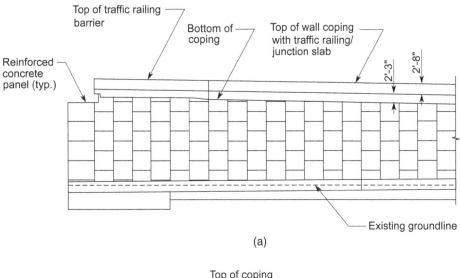

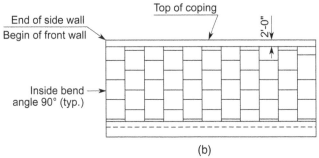

FIGURE 11-19 Typical MSE Wall Layout, (a) Side View, (b) Front View.

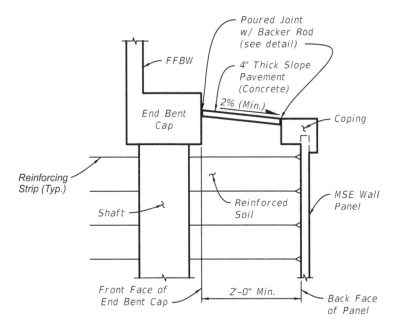

FIGURE 11-20 Typical Side View of End Bent and Front MSE Wall[11-7, 11-8].

11.4.2 ELASTOMERIC BEARINGS

11.4.2.1 General

The principal component of elastomeric bearings is a neoprene pad which distributes the loads from the super-structure to the substructure and accommodates the rotation and longitudinal movement of the super-structure. Neoprene pads are available with durometer hardness values ranging from 50 to 70. However, most current manufacturers are specifying a hardness value of 55 ± 5. Typical material properties are listed in Table 11-1. For the neoprene pad with a hardness value of 55, its shear modulus can be taken as 150 psi. There are two types of elastomeric bearings: plain and laminated. In concrete segmental bridges, steel laminated elastomeric bearings are typically used due to the bridges' higher loads and greater thermal expansions. Their maximum vertical capacities are related to their sizes and typically can range up to 580 kips[11-7]. The components of a typical steel laminated elastomeric bearing and its dimensions are illustrated in Fig. 11-21. For steel laminated elastomeric bearings, the ratio of the shape factor to the number of interior layers of the elastomeric bearing should meet

$$\frac{S_i^2}{n} < 22 \tag{11-1}$$

where

S_i = $\frac{LW}{2h_{ri}(L+W)}$ = shape factor of layer of rectangular elastomeric bearing
L = length of bearing perpendicular to bridge longitudinal direction
W = width of bearing parallel to bridge longitudinal direction
h_{ri} = thickness of ith elastomeric interior layer, typically constant for each layer of about 0.375 to 0.5 in.
n = number of interior layers of elastomer. If the thickness of the exterior layer of elastomer is equal to or greater than one-half the thickness of an interior layer, n can be increased by one-half for each such exterior layer.

In Fig. 11-21, the designations h_{se} and h_{is} represent the thicknesses of the exterior and interior steel plates, respectively. These values typically range from 3/32 to 3/16 in. for segmental bridges. The designation h_{re} indicates the thickness of the elastomer cover layer and should not be greater than $0.7h_{ri}$, i.e.,

$$h_{re} \leq 0.7h_{ri} \tag{11-2}$$

11.4.2.2 Design Requirements for Steel Laminated Elastomeric Bearings

AASHTO LRFD specifications provide two methods for designing the steel-reinforced elastomeric bearings: method A and method B[11-9]. The design requirements for method A are discussed in this section.

TABLE 11-1
Correlated Material Properties of Neoprene Pads

Items	Hardness (Shore A)		
	50	60	70
Shear modulus at 73°F (ksi) (G)	0.095–0.130	0.130–0.200	0.200–0.300
Creep deflection at 25 years divided by initial deflection (α_{cr})	0.25	0.35	0.45

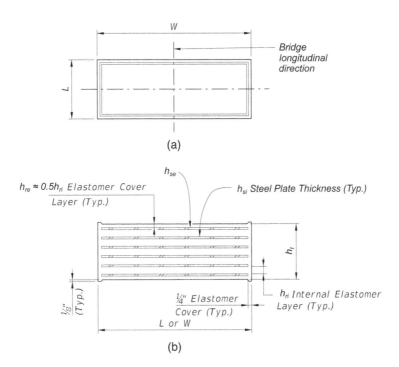

FIGURE 11-21   Elastomeric Bearing, (a) Plan View, (b) Elevation View[11-7].

11.4.2.2.1   Compressive Stress at Service Limit State

The maximum average stress f_s for steel-reinforced elastomeric bearings due to the total load from applicable service load combinations specified in Table 2-10 should meet the following equations:

$$f_s \leq 1.25GS_i \qquad (11\text{-}3a)$$

and

$$f_s \leq 1.25 \ ksi \qquad (11\text{-}3b)$$

where
   G = shear modulus of elastomeric bearing
   S_i = shape factor of interior layer of bearing

11.4.2.2.2   Compressive Deflection at Service Limit State

The compressive deflection Δv_i under live load and initial dead load of an internal layer of a steel-reinforced elastomeric bearing at the service limit state without impact should meet the following equation:

$$\Delta v_i \leq 0.09 h_{ri} \qquad (11\text{-}4)$$

where h_{ri} = thickness of an internal layer.
   The total deflection Δ_L due to live loading should be calculated as

$$\Delta_L = \sum \varepsilon_{Li} h_{ri} \qquad (11\text{-}5a)$$

where ε_{Li} = compressive strain due to instantaneous live load in ith elastomeric layer.

It is recommended that

$$\Delta_L \leq 0.125 \ in.$$ (11-5b)

The total deflection Δ_d due to initial dead load should be calculated as

$$\Delta_D = \sum \varepsilon_{di} h_{ri}$$ (11-5c)

where ε_{di} = compressive strain due to initial dead load in ith elastomeric layer.

The total long-term dead load deflection, including creep effects can be determined as

$$\Delta_{ltd} = \Delta_D \left(1 + \alpha_{cr}\right)$$ (11-6)

where α_{cr} = creep ratio (see Table 11-1).

The strains ε_{Li} and ε_{di} in Eqs. 11-5a and 11-5c, respectively, should be determined from tests or estimated by the following equation:

$$\varepsilon = \frac{f_s}{4.8 G S_i^2}$$ (11-7)

where
f_s = compressive stress
S_i = shape factor

AASHTO specifications also provide some stress-strain curves for determining the strain in an elastomeric layer based on the shape factor and durometer hardness of the steel-reinforced elastomeric bearings (see Fig. 11-22). Thus, the strains ε_{Li} and ε_{di} in Eqs. 11-5a and 11-5c, respectively, also can be determined based on Fig. 11-22.

11.4.2.2.3 Shear Deformation at Service Limit State

The maximum shear deformation Δ_s of a steel-reinforced elastomeric pad under the service limit load combination should meet the following:

$$\Delta_s \leq 0.5 h_{rt}$$ (11-8)

where h_{rt} = elastomeric thickness.

11.4.2.2.4 Stability

To ensure bearing pad stability, the total thickness of the elastomer should meet

$$h_{rt} \leq \frac{L}{3}$$ (11-9a)

and

$$h_{rt} \leq \frac{W}{3}$$ (11-9b)

11.4.2.2.5 Reinforcement Requirement

The minimum thickness h_s of the steel reinforcement should be

$$h_s \geq 0.0625 \ in.$$ (11-10a)

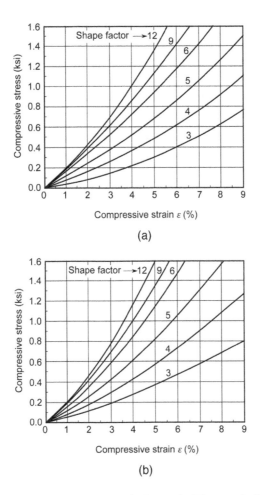

FIGURE 11-22   Stress-Strain Curves for Elastomeric Layers in Elastomeric Bearings, (a) 50 Durometer Reinforced Bearing, (b) 60 Durometer Reinforced Bearing[11-9].

The thickness h_s of steel reinforcement also should meet the following:

- At the service limit state:

$$h_s \geq \frac{3h_{ri}f_s}{f_y}$$
(11-10b)

- At the fatigue limit state:

$$h_s \geq \frac{2h_{ri}f_L}{\Delta F_{TH}}$$
(11-10c)

where
f_s = average compressive stress due to total service load combination
f_L = average stress at service limit state due to live load
f_y = yield stress of steel
ΔF_{TH} = constant amplitude fatigue threshold for category A = 24 ksi

Equations 11-10a and 11-10b are applicable for the reinforcement without holes in it. If there are holes in the reinforcement, the minimum thickness should be increased by a factor equal to twice the ratio of the gross width to net width.

11.4.2.2.6 Rotation

The design for rotation is implicit in the geometric and stress limits as previously discussed. The maximum rotation from applicable service load combinations specified in Table 2-10 plus an allowance of 0.005 rad should not be greater than the allowable rotation determined based on the maximum allowable average compressive stress. It should be mentioned that the rotation design in method A is implicit in Eq. 11-1.

11.4.2.2.7 Design Example of Steel-Reinforced Elastomeric Bearings

Take the bridge as shown in Fig. 5-38 for an example to show the design procedure for steel-reinforced elastomeric bearings. The five-span continuous segmental bridge is supported by one row of steel-reinforced elastomeric bearings on each of the tops of the six piers. The single row of bearings has two bearings in the bridge transverse direction. The end support bearings of the girder will be used for this example (see Fig. 11-23).

11.4.2.2.7.1 Super-Structure Reactions

The basic design requirements and analytical model for the super-structure are given in Sections 5.6.1.1 and 5.6.4.2, respectively. The reactions on the bearing are calculated through computer analysis and given as follows:

Maximum reaction at Service I limit state: $P_T = 1097$ kips
Maximum live load reaction at Service I limit state: $P_{LL} = 559$ kips
Maximum dead load at Service I limit state: $P_{DL} = 538$ kips
Longitudinal shear force for shear deformation due to temperature, creep, and shrinkage:
$F_L = 47.4$ kips
Preliminarily calculated longitudinal movement due to temperature, creep, and shrinkage:
$\Delta_{es} = 2.96$ in.

11.4.2.2.7.2 Material Properties

A. Elastomer
Hardness = 60
Minimum shear modulus: $G_{min} = 0.113$ ksi
Maximum shear modulus: $G_{max} = 0.165$ ksi
Creep ratio: $\alpha_{cr} = 0.25$

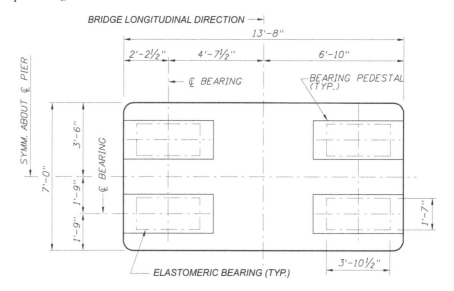

FIGURE 11-23 Layout of Elastomeric Bearings on Top of the Pier Cap.

B. Steel
Yield strength: $F_y = 36$ ksi
Constant amplitude fatigue threshold for category A: $\Delta F_{th} = 24$ ksi

11.4.2.2.7.3   Estimation of Preliminary Geometry of the Bearing
A. Plan Sizes
From Eq. 11-3b, the minimum area of the elastomeric bearing is

$$A_{es} = \frac{P_T}{1.25} = 877.6 \ in.^2$$

From the layout of the bearings on the top of the pier cap (see Fig. 11-23), we preliminarily select the length of the bearing as 19 in., i.e.,

$$L = 19 \ in.$$

Then, the minimum required width of the bearing is

$$W_{min} = \frac{A_{es}}{L} = 46.19 \ in.$$

Use $W = 46.5$ in.

B. Thickness
From Eq. 11-8, the minimum thickness of the elastomer is

$$h_{rt} = 2 \times \Delta_{es} = 5.92 \ in.$$

Assume the following:

Thickness of exterior elastomeric cover layers: $h_{re} = \frac{3}{16}$ in.
Thickness of interior steel laminate in the elastomeric bearing pad: $h_{si} = \frac{1}{16}$ in.
Thickness of exterior steel laminate in the elastomeric bearing pad: $h_{se} = \frac{3}{16}$ in.
Number of interior elastomeric layers: $n = 15$

Then, we can calculate the thickness of interior elastomeric layers:

$$h_{ri} = \frac{3}{8} \ in.$$

$$h_{re} < 0.7 h_{ri} = 0.263 \ in. \ (ok)$$

$$h_{rt} = n h_{ri} + 2 h_{re} = 6.00 \ in.$$

Total thickness of bearing pad: $h_t = n h_{ri} + 2(h_{re} + h_{se}) + (n-1) h_{si} = 7.25$.

11.4.2.2.7.4   Capacity Verification
11.4.2.2.7.4.1   General Geometry Check
From Eq. 11-1, the shape factor:

$$S_i = \frac{LW}{2 h_{ri} (L+W)} = \frac{19 \times 46.5}{2\frac{3}{8}(19 + 46.5)} = 17.99$$

$$\frac{S_i^2}{n} = \frac{17.99^2}{15} = 21.56 < 22 \ (ok)$$

11.4.2.2.7.4.2 Compressive Stress Check

The compressive stress due to the total service load is

$$f_s = \frac{P_T}{LW} = \frac{1097}{19 \times 46.5} = 1.24 \ ksi$$

From Eq. 11-3a, we can see

$$f_s \leq 1.25 \ ksi < 1.25 G S_i = 1.25 \times 0.113 \times 17.99 = 2.54 \ ksi \ (\text{ok})$$

11.4.2.2.7.4.3 Shear Deformation Check

We define the equivalent shear stiffness of the bearing as the shear force required to produce a unit longitudinal displacement, i.e.,

$$\text{Equivalent shear stiffness of bearing: } K_s = \frac{G_{min}LW}{h_{rt}} = \frac{0.113 \times 19 \times 46.5}{6} = 16.64 \left(\frac{kip}{in} \right)$$

Thus, the maximum shear deformation Δ_s of a steel-reinforced elastomeric pad under the service limit load combination can be determined as

$$\Delta_s = \frac{F_L}{K_s} = \frac{47.4}{16.64} = 2.85 \ in.$$

$$\Delta_s \leq 0.5 h_{rt} = 3.0 \ in. \ (\text{ok})$$

11.4.2.2.7.4.4 Compressive Deflection Calculations and Check

From Eq. 11-7, the equivalent modulus of elasticity of the bearing can be defined as

$$E_{eq} = \frac{f_s}{\varepsilon} = 4.8 G_{min} S_i^2 = 4.8 \times 0.113 \times 17.99^2 = 175.54 \ ksi$$

The compressive deflection due to live load in the ith elastomeric layer is

$$\Delta v_{Li} = h_{ri} \varepsilon_{Li} = \frac{h_{ri} P_{LL}}{LWE_{eq}} = \frac{\frac{3}{8} \times 559}{19 \times 46.5 \times 175.54} = 0.001 \ in.$$

The compressive deflection due to the dead load in the ith elastomeric layer is

$$\Delta v_{Di} = h_{ri} \varepsilon_{Di} = \frac{h_{ri} P_{DL}}{LWE_{eq}} = \frac{\frac{3}{8} \times 538}{19 \times 46.5 \times 175.54} = 0.001 \ in.$$

The total compressive deflection Δv_i under live load and initial dead load of an internal layer is

$$\Delta v_i = \Delta v_{Li} + \Delta v_{Di} = 0.002 \leq 0.09 h_{ri} = 0.034 \ in. \ \left(\text{meets Eq. 11-4, ok}\right)$$

The total deflection of the bearing due to live load Δ_L is

$$\Delta_L = \frac{h_{rt} P_{LL}}{LWE_{eq}} = \frac{6 \times 559}{19 \times 46.5 \times 175.54} = 0.022 \ in.$$

$$\Delta_L < 0.125 \ in. \ (\text{meet Eq. 11-5b, ok})$$

The total deflection Δ_D due to the initial dead load can be calculated as

$$\Delta_D = \frac{h_{rt}P_{DL}}{LWE_{eq}} = \frac{6\times538}{19\times46.5\times175.54} = 0.021 \ in.$$

The total long-term dead load deflection, including creep effects can be determined as

$$\Delta_{ltd} = \Delta_D\left(1+\alpha_{cr}\right) = 0.021(1+0.25) = 0.026 \ in.$$

11.4.2.2.7.4.5   Stability Check
From Eq. 11-9, we have

$$h_{rt} = 6 \ in. \le \frac{L}{3} = \frac{46.5}{3} = 15.5 \ in. \ (ok)$$

$$h_{rt} = 6 \ in. \le \frac{W}{3} = \frac{19}{3} = 6.3 \ in. \ (ok)$$

11.4.2.2.7.4.6   Reinforcement Check
The selected thickness h_s of the steel reinforcement meets the minimum thickness requirement, i.e.,

$$h_s = 0.0625 \ in.$$

$$h_s = 0.0625 \ge \frac{3h_{ri}f_s}{F_y} = \frac{3\times\frac{3}{8}\times1.24}{36} = 0.04 \ in. \ (ok) \ (see \ Eq. \ 11\text{-}10b)$$

$$h_s = 0.0625 \ge \frac{2h_{ri}f_L}{\Delta F_{TH}} = \frac{2\times\frac{3}{8}559}{24.0\times19\times46.5} = 0.02 \ in. \ (ok) \ (see \ Eq. \ 11\text{-}10c)$$

11.4.3   DISC BEARINGS
Disc bearings consist of unconfined elastomeric discs and shear restriction mechanisms to prevent their rotational and translational movements (see Fig. 11-24). The design methods for disc bearings can be found in AASHTO LRFD specifications chapter 14. Many manufacturers provide different types of disc bearings designed based on the specifications for bridge designers/contractors to select. There are typically three types of disc bearings: fixed, unidirectional, and multi-directional (see Fig. 11-24). For each type of bearing, the D.S. Brown Company provides two groups based on horizontal load capacities of 10% and 30% of their design vertical load capacities. The vertical capacity of disc bearings can range up to 3000 kips. Table 11-2 gives some design data for unidirectional disc bearings with horizontal capacities of 10% of their vertical capacities provided by D.S. Brown. In Table 11-2, the dimensional designations are shown in Fig. 11-24. For other types of disc bearings developed by other manufacturers, design information can be found in the related manufacturers' websites.

11.4.4   POT BEARINGS
The pot bearing may be one of the most often used bearings in segmental bridge constructions and mainly consists of a metal cylinder, elastomeric disc, metal piston, sealing rings, and top and bottom

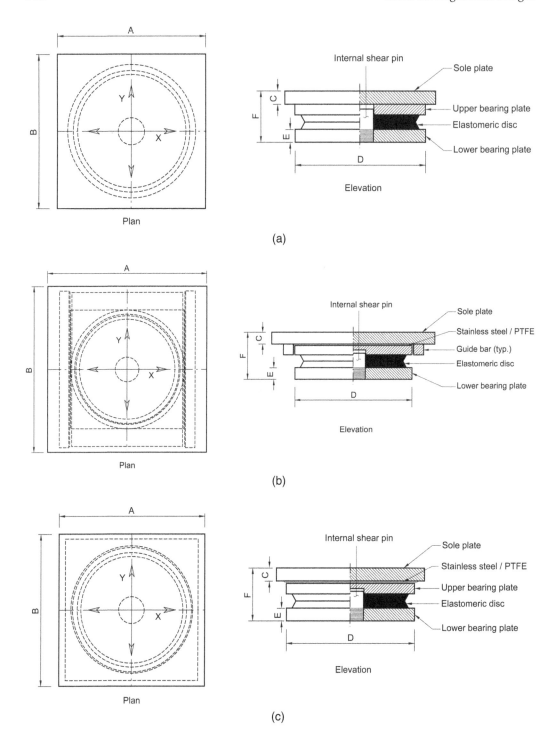

FIGURE 11-24 Typical Disc Bearings Provided by D.S. Brown, (a) Fixed Disc Bearing, (b) Unidirectional Disc Bearing, (c) Multi-Directional Disc Bearing[11-10] (Used with permission of D.S. Brown).

TABLE 11-2

Design Data for Unidirectional Disc Bearings Provided by D.S. Brown

Model Number	Vertical Capacity (kips)	Movement Y (in.)	A (in.)	B (in.)	C (in.)	D (in.)	E (in.)	F (in.)
DMG 400	400	3	19.00	17.50	1.000	12.500	0.750	4.750
DMG 500	500	3	20.75	18.75	1.000	13.875	0.750	5.000
DMG 600	600	3	22.25	20.25	1.000	15.000	0.875	5.375
DMG 700	700	3	23.50	21.25	1.000	16.125	1.000	5.625
DMG 800	800	3	24.75	22.50	1.000	17.125	1.000	5.875
DMG 900	900	3	25.75	23.50	1.000	18.000	1.000	6.000
DMG 1000	1000	3	27.00	24.50	1.000	19.000	1.125	6.375
DMG 1250	1250	4	29.50	27.75	1.000	21.000	1.125	6.750
DMG 1500	1500	4	31.75	29.75	1.000	22.875	1.375	7.250
DMG 1750	1750	4	34.00	31.50	1.000	24.625	1.500	7.750
DMG 2000	2000	4	36.00	33.25	1.000	26.125	1.500	8.000
DMG 2250	2250	6	37.75	37.00	1.000	27.750	1.625	8.250
DMG 2500	2500	6	39.75	38.50	1.000	29.250	1.750	9.000
DMG 2750	2750	6	41.50	40.00	1.000	30.625	1.750	9.250
DMG 3000	3000	6	43.00	41.50	1.000	32.125	1.875	9.625

Notes: Rotation: 0.02 rad. Horizontal capacity: 10% of vertical capacity. Movement: $X = \pm 0.063$ in. Steel strength: $F_y = 50$ ksi.

plates (see Fig. 11-25). The metal piston is supported on the elastomeric disc that is confined in the metal cylinder called a pot. This type of pot bearing can be fixed and provides for both rotational and horizontal movements. Pot bearings have high capacities and are more expensive than disc bearings. AASHTO LRFD specifications[11-9] require that the minimum vertical load on a pot bearing be not less than 20% of its vertical design capacity and that the depth of the elastomeric disc h_r should meet

$$h_r \geq 3.33 D_r \theta_u \tag{11-11}$$

where

D_r = internal diameter of pot (in.)

θ_u = design rotation angle (rad) calculated based on maximum strength limit state, which is sum of rotation from applicable strength load combination specified in Table 2-10 plus 0.01 rad (fabrication and installation tolerance 0.005 rad and an allowance of uncertainties 0.005 rad)

Currently, many manufacturers design their standard pot bearings based on AASHTO LRFD specifications for the bridge designers'/contractors' selection. Pot bearings with different vertical capacities up to 5000 kip are typically available. Three types of pot bearings, fixed, unidirectional, and multi-directional, can be chosen. For each type of bearing, D.S. Brown provides two groups of devices based on the horizontal capacities of 10% and 30% of their vertical design capacities. All three types of bearings can accommodate rotational movement of the bridge super-structure. The fixed bearing resists horizontal forces in any direction through the contact between the piston and the inside of the pot wall. The multi-directional bearing allows horizontal movement in any direction through a polytetrafluoroethylene (PTFE) and stainless-steel sliding surface. There are two types of unidirectional bearings: center-guided and edge-guided. Generally, the center-guided unidirectional bearing cannot be used where the horizontal force

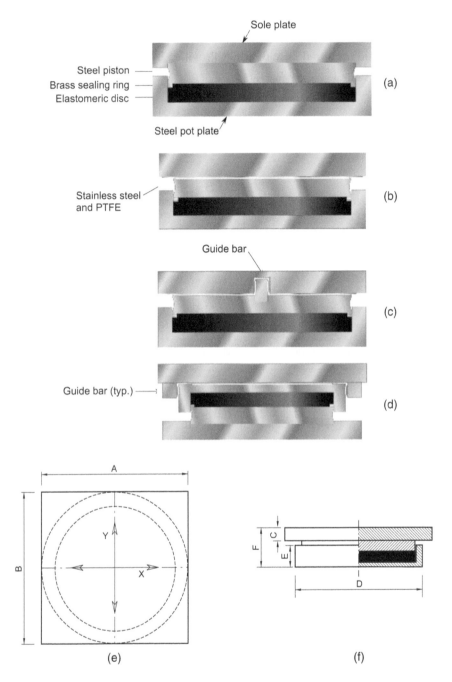

FIGURE 11-25   Typical Pot Bearings Provided by D.S. Brown, (a) Fixed, (b) Multi-Directional, (c) Center-Guided Unidirectional, (d) Edge-Guided Unidirectional, (e) Plan Dimensions, (f) Elevation Dimensions (Used with permission of D.S. Brown).

exceeds 30% of vertical force. Some design data for the fixed pot bearings with horizontal capacities of 10% of their vertical capacities, provided by D.S. Brown, are given in Table 11-3. The dimensional designations of the bearings are illustrated in Fig. 11-25e and f. For other types of pot bearings and those developed by other manufacturers, design information can be found in the related manufacturers' websites.

TABLE 11-3
Typical Design Data for Fixed Pot Bearings Developed by D.S. Brown

Model Number	Vertical Capacity (kips)	A, B, D (in.)	C (in.)	E (in.)	F (in.)
PF400	400	14.25	1.00	2.75	4.875
PF450	450	15.25	1.00	3.00	5.125
PF500	500	15.75	1.00	3.00	5.000
PF550	550	16.75	1.00	3.25	5.375
PF600	600	17.50	1.00	3.25	5.375
PF650	650	18.00	1.00	3.50	5.625
PF700	700	18.75	1.00	3.75	6.000
PF750	750	19.50	1.00	3.75	6.000
PF800	800	20.25	1.00	4.00	6.125
PF850	850	20.75	1.00	4.00	6.125
PF900	900	21.25	1.00	4.00	6.250
PF950	950	22.00	1.00	4.25	6.500
PF1000	1000	22.50	1.00	4.25	6.500
PF1100	1100	23.50	1.00	4.50	6.750
PF1200	1200	24.50	1.00	4.75	7.125
PF1300	1300	25.50	1.00	5.00	7.250
PF1400	1400	26.50	1.00	5.00	7.375
PF1500	1500	27.50	1.00	5.25	7.625
PF1600	1600	28.25	1.00	5.50	7.875
PF1700	1700	29.25	1.00	5.50	8.000
PF1800	1800	30.00	1.00	5.75	8.250
PF1900	1900	31.00	1.00	6.00	8.375
PF2000	2000	31.50	1.00	6.00	8.500
PF2250	2250	33.50	1.00	6.25	8.750
PF2500	2500	35.25	1.25	6.75	9.625
PF2750	2750	37.00	1.25	7.00	9.875
PF3000	3000	38.75	1.25	7.25	10.250
PF3500	3500	41.75	1.25	7.75	10.750
PF4000	4000	44.50	1.25	8.25	11.375
PF5000	5000	49.75	1.25	9.25	12.500

Notes: Rotation: 0.02 rad. Horizontal capacity: 10% of vertical capacity. Movement: $X = 0$, $Y = 0$.

11.4.5 ANCHOR DETAILS FOR POT AND DISC BEARINGS

In segmental bridge construction, the bearings are generally installed after all related segments are completely erected to provide a chance to adjust for girder geometry and internal forces, if necessary. Figure 11-26 illustrates a typical anchor detail for pot and disc bearings. The general installation procedures are as follows:

a. Install hex bolts in coupler on lower plate and tighten snug.
b. Thread swedge bolt into another end of the coupler.
c. Install hex bolts in top plate and tighten snug.
d. Thread headed stud into another end of the coupler.
e. Clean preformed holes and bearing surfaces.

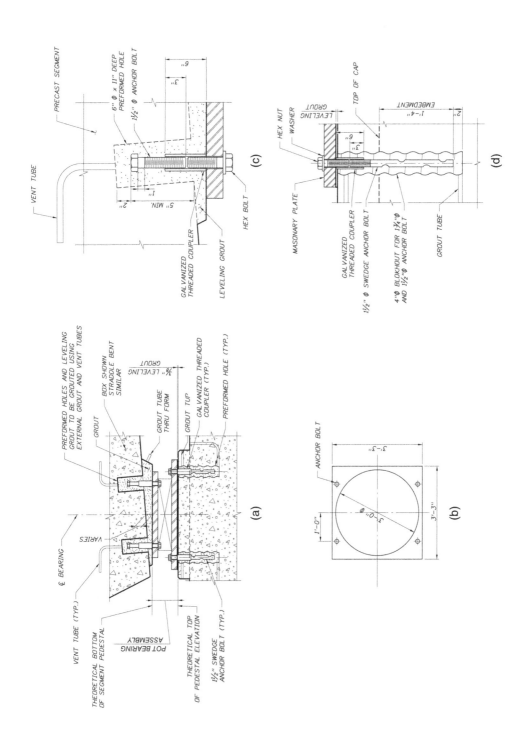

FIGURE 11-26 Typical Anchor Rod Details for Disc and Pot Bearings, (a) Elevation View, (b) Plan View of Anchor Plate, (c) Anchor Details of Top Plate, (d) Anchor Details of Bottom Plate[11-8].

f. Position bearing assembly in proper alignment and elevation on pedestal top to ensure there is no interference of anchor bolts in the preformed holes.
g. Place a leak-tight form frame on the pedestal top around the perimeter of the bearing.
h. Fill preformed holes and the area under the bearing with flowable, high-strength, non-shrink cementitious grout. Fill to the top of the form.
i. Verify the alignment and elevation.
j. Clean the top of the pedestal of excess grout.

11.5 ANALYSIS AND CAPACITY VERIFICATION OF SUBSTRUCTURES

Bridge substructures include piers and abutments. Their basic function is to transfer the loadings from super-structures to the foundation. The bridge abutments may be subjected to large earth pressures, in addition to the loadings from the super-structures. However, their basic analytical and design methods are similar. In this section, only the analytical and design methods of piers will be discussed.

11.5.1 PIER ANALYSIS

11.5.1.1 Loads on Piers

A bridge pier may carry the vertical forces, horizontal forces, bending moments, and torsion from its super-structures due to live loads, dead loads, temperature changes, wind loadings, shrinkage and creep, etc. (see Fig. 11-27). In addition to the loads from the super-structure, it is also subjected to the loads from wind; water; ice; and earthquake, vehicle, and ship collision forces, etc. The pier should be designed to meet all the related load combinations specified in Table 2-10.

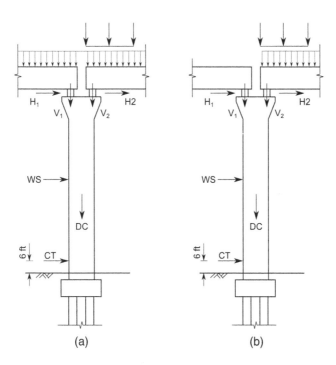

FIGURE 11-27 Loads on Pier, (a) Live Loading Position for Maximum Axial Force, (b) Live Loading Position for Maximum Moment[11-11].

11.5.1.2 Analysis of Flexible Piers with and without Neoprene Bearings

11.5.1.2.1 General

For short and rigid piers with slenderness ratios not greater than 22, their super-structure is typically designed to be simply supported on the piers to avoid large horizontal forces transferring to the foundations. For these cases, the super-structure and substructure can be analyzed separately. The nonlinear effects of the piers can be neglected, and the pier design is comparatively simple. For tall piers, the pier slenderness ratio in the completed structure may be over 70 and even higher during construction. The nonlinear effect should be considered. To increase pier stability, the super-structure is often hinged or partially hinged or fixed to the piers, allowing the transfer of horizontal forces or moments from the super-structure to the piers. In this case, the determination of the forces at the top of the pier should consider the interaction between the super-structure and substructure. Thus, the focus of this section is the analysis of flexible piers.

11.5.1.2.2 Analytical Models

An accurate analysis to account for the pier geometric non-linear effects can be performed by the non-linear finite-element method (refer to Section 4.3.7.2) and by modeling the super-structure and substructure as one structure through a computer program. However, for simplicity, the pier analysis is typically carried out through two separate models in practical bridge design. First, the reactions at the tops of piers due to the loads from the super-structure are determined by linear theory based on the entire bridge model as shown in Fig. 11-28a. Then, the individual pier is analyzed by a geometric non-linear method with the obtained reactions and additional loadings applied to it as shown in Fig. 11-28b. The results obtained by this simplified method are generally conservative. If the pier foundation needs

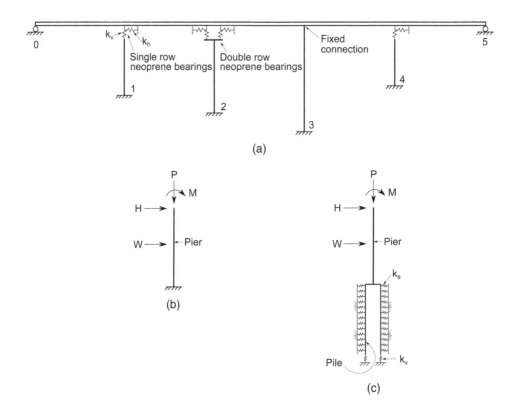

FIGURE 11-28 General Analytical Model for Substructures, (a) Analytical Model for Reactions at Pier Tops, (b) Analytical Model for Piers, (c) Analytical Model for Piers and Piles.

to be analyzed, a soil-structure interaction model may be used (see Fig. 11-28c). The footing is generally analyzed by a strut-and-tie model (refer to Section 3.8.2.3.3). If elastomeric bearings are used, they can be modeled as a number of springs (see Fig. 11-28a). For a single row of elastomeric bearings in a plane model, two springs can be used to simulate the elastomeric behaviors (see Fig. 11-28, piers 1 and 4). For a double row of elastomeric bearings in a plane model, four springs can be used to simulate the elastomeric bearing behaviors (see Fig. 11-28, pier 2). The elastomeric spring stiffness coefficients are

$$\text{Horizontal: } k_h = \frac{GA}{h_t} \tag{11-12a}$$

$$\text{Vertical: } k_v = \frac{EA}{h_t} \tag{11-12b}$$

where

G, E, A, h_t = shear modulus, modulus of elasticity, plan area, and total thickness of bearing, respectively

k_h, k_v = horizontal and vertical spring stiffness coefficients, respectively, representing forces required to produce unit displacements

Analytical results show that the effect of the vertical spring stiffness of the elastomeric bearings on the bridge internal forces is generally small and the bearing vertical support can be treated as rigid.

11.5.1.2.3 Geometric Nonlinear Analysis for Flexible Piers

11.5.1.2.3.1 Closed-Form Solutions for Cantilever Piers with a Constant Cross Section An approximate method for estimating the P-Δ effects of an axially loaded beam or column was discussed in Section 4.3.7.3. Now, let us examine a typical loading condition of a pier with its tip free from restraints (see Fig. 11-29) during construction. For ease of discussion, we first neglect the pier

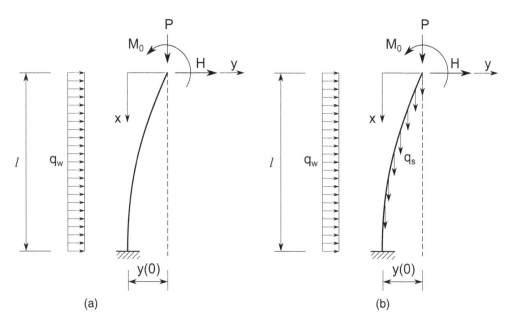

(a)   (b)

FIGURE 11-29 Typical Loading Case of Piers and Geometrical Nonlinear Effects, (a) Typical Loading Case without Self-Weight, (b) Analytical Model with Self-Weight[11-12].

self-weight and then extend the analytical results to a general case with self-weight loaded by the minimum energy theorem.

Assume the pier has a constant cross section and is loaded by uniform lateral loading q_w and vertical load P, horizontal load H, and bending moment M at its top. The horizontal displacement at the top due to these loadings is denoted as $y(0)$ (see Fig. 11-29a).

Taking a coordinate system as shown in Fig. 11-29, the linear moment at any location x due to external loads is

$$M_q = M_0 - Hx - \frac{q_w x^2}{2} \tag{11-13a}$$

The nonlinear moment due to lateral deformation $y(x)$ is

$$M_{ld} = P\big((y(x) - y(0))\big) \tag{11-13b}$$

The total moment at x is

$$M(x) = M_q + M_{ld} = M_0 - Hx - \frac{q_w x^2}{2} + P\big((y(x) - y(0))\big) \tag{11-14}$$

From Eqs. 11-14, 4-4, and 4-6, noting the coordinate system, we have

$$y'''' + \frac{P}{EI} y'' = \frac{q_w}{EI} \tag{11-15}$$

The solution of the differential equation (11-15) consists of two portions of general and partial solutions.

The general solution corresponding to the homogeneous equation can be written as

$$y_g = A \sin kx + B \cos kx + Cx + D \tag{11-16a}$$

where

$$k = \sqrt{\frac{P}{EI}}$$

A particular solution of the inhomogeneous equation can be assumed to be

$$y_p = cx^2 \tag{11-16b}$$

Substituting Eq. 11-16b into Eq. 11-15 yields

$$y_p = \frac{q_w}{2H} x^2 \tag{11-16c}$$

Thus, the general solution of Eq. 11-15 can be written as

$$y = y_g + y_p = A \sin kx + B \cos kx + Cx + D + \frac{q_w}{2H} x^2 \tag{11-17}$$

The integral constants of A, B, C, and D can be obtained using the following boundary conditions:

At $x = 0$:
$$M(0) = -EIy''(0) = M_0$$
$$V(0) = -EIy'''(0) = -H + Py'(0)$$
At $x = l$:
$$y(l) = 0, \; y'(l) = 0$$

And the constants of A, B, C, and D are solved as

$$A = \frac{1}{Pk\cos kl}\left[\left(M_0 + \frac{q_w}{P}EI\right)k\sin kl - H - q_w l\right]$$

$$B = \frac{1}{P}\left(M_0 + \frac{q_w}{P}EI\right)$$

$$C = \frac{H}{P}$$

$$D = \frac{\tan kl}{Pk}\left[\left(M_0 + \frac{q_w}{P}EI\right)k\sin kl - H - q_w l\right] - \frac{1}{P}\left(M_0 + \frac{q_w}{P}EI\right)\cos kl - \frac{Hl}{P} - \frac{q_w l^2}{2P}$$

Using Eqs. 4-4 and 11-17, we can obtain the moment at the bottom of the pier as

$$M(l) = \frac{1}{\cos kl}\left[M_0 + \frac{q_w}{P}EI - (H + q_w l)\frac{\sin kl}{k}\right] - \frac{q_w}{P}EI \tag{11-18a}$$

Using the relationship between moment and shear, we can obtain the shear at the bottom of the pier as

$$V(l) = -(H + q_w l) \tag{11-19}$$

If the effect of pier self-weight is considered in the nonlinear analysis, the pier lateral deformation can be obtained by the minimum energy theorem. The moment at the pier bottom with the effect of pier self-weight q_s can be written as

$$M(l) = \frac{1}{\cos kl}\left[M_0 + \frac{q_w}{\left(P + \frac{lq_s}{3}\right)}EI - (H + q_w l)\frac{\sin kl}{k}\right] - \frac{q_w}{\left(P + \frac{q_{sl}}{3}\right)}EI \tag{11-18b}$$

where q_s = pier self-weight per unit length.

Comparing Eq. 11-18a with Eq. 11-18b, it can be seen that the geometric nonlinear effect of the pier self-weight on the bottom moment can be calculated by applying a concentrated axial force of one-third of the pier's total self-weight at the top of the pier.

11.5.1.2.3.2 Approximate Method In many cases, it may be difficult to obtain closed-form solutions for the deflection and moment including the nonlinear effect. An approximate method similar to that discussed in Section 4.3.7.3 can be used, i.e.:

The pier lateral displacement at location x, including the nonlinear effect, can be written as

$$y(x) = \mu_0 y_I(x) \tag{11-20a}$$

where

$\mu_0 = \frac{1}{1-\frac{P}{P_e}}$ = theoretical magnifier of second-order effect

$P_e = \frac{\pi^2 EI}{(kl_u)^2}$ = buckling load of pier (see Eq. 4-167d for notation definitions)

$y_l(x)$ = pier lateral displacement determined based on linear theory

The total moment at location x, including the geometric effect, is

$$M(x) = M_I(x) + \mu_0 P y_l(x) \qquad (11\text{-}20b)$$

For the loading case as shown in Fig. 11-29a, the displacement at the pier top determined by linear theory can be written as

$$y_l(0) = \frac{l^2}{EI}\left(\frac{q_w l^2}{8} + \frac{Hl}{3} - \frac{M_0}{2}\right)$$

The buckling load of the pier is

$$P_e = \frac{\pi^2 EI}{(kl_u)^2} = \frac{\pi^2 EI}{(2l)^2} = \frac{\pi^2 EI}{4l^2}$$

Then, the deflection at the top of the pier including the nonlinear effect (second-order effect) can be written as

$$y(0) = \frac{l^2}{EI}\left(\frac{q_w l^2}{8} + \frac{Hl}{3} - \frac{M_0}{2}\right)\frac{1}{1-\frac{4Pl^2}{\pi^2 EI}} \qquad (11\text{-}21a)$$

The first-order moment at the bottom of the pier is

$$M_I = -\left(\frac{q_w l^2}{2} + Hl - M_0\right)$$

The total moment including the nonlinear effect at the bottom of the pier is

$$M_{tot} = M_I + Py(0) = -\left(\frac{q_w l^2}{2} + Hl - M_0 + Py(0)\right) \qquad (11\text{-}22a)$$

The difference of the results obtained based on Eq. 11-18a and Eq. (11-22a) is less than 0.5%.

During pier construction, the second-order effect of pier self-weight on the lateral displacement at pier top and the moment at pier bottom can be considered by applying one-third of its total self-weight at the top of the pier (see Fig. 11-30). Then, the second-order effect can be determined by Eqs. 11-21a and 11-22a with P replaced by $P + \frac{q_s l}{3}$, i.e.,

$$y(0) = \frac{l^2}{EI}\left(\frac{q_w l^2}{8} + \frac{Hl}{3} - \frac{M_0}{2}\right)\frac{1}{1-\frac{4\left(P+\frac{q_s l}{3}\right)l^2}{\pi^2 EI}} \qquad (11\text{-}21b)$$

$$M_{tot} = -\left(\frac{q_w l^2}{2} + Hl - M_0 + \left(P + \frac{q_s l}{3}\right)y(0)\right) \qquad (11\text{-}22b)$$

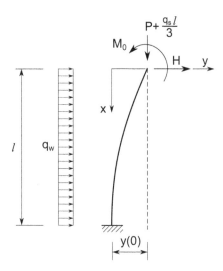

FIGURE 11-30 Analytical Model for Estimating Second-Order Effect at Bottom of Pier Due to Pier Self-Weight.

11.5.2 CAPACITY VERIFICATION OF PIERS

Bridge piers are typically biaxial flexural and compression members. Their capacities should be verified using Eq. 3-230a or 3-230b for applicable strength and extreme event limit states. For prestressed concrete piers, the concrete tension should be checked for Service IV limit state. For reinforced concrete piers, the concrete crack width should be checked to meet the requirement discussed in Section 3.7.4 and Service I limit state. For tall piers, the flexural and compressive strengths of bridge piers are generally controlled by Strength Limit States III, IV, or V and shear strength may be controlled by the extreme event limit state.

11.6 EXPANSION JOINTS

11.6.1 GENERAL

The lengths of bridge decks will change due to the effects of thermal changes and concrete shrinkage and creep. To reduce the effects of the bridge deck length changes on super-structures and substructures, expansion joints are typically provided at bridge ends and other locations along the bridge longitudinal direction as necessary. The main functions of the expansion joints are to:

1. Accommodate the translation and rotation of the super-structure at the joint.
2. Accommodate the movement of motorcycles, bicycles, and pedestrians without significantly affecting the riding characteristics of the roadway and causing damage to the vehicles.
3. Prevent damage to the bridge from water, deicing chemicals, and debris.

There are many types of expansion joints available and used in current bridge structures, such as compression seal joints, strip seal joints, modular joints, finger joints, and swivel joints. Currently, the most often used expansion joints in segmental bridges may be modular joints and finger joints, which can accommodate large movements. Strip seal joints may be also be considered for shorter bridges.

11.6.2 DESIGN CRITERIA

Expansion joints should be designed to satisfy the requirements of all strength, fatigue, fracture, and service limit states specified in Section 2.3.3.2. Joints and their supports should be designed to resist factored force effects over the range of movements for the applicable design limits.

The force effects and movements should consider the following factors:

- Temperature, creep, and shrinkage
- Construction sequences and tolerances
- Skew and curvature
- Substructure movements
- Structural restraints
- Static and dynamic loadings

The width of expansion joint gap W, measured in the direction of travel and determined by the appropriate strength load combination specified in Table 2-10, should not exceed:

- 4.0 in., for a single gap
- 3.0 in., for multiple modular gaps

The maximum open width between adjacent fingers on a finger plate should satisfy:

$W_f \leq 2.0$ in., for longitudinal opening greater than 8.0 in.
$W_f \leq 3.0$ in., for longitudinal opening less than or equal to 8.0 in.

The finger overlap at the maximum movement should not be less than 1.5 in. at the strength limit state. If pedestrians or bicycles are expected in the roadway, a special covering floor plate in shoulder areas should be provided.

11.6.3 SELECTION AND INSTALLATION DETAILS OF EXPANSION JOINTS USED IN SEGMENTAL BRIDGES

11.6.3.1 Typical Expansion Joints

For the selection of the type of expansion joints, Table 11-4 gives several types of expansion joints provided by D.S. Brown. The expansion joints are ranked by cost from inexpensive to more expensive. The expansion joints provided by other manufacturers can be found at their websites.

11.6.3.2 Strip Seal Expansion Joints

Strip seal expansion joint systems have been proven to have superior watertight performance and long service life. This type of joint consists of two proprietary components: steel rail profiles and a matching neoprene sealing element. The components and typical installation details are illustrated in Fig. 11-31.

11.6.3.3 Modular Expansion Joints

Modular expansion joint systems are highly engineered assemblies that consist of center beams and edge beams (see Fig. 11-32). The center beams and edge beams carry the dynamic wheel loads and are connected by a series of neoprene sealing strips that create a watertight joint. The center beams

TABLE 11-4
Expansion Joint Selection Guide (D.S. Brown)

Types of Joint	Strip Seal	Modular Joints	Finger Joints	Swivel Joints
Movement (in.)	≤ 4	≥ 4	≥ 4	≥ 4
Life expectancy (year)	20–25	20–25	20–25	20–25

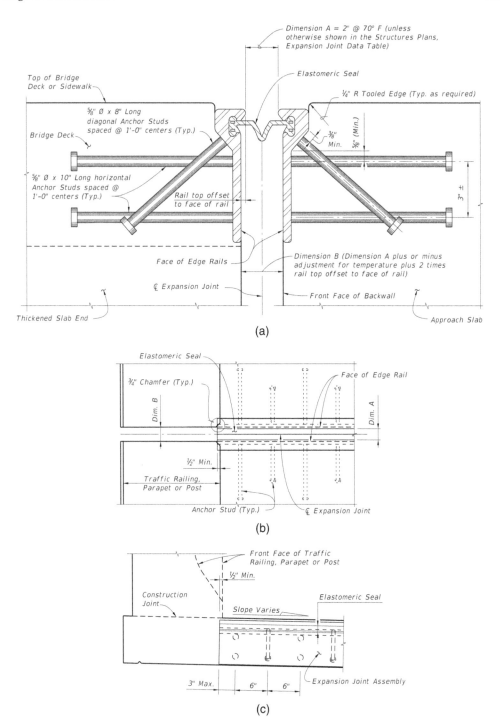

FIGURE 11-31 Typical Strip Seal Joint and Installation, (a) Strip Seal Joint Details, (b) Installation Plan View, (c) Installation Elevation[11-7].

are rigidly connected to the support bars which attach to the stainless-steel slide plates at each end of the support beams (see Fig. 11-32). The design data of the modular expansion joint systems provided by D.S. Brown are given in Table 11-5 for the bridge designer's selection.

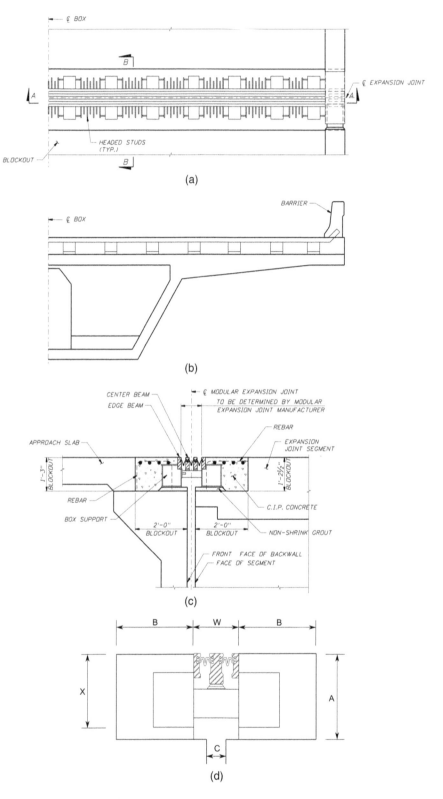

FIGURE 11-32 Typical Modular Joint and Installation Details, (a) Plan View, (b) Section *A-A* View, (c) Section *B-B* View, (d) Dimensional Designations.

TABLE 11-5

Design Data of Modular Expansion Joints

Joint Symbol	Model Number	Movement [in. (mm)]	Blockout Depth A [in. (mm)]	Blockout Width B [in. (mm)]	At Mid Temp C (in.)	At Mid Temp W (in.)	X (in.)
	D-160	6.30 (160)	14 (356)	14 (356)	3.35–8.17	8.17	12.2
	D-240	9.45 (240)	14 (356)	17 (432)	4.92–12.24	12.24	12.2
	D-320	12.60 (320)	14 (356)	20 (508)	6.50–16.32	16.32	12.2
	D-400	15.75 (400)	14 (356)	23 (584)	8.07–20.39	20.39	12.2
	D-480	18.90 (480)	14 (356)	27 (686)	9.65–24.47	24.47	12.2
	D-560	22.05(560)	14 (356)	30 (762)	11.22–28.54	28.54	12.2
	D-640	25.20 (640)	14.5(368)	33 (838)	12.80–32.62	32.62	12.5
	D-740	28.35(720)	15(381)	37 (940)	14.37–36.69	36.69	12.9

Source: Used with permission of D.S. Brown.

11.6.3.4   Finger Joints

Finger joints (see Fig. 11-33a) have proven long-term structural performance, are convenient to install, and require a shallow joint depth and/or staged construction. The typical components and installation details are illustrated in Fig. 11-33b to d.

11.7   DESIGN EXAMPLE VII: PIER

Take a pier modified from an existing bridge owned by FDOT (used with permission of FDOT) for a design example to show the basic procedure for pier design.

11.7.1   Design Information

The super-structure design information is the same as for those described in Section 5.6.1. The material properties used in the design example are as follows:

1. Concrete:
   $f'_c = 5.5$ ksi, $f'_{ci} = 4.4$ ksi
   $E_c = 33,000 \omega_c^{1.5} \sqrt{f'_c} = 4273$ ksi
2. Reinforcing steel:
   ASTM 615, Grade 60
   Yield stress: $f_y = 60$ ksi
   Modulus of elasticity: $E_s = 29,000$ ksi

11.7.2   Pier Geometry

Take pier 3 as shown in Fig. 11-34 for this design example. Its super-structure is a seven-span continuous segmental bridge constructed by the balanced cantilever method as discussed in Chapter 6. Two fixed multi-rotational pot bearings are located on two separate bearing pedestals with a height

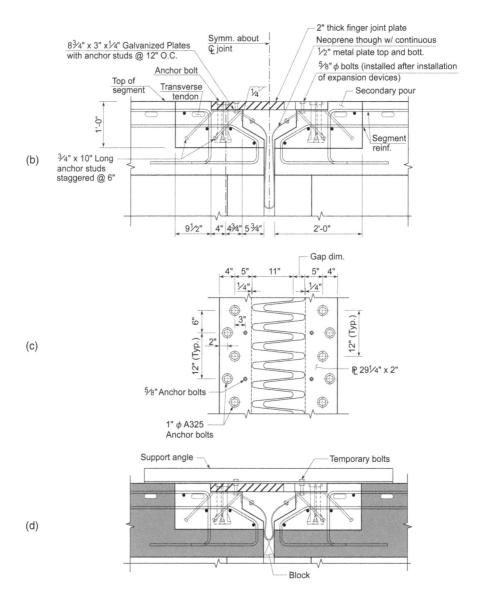

FIGURE 11-33 Finger Joint and Installation Details, (a) Overview of Finger Joint, (used with permission of D.S. Brown), (b) Elevation View, (c) Plan View, (d) Sketch of Installation.

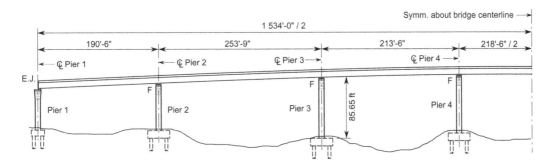

FIGURE 11-34 Super-Structure and Substructure Layout of Design Example VII.

of 4 in. (see Fig. 11-35). The total height of the pier is 85.65 ft. The cross section of the pier is a solid rectangle selected based on the consistency throughout the entire bridge project. The pier sizes are preliminarily determined with a length of 8 ft (in the bridge transverse direction) and a width of 7 ft (in the bridge longitudinal direction), based on the forces from the super-structure at the top of the pier. The pier is supported on a reinforced concrete footing supported by four 36-in.-diameter drilled shafts (see Fig. 11-34). The length of the cap is 13.67 ft.

11.7.3 Pier Analysis

For simplicity, the pier analysis is completed in two steps. First, the entire bridge super-structure and substructure are analyzed by neglecting the pier nonlinear effect. Then, the pier P-Δ effect is analyzed by an approximate method as discussed in Sections 4.3.7.3 and 11.5.1.2.3.2.

11.7.3.1 Linear Analysis of Entire Bridge Structure

The entire bridge is modeled as a three-dimensional frame structure and analyzed by the finite-element method using computer program LARSA 4D. The piers are treated approximately as fixed at the tops of related footings. All the related loadings are applied to produce the most unfavorable effects on the piers. In this example, we take load combinations for Service I and Strength V limit states to illustrate the pier capacity verification procedures. The factored loads for Service I and Strength V for this bridge can be written as follows (refer to Table 2-10):

Service I:

$$F_u = 1.0DC + 1.0LL\ (1 + IM) + 0.3WS + 1.0FR + 1.0WL + 1.0TU + 0.5TG \qquad (11\text{-}23)$$

Strength V:

$$F_u = 1.25DC + 1.35LL\ (1 + IM) + 0.4WS + 1.0FR + 1.0WL + 0.5TU \qquad (11\text{-}24)$$

Based on the above load combinations, the factored loadings at the bottom of the pier for Service I and Strength V are summarized in Table 11-6.

TABLE 11-6
Factored Forces at the Bottom of the Piers

Limit State	V_{ux} Shear in x-Direction (kips)	V_{uy} Shear in y-Direction (kips)	P_u Axial Force (kips)	M_{ux}^l Moment about x-Axis (kips-ft)	M_{uy}^l Moment about y-Axis (kips-ft)
Service I	123	110	5,450	9,666.7	4,825.0
Strength V	136	164	6,920	12,416.7	8,025.0

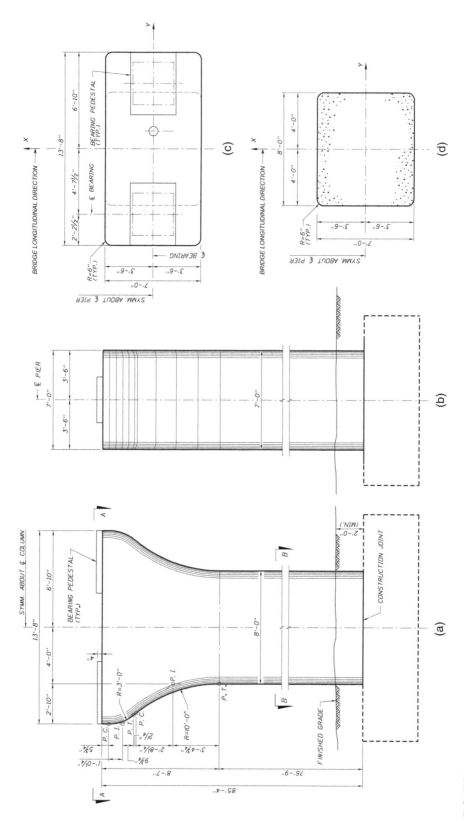

FIGURE 11-35   Principal Dimensions of Pier 3, (a) Elevation View, (b) Side View, (c) Cap Plan View, (d) Typical Section of Pier.

11.7.3.2 Nonlinear Analysis of the Pier

11.7.3.2.1 Determination of Pier Slenderness Ratios

Neglecting the effect of the pier cap, the cross section of the pier can be treated as a constant. Its cross-sectional area and moments of inertia about the bridge longitudinal and transverse directions can be approximated by using the concrete gross section as

$$A_g = 56 \ ft^2$$

$$I_{gx} = \frac{8^3 \times 7}{12} = 298.6 \ ft^4$$

$$I_{gy} = \frac{7^3 \times 8}{12} = 228.6 \ ft^4$$

The radii of gyration about bridge transverse and longitudinal directions are

$$r_x = \sqrt{\frac{I_{gx}}{A_g}} = \sqrt{\frac{298.6}{56}} = 2.309 \ ft$$

$$r_y = \sqrt{\frac{I_{gy}}{A_g}} = \sqrt{\frac{228.6}{56}} = 2.020 \ ft$$

From the super-structure layout and support conditions shown in Figs. 11-34 and 11-35, it is evident that the support at the top of the pier cannot be treated as either completely fixed nor completely free. The super-structure can provide a certain degree of restraint for the pier lateral and longitudinal deformations (see Fig. 11-36) as well as the rotation about the bridge longitudinal direction.

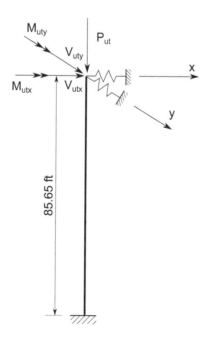

FIGURE 11-36 Analytical Models for P-Δ Effect on Pier.

In this design example, we conservatively take the effective factors of the pier in the transverse and longitudinal directions as

$$k_x = 1.0$$

$$k_y = 1.25$$

Then, the slenderness ratios of the pier about the bridge transverse and longitudinal directions are

$$i_x = \frac{k_x h}{r_x} = \frac{1.0 \times 85.65}{2.309} = 37.09 < 100$$

$$i_y = \frac{k_y h}{r_y} = \frac{1.25 \times 85.65}{2.02} = 53.00 < 100$$

Thus, the moment magnification method discussed in Section 4.3.7.3 can be used for estimating the P-Δ effect of the pier.

11.7.3.2.2 Pier P-Δ Effect

11.7.3.2.2.1 Determination of Equivalent Factored Axial Loading at Top of Pier From Fig. 11-35, the pier self-weight can be calculated as

$$W_{self} = 718.9 \ kips$$

In Table 11-6, the factored axial loadings at the bottom of the pier include the pier self-weight. The equivalent factored axial loadings acting at the top of the pier for Service I and Strength V are calculated as discussed in Section 11.5.1.2.3.2 as

$$\text{Service I: } P_{utI} = 5450 - \frac{2W_{self}}{3} = 4970.7 \ kips$$

$$\text{Strength V: } P_{utV} = 6920 - 1.25\frac{2W_{self}}{3} = 6320.9 \ kips$$

11.7.3.2.2.2 Determination of Effective Stiffness of the Pier The moment of inertia of reinforcement (see Fig. 11-37) can be calculated as

$$\text{About } x\text{-axis: } I_{sx} = 234575 \ in.^4$$

$$\text{About } y\text{-axis: } I_{sy} = 114658 \ in.^4$$

The moment due to permanent loads is zero. Thus the ratio of factored permanent moment to factored total moment is zero, i.e.,

$$\alpha_d = 0$$

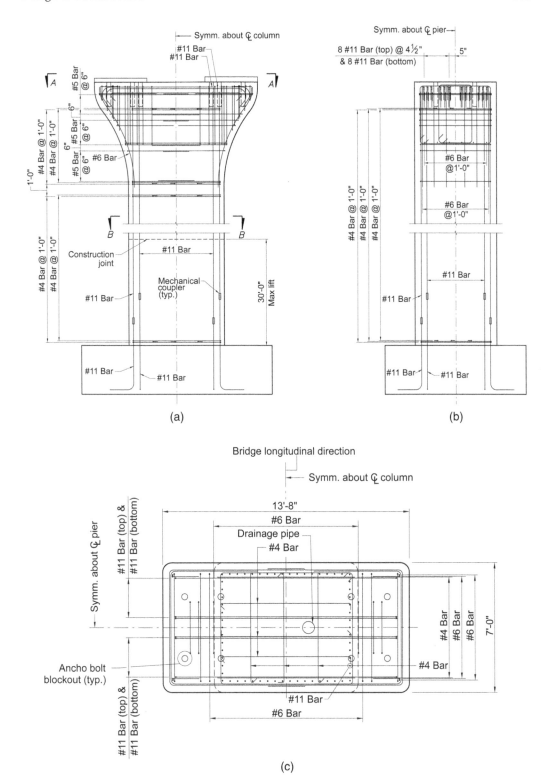

FIGURE 11-37 Reinforcement Layout of Pier, (a) Elevation View, (b) Side View, (c) Cap Plan View, (d) Typical Section of Pier.

(*Continued*)

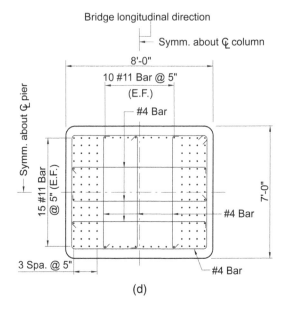

(d)

FIGURE 11-37 (Continued)

From Eqs. 4-168a and 4-168b, the effective stiffness of the pier in the bridge transverse direction should be taken as the larger of the following:

$$E_c I_{ex1} = \frac{\frac{E_c I_{gx}}{5} + E_s I_{sx}}{1 + \alpha_d} = \frac{\frac{4273 \times 298.6 \times 12^4}{5} + 30000 \times 234575}{1 + 0} = 1.2330 \times 10^{10} \text{ kips-in.}^2$$

$$E_c I_{ex2} = \frac{\frac{E_c I_{gx}}{2.5}}{1 + \alpha_d} = \frac{\frac{4273 \times 298.6 \times 12^4}{2.5}}{1 + 0} = 1.0585 \times 10^{10} \text{ kips-in.}^2$$

Thus, the effective stiffness of the pier in the bridge transverse direction is

$$EI_{ex} = 1.2330 \times 10^{10} \text{ kips-in.}^2$$

Similarly, we can obtain the effective stiffness of the pier in the bridge longitudinal direction as

$$EI_{ex} = 8.1044 \times 10^9 \text{ kips-in.}^2$$

11.7.3.2.2.3 Determination of Moment Magnification Factors From Eq. 4-167d, the Euler buckling loadings along the bridge transverse and longitudinal directions are:

Along the transverse direction: $P_{ex} = \dfrac{\pi^2 E_c I_{ex}}{(k_x h)^2} = \dfrac{\pi^2 \times 1.2330 \times 10^{10}}{(1.0 \times 85.65 \times 12)^2} = 115194.6 \; kip$

Along the longitudinal direction: $P_{ey} = \dfrac{\pi^2 E_c I_{ey}}{(k_y h)^2} = \dfrac{\pi^2 \times 8.1044 \times 10^9}{(1.25 \times 85.65 \times 12)^2} = 48457.6 \; kip$

The moment magnification factors along the bridge transverse and longitudinal directions can be calculated using Eqs. 4-167b and 4-167c as
Service I:

Along the transverse direction: $\mu_x = \dfrac{1}{1 - \frac{P_{utl}}{\phi_k P_{ex}}} = \dfrac{1}{1 - \frac{4970.7}{0.75 \times 115194.6}} = 1.061$

Along the longitudinal direction: $\mu_y = \dfrac{1}{1 - \frac{P_{utl}}{\phi_k P_{ey}}} = \dfrac{1}{1 - \frac{4970.7}{0.75 \times 48457.6}} = 1.158$

Strength V:

Along the transverse direction: $\mu_x = \dfrac{1}{1 - \frac{P_{utV}}{\phi_k P_{ex}}} = \dfrac{1}{1 - \frac{6320.9}{0.75 \times 115194.6}} = 1.079$

Along the transverse direction: $\mu_y = \dfrac{1}{1 - \frac{P_{utV}}{\phi_k P_{ey}}} = \dfrac{1}{1 - \frac{6320.9}{0.75 \times 48457.6}} = 1.211$

11.7.3.2.2.4 Calculation of the Moment at the Bottom of the Pier with P-Δ Effects The moments with P-Δ effects at the bottom of the pier can be obtained by multiplying the moments obtained based on linear analysis as shown in Table 11-6 by the above related moment magnification factors. The Service I moment at the bottom of the pier along the longitudinal direction can be determined as

$$M_y^{II} = \mu_y M_y^{I} = 1.158 \times 4825 = 5587.4 \ kips\text{-}ft \qquad (11\text{-}25)$$

The remaining moments with P-Δ effects at the bottom of the pier can be obtained in the same way and are shown in Table 11-7. In Eq. 11-25 and Table 11-7, superscripts I and II represent the moments determined by linear and nonlinear methods, respectively.

11.7.4 PRELIMINARY REINFORCEMENT LAYOUT

From the "try-and-revise" analysis of the pier internal forces, the required reinforcement of the pier can be estimated based on the methods discussed in Section 4.1.3.2. The preliminary reinforcement layout is shown in Fig. 11-37. From Fig. 11-37d, it can be seen that there are a total of 140 #11 bars in the pier vertical direction with a total area of 218.4 in.² and a bar spacing of 5 in. at the critical section.

TABLE 11-7
Forces at the Bottom of the Pier with P-Δ Effects

Limit State	V_{ux} Shear in x-Direction (kips)	V_{uy} Shear in y-Direction (kips)	P_u Axial Force (kips)	M_{ux}^{II} Moment about x-axis (kips-ft)	M_{uy}^{II} Moment about y-axis (kips-ft)
Service I	123	110	5,450	10,256.4	5,587.4
Strength V	136	164	6,920	1,3397.6	9,718.3

11.7.4.1 Minimum and Maximum Bar Spacing Check

The bar spacing should meet the requirements of the AASHTO specifications Article 5.10.3.1, i.e.:

$$1.5 \ times\ rebar\ diameter = 1.5 \times \frac{11}{8} = 2.06 \ in. < 5 \ in. \ (\text{ok})$$

$$1.5 \ times\ maximum\ coarse\ aggregate\ size = 1.5 \times 1.5 = 2.25 \ in. < 5 \ in. \ (\text{ok})$$

11.7.4.2 Minimum and Maximum Reinforcement Check

The reinforcement ratio of the column is

$$\rho_{cs} = \frac{218.4}{8 \times 7 \times 12 \times 12} = 0.027$$

From Section 3.7.4, Eqs. 3-231 and 3-233a, the maximum and minimum reinforcement ratios are:

Maximum reinforcement ratio:

$$\rho_{max} = 0.08 > \rho_{cs} = 0.027 \ (\text{ok})$$

Minimum reinforcement ratio:

$$\rho_{min} = 0.135 \frac{f_c'}{f_y} = 0.135 \frac{5.5}{60} = 0.0124 < \rho_{cs} = 0.027 \ (\text{ok})$$

Use #4 bars with a spacing of 12 in. for the ties, which meets AASHTO LRFD specifications Article 5.10.6.3 for the ties in compression members.

11.7.5 Pier Capacity Verifications

11.7.5.1 Strength V Capacity Check

11.7.5.1.1 Biaxial Flexure Capacity Check

11.7.5.1.1.1 Determination of Pier Capacity Check Equations From Section 2.3.3.5, the axial compressive resistance factor is

$$\phi_a = 0.75$$

$$P_{ca} = 0.1\phi_a f_c' A_g = 0.1 \times 0.75 \times 5.5 \times 56 \times 12^2 = 3326.4 \ kips < P_u = 6920$$

(see Table 11-7).

Thus, use Eq. 3-230a to check the pier capacity.

11.7.5.1.1.2 Determination of Factored Axial Resistance Nominal axial resistance with zero eccentricity can be calculated from Eq. 3-224 as

$$P_0 = k_c f_c' (A_g - A_s) + f_y A_s = 0.85 \times 5.5 (56 \times 12^2 - 218.4) + 60 \times 218.4 = 49782.2 \ kps$$

The factored axial resistances P_{rx} and P_{ry} can be determined based on the methods discussed in Sections 3.7.2 and 3.7.3.2 by using the following procedures:

1. Determine the eccentricities of the factored axial load:

$$e_x = \frac{M_{uy}^{II}}{P_u} = \frac{9718.3 \times 12}{6920} = 16.85 \ in.$$

$$e_y = \frac{M_{ux}^{II}}{P_u} = \frac{13397.6 \times 12}{6920} = 23.23 \ in.$$

2. Assume $e_x = 0$ and assume the location of neutral axis c_y in the bridge transverse direction.
3. Use Eqs. 3-229 and 3-164 to calculate P_{nx} and M_{nx} based on the assumed neutral axis.
4. Revise c_y until $\frac{M_{nx}}{P_{nx}} \approx e_y$.
5. Determine the net tensile strain ε_t under P_{nx} and M_{nx}.
6. Determine the resistance factor ϕ based on Eq. 2-53b.
7. Determine the factored axial strength:

$$P_{rx} = \phi P_{nx}$$

8. Using similar procedures as in steps 2 to 7, obtain P_{ry}.

In this design example, the factored strengths P_{rx} and P_{ry} are determined by using the Biaxial Column computer program developed by FDOT. The factored axial load–moment diagrams for Strength State V are illustrated in Fig. 11-38. From Fig. 11-38, $\frac{\phi M_{nx}}{\phi P_{nx}} = e_x = 16.85$ in., and $\frac{\phi M_{ny}}{\phi P_{ny}} = e_y = 23.23$ in., we can obtain

$$P_{rx} = \phi P_{nx} = 21428.6 \ kips$$

$$P_{ry} = \phi P_{ny} = 25142.9 \ kips$$

From Eq. 3-230a, the factored axial resistance of the pier can be written as

$$P_{rxy} = \frac{\phi_a P_{rx} P_{ry} P_0}{\phi_a P_{rx} P_0 + \phi_a P_{ry} P_0 - P_{rx} P_{ry}}$$

$$= \frac{0.75 \times 21428.6 \times 25142.9 \times 49782.2}{0.75 \times 21428.6 \times 49782.2 + 0.75 \times 25142.9 \times 49782.2 - 21428.6 \times 25142.9} = 16762.8 \ kips$$

11.7.5.1.1.3 Axial Flexural Capacity Check From Table 11-7, we have

$$P_u = 6920 < P_{rxy} = 16762.8 \text{ kips (ok)}$$

11.7.5.1.2 Shear Capacity Check

Shear and torsion of the pier for Service V limit state are small, in comparison with the pier section and, therefore, the capacity checks are omitted in this example.

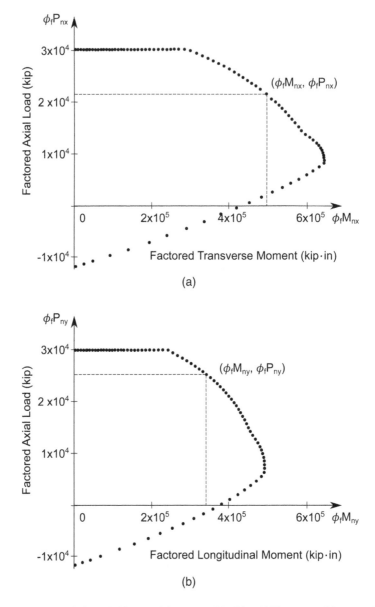

FIGURE 11-38 Factored Axial Load–Moment Diagrams of the Pier, (a) Transverse Moment, (b) Longitudinal Moment.

11.7.5.2   Service I Capacity Check

The Service I limit state for reinforced concrete piers is used to control cracking and to verify that the maximum rebar tensile stress does not exceed the allowable tensile stress of 24 ksi specified by FDOT or the rebar spacing meeting the requirement by Eq. 3-169. From Table 11-7, the resultant moment and axial load at the bottom of the pier at Service I limit state is

$$M_{ur} = \sqrt{M_{ux}^{II\,2} + M_{uy}^{II\,2}} = \sqrt{10256.4^2 + 5587.4^2} = 11679.6 \text{ kips-ft}$$

$$P_u = 5450 \text{ kips}$$

The angle between the resultant moment and bridge longitudinal direction (x-axis) is

$$\alpha_r = \tan^{-1} \frac{M_{ux}^{II}}{M_{ur}} = \tan^{-1} \frac{10256.4}{11679.6} = 61.42°$$

The pier section becomes an unsymmetrical cross section. Using the conditions of strain compatibility, $\Sigma P_z = 0$ and $\Sigma M_r = 0$, we can determine the locations of the neutral axis and the maximum stress in the reinforcing steel. The maximum stress in the outermost layer rebars under Service I loadings is determined through computer analysis as

$$f_s = 3.701 \ ksi \ll 24 \ ksi \ (ok!)$$

11.7.6 CAP REINFORCEMENT LAYOUT AND CAPACITY VERIFICATION

11.7.6.1 Development of Strut-and-Tie Model

The pier cap is subjected to two large concentrated loads and is typically analyzed by a strut-and-tie model. From Section 3.8.2.3.3 and Fig. 3-68c, the strut-and-tie model for this pier cap can be developed as shown in Fig. 11-39. Assume the distance from the centerline of the strut tie to the top of the pier d_t as 10.5 in., i.e.,

$$d_t = 10.5 \ in.$$

The angle between the tie and strut can be calculated as

$$\theta = \tan^{-1} \frac{71.5}{31.5} = 66.22°$$

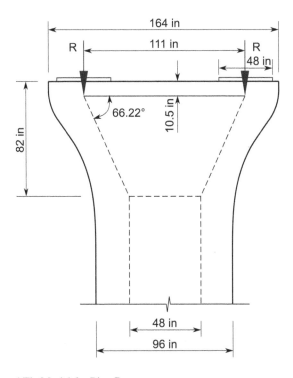

FIGURE 11-39   Strut-and-Tie Model for Pier Cap.

From Figs. 11-37 and 3-74b, the height of the tie-back face:

$$h_a = 2 \times d_t = 2 \times 10.5 = 21 \ in.$$

From Figs. 11-39 and 3-74b, the length l_b of the bearing plate is

$$l_b = 48 \ in.$$

The width of the strut is

$$w_{strut} = l_b \sin\theta + h_a \cos\theta = 48 \times \sin 66.22° + 21 \times \cos 66.22° = 52.39 \ in.$$

11.7.6.2   Capacity Check of the Struts and Ties for Strength V Limit State

From Table 11-7, we have

$$\text{Bearing reaction at Strength V: } R_V = \frac{6920}{2} = 3460 \ kips$$

$$\text{Strength V compressive force in strut: } F_{uV} = \frac{R_V}{\sin\theta} = 3781 \ kips$$

$$\text{Strength V tensile force in tie: } T_{uV} = \frac{R_V}{\tan\theta} = 1525 \ kips$$

Assuming that the steel tie is yielded before the compressive strut reaches its ultimate, we have the tensile strain as

$$\varepsilon_s = \frac{f_y}{E_s} = \frac{60}{30000} = 0.002$$

From Eq. 3-246c, we have

$$\varepsilon_I = \varepsilon_s + (\varepsilon_s + 0.002)(\cot\theta)^2 = 0.002777$$

From Eq. 3-246b, the limiting compressive strain of the strut is

$$f_{cu} = \frac{f_c'}{0.8 + 170\varepsilon_I} = 4.323 \ ksi < 0.85 f_c' \ (\text{ok})$$

From Section 2.3.3.5, the resistance factor for the strut-and-tie is $\phi_{snt} = 0.7$.
From Eq. 3-246a, the factored strength of the strut is

$$\phi_{snt} P_{nst} = 0.7 \times f_{cu} \times w_{strut} \times 7 \times 12 = 13317 \ kips > F_{uV} = 3781 \ kps \ (\text{ok})$$

From Section 3.9.4., the factored strength of the node is

$$\phi_{snt} P_{nno} = 0.7 \times 0.75 f_c' w_{strut} \times 7 \times 12 = 12707 \ kips > F_{uV} = 3781 \ kps \ (\text{ok})$$

The resistance factor for the tension member is $\phi_t = 0.9$.

From Fig. 11-37, we can see that 32 #11 bars are provided, and the total steel area (see Table 1-2) is

$$A_s = 49.92 \ in.^2$$

The factored steel tie resistance is

$$\phi_t P_{ns} = 0.9 f_y A_s = 2696 \ kips > T_{uV} = 1525 \ kips \ (\text{ok})$$

11.7.6.3 Crack Check for Service I Limit State

$$\text{Bearing reaction at Service I: } R_I = \frac{5450}{2} = 2725 \ kips$$

$$\text{Service I tensile force in tie: } T_{ul} = \frac{R_I}{\tan \theta} = 1201 \ kips$$

It is customary to control cracks in tension members by limiting the steel stress at the service limit to 24 ksi or less.

$$\text{Steel tie stress: } f_s = \frac{T_{ul}}{A_s} = \frac{1201}{49.92} \approx 24 \ ksi \ (ok)$$

11.7.7 FOOTING CAPACITY CHECK

The footing is a reinforced concrete member and can also be analyzed by the strut-and-tie model method. The footing capacity checks are similar to those for the cap discussed above, and its detailed analytical procedures are omitted in this section.

REFERENCES

11-1. Shao, X. D., and Gu, A. B., *Bridge Engineering*, China Communication Press, 3rd edition, Beijing, China, 2014.

11-2. Liu, L. J., and Xu. Y., *Bridge Engineering*, China Communication Press, 3rd edition, Beijing, China, 2016.

11-3. Podolny, W., and Muller, J. M., *Construction and Design of Prestressed Concrete Segmental Bridges*, A Wiley-Interscience Publication, John Wiley & Sons, New York, 1982.

11-4. AASHTO-PCI-ASBI, *Segmental Box Girder Standards for Span-by-Span and Balanced Cantilever Construction*, AASHTO, PCI, ASBI, 2000.

11-5. ASBI, *Guidelines for Construction of Concrete Segmental Bridges*, American Segmental Bridge Institute, Phoenix, AZ, 2004.

11-6. ASBI, *Construction Practices Handbook for Concrete Segmental and Cable-Supported Bridges*, American Segmental Bridge Institute, Buda, TX, 2008.

11-7. FDOT, *Standard Drawings for Road and Bridge Construction*, Tallahassee, FL, 2018.

11-8. FDOT, *Structural Detailing Manual*, Tallahassee, FL, 2018.

11-9. AASHTO, *LRFD Bridge Design Specifications*, 7th edition, Washington, D.C., 2014.

11-10. D.S. Brown, "Design Data," www.dsbrown.com, North Baltimore, OH.

11-11. Fan, L. C., *Bridge Engineering*, China Communication Press, Beijing, China, 1989.

11-12. Menn, C., *Prestressed Concrete Bridges*, Birkhauser Verlag AG Basel, 1990.

12 Segmental Bridge Construction

- Segment fabrication
- Geometric control
- Post-tensioning tendon installation and grouting

12.1 INTRODUCTION

Segmental construction features the use of a variety of construction equipment, procedures, and methods from which a specific construction procedure can be chosen to best fit the conditions of a specific project and to work within construction constraints. From the start, the design of a segmental bridge project depends heavily on the chosen construction method and needs to account for associated design loads and constraints. The selection and basic construction methods and procedures of the different types of segmental bridges have been discussed in Chapter 1 and Chapters 5 to 10. In this chapter, we will first discuss how the concrete segments are fabricated. Then, we will present how to control the bridge geometry during segment casting and erection. Finally, the post-tensioning tendon installation and grouting issues will be discussed.

12.2 SEGMENT FABRICATIONS

12.2.1 Segment Fabrication for Cast-in-Place Cantilever Construction

12.2.1.1 Typical Form Traveler

Form travelers are typically used in cast-in-place balanced cantilever segmental bridge construction. A typical form traveler is illustrated in Fig. 12-1[12-1]. The form traveler normally consists of eight systems: main structure, bottom platform, vertical suspension, interior formwork, cantilever formwork, launching system, tie-down, and work platform. The main structural system carries most of the weight from the newly cast segment. It typically consists of a front truss, a rear truss, and a main truss (see Fig. 12-1). The bottom platform system supports the concrete weight of the segment bottom slab and walls. The bottom platform system includes the bottom platform framework, side work platform, exterior wall formwork, front suspension bars, and rear tie-down device (see Fig. 12-1). The vertical suspension system supports the leading end of the bottom platform, interior soffit formwork, overhang, and wall formwork. The vertical suspension system typically consists of spreader beams and hanger channels pinned to front truss (see Fig. 12-1). The interior formwork system carries the concrete weight between the segment web walls. It consists of front suspension bars, interior framework rail and roller system, interior wall formwork, top formwork, and rear tie-down bars through concrete (see Fig. 12-1). The cantilever formwork system carries the deck overhang weight outside of each segment web and provides access for post-tensioning of transverse tendons (see Fig. 12-1a). It includes front suspension bars, supporting framework, overhang formwork, and a work platform (see Fig. 12-1a). The work platform provides access for construction, including segment bulkhead and post-tensioning. It consists of an upper work platform, a lower work platform, and a rear trailing platform (see Fig. 12-1). The launching system is used to move the complete traveler assembly forward. It includes main rails and rear tie-down bars, hydraulic launching cylinders, rear bogey assembly, and front bogey (see Fig. 12-1). The tie-down system is used to restrain the entire form traveler from tipping over during concrete placement. It consists of rear tie-down bars and pull-down rams to release tie-down bars (see Fig. 12-1a). As mentioned in Chapter 6, the weight of a typical form traveler including formwork ranges from 160 to 180 kips for a single-cell box and may reach 280 kips for twin-cell box.

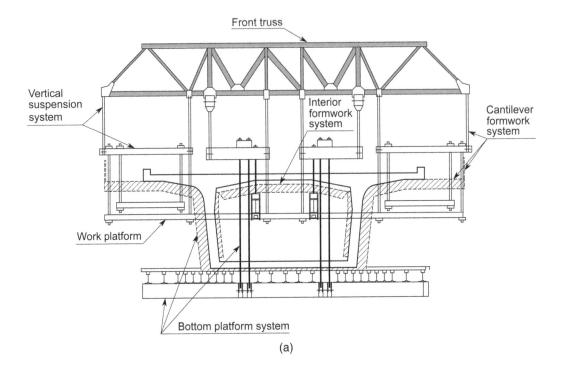

(a)

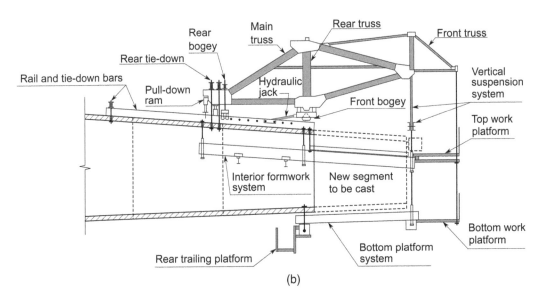

(b)

FIGURE 12-1   Typical Form Traveler, (a) Front View, (b) Side View[12-1].

12.2.1.2   Typical Procedure for Segment Fabrication

Typically, the basic procedure for fabricating the cast-in-place segments consists of:

    Step 1: Move and anchor the form traveler from the existing segment to the next segment posi-
        tion as designed (typically in one day).

    Step 2: Install bulkhead and rebars in the bottom slab as well as the webs (typically in one day).

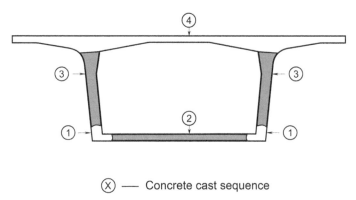

(X) — Concrete cast sequence

FIGURE 12-2 Typical Sequence of Concrete Placement.

Step 3: Move interior formwork to the next segment position as designed; install wall ties and rebars in top slab; install longitudinal and transverse tendon ducts as well as tendons (typically in one day).

Step 4: Survey final elevations, cast the new segment following the sequence shown in Fig. 12-2 for pouring the concrete, and cure the concrete to obtain sufficient strength for post-tensioning (typically in one day).

Step 5: Stress transverse and longitudinal tendons (typically in one day).

Step 6: Repeat steps 1 to 5 until all segments in the cantilever are installed.

12.2.1.3 Fabrication of the Pier Segment

The segment over each of the piers must first be completed before proceeding to the cantilever construction using the form traveler. The pier segment is also called a pier table and is normally built on a temporary platform anchored by prestressing bars at the top of the pier (see Fig. 12-3). The length of the pier segment is determined based on (1) providing enough length to ensure the stability of the future cantilever, and (2) allowing both travelers to be installed simultaneously. The minimum length of a pier segment is approximately 21 ft.

12.2.2 FABRICATION OF PRECAST SEGMENTS

The manufacturing methods for precasting segments can be classified into two categories: long-line casting and short-line casting. In the long-line casting method, all segments of a run are cast by moving the form from one segment to the next without moving any segments[12-1, 12-2]. In the short-line casting method, the segments are cast in a one-by-one process with the forms kept at a stationary position.

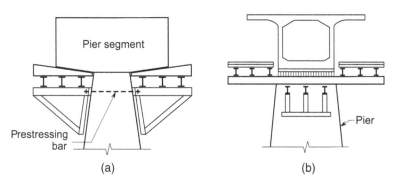

FIGURE 12-3 Typical Construction of Pier Segment for Cast-in-Place Cantilever Segmental Bridges, (a) Elevation View, (b) Side View.

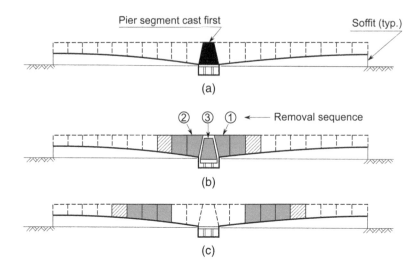

FIGURE 12-4 Typical Long-Line Casting Bed and Sequence, (a) Casting Pier Segment, (b) Removal Sequence, (c) Removing Center Part for Storage.

12.2.2.1 Long-Line Casting Method

The long-line casting method can be used for both span-by-span and balanced cantilever bridges. Figure 12-4 illustrates a typical long-line casting bed and cast sequences for typical balanced cantilever construction. The long-line casting bed typically consists of a bottom form or soffit, external forms, a core form, and a front bulkhead, similar to the systems shown in Figs 12-5 and 12-6 for the short-line casting method. The length of the soffit is the same as the entire cast run (see Fig. 12-4). The soffit is exactly set to the casting curve or cambered geometry. The casting form has one segment length and is moved along the run. It is essential that the ground support has enough stiffness to ensure that the settlements of any segment will not affect the desired geometry. For a balanced cantilever-constructed bridge, the pier segment is cast first (see Fig. 12-4a) and then the segments on either side of the pier segment (see Fig. 12-4 b). The initially cast segments may be removed for storage as segment casting progresses, leaving the center part of the casting bed free (see Fig. 12-4c). The removal sequence is illustrated in Fig. 12-4b.

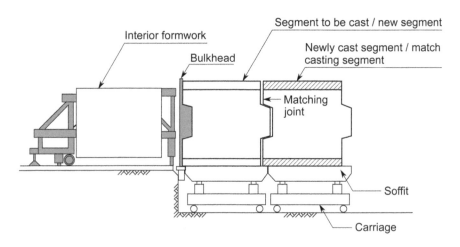

FIGURE 12-5 Typical Match-Casting System.

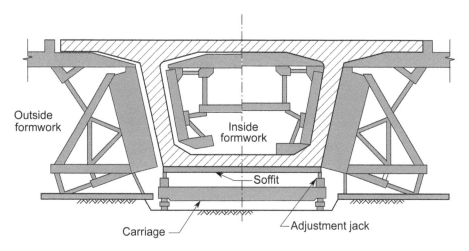

FIGURE 12-6   Typical Formwork for Short-Line Casting Method.

The main advantages of the long-line casting method are:

- It is comparatively easy to control and set the deck geometry.
- It is not necessary to move the segments to a storage area after form stripping.

The main disadvantages are:

- A large casting area is required.
- A firm nonsettling foundation for the casting bed is a must.
- It is difficult to reuse the forms from one project to another.

12.2.2.2   Short-Line Casting Method

12.2.2.2.1   Basic Concept of Short-Line Casting

To ensure a perfect fit between the ends of the segments, both long-line and short-line casting methods use the concept of match casting, in which a segment is cast against an existing segment to produce a matching joint. Then, these two segments are separated and reassembled in the final structure. In short-line casting, all segments are cast in the same place with stationary forms against the previously cast segment to create a match-cast joint. Figure 12-5 illustrates the schematic drawing of a typical match-casting system. All segments are cast in a level position against a stiff fixed bulkhead (see Fig. 12-5). After the newly cast segment reaches the required strength, the side forms are closed, and the inside form is rolled forward. Then the newly cast segment is moved out to form the end form of the next segment to be cast[12-2 to 12-5]. The newly cast segment is positioned and adjusted using different jacks (see Fig. 12-6) to develop anticipated horizontal, vertical, and twist angles in reference to the cast bed to ensure the correct geometric relationship between the two segments.

12.2.2.2.2   Realization of Bridge Geometry with Short-Line Casting Method

Usually, the segmental bridge deck is not level in both longitudinal and transverse directions. In many cases, the bridge is curved in both the vertical and horizontal planes (see Fig. 12-7). In short-line casting, the anticipated bridge geometry is realized to set the relative position between the new segment and the match-cast segment. The relative position is set by always keeping the new segment level and adjusting only the position of the match-cast segment.

Figure 12-8a shows a bridge vertical profile with a curved shape. For simplicity, we take three segments for an example and assume the relative angles between segments 1 and 2 as α_2 during erection. Assume that pier segment 1 is already cast and keeping the bulkhead in a fixed position. To ensure that the related angle between segments 1 and 2 is α_2, when segment 2 is installed, it is only

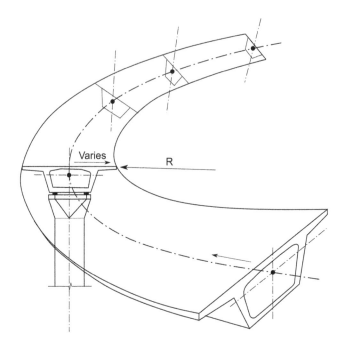

FIGURE 12-7   Typical Curved Segmental Bridge.

necessary to turn the march-casting segment (segment 1) by an angle α_2 through regulating the soffit during the adjustment operation by jacks (see Fig. 12-8b). Section 12.3 will show how to determine the angle α_2, which is related to both the designed deck geometry and the construction sequence.

A bridge horizontal curve can be obtained in the same fashion as that for a vertical curve discussed above. Assuming the relative angle between two segments 1 and 2 as β_2 (see Fig. 12-9a)

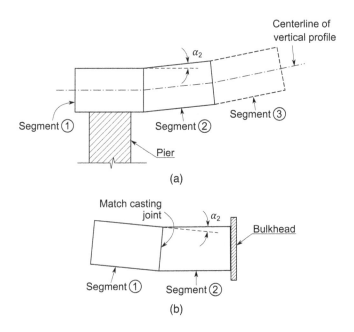

FIGURE 12-8   Realization of Vertical Curve with Short-Line Casting Method, (a) Bridge Vertical Curve and Segment Notations, (b) Position of Match-Cast Segment.

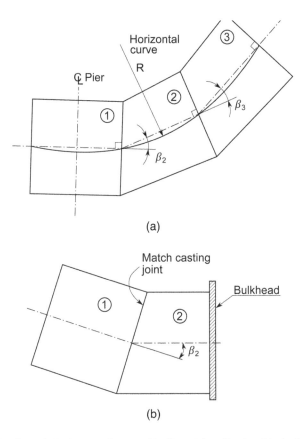

(a)

(b)

FIGURE 12-9   Realization of Horizontal Curve with Short-Line Casting Method, (a) Bridge Horizontal Curve and Segment Notations, (b) Position of Match-Cast Segment.

and keeping the bulkhead in a fixed position, the desired bridge curve can be realized by moving the match-cast segment (segment 1) by a pure transition following by an angle β_2 in plane through adjusting the soffit during the adjustment operation by jacks (see Fig. 12-9b).

For bridge sections with constant transverse slope, no special casting treatment for twisting angles is needed, because when the pier segments are set up at the correct cross slope angle, the following segments without casting twisting adjustment will automatically form a super-elevated cross section, no matter whether the bridge alignment is straight or curved. For a bridge with reverse curves and in transition areas between straight and curved alignments, a variable super-elevation is inevitable. The variable super-elevation should be accounted for in the casting operation and be achieved by rotating the match-cast segment by a small angle about the segment centerline as shown in Fig. 12-10. Assuming the difference of the deck transverse slopes between the ends of the segment is γ, the variable super-elevation can be realized through twisting the match-cast segment by an angle γ at the match-cast joint and 2γ at another end of the segment (see Fig. 12-10). It should be mentioned that using variable super-elevation in roadway bridge design should be avoided.

12.3   GEOMETRY CONTROL

12.3.1   INTRODUCTION

Geometry control of segmental bridges achieves the design roadway alignment and profile for service by properly setting up the casting geometry and erection geometry. With staged construction, geometry control in segmental construction needs to consider the target roadway geometry,

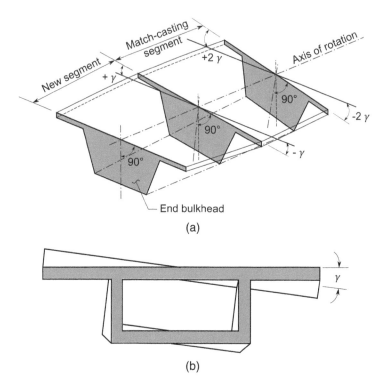

FIGURE 12-10 Realization of Variable Super-Elevation, (a) Isometric View of Segment Casting with Variable Super-Elevation, (b) End View of Twisted Segment.

instantaneous deformations due to segment self-weight and prestressing, deformations from additional dead load after the structure is completed, and long-term deformations.

In nonsegmental bridges, geometry control is achieved either by girder camber with variable haunch thickness in slab-on-girder bridges, or by adjusted soffit formwork elevations in cast-in-place concrete bridges. In segmental construction, the bridge is constructed with girder segments. Thus, the geometry control is mainly with that of each of the segments in the cast yard. Though geometry control methods during construction are available, they should be considered as the last resort because the internal force distribution either in global or in local scale will be adversely affected. The casting curve means the geometric profile with which the segments must be fabricated to obtain the theoretical bridge profile after all final structural and time-dependent deformations have taken place[12-1]. A scratch of the casting curve for a typical cantilever bridge is illustrated in Fig. 12-11.

The difficulties in geometry control of segmental construction include[12-6 to 12-8]:

*   In segmental cantilever construction, the geometry error at the beginning will be magnified as the construction proceeds because the cantilever length keeps increasing, and any angular error will result in more and more significant errors in bridge profiles.
*   The bridge geometry error in segmental construction can be observable only at very late stages. The measures available to correct the errors are very limited.
*   Even with the same mix design, the Young's modulus of concrete, especially of concrete at an early age, varies significantly and causes serious uncertainty in predicting the actual deflection during construction.

Though the principles of geometry control in short-line construction, long-line construction, and incremental launching are similar, the geometric control for the short-line construction is much more complex and is discussed in this section.

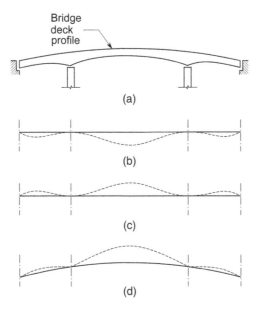

FIGURE 12-11 Typical Casting Curve for Balanced Cantilever Bridges, (a) Three-Span Cantilever Bridge, (b) Deflection due to Self-Weight, Post-Tensioning Force, Creep, Shrinkage, etc., (c) Camber, (d) Casting Curve Including Deck Profile and Camber.

12.3.2 Determination of Vertical Casting Angles for Match-Cast Segments

From Section 12.2.2.2.2, it is evident that the determination of horizontal and twisting angles of the match-cast segments is comparatively straightforward as the effect of the construction sequence on the bridge plan geometry generally can be neglected. In this section, we will focus on how to determine the casting angle of the match-cast segment which can achieve the required theoretical vertical casting curve. The basic concept of bridge camber has been discussed in Section 6.11.2. We take a simple example of a free cantilever beam as shown in Fig. 12-12 to illustrate the development of the vertical deflections and relative angles between two segments during stage construction.

The free cantilever beam is assumed to be assembled in four construction stages, and each stage installs one segment connecting to previously installed segments, as shown in Fig. 12-12b. For

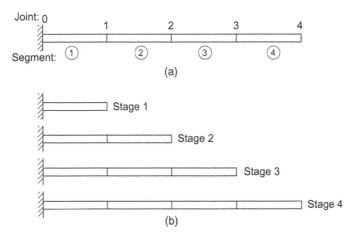

FIGURE 12-12 Cantilever Beam Example for Determination of Casting Angle between Adjacent Segments, (a) Cantilever Beam, (b) Assumed Construction Stage.

TABLE 12-1

Assumed Segment Deflection during Construction

Construction Stage	Deflection (in.)				
	Joint 0	Joint 1	Joint 2	Joint 3	Joint 4
1	0	2			
2	0	2	3		
3	0	3	7	10	
4	0	3	10	20	40
Total	0	10	20	30	40

simplicity, the theoretical target profile grade is assumed to be level as shown in Fig. 12-12a, and the total deflection curve is assumed to be a straight line as given in the last row in Table 12-1. Table 12-1 provides the assumed deflections at each of the segment joints in each construction stage. The total deflections shown in Table 12-1 are the total accumulated long-term deflections induced due to construction stages 1 to 4. For example, the total deflection at joint 2 is $3 + 7 + 10 = 20$ in. If we neglect the effect of creep and shrinkage, the line connecting these accumulation deflections at segment joints 0 to 4 represents the long-term deflection curve $OA_0'B_0'C_0'D_0'$ (see Fig. 12-13). The opposite of the long-term deflection curve is the casting curve of the free cantilever beam $OA_0B_0C_0D_0$ (see Fig. 12-13).

Let us examine the development of segment deflections during the stage erection and the determination of the required casting angles that can achieve the assumed level bridge profile.

Stage 1: Erect segment 1 by positioning its left end (joint 0) and right end (joint 1) at point O ($y_0 = 0$) and point A_0 ($y_A = 10$ in.), respectively, in the casting curve as shown in Fig. 12-13 (in thick dotted line) and Table 12-2. Point A_0 moves down by 2 in. to A_1 ($y_{A1} = 8$ in.) due to self-weight and post-tensioning force based on the assumed values shown in Table 12-1.

Stage 2: Erect segment 2 by positioning its left end (joint 1) at point A_1 ($y_{A1} = 8$ in.) and right end (joint 2) at point B_0 in the casting curve ($y_{B0} = 20$ in.) as shown in Fig. 12-13 (in thick dotted line) and Table 12-2. After the designed post-tensioning tendons are applied and its segment lifter

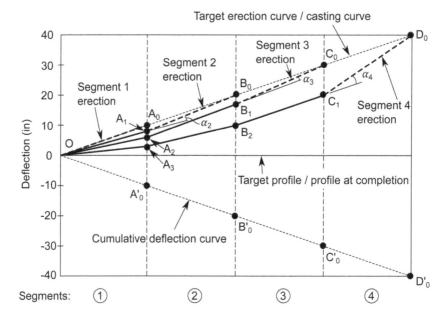

FIGURE 12-13   Vertical Casting Angles.

TABLE 12-2
Segment Positions during Construction

Construction Stage	Segment Erection Status	Location (y) of Segment Joints (in.)				
		Joint 0	Joint 1	Joint 2	Joint 3	Joint 4
1	Start	0	10			
	End	0	8			
2	Start	0	8	20		
	End	0	6	17		
3	Start	0	6	17	30	
	End	0	3	10	20	
4	Start	0	3	10	20	40
	End	0	0	0	0	0
Target		0	0	0	0	0

released, points A_1 and B_0 move down to A_2 (y_{A2} = 6 in.) and B_1 (y_{B1} = 17 in.), respectively, based on the assumption given in Table 12-1 (see Table 12-2 and Fig. 12-13).

Stage 3: Erect segment 3 by positioning its left end (joint 2) at point B_1 (y_{B1} = 17 in.) and right end (joint 3) at point C_0 in the casting curve (y_{C0} = 30 in.) as shown in Fig. 12-13 (in thick dotted line) and Table 12-2. After the designed post-tensioning tendons are applied and its segment lifter released, points A_2, B_1, and C_0 move down to A_3 (y_{A3} = 3 in.), B_2 (y_{B2} = 10 in.), and C_1 (y_{C1} = 20 in.) based on the assumed values given in Table 12-1, respectively (see Table 12-2 and Fig. 12-13).

Stage 4: Erect segment 4 by positioning its left end (joint 3) at point C_1 (y_{C1} = 20 in.) and right end (joint 4) at point D_0 in the casting curve (y_{D0} = 40 in.) as shown in Fig. 12-13 (in thick dotted line) and Table 12-2. After the designed post-tensioning tendons are applied and its segment lifter released, points A_3, B_2, C_1, and D_0 all move down to the bridge design level position based on the assumed value given in Table 12-1 (see Table 12-2 and Fig. 12-13).

From Fig. 12-13, we can see that the relative casting position between the new segment and the previous segment is not level; even the casting curve/target erection curve is a straight line to ensure the anticipated bridge deck profile is achieved. The rotation angles of the related match-cast segments for casting segments 2 to 4 are shown in Fig. 12-13 as α_2, α_3, and α_4, respectively. The rotation angle of the match-cast segment is typically referred to as the casting angle. The casting angle is equal to the sharp angle between the initial position (shown as a thick dotted line in Fig. 12-13) of the erected segment and the previously erected segment. For this example, the casting angles of the related match-cast segments for casting segments 2 to 4 can be calculated as

$$\alpha_2 = 180° - \angle OA_1B_0 \approx \tan^{-1}\frac{20-16}{120} = 1.909°$$

$$\alpha_3 = 180° - \angle A_2B_1C_0 \approx \tan^{-1}\frac{30-28}{120} = 0.955°$$

$$\alpha_4 = 180° - \angle B_2C_1D_0 \approx \tan^{-1}\frac{40-30}{120} = 4.764°$$

It should be mentioned that the casting curve has included the effect of the construction sequence. The elevation at the right end (tip of the cantilever) of the newly erected segment should always be set to the casting curve (see Fig. 12-13) in determining the rotation angle of the match-cast segment in the short-line casting method. The fundamental principle in geometry control of segmental bridges discussed above to ensure the final profile meets the target is applicable to either long-line or short-line casting.

12.3.3 Control Points of Segments and Coordinate Systems

12.3.3.1 Control Points of Segment

To mathematically describe the position of the segment during casting and erection, we need some reference points on the segment, which are often called control points. Typically, six points are used on the top of the segment to control the vertical and horizontal positions of the segment (see Fig. 12-14a). Although theoretically, three points on a rigid surface are enough to define a plane, for practical purposes, there are four level bolts A, B, C, D used for segment vertical control. The level bolts are also called elevation bolts and are set along the centerlines of webs to eliminate any vertical deflections occurring from transverse flexure or post-tensioning. Two centerline markers, E and F, are used for segment horizontal control along the segment centerline. The center markers are often called hairpins. In the bridge longitudinal direction, the control points should be located as close as possible to the edges of the segment and typically are about 2 in. (51 mm) from the edge to increase the level of survey precision. When the top of the slab is being finished, the elevation bolts and hairpins are inserted into

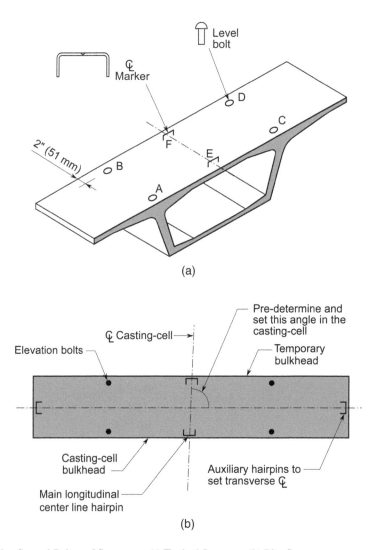

(a)

(b)

FIGURE 12-14   Control Points of Segments, (a) Typical Segment, (b) Pier Segment.

the concrete deck. The following morning, the elevations of the bolts are recorded, and the centerline of the segment is scribed onto the centerline markers.

The starting segment, typically a pier segment, is cast using the casting cell bulkhead and a temporary bulkhead. When put in the match-cast position, because it is typically short, the precision of horizontal position is harder to achieve as compared to typical segments. Auxiliary hairpins are usually installed at the tip of the segment overhangs to mark a line with a known angle to the centerline of the segment, as shown in Fig. 12-14b. These two auxiliary line markers have sufficient spacing to ensure the precision in positions for the short starting segment during casting and erection.

12.3.3.2 Coordinate Systems for Geometric Control

Four coordinate systems are typically used in geometry control of segmental bridges:

- The global coordinate system. Typically, for convenience in coordinating with roadway design, the global coordinate system is coincident with the northing (Y), easting (X), and elevation (Z) system of the project (see Fig. 12-15).
- The local coordinate system in the casting cell. This coordinate system is used to set up the target match-cast geometry and survey the actual match-cast geometry and is defined by the bulkhead and the cell centerline (xyz) as shown in Fig. 12-16.
- The local coordinate system aligned with the match-cast segment. This coordinate system is used to take the surveyed geometry relationship back to the global space for geometry adjustment for variations as shown in Fig. 12-16 ($x'y'z'$).
- The local coordinate system aligned with the segment being assembled. This coordinate system is used to facilitate the observation of target relative position of the previously erected segment with respect to the assembling segment. Theoretically, the casting cell coordinate system coincides with the assembling segment coordinate system with respect to the relative position and orientation in the cast segment (assembling segment). During casting, however, the segments may get slightly displaced, causing a misalignment of these two coordinate systems.

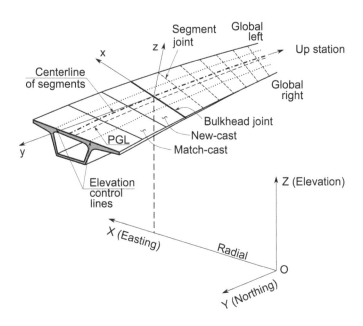

FIGURE 12-15   Global Coordinate System.

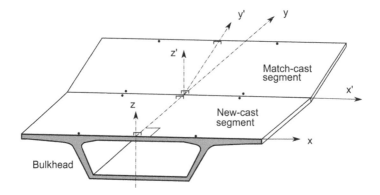

FIGURE 12-16　Local Coordinate Systems.

12.3.3.3　Coordinate Transformation

As stated in the aforementioned discussion, we typically use several different coordinate systems in bridge geometric control. In this section, we will briefly discuss how to transform the coordinates of a fixed point defined in one coordinate system to another coordinate system. As it is easy to move the origin of a coordinate system to another location, we assume two coordinate systems XYZ (global) and xyz (local) with the same origin at point O (see Fig. 12-17). The coordinates of a point P are $[x_P, y_P, z_P]$ in the coordinate system xyz, and are $[X_P, Y_P, Z_P]$ in the coordinate system XYZ. The local coordinates $[x_P, y_P, z_P]$ in coordinate system xyz can be transformed to the global coordinates $[X_P, Y_P, Z_P]$ in the coordinate system XYZ by the following equation:

$$\begin{Bmatrix} X_P \\ Y_P \\ Z_P \end{Bmatrix} = \begin{bmatrix} u_x & v_x & w_x \\ u_y & v_y & w_y \\ u_z & v_z & w_z \end{bmatrix} \begin{Bmatrix} x_P \\ y_P \\ z_P \end{Bmatrix} = [\lambda] \begin{Bmatrix} x_P \\ y_P \\ z_P \end{Bmatrix} \qquad (12\text{-}1)$$

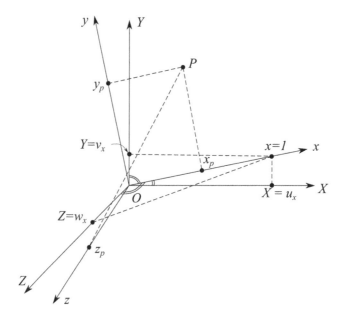

FIGURE 12-17　Coordinate Transformation.

where

$$\begin{Bmatrix} X_P \\ Y_P \\ Z_P \end{Bmatrix} = \text{vector of coordinates at point } P \text{ in coordinate system } XYZ$$

$$\begin{Bmatrix} x_P \\ y_P \\ z_P \end{Bmatrix} = \text{vector of coordinates at point } P \text{ in coordinate system } xyz$$

$$[\lambda] = \text{direction cosine matrix}$$

$$[\lambda] = \begin{bmatrix} u_x & v_x & w_x \\ u_y & v_y & w_y \\ u_z & v_z & w_z \end{bmatrix} \tag{12-2}$$

u_x, v_x, w_x = direction cosines of x-axis in X-axis, Y-axis, Z-axis directions (see Fig. 12-17)
u_y, v_y, w_y = direction cosines of y-axis in X-axis, Y-axis, Z-axis directions
u_z, v_z, w_z = direction cosines of z-axis in X-axis, Y-axis, Z-axis directions

From Eq. 12-1, we can transform the global coordinates at point P into the local coordinates by the following:

$$\begin{Bmatrix} x_P \\ y_P \\ z_P \end{Bmatrix} = [\lambda]^{-1} \begin{Bmatrix} X_P \\ Y_P \\ Z_P \end{Bmatrix} = \begin{bmatrix} u_x & u_y & u_z \\ v_x & v_y & v_z \\ w_x & w_y & w_z \end{bmatrix} \begin{Bmatrix} X_P \\ Y_P \\ Z_P \end{Bmatrix} \tag{12-3}$$

12.3.3.4 Procedure for Determining the Cast Geometry

As discussed in Section 12.2.2.2.1, the casting segment is always level and perpendicular to the bulk-head. The bridge vertical and horizontal curves are realized though adjusting the positions of the match-cast segment in reference to the casting segment. Thus, we need to determine the position of the match-cast segment in the local cell based on the predetermined casting curve. The position of a segment is mathematically described by its coordinates at the control points in the local coordinate system as shown in Fig. 12-18. Spatial coordinate transformation is typically used to determine the casting geometry. In practice, the coordinates at the control points are typically determined by computer algorithms or spreadsheets, as the same procedure is repeated for each cast segment during match-casting. In this section, one control point A in a match-cast segment (see Fig. 12-18) is taken as an example to show how to determine its coordinates in the local coordinate system. The position of the segment can be easily determined after the coordinates at all the control points have been calculated.

Step 1: Based on the design roadway profile, determine the global coordinates at the control points in the match-cast segment (previous segment) after it is erected and before the new segment (casting segment) is installed (see Fig. 12-15). This can be done by using the designed deck coordinate at the point of concern plus the deflection with opposite sign included after the match-cast segment is installed. For example, the global coordinates at point A in the match-cast segment can be written as:

$$X_{A.m} = X_{A.m.road} - \Delta X_{A.m.future} \tag{12-4a}$$

$$Y_{A.m} = Y_{A.m.road} - \Delta Y_{A.m.future} \tag{12-4b}$$

$$Z_{A.m} = Z_{A.m.road} - \Delta Z_{A.m.future} \tag{12-4c}$$

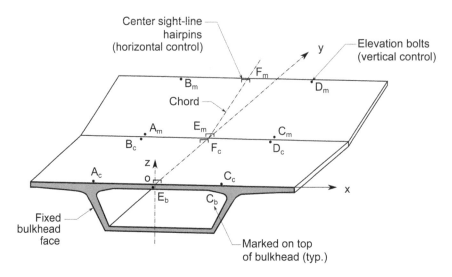

FIGURE 12-18   Coordinate System and Notations of Control Points in Local Cell[12-1].

where

$X_{A.m}, Y_{A.m}, Z_{A.m}$	= global coordinates at control point A in match-cast segment in X-, Y-, Z-directions, respectively
$X_{A.m.road}, Y_{A.m.road}, Z_{A.m.road}$	= global coordinates at control point A in match-cast segment on design roadway in X-, Y-, Z-directions, respectively
$\Delta X_{A.m.future}, \Delta Y_{A.m.future}, \Delta Z_{A.m.future}$	= total deflections at control point A in match-cast segment, occurring after the match-cast segment is installed (see Fig. 12-15) in X-, Y-, Z-directions, respectively. The sign of the value is in accordance with the reference coordinate system. For example, in Fig. 12-15, up is positive and down is negative.
	= -camber minus deflection immediately after the match-cast segment is installed

Step 2: Determine the target erection coordinates of the control points in the casting segment which are equal to the sum of the design roadway coordinate at the control point concerned and the camber which is the opposite value of the total accumulated deflection (see Fig. 12-11c). For example, the coordinates at control point A in the casting segment can be written as

$$X_{A.c} = X_{A.c.road} - \Delta X_{A.c.total} \tag{12-5a}$$

$$Y_{A.c} = Y_{A.c.road} - \Delta Y_{A.c.total} \tag{12-5b}$$

$$Z_{A.c} = Z_{A.c.road} - \Delta Z_{A.c.total} \tag{12-5c}$$

where

$X_{A.c}, Y_{A.c}, Z_{A.c}$	= global coordinates at control point A in casting segment in X-, Y-, Z-directions, respectively
$X_{A.c.road}, Y_{A.c.road}, Z_{A.c.road}$	= global coordinates at control point A in casting segment on design roadway in X-, Y-, Z-directions, respectively
$\Delta X_{A.c.total}, \Delta Y_{A.c.total}, \Delta Z_{A.c.total}$	= total accumulated deflections at control point A in casting segment (see Fig. 12-11b) in X-, Y-, Z-directions, respectively. The sign of the value is in accordance with the reference coordinate system. For example, in Fig. 12-15, up is positive and down is negative.

Step 3: Establish a local coordinate system with the origin at point E_b in the cast segment with the x-axis parallel to line AC and the y-axis parallel to line EF in the cast segment, as shown in Fig. 12-18.

Step 4: Determine the direction cosine vectors of the xyz coordinate system in the XYZ coordinate systems. From Eq. 12-2, we have

$$u_x = \frac{X_{C.c} - X_{A.c}}{|A_cC_c|}$$

$$u_y = \frac{Y_{C.c} - Y_{A.c}}{|A_cC_c|}$$

$$u_z = \frac{Z_{C.c} - Z_{A.c}}{|A_cC_c|}$$

$$v_x = \frac{X_{F.c} - X_{E.c}}{|E_cF_c|}$$

$$u_y = \frac{Y_{F.c} - Y_{E.c}}{|E_cF_c|}$$

$$u_z = \frac{Z_{F.c} - Z_{E.c}}{|E_cF_c|}$$

$$w = u \times v \begin{vmatrix} i & j & k \\ u_x & u_y & u_z \\ v_x & v_y & v_z \end{vmatrix}$$

$$w_x = u_y v_z - u_z v_y$$

$$w_y = u_z v_x - u_x v_z$$

$$w_z = u_x v_y - u_y v_x$$

Step 5: Convert all control point coordinates from the global coordinate system XYZ to the local coordinate system xyz by Eq. 12-3. For example, the local coordinates at point A in the match-cast segment can be written as

$$\begin{Bmatrix} x_{A.m} \\ y_{A.m} \\ z_{A.m} \end{Bmatrix} = \begin{bmatrix} u_x & u_y & u_z \\ v_x & v_y & v_z \\ w_x & w_y & w_z \end{bmatrix} \begin{Bmatrix} X_{A.m} - X_{E.c} \\ X_{A.m} - X_{E.c} \\ X_{A.m} - X_{E.c} \end{Bmatrix}$$

where vector $\begin{Bmatrix} X_{A.m} - X_{E.c} \\ X_{A.m} - X_{E.c} \\ X_{A.m} - X_{E.c} \end{Bmatrix}$ indicates moving the origin of the global coordinate system to that of the local coordinate system.

Step 6: After the coordinates of the six control points in the match-cast segment are determined, the position of the match-cast segment in the casting cell can be set up.

12.3.3.5 Geometric Control Procedure

Though the geometry control of precast segment bridges is obtained in the casting yard, it is still possible to correct some minor errors in bridge geometry during erection. In this section, we will first discuss the geometry control in the casting yard and then the erection geometry control.

12.3.3.5.1 Casting Geometry Control Setup and Procedure

In the short-line casting method, it is necessary to make very fine adjustments to each match-cast segment in the casting cell, and this requires precision geometry control. It is critical to obtain the

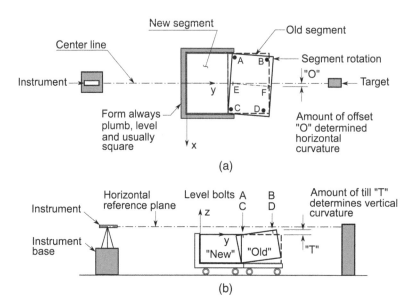

FIGURE 12-19   Cast Setup and Geometry Adjustment, (a) Plan View, (b) Elevation View.

required precision for the measurements of the relative as-cast position of the new segment with its match-cast segment. The casting cell has a permanent instrument base for survey equipment and a permanent target to establish the centerline of new segments, as shown in Fig. 12-19. The instrument and target should not be disturbed throughout the production. Otherwise, the control system must be reestablished. The casting and geometry control procedures are described as follows:

Step 1: Construct a permanent instrument base. Install the instrument, target, and casting cell, which should be plumb, level, and usually square (see Fig. 12-19a).

Step 2: Cast the first segment using the casting cell and temporarily bulkheads, and install four elevation control bolts (A, B, C, D) and two centerline control markers $(E$ and $F)$ after the top slab has been finished.

Step 3: Survey and record the elevations of the tops of the bolts, and scribe the centerline onto the centerline marks on the following day.

Step 4: Roll forward the first segment for match-casting, and reset the position of the old segment according to the theoretical local coordinates calculated by the method described in Section 12.3.3.4 based on the casting curve.

Step 5: Cast the new segment between the cast bulkhead and the match-cast segment as well as follow steps 2 and 3.

Step 6: The theoretical position of the match-cast segment is often slightly changed after the new segment has been cast due to the settlement of the soffit caused by the segment weight, vibration of the fresh concrete next to it, forces applied to it when closing the form, uneven concrete hydration heat, etc.[12-1]. It is important to correct the cast errors in the casting yard by bringing the casting variation to the global space and using the actual segment position instead of the theoretical cast curve in determining the geometries in subsequent match-casting. Before moving the match-cast segment, survey the as-cast segment geometry control points and transform the surveyed values into the casting cell geometry coordinate system. The surveyed values include the elevation readings on all eight bolts, offset readings on all four hairpins, and the auxiliary length measurements along the lines of the elevation bolts (see Fig. 12-20). All the observations should have an accuracy of 0.001 ft[12-1].

Step 7: Move the match-cast segment out for storage.

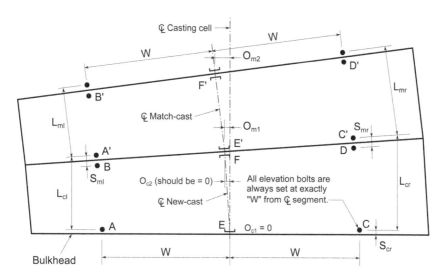

FIGURE 12-20   Observations for Geometry Control[12-1].

Step 8: Transform the casting segment geometry control point coordinates to the match-cast segment coordinate system. Note the match-cast segment coordinate system at this phase is based on survey, not on theory.

Step 9: Transform the segment geometry control point coordinates to the global coordinate system by Eq. 12-1, and determine the actual position of the new casting segment.

Step 10: Use the actual position of the casting segment instead of the theoretical geometry to determine the target casting geometry in the next match casting, in which the casting segment will become the match-cast segment.

Step 11: Repeat steps 5 to 10 until all segments are cast.

As the twist errors are more difficult to compensate for during erection than vertical and horizontal alignment errors, it is important to record and monitor the accumulative twist during the match casting. Priority should be given to casting adjustment to eliminate twisting tendencies[12-4].

The segment twist can be directly measured in the casting yard (see Fig. 12-21). To simplify the correction of the twist errors, we define the twist T as a unique linear dimension rather than twist angle γ (see Fig. 12-10) to express the effective surface warping, i.e.,

$$T = \gamma \cdot L_B \tag{12-6a}$$

where

T = twist (linear dimension)
γ = twist angle
L_B = horizontal distance between two level bolts on centers of two webs (see Fig. 12-21)

From Fig. 12-21, the twist angle γ can be written as

$$\gamma = \frac{S_C}{L_B} - \frac{S_{MC}}{L_B} \tag{12-7}$$

where S_C, S_{MC} = differences of vertical coordinates between control points A and D for casting segment and match-cast segment, respectively.

Substituting Eq. 12-7 into Eq. 12-6a yields another expression for the twist as

$$T = \gamma \cdot L_B = (A_C - D_C) - (A_{MC} - D_{MC}) \tag{12-6b}$$

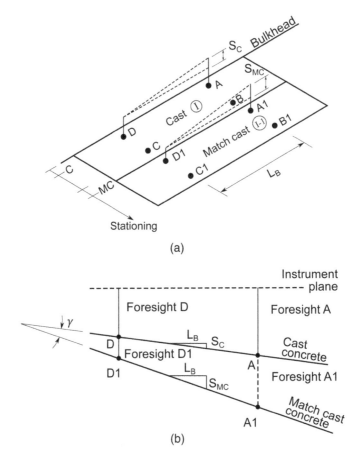

FIGURE 12-21 Twist Check in Casting Yard, (a) Deformations at Control Points in Casting and Match-Cast Segments, (b) Twist Angle.

where

A_C, D_C = vertical coordinates at control points A and D in casting segment, respectively
A_{MC}, D_{MC} = vertical coordinates at control points A and D in match-cast segment, respectively

The total twist after n cantilever segments is cast can be written as

$$T_{total} = \sum_{i}^{n} T_i \qquad (12\text{-}8)$$

As match-casting is carried on and if adjustment is needed during casting, proper counter-rotation can be taken to offset the accumulated twist.

Because of the nature of segmental construction, these small geometric tolerances can result in significant errors in bridges during erection. Therefore, rigorous survey measurements and immediate corrections in the production of the following segment are essential in the casting geometry. The survey should have an accuracy of 1/1000 ft, and two separate teams should conduct the observations and the results should check with each other.

12.3.3.5.2   Erection Geometry Control

The segment geometric relationships are set after casting, and erection geometry control is to implement the segment geometry, monitoring the actual profile, and intervene to correct geometry features if absolutely necessary.

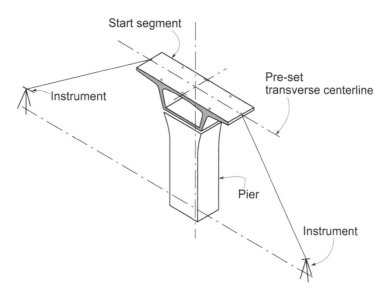

FIGURE 12-22 Setting of Pier Segments[12-1].

12.3.3.5.2.1 Erection Pier Segment/Start Segment Erection geometry control starts from erection of the first segment, typically a pier segment. It is extremely crucial to the geometry of the cantilever, because the increasing cantilever length will magnify any misalignment of a pier segment at the tip, causing difficulties in completing a span. Typically, horizontal and vertical alignment of pier segments should be 1/1000 of a foot maximum away from the theoretical values. The start segment is relatively short, resulting in difficulties in laying out the segment with sufficient precision as mentioned in Section 12.3.3.1. The solution is usually to establish a transverse reference line, typically the same transverse hairpin line during casting (see Fig. 12-14b), or an additional reference line observable to the survey stations as shown in Fig. 12-22. The transverse reference line is much longer than the segment length and, therefore, can help position the start segment with greater precision.

12.3.3.5.2.2 Surveyed Profiles Survey by any measure is the most crucial part in geometry control. At each segment erected, measurements should be taken and compared with expected values. A slight deviation is acceptable, but a record should be kept and updated to predict if the observed slight deviation can result in a substantial geometry error. If so, corrective measures should be assessed and determined, as it will be more difficult to correct the vertical, horizontal, and superelevation profiles as more segments are installed.

In balanced cantilever construction, the rotational component of surveyed deflections should be accounted for when determining if a correction is needed. Because of the flexibility of temporary supports at piers, the balanced cantilevers under construction may undergo a rigid body rotation about the centerline of the temporary support, causing the tips of the two cantilevers to deviate from the desired vertical profile in opposite directions. This can be corrected to the proper position by adjusting the jacks at the temporary support. The target correction is to equalize the profile difference at the tips of the two cantilevers.

The temperature gradient will cause additional deformation in the structures, which should be avoided in surveyed deflection (see Fig. 12-23). Typical surveyed deflections obtained before sunrise should be considered as "the real profile," upon which any evaluation of corrective methods should be based.

12.3.3.5.2.3 Corrective Methods There are several ways to correct bridge alignment errors during construction, e.g., through shims, strongbacks, lateral ties, counterweights, jacks, and wet joints.

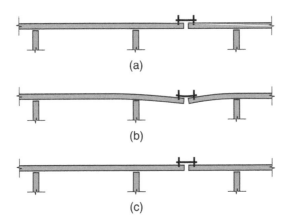

FIGURE 12-23 Temperature Gradient Effect on Deflection[12-1], (a) Before Sunrise, (b) About 3.00 p.m., (c) About 10 p.m.

12.3.3.5.2.3.1 Shims

Shimming during erection is typically achieved by inserting stainless-steel screen wire meshes or woven glass fiber matting, typically 1/16 inch (1.6 mm) thick, into the joint. The insertion position is between vertical faces of the adjacent segments to adjust vertical and horizontal alignments and is between shear key horizontal surfaces if rotational adjustment is needed, as shown in Fig. 12-24.

One issue with using shims is that it will affect the flow of epoxy across the joint, therefore risking the creation of hard points at the contact interface that can cause local damage to the segments. FDOT requires stainless-steel screen wire to be ASTM A240 Type 304 wire cloth (roving) with a maximum thickness of 1/8 in. (3.2 mm).

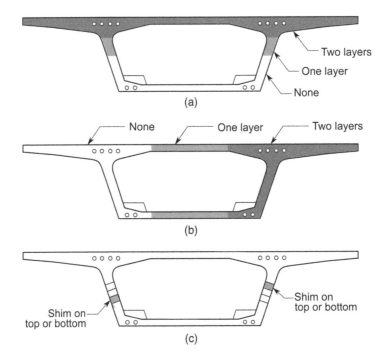

FIGURE 12-24 Shimming Joints in Geometry Control, (a) Vertical Adjusting Shim Layout, (b) Horizontal Adjusting Shim Layout, (c) Twist Adjusting Shim Layout[12-1].

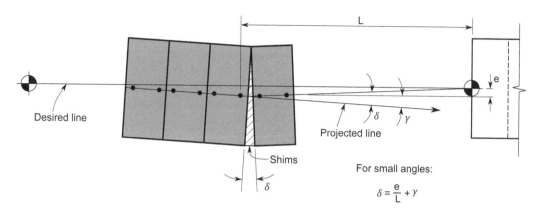

FIGURE 12-25   Correction of Alignment through Shimming between Segments[12-5].

Another issue is that the shimming effect will be magnified by the subsequent segments as shown in Fig. 12-25. Therefore, shimming should be used considering the effect on the geometry of the subsequent segments. If any shimming is needed early enough in the assembly process, the corrective horizontal angle should be projected several segments ahead so that the corrective angle is small in offset (see Fig. 12-25). Shimming should be considered as a last resort.

12.3.3.5.2.3.2   Strongbacks

Strongback frames are used to hold two cantilever tips in place to facilitate closure pour concrete casting and can transfer moment to the super-structure by using tension threaded rods[12-4]. Therefore, they have the capacity to align the two cantilever tips by force. The corrective capacity of this forcing method will depend on the capacities of the strongbacks and the bridge structure itself. Because the vertical rigidity of a cantilever beam is typically not substantial, strongbacks are sufficient for most bridge span closures. However, the effect on the capacities of the structure due to the additional forces induced during this process should be evaluated. Typical strongback frames are shown in Figs. 8-10 and 12-32.

12.3.3.5.2.3.3   Lateral Ties

The transverse stiffness of a segmental bridge is typically much greater than the vertical stiffness. If there is a horizontal misalignment between the two cantilever tips, for narrow bridges, a lateral strongback frame may be designable. However, in many practical cases, it will be difficult to develop anchorage into precast segments. In those cases, external adjustment methods, such as lateral ties, can be used. The method for adjusting lateral alignment by ties is a preferred way if the strongback method does not work[12-4].

12.3.3.5.2.3.4   Counterweights

When there is an elevation difference between two actual cantilever tips at the target closure position and the required closing force exceeds the capacity of the strongback frame, counterweights can be used to lower the cantilever tip and force the closure. Erection engineers must conduct a structure integrity check for the bridge during construction and evaluate the structural behavior during service with the resulting locked-in forces.

12.3.3.5.2.3.5   Jacks

Jacks at pier and abutment temporary supports can be used to rotate part of a bridge as a rigid body during construction, and therefore adjust the profile. Typically, jacks are used to equalize the elevation differences between cantilever tips.

12.3.3.5.2.3.6   Wet Joint

Using wet joints can correct geometry when shimming becomes excessive and exceeds the allowable limit. In this method, a short segment is cast in place between two match-cast segments to correct the geometry in subsequent segment assembly. The wet joint method is versatile and provides much higher geometric correction than shimming, but it is also very expensive. This method will affect the construction schedule and demands for temporary construction equipment and is not desirable.

12.4   CONSTRUCTION TOLERANCES

As previously mentioned, the majority of geometry control in segmental bridges essentially takes place in the casting yard. The ability to correct and control the geometry during assembly is limited and expensive. As very small angular errors between segments would result in large final offsets, the setting of the match-cast segment with respect to the cast form requires a high degree of precision. The American Segmental Bridge Institute recommends that all observations after casting should have an accuracy of 0.001 ft[12-1]. Bridges and Coulter[12-5] suggested the tolerances used for constructing a cable-stayed bridge with the center span length of 991 ft (299.0 m), total deck width of 80 ft, and segment length of 27 ft (see Fig. 12-26) as:

Maximum transverse offset at control point F (see Figs. 12-14a and 12-26) in match-cast segment: 0.003 ft (0.9 mm)
Maximum transverse offset at control point E (see Figs. 12-14a and 12-26) in match-cast segment due to horizontal rotation: 0.002 ft (0.6 mm)
Maximum vertical offset between control points A and B or C and D (see Figs. 12-14a and 12-26) in match-cast segment: 0.005 ft (±1.5 mm).

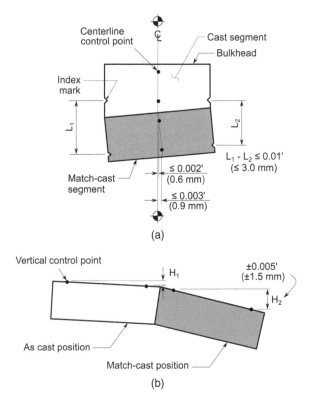

FIGURE 12-26   Match-Cast Tolerances, (a) Horizontal Tolerance, (b) Vertical Tolerance[12-5].

TABLE 12-3
Tolerances of Precast Segments

Super-Structure Box Segments	
Width of web	±1/4 in. (±6 mm)
Thickness of bottom slab	±3/16 in. (±5 mm)
Thickness of top slab	±3/16 in. (±5 mm)
Overall depth of segment	±3/16 in. (±5 mm)
Overall width of segment	±1/4 in. (±6 mm)
Length of segment	±3/8 in. (±10 mm)
Diaphragm dimensions	±3/8 in. (±10 mm)
Precast Box Pier Segments	
Height of segment	±1/4 in. (±6 mm)
Width and breadth of segment	±1/4 in. (±6 mm)
Wall thickness	±1/4 in. (±6 mm)
Warping	±1/4 in. per 20 ft with a maximum of 1/2 in. (±1/960 with a maximum of 13 mm)
Flatness	±0.025 in./ft with a maximum of 1/4 in. (±1/480 with a maximum of 6 mm)

To ensure the quality and serviceability of the final bridges, the state departments of transportation in the United States specify allowable tolerances for constructing their segmental bridges based on their construction practices. The tolerances for precasting and erection of the segments required by FDOT are given in Tables 12-3 and 12-4, respectively[12-9, 12-14].

Moretón[12-10] summarized the main reasons for rejected segments. The major issues are voids/honeycombing, displacement of post-tensioning ducts, conflict between reinforcing bars and post-tensioning, and weak forms. Other minor issues include dimensional tolerances, geometric alignment control, improper handling/storage, curing/thermal, concrete low strength, and damaged shear keys. A reasonable rejected segment rate should not exceed 0.5%.

12.5 HANDLING SEGMENTS

12.5.1 LIFTING OF SEGMENTS

Erection of segments typically uses lifting holes through the top slab near the inside or outside of the webs (see Fig. 12-27a and b). The lifting frame as shown in Fig. 12-27 is secured with post-tensioning

TABLE 12-4
Erection Tolerances of Segments

Super-Structure Box Segments	
Outside faces of adjacent segments	±3/16 in. (±5 mm)
Transverse angular deviation between adjacent joints	0.001 rad
Longitudinal angular deviation between two adjacent segments	0.003 rad
Vertical profile and horizontal alignment	1/1000
Precast Box Pier Segments	
Outside faces of adjacent segments	±3/16 in. (±5 mm)
Rotation about vertical axis	0.001 rad
Angular deviation from a vertical axis	0.003 rad for one segment 1/1200 for overall
Plan location	1/2 in. (13 mm)

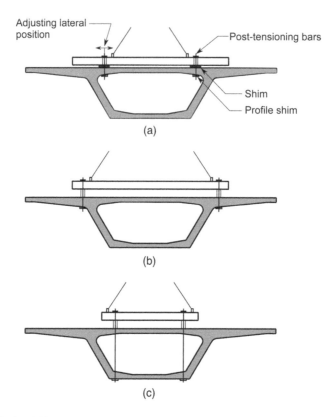

FIGURE 12-27 Typical Lifting Methods, (a) Anchored on the Top Slab between the Webs, (b) Anchored on the Top Slab in Cantilevers, (c) Anchored on the Bottom Slab[12-1].

bars through the hole in the top slab. This method allows the segment to hang at the crossfall required for erection by adjusting the lateral position of the frame where the bar is attached (see Fig. 12-27a). The post-tensioning bars can also be anchored in the bottom slab as shown in Fig. 12-27c. The bridge designer should generally provide the locations and sizes of the holes and bars as well as ensure there is enough capacity of the segment in both the casting yard and the erection site for the proposed lifting method.

12.5.2 STORING OF SEGMENTS

To avoid any warping developed in storage with time, segments should be stored using a three-point support system located under the webs (see Fig. 12-28a). The double stacking as shown in Fig. 12-28b is generally possible if the segments are supported on three points. To avoid any cracking, the designer should check the effects of localized loadings. Thus, double stacking requires the approval of the designer. The segments should also be periodically checked to see if there are any detrimental effects due to double stacking[12-1].

The segment transverse post-tensioning and grouting, clearing of joint faces, etc., can be done after the segment is placed in storage. Repairs on match-cast faces should not be made.

12.6 DETAILS DURING ERECTION

12.6.1 TEMPORARY POST-TENSIONING

As discussed in Section 5.2.3.5, it is common practice to use temporary post-tensioning bars to secure the erected segment prior to installing the permanent longitudinal post-tensioning tendons. The temporary individual bars or coupled bars can be overlapped and extend only one or a few

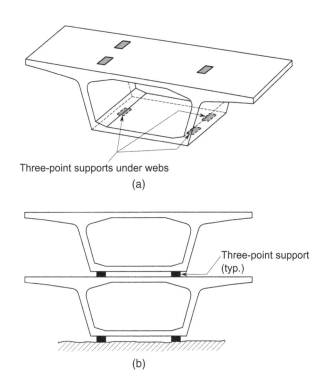

FIGURE 12-28   Segment Storage, (a) Single Segment, (b) Stacking of Segments[12-1].

segments (see Fig. 5-5). Temporary bars can also be continuously coupled throughout an entire cantilever or span (see Fig. 12-29). However, it is important to evaluate in advance the effect of cumulative bar extension and concrete shortening with the continuous coupling. The designed positions of the coupling points may be significantly altered and exceed the tolerance of the spaces within

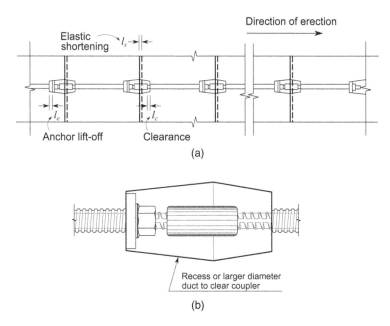

FIGURE 12-29   Typical Continuously Coupled Post-Tensioning Bars, (a) Continuously Coupled Bars, (b) Blockout for Coupler[12-1].

the blockouts, and there may be liftoff of previous anchorages, etc. Adequate clearance between the couplers and the edges of the blockouts should be provided.

As temporary bars and their anchors and couplers are expensive, it is desirable to reuse them to the maximum extent possible as mentioned in Chapter 5. For this reason, it is normal practice to stress the temporary bars to the limit of 50% of bar breaking strength. The average prestress on the match-cast epoxy joint provided by the temporary post-tensioning bars normally should be 30 to 50 psi. The normal stress on the epoxy joint can be calculated as

$$f_{epoxy} = \frac{My}{I} + \frac{P}{A} + \frac{Pey}{I} \tag{12-9}$$

where
M = moment due to dead and construction load
y = distance from section neutral axis
I = moment of inertia of section
A = area of section
P = force of post-tensioning bars
e = eccentricity of post-tensioning bars (above the neutral axis positive)

In Eq. 12-9, the sign convention and coordinate system are as same as those shown in Fig. 3-1. From Eq. 12-9, the required post-tensioning bars and their distribution on the section can be estimated.

12.6.2 Stabilizing Methods

12.6.2.1 Longitudinal Direction
As mentioned in Section 6.7, there are always unbalanced moments in balanced cantilever construction. It is important to maintain the structure stability during the construction by using the methods discussed in Section 6.7. If the bridge is comparatively straight and is erected by a gantry as discussed in Section 1.5.3, the stability of the cantilevers can be provided through the overhead launching gantry (see Fig. 12-30). The cantilevers are suspended from the gantry, and the hanger bars are always positioned next to the last erected segment.

12.6.2.2 Transverse Direction
In curved bridge construction, since the cantilevered portion also has a transverse eccentric centroid of gravity relative to the pier support, substantial transverse overturning moment can result with a tight curvature. Tie-downs may be needed in the temporary support system. When using temporary supports to resist this load effect is not economical or not convenient, a construction counterweight loaded close to the tip of the cantilever slab can be used to mitigate the uplift force in the temporary support, as shown in Fig. 12-31.

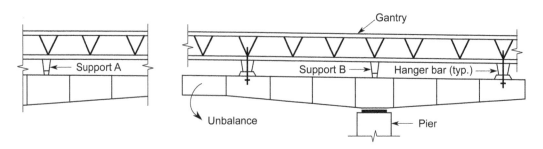

FIGURE 12-30 Stabilizing by Gantry.

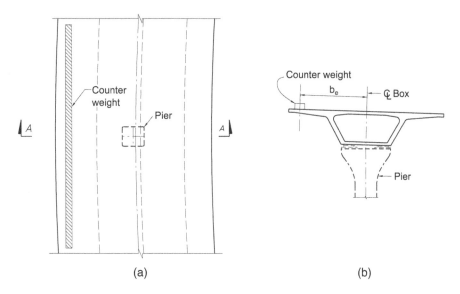

(a)   (b)

FIGURE 12-31   Counterweight in Balanced Cantilever Construction of Curved Bridges, (a) Plan View, (b) Section *A-A*.

12.6.3   MIDSPAN CLOSURE CONSTRUCTION IN BALANCED CANTILEVER SEGMENTAL BRIDGES

The lengths of midspan closures in balanced cantilever bridges typically range from 2 to 5 ft, although they can be as long as a segment length. Before connecting the cantilevers, the cantilevers are comparatively flexible. The deformation at the cantilever tip may be significantly affected by the loads applied at this location and by temperature (see Fig. 12-23). Thus, it is necessary to secure the newly built cantilever to the previously built cantilever to construct the closure. Typically, strongbacks stressed with post-tensioning bars to the precast segments (see Fig. 8-10) are employed to accomplish this purpose (see Fig. 12-32a)[12-1]. The strongbacks should have enough capacity to resist the following loadings:

- Weight of the closure pour concrete
- Remaining unbalanced forces
- Forces applied to pull the cantilever tips up and down to level
- Forces induced by temperature, such as the friction force on the adjacent piers
- Other forces

The midspan closure tends to deflect and rotate due to self-weight and temperature gradient. To prevent any cracks at the bottom of the joint, the concrete should be poured from one end of the segment to the other. If large deflection and rotation at the closure are expected, the bottom slab can be cast first and then a normal post-tensioning force can be applied prior to casting the remaining concrete. If a temporary stability tower is used to resist the unbalanced moment as shown in Fig. 12-32a, the cast-in-place concrete will induce more unbalanced moment on the cantilevers and cause the tower to deflect under the additional load. This deflection will reduce the elevation at the tip of the cantilever where the cast-in-place splice is poured. This effect should be carefully calculated and considered.

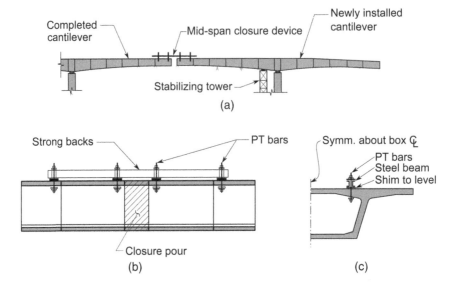

FIGURE 12-32   Construction of Midspan Closure, (a) Sketch of Midspan Closure Joint Device, (b) Elevation View of Strongback Connection, (c) Side View of Strongback Connection.

12.7   POST-TENSIONING TENDON INSTALLATION AND GROUTING

12.7.1   GENERAL

Post-tensioning tendon installation and grouting are key issues for successfully constructing a concrete segmental bridge. Generally, the designers determine and show the type, size, location, and number of required tendons in the contract plans. The contractors select the post-tensioning systems. Shop drawings are required for all post-tensioning systems selected and should be approved by the bridge designer or construction inspector before being used. The shop drawings from the manufacturer of the selected post-tensioning system typically includes:[12-11, 12-12]:

   a. Dimensions, details, and materials of the components
   b. Dimensions and details of anchors, wedge-plates, and wedges for the tendons used
   c. Dimensions and details of anchor plates, anchor nuts, bars, and couplers for the post-tensioning bars used
   d. Details of grout/filler inlets and outlets at the anchorages selected
   e. Size, type of connection, and sealing details of grout caps
   f. Details, type of material, duct connectors, and methods of connecting ducts to anchor cones
   g. Segmental duct couplers used
   h. Details for attaching intermediate grout/filler inlets and outlets to the ducts, including sizes of grout/filler pipes, materials, and shut-off valves
   i. Dimensions, clearances, force, and stroke of stressing jacks for post-tensioning bars and strands
   j. Details of ancillary equipment such as power source, hydraulic lines, and pressure gages for use with the stressing jacks
   k. Jack calibration charts to show the relationship between dial gage pressure and force delivered, etc.

Additional shop drawings from the contractor include other information, such as:

 a. Duct profile with minimum clearances
 b. Details, types, and locations of duct supports, connections to temporary bulkheads, etc.
 c. Methods for installing strands, individually or in a complete bundle for tendons.
 d. Segmental duct couplers
 e. Sequence for stressing tendons
 f. Tendon stressing ends
 g. Assumed coefficient of friction and wobble coefficient
 h. Estimated anchor set or seating loss
 i. Estimated elongation and maximum jacking force for tendons
 j. Sequence and force to which each of the proposed temporary post-tensioning bars should be coupled and stressed around the cross section.
 k. Sequence and means for de-tensioning and removing the temporary post-tensioning bars or strand tendons
 l. Locations of grout inlets and outlets, details, direction of grouting, and sequence in which tendons are grouted

12.7.2 DUCT AND TENDON INSTALLATION

12.7.2.1 Duct Installation

As discussed in Section 1.4.3, tendons are housed in the ducts that are used for installation and protection of corrosion. It is especially important to ensure the correct duct alignment for the proper functioning of a post-tensioning tendon. Ducts tend to float, as they are hollow, and should be properly secured to prevent them from displacement and resulting in excessive wobble. For this purpose, the tendon ducts are typically supported by tie-wires, straight, L-, U-, or Z-shape reinforcing bars, etc. The intervals of the supports generally should not exceed 48 in. for steel pipes or round galvanized metal ducts or 24 in. for round and flat plastic ducts. To prevent the ducts from excessive misalignment, attention should be paid to the following issues:

 a. When there is a conflict between the duct and the rebars as shown in Fig. 12-33a, relocate the rebars (see Fig. 12-33b) and add additional rebars as necessary.
 b. Significant pressure and local forces can be exerted on the reinforcing cages and tendon ducts when wet concrete is discharged into the form and consolidated by vibration. The reinforcing cages should be firmly tied and held in place by spacer blocks or chairs and adequate concrete cover. The ducts should be tightly supported and attached to the reinforcing cage.
 c. Use the concrete pour sequence as shown in Fig. 12-34 to reduce the concrete flow pressure and the displacement of ducts.
 d. Cover or temporarily plug open ends of ducts to prevent debris, water, and vermin from entering.

12.7.2.2 Tendon Installation

A post-tensioning tendon may be pulled through the duct using a special steel wire sock or other device securely attached to the end of the bundle. The post-tensioning strands in a tendon also can be individually pushed or pulled through the duct and then are formed into a tendon. To avoid any strands from getting caught or damaging the duct, each of the strands should have the protective plastic or metal caps provided by the post-tensioning system supplier before pushing or pulling. For flat ducts, it is typical practice to place the strands in the ducts prior to installation to provide required stiffness and maintain alignment during the concrete pour.

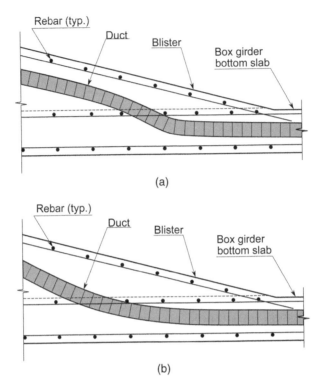

FIGURE 12-33 Excessive Duct Wobble Due to Rebar and Duct Conflict, (a) Unacceptable Misalignment, (b) Good Duct Alignment by Repositioning the Rebar.

More detailed information regarding tendon installation, including the tendon stressing equipment, jacking methods, and calibration can be found in the FHWA *Post-Tensioning Installation and Grouting Manual*[12-11].

12.7.3 GROUTING

12.7.3.1 Introduction

As discussed in Section 1.8, there are two types of tendon duct filler materials: cement grout and flexible filler. The cement grout or flexible filler is often the last line of defense against corrosion of the post-tensioning tendons. It is often difficult to inspect the tendons, and severe corrosion may be undetected for a long time before failure occurs. As the post-tensioning tendons consist of many

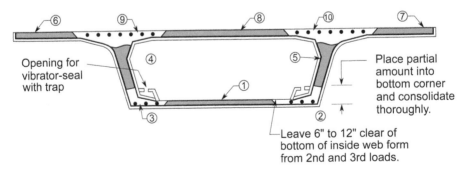

FIGURE 12-34 Concrete Pour Sequence of Box Section for Reducing the Effect of Concrete Flow Pressure and Internal Vibration[12-11].

steel wires with small diameter and large stresses, they are highly susceptible to corrosion damage. It is especially important that the ducts are completely filled with qualified grout or filler material without any voids. In this section, we will mainly discuss cement grouting issues. In general, the contractor should develop a grouting plan based on the requirements of the project specifications for post-tensioning and grouting. The grouting plan should typically include the following[12-11]:

1. Grouting procedures
2. Qualifications and certifications of grouting personnel
3. Proposed grout material in a prequalified or approved product list or with its laboratory qualification tests
4. Storage and protection of all grout material
5. A source of potable water
6. Means of measuring correct quantities of grout, water, and additives
7. Equipment for mixing and testing daily grout production
8. Sequence of injecting and evacuating grout for each type of tendon
9. Location for injecting grout at the low point of each tendon profile
10. Direction of grout injection and sequence of closing
11. Methods of inspecting to ensure all tendons are completely filled with grout
12. Methods, means, and details for sealing grout inlets, vents, and drains
13. A procedure for secondary grouting using vacuum grouting techniques
14. Forms or other means of keeping records of the grouting operations
15. Procedures to ensure that all internal ducts used for temporary post-tensioning for any purpose are fully grouted at the end of erection

FDOT[12-12] and FHWA[12-11] provide detailed requirements regarding the grouting material, equipment, testing methods, and grouting procedures. Some of the typical cement grouting procedures are briefly discussed below.

12.7.3.2 Grouting Procedures for Precast Segmental Span-by-Span Bridges

12.7.3.2.1 Grouting Procedures for End Spans

Figure 12-35a shows a typical end span of a span-by-span bridge. The following procedures may be used for grouting the tendons:

Step 1: Arrange grout outlet vents at end anchors A and E, the highest point D of the tendon, and at point C if the tendon is longer than 150 ft as well as provide grout injection ports at the low point B of the tendon profile.

Step 2: Open all vents and drains at points A through E, and then blow out with oil-free air if there is any free water in the ducts.

Step 3: Inject qualified grout at a steady rate in accordance with the specifications at the lowest point B of the duct system.

Step 4: Allow air, excess water, and grout to flow freely from the vent at point C at a satisfactory consistency; then close the vent at point C and continue pumping at a steady rate. Allow air, excess water, and grout to flow freely from the end anchor vent at point A until consistency is satisfactory; then close the vent at point A and continue pumping at a steady rate. Allow air, water, and grout to flow from the vent at point D until consistency is satisfactory; then close the vent at point D and continue pumping at a steady rate. Allow air, water, and grout to flow from the anchor vent at point E until consistency is satisfactory, and then close the vent at point E.

Step 5: Pump to a pressure of 75 psi; then hold for 2 min and check for grout leaks. If no leaks are found, reduce pressure to 5 psi for 10 min to allow entrapped air to flow to the high points.

Step 6: Open the vent at point A to release any accumulated air or bleed water, and pump grout in again as necessary until the grout flows consistently from the vent at point A. Then close

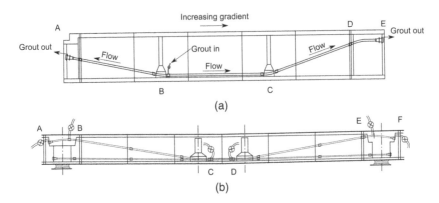

FIGURE 12-35 Grouting Procedure for Span-by-Span Bridges and Similar Tendon Profiles, (a) End Span, (b) Interior Span.

the vent at point A. The maximum grouting pressure is 150 psi and normally should be approximately 80 to 100 psi.

Step 7: Repeat the same operation as described in step 6 for the vents at points C, D, and E.

Step 8: Pump up the pressure, lock off at 30 psi, and allow the grout to take an initial set.

Step 9: Probe each vent for any vents not completely filled, and perform secondary grouting of unfilled zones as necessary.

Step 10: Seal all grout injection ports, grout outlet vents, and drain vents after completion of the grouting.

12.7.3.2.2 Grouting Procedure for Interior Spans

Figure 12-35b illustrates a typical interior span of a span-by-span segmental bridge with external tendons. If the system consists of sealed plastic pipes connected to steel pipes at diaphragms or deviators fitted with injection ports, vents, and drains, the following procedures can be used for grouting the tendons:

Step 1: Provide outlet vents at the end anchors at points A and F, at the highest points B and E if the anchors at points A and F are not the highest points, and at point D if the tendon is longer than 150 ft.

Step 2: Provide grout injection ports at the low point C of the tendon profile.

Step 3: Use similar grouting procedures as described in Section 12.7.3.2.1 (steps 2 to 10) to grout the tendon. The sequence of closing vents is D, A, B, E, F, and C.

12.7.3.3 Grouting Procedure for Precast Segmental Balanced Cantilever Bridges

12.7.3.3.1 Grouting Procedure for Top Slab Cantilever and Continuous Tendons

Take a typical cantilever tendon on a rising longitudinal gradient as an example to illustrate the grouting procedure (see Fig. 12-36a).

Step 1: Consider the longitudinal gradient, and establish the intended direction of grouting from points A to D as shown by an arrow in Fig. 12-36a.

Step 2: Provide a grout injection port at the lowest end anchor at point A.

Step 3: Provide grout outlets at the highest point C of the tendon profile, at the end anchor at point D, and at point B around the mid-length of the tendon if the tendon is longer than 150 ft.

Step 4: Use a similar grouting procedure as described in Section 12.7.3.2.1 (steps 2 to 10) to complete the tendon grouting. The sequence of closing vents is B, D, C, and A.

The above grouting procedures are also applicable to the top continuous tendons located in the midspan range.

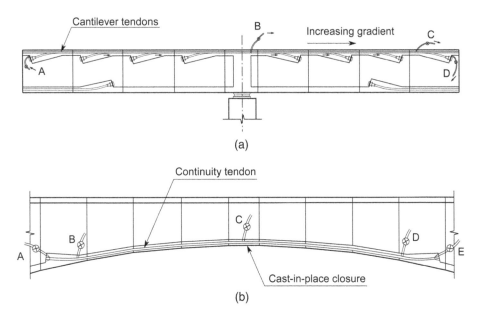

FIGURE 12-36   Tendon Grouting for Precast Segmental Balanced Cantilever Bridges and Similar Tendon Profiles, (a) Grouting Cantilever Tendons, (b) Grouting Bottom Continuity Tendons[12-12].

12.7.3.3.2   Grouting Procedures for Bottom Continuity Tendons

Take a typical bottom continuity tendon in an interior span with variable girder depth as an example to illustrate the grouting procedures (see Fig. 12-36b).

Step 1: Consider the longitudinal gradient, and establish the intended direction of grouting.

Step 2: Provide grout outlet vents at the end anchors at points A and E, at the highest point C of the tendon profile, or near the mid-length of the tendon if the tendon is longer than 150 ft.

Step 3: Provide a grout injection port at the low port at point B and another injection port at point D if the tendon profile at this point is more than 20 in. lower than the intermediate vent at point C, and at the end anchor at point E (see Fig. 12-36b).

Step 4: Use a similar grouting procedure as described in Section 12.7.3.2.1 (steps 2 to 10) to complete the tendon grouting. The sequence of closing vents is A, C, D, E, and B.

The above grouting procedure may be applicable to a structure of constant depth with a rising longitudinal grade.

12.7.3.4   Grouting Procedure for Spliced I-Girder Bridges

Take a typical tendon in a three-span variable-depth continuous cantilever and drop-in spliced I-girder bridge as an example to illustrate the grouting procedure (see Fig. 12-37). As the change in the depth of the spliced I-girder is significant, grout must be injected from the low point. The following procedure can be used to grout the tendons:

Step 1: Determine the lowest points of profile (points B and F), and then select one (say point B) as the injection point (see Fig. 12-37).

Step 2: Arrange the end anchors at points A and L so that grout vents are at the top.

Step 3: Provide grout vents at crest points D and I; at points C, E, H, and J, which are about 3 to 6 ft from the crests in both directions; and at the end anchor at point L.

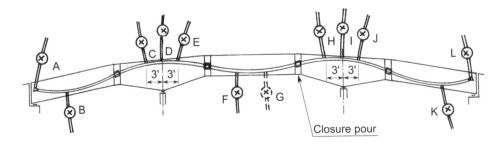

FIGURE 12-37 Tendon Grouting for Spliced I-Girder Bridges and Similar Tendon Profiles[12-12].

Step 4: Provide drain vents at all other low points *B*, *F*, *G*, and *K*.
Step 5: Use a similar grouting procedure as described in Section 12.7.3.2.1 (steps 2 to 10) to complete the tendon grouting. The sequence of closing vents is *A*, *C*, *F*, *G*, *E*, *D*, *H*, *K*, *J*, *I*, *L*, and *B*.

12.7.4 FLEXIBLE FILLER

For an easier replacement of tendons and greater effectiveness for protecting the tendons from corrosion, flexible fillers are currently being used more often in segmental bridge construction. FDOT requires most longitudinal tendons to be designed as unbonded tendons with flexible filler in the tendon ducts[12-9]. The locations of inlets and outlets as well as the flow direction of flexible filler for several typical tendon layouts are illustrated in Fig. 12-38[12-13]. The basic filling procedure for the flexible filler is generally similar to the procedures described in Section 12.7.3.

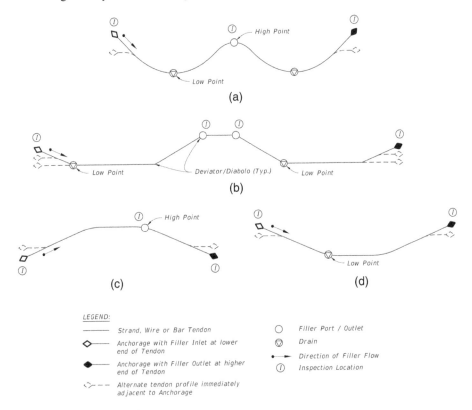

FIGURE 12-38 Flexible Filler Installation for Typical Tendon Profiles, (a) Tendon Profile One, (b) Tendon Profile Two, (c) Tendon Profile Three, (d) Tendon Profile Four[12-13].

REFERENCES

12-1. American Segmental Bridge Institute, *Construction Practices Handbook for Concrete Segmental and Cable-Supported Bridges*, 2nd edition, June 2008.

12-2. Podolny, W., and Muller, J. M., *Construction and Design of Prestressed Concrete Segmental Bridges*, A Wiley-Interscience Publication, John Wiley & Sons, New York, 1982.

12-3. Bender, B. F., and Janssen, H. H., "Geometry Control of Precast Segmental Concrete Bridges," *Journal of Prestressed Concrete Institute*, Vol. 27, No. 4, 1982, pp. 72–86.

12-4. Breen, J. E., "Controlling Twist in Precast Segmental Concrete Bridges," *Prestressed Concrete Institute Journal*, Vol. 4, 1985, pp. 15–17.

12-5. Bridges, C. P., and Clifford S. C., "Geometry Control for the Intercity Bridge," *Precast/Prestressed Concrete Institute Journal*, Vol. 24, No. 3, 1979.

12-6. Dai, X. H., Wu, X. F., and Zhang, T., "Research on the Precision Control Technology of Short-Line Segmental Prefabricated Assembly Bridge." In IOP Conference Series: *Earth and Environmental Science*, Vol. 128, No. 1, pp. 012088, IOP Publishing, 2018.

12-7. Hitchcox, J., and Mueller, D., "Geometry Control of Precast Segmental Balanced Cantilever Bridges on the Hunter Expressway." In *Australasian Structural Engineering Conference*: ASEC 2016, p. 175, Engineers Australia, 2016.

12-8. Song, X. M., Hani, M., Li, J., Xu, Q. Y., and Cheng, L. J., "Effects of Solar Temperature Gradient on Long-Span Concrete Box Girder during Cantilever Construction," *Journal of Bridge Engineering*, Vol. 21, No. 3, 2016, pp. 04015061.

12-9. FDOT, *Structures Design Guidelines*, Florida Department of Transportation, Tallahassee, FL, 2019.

12-10. Moretón, A. J. "Segmental Bridge Construction in Florida: A Review and Perspective," *Proceedings, Institution of Civil Engineers*, Pt 1, Vol. 88, 1990.

12-11. Corven, J., and Moretón, A., Post-Tensioning Tendon Installation and Grouting Manual, Federal Highway Administration, U.S. Department of Transportation, Version 2.0, FHWA-NHI-13-026, 2013.

12-12. Corven Engineering, Inc., *New Directions for Florida Post-Tensioned Bridges*, Florida Department of Transportation, Tallahassee, FL, 2002.

12-13. FDOT, *Standard Plans for Road and Bridge Construction*, FY2019-20, Florida Department of Transportation, Tallahassee, Florida, 2019.

12-14. FDOT, *Standard Specifications for Road and Bridge Construction*, Florida Department of Transportation, Tallahassee, Florida, 2019.

Appendix A*:
AASHTO-PCI-ASBI Segmental Box Girder Standards for Span-by-Span and Balanced Cantilever Construction

Span Length Range: 100 ft to 150 ft for Span-by-Span
                               100 ft to 200 ft for Balanced Cantilever
Deck Width Range: 28 ft to 45 ft

* Used with permission by ASBI

Purpose

The standards shown on these sheets have been developed to establish a limited number of practical sections leading to uniformity and simplicity of forming and production methods. These standards are applicable to most conditions of highway bridge loading and usage within the approximate span limits indicated for the sections, and the design loads specified in these General Notes.

Span Limits

The span limits shown on these sheets are approximate only and are not mandatory at either limit. The span limits shown contemplate the use of concrete weighing 155 pcf (including rebar) and concrete strength of not less than 5000 psi. It is intended that the segment depth should generally increase in 1'-0" increments for each 20' increase in span above the minimum span of 100'.

Web Thickness

Web thickness for balanced cantilever construction is based on use of 100% internal tendons in top and bottom slabs (no draped internal or external tendons). The web thickness for balanced cantilever bridges with 100% straight internal tendons may be reduced for segments in the interior 60% of spans in accordance with shear requirements and other provisions of the "AASHTO Guide Specifications for Design and Construction of Segmental Concrete Bridges". Reductions in shear and web thickness requirements for balanced cantilever construction may also be achieved by use of draped external tendons in the box cells in conjunction with straight internal tendons.

Precast Concrete

Recommended minimum strength of concrete is 5000 psi. Concrete of greater compressive strength may be used, and may be required for structural considerations, in which case limiting stresses will be based on the concrete specifications for the actual project.

Segment Lengths

Maximum Segment Length using these standards is 10'-0". In curved alignments, the segment length should be kept as close to the Maximum as possible.

Post-Tensioning Steel

Post-Tensioning steel shall be 7-wire, 1/2 inch or 0.6 inch diameter strands, conforming to ASTM A416 (AASHTO M203). Grade 270. The maximum internal tendon size used for balanced cantilever construction under these standards shall not exceed 15-1/2 inch, or 12-0.6 inch diameter Grade 270 low relaxation strands. Unless otherwise stated in the contract special provisions, other aspects of furnishing, installing and grouting of prestressing steel shall be in accordance with the details shown on the plans, and the "Recommended Contract Administration Guidelines for Design and Construction of Segmental Concrete Bridges", March, 1995, American Segmental Bridge Institute.

Reinforcing Steel

All reinforcing steel shall conform to the requirements of the AASHTO Standard Specifications, and shall be ASTM A615, Grade 60, or ASTM A706. When permitted welded grillages shall be shop prepared. Field welding of reinforcing steel will be permitted at the discretion of the engineer.

Shop Drawing Requirements

Shop Drawing Requirements shall be in accordance with the "Recommended Contract Administration Guidelines for Design and Construction of Segmental Concrete Bridges" published by the American Segmental Bridge Institute, March, 1995, unless other provisions are stated in the Contract Special Provisions.

Fabrication, Formwork, Handling, Storage, Shipment and Erection

Fabrication, formwork, handling, storage, shipment and erection of precast segments shall be in accordance with the "Recommended Contract Administration Guidelines for Design and Construction of Segmental Concrete Bridges", March, 1995, American Segmental Bridge Institute, unless other requirements are specified in the Contract Special Provisions. Angular intersections of formwork shall have a minimum radius of 2". Slab and box edges shall have a minimum chamfer of 3/4".

Epoxy Joining of Precast Concrete Segments

When required by the Contract Drawings, epoxy joining of precast segments shall be in accordance with the Recommended Contract Administration Guidelines for Design and Construction of Segmental Concrete Bridges, March, 1995, American Segmental Bridge Institute, unless other requirements are specified in the Contract Special Provisions.

Temporary Post-Tensioning

Temporary Post-tensioning required for construction of span-by-span, or balanced cantilever bridges using these standard segments shall be internal bars or tendons in top and bottom slabs unless specifically detailed otherwise in the contract drawings.

Camber Diagrams

For span-by-span construction, a final, long-term camber diagram that compensates for deflections in accordance with the assumed material properties shall be provided by the designer. For balanced cantilever construction, camber diagrams shall be prepared by the contractor and reviewed by the designer.

Crown Roadway Cross Sections

Crown roadways should be accommodated by rotating the cantilever wings downward and building up the top slab between the webs. The shape of the inside void shall remain unchanged.

Wearing Surfaces

For those regions in which deicing chemicals are used on roadways, a sacrificial wearing surface is recommended to protect the structural deck and thereby enhance the life of the structure. In regions where deicing chemicals are not used, as-cast riding surfaces without wearing surfaces may be used.

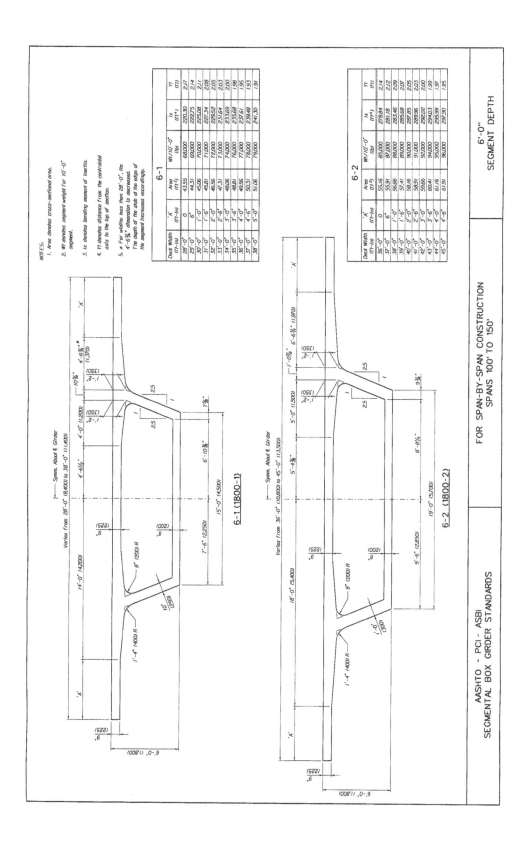

NOTES:
1. Area denotes cross-sectional area.
2. Wt denotes segment weight for 10'-0" segment.
3. Ix denotes bending moment of inertia.
4. Yt denotes distance from the centroidal axis to the top of section.
5. # For widths less than 28'-0", the 4'-6¾" dimension is decreased. The depth of the slab at the edge of the segment increases accordingly.

6-1

Deck Width (ft-in)	"A" (ft-in)	Area (ft²)	Wt/10'-0" (lb)	Ix (ft⁴)	Yt (ft)
28'-0"	0	43.55	68,000	220.30	2.17
29'-0"	6"	44.31	69,000	222.75	2.14
30'-0"	1'-0"	45.06	70,000	225.08	2.11
31'-0"	1'-6"	45.81	71,000	227.34	2.08
32'-0"	2'-0"	46.56	72,000	229.52	2.05
33'-0"	2'-6"	47.31	73,000	231.64	2.03
34'-0"	3'-0"	48.06	74,000	233.69	2.00
35'-0"	3'-6"	48.81	76,000	235.68	1.98
36'-0"	4'-0"	49.56	77,000	237.61	1.95
37'-0"	4'-6"	50.31	78,000	239.48	1.93
38'-0"	5'-0"	51.06	79,000	241.30	1.91

6-2

Deck Width (ft-in)	"A" (ft-in)	Area (ft²)	Wt/10'-0" (lb)	Ix (ft⁴)	Yt (ft)
36'-0"	0	55.16	85,000	278.84	2.14
37'-0"	6"	55.91	87,000	281.18	2.12
38'-0"	1'-0"	56.66	88,000	283.46	2.09
39'-0"	1'-6"	57.41	89,000	285.68	2.07
40'-0"	2'-0"	58.16	90,000	287.85	2.05
41'-0"	2'-6"	58.91	91,000	289.96	2.03
42'-0"	3'-0"	59.66	92,000	292.02	2.00
43'-0"	3'-6"	60.41	94,000	294.03	1.99
44'-0"	4'-0"	61.16	95,000	295.99	1.97
45'-0"	4'-6"	61.91	96,000	297.90	1.95

6-1 (1800-1)

6-2 (1800-2)

AASHTO - PCI - ASBI
SEGMENTAL BOX GIRDER STANDARDS

FOR SPAN-BY-SPAN CONSTRUCTION
SPANS 100' TO 150'

6'-0"
SEGMENT DEPTH

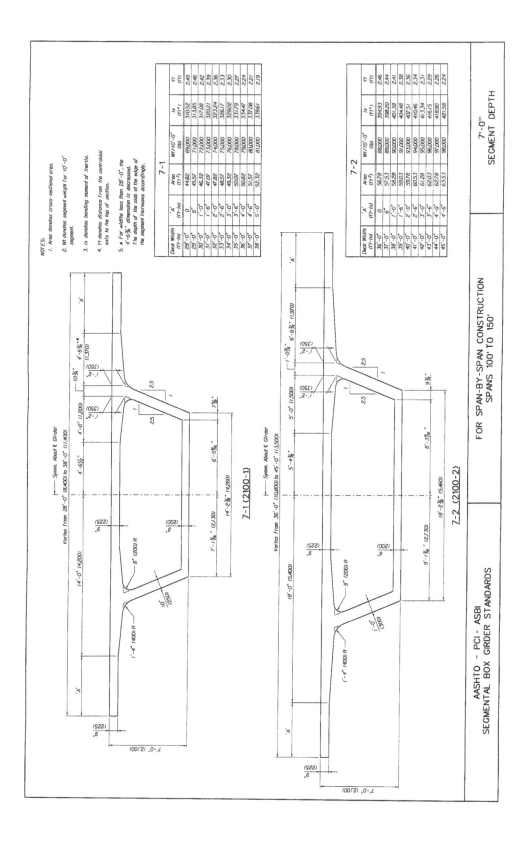

NOTES:
1. Area denotes cross-sectional area.
2. Wt denotes segment weight for 10'-0" segment.
3. Ix denotes bending moment of inertia.
4. Yt denotes distance from the centroidal axis to the top of section.
5. * For widths less than 28'-0", the 4'-6¾" dimension is decreased. The depth of the slab at the edge of the segment increases accordingly.

7-1

Deck Width (ft-in)	'A' (ft-in)	Area (ft²)	Wt/10'-0" (lb)	Ix (ft⁴)	Yt (ft)
28'-0"	0	44.82	69,000	310.52	2.49
29'-0"	6"	45.57	71,000	313.85	2.46
30'-0"	1'-0"	46.32	72,000	317.08	2.42
31'-0"	1'-6"	47.09	73,000	320.21	2.39
32'-0"	2'-0"	47.82	74,000	323.24	2.36
33'-0"	2'-6"	48.57	75,000	326.17	2.33
34'-0"	3'-0"	49.32	76,000	329.02	2.30
35'-0"	3'-6"	50.07	78,000	331.79	2.27
36'-0"	4'-0"	50.82	79,000	334.47	2.24
37'-0"	4'-6"	51.57	80,000	337.08	2.21
38'-0"	5'-0"	52.32	81,000	339.61	2.19

7-2

Deck Width (ft-in)	'A' (ft-in)	Area (ft²)	Wt/10'-0" (lb)	Ix (ft⁴)	Yt (ft)
36'-0"	0	56.78	88,000	394.93	2.46
37'-0"	6"	57.53	89,000	398.20	2.44
38'-0"	1'-0"	58.28	90,000	401.38	2.41
39'-0"	1'-6"	59.03	91,000	404.48	2.38
40'-0"	2'-0"	59.78	93,000	407.51	2.36
41'-0"	2'-6"	60.53	94,000	410.46	2.34
42'-0"	3'-0"	61.28	95,000	413.34	2.31
43'-0"	3'-6"	62.03	96,000	416.15	2.29
44'-0"	4'-0"	62.78	97,000	418.90	2.26
45'-0"	4'-6"	63.53	98,000	421.58	2.24

7-1 (2100-1)

7-2 (2100-2)

AASHTO - PCI - ASBI
SEGMENTAL BOX GIRDER STANDARDS

FOR SPAN-BY-SPAN CONSTRUCTION
SPANS 100' TO 150'

7'-0"
SEGMENT DEPTH

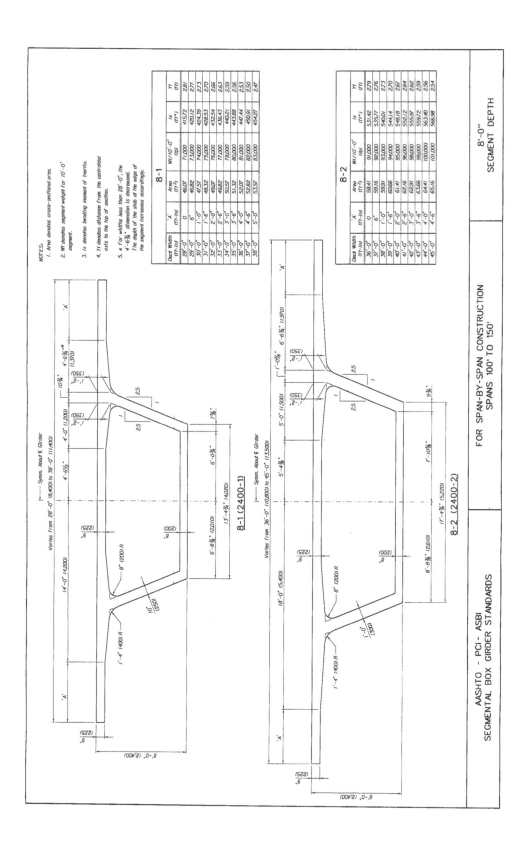

NOTES:

1. Area denotes cross-sectional area.

2. Wt denotes segment weight for 10'-0" segment.

3. Ix denotes bending moment of inertia.

4. Y1 denotes distance from the controlled axis to the top of section.

5. * For widths less than 28'-0", the 4'-6¾" dimension is decreased. The depth of the slab at the edge of the segment increases accordingly.

8-1

Deck Width (ft-in)	'A' (ft-in)	Area (ft²)	Wt/10'-0" (lb)	Ix (ft⁴)	Y1 (ft)
28'-0"	0	46.07	71,000	415.72	2.81
29'-0"	6"	46.82	73,000	420.12	2.77
30'-0"	1'-0"	47.57	74,000	424.39	2.73
31'-0"	1'-6"	48.32	75,000	428.53	2.70
32'-0"	2'-0"	49.07	76,000	432.54	2.66
33'-0"	2'-6"	49.82	77,000	436.43	2.63
34'-0"	3'-0"	50.57	78,000	440.21	2.59
35'-0"	3'-6"	51.32	80,000	443.88	2.56
36'-0"	4'-0"	52.07	81,000	447.44	2.53
37'-0"	4'-6"	52.82	82,000	450.91	2.50
38'-0"	5'-0"	53.57	83,000	454.27	2.47

8-2

Deck Width (ft-in)	'A' (ft-in)	Area (ft²)	Wt/10'-0" (lb)	Ix (ft⁴)	Y1 (ft)
36'-0"	0	58.41	91,000	531.42	2.79
37'-0"	6"	59.16	92,000	535.77	2.76
38'-0"	1'-0"	59.91	93,000	540.01	2.73
39'-0"	1'-6"	60.66	94,000	544.14	2.70
40'-0"	2'-0"	61.41	95,000	548.18	2.67
41'-0"	2'-6"	62.16	96,000	552.12	2.64
42'-0"	3'-0"	62.91	98,000	555.97	2.62
43'-0"	3'-6"	63.66	99,000	559.72	2.59
44'-0"	4'-0"	64.41	100,000	563.40	2.56
45'-0"	4'-6"	65.16	101,000	566.98	2.54

8-1 (2400-1)

8-2 (2400-2)

AASHTO - PCI - ASBI
SEGMENTAL BOX GIRDER STANDARDS

FOR SPAN-BY-SPAN CONSTRUCTION
SPANS 100' TO 150'

8'-0"
SEGMENT DEPTH

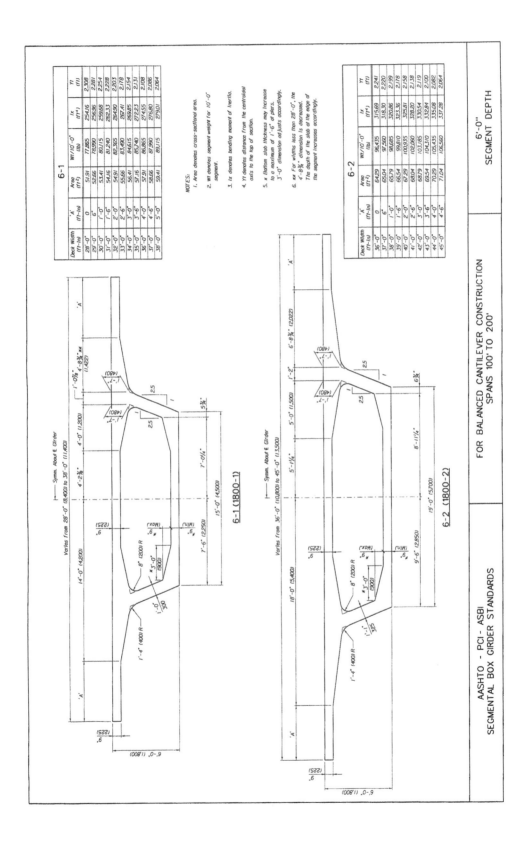

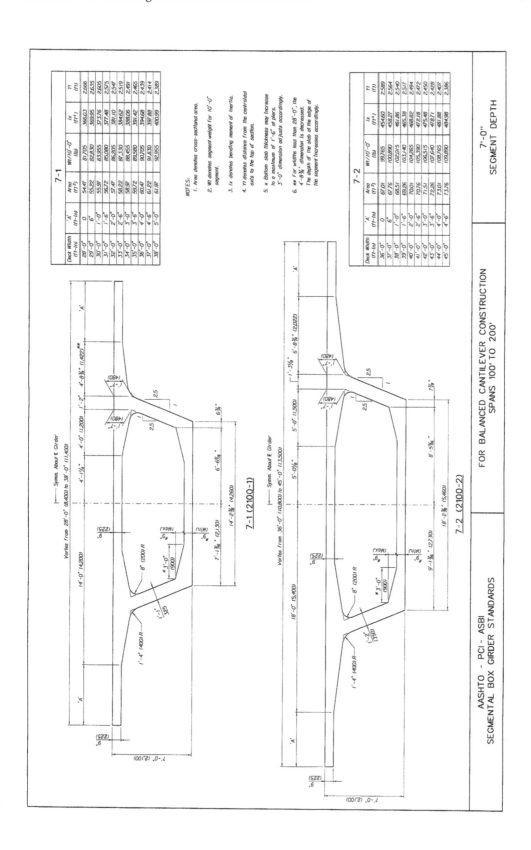

7-1

Deck Width (ft-in)	'A' (ft-in)	Area (ft²)	Wt/10'-0" (lb)	Ix (ft⁴)	Yt (ft)
28'-0"	0	54.47	81,705	366.03	2.566
29'-0"	6"	55.22	82,830	369.95	2.635
30'-0"	1'-0"	55.97	83,955	373.76	2.605
31'-0"	1'-6"	56.72	85,080	377.48	2.575
32'-0"	2'-0"	57.47	86,205	381.10	2.547
33'-0"	2'-6"	58.22	87,330	384.62	2.519
34'-0"	3'-0"	58.97	88,455	388.06	2.491
35'-0"	3'-6"	59.72	89,580	391.42	2.465
36'-0"	4'-0"	60.47	90,705	394.68	2.439
37'-0"	4'-6"	61.22	91,830	397.88	2.414
38'-0"	5'-0"	61.97	92,955	400.99	2.389

7-2

Deck Width (ft-in)	'A' (ft-in)	Area (ft²)	Wt/10'-0" (lb)	Ix (ft⁴)	Yt (ft)
36'-0"	0	67.01	99,765	454.60	2.589
37'-0"	6"	67.76	100,890	458.27	2.564
38'-0"	1'-0"	68.51	102,015	461.86	2.540
39'-0"	1'-6"	69.26	103,140	465.38	2.517
40'-0"	2'-0"	70.01	104,265	468.82	2.494
41'-0"	2'-6"	70.76	105,390	472.18	2.472
42'-0"	3'-0"	71.51	106,515	475.48	2.450
43'-0"	3'-6"	72.26	107,640	478.71	2.428
44'-0"	4'-0"	73.01	108,765	481.88	2.407
45'-0"	4'-6"	73.76	109,890	484.98	2.386

NOTES:

1. Area denotes cross-sectional area.

2. Wt denotes segment weight for 10'-0" segment.

3. Ix denotes bending moment of inertia.

4. Yt denotes distance from the centroidal axis to the top of the section.

5. * Bottom slab thickness may increase to a maximum of 1'-6" at piers. 3'-0" dimension adjusts accordingly.

6. ** For widths less than 28'-0", the 4'-8¾" dimension is decreased. The depth of the slab at the edge of the segment increases accordingly.

7-1 (2100-1)

7-2 (2100-2)

7'-0" SEGMENT DEPTH

FOR BALANCED CANTILEVER CONSTRUCTION SPANS 100' TO 200'

AASHTO - PCI - ASBI SEGMENTAL BOX GIRDER STANDARDS

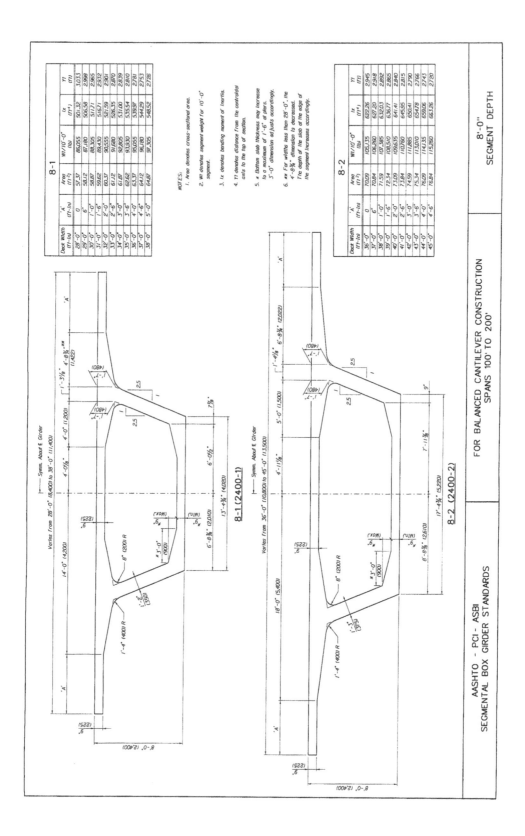

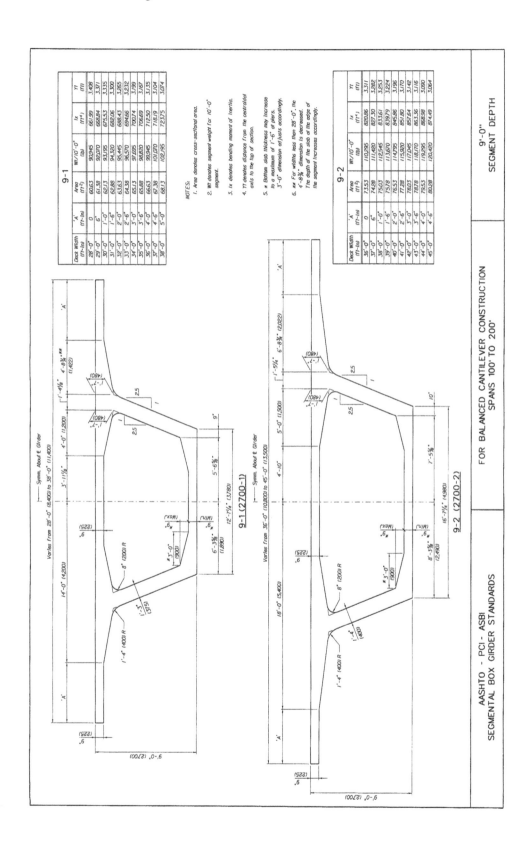

9-1

Deck Width (ft-in)	'A' (ft-in)	Area (ft²)	Wt/10'-0" (lb)	Ix (ft⁴)	YT (ft)
28'-0"	0	60.63	90,945	661.99	3,408
29'-0"	6"	61.38	92,070	668.84	3,571
30'-0"	1'-0"	62.13	93,195	675.53	3,335
31'-0"	1'-6"	62.88	94,320	682.06	3,300
32'-0"	2'-0"	63.63	95,445	688.43	3,265
33'-0"	2'-6"	64.38	96,570	694.66	3,232
34'-0"	3'-0"	65.13	97,695	700.74	3,199
35'-0"	3'-6"	65.88	98,820	706.69	3,167
36'-0"	4'-0"	66.63	99,945	712.50	3,135
37'-0"	4'-6"	67.38	101,070	718.19	3,104
38'-0"	5'-0"	68.13	102,195	723.75	3,074

NOTES:

1. Area denotes cross-sectional area.

2. Wt denotes segment weight for 10'-0" segment.

3. Ix denotes bending moment of inertia.

4. YT denotes distance from the centroidal axis to the top of section.

5. * Bottom slab thickness may increase to a maximum of 1'-6" at piers. 3'-0" dimension of justs accordingly.

6. ** For widths less than 28'-0", the 4'-8¾" dimension is decreased. The depth of the slab at the edge of the segment increases accordingly.

9-2

Deck Width (ft-in)	'A' (ft-in)	Area (ft²)	Wt/10'-0" (lb)	Ix (ft⁴)	YT (ft)
36'-0"	0	73.53	110,295	820.86	3,311
37'-0"	6"	74.28	111,420	827.30	3,282
38'-0"	1'-0"	75.03	112,545	833.61	3,253
39'-0"	1'-6"	75.78	113,670	839.79	3,224
40'-0"	2'-0"	76.53	114,795	845.86	3,196
41'-0"	2'-6"	77.28	115,920	851.80	3,170
42'-0"	3'-0"	78.03	117,045	857.64	3,142
43'-0"	3'-6"	78.78	118,170	863.36	3,116
44'-0"	4'-0"	79.53	119,295	868.98	3,090
45'-0"	4'-6"	80.28	120,420	874.49	3,064

9-1 (2700-1)

9-2 (2700-2)

AASHTO - PCI - ASBI
SEGMENTAL BOX GIRDER STANDARDS

FOR BALANCED CANTILEVER CONSTRUCTION
SPANS 100' TO 200'

9'-0"
SEGMENT DEPTH

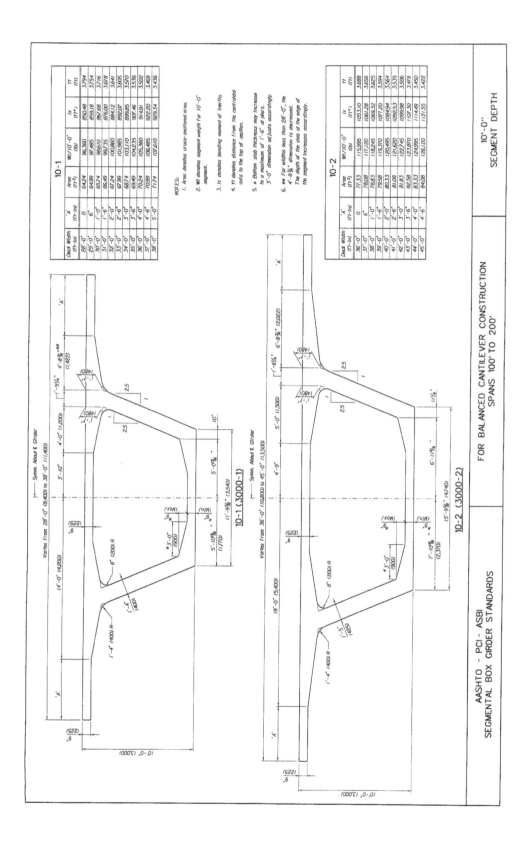

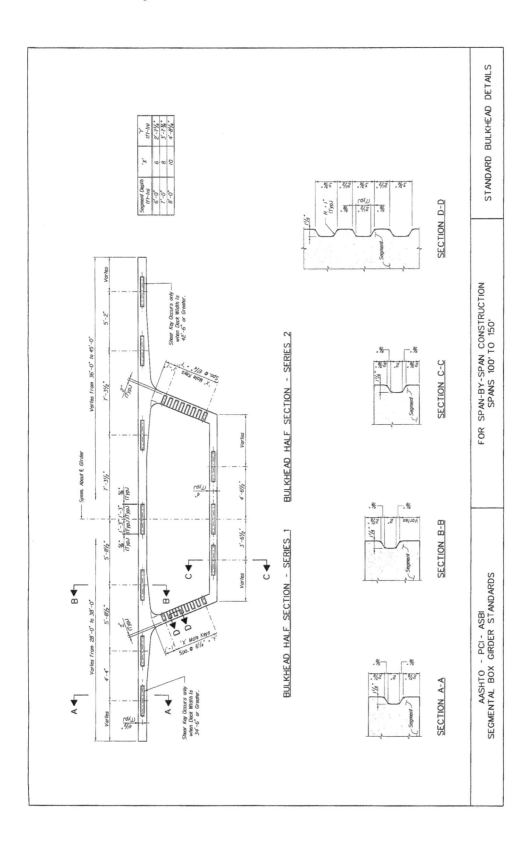

Segment Depth (ft-In)	'X'	'Y' (ft-In)
6'-0"	6	2'-7¼"
7'-0"	8	3'-7⅜"
8'-0"	10	4'-8¼"

SECTION D-D

BULKHEAD HALF SECTION - SERIES 2

BULKHEAD HALF SECTION - SERIES 1

SECTION C-C

SECTION B-B

SECTION A-A

STANDARD BULKHEAD DETAILS

FOR SPAN-BY-SPAN CONSTRUCTION
SPANS 100' TO 150'

AASHTO - PCI - ASBI
SEGMENTAL BOX GIRDER STANDARDS

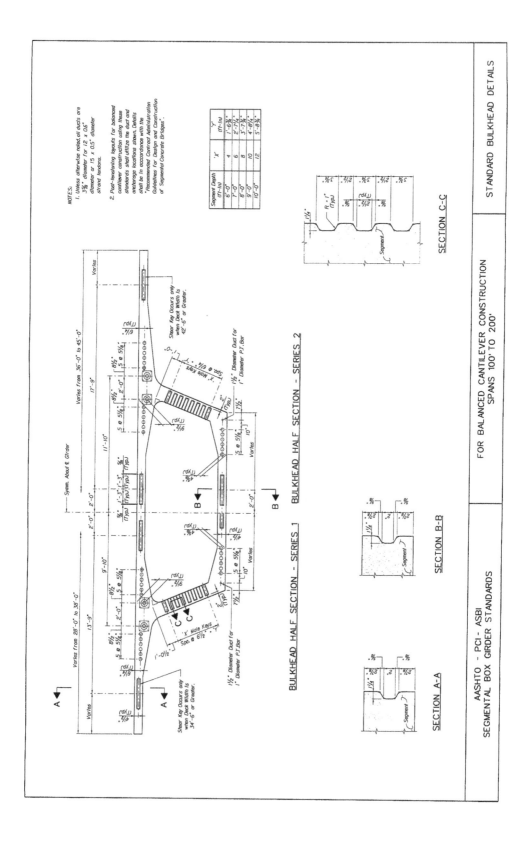

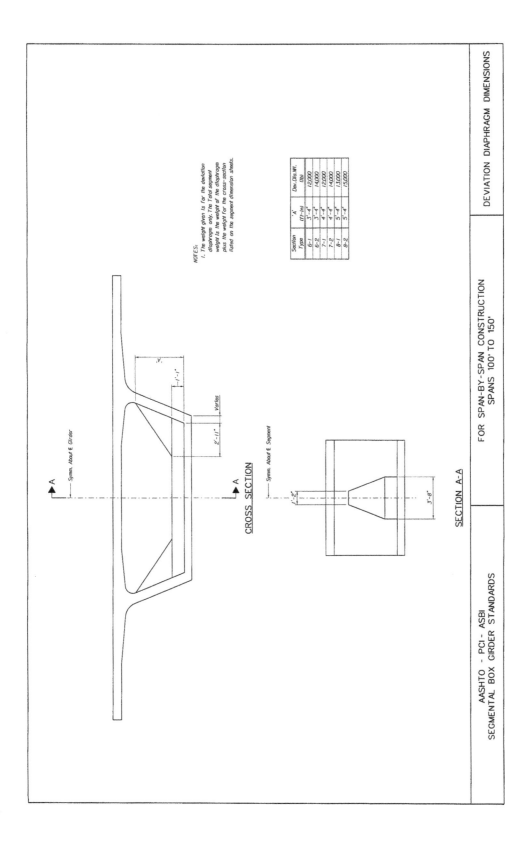

NOTES:
1. The weight given is for the deviation diaphragm only. The Total segment weight is the weight of the diaphragm plus the weight for the cross-section listed on the segment dimension sheets.

Section Type	"A" (ft-in)	Dev.Dia.Wt. (lb)
6-1	3'-4"	12,000
6-2	3'-4"	14,000
7-1	4'-4"	12,000
7-2	4'-4"	14,000
8-1	5'-4"	13,000
8-2	5'-4"	15,000

Symm. About ℄ Girder

CROSS SECTION

Symm. About ℄ Segment

SECTION A-A

1'-2"

3'-6"

"V"

1'-1"

Varies

2'-11"

FOR SPAN-BY-SPAN CONSTRUCTION
SPANS 100' TO 150'

DEVIATION DIAPHRAGM DIMENSIONS

AASHTO - PCI - ASBI
SEGMENTAL BOX GIRDER STANDARDS

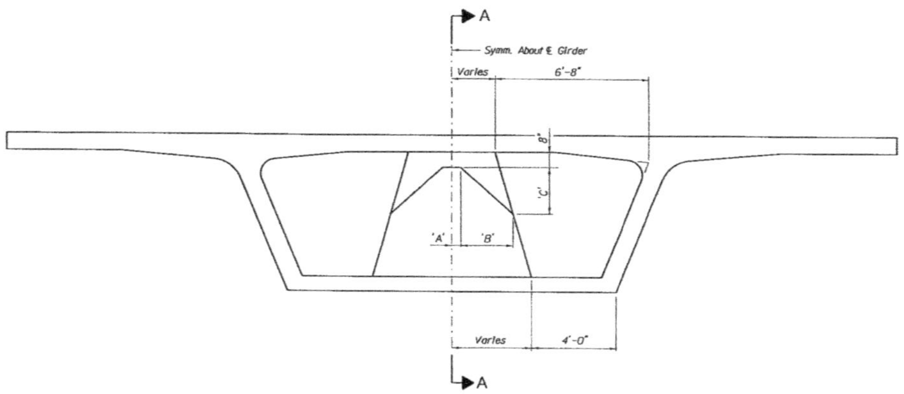

CROSS SECTION

SECTION A-A

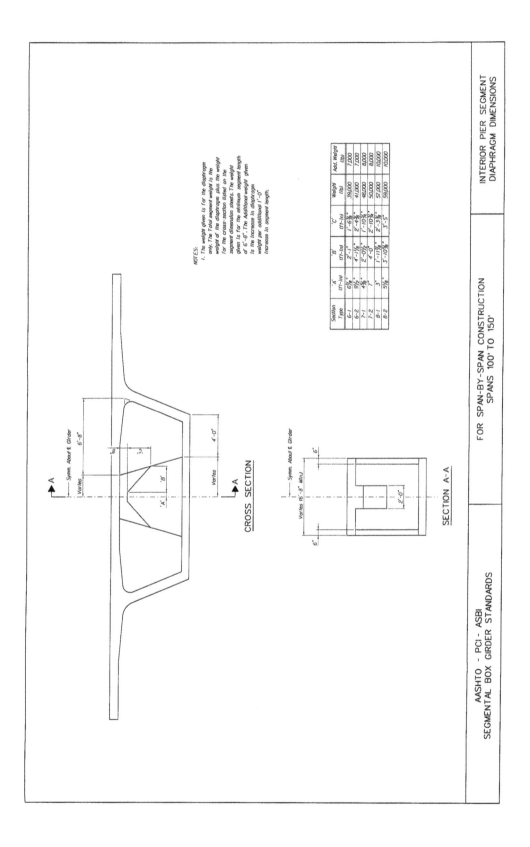

NOTES:

1. The weight given is for the diaphragm only. The Total segment weight is the weight of the diaphragm plus the weight for the cross-section listed on the segment dimension sheets. The weight given is for the minimum segment length of 6'-8". The Additional weight given is the increase in diaphragm weight per additional 1'-0" increase in segment length.

Section Type	'A' (ft-in)	'B' (ft-in)	'C' (ft-in)	Weight (lb)	Add. Weight (lb)
6-1	6⅞"	2'-1"	1'-6⅜"	39,000	7,000
6-2	9½"	4'-1⅛"	2'-4⅜"	41,000	7,000
7-1	4⅝"	2'-0½"	1'-10¾"	48,000	8,000
7-2	7"	4'-0"	2'-10¾"	50,000	8,000
8-1	3"	1'-11⅞"	2'-3⅜"	57,000	10,000
8-2	5⅛"	3'-10⅝"	3'-5"	59,000	10,000

AASHTO - PCI - ASBI SEGMENTAL BOX GIRDER STANDARDS	FOR SPAN-BY-SPAN CONSTRUCTION SPANS 100' TO 150'	INTERIOR PIER SEGMENT DIAPHRAGM DIMENSIONS

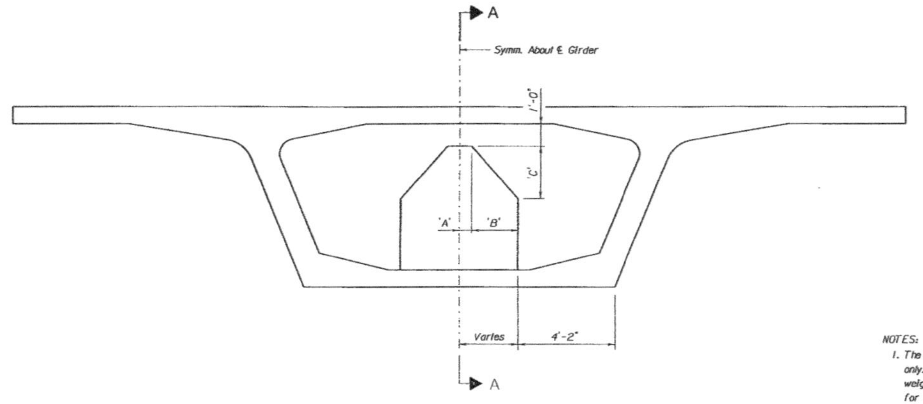

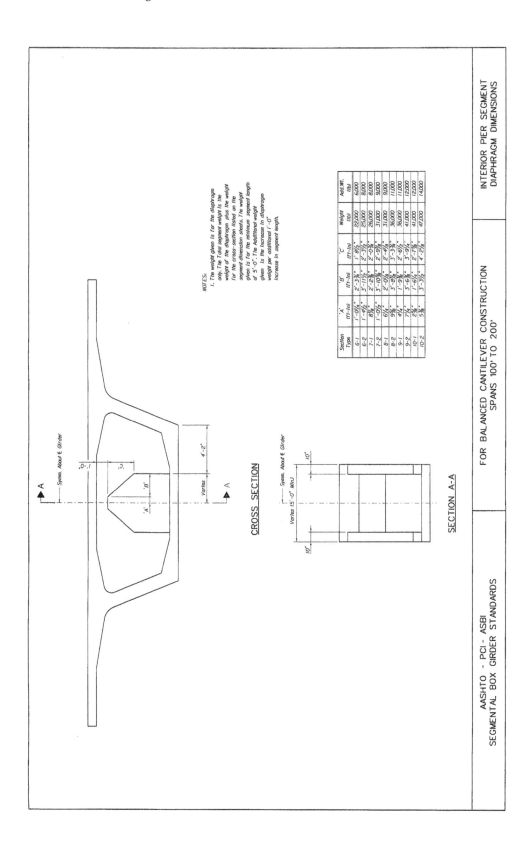

CROSS SECTION

SECTION A-A

NOTES:

1. The weight given is for the diaphragm only. The Total segment weight is the weight of the diaphragm plus the weight for the cross-section listed on the segment dimension sheets. The weight given is for the minimum segment length of 5'-0". The Additional weight given is the increase in diaphragm weight per additional 1'-0" increase in segment length.

Section Type	'A' (ft-in)	'B' (ft-in)	'C' (ft-in)	Weight (lb)	Add.Wt. (lb)
6-1	1'-0¼"	2'-3¾"	1' 8½"	22,000	6,000
6-2	1'-4½"	3'-11½"	2'-3½"	25,000	8,000
7-1	8⅞"	2'-2⅜"	2'-0¾"	26,000	8,000
7-2	1'-0½"	3'-10⅝"	2'-9⅞"	31,000	9,000
8-1	6¼"	2'-0⅛"	2'-4⅛"	31,000	9,000
8-2	9⅝"	3'-8⅞"	3'-3¾"	36,000	11,000
9-1	4¼"	1'-9⅜"	2'-6½"	36,000	11,000
9-2	7¼"	3'-6⅜"	3'-9¼'	41,000	12,000
10-1	2⅝"	1'-6¼"	2'-7⅝"	41,000	12,000
10-2	5⅜"	3'-3½"	4'-2⅛"	47,000	14,000

AASHTO - PCI - ASBI SEGMENTAL BOX GIRDER STANDARDS	FOR BALANCED CANTILEVER CONSTRUCTION SPANS 100' TO 200'	INTERIOR PIER SEGMENT DIAPHRAGM DIMENSIONS

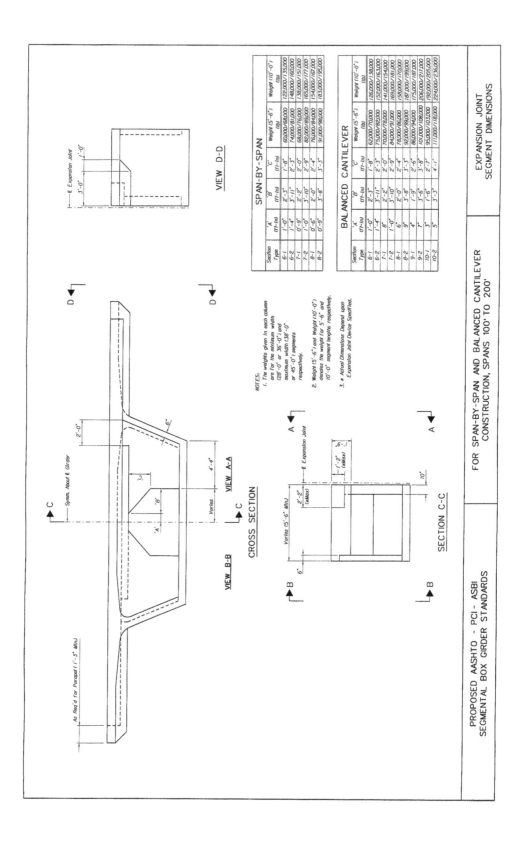

SPAN-BY-SPAN

Section Type	'A' (ft-in)	'B' (ft-in)	'C' (ft-in)	Weight (5'-6") (lb)	Weight (10'-0") (lb)
6-1	1'-0"	2'-3"	1'-6"	69,000/65,000	122,000/135,000
6-2	1'-4"	3'-11"	2'-3"	74,000/81,000	148,000/160,000
7-1	0'-9"	2'-2"	2'-0"	68,000/76,000	138,000/151,000
7-2	1'-0"	3'-10"	2'-9"	82,000/89,000	165,000/177,000
8-1	0'-6"	2'-0"	2'-4"	76,000/84,000	154,000/167,000
8-2	0'-9"	3'-8"	3'-3"	91,000/98,000	183,000/195,000

BALANCED CANTILEVER

Section Type	'A' (ft-in)	'B' (ft-in)	'C' (ft-in)	Weight (5'-6") (lb)	Weight (10'-0") (lb)
6-1	1'-0"	2'-3"	1'-8"	62,000/70,000	126,000/138,000
6-2	1'-4"	3'-11"	2'-3"	75,000/82,000	152,000/163,000
7-1	8"	2'-2"	2'-0"	70,000/78,000	141,000/154,000
7-2	1'-0"	3'-10"	2'-9"	84,000/91,000	169,000/181,000
8-1	6"	2'-0"	2'-4"	78,000/86,000	158,000/170,000
8-2	9"	3'-8"	3'-3"	92,000/99,000	187,000/199,000
9-1	4"	1'-9"	2'-6"	86,000/94,000	175,000/187,000
9-2	7"	3'-6"	3'-8"	101,000/108,000	206,000/217,000
10-1	3"	1'-6"	2'-7"	95,000/103,000	192,000/205,000
10-2	5"	3'-3"	4'-1"	111,000/118,000	224,000/236,000

VIEW D-D

VIEW A-A

VIEW B-B

CROSS SECTION

SECTION C-C

NOTES:
1. The weights given in each column are for the minimum width (28'-0" or 36'-0") and maximum width (38'-0" or 45'-0") segments respectively.

2. Weight (5'-6") and Weight (10'-0") denotes the weight for 5'-6" and 10'-0" segment lengths respectively.

3. * Actual Dimensions Depend upon Expansion Joint Device Specified.

EXPANSION JOINT SEGMENT DIMENSIONS

FOR SPAN-BY-SPAN AND BALANCED CANTILEVER CONSTRUCTION, SPANS 100' TO 200'

PROPOSED AASHTO - PCI - ASBI SEGMENTAL BOX GIRDER STANDARDS

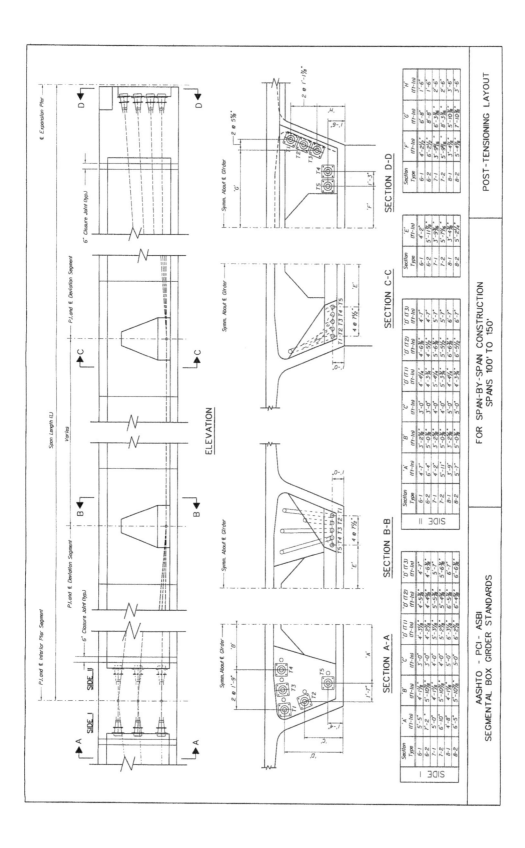

POST-TENSIONING LAYOUT

FOR SPAN-BY-SPAN CONSTRUCTION
SPANS 100' TO 150'

AASHTO - PCI - ASBI
SEGMENTAL BOX GIRDER STANDARDS

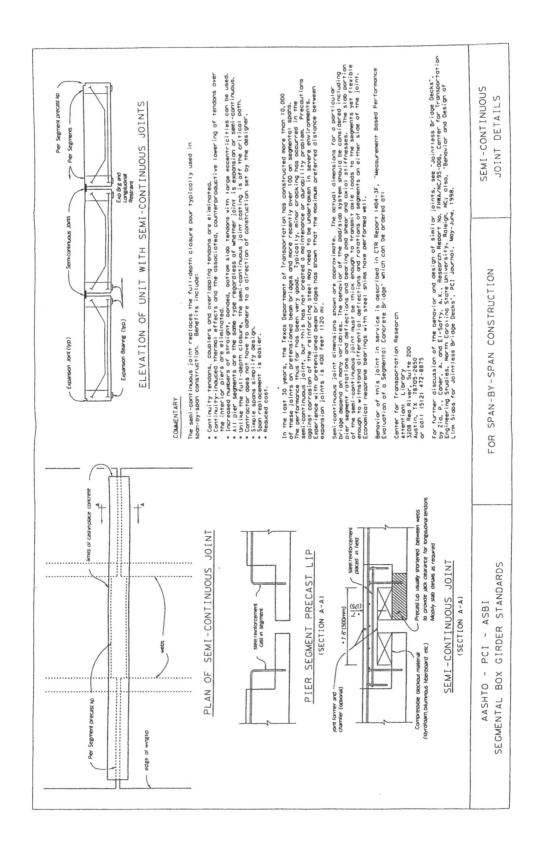

ELEVATION OF UNIT WITH SEMI-CONTINUOUS JOINTS

Per Segment precast lip
Per Segments
Exp Brg and Longitudinal Restraint
Semi-continuous Joints
Expansion Joint (typ.)
Expansion Bearing (typ.)

COMMENTARY

The semi-continuous joint replaces the full-depth closure pour typically used in span-by-span construction. Benefits include:

- Continuity tendons, couplers and overlapping tendons are eliminated.
- Continuity-induced thermal effects and the associated, counterproductive lowering of tendons over the interior piers are eliminated.
- Increased numbers are straight, bonded, bottom slab tendons with large eccentricities can be used.
- All pier segments are the same type regardless of whether joint is expansion or semi-continuous.
- Unlike the full-depth closure, the semi-continuous joint casting is off the critical path.
- Contractor does not have to adhere to a direction of construction set by the designer.
- Simple spans simplify design.
- Span replacement is easier.
- Reduced cost.

In the last 30 years, the Texas Department of Transportation has constructed more than 10,000 of these joints on pretensioned beam bridges and more recently over 100 on segmental spans. Their performance thus far has been very good. Typically minor cracking has occurred in the semi-continuous joint, but this has not created a maintenance or durability problem. Precautions against corrosion of the reinforcing steel may need to be undertaken in severe environments. Experience with pretensioned beam bridges has shown that the maximum preferred distance between expansion joints is 400 feet (120 m).

Semi-continuous joint dimensions shown are approximate. The actual dimensions for a particular bridge depend on many variables. The behavior of the pad/slab system should be considered including the pier segment rotations and deflections, and the bearing pad shear and axial stiffnesses. The slab portion of the semi-continuous joint must be thick enough to transmit axle loads to the segments yet flexible enough to withstand differential deflections and rotations of segments on either side of the joint. Economical neoprene bearings with steel shims have performed well.

Behavior of this joint in service is described in CTR Report 1404-3F, "Measurement Based Performance Evaluation of a Segmental Concrete Bridge" which can be ordered at:

Center for Transportation Research
Attention: Library
3208 Red River, Suite 200
Austin, TX 78705-2650
or call (512) 472-8875

For further discussion of the behavior and design of similar joints, see "Jointless Bridge Decks", by Zia, P., Caner, A. and El-Safty, A.K., Research Report No. FHWA/NC/95-006, Center for Transportation Engineering Studies, North Carolina State University, Raleigh, NC; also, "Behavior and Design of Link Slabs for Jointless Bridge Decks", PCI Journal, May-June, 1998.

SEMI-CONTINUOUS JOINT DETAILS

FOR SPAN-BY-SPAN CONSTRUCTION

PLAN OF SEMI-CONTINUOUS JOINT

edge of wing(s)
Pier Segment precast lip
sides of cast-in-place concrete
steel reinforcement cast in segment
webs

PIER SEGMENT PRECAST LIP
(SECTION A-A)

steel reinforcement cast in segment
steel reinforcement placed in field

SEMI-CONTINUOUS JOINT
(SECTION A-A)

Joint former and chamfer (optional)
Compressible backout material (styrofoam, bituminous fiberboard etc.)
Precast lip usually shortened between webs to provide jack clearance for longitudinal tendons Modify slab details as required
~ 1'6" (500mm)
~7" (175)

AASHTO - PCI - ASBI
SEGMENTAL BOX GIRDER STANDARDS

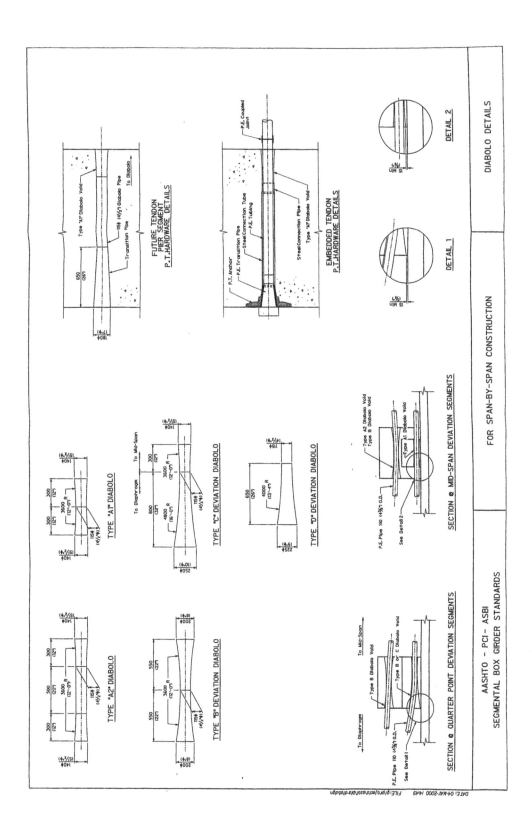

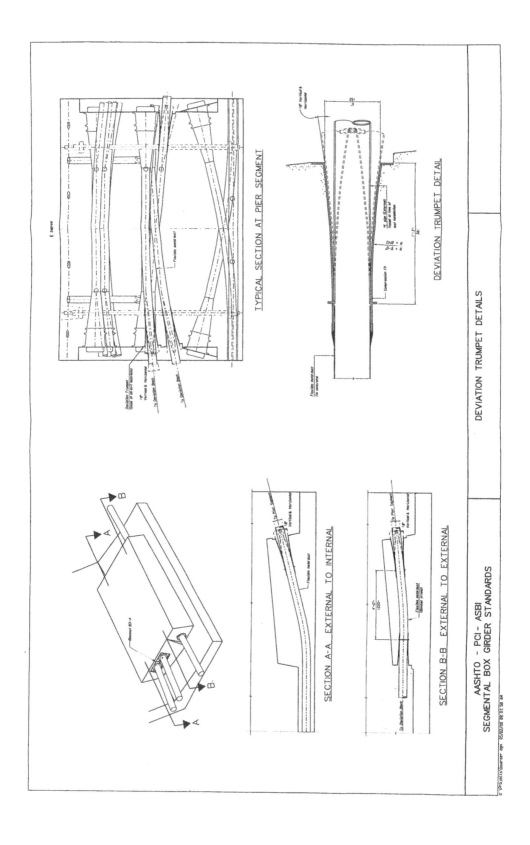

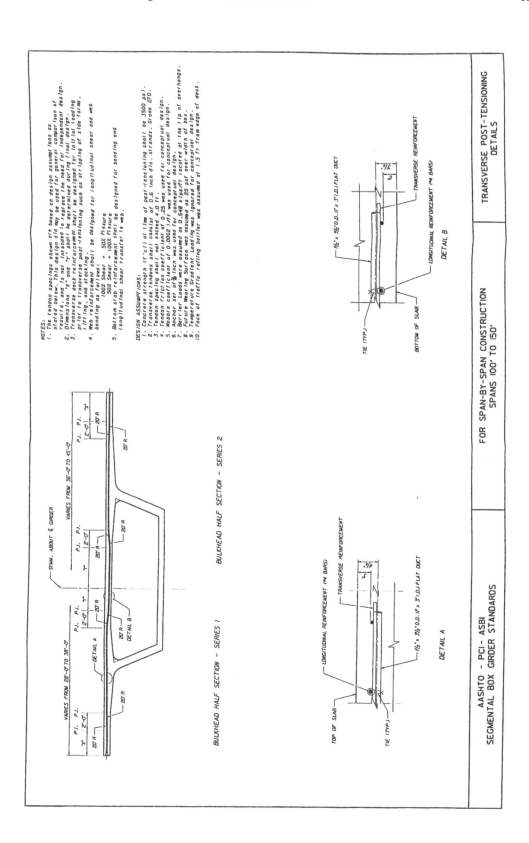

NOTES:
1. The tendon spacings shown are based on design assumptions as stated below. This design old may be used for general comparison of results, and is not intended to replace the need for independent design.
2. Dimensions 'x' and 'y' shall be determined during final design.
3. Transverse deck reinforcement shall be designed for initial loading prior to transverse post-tensioning such as stripping of side forms, lifting, and stacking.
4. Web reinforcement shall be designed for longitudinal shear and web bending as follows:
   100% Shear + 50% Flexure
   50% Shear + 100% Flexure
5. Bottom slab reinforcement shall be designed for bending and longitudinal shear transfer in web.

DESIGN ASSUMPTIONS:
1. Concrete strength (f'c) at time of post-tensioning shall be 3500 psi.
2. Transverse tendons shall consist of 0.6 inch dia. strands. Grade 270.
3. Tendon spacing shall not exceed 4.0 ft.
4. Tendon friction coefficient of 0.25 was used for conceptual design.
5. Wobble coefficient of 0.0002 /ft was used for conceptual design.
6. Anchor set of 3/8 inch was used for conceptual design.
7. Barrier loads were assumed as 0.548 k/ft located at the tip of overhangs.
8. Future Wearing Surface was assumed as 25 psf over width of box.
9. Temperature Gradient Loading was ignored for conceptual design.
10. Face of traffic railing barrier was assumed at 1.5 ft from edge of deck.

BULKHEAD HALF SECTION – SERIES 1

BULKHEAD HALF SECTION – SERIES 2

DETAIL A

DETAIL B

AASHTO - PCI - ASBI
SEGMENTAL BOX GIRDER STANDARDS

FOR SPAN-BY-SPAN CONSTRUCTION
SPANS 100' TO 150'

TRANSVERSE POST-TENSIONING
DETAILS

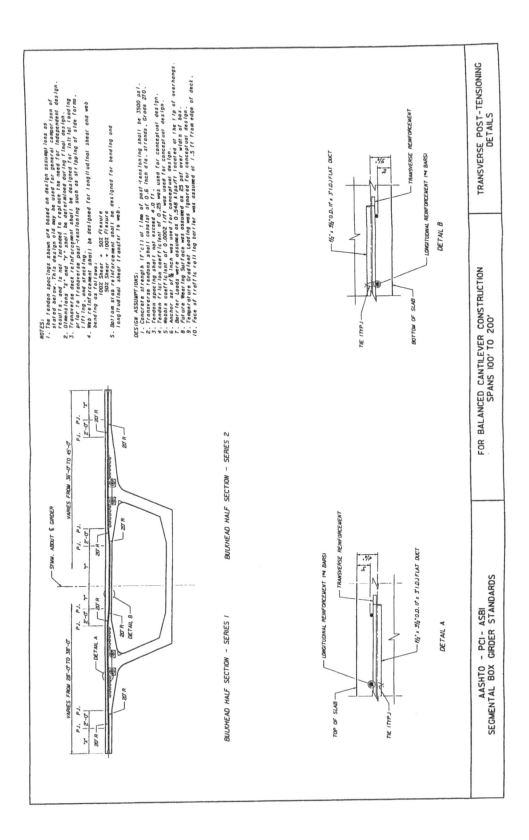

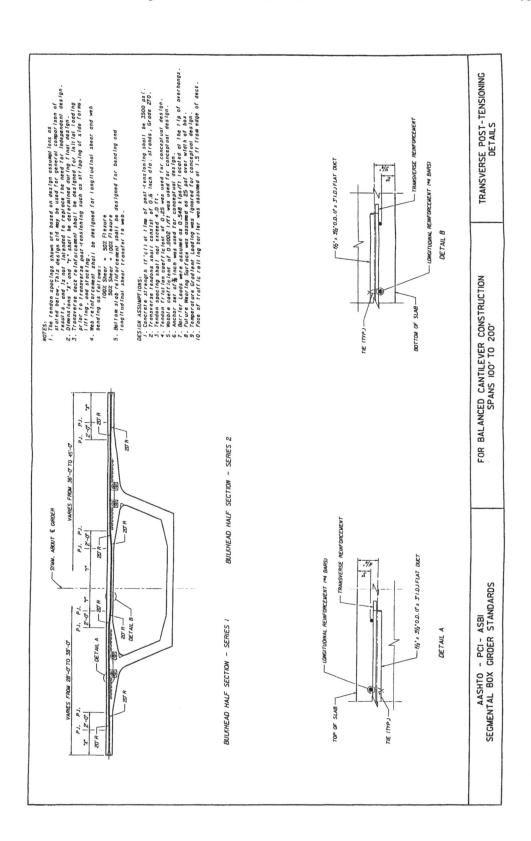

NOTES:
1. The tendon spacings shown are based on design assumptions as stated below. This design old may be used for general comparison of results, and is not intended to replace the need for independent design.
2. Dimensions "X" and "Y" shall be determined during final design.
3. Transverse deck reinforcement shall be designed for initial loading prior to transverse post-tensioning such as stripping of side forms, lifting, and stacking.
4. Web reinforcement shall be designed for longitudinal shear and web bending as follows:
   100% Shear + 50% Flexure
   50% Shear + 100% Flexure
5. Bottom slab reinforcement shall be designed for bending and longitudinal shear transfer to web.

DESIGN ASSUMPTIONS:
1. Concrete strength (f'ci) at time of post-tensioning shall be 3500 psi.
2. Transverse tendons shall consist of 0.6 inch dia. strands. Grade 270.
3. Tendon spacing shall not exceed 4.0 ft.
4. Tendon friction coefficient of 0.25 was used for conceptual design.
5. Wobble coefficient of 0.0002 1/ft was used for conceptual design.
6. Anchor set of 3/8 inch was used for conceptual design.
7. Barrier loads were assumed as 0.548 kips/ft located at the tip of overhangs.
8. Future Wearing Surfaces assumed as 25 psf over width of box.
9. Temperature Gradient Loading was ignored for conceptual design.
10. Face of traffic railing barrier was assumed at 1.5 ft from edge of deck.

BULKHEAD HALF SECTION – SERIES 1   BULKHEAD HALF SECTION – SERIES 2

DETAIL A

DETAIL B

AASHTO – PCI – ASBI SEGMENTAL BOX GIRDER STANDARDS | FOR BALANCED CANTILEVER CONSTRUCTION SPANS 100' TO 200' | TRANSVERSE POST-TENSIONING DETAILS

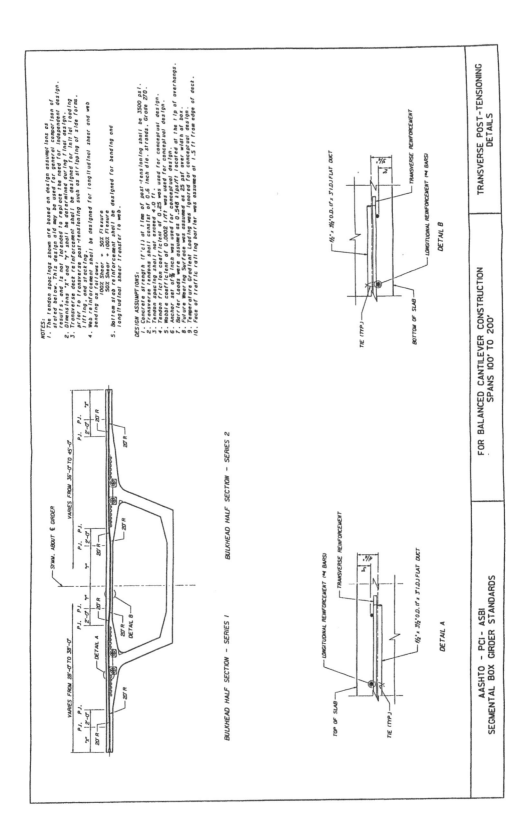

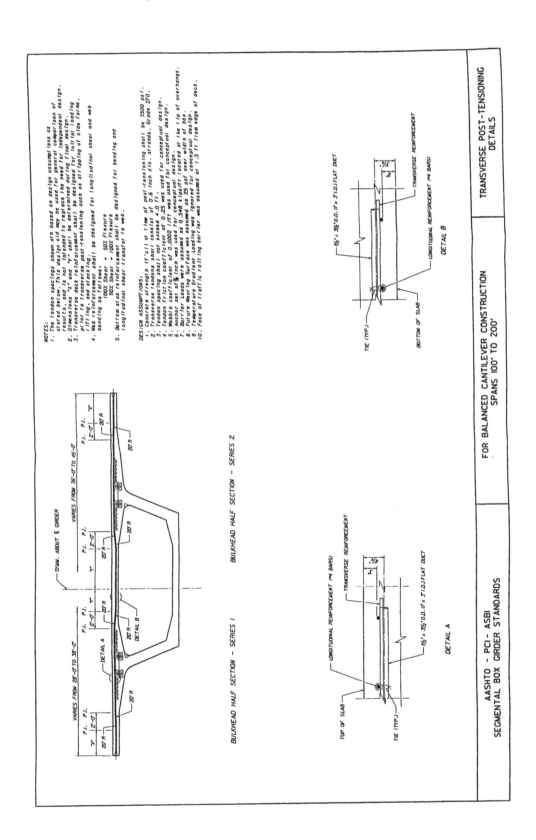

NOTES:
1. The tendon spacings shown are based on design assumptions as stated below. This design aid may be used for general comparison of results, and is not intended to replace the need for independent design.
2. Dimensions "x" and "y" shall be determined during final design.
3. Transverse deck reinforcement shall be designed for initial loading prior to transverse post-tensioning such as stripping of side forms, lifting, and stacking.
4. Web reinforcement shall be designed for longitudinal shear and web bending as follows:
   100% Shear + 50% Flexure
   50% Shear + 100% Flexure
5. Bottom slab reinforcement shall be designed for bending and longitudinal shear transfer to web.

DESIGN ASSUMPTIONS:
1. Concrete strength (f'c) at time of post-tensioning shall be 3500 psi.
2. Transverse tendons shall consist of 0.6 inch dia. strands, Grade 270.
3. Tendon spacing shall not exceed 4.0 ft.
4. Tendon friction coefficient of 0.25 was used for conceptual design.
5. Wobble coefficient of 0.0002 l/ft was used for conceptual design.
6. Anchor set of ¼ inch was used for conceptual design.
7. Barrier Loads were assumed as 0.5#/ft part located at the tip of overhangs.
8. Future Wearing Surface was assumed as 25 psf over width of bar.
9. Temperature Gradient Loading was ignored for conceptual design.
10. Face of traffic railing barrier was assumed at 1.5 ft from edge of deck.

VARIES FROM 28'-0 TO 38'-0

P.J. P.J.
2'-0

20' R

SYMM. ABOUT ℄ GIRDER

VARIES FROM 36'-0 TO 45'-0

P.J. P.J.
2'-0

20' R

DETAIL A

DETAIL B

20' R

BULKHEAD HALF SECTION - SERIES 1   BULKHEAD HALF SECTION - SERIES 2

LONGITUDINAL REINFORCEMENT (#4 BARS)

TRANSVERSE REINFORCEMENT

½" × 3½"O.D.(2 × 5"I.D.)FLAT DUCT

TOP OF SLAB

TIE (TYP.)

DETAIL A

TRANSVERSE REINFORCEMENT (#4 BARS)

½" × 3½"O.D.(2 × 5"I.D.)FLAT DUCT

LONGITUDINAL REINFORCEMENT

TIE (TYP.)

BOTTOM OF SLAB

DETAIL B

AASHTO - PCI - ASBI
SEGMENTAL BOX GIRDER STANDARDS

FOR BALANCED CANTILEVER CONSTRUCTION
SPANS 100' TO 200'

TRANSVERSE POST-TENSIONING
DETAILS

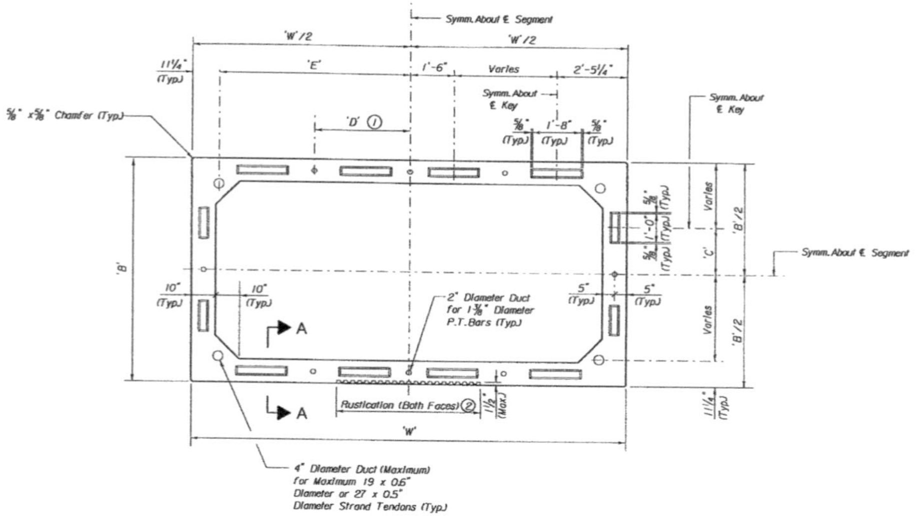

PLAN VIEW

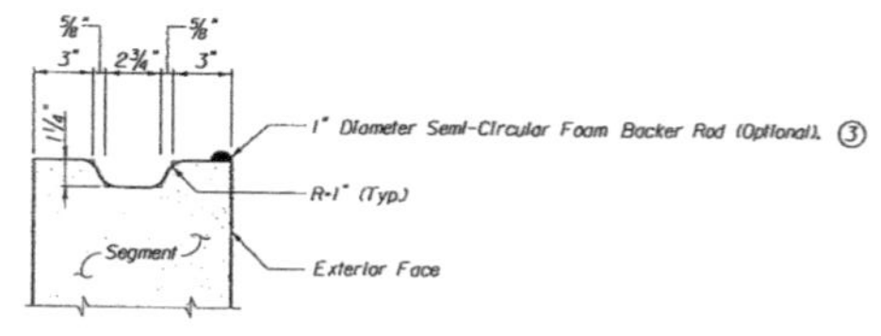

SECTION A-A

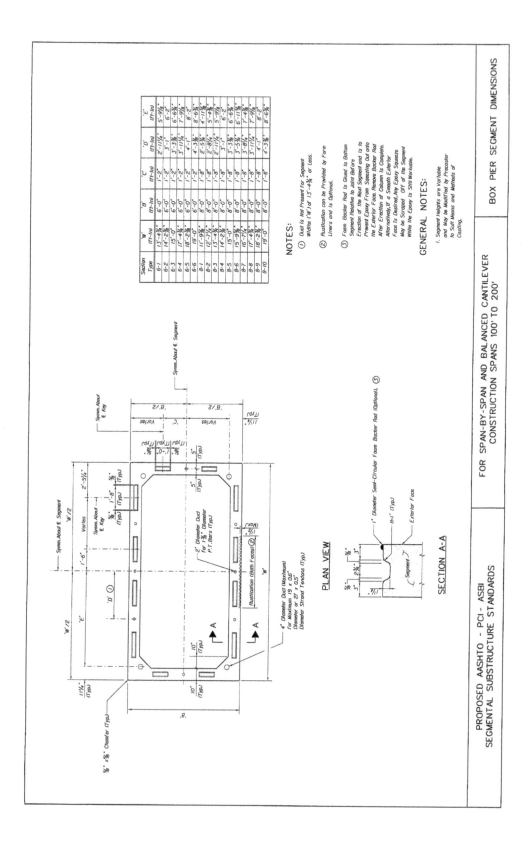

Section Type	'W' (ft-In)	'B' (ft-In)	'C' (ft-In)	'D' (ft-In)	'E' (ft-In)
6-1	13'-4¾"	6'-0"	1'-2"	2'-11¼"	5'-9⅛"
6-2	14'-2⅜"	6'-0"	1'-2"	3'-1"	6'-2"
6-3	15'-0"	6'-0"	1'-2"	3'-3⅜"	6'-6¾"
6-4	17'-4¾"	6'-0"	1'-2"	3'-11¼"	7'-9⅛"
6-5	18'-2⅜"	6'-0"	1'-2"	4'-1"	8'-2"
6-6	19'-0"	6'-0"	1'-2"	4'-3⅜"	8'-6¾"
8-1	11'-9⅝"	8'-0"	1'-8"	2'-5¾"	4'-11⅝"
8-2	12'-7¼"	8'-0"	1'-8"	2'-8¼"	5'-4⅜"
8-3	13'-4¾"	8'-0"	1'-8"	2'-11¼"	5'-9⅛"
8-4	14'-2⅜"	8'-0"	1'-8"	3'-1"	6'-2"
8-5	15'-0"	8'-0"	1'-8"	3'-3⅜"	6'-6¾"
8-6	15'-9⅝"	8'-0"	1'-8"	3'-5¾"	6'-11⅝"
8-7	16'-7¼"	8'-0"	1'-8"	3'-8¼"	7'-4⅜"
8-8	17'-4¾"	8'-0"	1'-8"	3'-11¼"	7'-9⅛"
8-9	18'-2⅜"	8'-0"	1'-8"	4'-1"	8'-2"
8-10	19'-0"	8'-0"	1'-8"	4'-3⅜"	8'-6¾"

NOTES:

① Duct Is Not Present for Segment Widths ('W') of 13'-4¾" or Less.

② Rustication can be Provided by Form Liners and Is Optional.

③ Foam Backer Rod Is Glued to Bottom Segment Relative to Joint Before Erection of the Next Segment and Is to Prevent Epoxy From Sqeezing Out onto the Exterior Face. Remove Backer Rod After Erection of Column Is Complete. Alternatively, If a Smooth Exterior Face Is Desired, Any Epoxy Squeeze May be Scraped Off of the Segment While the Epoxy Is Still Workable.

GENERAL NOTES:

1. Segment Heights are Variable and May be Modified by Precaster to Suit Means and Methods of Casting.

PROPOSED AASHTO - PCI - ASBI SEGMENTAL SUBSTRUCTURE STANDARDS	FOR SPAN-BY-SPAN AND BALANCED CANTILEVER CONSTRUCTION SPANS 100' TO 200'	BOX PIER SEGMENT DIMENSIONS

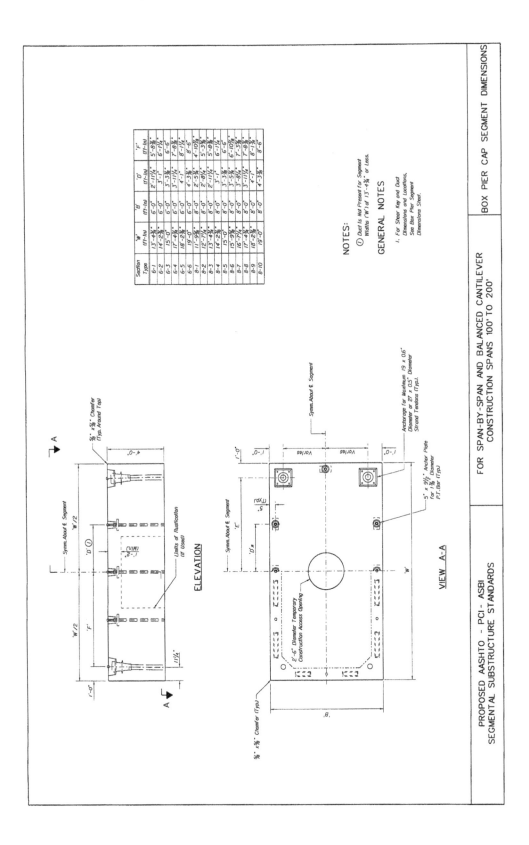

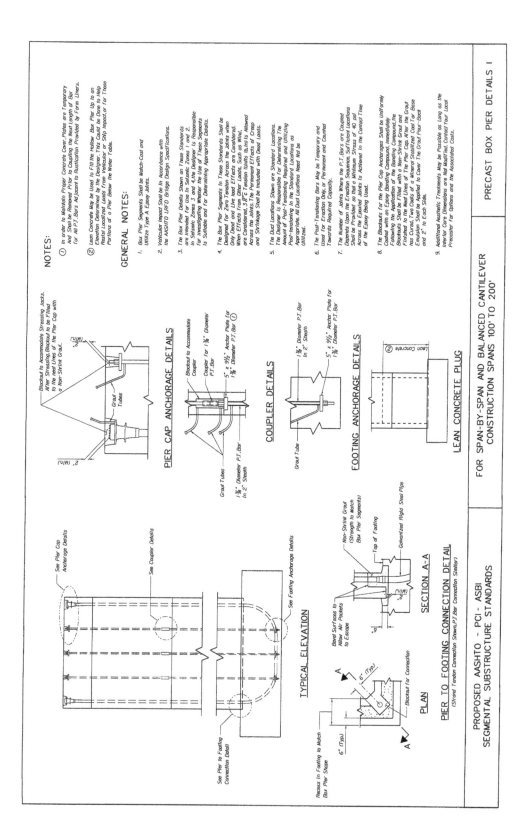

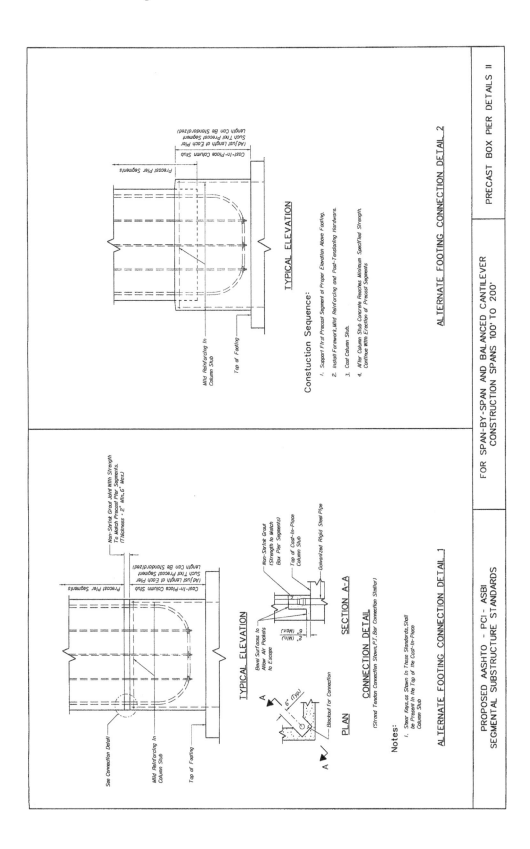

PRECAST BOX PIER DETAILS II

FOR SPAN-BY-SPAN AND BALANCED CANTILEVER
CONSTRUCTION SPANS 100' TO 200'

PROPOSED AASHTO - PCI - ASBI
SEGMENTAL SUBSTRUCTURE STANDARDS

TYPICAL ELEVATION (upper figure)

Cast-In-Place Column Stub
(Adjust Length at Each Pier
Such That Precast Segment
Length Can Be Standardized)

Precast Pier Segments

Mild Reinforcing In
Column Stub

Top of Footing

Construction Sequence:

1. Support First Precast Segment at Proper Elevation Above Footing.

2. Install Formwork, Mild Reinforcing and Post-Tensioning Hardware.

3. Cast Column Stub.

4. After Column Stub Concrete Reaches Minimum Specified Strength,
   Continue With Erection of Precast Segments

ALTERNATE FOOTING CONNECTION DETAIL 2

TYPICAL ELEVATION (lower figure)

Non-Shrink Grout Joint With Strength
To Match Precast Pier Segments.
(Thickness - 2" Min., 6" Max.)

Cast-In-Place Column Stub
(Adjust Length at Each Pier
Such That Precast Segment
Length Can Be Standardized)

Precast Pier Segments

See Connection Detail

Mild Reinforcing In
Column Stub

Top of Footing

Bevel Surfaces to
Allow Air Pockets
to Escape

Non-Shrink Grout
(Strength to Match
Box Pier Segments)

Top of Cast-In-Place
Column Stub

Galvanized Rigid Steel Pipe

2" (Min.)   6" (Max.)

SECTION A-A

A

6" (Typ.)

A

Blockout for Connection

PLAN

CONNECTION DETAIL

(Strand Tendon Connection Shown, P.T. Bar Connection Similar)

Notes:

1. Shear Keys, as Shown In These Standards, Shall
   be Present In the Top of the Cast-In-Place
   Column Stub

ALTERNATE FOOTING CONNECTION DETAIL 1

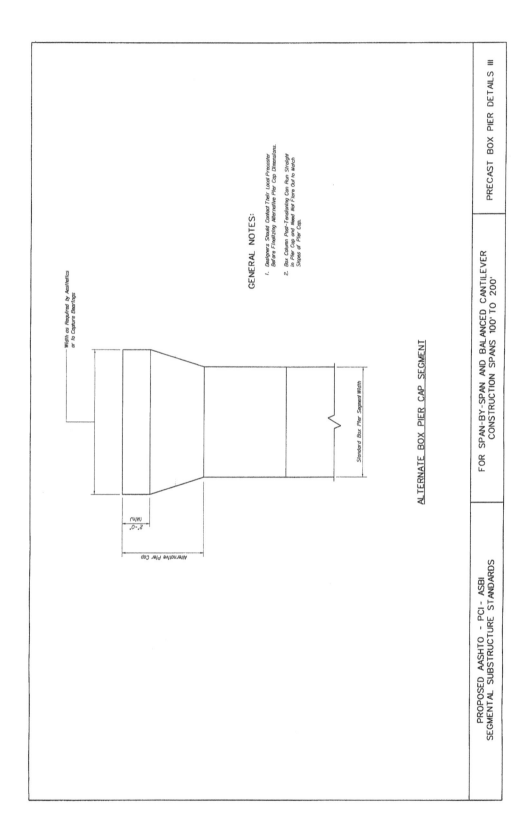

GENERAL NOTES:

1. Designers Should Contact Their Local Precaster Before Finalizing Alternative Pier Cap Dimensions.

2. Box Column Post-Tensioning Can Run Straight In Pier Cap and Need Not Flare Out to Match Slopes of Pier Cap.

Width as Required by Aesthetics or to Capture Bearings

Standard Box Pier Segment Width

2'-0" (Min)

Alternative Pier Cap

ALTERNATE BOX PIER CAP SEGMENT

PRECAST BOX PIER DETAILS III

FOR SPAN-BY-SPAN AND BALANCED CANTILEVER CONSTRUCTION SPANS 100' TO 200'

PROPOSED AASHTO - PCI - ASBI SEGMENTAL SUBSTRUCTURE STANDARDS

Appendix B: Typical Prestressing and Stay Cable Systems Provided by Freyssinet and TENSA

BI PART OF FREYSSINET PRESTRESSING AND STAY CABLE SYSTEMS

The figures and tables in Section BI are used with permission by Freyssinet. For more information, see www.freyssinet.com.

BI.1 PRESTRESSING SYSTEMS

BI.1.1 ANCHORAGES

Freyssinet prestressing anchorages include two ranges: C range and F range. C range can hold 3 to 55 strand tendons, and F range can hold 1 to 4 strand tendons.

BI.1.1.1 C Range Anchors

BI.1.1.1.1 Introduction

The C range prestressing system can be applicable for the following:

- 13-mm (0.5 in.), 15-mm (0.6 in.) strands of all grades
- up to 55 strands
- bonded or unbonded
- retensioning possible
- replaceable, adjustable
- detensioning possible

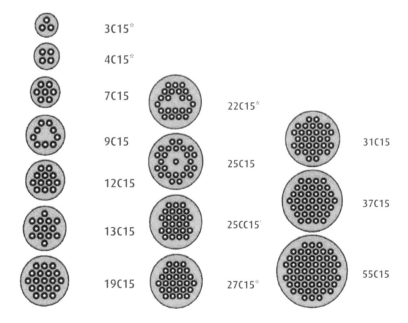

3C15*

4C15*

7C15

22C15*

9C15

25C15

12C15

13C15   25CC15

31C15

37C15

19C15   27C15*   55C15

Anchor Units (Courtesy of Soletanche Freyssinet Photo Library)

BI.1.1.1.2   Dimensions of Typical Freyssinet Anchorages
BI.1.1.1.2.1   General Dimensions

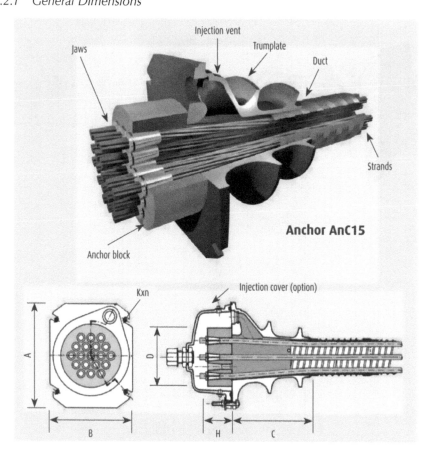

Units	A (mm)	B (mm)	C (mm)	D (mm)	H (mm)	Kxn (mm)
3C15	150	110	120	85	50	M10x2
4C15	150	120	125	95	50	M10x2
7C15	180	150	186	110	55	M12x2
9C15	225	185	260	150	55	M12x4
12C15	240	200	165	150	65	M12x4
13C15	250	210	246	160	70	M12x4
19C15	300	250	256	185	80	M12x4
22C15	330	275	430	220	90	M12x4
25C15	360	300	400	230	95	M16x4
25CC15	350	290	360	220	95	M16x4
27C15	350	290	360	220	100	M16x4
31C15	385	320	346	230	105	M16x4
37C15	420	350	466	255	110	M16x4
55C15	510	420	516	300	145	M20x4

(courtesy of) Soletanche Freyssinet Photo Library

BI.1.1.1.2.2   Applications
- For bonded internal prestressing with bare strands with cement grouting

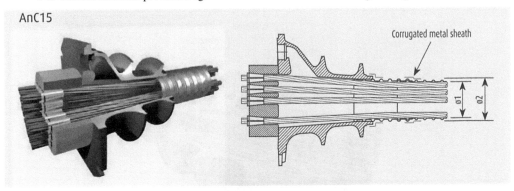

AnC15

Units	Ø1* (mm)	Ø2** (mm)
3C15	40	45
4C15	45	50
7C15	60	65
9C15	65	70
12C15	80	85
13C15	80	85
19C15	95	100
22C15	105	110
25C15	110	115
25CC15	110	115
27C15	115	120
31C15	120	125
37C15	130	135
55C15	160	165

(courtesy of) Soletanche Freyssinet Photo Library

*/**: refer to manufacturers' websites

- For unbonded internal prestressing with greased sheathed strands with cement grouting

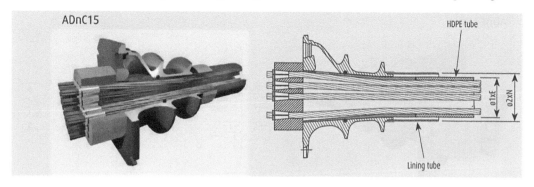

Units	Ø1* (mm)	E (mm)	Ø2** (mm)	N (mm)
3C15	50	3.7	70	2.9
4C15	63	4.7	82.5	3.2
7C15	63	4.7	82.5	3.2
9C15	75	5.5	101.6	5
12C15	90	6.6	114.3	3.6
13C15	90	6.6	114.3	3.6
19C15	110	5.3	133	4
22C15	110	5.3	139.7	4
25C15	125	6	152.4	4.5
25CC15	125	6	152.4	4.5
27C15	125	6	152.4	4.5
31C15	140	6.7	177.8	5
37C15	140	6.7	177.8	5

(courtesy of) Soletanche Freyssinet Photo Library

*/**: refer to manufacturers' websites

- For unbonded external prestressing with greased sheathed strands with cement grouting

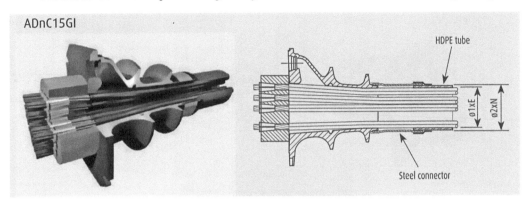

ADnC15GI

Units	Ø1* (mm)	E (mm)	Ø2** (mm)	N (mm)
3C15	70	2.9	63	4.7
4C15	82.5	3.2	75	5.5
7C15	82.5	3.2	90	6.6
9C15	101.6	5	90	6.6
12C15	114.3	3.6	110	5.3
13C15	114.3	3.6	110	5.3
19C15	133	4	125	6
22C15	139.7	4	125	6
25C15	152.4	4.5	140	6.7
25CC15	152.4	4.5	140	6.7
27C15	152.4	4.5	140	6.7
31C15	177.8	5	160	7.7
37C15	177.8	5	160	7.7
55C15	219.1	6.3	200	9.6

(courtesy of) Soletanche Freyssinet Photo Library

*/**: refer to manufacturers' websites

- For unbonded external prestressing with bare strands with cement grouting

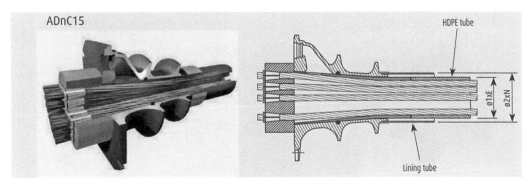

Units	Ø1* (mm)	E (mm)	Ø2** (mm)	N (mm)
3C15	50	3.7	70	2.9
4C15	63	4.7	82.5	3.2
7C15	63	4.7	82.5	3.2
9C15	75	5.5	101.6	5
12C15	90	6.6	114.3	3.6
13C15	90	6.6	114.3	3.6
19C15	110	5.3	133	4
22C15	110	5.3	139.7	4
25C15	125	6	152.4	4.5
25CC15	125	6	152.4	4.5
27C15	125	6	152.4	4.5
31C15	140	6.7	177.8	5
37C15	140	6.7	177.8	5

(courtesy of) Soletanche Freyssinet Photo Library

*/**: refer to manufacturers' websites

- For unbonded external prestressing with bare strands with injection of flexible product

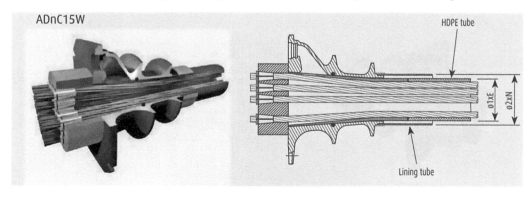

Units	Ø1* (mm)	E (mm)	Ø2** (mm)	N (mm)
3C15	50	3.7	70	2.9
4C15	63	4.7	82.5	3.2
7C15	63	4.7	82.5	3.2
9C15	75	5.5	101.6	5
12C15	90	6.6	114.3	3.6
13C15	90	6.6	114.3	3.6
19C15	110	8.1	133	4
22C15	110	8.1	139.7	4
25C15	125	9.2	152.4	4.5
25CC15	125	9.2	152.4	4.5
27C15	125	9.2	152.4	4.5
31C15	140	10.3	177.8	5
37C15	140	10.3	177.8	5

(courtesy of) Soletanche Freyssinet Photo Library

*/**: refer to manufacturers' websites

BI.1.1.1.3   Reinforcement at Anchorages

The anchorage anti-burst reinforcement can be used with either crossed hoops (stirrups) or spiral reinforcement. The spiral reinforment requirements are shown below.

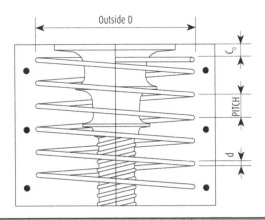

Units	Spiral reinforcement (Fy 235)					(Fy500) Additional reinforcements (stirrups)		
	Pitch (mm)	Diameter d (mm)	Number	Co (mm)	Outside diameter D (mm)	Pitch (mm)	Diameter d (mm)	Number
3C15	50	8	5	40	160	110	8	3
4C15	60	10	5	40	190	115	10	3
7C15	60	14	6	40	270	120	10	4
9C15	70	14	6	40	320	125	12	4
12C15	70	14	7	40	370	140	16	4
13C15	70	14	7	40	390	130	16	4
19C15	60	16	8	40	470	180	20	4
22C15	70	16	8	40	510	130	20	5
25C15	80	20	7	40	550	150	20	5
27C15	80	20	7	40	570	160	20	5
31C15	80	20	7	40	600	140	20	6
37C15	90	20	7	40	660	130	25	5
55C15	100	25	9	40	930	200	20	6

$f_{cm,0} = 24$ MPa

Units	Spiral reinforcement (Fy 235)					(Fy500) Additional reinforcements (stirrups)		
	Pitch (mm)	Diameter d (mm)	Number	Co (mm)	Outside diameter D (mm)	Pitch (mm)	Diameter d (mm)	Number
3C15	50	8	5	40	150	150	8	2
4C15	60	10	5	40	160	250	8	3
7C15	60	12	6	40	200	140	10	4
9C15	70	14	6	40	250	150	12	3
12C15	50	14	7	40	260	240	14	3
13C15	70	14	7	40	290	120	14	4
19C15	60	16	8	40	320	200	16	3
22C15	70	16	8	40	350	160	14	4
25C15	80	20	7	40	380	165	16	3
27C15	80	20	7	40	400	165	16	3
31C15	80	20	8	40	420	210	16	3
37C15	90	20	9	40	520	210	20	4
55C15	100	25	10	40	650	250	20	3

$f_{cm,0} = 44$ MPa

(courtesy of) Soletanche Freyssinet Photo Library

Bl.1.1.1.4   Tendon Coupling
- CU fixed multi-strand coupler

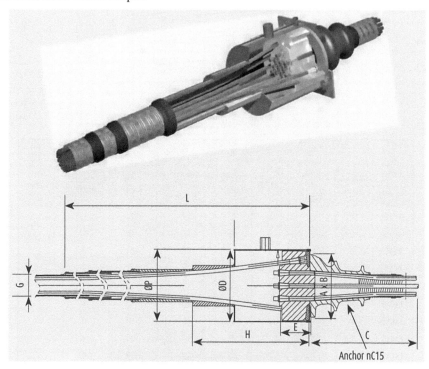

Units	A (mm)	B (mm)	C (mm)	G (mm)	ØD (mm)	E (mm)	L (mm)	H (mm)	ØP (mm)
CU 3C15	150	110	120	40	140	120	150	150	150
CU 4C15	150	120	125	45	150	127	150	150	150
CU 7C15	180	150	186	60	200	120	180	180	180
CU 9C15	225	185	260	65	255	122	225	225	225
CU 12C15	240	200	165	80	265	130	240	240	240
CU 13C15	250	210	246	80	276	130	250	250	250
CU 19C15	300	250	256	95	306	140	300	300	300
CU 22C15	330	275	430	105	335	145	330	330	330
CU 25C15	360	300	400	110	346	145	360	360	360
CU 25CC15	350	290	360	110	354	150	350	350	350
CU 27C15	350	290	360	115	354	150	350	350	350
CU 31C15	385	320	346	120	356	150	385	385	385
CU 37C15	420	350	466	130	386	156	420	420	420

(courtesy of) Soletanche Freyssinet Photo Library

- CM mobile multi-strand coupler for coupling for untensioned tendons

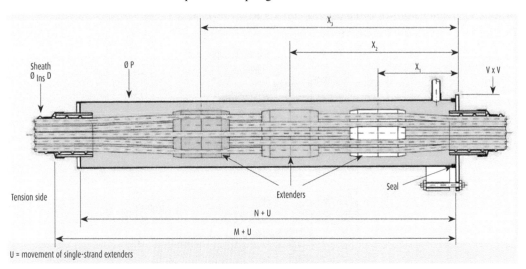

Units	D (mm)	M (mm)	N (mm)	P (mm)	X_1 (mm)	X_2 (mm)	X_3 (mm)	V (mm)
CM 3C15	40	1,050	1,000	102	250	500	750	130
CM 4C15	45	1,050	1,000	108	250	500	750	140
CM 7C15	60	1,050	1,000	114	250	500	750	150
CM 9C15	65	1,100	1,050	159	300	550	800	200
CM 12C15	80	1,150	1,100	159	300	550	800	200
CM 13C15	80	1,200	1,150	168	300	550	800	200
CM 19C15	95	1,200	1,150	194	300	550	800	230
CM 22C15	105	1,250	1,200	219	350	600	800	230
CM 25C15	110	1,250	1,200	219	350	600	850	250
CMI 27C15	115	1,300	1,250	219	350	600	850	250
CM 31C15	120	1,350	1,300	244	400	650	900	280
CM 37C15	130	1,530	1,480	273	400	650	900	310

(courtesy of) Soletanche Freyssinet Photo Library

BI.1.1.2   F13/F15 Anchorages

BI.1.1.2.1   Dimensions of Multi-Strand Units 3 to 5 F13/F15

BI.1.1.2.1.1   Bonded Internal Prestressing

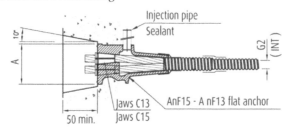

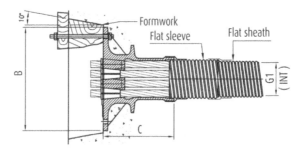

Units	A (mm)	B (mm)	C (mm)	G1 x G2 (mm²)	G (mm)	H (mm)
A3 F13/15	85	190	163	58 x 21	95	200
A4 F13/15	90	230	163	75 x 21	100	240
A5 F13/15	90	270	163	90 x 21	100	280

(courtesy of) Soletanche Freyssinet Photo Library

BI.1.1.2.1.2   Unbonded Internal Prestressing with Greased Sheathed Stands

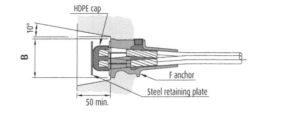

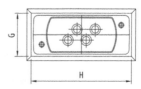

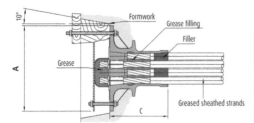

Units	A (mm)	B (mm)	C (mm)	G (mm)	H (mm)
A 3F 13/15	190	85	163	95	200
A 4F 13/15	230	90	163	100	240
A 5F 13/15	270	90	163	100	280

(courtesy of) Soletanche Freyssinet Photo Library

BI.1.1.2.2   Reinforcement for Multi-Strand Units (3 to 5 F13/F15)

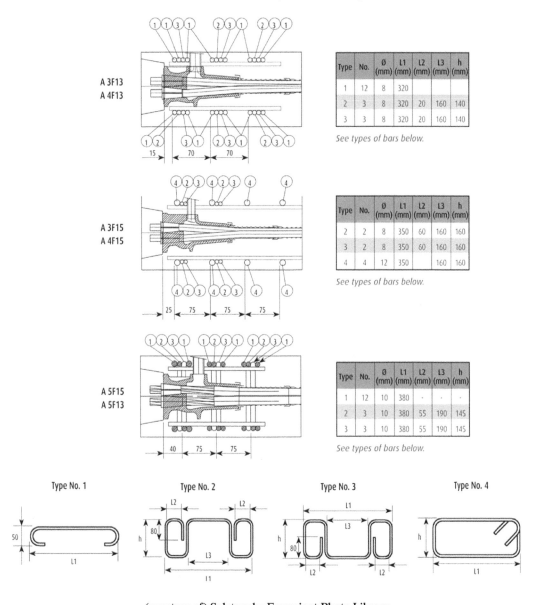

A 3F13
A 4F13

Type	No.	Ø (mm)	L1 (mm)	L2 (mm)	L3 (mm)	h (mm)
1	12	8	320			
2	3	8	320	20	160	140
3	3	8	320	20	160	140

See types of bars below.

A 3F15
A 4F15

Type	No.	Ø (mm)	L1 (mm)	L2 (mm)	L3 (mm)	h (mm)
2	2	8	350	60	160	160
3	2	8	350	60	160	160
4	4	12	350		160	160

See types of bars below.

A 5F15
A 5F13

Type	No.	Ø (mm)	L1 (mm)	L2 (mm)	L3 (mm)	h (mm)
1	12	10	380	-	-	-
2	3	10	380	55	190	145
3	3	10	380	55	190	145

See types of bars below.

Type No. 1   Type No. 2   Type No. 3   Type No. 4

(courtesy of) Soletanche Freyssinet Photo Library

BI.1.2 Dimensions of Type CC Jacks for C Range

Freyssinet provides different jacks for C range and F range. The dimensions of Type CC jacks for C range are shown below.

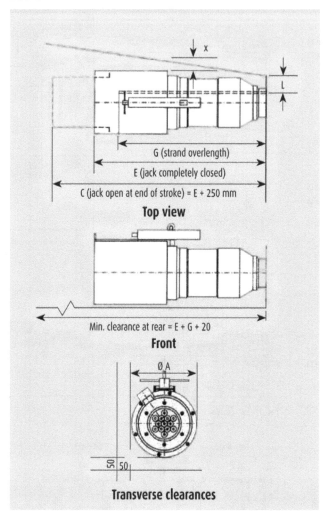

(courtesy of) Soletanche Freyssinet Photo Library

Outside dimensions of CC jacks

Jacks	Units	ØA (mm)	E (mm)	G (mm)	L (mm)	α for x ≈ 50	Stroke (mm)
CC 350	7C15	360	1,105	690	120	11°	250
	9C15		1,105	690	150	8°	
	12C15		1,115	700	150	8°	
	13C15		1,074	660	150	9°	
CC 500	7C15	438	1,085	688	120	15°	250
	9C15		1,085	688	150	13°	
	12C15		1,095	698	150	13°	
	13C15		1,100	703	150	12°	
	19C15		1,071	674	170	11°	
CC 1000	19C15	593	1,160	723	170	16°	250
	22C15		1,170	733	210	13°	
	25C15		1,175	738	210	13°	
	25C15P		1,175	738	210	13°	
	27C15		1,180	743	210	13°	
	31C15		1,146	709	210	13°	
	37C15		1,151	714	240	10°	
CC 1500	37C15	722	1,550	770	240	9°	350
	55C15		1,986	700	280	8°	

(courtesy of) Soletanche Freyssinet Photo Library

BI.2   STAY CABLES SYSTEM

BI.2.1   THE FREYSSINET STAY CABLE ANCHORAGES

BI.2.1.1   Components and Dimensions

- Fixed anchorage

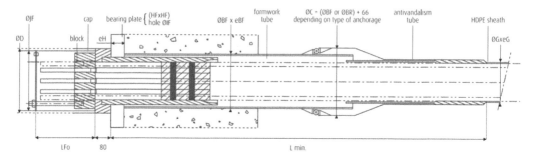

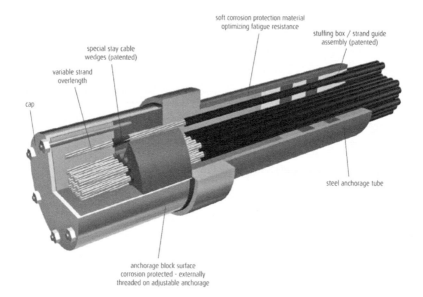

(courtesy of) Soletanche Freyssinet Photo Library

- Adjustable anchorage

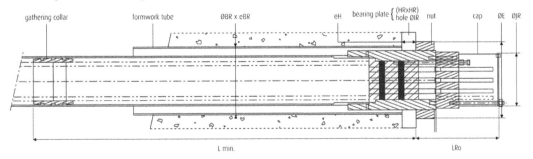

	Formwork tube				Flange/Nut		Outer pipe		Bearing plate*					Cap				Gathering collar
Type	ØBF	eBF	ØBR	eBR	ØD	ØE	ØG	eG	HF	HR	eH*	ØIF	ØIR	ØJF	ØJR	LFo	LRo	L min
12	177.8	6.3	219.1	6.3	210	235	125	6	275	300	50	151	192	200	160	275	346	1 200
19	219.1	6.3	244.5	6.3	250	284	140	6	340	350	50	186	230	240	194	275	356	1 400
27	244.5	6.3	298.5	8	280	336	160	6	400	420	60	212	260	270	222	285	376	1 750
31	244.5	6.3	298.5	8	290	346	160	6	420	440	60	221	270	280	233	290	386	1 750
37	273	6.3	323.9	8	320	368	180	6	460	470	70	239	290	300	252	305	411	1 900
48	323,9	8	368	8	356	415	200	6,2	520	540	80	273	330	345	291	315	434	2 100
55	323.9	8	368	8	370	438	200	6.2	550	570	80	285	350	360	304	320	446	2 200
61	355.6	8.8	406.4	8.8	405	460	225	6.9	600	610	90	318	375	395	336	330	466	2 400
75	368	8.8	445	10	433	506	250	7.7	640	670	100	342	405	423	368	340	481	2 500
91	419	10	482.6	11	480	546	280	8.6	720	750	110	374	450	470	410	360	524	2 850
109	431.8	10	530	12.5	500	600	315	9.7	770	815	120	386	480	490	435	380	560	3 100
127	457.2	10	558.8	12.5	545	640	315	9.7	810	850	130	424	525	535	478	400	600	3 250
169	530	12.5	635	12.5	625	740	355	10.9	950	980	140	490	605	615	555	430	660	3 700

(courtesy of) Soletanche Freyssinet Photo Library

*: refer to manufacturers' websites

Bl.2.1.2   Adjustable Dimensions

Stay cables	ØA	H1	H2	C1	C2
12H15	518	370	470	140	240
19H15	518	370	470	130	230
27H15	598	400	500	150	250
31H15	598	400	500	145	245
37H15	598	400	500	135	235
55H15	700	500	600	150	250
61H15	740	510	610	150	250
75H15	740	510	610	150	250
91H15	878	560	660	180	280
109H15	878	560	660	160	260
127H15	938	610	710	185	285

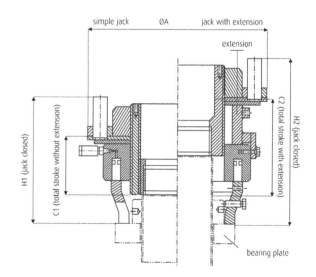

(courtesy of) Soletanche Freyssinet Photo Library

Bl.2.2   AERODYNAMIC STABILITY

Freyssinet provides a wide range of vibration-resistant devices adaptable to each project, including

   a. Double helical ribs on outer sheathes
   b. Internal damper
   c. External damper
   d. Cross ties

(courtesy of) Soletanche Freyssinet Photo Library

BII   PART OF TENSA PRESTRESSING AND STAY CABLES SYSTEMS

The figures and tables in Section BII are used with permission by TENSA (www.tensainternational.com).

BII.1   PRESTRESSING SYSTEMS

BII.1.1   MULTI-STRAND POST-TENSIONING SYSTEMS

BII.1.1.1   Introduction

Multi-strand systems are provided with a wide range of anchorages and solutions for different construction needs, such as internal MTAI live anchorage (Fig. BII-1a), internal MTAIM dead anchorage (Fig. BII-1b), MTRN adjustable anchorage (Fig. BII-1c), PTS flat anchorage (Fig. BII-1d), internal PTS flat anchorage (Fig. BII-1e), external MTAIE anchorage (Fig. BII-1f), MTG coupler anchorage (Fig. BII-1g), CU coupler system (Fig. BII-1h).

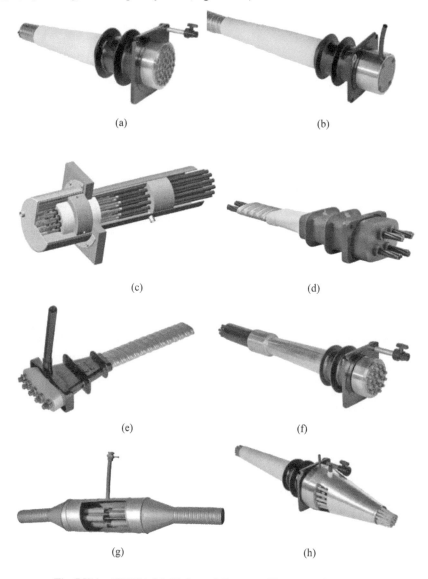

(a)   (b)

(c)   (d)

(e)   (f)

(g)   (h)

Fig. BII-1   TENSA Multi-Strand Systems (Courtesy of TENSA)

BII.1.1.2　Multi-Strand Post-Tensioning Anchorages

BII.1.1.2.1　MTAI and MTAIM Systems

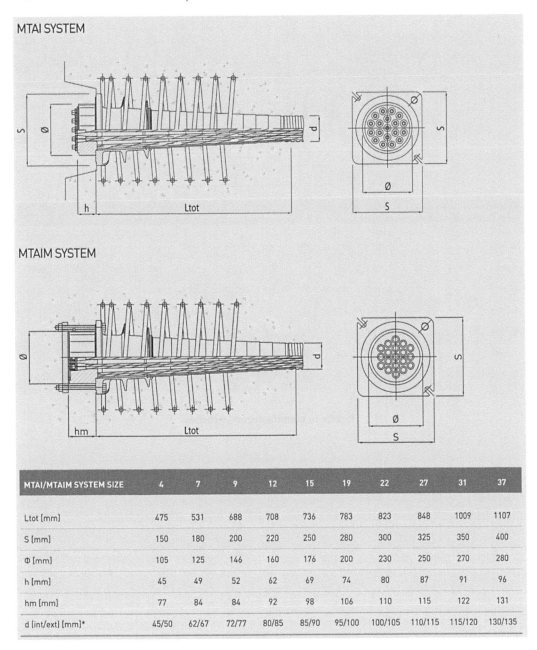

MTAI/MTAIM SYSTEM SIZE	4	7	9	12	15	19	22	27	31	37
Ltot [mm]	475	531	688	708	736	783	823	848	1009	1107
S [mm]	150	180	200	220	250	280	300	325	350	400
Φ [mm]	105	125	146	160	176	200	230	250	270	280
h [mm]	45	49	52	62	69	74	80	87	91	96
hm [mm]	77	84	84	92	98	106	110	115	122	131
d (int/ext) [mm]*	45/50	62/67	72/77	80/85	85/90	95/100	100/105	110/115	115/120	130/135

*: refer to manufacturers' websites

BII.1.1.2.2   MTAID System

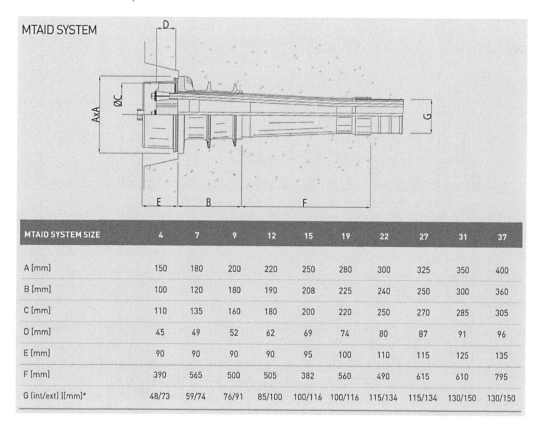

MTAID SYSTEM SIZE	4	7	9	12	15	19	22	27	31	37
A [mm]	150	180	200	220	250	280	300	325	350	400
B [mm]	100	120	180	190	208	225	240	250	300	360
C [mm]	110	135	160	180	200	220	250	270	285	305
D [mm]	45	49	52	62	69	74	80	87	91	96
E [mm]	90	90	90	90	95	100	110	115	125	135
F [mm]	390	565	500	505	382	560	490	615	610	795
G (int/ext)][mm]*	48/73	59/74	76/91	85/100	100/116	100/116	115/134	115/134	130/150	130/150

*: refer to manufacturers' websites

BII.1.1.2.3   MTAIE System

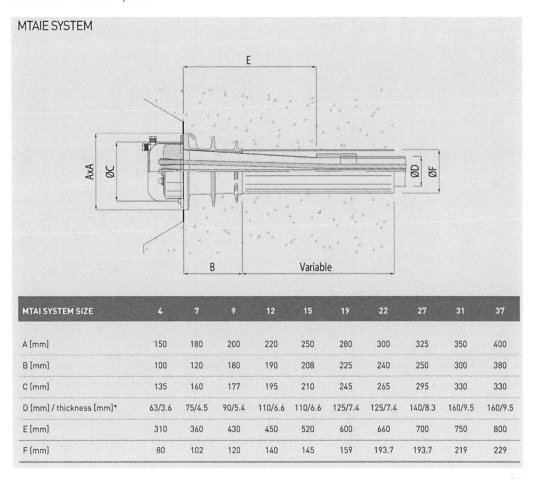

MTAI SYSTEM SIZE	4	7	9	12	15	19	22	27	31	37
A [mm]	150	180	200	220	250	280	300	325	350	400
B [mm]	100	120	180	190	208	225	240	250	300	380
C [mm]	135	160	177	195	210	245	265	295	330	330
D [mm] / thickness [mm]*	63/3.6	75/4.5	90/5.4	110/6.6	110/6.6	125/7.4	125/7.4	140/8.3	160/9.5	160/9.5
E [mm]	310	360	430	450	520	600	660	700	750	800
F [mm]	80	102	120	140	145	159	193.7	193.7	219	229

*: refer to manufacturers' websites

BII.1.1.2.4   PTS System

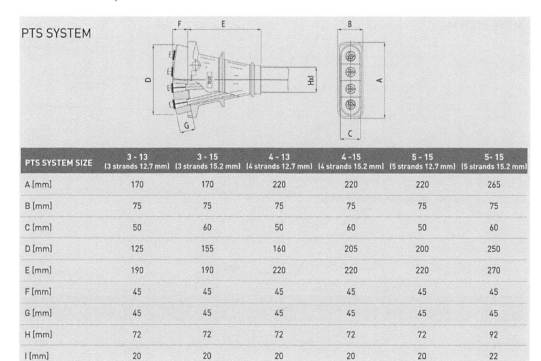

PTS SYSTEM SIZE	3 - 13 (3 strands 12.7 mm)	3 - 15 (3 strands 15.2 mm)	4 - 13 (4 strands 12.7 mm)	4 -15 (4 strands 15.2 mm)	5 - 15 (5 strands 12.7 mm)	5- 15 (5 strands 15.2 mm)
A [mm]	170	170	220	220	220	265
B [mm]	75	75	75	75	75	75
C [mm]	50	60	50	60	50	60
D [mm]	125	155	160	205	200	250
E [mm]	190	190	220	220	220	270
F [mm]	45	45	45	45	45	45
G [mm]	45	45	45	45	45	45
H [mm]	72	72	72	72	72	92
I [mm]	20	20	20	20	20	22

BII.1.1.3 Confinement and Busting Reinforcement

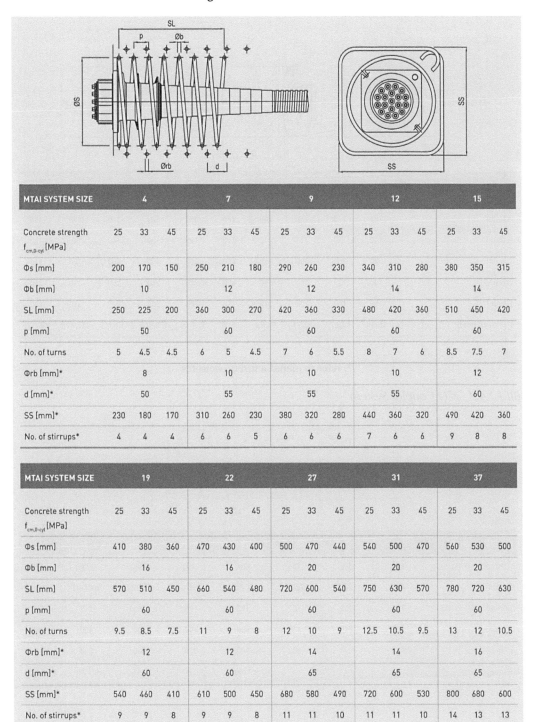

MTAI SYSTEM SIZE	4			7			9			12			15		
Concrete strength $f_{cm,0-cyl}$ [MPa]	25	33	45	25	33	45	25	33	45	25	33	45	25	33	45
Φs [mm]	200	170	150	250	210	180	290	260	230	340	310	280	380	350	315
Φb [mm]		10			12			12			14			14	
SL [mm]	250	225	200	360	300	270	420	360	330	480	420	360	510	450	420
p [mm]		50			60			60			60			60	
No. of turns	5	4.5	4.5	6	5	4.5	7	6	5.5	8	7	6	8.5	7.5	7
Φrb [mm]*		8			10			10			10			12	
d [mm]*		50			55			55			55			60	
SS [mm]*	230	180	170	310	260	230	380	320	280	440	360	320	490	420	360
No. of stirrups*	4	4	4	6	6	5	6	6	6	7	6	6	9	8	8

MTAI SYSTEM SIZE	19			22			27			31			37		
Concrete strength $f_{cm,0-cyl}$ [MPa]	25	33	45	25	33	45	25	33	45	25	33	45	25	33	45
Φs [mm]	410	380	360	470	430	400	500	470	440	540	500	470	560	530	500
Φb [mm]		16			16			20			20			20	
SL [mm]	570	510	450	660	540	480	720	600	540	750	630	570	780	720	630
p [mm]		60			60			60			60			60	
No. of turns	9.5	8.5	7.5	11	9	8	12	10	9	12.5	10.5	9.5	13	12	10.5
Φrb [mm]*		12			12			14			14			16	
d [mm]*		60			60			65			65			65	
SS [mm]*	540	460	410	610	500	450	680	580	490	720	600	530	800	680	600
No. of stirrups*	9	9	8	9	9	8	11	11	10	11	11	10	14	13	13

*: refer to manufacturers' websites

BII.1.1.4   Couplers

BII.1.1.4.1   MTG Coupler System

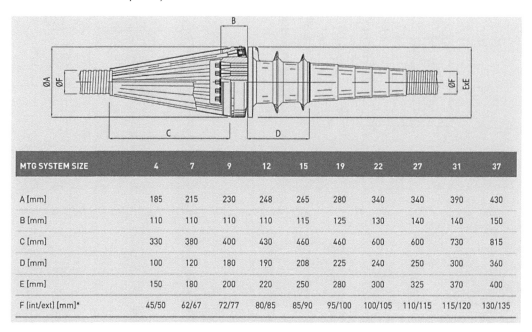

MTG SYSTEM SIZE	4	7	9	12	15	19	22	27	31	37
A [mm]	185	215	230	248	265	280	340	340	390	430
B [mm]	110	110	110	110	115	125	130	140	140	150
C [mm]	330	380	400	430	460	460	600	600	730	815
D [mm]	100	120	180	190	208	225	240	250	300	360
E [mm]	150	180	200	220	250	280	300	325	370	400
F (int/ext) [mm]*	45/50	62/67	72/77	80/85	85/90	95/100	100/105	110/115	115/120	130/135

*: refer to manufacturers' websites

BII.1.1.4.2   CU Coupler System

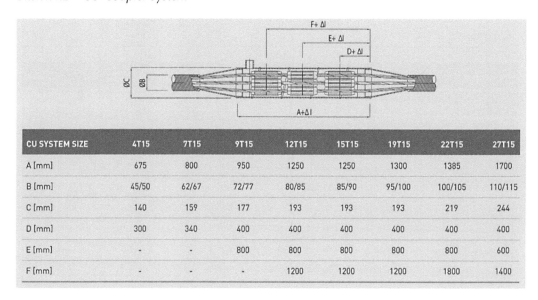

CU SYSTEM SIZE	4T15	7T15	9T15	12T15	15T15	19T15	22T15	27T15
A [mm]	675	800	950	1250	1250	1300	1385	1700
B [mm]	45/50	62/67	72/77	80/85	85/90	95/100	100/105	110/115
C [mm]	140	159	177	193	193	193	219	244
D [mm]	300	340	400	400	400	400	400	400
E [mm]	-	-	800	800	800	800	800	600
F [mm]	-	-	-	1200	1200	1200	1800	1400

BII.1.2   STRESSING JACKS

BII.1.2.1   Introduction

TENSA manufactures several types of "MT" stressing jacks, such as the MT series (see Fig. BII-2a) and MTA series (see Fig. BII-2b). They capacities range from 1000 to 10,000 kN.

(a)   (b)

Fig. BII-2   Typical Stressing Jacks (Courtesy of TENSA), (a) MT Series, (b) MTA Series

BII.1.2.2 Multi-Stand Stressing Jacks

BII.1.2.2.1 MT Series

MT SERIES

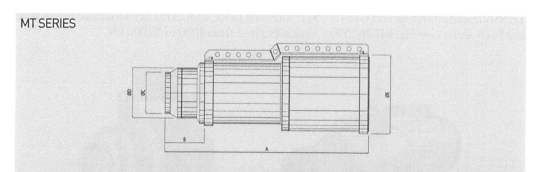

TYPE OF JACK	MT1000kN	MT1500kN	MT2500kN	MT3000kN	MT3500kN	MT4500kN	MT6000kN	MT9000kN
Capacity [kN]	1000	1500	2500	3000	3500	4500	6000	9000
Stroke [mm]	250	250	250	250	250	250	300	150
Weight [kg]	100	180	290	350	420	600	1000	1250
Tensioning section [cm²]	155.51	302.18	361.00	400.55	492.44	725.71	879.60	1625.00
Max. tensioning pressure [bar]	600	500	700	700	700	700	700	700
Max. return pressure [bar]	180	180	180	180	180	180	180	180
Max. locking pressure [bar]	165	165	165	165	165	165	165	165
Tensioning over length with lock-off [cm]	35	37	37	38	38	45	51	52
A [mm]	950	931	951	984	970	1107	1237	1016
B [mm]	155	130	150	154	147	200	207	191
C [mm]	137	152	173	195	214	243	295	322
D [mm]	162	185	213	236	252	310	380	407
E [mm]	248	310	339	370	415	512	615	714

BII.1.2.2.2 MTA Series

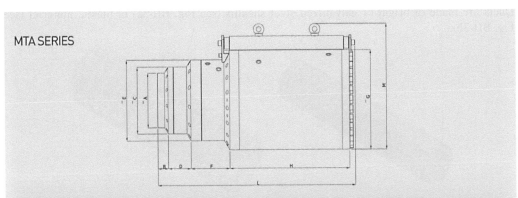

MTA SERIES

TYPE OF JACK	MTA 950kN	MTA 1700kN	MTA 2200kN	MTA 2900kN	MTA 3600kN	MTA 4600kN	MTA 5300kN	MTA 6500kN	MTA 7400kN	MTA 8800kN	MTA 10000kN
Capacity [kN]	950	1700	2200	2900	3600	4600	5300	6500	7400	8800	10000
Stroke [mm]	250	250	250	250	250	250	300	300	300	300	300
Weight [kg]	150	250	450	545	610	670	980	1055	1250	1400	1550
Tensioning section [cm²]	173.72	317.42	404.06	520.72	703.72	841.16	989.6	1193.8	1353.17	1643.11	1836.62
Max. tensioning pressure [bar]	550	550	550	550	550	550	550	550	550	550	550
Max. return pressure [bar]	110	110	110	110	110	110	110	110	110	110	110
Max. locking pressure [bar]	130	130	130	130	130	130	130	130	130	130	130
Tensioning over length with lock-off [cm]	70	70	70	70	70	70	75	75	75	75	75
A [mm]	137	168	189	203	225	258	276	293	333	363	400
B [mm]	33	36	39	42	45	50	55	65	70	75	82
C [mm]	195	230	250	270	285	315	345	360	410	450	490
D [mm]	90	90	100	108	110	111	163	108	110	115	120
E [mm]	270	300	320	337	352	382	412	427	470	500	540
F [mm]	160	165	170	176	181	187	187	220	235	250	270
G [mm]	320	340	360	385	420	470	520	545	565	595	630
H [mm]	517	559	591	591	585	580	639	646	665	710	728
L [mm]	800	850	900	945	950	957	1072	1072	1080	1150	1200
M [mm]	450	470	490	512	545	595	660	678	705	735	770

BII.1.3 Ducts

Ducts are made of bright or galvanized steel sheaths (see Fig. BII-3a) or plastic material (see Fig. BII-3b).

(a)   (b)

Fig. BII-3   Typical Ducts, (a) Steel, (b) Plastic (Courtesy of TENSA)

STRAND NO.	4	7	9	12	15	19	22	27	31	37
Internal Ø [mm]	45	62	72	80	85	95	100	110	115	130
Grout requirement [l/m]	1.2	2.3	2.8	3.6	3.8	4.7	5.2	6.2	6.9	8.6
Cement [kg/m]	1.9	3.6	4.5	5.8	6.1	7.5	8.4	9.9	10.8	13.8

BII.2 PARALLEL STRANDS STAY CABLE SYSTEM

BII.2.1 Introduction

There are two types of cable stay systems: TSR and TSRF.

The TSR stay cable system consists of a compact bundle of parallel seven-wire steel strands enclosed in a co-extruded high-density polyethylene circular duct (see Fig. BII-4a). The TSRF stay cable system features all the main advantages of the TSR system, and it is provided with a fork and pin connection that links to a clevis plate on the structure (see Fig. BII-4b).

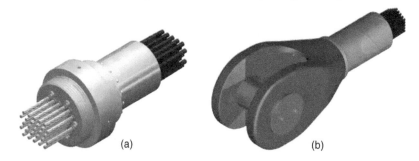

(a)   (b)

Fig. BII-4   TENSA Stay Cable Systems, (a) TSR, (b) TSRF (Courtesy of TENSA)

BII.2.2 TSR Stay Cable System

BII.2.2.1 Anchorage Components

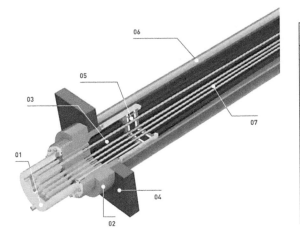

PART	NAME
01	PROTECTION CAP
02	ADJUSTABLE ANCHORAGE (TYPE TSRA)
03	ANTICORROSIVE COMPOUND
04	BEARING PLATE
05	WAX BOX SYSTEM
06	FORM PIPE
07	GALVANIZED, WAXED AND HDPE COATED STRAND
08	DAMPER SYSTEM/DEVIATION SYSTEM
09	ANTIVANDALISM/TELESCOPIC TUBE
10	EXTERNAL HDPE PIPE

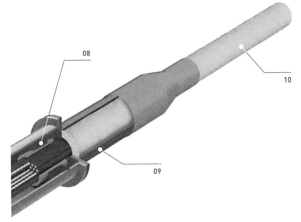

BII.2.2.2   System Properties and Dimensions

BII.2.2.2.1   Parallel Strands Stay Cables System Main Characteristics

Main Characteristics of TENSA Parallel Strands Stay Cables System

N° of STRANDS	STEEL NOMINAL CROSS SECTION [1] A_P [mm²]	STEEL NOMINAL MASS [1] M [kg/m]	STEEL NOMINAL BREAKING LOAD [1] F_m [kN]	MAXIMUM WORKING LOAD [2] 50% F_m [kN]	MAXIMUM WORKING LOAD [3] 60% F_m [kN]
2	300	2.34	558	279	335
4	600	4.69	1 116	558	670
7	1 050	8.20	1 953	977	1 172
12	1 800	14.06	3 348	1 674	2 009
19	2 850	22.27	5 301	2 651	3 181
27	4 050	31.64	7 533	3 767	4 520
31	4 650	36.33	8 649	4 325	5 189
37	5 550	43.36	10 323	5 162	6 194
43	6 450	50.40	11 997	5 999	7 198
55	8 250	64.46	15 345	7 673	9 207
61	9 150	71.49	17 019	8 510	10 211
73	10 950	85.56	20 367	10 184	12 220
91	13 650	106.65	25 389	12 695	15 233
109	16 350	127.75	30 411	15 206	18 247
127	19 050	148.84	35 433	17 717	21 260
169	25 350	198.07	47 151	23 576	28 291

BII.2.2.2.2   TSR System
BII.2.2.2.2.1   Deck Connection with Fixed Anchorage

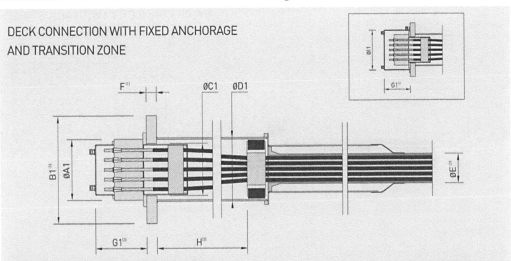

DECK CONNECTION WITH FIXED ANCHORAGE AND TRANSITION ZONE

Main dimensions (using steel strand diameter 15.7 mm and grade 1 860 MPa)

N° of STRANDS	ØA1	B1[1]	ØC1	ØD1	ØE STANDARD	ØE SLIM	F[1]	G1[2]	H STANDARD	H SLIM	ØI1
	[mm]	[mm]	[mm]	[mm]	[mm]	[MM]	[mm]	[mm]	[mm]	[mm]	[mm]
4	130	280	100	127	63	63	20	285	460	610	190
7	150	300	122	152.4	75	63	30	295	580	790	210
12	190	375	160	193.7	110	110	40	300	910	1 250	255
19	225	390	180	219.1	125	110	50	320	1 010	1 420	290
27	260	410	217	254	160	140	60	390	1 330	1 860	330
31	275	415	230	267	160	140	70	400	1 330	1 860	345
37	280	430	237	273	180	160	80	410	1 460	2 070	355
43	320	475	267	305	200	180	80	425	1 660	2 360	400
55	335	475	282	323.9	200	180	90	445	1 770	2 490	425
61	360	550	305	355.6	225	200	100	475	1 920	2 730	445
73	390	590	325	368	250	225	100	525	2 080	2 950	475
91	425	650	365	419	280	250	120	555	2 330	3 340	525
109	450	700	380	431.8	280	250	125	585	2 500	3 560	550
127	500	750	425	482.6	315	280	130	615	2 800	4 010	600
169	570	900	485	558.8	400	315	145	655	3 220	4 620	680

BII.2.2.2.2.2   Pylon Connection with Adjustable Anchorage

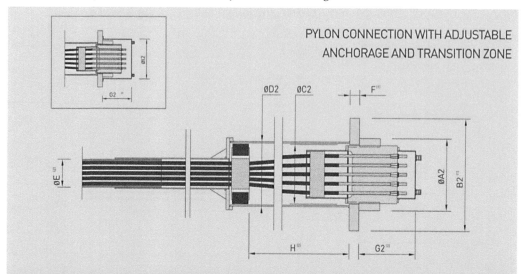

PYLON CONNECTION WITH ADJUSTABLE
ANCHORAGE AND TRANSITION ZONE

N° of STRANDS	ØA2	B2[1]	ØC2	ØD2	E[2]		F[1]	G2[2]	H		Øl2
					STANDARD	SLIM			STANDARD	SLIM	
	[mm]	[mm]	[mm]	[mm]	[mm]	[mm]	[mm]	[mm]	[mm]	[mm]	[mm]
4	160	300	140	168.3	63	63	20	325	430	580	190
7	180	340	160	193.7	75	63	30	345	550	760	210
12	220	440	200	229	110	110	40	345	860	1 200	250
19	280	450	235	267	125	110	50	360	960	1 370	310
27	320	480	270	305	160	140	60	380	1 280	1 810	350
31	330	500	285	323.9	160	140	70	390	1 280	1 810	360
37	345	500	290	323.9	180	160	80	400	1 410	2 020	375
43	390	560	330	368	200	180	80	435	1 610	2 310	420
55	410	560	345	394	200	180	90	450	1 720	2 440	445
61	440	610	370	419	225	200	100	450	1 870	2 680	475
73	475	650	400	445	250	225	100	470	2 030	2 900	510
91	520	700	435	482.6	280	250	120	505	2 280	3 290	555
109	545	740	460	508	280	250	125	540	2 400	3 460	580
127	600	800	510	558.8	315	280	130	580	2 700	3 910	640
169	680	900	580	635	400	315	145	670	3 120	4 520	720

BII.2.3 SADDLE SYSTEM

There are several different saddles, such as TSS-T (see Fig. BII-5a), TSS-B, and TSS-ST.

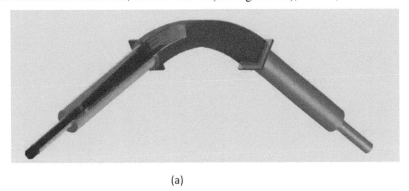

(a)

(b)

BII-5   Typical Saddle, (a) Type TSS-T, (b) Application on Bridge (Courtesy of TENSA)

Index